Lecture Notes in Physics

The Lecture Notes in Physics

The series Lecture Notes in Physics (LNP), founded in 1969, reports new developments in physics research and teaching – quickly and informally, but with a high quality and the explicit aim to summarize and communicate current knowledge in an accessible way. Books published in this series are conceived as bridging material between advanced graduate textbooks and the forefront of research to serve the following purposes:

• to be a compact and modern up-to-date source of reference on a well-defined topic;

• to serve as an accessible introduction to the field to postgraduate students and nonspecialist researchers from related areas;

• to be a source of advanced teaching material for specialized seminars, courses and schools.

Both monographs and multi-author volumes will be considered for publication. Edited volumes should, however, consist of a very limited number of contributions only. Proceedings will not be considered for LNP.

Volumes published in LNP are disseminated both in print and in electronic formats, the electronic archive is available at springerlink.com. The series content is indexed, abstracted and referenced by many abstracting and information services, bibliographic networks, subscription agencies, library networks, and consortia.

Proposals should be sent to a member of the Editorial Board, or directly to the managing editor at Springer:

Dr. Christian Caron
Springer Heidelberg
Physics Editorial Department I
Tiergartenstrasse 17
69121 Heidelberg/Germany
christian.caron@springer.com

Mauro Ferrario   Giovanni Ciccotti   Kurt Binder

Computer Simulations in Condensed Matter Systems: From Materials to Chemical Biology Volume 1

 Springer

Editors

Professor Mauro Ferrario
Dipartimento di Fisica
Università di Modena e Reggio Emilia
Via Campi, 213/A
41100 Modena, Italy
E-mail: mauro.ferrario@unimore.it

Professor Kurt Binder
Institut für Physik
Universität Mainz
Staudinger Weg 7
55128 Mainz, Germany
E-mail: kurt.binder@uni-mainz.de

Professor Giovanni Ciccotti
Dipartimento di Fisica, INFN
Università di Roma La Sapienza
Piazzale Aldo Moro 2
00185 Roma, Italy
E-mail: giovanni.ciccotti@roma1.infn.it

M. Ferrario et al., *Computer Simulations in Condensed Matter Systems: From Materials to Chemical Biology Volume 1*, Lect. Notes Phys. 703 (Springer, Berlin Heidelberg 2006), DOI 10.1007/b11604457

Library of Congress Control Number: 2006927291

ISSN 0075-8450
ISBN-10 3-540-35270-8 Springer Berlin Heidelberg New York
ISBN-13 978-3-540-35270-9 Springer Berlin Heidelberg New York

Springer is a part of Springer Science+Business Media
springer.com
© Springer-Verlag Berlin Heidelberg 2006

Typesetting: by the authors and techbooks using a Springer LaTeX macro package
Cover design: WMXDesign GmbH, Heidelberg

Printed on acid-free paper   SPIN: 11604457   54/techbooks   5 4 3 2 1 0

Preface

The school that was held at the Ettore Majorana Foundation and Center for Scientific Culture (EMFCSC), Erice (Sicily), in July 2005, aimed to provide an up-to-date overview of almost all technical advances of computer simulation in statistical mechanics, giving a fair glimpse of the domains of interesting applications. Full details on the school programme and participants, plus some additional material, are available at its Web site, `http://cscm2005.unimore.it`

Computer simulation is now a very well established and active field, and its applications are far too numerous and widespread to be covered in a single school lasting less than 2 weeks. Thus, a selection of topics was required, and it was decided to focus on perspectives in the celebration of the 65th birthday of Mike Klein, whose research has significantly pushed forward the frontiers of computer simulation applications in a broad range, from materials science to chemical biology. Prof. M. L. Klein (Dept. Chem., Univ. Pennsylvania, Philadelphia, USA) is internationally recognized as a pioneer in this field; he is the winner of both the prestigious Aneesur Rahman Prize for Computational Physics awarded by the American Physical Society, and its European counterpart, the Berni J. Alder CECAM Prize, given jointly with the European Physical Society. The festive session held on July 23rd, 2005, highlighting these achievements, has been a particular focus in this school. In the framework of the EMFCSC International School of Solid State Physics Series, the present school was the 34th course of its kind.

However, this school can be considered as being the third (and perhaps last?) event in a series of comprehensive schools on computer simulation, 10 years after the COMO Euroconference on "Monte Carlo and Molecular Dynamics of Condensed Matter systems," and 20 years after the VARENNA Enrico Fermi Summer School on "Molecular Dynamics of Statistical Mechanical Systems." Comparing the topics emphasized upon in these schools, both the progress in achieving pioneering applications to problems of increasing complexity, and the impressive number of new methodological developments are evident. While the focus of the Varenna School was mostly on Molecular Dynamics (MD) and its applications from simple to complex fluids, the Como school included both Monte Carlo (MC) simulations of lattice systems (from quantum problems to the advanced analysis of critical phenomena in classical systems like the simple Ising model), and the density functional theory of electronic structure up to the Car-Parrinello ab initio Molecular Dynamics techniques (CPMD). At the Erice school, a new focus was put on the paradigma of "Multiscale Simulation", i.e. the idea to combine different methods of simulation on different scales of length and time in a coherent fashion. This method allow us to clarify the properties of complex materials or biosystems where a single technique (like CPMD or MD or MC etc.) due to excessive needs of computer resources is bound to fail. Good examples presented at this school for such multiscale simulation approaches included MD studies of polymers coupled with a solvent, which is described only in a coarse-grained fashion by the lattice Boltzmann technique and hybrid quantum mechanical/molecular mechanics (QM/MM) methods for CPMD simulations of biomolecules, etc.

As a second "leitmotif," emphasis has been put on rapidly emerging novel simulation techniques. Techniques that have been dealt with at this school include the methods of "transition path sampling" (i.e. a Monte Carlo sampling not intending to clarify the properties of a state in the space of thermodynamic variables, but the properties of the dominating paths that lead "in the course of a transition" from one stable state to another), density of state methods (like Wang-Landau sampling and multicanonical Monte Carlo, allowing an elegant assessment of free energy differences and free energy barriers, etc.) and so on. These techniques promise substantial progress with famous "grand challenge problems" like the kinetics of protein folding, as well as with classical ubiquitous problems like the theory of nucleation phenomena. Other subjects where significant progress in methodological aspects was made included cluster algorithms for off-lattice systems, evolutionary design in biomedical physics, construction of coarse-grained models describing the self-assembly and properties of lipid layers or of liquid crystals under confinement and/or shear, glass simulations, novel approaches to quantum chemistry, formulation of models to correctly describe the essence of dry friction and lubrication, rare event sampling, quantum Monte Carlo methods, etc. The diversity of this list vividly illustrates the breadth and impact that simulation methods have today.

While the most simple MC and MD methods have been invented about 50 years ago (the celebration of the 50th anniversary of the Metropolis algorithm was held in 2003, the 50th anniversary of the Alder-Wainwright spectacular first discovery by MD of the (then unexpected) phase transition in the hard sphere fluid is due in 2007), even the "second generation" of scientists, who started out 30-40 years ago as "simulators" are now already the "old horses" of the field, either close to the end of their scientific career, or, in the best case, near it. Thus, we can clearly observe that the task of developing the computer simulation methodology is further taken over with vigor by the "third generation" of well-established younger scientists who have emerged in the field. Because two of the organizers of the school (KB, GC) do belong to the "old horse" category, it was clearly necessary to get an energetic younger co-organizer involved (MF), and we also felt it was the appropriate time that the most senior experts need not give the main lectures of the school, but rather the younger generation who are now most actively driving forward the frontier of research. Of course, it was crucial to involve the very valuable experience and knowledge of our senior colleagues into the school as well, and we are very glad that so many of them have accepted our invitation to give one-hour seminars providing tutorial introductions to various advanced research topics, which is at the heart of the research interests of the speakers. In this way, it was possible to produce an exciting event on the forefront of research on computer simulation in condensed matter, in a very stimulating and interactive atmosphere, with plenty of fruitful discussions.

It is with great pleasure that we end this preface with several acknowledgments. This school, of which the lecture notes are collected here, could not have taken place without the generous support of the European Community under the Marie Curie Conference and Training Courses, Contract No. MSCF-CT-2003-503840. We are grateful to the coordinators of this program, Michel Mareschal and Berend Smit, for their help in securing this support. We also wish to thank the CECAM secretaries, Emmanuelle Crespeau and Emilie Bernard.

We thank the Ettore Majorana Foundation and Centre for Scientific Culture in Erice, Sicily, for providing their excellent facilities to hold this school, and also Giorgio Benedek, Director of the International School of Solid State Physics, for the opportunity to hold our school as its 34th course: for his enthusiastic support during the school, and for his personal scientific participation. We are particularly grateful to him for providing the beautiful facilities of Erice.

MF thanks Davide Calanca, INFM-S3, Modena, for his valuable help in setting up the Web site of the school.

We thank the director of the physics department of the University of Rome "La Sapienza", Guido Martinelli, and the Administrative Secretary of the Department, Mrs. Maria Vittoria Marchet and her assistant, Mrs. Maria Proietto, for helping us in the difficult duty of managing all the financial matters. Mrs. Fernanda Lupinacci deserves grateful appreciation for her devoted and untiring presence and skillful help in overcoming all practical difficulties related to the organizational needs, and for providing a hospitable atmosphere to all the participants.

We are very grateful to Daan Frenkel, Mike Klein, and Peter Nielaba for their very valuable input when setting up the scientific program of the school, to all the lecturers, for their willingness to engage in the endeavor, and to all the participants, for their engagement and enthusiasm.

May 2006

Mauro Ferrario
Giovanni Ciccotti
Kurt Binder

Contents

Contents
LNP 704 Computer Simulations in Condensed Matter Systems: From Materials to Chemical Biology Volume 2

List of Contributors

Jerry B. Abrams
Dept. of Chemistry
New York University
New York, NY 10003, U.S.A.
jerry.abrams@nyu.edu

Ali Alavi
University of Cambridge
Chemistry Department
Lensfield Road
Cambridge CB2 1EW, U.K.
asa10@cam.ac.uk

Axel Arnold
Max-Planck-Institut
für Polymerforschung
Ackermannweg 10
55128 Mainz, Germany
and
Frankfurt Institute for
Advanced Studies (FIAS)
Johann Wolfgang Goethe-Universität
Frankfurt/Main, Germany
arnolda@mpip-mainz.mpg.de

Kurt Binder
Institut fuer Physik
Universitaet Mainz
Staudinger Weg 7
D-55099 Mainz-Germany
Kurt.Binder@uni-mainz.de

Peter G. Bolhuis
van 't Hoff Institute for Molecular
Sciences
University of Amsterdam
Nieuwe Achtergracht 166
1018 WV Amsterdam
The Netherlands
bolhuis@science.uva.nl

Sara Bonella
NEST Scuola Normale Superiore
Piazza dei Cavalieri 7
It-56126 Pisa
s.bonella@sns.it

David M. Ceperley
Department of Physics and NCSA
University of Illinois at
Urbana-Champaign
Urbana IL 61801, U.S.A.
david.ceperley@uiuc.edu

Giovanni Ciccotti
Dipartimento di Fisica
Università "La Sapienza"
Piazzale Aldo Moro
2 00185 Roma, Italy
giovanni.ciccotti@roma1.infn.it

David F. Coker
Department of Chemistry
Boston University
590 Commonwealth Avenue
Boston, MA 02215, U.S.A.
coker@bu.edu

Christoph Dellago
Faculty of Physics
University of Vienna
Boltzmanngasse 5
1090 Wien, Austria
Christoph.Dellago@univie.ac.at

Evelyn Dittmer
Institut für Mathematik II
Freie Universität Berlin
Arnimallee 2–6
14195 Berlin, Germany
dittmer@math.fu-berlin.de

Ron Elber
Department of Computer Science
Cornell University
4130 Upson Hall
Ithaca NY 14853
ron@cs.cornell.edu

Mauro Ferrario
Dipartimento di Fisica
Università di Modena e
Reggio Emilia
Via G. Campi 213/A
I-41100 Modena-Italy
Mauro.Ferrario@unimore.it

Daan Frenkel
FOM Institute for Atomic
and Molecular Physics (AMOLF)
Kruislaan 407
1098 SJ Amsterdam
The Netherlands
frenkel@amolf.nl

Phillip L. Geissler
Department of Chemistry
University of California
at Berkeley
94720 Berkeley, CA, U.S.A.
geissler@cchem.berkeley.edu

Christian Holm
Max-Planck-Institut für
Polymerforschung
Ackermannweg 10
55128 Mainz, Germany
and
Frankfurt Institute for
Advanced Studies (FIAS)
Johann Wolfgang Goethe-Universität
Frankfurt/Main, Germany
c.holm@fias.uni-frankfurt.de

Illia Horenko
Institut für Mathematik II
Freie Universität Berlin
Arnimallee 2–6
14195 Berlin, Germany
horenko@math.fu-berlin.de

Jürg Hutter
Physical Chemistry Institute
University of Zurich
Winterthurerstrasse 190
8057 Zurich, Switzerland
hutter@pci.unizh.ch

Marcella Iannuzzi
Physical Chemistry Institute
University of Zurich
Winterthurerstrasse 190
8057 Zurich, Switzerland
marcella@pci.unizh.ch

Raymond Kapral
Chemical Physics Theory Group
Department of Chemistry
University of Toronto
Toronto, ON M5S 3H6, Canada
rkapral@chem.utoronto.ca

Alessandro Laio
Statistical and Biological Physics
SISSA – International School for
Advanced Studies
Via Beirut 2
34100, Trieste, Italy
Laio@sissa.it

Erik Luijten
Department of Materials Science
and Engineering
Frederick Seitz Materials
Research Laboratory
University of Illinois
at Urbana-Champaign
Urbana, Illinois 61801, U.S.A.
luijten@uiuc.edu

Bernward A.F. Mann
Max-Planck-Institut für
Polymerforschung
Ackermannweg 10
55128 Mainz, Germany
mann@mpip-mainz.mpg.de

Glenn J. Martyna
T.J. Watson Research Center
International Business Corporation
P.O. Box 218, Yorktown Heights
NY 10598, USA
martyna@us.ibm.com

Eike Meerbach
Institut für Mathematik II
Freie Universität Berlin
Arnimallee 2–6
14195 Berlin, Germany
meerbach@math.fu-berlin.de

Marcus Müller
Institut für Theoretische Physik
Georg-August-Universität
37077 Göttingen, Germany
mmueller@theorie.physik.uni-
goettingen.de

Michele Parrinello
Computational Science
Dept. of Chemistry and Applied
Biosciences
ETH Zurich. c/o USI Campus
Via Buffi 13
6900 Lugano, Switzerland
parrinello@phys.chem.ethz.ch

Juan J. de Pablo
Department of Chemical
and Biological Engineering
University of Wisconsin-Madison
Madison Wisconsin 53706-1691,
U.S.A.
depablo@engr.wisc.edu

Carlo Pierleoni
Department of Physics
University of L'Aquila
Polo di Coppito, Via Vetoio
L'Aquila, 67010, Italy
carlo.pierleoni@aquila.infn.it

Christof Schütte
Institut für Mathematik II
Freie Universität Berlin
Arnimallee 2–6
14195 Berlin, Germany
schuette@math.fu-berlin.de

Alex J.W. Thom
University of Cambridge
Chemistry Department
Lensfield Road
Cambridge CB2 1EW, U.K.
ajwt3@cam.ac.uk

Simon Trebst
Microsoft Research
and Kavli Institute
for Theoretical Physics
University of California
Santa Barbara, CA 93106, U.S.A.
trebst@kitp.ucsb.edu

Matthias Troyer
Theoretische Physik
ETH Zürich
CH-8093 Zürich, Switzerland
troyer@comp-phys.org

Mark E. Tuckerman
Courant Institute
of Mathematical Sciences
New York University
New York, NY 10003, U.S.A.
mark.tuckerman@nyu.edu

Eric Vanden-Eijnden
Courant Institute
of Mathematical Sciences
New York University
New York, NY 10012
eve2@cims.nyu.edu

Joost VandeVondele
Department of Chemistry
University of Cambridge
Lensfield Road
Cambridge CB2 1EW, U.K.
jv244@hermes.cam.ac.uk

Rodolphe Vuilleumier
Laboratoire de Physique
Théorique de la Matière
Condensée, Tour 24-25
2ème étage c.c. 121
Universitè Pierre
et Marie Curie
4 place Jussieu
F-75005 Paris, France
rodolphe.vuilleumier-
@lptmc.jussieu.fr

N.B. Wilding
Department of Physics
University of Bath
Bath, BA2 7AY, United Kingdom
N.B.Wilding@Bath.ac.uk

Introduction: Condensed Matter Theory by Computer Simulation

G. Ciccotti[1], K. Binder[2], and M. Ferrario[3]

[1] Dipartimento di Fisica, Università di Roma "La Sapienza", P.le A. Moro 5,
00185 Rome-Italy
Giovanni.Ciccotti@roma1.infn.it
[2] Institut fuer Physik, Universitaet Mainz, Staudinger Weg 7, 55099
Mainz-Germany
Kurt.Binder@uni-mainz.de
[3] Dipartimento di Fisica, Università di Modena e Reggio Emilia, Via G. Campi
213/A, 41100 Modena-Italy
Mauro.Ferrario@unimore.it

Kurt Binder, Mauro Ferrario and Giovanni Ciccotti

G. Ciccotti et al.: *Introduction: Condensed Matter Theory by Computer Simulation*, Lect.
Notes Phys. **703**, 1–11 (2006)
DOI 10.1007/3-540-35273-2_0

The basic laws of physics that govern the phenomena on the scales of length and energy relevant for condensed matter systems, ranging from simple fluids and solids to complex multicomponent materials and even problems of chemical biology, are well known and understood: one just deals with the Schrödinger equation for the quantum many-body problem of the nuclei and electrons interacting with Coulomb potentials (for simplicity, we disregard, here throughout, the need for relativistic corrections in electronic structure calculations of matter containing heavy atoms). Statistical mechanics then supplies the framework to extend this quantum many-body theory to provide a statistical description in terms of averages taken at nonzero temperature.

However, it is also well-known that one cannot carry out this program with any mathematical rigor. Even the problem of one nucleus (or a few nuclei) with the associated electrons is still a challenge for the methods of quantum chemistry. Dealing with the quantum-many-body problem in terms of approximations such as the Hartree-Fock method, which tries effectively to reduce the many-body problem to a single electron problem, introduces errors that at least for excited states, cannot be controlled. Similarly, statistical mechanics, as founded by Boltzmann, Maxwell, Gibbs and others more than one-hundred years ago, can only make analytically precise predictions for problems of a type where the many-body problem can be reduced to a system of independent particles or quasiparticles. Such problems are, for instance, the ideal gas, the ideal paramagnet, or the multidimensional harmonic oscillator describing phonons in perfectly harmonic crystals. Of course, these problems are useful and nicely illustrate the spirit of the general theoretical framework and hence, we all teach them to our students. But we should not fail to admit that the predictive power that emerges from these few problems is very scarce. One has to be very careful about concluding anything about the problems of real matter as it occurs in nature or in the experimentalist's laboratory. For instance, in gases and paramagnets the degrees of freedom considered almost never do not interact at all; real solids show thermal expansion and finite lifetime of the phonon excitations, unlike strictly harmonic crystals; etc.

It is true that one can try to account for those neglected interactions either by systematic expansions, e.g. dealing with anharmonic terms in crystals via perturbative methods, or by closed-form approximations, e.g. exchange interactions among the magnetic moments in a crystal may be treated within molecular field theory. But the parameter range over which the systematic expansions are accurate is often doubtful; carrying them to high enough order often requires extremely heavy use of very sophisticated computer programs, which typically give relatively little reward in the form of physical insight. A characteristic example is the study of critical phenomena in systems of interacting spins with the high temperature series expansion method. This method indeed can give very good estimates for critical exponents at second order transitions, but still may suffer from the problem that the answers gotten are misleading: if the model system studied exhibits a phase transition that is weakly of first order, rather than second order, it has no critical exponents at all. In addition, it is hard to improve such methods systematically by going to still higher order, since the effort then typically increases exponentially fast.

The situation with closed-form approximations is even worse, since these are typically uncontrolled in a mathematical sense and often lead to very bad and misleading results. Although the simplest of these approximations are standard material of university courses in statistical thermodynamics, one must not sweep under the rug that approximations such as the molecular field theory of magnetism erroneously predict long range order in a one-dimensional chain although there is none; the van der Waals theory of the liquid-gas transition produces isotherms with spurious loops, and none of these approximations can describe the critical behavior near second order phase transitions correctly. The reason of these shortcomings is that nontrivial correlations between the degrees of freedom of the many-body system arise. These correlations cannot be dealt with appropriately by these approximations, which always involve unjustifiable factorization of such correlations in one way or another. For more complicated problems, even such a mean-field like factorization requires very heavy and technically demanding computer use. For instance when one deals with quantum-many-body problems by Hartree-Fock techniques and their extensions, or when one deals with the glass transition of supercooled fluids by the mode-coupling theory beyond schematic models, cumbersome numerics is required. In such cases it is particularly difficult to justify which steps of the approximate theoretical treatment are accurate. Often direct comparison to experiment may be misleading, too, since the simplified model on which the theory is based does not correspond to a real system in sufficient detail. Hence discrepancies between "analytical theory" and experiment can be attributed to the choice of an inappropriate model, an inappropriate approximation, or both. Conversely, sometimes agreement between experiment and theory is claimed which is completely spurious because inadequacies of a model somehow are effectively more or less compensated by wrong approximations. An example of such spurious agreement are fits of the Flory-Huggins equation of state to phase diagrams of polymer mixtures.

A long list of theories to which these criticisms apply could be compiled, notwithstanding the fact that there are some special models, for instance, lattice models like the one- and two-dimensional Ising and Potts models with nearest neighbor interactions, which can be solved exactly by analytical methods. While one has certainly learned a lot of physics from the results of these exceptional nontrivial models that were exactly soluble, usually the method of solution is fairly special – if not tricky – and not illuminating the physics of the problem. And, in addition, the overwhelming majority of problems that one encounters in the physics of condensed matter does not fall in this category of solvable problems. While very respectable research on mathematical statistical mechanics is still going on, it is not likely that it will change this situation.

Hence, until about 50 years ago, condensed matter theory was in a very unsatisfactory status: although a formal framework for the theoretical description of the problems in terms of quantum and classical statistical mechanics did exist, a reliable set of tools, which would allow to make valuable predictions for static and dynamic properties and phenomena of most condensed-matter systems, was missing.

This situation has changed fundamentally through the invention of computer simulation, which provides a different paradigm towards all these problems. This novel methodic route towards the treatment of correlations in many-body systems started with the introduction of Monte Carlo and Molecular Dynamics methods about 50 years ago, and since then the scope of the approach has been – and still is – expanding systematically. The first studies were mostly concerned with the classical statistical mechanics of liquid matter, in thermal equilibrium, dealing with simple models for the interactions such as hard disks, hard spheres, or atoms interacting with effective forces described by Lennard-Jones potentials. Although these pioneering studies, due to the extremely limited computer resources then available, were limited to very small numbers of particles (a few hundred or even less), they still made a great step forward toward understanding the liquid state of matter, and moreover provided ideas to develop a theory of liquids. Despite the limited computer speed and storage capacity, the fact that only a few practitioners had access to these rare computing facilities and that the appropriate algorithms needed to be formulated and tested, impressive discoveries were already made by the first generation of computer simulation pioneers. Examples of important discoveries from that time include the surprises that hard spheres crystallize at a density long before close packing is achieved, and that dynamic correlation functions in fluids exhibit long time tails. Both discoveries have been a strong boost for the search of more traditional theoretical methods to reproduce these findings, such as an improved kinetic theory of dense gases, or density functional theories of liquids, etc. In view of these successes, the complaint of Nobel Laureate P.W. Anderson that never has anything of interest been discovered by a computer simulation on a supercomputer[1] is rather misleading: Of course, for the standards of the 50ies and 60ies of the 20th century, the computers at Los Alamos and Livermore were the "supercomputers" of the time.

Since the days of the pioneers, the field has greatly expanded and still continues to expand at a breathtaking speed, and more and more scientists are attracted by the fascination of what computer simulation methods have to offer. One line of expansion that was fairly obvious from the beginning is directed towards increasing the complexity of the systems that can be studied: the first studies of "Lennard Jonesium" did already allow very useful comparisons with experimental neutron scattering studies on fluid argon and other rare gases. But there was a clear need to extend the methodology as a next step to fluids formed from small molecules, and then to more and more

---

[1] See for example "Superficial Solution", Physics World, vol. **4**, August 1991, page 20.

complicated molecules, up to synthetic and biological macromolecules. Testing effective potentials (within the framework of a description within classical mechanics) is an ongoing challenge, driven by the desire to be able to compare the results to experiment; but at the same time the methodology needed to be refined in order to make a most efficient use of the always limited computer resources: in a molecular dynamics simulation of almost rigid molecules, it is preferable to treat them as perfectly rigid in order to avoid a too small integration time step. This needed a reformulation of the standard Molecular Dynamics methods (where Newton's equation of motion is integrated numerically, e.g. by the Verlet algorithm) to take the constraints correctly into account (SHAKE algorithm). Other methodic developments concerned methods to deal with long range interaction, implementation of the different ensembles of statistical mechanics, implementation of concepts from linear response theory to extract transport coefficients from Kubo formulae, and techniques to deal with driven systems in which currents far from equilibrium are maintained ("Non-Equilibrium Molecular Dynamics", NEMD).

Of course, this evolution of Molecular Dynamics (MD) was paralleled by various methodic developments dealing with Monte Carlo (MC) simulations. While MD realizes "Boltzmann's dream" to generate averages along a trajectory through phase space that follows from Newton's equations, MC realizes the Gibbs approach to statistical mechanics, sampling the appropriate probability distribution. A rather straightforward step was again the extension from the original NVT ensemble, for which the Metropolis algorithm was formulated, to other statistical ensembles, including grandcanonical ensembles (and semi-grandcanonical ones, in the case of mixtures). Also virtual particle insertion/deletion steps were invented to "measure" the chemical potential in the canonical ensemble in such a "computer experiment": for the theoretical physicist, computer simulation is like a laboratory where he can execute "thought experiments", and bring many concepts of statistical mechanics to work.

The MC method is also very useful to deal with problems on rigid lattices, from the self-avoiding walk problem, which can be viewed as a coarse-grained description for the statistical properties of long flexible macromolecules in a good solvent, to Ising- and Heisenberg models of ferro- and antiferromagnets, and models for order-disorder problems in crystalline alloys, adsorbed monolayers on crystalline surfaces, etc. Not only is there a rich variety of physical problems that could be dealt with, but also the great flexibility of the MC method is an asset: for the problem at hand one can taylor the stochastic move from one microstate of the model system to the next one in such a way that one greatly improves the efficiency of the algorithm. A good example is the Swendsen-Wang-Wolff cluster algorithm for the Ising model, where near the criticality, a large cluster of spins, rather than a single spin, is overturned so that the critical slowing down is reduced. Another example is the pivot algorithm for a self-avoiding walk, producing a much larger change of the configuration rather than local crankshaft motions, etc. Finding better and better algorithms where one moves faster through the phase space of the model is still a very active line of research, in particular for models of macro-molecules, where the large number of degrees of freedom of a single polymer is already a challenge.

Apart from the development of more efficient algorithms, also the improved understanding of limitations of simulation methods, and of how to best ex-tract the desired information on physical properties from the simulations, has been – and still is – a very active line of research. Although the MC method was invented in the context of equilibrium statistical mechanics, one can give the Metropolis algorithm a dynamic interpretation in terms of a Markovian master equation. This approach not only is helpful to understand what ini-tially was called "statistical inefficiency", i.e. dynamic correlation controlling statistical errors, but also forms the basis for many studies of dynamic prob-lems in condensed matter. These include problems from the critical dynamics of anisotropic magnets to the kinetics of phase separation in alloys, and non-equilibrium Monte Carlo (NEMC) methods where gradients together with pe-riodic boundary conditions maintain stationary fluxes, analogous to NEMD methods. But also irreversible growth phenomena can be studied from prob-lems such as diffusion limited aggregation (DLA), a problem that was first formulated as a computer simulation algorithm, to the chemical kinetics of polymerization.

A very important development of MC methods, but also easily imple-mentable in MD, addressed the problem of estimating free energies. Some of these methods originate from the idea of single histogram reweighting, and now include multiple histogram extrapolation, and expanded ensemble meth-ods, from "umbrella sampling" to the so-called "multicanonical Monte Carlo" and "Wang-Landau sampling" of the energy density of states. This is still an extremely active area of research and will be thoroughly dealt with in the book. These methods are crucial for making full use of the information that is generated by simulations.

A very important limitation of simulations, that could be turned into the most powerful theoretical method to explore phase transition phenomena, is related to the fact that simulations usually deal with systems of rather small size, and these restricted linear dimensions may impose important constraints on the phenomena one wishes to observe. E.g., near a second order phase tran-sition the growth of the correlation length of the order parameter fluctuations is limited, and critical singularities are rounded off and shifted away from the parameter values where they occur in the thermodynamic limit. Similar finite size rounding occurs also for first order transitions, where the delta function singularity of the specific heat representing the latent heat of the transition is smeared into a broad peak by the finite size of the simulated system. How-ever, such phenomena are meanwhile rather well-understood by the so-called "finite size scaling" theories, which have been thoroughly tested and explored by simulations, the last having also greatly inspired their development into a powerful analysis tool. Of course, these finite size scaling techniques can be used in conjunction with MD methods as well.

As a last comment on this overview on the role of simulation in classical statistical mechanics, we emphasize that, by suitable preparation of a system in terms of boundary conditions and/or initial conditions, one can extend the study from phenomena in bulk phases to phase coexistence and interfacial phenomena, be it flat interfaces or nanoscopic droplets or crystallites. One can derive methods to estimate all kinds of surface excess free energies, interface and line tensions, etc. Again all the concepts of statistical physics can be put to work in a very immediate and efficient way, in particular since the basic idea of dividing a system into a subsystem and its environment (with which it may or may not exchange, particles, etc.) can be very directly implemented. Any desired distribution function (also for quantities that are not accessible directly in any real experiment) can be sampled. One is not restricted to work only in the experimentally accessible dimensionalities of space $d = 1, 2, 3$, but also in higher dimensionalities, if this helps to elucidate the theoretical questions. So, it again emerges very clearly, that within the framework of classical statistical mechanics, computer simulation has become the method for most theoretical problems of interest. For its ability to test whether a considered model describes a real system well enough, and to test analytical approximations in a decisive way (since one implements exactly the same model as the one used by the theory, one can test the approximations involved in the latter step by step, computing quantities that may not be accessible in an experiment at all), simulational approach in the early days has been sometimes referred to as the "third branch" of science, intermediate between experiment and theory. Now, on the contrary, in classical statistical mechanics computer simulation is the basic theoretical method that fills the wide gap between the foundations of statistical mechanics, based on classical mechanics and probability theory, and the phenomena to which classical statistical mechanics can be applied.

It is important to realize, of course, that in general condensed matter theory needs to be based on a treatment of nuclei and electrons by quantum statistical mechanics rather than classical statistical mechanics. Unfortunately, we are still far from the possibility to solve the many-body Schrödinger equation, by brute force computer simulation, for a sufficiently large number of electrons and nuclei. However, as far as equilibrium properties are concerned, the Feynman path integral representation of the partition function of a quantum particle in imaginary time provides a mapping on a problem again in classical statistical mechanics, namely a "ring polymer" with P segments, the gyration radius of the ring polymer being of the order of the thermal De Broglie wavelength. The approach in principle can be generalized to many-body systems, although for fermions the antisymmetry of the many-body wave function gives rise to the famous "minus sign problem", i.e. one can no longer interpret the quantity that needs to be sampled as a positive definite probability, and hence a straightforward application of the Metropolis algorithm of classical statistical mechanics is no longer possible. Although many ideas have been developed to deal with this problem, in the general case it is still unsolved, and, hence, important problems of solid state physics involving electrons and their correlations (e.g. high temperature superconductivity) are very difficult to deal with. So several strategies are followed: Either one approximates the problem by fixing the nuclei (and part of the electrons) on rigid lattice sites and tries to deal only with the remaining electrons to explain specific properties, e.g. magnetic properties of solids, by the methods of quantum statistical mechanics. Or else, alternatively, electrons are not explicitly considered at all by working with phenomenological effective potentials between the atoms, while low temperature properties of matter are accessible via a quantum statistical mechanics treatment by path integral formalism. A further, more ambitious, step leads beyond the use of effective potentials, carrying out a separation of electrons and nuclei via the Born-Oppenheimer approximation, and assuming that the electrons remain in their ground state. We shall discuss all these approaches in turn. But also when one deals with electrons on rigid lattice sites, or when one deals with fermionic atoms interacting with effective potentials (e.g. ^{3}He), the minus sign problem is a major hurdle. For bosonic systems, however, the situation is better, and promising progress with nontrivial problems such as the suprafluidity of liquid Helium and Bose condensation of weakly interacting atoms in dilute ultra cold quantum gases has been made. Note that for propagation of the effective classical particles of the "ring polymer" through their phase space one is not restricted to MC, but often it is preferable to use MD, and hence both "Path Integral Molecular Dynamics" (PIMD) and "Path Integral Monte Carlo" (PIMC) techniques are available and very useful. This is also true of quantum effects relating to the nuclei in crystals, where the quantum-mechanical indistinguishability of the particles does not play any role, and hence quantities such as the thermal expansion coefficient of crystals at low temperature, their elastic constants, etc. can be reliably computed, if good enough effective potentials describing the forces among the atoms are available.

Of course, obtaining such effective potentials that are accurate enough is a problem that in general is not solved, and it would be preferable to include both electrons and nuclei in the path integral treatment. However, the disparity in their masses and the corresponding energy scales is a serious problem, and hence such a full path integral treatment is rarely attempted. So, in solid state physics one often works with a very reduced description only, dealing with quantum mechanical models such as the quantum-mechanical version of the Heisenberg model of magnetism, or the Ising model in transverse magnetic field, or the Hubbard model, etc. In all these models, only a subset of the degrees of freedoms is considered, just in the same spirit as for the Ising and Potts models and other lattice models of classical statistical mechanics, where all other degrees of freedom of the quantum-many body system are thought to be simply responsible for the formation of a rigid lattice, on the sites of which the few remaining degrees of freedom onto which attention is focused (e.g., the spin variables, or electrons that may "hop" from lattice site to lattice site) "live". While this reduction of the problem is clearly highly phenomenological, as it was in the case of the mentioned classical models, it still has given very valuable insight into the phenomena of condensed matter physics that can be modelled in this way, such as magnetism, ferroelectricity, and electronic conduction in systems with highly correlated electrons, which cannot at all be described by the standard reciprocal space one-electron picture. As will be also mentioned in the present book, the treatment of these problems by versions of PIMC such as Quantum Monte Carlo (QMC) "world line algorithms" and their efficient implementation (via "Loop Algorithms" complementing local updates, or cluster algorithms similar to those used for classical spin models, etc.) is an active area of research. Note also, that the techniques to sample free energies (such as multihistogram algorithms, Wang-Landau-sampling, etc., that were mentioned in the context of classical statistical mechanics) are useful here, as well as the finite size scaling techniques, if phase transitions need to be explored. For some problems involving highly correlated electrons at lattices an approximate but useful variant of QMC is the combination of the "Dynamical Mean-Field-Theory" with the local density approximation of density functional theory.

Another reduction is possible when one wishes to treat only one or a few degrees of freedom quantum-mechanically while the rest of the system can be treated still in a classical way. First pioneering studies along such lines treated the problem of electron solvation in molten salts and liquid ammonia. But, it must be noted that, when one studies the dynamics of quantum degrees of freedom coupled to a classical environment, particular care is required: This "mixed" quantum-classical dynamics has subtle features, and is still an active area of research.

While all the work alluded to above still involves model potential functions, it is desirable to have a much more systematic approach, where one combines a quantum-mechanical calculation of the total energy in the course ("on the fly") of an MD simulation. This fundamentally important approach is termed "ab initio Molecular Dynamics", often also referred to (in the specific implementation developed by Car and Parrinello) as "Car-Parrinello Molecular Dynamics" (CPMD). This scheme is based on (i) density functional theory (DFT) and (ii) the Born-Oppenheimer approximation. Because the masses of the nuclei are many orders of magnitude larger than the electron mass one can treat the positions of the nuclei as parameters for the quantum problem of the electrons and invoke the Hohenberg-Kohn theorem, which states that the energy of a quantum system can be expressed exactly in terms of a functional of the electronic density, to compute the Hellman-Feynman forces between the nuclei. The CPMD method is then based on a Lagrangian with a fictitious dynamics for the electronic degrees of freedom to generate equations of motion that can be solved by classical MD and determine the potential energy surface. This method has been extremely powerful to deal with systems where covalent chemical bonds can form and break, for instance semiconductors, oxides, silicates, etc. Although one should admit that even beyond the Born-Oppenheimer approximation further approximations are required, since the exchange-correlation part of the energy functional is not known explicitly, a serious limitation if one deals with van der Waals' like interactions, the CPMD method can be very successfully used for an extremely wide range of problems. Also extensions have been developed where one goes beyond the adiabatic limit, applying techniques related to those of the "mixed" quantum-classical dynamics, alluded to above. A particular ambitious extension involves a combination of CPMD with PIMD to treat the nuclei quantum-mechanically: such techniques are needed, for instance, to understand proton transfer in liquid water and ice. Another very interesting development is a coupled electron-ion Monte Carlo method where ions and electrons are treated quantum mechanically (the electrons being always in their ground state).

The CPMD approach is limited to a few hundred nuclei and the time scale that is accessed is in the picosecond range, while with classical MD one can either perform very long runs (i.e., up to about one microsecond) for a medium number of particles ($< 10^4$) or distinctly shorter runs for very large systems (10^6 particles or more). For many problems of interest there is the need to cover much longer scales of time, or much larger scales of length, or both (the latter is true e.g. in the case of biomolecular simulations) than is possible with CPMD. If the problem is a matter of enhancing the accessible time scales, one may combine CPMD with techniques to sample rare events that are useful already in a classical context, e.g., classical force fields can be used as bias potentials for an enhanced sampling of conformational transitions, and one can also combine CPMD with advanced techniques to sample "reaction rates" such as transition path sampling. The latter technique, and related ones are of great interest also in the context of classical statistical mechanics (e.g. to solve problems such as estimating nucleation rates at first order phase transitions, etc.), and are rapidly developing at present. They will be dealt with in the book from several points of view. In order to solve the length scale problem, one needs to develop a mixed CPMD/MD approach, where a subsystem is treated by CPMD, and coupled to a classical environment via a transition region where a proper "handshaking" between both types of simulation is provided. This is a case of multiscale simulation, a very important problem of computer simulation, since the need to combine in a coherent fashion different methods of simulation on different scales of length and time arises in many contexts. E.g., for complex materials, in particular in biological contexts (e.g., biomembranes), a molecular simulation may not be feasible due to excessive needs for computer resources, and one may have to resort to highly coarse-grained models. However, then the need arises to validate the coarse-graining by a "mapping" to an atomistic model on smaller scales, and, eventually, also to reintroduce atomistic detail in coarse-grained models ("inverse mapping"). Such techniques have found much attention in the context of polymer melts, for instance, but it is clear that the problem in general is far from solved. Again, topics belonging into this class are emphasized in the book. Clearly, the ultimate goal must be to have simulation approaches for complex materials, including biological molecules, that successfully bridge the gaps from the small scales of electronic structure calculations to the mesoscopic scales of coarse-grained models such as dissipative particle dynamics (DPD), multiparticle collision dynamics (MPCD) and other popular mesoscopic simulation methods.

While these problems, starting from the many-body Schrödinger equation, and ranging to pattern formation in driven complex fluids and chemical reactions at the biomolecule-membrane interface in aqueous solution, clearly are not yet solved and will remain at the forefront of challenges for many years to come, it must be emphasized that steady and important progress towards reaching these goals can be anticipated. Thus, we can expect that the role of computer simulation as the central and basic methodic approach of condensed matter theory will become even much more important in the future.

Introduction to Cluster Monte Carlo Algorithms

E. Luijten

Department of Materials Science and Engineering, Frederick Seitz Materials Research Laboratory, University of Illinois at Urbana-Champaign, Urbana, Illinois 61801, U.S.A.
luijten@uiuc.edu

Erik Luijten

E. Luijten: *Introduction to Cluster Monte Carlo Algorithms*, Lect. Notes Phys. **703**, 13–38 (2006)
DOI 10.1007/3-540-35273-2_1

This chapter provides an introduction to cluster Monte Carlo algorithms for classical statistical-mechanical systems. A brief review of the conventional Metropolis algorithm is given, followed by a detailed discussion of the lattice cluster algorithm developed by Swendsen and Wang and the single-cluster variant introduced by Wolff. For continuum systems, the geometric cluster algorithm of Dress and Krauth is described. It is shown how their geometric approach can be generalized to incorporate particle interactions beyond hard-core repulsions, thus forging a connection between the lattice and continuum approaches. Several illustrative examples are discussed.

1 Introduction

The Monte Carlo method is applied in wide areas of science and engineering. This chapter specifically focuses on its use in equilibrium statistical mechanics of classical systems of particles or spins. To set the stage, I first review the fundamental concepts of Monte Carlo simulation and discuss the importance sampling method introduced by Metropolis et al. However, the main emphasis in these notes lies on so-called cluster methods that have been developed over the last two decades. These are collective-update schemes that are capable of generating independent particle configurations in a very efficient manner. While originally believed to be only applicable to lattice-based systems, they now have been extended to large classes of off-lattice models. Their wide range of applicability is illustrated by means of several examples.

2 Local Monte Carlo Simulations

2.1 Importance Sampling and the Metropolis Method

It is one of the fundamental results of statistical mechanics that a thermodynamic system is described by its *partition function*,

$$Z = \sum_{\{s\}} \exp(-\beta E_s) \,, \tag{1}$$

where the sum runs over all possible states of the system, E_s denotes the energy of state s, and $\beta = 1/(k_B T)$, with k_B Boltzmann's constant and T the absolute temperature. Thermodynamic properties can be computed as ensemble averages,

$$\langle A \rangle = \frac{1}{Z} \sum_{\{s\}} A_s \exp(-\beta E_s) \,, \tag{2}$$

where $\langle A \rangle$ is the expectation value of an observable and A_s is the value of this observable if the system resides in state s.

The integrals (summations) in (1) and (2) are taken over phase space, which spans $2dN$ dimensions for N particles in a system of dimensionality d. If a conventional numerical integration method would be applied, a prohibitive computational effort would be required to obtain an acceptable accuracy. An alternative approach is *simple sampling*. The integrand is evaluated for a set of randomly chosen states ("samples"), and the mean of these individual evaluations is an estimate of the integral. While this indeed works for smoothly varying functions, the Boltzmann factor $\exp(-\beta E_s)$ is vanishingly small for most samples, making the statistical uncertainty in the integral very large. In general, this problem can be resolved via *importance sampling*, in which the samples are chosen according to a probability distribution. By preferentially sampling states that strongly contribute to the integral in (2) the variance in the estimate of this integral is greatly reduced. Specifically, we desire to sample the states with a probability distribution $\exp(-\beta E_s)/Z$.

However, even though we can compute the *relative* probability with which two specific states should occur in a set of samples, we cannot compute their *absolute* probability, since we do not know the normalization constant Z. It is the accomplishment of Metropolis et al. [1] to have found a way to calculate expectation values (2) *without* evaluating the partition function. The basic idea is to create a Markov chain of states, i.e., a sequence of states in which each state only depends on the state immediately preceding it. One starts from a configuration s_i that has a nonvanishing Boltzmann factor p_i. This is the first member of the Markov chain. From this configuration, a new *trial* configuration s_j is created, which has a Boltzmann factor p_j. The trial configuration is either accepted or rejected. If it is accepted, it is the next member of the Markov chain. If it is rejected, then the next member of the Markov chain is again s_i. This process is repeated iteratively to generate a sequence of configurations. It is emphasized that each trial configuration is created from the previous state in the Markov chain and accepted or rejected only based upon a comparison with this previous state. There is thus a *transition probability* from each state s_i to each state s_j, represented by a *transition matrix* π_{ij}. It is our goal to find a transition matrix that yields the equilibrium distribution p_j. Evidently, this matrix must satisfy the condition

$$\sum_i p_i \pi_{ij} = p_j \ . \tag{3}$$

Finding a solution π_{ij} of this equation is greatly simplified by imposing the condition of *microscopic reversibility* or *detailed balance*, i.e., on average the number of transitions from a state i to a state j is balanced by the number of transitions from state j to state i.

$$p_i \pi_{ij} = p_j \pi_{ji} \ . \tag{4}$$

Indeed, summation over all states s_i reduces (4) to (3):

$$\sum_i p_i \pi_{ij} = \sum_i p_j \pi_{ji} = p_j \sum_i \pi_{ji} = p_j \ , \tag{5}$$

where we have used that

$$\sum_i \pi_{ji} = 1 \ . \tag{6}$$

The matrix elements π_{ij} are the product of two factors, namely an *a priori probability* α_{ij} of generating a trial configuration s_j from a configuration s_i and an *acceptance probability* P_{ij} of accepting the trial configuration as the new state. The detailed balance condition can thus be written as

$$p_i \alpha_{ij} P_{ij} = p_j \alpha_{ji} P_{ji} \ . \tag{7}$$

In the simplest scheme, α_{ij} is symmetric and the condition reduces to

$$p_i P_{ij} = p_j P_{ji} \ , \tag{8}$$

which can be rewritten as

$$\frac{P_{ij}}{P_{ji}} = \exp\left[-\beta(E_j - E_i)\right] \ . \tag{9}$$

The acceptance probability is not uniquely defined by this equation. Metropolis et al. [1] proposed the solution

$$P_{ij} = \begin{cases} \exp\left[-\beta(E_j - E_i)\right] & \text{if } E_j > E_i \\ 1 & \text{if } E_j \leq E_i \end{cases} , \tag{10}$$

which is sometimes summarized as

$$P_{ij} = \min\left[\exp\left(-\beta\Delta_{ij}\right), 1\right] \ , \tag{11}$$

with $\Delta_{ij} = E_j - E_i$.

The trial configuration s_j is generated via a so-called *trial move*. For example, if we consider an assembly of N particles, then a trial move can consist of a small displacement of one particle, in a random direction. If the resulting configuration has an energy that is lower than the original configuration, the new state is always accepted. If the trial configuration has a higher energy, it is only accepted with a probability equal to the ratio of the Boltzmann factor of the new configuration and the Boltzmann factor of the original configuration. In practice, this is realized by generating a random number $0 \leq r < 1$ and accepting the trial configuration only if $r < P_{ij}$.

The expectation value of a thermodynamic property A is calculated as follows. A sequence $\{s_1, \ldots, s_M\}$ of M configurations is generated, and for each configuration s_n the property A_n is sampled. The thermodynamic average (2) is then estimated as a *simple average*,

$$\langle A \rangle \approx \frac{1}{M} \sum_{n=1}^{M} A_n \ . \tag{12}$$

2.2 Elementary Moves and Ergodicity

The *trial moves* or *elementary moves* that are used to generate a trial configuration depend on the nature of the system. A simple model for a fluid is the
above-mentioned example of an assembly of N spherical particles, confined to
a certain volume. In an elementary move, one randomly selected particle with
position $\mathbf{r}$ is displaced to a new position $\mathbf{r}' = \mathbf{r} + \delta\mathbf{r}$. The displacement $\delta\mathbf{r}$ can
be chosen as a randomly oriented vector on a sphere with radius $0 < |\delta\mathbf{r}| < \ell$.
However, it is computationally simpler to choose the new position $\mathbf{r}'$ within
a cube of linear dimension ℓ, centered around the original position $\mathbf{r}$. In either case (sphere or cube) the a priori probability α_{ij} is symmetric, i.e., the
probability to generate the trial configuration from the original configuration
is identical to the probability of the reverse process. As will be discussed in
Sect. 2.3, the parameter ℓ permits control over the efficiency of the simulation.
Monte Carlo algorithms with trial moves that involve small displacements of
individual particles are also called *local-update algorithms*.

A valid Monte Carlo scheme must not only obey detailed balance, but must
also be ergodic. This means that there is a path in phase space from every
state to every other state, via a succession of trial moves. Clearly, if the trial
states are chosen in such a manner that certain states can never be reached,
then the estimator for a thermodynamic observable can differ severely from
the correct expectation value.

2.3 Efficiency Considerations

The statistical quality of the estimate (12) depends on the number of *independent* samples in the sequence of configurations, and it is therefore the objective
of a simulation to maximize the rate at which independent configurations are
generated. If a trial configuration is generated via a small change to the previous configuration, the energy difference Δ_{ij} will typically be small and the
acceptance ratio will be large. However, many configurations may have to be
generated before an *independent* configuration results. Conversely, if a trial
configuration is generated via a big change to the previous configuration then
the sequence of configurations would decorrelate quickly, were it not that the
typical energy difference will be large and the acceptance probability thus
very small. For the elementary moves described in Sect. 2.2, the parameter
ℓ controls the maximum displacement of a particle and thus the acceptance
ratio.

It follows from these considerations that it is not always desirable to sample
the property A_n in (12) in each successive configuration, in particular not if
the calculation of A_n is computationally expensive. However, it is crucial that
the sampling takes place at a *regular* interval. A typical mistake in Monte
Carlo calculations is that a sample only is taken after a fixed number of trial
configurations have been *accepted*. This is wrong, as can also easily be seen
from the following example: At low temperatures a configuration with a low

energy is very unlikely to make a transition to a state with a higher energy. But this precisely borne out by (2): Low-energy states contribute more strongly to the expectation value and hence should be sampled more frequently than high-energy states.

In addition to variation of the maximum single-particle displacement, one may also attempt to increase the rate at which configurations evolve by moving several particles at a time. However, if the moves of these particles are independent, this is less efficient than a sequence of single-particle moves [2].

2.4 Ising Model

In order to describe the collective-update schemes that are the focus of this chapter, it is necessary to introduce the Ising model. This model is defined on a d-dimensional lattice of linear size L (a square lattice in $d = 2$ and a cubic lattice in $d = 3$) with, on each vertex of the lattice, a one-component spin of fixed magnitude that can point up or down. This system is described by the Hamiltonian,

$$\mathcal{H}_{\text{Ising}} = -J \sum_{\langle ij \rangle} s_i s_j \ . \tag{13}$$

The spins s take values ± 1. The sum runs over all pairs of nearest neighbors, which are coupled via a ferromagnetic coupling with strength $J > 0$. The Metropolis algorithm can be applied directly to this system. Local trial moves amount to the inversion of a single spin and are accepted or rejected on the basis of the change in coupling energy.

3 Lattice Cluster Algorithms

3.1 Swendsen–Wang Algorithm

In 1987, Swendsen and Wang (SW) [3] introduced a new Monte Carlo algorithm for the Ising spin model, which constituted a radical departure from the Metropolis or "single-spin flip" method used until then. Since the "recipe" is relatively straightforward, it is instructive to begin with a description of this algorithm.

Starting from a given configuration of spins, the SW algorithm proceeds as follows:

1. A "bond" is formed between every pair of nearest neighbors that are aligned, with a probability $p_{ij} = 1 - \exp(-2\beta J)$, where J is the coupling constant [cf. (13)].
2. All spins that are connected, directly or indirectly, via bonds belong to a single cluster. Thus, the bond assignment procedure divides the system into clusters of parallel spins (a so-called *cluster decomposition*). Note how the bond probability (and hence the typical cluster size) grows with

increasing coupling strength βJ (decreasing temperature). For finite βJ, $p_{ij} < 1$ and hence a cluster is generally a *subset* of all spins of a given sign – in other words, two spins of the same sign need not belong to the same cluster, even if these spins are adjacent on the lattice.

3. All spins in each cluster are flipped *collectively* with a probability $\frac{1}{2}$. I.e., for each cluster of spins a spin value ± 1 is chosen and this value is assigned to all spins that belong to the cluster.

4. All bonds are erased and the "cluster move" is complete; a new spin configuration has been created. The algorithm restarts at step (1).

Step (3) is the crucial one. It is made possible by the so-called Fortuin–Kasteleyn mapping [4, 5] of the Ising model on the random-cluster model.[1] This mathematical result essentially shows that the partition function of the Potts model can be written as a sum over all possible clusters, or "graphs," on the lattice. Consequently, all spins in a cluster (a "connected component" in the random-cluster model) are uncorrelated with all spins that belong to other clusters and can be independently assigned a new spin value. Here, we use the word *cluster* in a general sense: A cluster may also consist of a single, isolated spin. The independence of clusters also implies that the cluster-flip probability in step (3) can be chosen at will. Evidently, if this probability is very small the configuration does not change much. On the other hand, flipping the spins in *all* clusters amounts to an inversion of the entire sample and therefore does not accomplish anything. Thus, a probability halfway these two extremes is typically chosen in order to maximize the rate at which the system evolves.

One remarkable aspect of this algorithm is that it is *rejection free*.[2] Indeed, the assignment of bonds involves specific probabilities, but once the clusters have been formed each of them can be flipped independently without imposing an acceptance criterion that involves the energy change induced by such a collective spin-reversal operation. We note that the absence of an acceptance criterion does *not* imply that a cluster flip does not entail an energy difference! Indeed, there is nothing in the algorithm that would guarantee this property (which would require that the boundary of the cluster cuts through an equal number of parallel and antiparallel pairs of interacting spins). Furthermore, this peculiar property would cause the system to move over a constant-energy surface in phase space, which is certainly not what one desires for a simulation that operates in the canonical ensemble. In contrast, the cluster flips *do* result in a sequence of configurations with different energies, in such a way that they appear exactly according to the Boltzmann distribution.

The aspect that made the SW algorithm very popular is its ability to strongly suppress dynamic slowing down near a critical point. This can be

---

[1] Strictly, the Fortuin–Kasteleyn mapping applies to the q-state Potts model, but the Ising model is equivalent to a Potts model with $q = 2$.

[2] The selection of clusters that are flipped and clusters that are not flipped is considered as part of the move, and not as a "rejection."

explained by a brief digression into the field of critical phenomena. For a substance near a critical point (a continuous phase transition), the relaxation time of thermodynamic properties depends as a power law on the correlation length ξ,

$$\tau \propto \xi^z \, , \tag{14}$$

where $z \approx 2$ is the so-called dynamical critical exponent [6]. The correlation length itself diverges as a power law of the difference between the temperature T of the substance and the critical temperature T_c,

$$\xi \propto |T - T_c|^{-\nu} \, , \tag{15}$$

where ν is a positive exponent. In simulations of finite systems, e.g., a d-dimensional "hypercube" of volume L^d, the correlation length is bounded by the linear system size L. Thus, if the temperature approaches T_c, ξ grows according to (15) until it reaches a maximum value $\xi_{max} \propto L$, and for temperatures sufficiently close to the critical temperature, (14) is replaced by

$$\tau \propto L^z \, . \tag{16}$$

We thus encounter a phenomenon called *critical slowing down*. If a system becomes larger, the correlation time grows very rapidly and it becomes increasingly difficult to generate statistically independent configurations. However, the clusters created in the Swendsen–Wang algorithm have a structure that is very efficient at destroying nonlocal correlations. As a result, the dynamical critical exponent z is lowered to a much smaller value and independent configurations can be generated at a much faster rate than with a single-spin flip algorithm. This advantage only holds in the vicinity of the critical temperature, which also happens to be the most interesting point in the study of lattice spin models such as the Ising model.

3.2 Wolff or Single-Cluster Algorithm

Soon after the SW algorithm described in Sect. 3.1 had been developed, Wolff [7] introduced a so-called single-cluster variant of this algorithm. In the SW algorithm, small and large clusters are created. While the destruction of critical correlations is predominantly due to the large clusters, a considerable amount of effort is spent on constructing the smaller clusters. In Wolff's implementation, no decomposition of the entire spin configuration into clusters takes place. Instead, only a single cluster is formed, which is then always flipped. If this cluster turns out to be large, correlations are destroyed as effectively as by means of the large clusters in the SW algorithm, without the effort of creating the smaller clusters that fill up the remainder of the system. If the Wolff cluster turns out to be small, then not much is gained, but also not much computational effort is required. As a result, critical slowing down is suppressed even more strongly than in the SW algorithm, and the dynamical critical exponent z [see (16)] is even smaller.

A convenient side effect, which has certainly contributed to the popularity of the Wolff algorithm, is that it is exceedingly simple to implement. The prescription is as follows:

1. A spin i is selected at random.
2. All nearest neighbors j of this spin are added to the cluster with a probability $p_{ij} = 1 - \exp(-2\beta J)$, provided spins i and j are parallel and the bond between i and j has not been considered before.
3. Each spin j that is indeed added to the cluster is also placed on the stack. Once all neighbors of i have been considered for inclusion in the cluster, a spin is retrieved from the stack and all its neighbors are considered in turn for inclusion in the cluster as well, following step (2).
4. Steps (2) and (3) are repeated iteratively until the stack is empty.
5. Once the cluster has been completed, all spins that belong to the cluster are inverted.

Again, this is a rejection-free algorithm, in the sense that the cluster is always flipped. Just as in the SW algorithm, the cluster-construction process is probabilistic, but the probabilities p_{ij} involve energies of individual spin pairs in contrast with an acceptance criterion that involves the *total* energy change induced by a cluster flip. The implementation can be simplified by a small trick: In step (2), each spin j that is added to the cluster can immediately be inverted. This guarantees that a spin is never added twice. Step (5) can then be eliminated.[3]

3.3 Cluster Algorithms for Other Lattice Models

The algorithms described here are not restricted to the Ising model, but can be applied to a number of other problems. (i) For multicomponent spins (such as the XY or planar model), Wolff [7] replaced the spin-inversion operation by a reflection operation in which only the component of a spin is reversed that is orthogonal to a randomly oriented plane. For each cluster, a new orientation of the plane is chosen. (ii) The original Fortuin–Kasteleyn mapping is valid for q-state Potts models in which each lattice site corresponds to a variable that can take q different, equivalent states. In the SW algorithm, after the cluster decomposition, one of these q values is assigned with probability $1/q$ to each cluster and all variables in the cluster take this new value. (iii) The Fortuin–Kasteleyn mapping can also be applied to systems in which each spin interacts not only with its nearest neighbors, but also with other spins [8]. In particular, the coupling strength can be different for different spin pairs, leading to a probability p_{ij} that is, e.g., dependent on the separation between spins i and j. While this permits the direct formulation of a cluster Monte

---

[3] To formulate this more precisely, also spin i should be inverted upon selection in step (1) and in step (2) the spins j must be considered for addition to the cluster if their sign is the same as the *original* sign of spin i.

Carlo algorithm for these systems, the cluster addition probability becomes very small if the interaction is long-ranged. As shown by Luijten and Blöte [9], this can be resolved by reformulating the algorithm, allowing the study of spin systems with medium- [10] and long-range [11,12] ferromagnetic interactions, as well as dipolar [13] interactions.

4 Cluster Algorithms for Continuum Systems

4.1 Geometric Cluster Algorithm for Hard-Sphere Mixtures

The advantages brought by the cluster algorithms described in Sect. 3, in particular the suppression of critical slowing down, made it a widely pursued goal to generalize the SW and Wolff algorithms to fluid systems in which particles are not confined to lattice sites but can take arbitrary positions in continuum space. Unfortunately, the absence of a lattice structure breaks a fundamental symmetry, rendering such attempts largely unsuccessful. An Ising model can be interpreted as a so-called *lattice gas*, where a spin +1 corresponds to a particle and a spin −1 corresponds to an empty site. Accordingly, a spin-inversion operation corresponds to a particle being inserted into or removed from the system. This "particle–hole symmetry" is absent in off-lattice (continuum) systems. While a particle in a fluid configuration can straightforwardly be deleted, there is no unambiguous prescription on how to transform empty space into a particle. More precisely, in the lattice cluster algorithms the operation performed on every spin is self-inverse. This requirement is not fulfilled for off-lattice fluids.

Independently of these efforts, in 1995 Dress and Krauth [14] proposed a method to efficiently generate particle configurations for a hard-sphere liquid. In this system, particles are represented by impenetrable spheres (or disks, in the two-dimensional variant) that have no interaction as long as they do not overlap. Because of the *hard-core repulsion*, a Monte Carlo algorithm involving local moves is relatively inefficient, since any move that generates a particle overlap is rejected. Instead, the *geometric cluster algorithm* (GCA) [14] is designed to avoid such overlaps while generating a new configuration, by proceeding as follows (cf. Fig. 1).

1. In a given configuration C of particles, a "pivot" is chosen at random.
2. A configuration $\tilde{C}$ is now generated by carrying out a point reflection for all particles in C with respect to the pivot.[4]
3. The configuration C and its transformed counterpart $\tilde{C}$ are superimposed, which leads to groups of overlapping particles. The groups generally come

---

[4] In the original algorithm, a π rotation with respect to the pivot was performed for all particles. In the two-dimensional example of Fig. 1 this is equivalent to a point reflection, but the reflection is more suitable for generalization to higher dimensions.

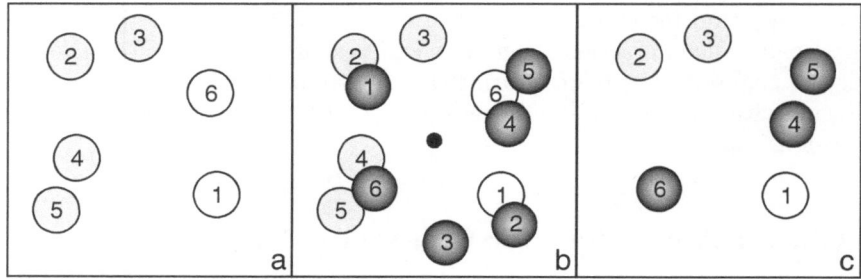

Fig. 1. Illustration of the geometric cluster algorithm for hard disks [14]. (**a**) Original configuration. (**b**) A new configuration (*shaded circles*) is created by means of a point reflection of all particles with respect to a randomly chosen pivot point (*small filled disk*). The superposition of the original and the new configuration leads to groups of overlapping particles. In this example, there are three *pairs* of groups ({1, 2}, {3}, {4, 5, 6}). Each pair is denoted a *cluster*. The particles in any one of these clusters can be point-reflected with respect to the pivot without affecting the other two clusters. This can be used to carry out the point reflection for every cluster with a pre-set probability. (**c**) Final configuration that results if, starting from the original configuration, only the particles in the third cluster {4, 5, 6} are point-reflected. This approach guarantees that every generated configuration will be free of overlaps. Note that the pivot will generally not be placed in the center of the cell, and that the periodic boundary conditions indeed permit any position. Reprinted figure with permission from [16]. Copyright 2005 by the American Physical Society

in pairs, except possibly for a single group that is symmetric with respect to the pivot. Each pair is denoted a "cluster."

4. For each cluster, all particles can be exchanged between C and $\tilde{C}$ without affecting particles belonging to other clusters. This exchange is performed for each cluster independently with a probability $\frac{1}{2}$. Thus, if the superposition of C and $\tilde{C}$ is decomposed into N clusters, there are 2^N possible new configurations. The configurations that are actually realized are denoted C' and $\tilde{C}'$, i.e., the original configuration C is transformed into C' and its point-reflected counterpart $\tilde{C}$ is transformed into $\tilde{C}'$.

5. The configuration $\tilde{C}'$ is discarded and C' is the new configuration, serving as the starting point for the next iteration of the algorithm. Note that a new pivot is chosen in every iteration.

Observe that periodic boundary conditions must be employed, such that an arbitrary placement of the pivot is possible. Other self-inverse operations are permissible, such as a reflection in a plane [15], in which case various orientations of the plane must be chosen in order to satisfy ergodicity.

Comparison to the lattice cluster algorithms of Sect. 3 shows that the SW and Wolff algorithms operate in the grand-canonical ensemble, in which the cluster moves do not conserve the magnetization (or the number of particles, in the lattice-gas interpretation), whereas the geometric cluster algorithm operates in the canonical ensemble. Nevertheless, this prescription bears a remarkable resemblance to the SW algorithm. The original configuration is decomposed into clusters by exploiting *a symmetry operation that leaves the Hamiltonian invariant if applied to the entire configuration;* in the SW algorithm this is the spin-inversion operation and in the geometric cluster algorithm it is a geometric symmetry operation. Subsequently, a new configuration is created by moving each cluster independently with a certain probability.

This approach is very general. For example, it is not restricted to monodisperse systems, and Krauth and co-workers have applied it successfully to binary [17] and polydisperse [18] mixtures. Indeed, conventional simulations of size-asymmetric mixtures typically suffer from jamming problems, in which a very large fraction of all trial moves is rejected because of particle overlaps. In the geometric cluster algorithm particles are moved in a nonlocal fashion, yet overlaps are avoided.

The most important limitation of the GCA is the fact that the average cluster size increases very rapidly for systems with a density that exceeds the percolation threshold of the combined system containing the superposition of the configurations C and $\tilde{C}$. Once the clusters span the entire system, the algorithm is clearly no longer ergodic.

In order to emphasize the analogy with the lattice cluster algorithms, we can also formulate a single-cluster (Wolff) variant of the geometric cluster algorithm [15, 19].

1. In a given configuration C, a "pivot" is chosen at random.
2. A particle i is selected as the first particle that belongs to the cluster. This particle is moved via a point reflection with respect to the pivot. In its new position, the particle is referred to as i'.
3. The point reflection in step 2 is repeated *iteratively* for each particle j that overlaps with i'. Thus, if the (moved) particle j' overlaps with another particle k, particle k is moved as well. Note that all translations involve the same pivot.
4. Once all overlaps have been resolved, the cluster move is complete.

As in the SW-like prescription, a new pivot is chosen for each cluster that is constructed.

4.2 Generalized Geometric Cluster Algorithm for Interacting Particles

The geometric cluster algorithm described in the previous section is formulated for particles that interact via hard-core repulsions only. Clearly, in order to make this approach widely applicable, a generalization to other types of pair potentials must be found. Thus, Dress and Krauth [14] suggested to impose a Metropolis-type acceptance criterion, based upon the energy difference induced by the cluster move. Indeed, if a pair potential consists of a hard-core contribution supplemented by an attractive or repulsive tail, such as a

Yukawa potential, the cluster-construction procedure takes into account the excluded-volume contribution, guaranteeing that no overlaps are generated, and the acceptance criterion takes into account the tail of the interactions. For "soft-core" potentials, such as a Lennard-Jones interaction, the situation becomes already somewhat more complicated, since an arbitrary excluded-volume distance must be chosen in the cluster construction. As the algorithm will not generate configurations in which the separation (center-to-center distance) between a pair of particles is less than this distance (i.e., the particle "diameter," in the case of monodisperse systems), it must be set to a value that is smaller than any separation that would typically occur. It is important to recognize that both for hard-core particles with an additional tail and for soft-core particles the clusters are now constructed on the basis of only a part of the pair potential, and the evaluation of a part of the energy change resulting from a cluster move is deferred until the acceptance step. As a result, the computational efficiency can decrease significantly, since – similar to the situation for regular multiple-particle moves – rejection of a cluster move is quite likely. To make things worse, every rejection leads to a considerable waste of computational effort spent on the construction of the cluster and the evaluation of the corresponding energy change. Nevertheless, this approach certainly works in principle, as shown by a study of Yukawa mixtures with moderate size asymmetry (diameter ratio ≤ 5) [20].

However, an entirely different approach is possible, by carrying the analogy with the lattice cluster algorithms further. The probability p_{ij} to add a spin j (which is neighboring a spin i) to the cluster in the SW algorithm can be phrased in terms of the corresponding energy difference. Two different situations can be discerned that lead to a change in the relative energy Δ_{ij}^{SW} between a spin i that belongs to the cluster and a spin j that does not yet belong to the cluster. If i and j are initially *antiparallel*, j will never be added to the cluster and only spin i will be inverted, yielding an energy change $\Delta_{ij}^{\text{SW}} = -2J < 0$ that occurs with probability unity. If i and j are initially *parallel* and j is not added to the cluster, the resulting change in the pair energy equals $\Delta_{ij}^{\text{SW}} = +2J > 0$. This occurs with a probability $\exp(-2\beta J) < 1$. These two situations can be summarized as

$$1 - p_{ij} = \min\left[\exp(-\beta\Delta_{ij}^{\text{SW}}), 1\right] , \qquad (17)$$

so that the probability of adding spin j to the cluster can be written as $p_{ij} = \max[1 - \exp(-\beta\Delta_{ij}^{\text{SW}}), 0]$. The GCA, although formulated in continuum space rather than on a lattice, can now be interpreted as special situation in which either $\Delta_{ij} = 0$ (after reflection of particle i, there is no overlap between particles i and j), leading to $p_{ij} = 0$, or $\Delta_{ij} = \infty$ (after point reflection, particle i overlaps with particle j), leading to $p_{ij} = 1$.

A generalization of the GCA to general pair potentials then follows in a natural way [19]. All interactions are treated in a unified manner, so that there is no technical distinction between attractive and repulsive interactions or between hard-core and soft-core potentials. This *generalized GCA* is most

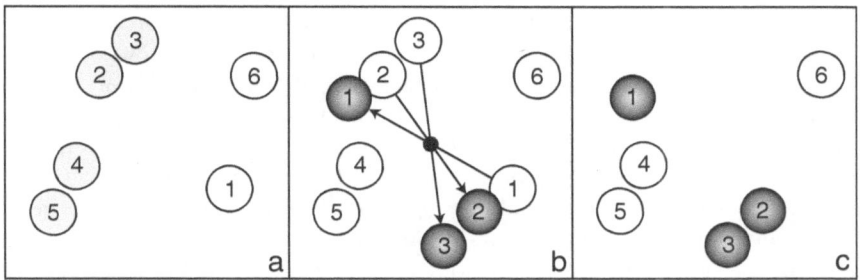

Fig. 2. Two-dimensional illustration of the interacting geometric cluster algorithm. Like in Fig. 1, open and shaded disks denote the particles before and after the geometric operation, respectively, and the small disk denotes the pivot. However, in the *generalized* GCA a single cluster is constructed, to which particles are added with an interaction-dependent probability. (**a**) Original configuration. (**b**) A cluster is constructed as follows. Particle 1 is point-reflected with respect to the pivot. If, in its new position, it has a repulsive interaction with particle 2, the latter has a certain probability to be point-reflected as well, with respect to the same pivot. Assuming an attractive interaction between particles 2 and 3, particle 3 is translated as well, but only with a certain probability. If particles 4–6 are not affected by these point reflections, the cluster construction terminates. (**c**) The new configuration consists of particles 1–3 in their new positions and particles 4–6 in the original positions. A new pivot is chosen and the procedure is repeated. Reprinted figure with permission from [16]. Copyright 2005 by the American Physical Society

easily described as a combination of the single-cluster methods formulated in Sect. 3.2 and Sect. 4.1. We assume a general pair potential $V_{ij}(\mathbf{r}_{ij})$ that does not have to be identical for all pairs (i, j) (see Fig. 2).

1. In a given configuration C, a "pivot" is chosen at random.
2. A particle i at position $\mathbf{r}_i$ is selected as the first particle that belongs to the cluster. This particle is moved via a point reflection with respect to the pivot. In its new position, the particle is referred to as i', at position $\mathbf{r}'_i$.
3. Each particle j that interacts with i or i' is now considered for addition to the cluster. A particle j that interacts with i both in its old and in its new position is nevertheless treated once. Unlike the first particle, particle j is point-reflected with respect to the pivot only with a probability $p_{ij} = \max[1 - \exp(-\beta\Delta_{ij}), 0]$, where $\Delta_{ij} = V(|\mathbf{r}'_i - \mathbf{r}_j|) - V(|\mathbf{r}_i - \mathbf{r}_j|)$.
4. Each particle j that is indeed added to the cluster (i.e., moved) is also placed on the stack. Once all particles interacting with i or i' have been considered, a particle is retrieved from the stack and all its neighbors that are not yet part of the cluster are considered in turn for inclusion in the cluster as well, following step 3.
5. Steps 3 and 4 are repeated iteratively until the stack is empty. The cluster move is now complete.

If a particle interacts with multiple other particles that have been added to the cluster, it can thus be considered multiple times for inclusion. However, once it has been added to the cluster, it cannot be removed. This is an important point in practice, since particles undergo a point reflection already during the cluster construction process (and thus need to be tagged, in order to prevent them from being returned to their original position by a second point reflection). A crucial aspect is that the probability p_{ij} *only* depends on the change in *pair energy* between i and j that occurs if particle i is point-reflected with respect to the pivot, but particle j is *not* added to the cluster (and hence not point-reflected). This happens with a probability $1 - p_{ij} = \min[\exp(-\beta\Delta_{ij}), 1]$, just as we found for the SW algorithm in (17). The similarity of this probability to the Metropolis acceptance criterion is deceptive (and merely reflects the fact that both algorithms aim to generate configurations according to the Boltzmann distribution), since Δ_{ij} does not represent the *total* energy change resulting from the translation of particle i. Instead, other energy changes are taken into account via the iterative nature of the algorithm.

It is interesting to note that the GCA can also be applied to lattice-based models. This was first done for the Ising model by Heringa and Blöte [21, 22], who also devised a way to take into account the nearest-neighbor interactions between spins already during the cluster construction. While this lattice model can obviously be simulated by the SW and Wolff algorithms, their approach permits simulation in the constant-magnetization ensemble. Since the geometric operations employed map the spin lattice onto itself, excluded-volume conditions are satisfied automatically: Every spin move amounts to an *exchange* of spins. For every spin pair (i, i') that is exchanged, each of its nearest-neighbor pairs (k, k') is exchanged with a probability that depends on the change in pair energy, $\Delta = (E_{ik} + E_{i'k'}) - (E_{ik'} + E_{i'k})$. This procedure is then again performed iteratively for the neighbors of all spin pairs that are exchanged. This is similar to the generalized GCA discussed above, although, in the absence of a lattice, particles are added to the cluster on an individual basis rather than in pairs.

In order to establish the correctness of the generalized GCA, we need to prove ergodicity as well as detailed balance. The ergodicity of this algorithm follows from the fact that there is a nonvanishing probability that a cluster consists of only one particle, which can be moved over an arbitrarily small distance, since the location of the pivot is chosen at random. This obviously requires that not all particles are part of the cluster, a condition that is violated at high packing fractions and, depending on the nature of the interactions, at very strong coupling strengths.

Detailed balance is proven as follows. We consider a configuration X that is transformed into a configuration Y by means of a cluster move. All particles included in the cluster maintain their relative separation; as noted above, an energy change arises if a particle is *not* included in the cluster, but interacts with a particle that does belong to the cluster. Following Wolff [7] we denote each of these interactions as a "broken bond." A broken bond k that corresponds to an energy change Δ_k occurs with a probability $1 - p_k = 1$ if $\Delta_k \leq 0$ and a probability $1 - p_k = \exp(-\beta\Delta_k)$ if $\Delta_k > 0$. The formation of an entire cluster corresponds to the breaking of a set $\{k\}$ of bonds, which has a probability P. This set is comprised of the subset $\{l\}$ of broken bonds l that lead to an increase in pair energy and the subset $\{m\}$ of broken bonds that lead to a decrease in pair energy, such that

$$P = \prod_k (1 - p_k) = \exp\left[-\beta \sum_l \Delta_l\right] . \tag{18}$$

The transition probability from configuration X to configuration Y is proportional to the cluster formation probability,

$$T(X \rightarrow Y) = C \exp\left[-\beta \sum_l \Delta_l\right] , \tag{19}$$

where the factor C accounts for the fact that various arrangements of bonds within the cluster ("internal bonds") correspond to the same set of broken bonds. In addition, it incorporates the probability of choosing a particular pivot and a specific particle as the starting point for the cluster.

If we now consider the reverse transition $Y \rightarrow X$, we observe that this again involves the set $\{k\}$, but all the energy differences change sign compared to the forward move. Consequently, the subset $\{l\}$ in (19) is replaced by its complement $\{m\}$ and the transition probability is given by

$$T(Y \rightarrow X) = C \exp\left[+\beta \sum_m \Delta_m\right] , \tag{20}$$

where the factor C is identical to the prefactor in (19). Since we require the geometric operation to be self-inverse we thus find that the cluster move satisfies detailed balance at an acceptance ratio of unity,

$$\frac{T(X \rightarrow Y)}{T(Y \rightarrow X)} = \frac{\exp\left[-\beta \sum_l \Delta_l\right]}{\exp\left[+\beta \sum_m \Delta_m\right]} = \exp\left[-\beta \sum_k \Delta_k\right]$$

$$= \exp\left[-\beta(E_Y - E_X)\right] = \frac{\exp(-\beta E_Y)}{\exp(-\beta E_X)} , \tag{21}$$

where E_X and E_Y are the internal energies of configurations X and Y, respectively. That is, the ratio of the forward and reverse transition probabilities is equal to the inverse ratio of the Boltzmann factors, so that we indeed have created a rejection-free algorithm. This is obscured to some extent by the fact that in our prescription the cluster is moved while it is being constructed, similar to the Wolff algorithm in Sect. 3.2. The central point, however, is that the construction solely involves single-particle energies, whereas a Metropolis-type approach only evaluates the total energy change induced by a multi-particle move and then frequently rejects this move. By contrast, the GCA avoids large energy differences by incorporating "offending" particles into the cluster with a high probability (i.e., strong bonds are unlikely to be broken).

4.3 Generalized Geometric Cluster Algorithm: Full Cluster Decomposition

It is instructive to also formulate a SW version of the generalized GCA, based upon the single-cluster version described in the previous section. This demonstrates that the generalized GCA is a true off-lattice counterpart of the cluster algorithms of Sect. 3. Furthermore, it is of conceptual interest, as this algorithm decomposes a continuum fluid configuration into *stochastically independent clusters*. This implies an interesting and remarkable analogy with the Ising model. As observed by Coniglio and Klein [23] for the two-dimensional Ising model at its critical point, the clusters created according to the prescription in Sect. 3 are just the so-called "Fisher droplets" [24]. Indeed, these "Coniglio–Klein clusters" are implied by the Fortuin–Kasteleyn mapping of the Potts model onto the random-cluster model [5], which in turn constitutes the basis for the Swendsen–Wang approach [3]. The clusters generated by the GCA do not have an immediate physical interpretation, as they typically consist of two spatially disconnected parts. However, just like the Ising clusters can be inverted at random, each cluster of fluid particles can be moved independently with respect to the remainder of the system. As such, the generalized GCA can be viewed as a continuum version of the Fortuin–Kasteleyn mapping.

The cluster decomposition of a configuration proceeds as follows. First, a cluster is constructed according to the single-cluster algorithm of Sect. 4.2, with the exception that the cluster is only *identified*; particles belonging to the cluster are marked but not actually moved. The pivot employed will also be used for the construction of all subsequent clusters in this decomposition. These subsequent clusters are built just like the first cluster, except that particles that are already part of an earlier cluster will never be considered for a new cluster. Once each particle is part of exactly one cluster the decomposition is completed. Like in the SW algorithm, every cluster can then be moved (i.e., all particles belonging to it are translated via a point reflection) independently, e.g., with a probability f. Despite the fact that all clusters except the first are built in a restricted fashion, each individual cluster is constructed according to the rules of the Wolff formulation of Sect. 4.2. The exclusion of particles that are already part of another cluster simply corresponds to the fact that every bond should be considered only once. If a bond is broken during the construction of an earlier cluster it should not be re-established during the construction of a subsequent cluster. The cluster decomposition thus obtained is not unique, as it depends on the placement of the pivot and the choice of the first particle. Evidently, this also holds for the SW algorithm.

In order to establish that this prescription is a true equivalent of the SW algorithm, we prove that each cluster can be moved (reflected) independently while preserving detailed balance. If only a single cluster is actually moved, this essentially corresponds to the Wolff version of the GCA, since each cluster is built according to the GCA prescription. The same holds true if several clusters are moved and no interactions are present between particles that belong to different clusters (the hard-sphere algorithm is a particular realization of this situation). If two or more clusters are moved and *broken* bonds exist between these clusters, i.e., a nonvanishing interaction exists between particles that belong to disparate (moving) clusters, then the shared broken bonds are actually preserved and the proof of detailed balance provided in the previous section no longer applies in its original form. However, since these bonds are identical in the forward and the reverse move, the corresponding factors cancel out. This is illustrated for the situation of two clusters whose construction involves, respectively, two sets of broken bonds $\{k_1\}$ and $\{k_2\}$. Each set comprises bonds l ($\{l_1\}$ and $\{l_2\}$, respectively) that lead to an *increase* in pair energy and bonds m ($\{m_1\}$ and $\{m_2\}$, respectively) that lead to a *decrease* in pair energy. We further subdivide these sets into *external* bonds that connect cluster 1 or 2 with the remainder of the system and *joint* bonds that connect cluster 1 with cluster 2. Accordingly, the probability of creating cluster 1 is given by

$$C_1 \prod_{i\in\{k_1\}} (1-p_i) = C_1 \prod_{i\in\{l_1\}} (1-p_i) = C_1 \prod_{i\in\{l_1^{\text{ext}}\}} (1-p_i) \prod_{j\in\{l_1^{\text{joint}}\}} (1-p_j) . \quad (22)$$

Upon construction of the first cluster, the creation of the second cluster has a probability

$$C_2 \prod_{i\in\{l_2^{\text{ext}}\}} (1 - p_i) , \quad (23)$$

since all joint bonds in $\{l_2^{\text{joint}}\} = \{l_1^{\text{joint}}\}$ already have been broken. The factors C_1 and C_2 refer to the probability of realizing a particular arrangement of internal bonds in clusters 1 and 2, respectively (cf. Sect. 4.2). Hence, the total transition probability of moving *both* clusters is given by

$$T_{12}(X \rightarrow Y) = C_1 C_2 \exp\left[-\beta \sum_{i\in\{l_1^{\text{ext}}\}} \Delta_i - \beta \sum_{j\in\{l_2^{\text{ext}}\}} \Delta_j - \beta \sum_{n\in\{l_1^{\text{joint}}\}} \Delta_n\right] .$$

$$(24)$$

In the reverse move, the energy differences for all external broken bonds have changed sign, but the energy differences for the joint bonds connecting cluster 1 and 2 are the same as in the forward move. Thus, cluster 1 is created with probability

$$C_1 \prod_{i\in\{m_1^{\text{ext}}\}} (1-\bar{p}_i) \prod_{j\in\{l_1^{\text{joint}}\}} (1-p_j) = C_1 \prod_{i\in\{m_1^{\text{ext}}\}} \exp[+\beta\Delta_i] \prod_{j\in\{l_1^{\text{joint}}\}} \exp[-\beta\Delta_j] ,$$

$$(25)$$

where the $\bar{p}_i$ reflects the sign change of the energy differences compared to the forward move and the product over the external bonds involves the complement of the set $\{l_1^{\text{ext}}\}$. The creation probability for the second cluster is

$$C_2 \prod_{i\in\{m_2^{\text{ext}}\}} (1 - \bar{p}_i) = C_2 \prod_{i\in\{m_2^{\text{ext}}\}} \exp[+\beta\Delta_i] \qquad (26)$$

and the total transition probability for the reverse move is

$$T_{12}(Y \to X) = C_1 C_2 \exp\left[+\beta \sum_{i\in\{m_1^{\text{ext}}\}} \Delta_i + \beta \sum_{j\in\{m_2^{\text{ext}}\}} \Delta_j - \beta \sum_{n\in\{l_1^{\text{joint}}\}} \Delta_n\right] .$$

$$(27)$$

Accordingly, detailed balance is still fulfilled with an acceptance ratio of unity,

$$\frac{T_{12}(X \to Y)}{T_{12}(Y \to X)} = \exp\left[-\beta \sum_{i\in\{k_1^{\text{ext}}\}} \Delta_i - \beta \sum_{j\in\{k_2^{\text{ext}}\}} \Delta_j\right] = \exp\left[-\beta(E_Y - E_X)\right] ,$$

$$(28)$$

in which $\{k_1^{\text{ext}}\} = \{l_1^{\text{ext}}\}\cup\{m_1^{\text{ext}}\}$ and $\{k_2^{\text{ext}}\} = \{l_2^{\text{ext}}\}\cup\{m_2^{\text{ext}}\}$ and E_X and E_Y refer to the internal energy of the system before and after the move, respectively. This treatment applies to any simultaneous move of clusters, so that *each cluster in the decomposition indeed can be moved independently* without violating detailed balance. This completes the proof of the multiple-cluster version of the GCA. It is noteworthy that the probabilities for breaking joint bonds in the forward and reverse moves cancel only because the probability in the cluster construction factorizes into individual probabilities.

4.4 Implementation Issues

The actual implementation of the generalized GCA involves a variety of issues. The point reflection with respect to the pivot requires careful consideration of the periodic boundary conditions. Furthermore, as mentioned above, particles that have been translated via a point reflection must not be translated again within the same cluster move, and particles that interact with a given cluster particle both before and after the translation of that cluster particle must be considered only once, on the basis of the difference in pair potential. One way to account for all interacting pairs in an efficient manner is the use of the cell index method [25]. For mixtures with large size asymmetries (the situation where the generalized GCA excels), it is natural to set up different cell structures, with cell lengths based upon the cutoffs of the various particle interactions. For example, in the case of a binary mixture of two species with very different sizes and cutoff radii ($r_{\text{cut}}^{\text{large}}$ and $r_{\text{cut}}^{\text{small}}$, respectively), the use of a single cell structure with a cell size that is determined by the large particles would be highly inefficient for the smaller particles. Thus, two cell structures are constructed in this case (with cell sizes l_{large} and l_{small}, respectively) and each particle is stored in the appropriate cell of the structure belonging to its species, and incorporated in the corresponding linked list, following the standard approach [25]. However, in order to efficiently deal with interactions between unlike species (which have a cutoff $r_{\text{cut}}^{\text{ls}}$), a mapping between the two cell structures is required. If all small particles that interact with a given large particle must be located, one proceeds as follows. First, the small cell **c** is identified in which the center of the large particle resides. Subsequently, the interacting particles are located by scanning over all small cells within a cubic box with linear size $2r_{\text{cut}}^{\text{ls}}$, centered around **c**. This set of cells is predetermined at the beginning of a run and their indices are stored in an array. Each set contains approximately $N_{\text{cell}} = (2r_{\text{cut}}^{\text{ls}}/l_{\text{small}})^3$ members. In an efficient implementation, l_{small} is not much larger than $r_{\text{cut}}^{\text{small}}$, which for short-range interactions is of the order of the size of a small particle. Likewise, $r_{\text{cut}}^{\text{ls}}$ is typically of the order of the size of the large particle, so that $N_{\text{cell}} = \mathcal{O}(\alpha^3)$, where $\alpha > 1$ denotes the size asymmetry between the two species. Since N_{cell} indices must be stored for each large cell, the memory requirements become very large for cases with large size asymmetry, such as the suspension of colloids and nanoparticles (size asymmetry $\alpha = 100$) studied in [26].

4.5 Illustration 1: Efficiency of the Generalized Geometric Cluster Algorithm

Probably the most important feature of the generalized GCA for practical applications is the efficiency with which it generates uncorrelated configurations for size-asymmetric mixtures. This performance directly derives from the non-local character of the point reflection employed. In general, the translation of a single particle over large distances has a very low acceptance ratio in conventional Monte Carlo simulations, except in extremely dilute conditions. The situation only deteriorates for multiple-particle moves, unless the particles involved in the move are selected in a very specific manner. The generalized GCA makes nonlocal collective moves possible, without any negative consequences in the acceptance ratio. The resulting efficiency gain is illustrated by means of an example taken from [19], namely a simple binary mixture containing 150 large particles of size σ_{22}, at fixed volume fraction $\phi_2 = 0.1$, and N_1 small particles, also at fixed volume fraction $\phi_1 = 0.1$. The efficiency is determined through the autocorrelation time, as a function of size asymmetry. As the size σ_{11} of these small particles is varied from $\sigma_{22}/2$ to $\sigma_{22}/15$ (i.e., the size ratio $\alpha = \sigma_{22}/\sigma_{11}$ is increased from 2 to 15), their number increases from $N_1 = 1\,200$ to $506\,250$.

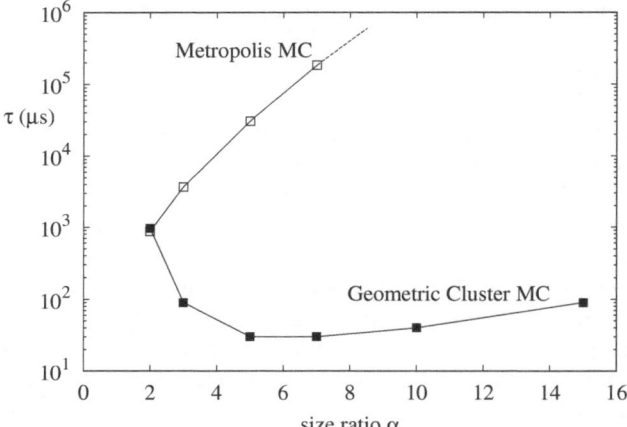

Fig. 3. Efficiency comparison between a conventional local update algorithm (*open symbols*) and the generalized geometric cluster algorithm (*closed symbols*), for a binary mixture (see text) with size ratio α. Whereas the autocorrelation time per particle (expressed in μs of CPU time per particle move) rapidly increases with size ratio, the GCA features only a weak dependence on α. Reprinted figure with permission from [19]. Copyright 2004 by the American Physical Society

Pairs of small particles and pairs involving a large and a small particle act like hard spheres. However, in order to prevent depletion-driven aggregation of the large particles [27], a short-ranged Yukawa repulsion is introduced,

$$
U_{22}(r) = \begin{cases} +\infty & r \leq \sigma_{22} \\ \frac{\sigma_{22}}{r}\varepsilon \exp[-\kappa(r - \sigma_{22})] & r > \sigma_{22} \end{cases} , \tag{29}
$$

where $\beta\varepsilon = 3.0$ and the screening length $\kappa^{-1} = \sigma_{11}$. In the simulation, the exponential tail is cut off at $3\sigma_{22}$.

The additional Yukawa interactions also lead to a fluctuating internal energy $E(t)$ that makes it possible to determine the rate at which the large (and slower) particles decorrelate. We consider the integrated autocorrelation time τ obtained from the energy autocorrelation function [28],

$$
C(t) = \frac{\langle E(0)E(t)\rangle - \langle E(0)\rangle^2}{\langle E(0)^2\rangle - \langle E(0)\rangle^2} , \tag{30}
$$

and compare τ for a conventional (Metropolis) MC algorithm and the generalized GCA, see Fig. 3. In order to avoid arbitrariness resulting from the computational cost involved with a single sweep or the construction of a cluster, we assume that both methodologies have been programmed in an efficient manner and express τ in actual CPU time. Furthermore, τ is normalized by the total number of particles in the system, to account for the variation in N_1 as the size ratio α is increased. The autocorrelation time for the conventional

MC calculations, τ_{MC}, rapidly increases with increasing α, because the large particles tend to get trapped by the small particles. Indeed, already for $\alpha > 7$ it is not feasible to obtain an accurate estimate for τ_{MC}. By contrast, τ_{GCA} exhibits a very different dependence on α. At $\alpha = 2$ both algorithms require virtually identical simulation time, which establishes that the GCA does not involve considerable overhead compared to standard algorithms (if any, it is mitigated by the fact that all moves are accepted). Upon increase of α, τ_{GCA} initially *decreases* until it starts to increase weakly. The nonmonotonic variation of τ_{GCA} results from the changing ratio N_2/N_1 which causes the cluster composition to vary with α. The main points to note are: (i) the GCA greatly suppresses the autocorrelation time, $\tau_{GCA} \ll \tau_{MC}$ for $\alpha > 2$, with an efficiency increase that amounts to more than three orders of magnitude already for $\alpha = 7$; (ii) the increase of the autocorrelation time with α is much slower for the GCA than for a local-move MC algorithm, making the GCA increasingly advantageous with increasing size asymmetry.

4.6 Illustration 2: Temperature and Cluster Size

The cluster size clearly has a crucial influence on the performance of the GCA. If a cluster contains more than 50% of all particles, an equivalent change to the system could have been made by moving its complement; unfortunately it is unclear how to determine this complement without constructing the cluster. Nevertheless, it is found that the algorithm can operate in a comparatively efficient manner for average relative cluster sizes as large as 90% or more. Once the total packing fraction of the system exceeds a certain value, the original hard-core GCA breaks down because each cluster occupies the entire system. The same phenomenon occurs in the generalized GCA, but in addition the cluster size can saturate because of strong interactions. Thus, the maximum accessible volume fraction depends on a considerable number of parameters, including the range of the potentials and the temperature. For multi-component mixtures, size asymmetry and relative abundance of the components are of importance as well, and the situation can be complicated further by the presence of competing interactions.

As an illustration, we consider the cluster-size distribution for a monodisperse Lennard-Jones fluid (particle diameter σ, interaction cut-off 2.5σ) at a rather arbitrary density $0.16\sigma^{-3}$, for a range of temperatures, see Fig. 4. Whereas the distribution is a monotonously decreasing function of cluster size at high temperatures, it becomes bimodal at temperatures around 25% above the critical temperature. The bimodal form is indicative of the formation of large clusters.

It turns out to be possible to influence the cluster-size distribution by placing the pivot in a biased manner. Rather than first choosing the pivot location, a particle is selected that will become the first member of the cluster. Subsequently, the pivot is placed at random within a cubic box of linear size δ, centered around the position of this particle. By decreasing δ, the

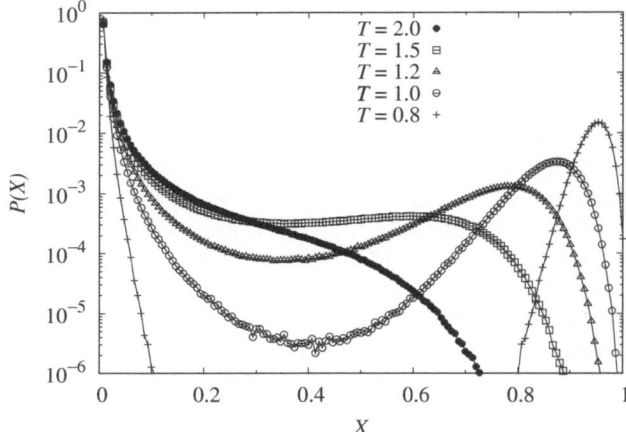

Fig. 4. Cluster-size distributions as a function of relative cluster size X, for a monodisperse Lennard-Jones fluid. The number density $\rho = 0.16\sigma^{-3}$ is set to 50% of the critical density. For low temperatures, the cluster-size distribution becomes bimodal. At higher temperatures, it decreases monotonically. All temperatures are indicated in units of ε/k_B. Reprinted figure with permission from [16]. Copyright 2005 by the American Physical Society

displacement of the first particle is decreased, as well as the number of other particles affected by this displacement. As a consequence, the average cluster size decreases, and higher volume fractions can be reached. Ultimately, the cluster size will still occupy the entire system (making the algorithm no longer ergodic), but it has been found that the maximum accessible volume fraction can be increased from approximately 0.23 to a value close to 0.34. This value indeed corresponds to the percolation threshold for hard spheres. Note that the proof of detailed balance is not affected by this modification.

Summary and Conclusions

In this chapter, I have presented a detailed discussion of cluster Monte Carlo algorithms for off-lattice systems. In order to emphasize the efficiency of these algorithms as well as their connection to methods developed earlier, I have first introduced the basic ingredients of a conventional (Metropolis-type) Monte Carlo algorithm. Subsequently, the Swendsen–Wang and Wolff algorithms for Ising and q-state Potts models have been discussed, which are striking because of their rejection-free character. The off-lattice cluster algorithms, which are based upon geometric symmetry operations, are a direct generalization of these lattice cluster methods, and illustrate that rejection-free algorithms are by no means rare exceptions, but can actually be phrased for large classes of systems.

Acknowledgments

This work is supported by the National Science Foundation under CAREER Award No. DMR-0346914 and by the U.S. Department of Energy, Division of Materials Sciences under Award No. DEFG02-91ER45439, through the Frederick Seitz Materials Research Laboratory at the University of Illinois at Urbana-Champaign.

References

1. N. Metropolis, A. W. Rosenbluth, M. N. Rosenbluth, A. H. Teller, and E. Teller (1953) Equation of state calculations by fast computing machines. *J. Chem. Phys.* **21**, pp. 1087–1092
2. D. Frenkel and B. Smit (2002) *Understanding Molecular Simulation.* San Diego: Academic, 2nd ed.
3. R. H. Swendsen and J.-S. Wang (1987) Nonuniversal critical dynamics in Monte Carlo simulations. *Phys. Rev. Lett.* **58**(2), pp. 86–88
4. P. W. Kasteleyn and C. M. Fortuin (1969) Phase transitions in lattice systems with random local properties. *J. Phys. Soc. Jpn. Suppl.* **26s**, pp. 11–14
5. C. M. Fortuin and P. W. Kasteleyn (1972) On the random-cluster model. I. Introduction and relation to other models. *Physica* **57**, pp. 536–564
6. S.-K. Ma (1976) *Modern Theory of Critical Phenomena.* Redwood City, Calif.: Addison-Wesley
7. U. Wolff (1989) Collective Monte Carlo updating for spin systems. *Phys. Rev. Lett.* **62**(4), pp. 361–364
8. R. J. Baxter, S. B. Kelland, and F. Y. Wu (1976) Equivalence of the Potts model or Whitney polynomial with an ice-type model. *J. Phys. A* **9**, pp. 397–406
9. E. Luijten and H. W. J. Blöte (1995) Monte Carlo method for spin models with long-range interactions. *Int. J. Mod. Phys. C* **6**, pp. 359–370
10. E. Luijten, H. W. J. Blöte, and K. Binder (1996) Crossover scaling in two dimensions. *Phys. Rev. E* **54**, pp. 4626–4636
11. E. Luijten and H. W. J. Blöte (1997) Classical critical behavior of spin models with long-range interactions. *Phys. Rev. B* **56**, pp. 8945–8958
12. E. Luijten and H. Meßingfeld (2001) Criticality in one dimension with inverse square-law potentials. *Phys. Rev. Lett.* **86**, pp. 5305–5308
13. A. Hucht (2002) On the symmetry of universal finite-size scaling functions in anisotropic systems. *J. Phys. A* **35**, pp. L481–L487
14. C. Dress and W. Krauth (1995) Cluster algorithm for hard spheres and related systems. *J. Phys. A* **28**, pp. L597–L601
15. J. R. Heringa and H. W. J. Blöte (1996) The simple-cubic lattice gas with nearest-neighbour exclusion: Ising universality. *Physica A* **232**, pp. 369–374
16. J. Liu and E. Luijten (2005) Generalized geometric cluster algorithm for fluid simulation. *Phys. Rev. E* **71**, p. 066701
17. A. Buhot and W. Krauth (1998) Numerical solution of hard-core mixtures. *Phys. Rev. Lett.* **80**, pp. 3787–3790
18. L. Santen and W. Krauth (2000) Absence of thermodynamic phase transition in a model glass former. *Nature* **405**, pp. 550–551

19. J. Liu and E. Luijten (2004) Rejection-free geometric cluster algorithm for complex fluids. *Phys. Rev. Lett.* **92**(3), p. 035504
20. J. G. Malherbe and S. Amokrane (1999) Asymmetric mixture of hard particles with Yukawa attraction between unlike ones: a cluster algorithm simulation study. *Mol. Phys.* **97**, pp. 677–683
21. J. R. Heringa and H. W. J. Blöte (1998) Geometric cluster Monte Carlo simulation. *Phys. Rev. E* **57**(5), pp. 4976–4978
22. J. R. Heringa and H. W. J. Blöte (1998) Geometric symmetries and cluster simulations. *Physica A* **254**, pp. 156–163
23. A. Coniglio and W. Klein (1980) Clusters and Ising critical droplets: a renormalisation group approach. *J. Phys. A* **13**, pp. 2775–2780
24. M. E. Fisher (1967) The theory of condensation and the critical point. *Physics* **3**, pp. 255–283
25. M. P. Allen and D. J. Tildesley (1987) *Computer Simulation of Liquids.* Oxford: Clarendon
26. J. Liu and E. Luijten (2004) Stabilization of colloidal suspensions by means of highly charged nanoparticles. *Phys. Rev. Lett.* **93**, p. 247802
27. S. Asakura and F. Oosawa (1954) On interaction between two bodies immersed in a solution of macromolecules. *J. Chem. Phys.* **22**, pp. 1255–1256
28. K. Binder and E. Luijten (2001) Monte Carlo tests of renormalization-group predictions for critical phenomena in Ising models. *Phys. Rep.* **344**, pp. 179–253

Generic Sampling Strategies
for Monte Carlo Simulation of Phase Behaviour

N.B. Wilding

Department of Physics, University of Bath, Bath, BA2 7AY, U.K.
N.B.Wilding@Bath.ac.uk

Nigel B. Wilding

N.B. Wilding: *Generic Sampling Strategies for Monte Carlo Simulation of Phase Behaviour*,
Lect. Notes Phys. **703**, 39–66 (2006)
DOI 10.1007/3-540-35273-2_2   © Springer-Verlag Berlin Heidelberg 2006

1 Introduction

The phenomenon of phase behavior is the organization of many-body systems into forms which reflect the interplay between constraints imposed macroscopically (through the prevailing external conditions) and microscopically (through the interactions between the elementary constituents). In this article we focus on *generic* computational strategies needed to address the problems of phase behavior, or more specifically the task of mapping equilibrium phase boundaries.

The physical context we shall explore will not extend beyond the structural organization of the elementary phases (liquid, vapor, crystalline) of matter, although the strategies are much more widely applicable than this. The problem may then be formulated (and in principle solved) entirely within the framework of the equilibrium statistical mechanics of the competing phases. Since there is no *need* to deal authentically with the phase transformation process itself, the natural computational tool becomes Monte Carlo (MC) rather than MD. The strategic advantage of this choice is the range of ways in which MC algorithms may be engineered to move around configuration space.

Rather than attempt to perform an exhaustive description of the large number of contemporary MC methods for dealing with the problem of phase behaviour, we shall instead attempt to survey the rather more limited portfolio of inter-phase pathways and sampling techniques on which most methods are based. A few illustrative examples are provided, taken mainly from the author's own work.

2 Basic Equipment

2.1 Statistical Mechanics

Consider a system of structure-less, classical particles, characterized macroscopically by a set of thermodynamic coordinates (such as the temperature T) and microscopically by a set of model parameters which prescribe their interactions. The two sets of parameters play a strategically similar role; it is therefore convenient to denote them, collectively, by a single label, c (for "conditions or "constraints" or "control parameters" in thermodynamic-and-model space).

The statistical behavior of the system is rooted in the position coordinates $r_i, i = 1, N$ which are the principal members of a set of *generalized coordinates* $\{q\}$ locating the system in its *configuration space*. In some instances it is advantageous to work with ensembles in which particle number N or system volume V is free to fluctuate; the coordinate set $\{q\}$ is then extended accordingly (to include N or V) and the control parameters c extended to include the corresponding fields (chemical potential μ or pressure P).

The statistical behavior of interest is encapsulated in the equilibrium probability density function $P_0(\{q\}|c)$. This PDF is determined by an appropriate dimensionless [1] configurational energy $\mathcal{E}(\{q\}, c)$. The relationship takes the form

$$P_0(\{q\}|c) = \frac{1}{Z(c)} e^{-\mathcal{E}(\{q\}, c)} \tag{1}$$

where the normalizing prefactor (the partition function) is defined by

$$Z(c) \equiv \int \prod_i dq_i e^{-\mathcal{E}(\{q\}, c)} \tag{2}$$

The different phases which the system displays will in general be distinguished by the values of some *order parameter*. Thus, for example, the density serves to distinguish a liquid from a vapor; a structure factor distinguishes a liquid from a crystalline solid. The order parameter, M, allows us to associate with each phase, α, a corresponding portion $\{q\}_\alpha$ of $\{q\}$-space:

$$\{q\} \in \{q\}_\alpha \quad \text{iff} \quad M(\{q\}) \in [M]_\alpha \tag{3}$$

where $[M]_\alpha$ is the set of order parameter values consistent with phase α. The partitioning of $\{q\}$-space into distinct regions is the key feature of the core problem.

The equilibrium properties of a particular phase α follow from the conditional counterpart of (1)

$$P_0(\{q\}|\alpha, c) = \begin{cases} \frac{1}{Z_\alpha(c)} e^{-\mathcal{E}(\{q\}, c)} & \{q\} \in \{q\}_\alpha \\ 0 & \text{otherwise} \end{cases} \tag{4}$$

with

$$Z_\alpha(c) \equiv \int \prod_i dq_i \Delta_\alpha[M(\{q\})] e^{-\mathcal{E}(\{q\}, c)} \equiv e^{-\mathcal{F}_\alpha(c)} \tag{5}$$

The last equation defines $\mathcal{F}_\alpha(c)$, the *free energy* of phase α, while

$$\Delta_\alpha[M] \equiv \begin{cases} 1 & M \in [M]_\alpha \\ 0 & \text{otherwise} \end{cases} \tag{6}$$

so that the integral is effectively confined to the set of configurations $\{q\}_\alpha$ associated with phase α.

Now referring back to (1) we may write

$$Z_\alpha(c) = \int \prod_i dq_i \Delta_\alpha[M] e^{-\mathcal{E}(\{q\}, c)} = Z(c) \int \prod_i dq_i \Delta_\alpha[M] P_0(\{q\}|c)$$

The a priori probability of phase α may thus be related to its free energy by

$$P_0(\alpha|c) \equiv \int \prod_i dq_i \Delta_\alpha[M] P_0(\{q\}|c) = \frac{Z_\alpha(c)}{Z(c)} = \frac{e^{-\mathcal{F}_\alpha(c)}}{Z(c)} \tag{7}$$

For two phases, α and $\tilde{\alpha}$ say, it then follows that

$$\Delta \mathcal{F}_{\alpha\tilde{\alpha}}(c) \equiv \mathcal{F}_\alpha(c) - \mathcal{F}_{\tilde{\alpha}}(c) = \ln \frac{Z_{\tilde{\alpha}}(c)}{Z_\alpha(c)} = \ln \frac{P_0(\tilde{\alpha}|c)}{P_0(\alpha|c)} \tag{8}$$

This is a key equation in several respects: it is conceptually helpful; it is cautionary; and it is suggestive, strategically.

At a conceptual level, (8) provides a helpful link between the languages of thermodynamics and statistical mechanics. According to the familiar mantra of thermodynamics the favored phase will be that of *minimal free energy*; from a statistical mechanics perspective the favored phase is the one of *maximal probability*, given the probability partitioning implied by (1).

In general, however, (1) presupposes ergodicity on the space $\{q\}$. The framework can thus be trusted to tell us what we will "see" for some given c (the "favored phase") only to the extent that appropriate kinetic pathways exist to allow sampling (ultimately, *comparison*) of the distinct regions of configuration space associated with the different phases. In the context of laboratory experiments on real systems the relevant pathways typically entail the nucleation and growth of droplets of one phase embedded in the other; the associated time scales are long; and (8) will be relevant only if the measurements extend over correspondingly long times. The fact that they frequently do not is signaled in the phenomena of metastability and hysteresis.

Finally, (8) helps to shape strategic thinking on how to broach the problem computationally. It reminds us that what is relevant here is the *difference* between free energies of two competing phases and that this free energy difference is a *ratio* of the a priori probabilities of the two phases. It implies that the phase boundary may be identified as the locus of points of equal a priori probability of the two phases, and that such points are in principle identifiable through the condition that the order parameter distribution will have equal integrated weights (areas) in the two phases. The pathways by which our simulated system passes between the two regions of configuration space associated with the two phases will, additionally, play a strategically crucial role. While, for the reasons just discussed, the details of those pathways are essential to the *physical applicability* of (8) they are *irrelevant* to the *values* of the quantities it defines; we are thus free to engineer whatever pathways we may wish. These considerations lead one naturally to the Monte Carlo toolkit.

2.2 Tools: Elements of Monte Carlo

The MC method [2–4] generates a sequence (Markov chain) of configurations in $\{q\}$-space. The procedure can be constructed to ensure that, in the "long-enough-term", configurations will appear in that chain with *any* probability density, $P_S(\{q\})$ (the "S" stands for "sampling") we care to nominate. The key requirement (it is not strictly necessary [5]; and-as we shall see – it is not always sufficient) is that the transitions, from one configuration $\{q\}$ to another $\{q'\}$, should respect the detailed balance condition

$$P_S(\{q\})P_S(\{q\} \to \{q'\}) = P_S(\{q'\})P_S(\{q'\} \to \{q\}) \tag{9}$$

where $P_S(\{q\} \to \{q'\})$ is the transition probability, the probability density of configuration $\{q'\}$ at Markov chain step $t+1$ given configuration $\{q\}$ at time t. (We have added a subscript to emphasize that its form is circumscribed by the choice of sampling density, through (9).) MC transitions satisfying this constraint are realized in a two-stage process. In the first stage, one generates a trial configuration $\{q'\} = T\{q\}$, where T is some generally stochastic selection procedure; the probability density of a trial configuration $\{q'\}$ given $\{q\}$ is of the form

$$P_T(\{q'\} \mid \{q\}) = \langle\delta(\{q'\} - T\{q\})\rangle_T \tag{10}$$

where $\langle\cdot\rangle_T$ represents an average with respect to the stochastic variables implicit in the procedure T. In the second stage the "trial" configuration is accepted (the system "moves" from $\{q\}$ to $\{q'\}$ in configuration space) with probability P_A, and is otherwise rejected (so the system "stays" at $\{q\}$); the form of the acceptance probability is prescribed by our choices for P_S and P_T since

$$P_S(\{q\} \to \{q'\}) = P_T(\{q'\} \mid \{q\})P_A(\{q\} \to \{q'\})$$

It is then easy to verify that the detailed balance condition (9) is satisfied, if the acceptance probability is chosen as

$$P_A(\{q\} \to \{q'\}) = \min\left\{1, \frac{P_S(\{q'\})P_T(\{q\} \mid \{q'\})}{P_S(\{q\})P_T(\{q'\} \mid \{q\})}\right\} \tag{11}$$

Suppose that, in this way, we build a Markov chain comprising a total of t_T steps; we set aside the first t_E configurations visited; we denote by $\{q\}^{(t)}$ $(t = 1, \ldots t_U)$ the configurations associated with the subsequent $t_U \equiv t_T - t_E$ steps. The promise on the MC package is that the expectation value $\langle Q \rangle_S$ of some observable $Q = Q(\{q\})$ *defined* by

$$\langle Q \rangle_S = \int \prod_i dq_i P_S(\{q\})Q(\{q\}) \tag{12}$$

may be *estimated* by the sample average

$$\langle Q \rangle_S \overset{eb}{=} \frac{1}{t_U} \sum_{t=1}^{t_U} Q(\{q\}^{(t)}) \tag{13}$$

Now we must consider the choices of P_S and P_T. Tailoring those choices to whatever task one has in hand provides potentially limitless opportunity for ingenuity. But at this point we consider only the simplest possibilities. The sampling distribution $P_S(\{q\})$ is chosen to be the appropriate *equilibrium* distribution $P_0(\{q\} \mid c)$ (1) so that the configurations visited are representative of a "real" system, even though their sequence is not an authentic representation of the "real" dynamics. We shall refer to this form of sampling distribution as *canonical*. The trial-coordinate selection procedure $\mathcal{T}$ is chosen to comprise some *small* change of *one* coordinate; the change is chosen to be small enough to guarantee a reasonable acceptance probability (11), but no smaller, or the Markov chain will wander unnecessarily slowly through the configuration space. We shall refer to this form of selection procedure as *local*. For such schemes (and sometimes for others) the selection probability density typically has the symmetry:

$$P_{\mathcal{T}}(\{q\} \to \{q'\}) = P_{\mathcal{T}}(\{q'\} \to \{q\}) \tag{14}$$

With these choices, (11) becomes

$$P_{\mathcal{A}}(\{q\} \to \{q'\}) = \min\{1, \exp[-\mathcal{E}(\{q'\}, c) - \mathcal{E}(\{q\}, c)]\} \tag{15}$$

which defines the Metropolis acceptance function.

These choices are not only the simplest, they are also the most frequent: the local-canonical strategy is the staple Monte Carlo method. However, it is in general not equal to the task of determining phase boundaries. The reason for this is that a simulation initiated in one phase will generally be confined on simulation timescales to the microstates of that phase and will not make excursion to the other phase. The observations thus provide no basis for assigning a value to the relative probabilities of the two phases, and thus of estimating the location of the phase boundary. This reluctance to explore outside the phase in which the system is initialized is a symptom of the high probability barrier separating one phase from the other, the origins of which are traceable to the surface tension (free energy cost) of forming *mixed* phase configurations.

To determine the relative stability of two phases under conditions c thus requires a MC framework that, in some sense, does *more* than sample the equilibrium configurations appropriate to the (two) c-macrostates. We have seen where there is room for maneuver – in the choices we make in regard to $P_{\mathcal{S}}$ and $P_{\mathcal{T}}$. The possibilities inherent in the latter are intuitively obvious: better to find ways of bounding or leaping through configuration space than be limited to the shuffle of local-updating. The fact that we have flexibility in regard to the choice of sampling distribution is perhaps less obvious so it is worth recording the simple result which shows us that we do.

Let $P_{\mathcal{S}}$ and $P_{\mathcal{S}'}$ be two arbitrary distributions of the coordinates $\{q\}$. Then the expectation values of some arbitrary observable Q with respect to the two distributions are formally related by the identity

$$\langle Q \rangle_{\mathcal{S}'} = \int \prod_i dq_i P_{\mathcal{S}'}(\{q\}) Q(\{q\}) = \int \prod_i dq_i P_{\mathcal{S}}(\{q\}) Q(\{q\}) \frac{P_{\mathcal{S}'}(\{q\})}{P_{\mathcal{S}}(\{q\})}$$

$$= \left\langle \frac{P_{\mathcal{S}'}}{P_{\mathcal{S}}} Q \right\rangle_{\mathcal{S}} \tag{16}$$

Thus, in particular, we can –in principle– determine *canonical* expectation values from an ensemble defined by an *arbitrary* sampling distribution through the relationship

$$\langle Q \rangle_0 = \left\langle \frac{P_0}{P_S} Q \right\rangle_S \tag{17}$$

We do not *have* to make the choice $P_S = P_0$. In many cases the "obvious" choice $P_S = P_0$, though not strictly optimal, is adequate. But for some observables the choice of a canonical sampling distribution is *so* "sub-optimal" as to be useless [6]. The problem of determining phase coexistence is a case in point.

3 Paths

There are many ways of motivating, constructing and describing the kind of MC sampling strategy we need; the core idea we shall appeal to here to structure our discussion is that of a *path*.

3.1 Meaning and Specification

For our purposes a *path* comprises a sequence of contiguous macrostates, $\mathcal{C}_1, \mathcal{C}_2 ... \mathcal{C}_\Omega \equiv \{\mathcal{C}\}$ [7]. By "contiguous" we mean that each adjacent pair in the sequence ($\mathcal{C}_j, \mathcal{C}_{j+1}$ say) have some configurations in common (or that a configuration of one lies arbitrarily close to a configuration of the other). A path thus comprises a quasi-continuous band through configuration space.

The physical quantities that distinguish the macrostates from one another will fall into one or other of two categories, which we shall loosely refer to as *fields*, and *macrovariables*. In the former category we include thermodynamic fields, model parameters and, indeed, the conjugate [8] of any "macrovariable". By "macrovariable" [9] we mean any collective property, aggregating contributions from all or large numbers of the constituent particles, free to fluctuate in the chosen ensemble, but in general sharply-prescribed, in accordance with the Central Limit Theorem. Note that we do not restrict ourselves to quantities that feature on the map of thermodynamics, nor to the parameter space of the physical system itself: with simulation tools at our disposal there are limitless varieties of parameters to vary and properties to observe.

3.2 Generic Routes

It is perhaps not surprising that the extended MC framework needed to solve the phase-equilibrium problem entails exploration of a path that *links* the macrostates of the two competing phases, for the desired physical conditions $\mathcal{C}$. The generic choices here are distinguished by the way in which the path is *routed* in relation to the two-phase region which separates the macrostates of the two phases, and which confers on them their (at least meta-) stability. Figure 1 depicts four conceptually different possibilities.

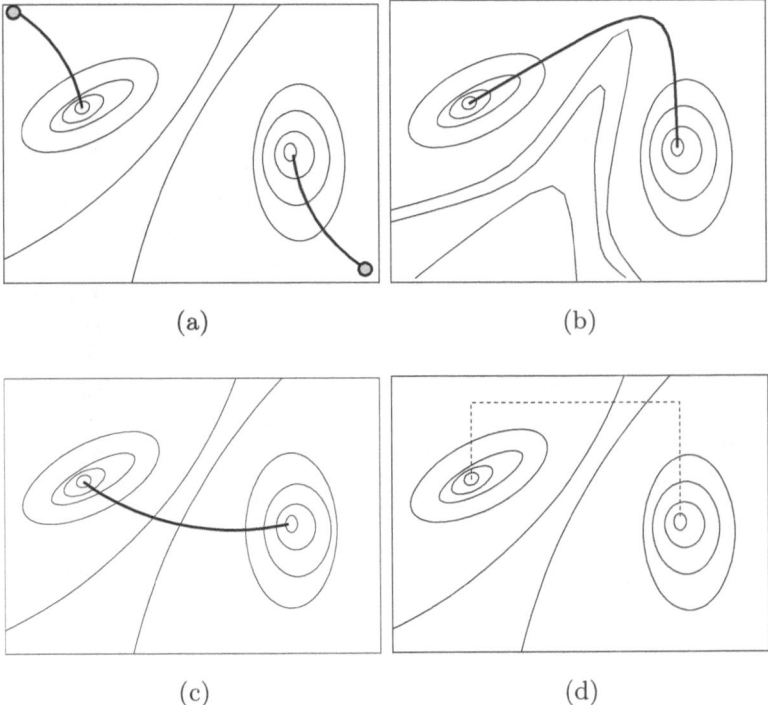

Fig. 1. Schematic representation of the four conceptually-different paths (the heavy lines) one may utilize to attack the phase-coexistence problem. Each figure depicts a configuration space spanned by two macroscopic properties (such as energy, density ...); the contours link macrostates of equal probability, for some given conditions c (such as temperature, pressure ...). The two mountain-tops locate the equilibrium macrostates associated with the two competing phases, under these conditions. They are separated by a probability-ravine (free-energy-barrier). In case (**a**) the path comprises two disjoint sections confined to each of the two phases and terminating in appropriate reference macrostates. In (**b**) the path skirts the ravine. In (**c**) it passes through the ravine. In (**d**) it leaps the ravine

First (Fig. 1(a)) the route may comprise two distinct sections, neither encroaching into the two-phase region, and each terminating in a *reference* macrostate. By *reference* macrostate we mean one whose partition function (and thus free energy) is already known – on the basis of exact calculation or previous measurement. The information accessible through MC study of the two sections of such a path has to be combined with the established properties of the reference macrostates to provide the desired link between the two equilibrium macrostates of interest. This is the traditional strategy for addressing the phase-coexistence problem.

In the case of a liquid-gas phase boundary it is possible to choose a path (Fig. 1(b)) that links the macrostates of the two phases *without* passing through the two-phase region: one simply has to sail around the critical point at which the phase-boundary terminates. The dependence on the existence of an adjacent critical point limits the applicability of this strategy.

The next option is a route which negotiates the probability ravine (free-energy barrier) presented by the two-phase region (Fig. 1(c)). The extended sampling tools we discuss in the next section are essential here; with their refinement in recent years this route has become increasingly attractive.

If either of the phases involved is a *solid*, there are *additional* reasons to avoid the ravine, over and above the low canonical probability of the macrostates that lie there. It is possible to do so. The necessary strategy is depicted in Fig. 1(d). As in (a) the path comprises two segments, each of which lies within a single phase region. But in contrast to (a) the special macrostates to which these segments lead are not of the traditional reference form: the defining characteristic of *these* macrostates is that they they should act as the ends of a "wormhole" along which the system may be transported by a collective Monte Carlo move, from one to phase to the other. In this case extended sampling methods are used to locate the wormhole ends.

4 Generic Sampling Strategies

The task of exploring (sampling) the macrostates which form a path can be accomplished in a number of conceptually different ways; we identify them here in a rudimentary way.

4.1 Serial Sampling

The most obvious way of gathering information about the set of macrostates $\{C\}$ is to use canonical Boltzmann sampling to explore them one at a time, with a sampling distribution set to

$$P_S(\{q\}) = P_0(\{q\}|C_j) \ \ (j = 1 \ldots \Omega) \tag{18}$$

in turn. One must then combine the information generated in these Ω independent simulations. The traditional approach to the problem ("integration methods" [10]) employs this strategy and is often applied to sample a path in the manner depicted in Fig. 1(a); its basic merit is that it is simple.

4.2 Parallel Sampling

Instead of exploring the path-macrostates serially we may choose to explore them in parallel. In this framework we simulate a set of Ω replicas of the physical system, with the j-th member chosen to realize the conditions C_j.

The sampling distribution in the composite configuration space spanned by $\{q\}^{(1)} \ldots \{q\}^{(\Omega)}$ is

$$P_{\mathcal{S}}(\{q\}^{(1)} \ldots \{q\}^{(\Omega)}) = \prod_{j=1}^{\Omega} P_0(\{q\}^{(j)}|\mathcal{C}_j) \qquad (19)$$

This is not simply a way of exploiting the availability of parallel computing architectures to treat Ω tasks at the same time; more significantly it provides a way of breaking out of the straight-jacket of local-update algorithms. The composite ensemble can be updated through interchanges of the coordinate sets associated with adjacent macrostates (j and $j+1$ say) to give updated coordinates

$$\{q'\}^{(j)} = \{q\}^{(j+1)} \quad \text{and} \quad \{q'\}^{(j+1)} = \{q\}^{(j)} \qquad (20)$$

The chosen sampling distribution (19) is realized if such configuration-exchanges are accepted with the appropriate probability (11, 15) reflecting the change incurred in the total energy of the composite system. This change is nominally "macroscopic" (scales with the size of the system) and so the acceptance probability will remain workably large only if the parameters of adjacent macrostates are chosen sufficiently close to one another. In practice this means utilizing a number Ω of macrostates that is proportional to $\sqrt{N}$ [11].

Interchanging the configurations of adjacent replicas is one instance (see Sect. 7.3 for another) of a *global-update* in which all the coordinates evolve simultaneously. The pay-off, potentially, is a more rapid evolution around coordinate space – more formally a stronger mixing of the Markov chain. This is of course a general desideratum of any MC framework. Thus it is not surprising that algorithms of this kind have been independently devised in a wide variety of disciplines, applied in a correspondingly wide set of contexts ... and given a whole set of different names [11]. These include Metropolis Coupled Chain [12], Exchange Monte Carlo [13], and Parallel Tempering [14].

4.3 Extended Sampling

We will use the term *extended sampling* (ES) to refer to an algorithm that allows exploration of a region of configuration space which is "extended" with respect to the range spanned by canonical Boltzmann sampling – specifically, one which assembles the statistical properties of our set of macrostates $\mathcal{C}_1 \ldots \mathcal{C}_\Omega$ within a *single* simulation. Again it is straightforward to write down the generic form of a sampling distribution that will achieve this end; we need only a "superposition" of the canonical sampling distributions for the set of macrostates.

$$P_{\mathcal{S}}(\{q\}) = W_0 \sum_{j=1}^{\Omega} P_0(\{q\}|\mathcal{C}_j) \qquad (21)$$

where W_0 is a normalization constant. The superficial similarity between this form and those prescribed in (18) and (19) camouflages a crucial difference. Each of the distributions $P_0(\{q\}|\mathcal{C}_j)$ involves a normalization constant identifiable as (1, 5)

$$w_j = Z(\mathcal{C}_j)^{-1} \tag{22}$$

We do not need to know the set of normalization constants $\{w\}$ to implement *serial* sampling (18) since each features *only* as the normalization constant for a *sampling* distribution, which the MC framework does not require. Nor do we need these constants in implementing *parallel* sampling (19) since in this case they feature (through their product) only in the one overall normalization constant for the sampling distribution. But (21) is different. Writing it out more explicitly

$$P_{\mathcal{S}}(\{q\}) = W_0 \sum_{j=1}^{\Omega} w_j e^{-\mathcal{E}(\{q\}, \mathcal{C}_j)} \tag{23}$$

we see that the weights $\{w\}$ control the relative contributions which the macrostates make to the sampling distribution. While we are in principle at liberty to choose whatever mixture we please (we do not *have* to make the assignment prescribed by (22)) it should be clear intuitively that the choice should confer *roughly* equal probabilities on each macrostate, so that all are well-sampled. It is not hard to see that the weight-assignment made in (22) is in fact what we need to fulfill this requirement. Evidently, to make *extended* sampling work we *do* need to "know" the weights $w_j = Z(\mathcal{C}_j)^{-1}$. There is an element of circularity here which needs to be recognized. Our prime objective is to determine the (relative) configurational weights of *two* macrostates (those associated with two different phases, under the same physical conditions [15]); to do so (somehow or other – we haven't yet said how) by extended sampling requires knowledge of the configurational weights of a *whole path's-worth* of macrostates. There *is* progress here nevertheless. While the two macrostates of interest are remote from one another, the path (by construction) comprises macrostates which are contiguous; it is relatively easy to determine the relative weights of pairs of contiguous macrostates, and thence the relative weights of all in the set. In effect the extended sampling framework allows us to replace one hard problem with a large number of somewhat easier problems.

The machinery needed to turn this general strategy into a practical method ("building the extended sampling distribution" or "determining the macrostate weights') has evolved over the years from a process of trial and error to algorithms that are systematic and to some extent self-monitoring. The workings of the machinery is more interesting than it might sound; we will discuss some aspects of what is involved in Sects. 7.2 and 7.3. But we relegate more technical discussion (focused on recent advances) to Sect. 5. Here we continue with a broader brush.

It would be hard to write a definitive account of the development of extended sampling methods; we will not attempt to do so. The seminal ideas are probably correctly attributed to Torrie and Valleau [16] who coined the terminology *umbrella sampling*. The huge literature of subsequent advances and rediscoveries may be rationalized a little by dividing it into two, according to how the macrostates to be weighted are defined.

If the macrostates are defined by a set of values $[\lambda]$ of some generalized "field" λ, the sampling distribution is of the form

$$P_S(\{q\}) = W_0 \sum_{j=1}^{\Omega} w_j e^{-\mathcal{E}(\{q\}, \lambda_j)} \tag{24}$$

Extended sampling strategies utilizing this kind of representation feature in the literature with a variety of titles: *expanded ensemble* [17], *simulated tempering* [18], *temperature scaling* [19].

On the other hand, if the macrostates are defined on some "macrovariable" M the sampling distribution is of the form

$$P_S(\{q\}) = W_0 \sum_{j=1}^{\Omega} w_j e^{-\mathcal{E}(\{q\})} \Delta_j[M] \tag{25}$$

where
$$\Delta_j[M] \equiv \begin{cases} 1 & M \in \text{range associated with } \mathcal{C}_j \\ 0 & \text{otherwise} \end{cases} \tag{26}$$

Realizations of this formalism go under the names *adaptive umbrella sampling* [20] and the *multicanonical ensemble* introduced by Berg and co-workers [21]. It seems right to attribute the recent revival in interest in extended sampling to the latter work.

In Sects. 7.2 and 7.3 we shall see that extended sampling strategies provide a rich variety of ways of tackling the phase coexistence problem, including the distinctive problems arising when one of the phases is of solid form. Before considering these applications, however, we must address the matter of how to obtain a suitable form of the extended sampling distribution.

5 Building Extended Sampling Distributions

In contrast to canonical sampling distributions whose form can be *written down* (1) the Extended Sampling (ES) distributions discussed in Sect. 4.3 have to be *built*. There is a large literature devoted to the building techniques, extending back at least as far as reference [16]. We restrict our attention to relatively recent developments (those that seem to be reflected in current practices); and we shall focus on those aspects which are most relevant to ES distributions facilitating two-phase sampling.

In the broad-brush classification scheme offered in Sect. 4.3 the domain of an ES distribution may be prescribed by a range of values of one or more *fields*, or one or more *macrovariables*. We shall focus on the latter representation which seems simpler to manage. The generic task then is to construct a ("multicanonical") sampling distribution which will visit all (equal-sized) intervals within a chosen range of the nominated macrovariable(s) with roughly equal probability: the multicanonical distribution of the macrovariable(s) is essentially flat over the chosen range.

In formulating a strategy for building such a distribution, most authors have chosen to consider the particular case in which the macrovariable space is one-dimensional, and is spanned by the configurational *energy*, E [22]. The choice is motivated by the fact that a distribution that is multicanonical in E samples configurations typical of a range of *temperatures*, providing access to the simple (reference-state) behavior that often sets in at high or low temperatures. For the purposes of two-phase sampling we typically need to track a path defined on some macrovariable other than the energy (ideally, in *addition* to it: we will come back to this). The hallmarks of a "good" choice are that in some region of the chosen variable (inevitably one with intrinsically low equilibrium probability) the system may pass (has a workably-large chance of passing) from one phase to the other. In discussing the key issues, then, we shall have in mind this kind of quantity; we shall continue to refer to it as an order parameter, and denote it by M.

One can easily identify the generic structure of the sampling distribution we require. It must be of the form

$$P_S(\{q\}) \doteq \frac{P_0(\{q\} \mid c)}{\hat{P}_0(M(\{q\}))} \tag{27}$$

Here $\hat{P}_0(M)$ is an *estimate* of the true canonical M-distribution. Appealing to the sampling identity (17) the resulting M-distribution is

$$P_S(M) = \langle \delta[M - M(\{q\})]\rangle_S \doteq \left\langle \frac{P_S}{P_0} \delta[M - M(\{q\})] \right\rangle_0 \doteq \frac{P_0(M \mid c)}{\hat{P}_0(M)} \tag{28}$$

and is multicanonical (flat) to the extent that our estimate of the canonical M-distribution is a good one.

We can also immediately write down the prescription for generating the ensemble of configurations defined by the chosen sampling distribution. We need a simple MC procedure with acceptance probability ((11), with the presumption of (14))

$$P_A(\{q\} \to \{q'\}) = \min\left\{1, \frac{P_S(\{q'\})}{P_S(\{q\})}\right\} \tag{29}$$

In turning this skeleton framework into a working technique one must make choices in regard to three key issues:

1. How to *parameterize* the estimator $\hat{P}_0$.
2. What *statistics* of the ensemble to use to guide the update of $\hat{P}_0$.
3. What *algorithm* to use in updating $\hat{P}_0$.

The second and third issues are the ones of real substance; the issue of parameterization is important only because the proliferation of different choices that have been made here may give the impression that there are more techniques available than is actually the case. That proliferation is due, in some measure, to the preoccupation with building ES distributions for the *energy*, E. There are as many "natural parameterizations" here as there are ways in which E appears in canonical sampling. Thus Berg and Neuhaus [21] employ an E-dependent *effective temperature*; Lee [23] utilizes a *microcanonical entropy* function; Wang and Landau [24] focus on a *density of states* function. Given our concern with macrovariables other than E the most appropriate parameterization of the sampling distribution here is through a *multicanonical weight function* $\hat{\eta}(M)$, in practice represented by a discrete set of multicanonical weights $\{\hat{\eta}\}$. Thus we write [25]

$$\hat{P}_0(M \mid c) \doteq e^{-\hat{\eta}(M)} \tag{30}$$

implying (through (27)) a sampling distribution

$$P_S(\{q\}) \doteq \sum_{j=1}^{\Omega} e^{-\mathcal{E}(\{q\})-\hat{\eta}_j} \Delta_j[M] \doteq e^{-\beta E(\{q\})+\hat{\eta}[M(\{q\})]} \tag{31}$$

which is of the general form of (25), with $w_j \equiv e^{\hat{\eta}_j}$.

There are broadly two strategic responses to the second of the issues raised above: to drive the estimator in the right direction one may appeal to the *statistics of visits to macrostates* or to the *statistics of transitions between macrostates*. We divide our discussion accordingly.

5.1 Statistics of Visits to Macrostates

The extent to which any chosen sampling distribution (weight function) meets our requirements is reflected most directly in the M-distribution it implies. One can estimate that distribution from a histogram $H(M)$ of the macrostates visited in the course of a set of MC observations. One can then use this information to refine the sampling distribution to be used in the next set of MC observations. The simplest update algorithm is of the form [26]

$$\hat{\eta}(M) \longrightarrow \hat{\eta}(M) - \ln[H(M) + 1] + k \tag{32}$$

The form of the logarithmic term serves to ensure that macrostates registering no counts (of which there will usually be many) have their weights incremented by the same finite amount (the positive constant k [27]); macrostates which *have* been visited are (comparatively) down-weighted.

Each successive iteration comprises a fresh simulation, performed using the weight function yielded by its predecessor; since the weights attached to unvisited macrostates is enhanced (by k) at every iteration which fails to reach them, the algorithm plumbs a depth of probability that grows exponentially with the iteration number. The iterations proceed until the sampling distribution is roughly flat over the entire range of interest.

There are many tricks of the trade here. One must recognize the interplay between signal and noise in the histograms: the algorithm will converge only as long as the signal is clear. To promote faster convergence one can perform a linear extrapolation from the sampled into the unsampled region. One may bootstrap the process by choosing an initial setting for the weight function on the basis of results established on a smaller (computationally-less-demanding) system. To avoid spending excessive time sampling regions in which the weight function has already been reliably determined, one can adopt a multistage approach. Here one determines the weight function separately within slightly overlapping windows of the macrovariable. The individual parts of the weight function are then synthesized using multi-histogram re-weighting (Sect. 6.2) to obtain the full weight function. For further details the reader is referred to [28].

The strategy we have discussed is generally attributed to Berg and Neuhaus (BN) [21]. Wang and Landau (WL) have offered an alternative formulation [24]. To expose what is different, and what is not, it is helpful to consider first the case in which the macrovariable is the energy, E. Appealing to what one knows *a priori* about the canonical energy distribution, the obvious parameterization is

$$\hat{P}_0(E) \doteq \hat{G}(E)e^{-\beta E} \tag{33}$$

where $\hat{G}(E)$ is an estimator of the *density of states* function $G(E)$. Matching this parameterization to the multicanonical weight function $\hat{\eta}(E)$ implied by choosing $M = E$ in (30) one obtains the correspondence

$$\hat{\eta}(E) = \beta E - \ln \hat{G}(E) \tag{34}$$

There is thus no major difference here. The differences between the two strategies reside rather in the procedure by which the parameters of the sampling distribution are *updated*, and the point at which that procedure is *terminated*.

Like BN, WL monitors visits to macrostates. But, while BN updates the weights of *all* macrostates after *many* MC steps, WL updates its "density of states" for *the current* macrostate after *every* step. The update prescription is

$$\hat{G}(E) \longrightarrow f\hat{G}(E) \tag{35}$$

where f is a constant, greater than unity. As in BN a visit to a given macrostate tends to reduce the probability of further visits. But in WL this change takes place *immediately* so the sampling distribution evolves on the basic timescale of the simulation. As the simulation proceeds, the evolution in the sampling distribution irons out large differences in the sampling probability across E-space, which is monitored through a histogram $H(E)$. When that histogram satisfies a nominated "flatness criterion" the entire process is repeated (starting from the current $\hat{G}(E)$, but zeroing $H(E)$) with a smaller value of the weight-modification factor, f.

Like BN, then, the WL strategy entails a two-time scale iterative process. But in BN the aim is only to generate a set of weights that can be utilized in a further, final multicanonical sampling process; the iterative procedure is terminated when the weights are sufficiently good to allow this. In contrast, in WL the iterative procedure is pursued further – to a point [29] where $\hat{G}(E)$ may be regarded as a definitive approximation to $G(E)$, which can be used to compute any (single phase) thermal property at any temperature through the partition function

$$Z(\beta) = \int dE\, G(E)e^{-\beta E} \qquad (36)$$

In the context of energy sampling, then, BN and WL achieve essentially the same ends, by algorithmically-different routes. Both entail choices (in regard to their update schedule) which have to rest on experience rather than any deep understanding. WL seems closer to the self-monitoring ideal, and may scale more favorably with system size.

The WL procedure *can* be applied to any chosen macrovariable, M. But while a good estimate $\hat{G}(E)$ is sufficient to allow multicanonical sampling in E (and a definitive one is enough to determine $Z(\beta)$, (36)) the M-density of states does not itself deliver the desired analogues: we need, rather, the *joint* density of states $G(E, M)$ which determines the restricted, single-phase partition functions through

$$Z_\alpha(\beta) = \int dE \int dM\, G(E, M)e^{-\beta E}\Delta_\alpha[M] \qquad (37)$$

The WL strategy *does* readily generalize to a 2D macrovariable space. The substantially greater investment of computational resources is offset by the fact that the relative weights of the two phases can be determined at *any* temperature. Reference [30] provides one illustration of this strategy, which seems simple, powerful and general.

5.2 Statistics of Transitions Between Macrostates

The principal general feature of the algorithms based on visited macrostates is that the domain of the macrovariable they explore expands relatively slowly into the regions of interest. The algorithms we now discuss offer significant improvement in this respect. Although (inevitably, it seems) they exist in a variety of guises, they have a common core which is easily established. We take the general detailed balance condition (9) and sum over configurations $\{q\}$ and $\{q'\}$ that contribute (respectively) to the macrostates M_i and M_j of some chosen macrovariable M. We obtain immediately

$$P_{\mathcal{S}}(M_i)\overline{P_{\mathcal{S}}(M_i \to M_j)} = P_{\mathcal{S}}(M_j)\overline{P_{\mathcal{S}}(M_j \to M_i)} \tag{38}$$

The terms with over-bars are macrostate transition probabilities (TP). Specifically $\overline{P_{\mathcal{S}}(M_i \to M_j)}$ is the probability (per unit time, say) of a transition from some nominated configuration in M_i to *any* configuration in M_j, ensemble-averaged over the configuration in M_i. Adopting a more concise (and suggestive) matrix notation:

$$p_{\mathcal{S}}^M[i]\rho_{\mathcal{S}}^M[ij] = p_{\mathcal{S}}^M[j]\rho_{\mathcal{S}}^M[ji] \tag{39}$$

This is a not-so-detailed balance condition; it holds for any sampling distribution and any macrovariable [26]. The components of the eigenvector of the TP matrix (of eigenvalue unity) thus identify the macrostate probabilities. This is more useful than it might seem. One can build up an approximation of the transition matrix by monitoring the transitions which follow when a simulation is launched from an *arbitrary* point in configuration space. The "arbitrary" point can be judiciously sited in the heart of the interesting region; the subsequent simulations then carry the macrovariable right through the chosen region, allowing one to accumulate information about it from the outset. With a sampling distribution parameterized as in (31), the update scheme is simply

$$\hat{\eta}(M) \longrightarrow \hat{\eta}(M) - \ln[\hat{P_{\mathcal{S}}}(M)] + k \tag{40}$$

where $\hat{P_{\mathcal{S}}}(M)$ is the estimate of the sampling distribution that is *deduced* from the measured TP matrix [31]. Reference [32] describes the application of this technique to a structural phase transition.

One particular case of (38) has attracted considerable attention. If one sets $M = E$, and considers the infinite temperature limit, the probabilities of the macrostates E_i and E_j can be replaced by the associated values of the density of states function $G(E_i)$ and $G(E_j)$. The resulting equation has been christened *the broad-histogram relation* [33]; it forms the core of extensive studies of transition probability methods referred to variously as "flat histogram" [34] and "transition matrix" [35]. Applications of these formulations seem to have been restricted to the situation where the energy is the macrovariable, and the energy spectrum is discrete.

Methods utilizing macrostate transitions do have one notable advantage with respect to those that rely on histograms of macrostate-visits. In transition-methods the results of separate simulation runs (possibly initiated from different points in macrovariable space) can be straightforwardly combined: one simply aggregates the contributions to the transition-count matrix [26]. Synthesizing the information in separate histograms is less straightforward. The easy synthesis of data sets makes the TP method ideally suited for implementation in parallel architectures. Whether these advantages are sufficient to offset the the fact that TP methods are somewhat more complicated to implement is, perhaps, a matter of individual taste.

6 Tracking the Phase Boundary

6.1 Gibbs-Duhem Integration

In principle knowledge of a *single point* on the coexistence curve permits the *entire curve* to be traced without further calculation of free energies. The key result is that the *slope* of the coexistence curve $\Delta \mathcal{F}_{\alpha\tilde{\alpha}}(\{\lambda\}_x) = 0$, is given by

$$\frac{d\lambda_1}{d\lambda_2} = -\frac{\langle M_2 \rangle_\alpha - \langle M_2 \rangle_{\tilde{\alpha}}}{\langle M_1 \rangle_\alpha - \langle M_1 \rangle_{\tilde{\alpha}}} \qquad (41)$$

This is the generalized Clausius-Clapeyron equation [36]. It expresses the slope entirely in terms of single-phase averages; the slope can be employed (in a predictor-corrector scheme) to estimate a nearby coexistence point. Fresh simulations performed at this new point yield the phase boundary gradient there, allowing further extrapolation to be made, and so on. In this manner one can in principle track the whole coexistence curve. This strategy is widely known as Gibbs-Duhem integration (GDI) [37].

GDI has been used effectively in a number of studies, most notably in the context of freezing of hard and soft spheres [38]. Its distinctive feature is simultaneously its strength and its weakness: once boot-strapped by knowledge of one point on the coexistence curve it subsequently requires only *single-phase* averages. This is clearly a virtue since the elaborate machinery needed for two-phase sampling is not to be unleashed lightly. But *without* any "reconnection" of the two configuration-spaces at subsequent simulation state points, the GDI approach offers no feedback on integration errors. Since there will generally exist a *band* of practically stable states on each side of the phase boundary, it is possible for the integration to wander significantly from the true boundary with no indication that anything is wrong.

6.2 Histogram Reweighting

A more robust (though computationally more intensive) alternative to GDI is provided by a synthesis of extended (multicanonical) sampling and histogram re-weighting techniques [39]. The method is boot-strapped by an ES measurement of the full canonical distribution of a suitable order parameter, at some point on the coexistence curve (identified by the equal areas criterion specified in (8)). HR techniques then allow one to map a region of the phase boundary close to this point. The range over which such extrapolations are reliable is limited and it is not possible to extrapolate arbitrarily far along the phase boundary: further multicanonical simulations will be needed at points that lie at the extremes of the range of reliable extrapolation. But there is no need to determine a new set of weights (a new extended sampling distribution) from *scratch* for these new simulations. HR allows one to generate a *rough estimate* of the equilibrium order parameter (at these points) by extrapolation from the original measured distribution. The "rough estimate" is enough to furnish a usable set of weights for the new multicanonical simulations. Repeating the combined procedure (multicanonical simulation followed by histogram extrapolation) one can track along the coexistence curve. The data from the separate histograms can subsequently be combined self consistently (through multihistogram extrapolation, as discussed in Sect. 6.2) to yield the whole phase boundary. If one wishes to implement this procedure for a phase boundary that terminates in a critical point it is advisable to start the tracking procedure nearby. At such a point the ergodic block presented to inter-phase traverses is relatively small (the canonical order parameter distribution is relatively weakly doubly-peaked); and so the multicanonical distribution (weights) required to initiate the whole process can be determined without extensive (perhaps without any) iterative procedures [40].

7 Examples

7.1 Parallel Tempering: Around the Critical Point

When the coexistence line of interest terminates in a critical point the two phases can be linked by a single continuous path (Fig. 1(b)) which loops around the critical point, eliminating the need for reference macrostates, while still avoiding the inter-phase region. In principle it is possible to establish the location of such a coexistence curve by integration along this route. But the techniques of parallel sampling (Sect. 4.2) provide a substantially more elegant way of exploiting such a path, in a technique known as (hyper) parallel tempering (HPT) [41].

Studies of liquid-vapor coexistence are, generally, best addressed in the framework of an open ensemble; thus the state variables here comprise both the particle coordinates $\{r\}$ and the particle number N. A path with the appropriate credentials can be constructed by identifying pairs of values of the chemical potential μ and the temperature T which trace out some rough *approximation* to the coexistence curve in the $\mu - T$ plane, but extend into the one-phase region beyond the critical point. Once again there is some circularity here to which we shall return. Making the relevant variables explicit, the sampling distribution (19) takes the form

$$P_{\mathcal{S}}(\{r\}^{(1)}, N^{(1)} \ldots \{r\}^{(\Omega)}, N^{(\Omega)}) = \prod_{j=1}^{\Omega} P_0(\{r\}^{(j)}, N^{(j)} | \mu_j, T_j) \qquad (42)$$

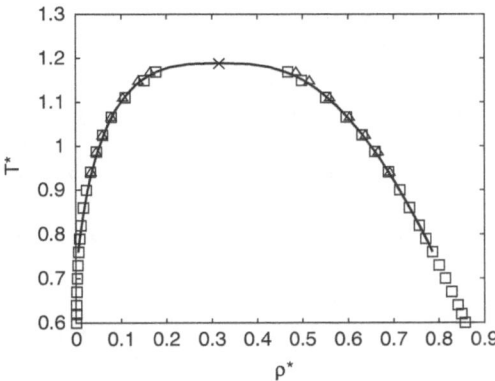

Fig. 2. Phase diagram for the LJ fluid. The *squares* show results obtained by the HPT method described in Sect. 7.1. The *triangles* show results obtained by the ESIT strategy described in Sect. 7.2 and illustrated in Fig. 3. Taken from Fig. 3 of [41]

In the context of liquid-vapor coexistence the particle number N (or equivalently the number density $\rho = N/V$) plays the role of an order parameter. Estimates of the distribution $P_0(N|\mu, T)$ are available from the simulation for all the points chosen to define the path. One may then identify the free energy difference from the integrated areas of the branches of this distribution associated with each phase and proceed to search for coexistence using the criterion that these integrated areas should be equal ((8) et seq.). Figure 2 show some explicit results [41] for a Lennard-Jones fluid.

7.2 Extended Sampling: Traversing the Barrier

Viewed from the perspectives of configuration space provided by the caricature in Fig. 1 the most direct approach to the phase-coexistence problem calls for a full frontal assault on the ergodic barrier that separates the two phases. The extended sampling strategies discussed in Sect. 4.3 make that possible. The framework we need is a synthesis of (8) and (25). We will refer to it generically as Extended Sampling Interface Traverse (ESIT).

Equation (8) shows that we can always accomplish our objective if we can measure the full canonical distribution of an appropriate order parameter. By "full" we mean that the contributions of both phases must be established *and calibrated on the same scale*. Of course it is the last bit that is the problem. (It is always straightforward to determine the two *separately* normalized distributions associated with the two phases, by conventional sampling in each phase in turn.) The reason that it is a problem is that the "full canonical" distribution of the (an) "order parameter" is typically vanishingly small at values intermediate between those characteristic of the two individual phases. The vanishingly small values provide a real, even quantitative, measure of the ergodic barrier between the phases. If the "full" order parameter distribution is to be determined by a "direct" approach (as distinct from the circuitous approach of Sect. 7.1, or the "off the map" approach to be discussed in Sect. 7.3) these low-probability macrostates *must* be visited.

Equation (25) shows how. We need to build a sampling distribution that "extends" along the path of M-macrostates running between the two phases. To do its job that sampling distribution must (Sect. 4.3) assign "roughly equal" values to the probabilities of the different macrostates. More explicitly the resulting measured distribution of M-values (following [21] we shall call it multicanonical)

$$P_\mathcal{S}(M_j) \equiv \int \prod_i dq_i P_\mathcal{S}(\{q\}) \Delta_j [M(\{q\})] \qquad (43)$$

should be "roughly flat". It needs to be "roughly flat" because the macrostate of lowest probability sets the size of the bottleneck through which inter-phase traverses must pass. It needs to be *no better* than "roughly" flat because of the way in which (ultimately) it is *used*. It is used to estimate the true canonical distribution $P_0(M)$. The two distributions are simply related by

$$P_0(M_j) \doteq w_j^{-1} P_\mathcal{S}(M_j) \qquad (44)$$

where $\{w\}$ are the multicanonical weights that define the chosen sampling distribution (25) and $\doteq$ means equality to within an overall normalization constant. The procedure by which one uses this equation to estimate the canonical distribution from the measured distribution is variously referred to as "unfolding the weights" or "re-weighting"; it is simply one realization of the identity given in (17). The procedure eliminates any *explicit* dependence on the weights (hence the looseness of the criteria by which they are specified); but it leaves the desired legacy: the relative sizes of the two branches of the *canonical* distribution are determined with a statistical quality that reflects the number of inter-phase traverses in the *multicanonical* ensemble.

This strategy has been applied to the study of a range of coexistence problems, initially focused on lattice models in magnetism [42] and particle physics [43]. Figure 3 [44, 45] shows the results of an application to liquid-vapor coexistence in a Lennard-Jones system with the particle number density chosen as an order parameter.

7.3 From Paths to Wormholes: Phase Switch Monte Carlo

Phase Switch Monte Carlo is a method for traversing an inter-phase path which avoids mixed phase stats by leaping directly from configurations of one pure phase to those of another pure phase (cf. Fig. 1(d)). The leap is implemented as a Monte Carlo move. It takes as its starting point the specification of a reference configuration $\{\boldsymbol{R}^\alpha\}$ for each of the phases α coexisting at the phase boundary. The specific choice of $\{\boldsymbol{R}^\alpha\}$ is arbitrary, the only condition

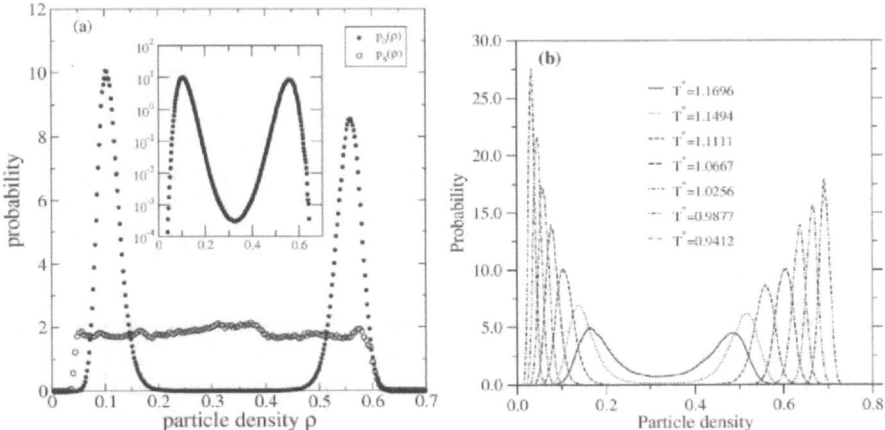

Fig. 3. (a) Results from a multicanonical simulation of the 3D Lennard-Jones fluid at a point on the coexistence curve. The figure shows both the multicanonical sampling distribution $P_S(\rho)$ (symbols: o) and the corresponding estimate of the equilibrium distribution $P_0(\rho)$ with $\rho = N/V$ the number density. The inset shows the value of the equilibrium distribution in the interfacial region [44]. (b) Coexistence density distributions for a selection of temperatures. The coexistence densities can be simply read off from the peak positions.

being that it should be a member of the set of pure phase configurations identifiable as "belonging" to phase α. For a crystalline phase, a suitably simple choice of $\{R^\alpha\}$ is the set of lattice sites. For a liquid phase, a suitable choice of reference state is a configuration selected randomly in the course of a short simulation of the liquid.

The next step is express the coordinates of each particle in phase α in terms of the displacement from its lattice site, i.e.

$$r_i^\alpha = R_i^\alpha + u_i \, . \tag{45}$$

Now, clearly one can reversibly map any configuration $\{r^\alpha\}$ of phase α onto a configuration of another phase $\tilde{\alpha}$ simply by *switching* the set of reference sites $\{R^\alpha\} \to \{R^{\tilde{\alpha}}\}$, while holding the set of displacements $\{u\}$ *constant*. This switch, which forms the heart of the method, can be incorporated in a global MC move. A complication arises however, because the displacements $\{u\}$ typical for phase α will not, in general, be typical for phase $\tilde{\alpha}$. Thus the switch operation will mainly propose high energy configurations of phase $\tilde{\alpha}$ which are unlikely to be accepted as a Metropolis update. This problem can be circumvented by employing extended sampling (biasing) techniques to seek out those displacements $\{u\}$ for which the switch operation *is* energetically favorable. We refer to such states as gateway configurations.

The bias necessary to promote sampling of the gateway states is administered with respect to an "order parameter" – some suitable chosen macrovariable M of the system. The freedom in selecting this order parameter is circumscribed by the requirement that the associated microstates form a contiguous path through phase space linking the large number of equilibrium (typical) microstates to the relatively small number of gateway states. In many cases a suitable order parameter is simply the energy cost associated with the switch.

To enhance the acceptance rate of phase switch attempts, biased sampling techniques are employed to extend the range of M values explored in each phase in order to encompass the gateway states (of low M). Operationally, this is achieved by incorporating a weight function $\eta(M)$ in the effective Hamiltonian (in the usual manner of extended sampling [21]). A suitable weight function is obtainable via a variety of methods, but we have found adaptive methods such as Transition Matrix Monte Carlo [35] to operate most effectively. It should be noted that the biasing procedure automatically seeks out those configurations $\{u\}$ for which the switch is energetically favorable – it is not necessary to specify such gateway configurations in advance. One discovers (as seems reasonable *a postiori*) that gateway states correspond to configurations in which the particles lie close to their reference sites.

Once a suitable weight function has been determined, a long simulation is performed in the course of which both phases are visited many times. During this run, the biased form of the order parameter distribution $P(M|\eta)$ is accumulated in the form of a histogram. The applied bias can be unfolded from this distribution in the usual fashion to yield the true equilibrium distribution $P(M)$, from which the relative free energy difference of the two phases can be read off as the logarithm of the ratio of peak weights in $P(M)$.

When the phase switch strategy is applied to the freezing transition the liquid and solid phases generally have significantly different densities and the simulation must be conducted at constant pressure. The coordinate set $\{u\}$ then contains the system volume and the switch must accommodate an appropriate dilation (and can do so easily through the specification of the volumes implicit in the reference configurations). While an order parameter based on the energy cost (or for hard spheres, the degree of overlap) of the switch remains appropriate for simulations conducted in the solid phase, in the liquid phase it is necessary to engineer something a little more elaborate to account for the fact that the particles are not spatially localized. Such considerations also lead to some relatively subtle but significant finite-size effects. Figure 4 shows some results locating the freezing pressure of hard spheres this way [46].

8 Discussion

The large number of contemporary MC approaches for dealing with the problem of phase behaviour can usefully be categorized in terms of the the interphase path they utilize to connect the phases and the sampling strategy adopted to traverse this path. We have exemplified some of the principal concepts within the limited context of the phase behaviour of simple particle

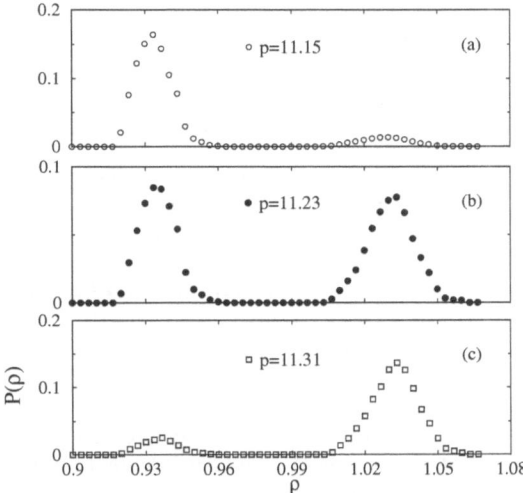

Fig. 4. The distribution of the density of a system of $N = 256$ hard particles in crystalline and liquid phases, as determined by PSMC. The three pressures are (**a**) just below, (**b**) at and (**c**) just above coexistence for this N. Taken from [46]

models. However, it is probably wise to have some idea of where the ultimate destination is, and the strategies that are most likely to take us there.

The long term goal is a computational framework that will be grounded in electronic structure as distinct from phenomenological particle potentials; that will predict global phase behavior a priori, rather than simply decide between two nominated candidate phases; and that will handle quantum behavior, in contrast to the essentially classical framework on which we have focused here. That goal is distant but not altogether out of sight. Integrating *ab-initio* electronic structure calculations with the statistical mechanics of phase behavior has already received some attention [47,48]. The WL algorithm ([24], Sect. 5) offers a glimpse of the kind of self-monitoring configuration-space search algorithm that one needs to make automated a priori predictions of phase behavior possible. And folding in quantum mechanics requires only a dimensionality upgrade [49].

Acknowledgement

The author thanks A.D. Bruce for a close collaboration from which many of the perspectives developed in this article have grown.

References

1. In the arguments of exponentials we shall generally absorb factors of $kT \equiv 1/\beta$ into effective energy functions
2. K. Binder and D. W. Heermann (1998) *Monte Carlo Simulation in Statistical Physics*. Springer, Berlin Heidelberg
3. D. P. Landau and K. Binder (2000) *A Guide to Monte Carlo Simulations in Statistical Physics*. Cambridge University Press
4. D. Frenkel and B. Smit (1996) *Understanding Molecular Simulation*. Academic Press
5. O. Narayan and A. P. Young (2001) Convergence of Monte Carlo simulations to equilibrium. *Phys. Rev. E* **64**, 021104
6. In general the expectation value of an observable which is an *exponential* function of some *extensive* ("macroscopic") property cannot be reliably estimated from its sample average if one chooses $P_S = P_0$
7. Throughout this section we use the notation in which the phase label is absorbed into an extended macrostate label $\mathcal{C}$. The path defined by the set of macrostates $\{\mathcal{C}\}$ may (and sometimes will) extend from one phase to another
8. We shall refer to a field λ and a macrovariable M as conjugate if the λ-dependence of the configurational energy is of the form $\mathcal{E}(\{q\}, c, \lambda) = \mathcal{E}(\{q\}, c) - \lambda M(\{q\})$
9. It would be more conventional to refer to "densities", which are *intensive*. But we prefer to deal with their *extensive* counterparts, which we shall refer to as *macrovariables*. The preference reflects our focus on simulation studies which are necessarily conducted on systems of some given (finite) size
10. D. Frenkel and A. J. C. Ladd (1984) New Monte Carlo method to compute the free energy of arbitrary solids. Application to the fcc and hcp phases of hard spheres. *J. Chem. Phys.* **81**, p. 3188
11. Y. Iba (2001) Extended ensemble Monte Carlo. *Int. J. Mod. Phys. C* **12**, p. 623
12. C. J. Geyer and E. A. Thompson (1992) Constrained Monte-Carlo Maximum-Likelihood For Dependent Data. *J. R. Statist. Soc. B* **54**, p. 657
13. K. Hukushima, H. Takayama and K. Nemoto (1996) Application of an extended ensemble method to spin glasses. *Int. J. Mod. Phys. C.* **3**, p. 337
14. E. Marinari (1997) Optimized Monte Carlo Methods in *Advances in Computer Simulation* ed J. Kertesz and I. Kondor. Springer Verlag
15. More explicitly $\mathcal{C}_1 \equiv \alpha, c$ and $\mathcal{C}_2 \equiv \tilde{\alpha}, c$
16. G. M. Torrie and J. P. Valleau (1977) Monte-Carlo Free-Energy Estimates Using Non-Boltzmann Sampling – Application to Subcritical Lennard-Jones Fluid. *Chem. Phys. Lett.* **28**, p. 578; *ibid.* (1974) Non-Physical Sampling Distributions in Monte-Carlo Free-Energy Estimation – Umbrella Sampling. *J. Comp. Phys.* **23**, p. 187
17. A. P. Lyubartsev, A. A. Martsinovski, S. V. Shevkunov and P. N. Vorontsov-Velyaminov (1992) New Approach to Monte-Carlo Calculation of the Free-Energy – Method of Expanded Ensembles. *J. Chem. Phys.* **96**, p. 1776
18. E. Marinari and G. Parisi (1992) Simulated Tempering – A New Monte-Carlo Scheme. *Europhysics Lett.* **19**, p. 451
19. J. P. Valleau (1999) Thermodynamic-Scaling Methods in Monte Carlo and their application to phase equilibria. *Adv. Chem. Phys.* **105**, p. 369
20. M. Mezei (1987) Adaptive Umbrella Sampling – Self-Consistent Determination of the Non-Boltzmann Bias. *J. Comp. Phys.* **68**, p. 237

21. B. A. Berg and T. Neuhaus (1991) Multicanonical Algorithms for 1st Order Phase-Transitions. *Phys. Lett. B* **267**, p. 249; *ibid.* (1992) Multicanonical Ensemble – A New Approach to Simulate 1st-Order Phase-Transitions. *Phys. Rev. Lett.* **68**, p. 9

22. In this section we refer to E rather than its dimensionless counterpart $\mathcal{E} \equiv \beta E$, since it seems advisable to make the measure of temperature $\beta = 1/kT$ explicit

23. J. Lee (1993) New Monte-Carlo Algorithm – Entropic Sampling. *Phys. Rev. Lett.* **71**, p. 211; Erratum: *ibid.* **71**, p. 2353

24. F. Wang and D. P. Landau (2001) Efficient, multiple-range random walk algorithm to calculate the density of states. *Phys. Rev. Lett.* **86**, p. 2050; *ibid.* (2001) Determining the density of states for classical statistical models: A random walk algorithm to produce a flat histogram. *Phys. Rev. E* **64**, 056101

25. Different authors have made different choices of notation and adopted different sign conventions at this point. We adopt those of [26]

26. G. R. Smith and A. D. Bruce (1995) A study of the multi-canonical Monte Carlo method. *J. Phys. A* **28**, p. 6623

27. The constant k merely serves to assign a convenient lower (or upper) bound to the weights; the value chosen is absorbed in the normalization of the sampling distribution

28. B. A. Berg (1993) Multicanonical Monte-Carlo Simulations. *Int. J. Mod. Phys. C* **4**, p. 249; *ibid.* (2000) Introduction to Multicanonical Monte Carlo Simulation. *Fields Inst. Commun.* **26**, p. 1

29. The WL scheme is terminated when the histogram is "sufficiently" flat *and* the weight-modification factor f is "sufficiently" close to unity.

30. Q. Yan, R. Faller and J. J. de Pablo (2002) Density-of-states Monte Carlo method for simulation of fluids. *J. Chem. Phys.* **116**, p. 8745

31. There is a caveat here: the traverse through the interesting region needs to be slow enough to allow a rough local equilibrium in each macrostate to be established; this condition is satisfied relatively poorly at the outset (where the MC trajectory heads rapidly for the equilibrium states); but it gets better the closer the sampling distribution comes to the multicanonical limit; one requires no more than this

32. G. R. Smith & A. D. Bruce (1996) Multicanonical Monte Carlo study of solid-solid phase coexistence in a model colloid. *Phys. Rev. E* **53**, p. 6530

33. P. M. C. de Oliveira, T. J. P. Penna and H. J. Herrmann (1996) Broad histogram Monte Carlo. *Braz. J. Phys.* **26**, p. 677; *ibid.* (1998) Broad histogram Monte Carlo. *Eur. Phys. J. B* **1**, p. 205; P. M. C. de Oliveira (1998) Broad histogram relation is exact. *Eur. Phys. J. B* **6**, p. 111

34. J. S. Wang and L. W. Lee (2000) Monte Carlo algorithms based on the number of potential moves. *Comput. Phys. Commun.* **127**, p. 131

35. J. S. Wang, T. K. Tay, and R. H. Swendsen (1999) Transition matrix Monte Carlo reweighting and dynamics. *Phys. Rev. Lett.* **82**, p. 476; J. S. Wang and R. H. Swendsen (2002) Transition matrix Monte Carlo method. *J. Stat. Phys.* **106**, p. 245

36. It reduces to the familiar form with the macrovariables M_1 and M_2 chosen as respectively the enthalpy and the volume whose conjugate fields are $1/T$ and P/T

37. D. A. Kofke (1993) Direct Evaluation of Phase Coexistence by Molecular Simulation via Integration along the Saturation Line. *J. Chem. Phys.* **98**, p. 4149;

R. Agrawal and D. A. Kofke (1995) Solid-Fluid Coexistence for Inverse-Power Potentials. *Phys. Rev. Lett.* **74**, p. 122

38. P. G. Bolhuis and D. A. Kofke (1996) Monte carlo study of freezing of polydisperse hard spheres. *Phys. Rev. E* **54**, p. 634; D. A. Kofke and P. G. Bolhuis (1999) Freezing of polydisperse hard spheres. *Phys. Rev. E* **59**, p. 618; M. Lisal and V. Vacek (1996) Direct evaluation of vapour-liquid equilibria of mixtures by molecular dynamics using Gibbs-Duhem integration. *Molecular Simulation* **18**, p. 75; F. A. Escobedo and J. J. de Pablo (1997) Pseudo-ensemble simulations and Gibbs-Duhem integrations for polymers. *J. Chem. Phys.* **106**, p. 2911

39. A. M. Ferrenberg and R. H. Swendsen (1989) New Monte-Carlo Technique for Studying Phase-Transitions. *Phys. Rev. Lett.* **61**, p. 2635; *ibid.* (1989) Optimized Monte-Carlo Data-Analysis. *Phys. Rev. Lett.* **63**, p. 1195; R. H. Swendsen (1993) Modern Methods of Analyzing Monte-Carlo Computer-Simulations. *Physica A* **194**, p. 53

40. Recall that knowing the weights that define a multicanonical distribution of some quantity M over some range is equivalent to knowing the true canonical M-distribution *throughout the chosen range*

41. Q. Yan and J. J. de Pablo (1999) Hyper-parallel tempering Monte Carlo: Application to the Lennard-Jones fluid and the restricted primitive model. *J. Chem. Phys.* **111**, p. 9509

42. B. A. Berg, U. H. E. Hansmann and T. Neuhaus (1993) Properties of Interfaces in the 2 and 3-Dimensional Ising-Model. *Z. Phys. B* **90**, p. 229

43. B. Grossmann, M. L. Laursen, T. Trappenberg and U. J. Wiese (1992) A Multicanonical Algorithm for Su (3) Pure Gauge-Theory. *Phys. Lett. B* **293**, p. 175

44. Reference [45] and N. B. Wilding (unpublished)

45. N. B. Wilding (1995) Critical-Point and Coexistence-Curve Properties of the Lennard-Jones Fluid – A Finite-Size-Scaling Study. *Phys. Rev. E* **52**, p. 602

46. N. B. Wilding and A. D. Bruce (2000) Freezing by Monte Carlo phase switch. *Phys. Rev. Lett.* **85**, p. 5138

47. D. Alfe, G. A. De Wijs, G. Kresse, M. J. Gillan (2000) Recent developments in ab initio thermodynamics. *Int. J. Quant. Chem.* **77**, p. 871

48. G. J. Ackland (2002) Calculation of free energies from ab initio calculation. *J. Phys. Condens. Mat.* **14**, p. 2975

49. F. H. Zong, D. M. Ceperley (1998) Path integral Monte Carlo calculation of electronic forces. *Phys. Rev. E* **58**, p. 5123; C. Rickwardt, P. Nielaba, M. H. Muser and K. Binder (2001) Path integral Monte Carlo simulations of silicates. *Phys. Rev. B* **63**, 045204

Simulation Techniques
for Calculating Free Energies

M. Müller[1] and J.J. de Pablo[2]

[1] Institut für Theoretische Physik, Georg-August- Universität, 37077 Göttingen,
Germany
`mmueller@theorie.physik.uni-goettingen.de`
[2] Department of Chemical and Biological Engineering, University of
Wisconsin-Madison, Madison Wisconsin 53706-1691, USA
`depablo@engr.wisc.edu`

Marcus Müller and Juan J. de Pablo

M. Müller and J.J. de Pablo: *Simulation Techniques for Calculating Free Energies*, Lect. Notes
Phys. **703**, 67–126 (2006)
DOI 10.1007/3-540-35273-2_3

1 Introduction

The study of phase transitions has played a central role in the study of condensed matter. Since the first applications of molecular simulations, which provided some of the first evidence in support of a freezing transition in hard-sphere systems, to contemporary research on complex systems, including polymers, proteins, or liquid crystals, to name a few, molecular simulations are increasingly providing a standard against which to measure the validity of theoretical predictions or phenomenological explanations of experimentally observed phenomena.

This is partly due to significant methodological advances that, over the past decade, have permitted study of systems and problems of considerable complexity. The aim of this chapter is to describe some of these advances in the context of examples taken from our own research.

The application of Monte Carlo simulations for the study of phase behavior in fluids attracted considerable interest in the early 90's, largely as a result of Monte Carlo methods, such as the Gibbs ensemble technique [1], which permitted direct calculation of coexistence properties (e.g., the orthobaric densities of a liquid and vapor phases at a given temperature) in a single simulation, without a need of costly thermodynamic integrations for the calculation of chemical potentials. Perhaps somewhat ironically, some of the latest and most powerful methods for the study of phase behavior have reverted back to the use of thermodynamic integration, albeit in a different form from that employed before the advent of the Gibbs ensemble technique. One of the central concepts that paved the way for the widespread use of free-energy based methods in simulations of phase behavior was the conceptually simple, but highly consequential realization that histograms of data generated at one particular set of thermodynamic conditions could be used to make predictions about the behavior of the system over a wide range of conditions [2, 3]. The so-called weighted histogram analysis or histogram-reweighing technique constitutes an essential component of the methods presented in this chapter, and we therefore begin with a brief description of its implementation. We then discuss multicanonical simulation techniques, Wang-Landau sampling and extensions, and successive umbrella sampling. We close this section on Methods with a discussion of the configurations that one encounters when one uses the order parameter of a first order phase transition as a reaction coordinate between the two coexisting phases. The methodological section is followed by several applications that illustrate the computational methods in the context of examples drawn from our own research. The chapter closes with a brief look ahead.

2 Methods

2.1 Weighted Histogram Analysis

In the simplest implementation of a molecular simulation, data corresponding to a specific set of thermodynamic conditions (e.g., number of particles, n, volume, V, and temperature, T, for a canonical ensemble) are used to estimate the ensemble or time average of a property of interest. For a canonical ensemble, for example, the outcome of such a simulation would be a pressure p corresponding to n, V, and T. Histogram reweighting techniques permit calculation of ensemble averages over a range of thermodynamic conditions without having to perform additional simulations. The underlying ideas, introduced by Ferrenberg and Swendsen in the late 80's [2, 3], provide a simple means for extrapolating data generated at one set of conditions to nearby points in thermodynamic space. For a canonical ensemble, the probability of observing a configuration having energy E at an inverse temperature $\beta = 1/k_B T$ (where k_B is Boltzmann's constant) takes the form

$$P_\beta(E) = \frac{\Omega(E) \exp(-\beta E)}{Z(\beta)} , \tag{1}$$

where $\Omega(E)$ is the density of states of the system and $Z(\beta)$ is the canonical partition function, given by

$$Z(\beta) = \sum_E \Omega(E) \exp(-\beta E) . \tag{2}$$

The probability distribution at a nearby inverse temperature, denoted by β', can be expressed in terms of the distribution at β according to

$$P_{\beta'}(E) = \frac{P_\beta(E) \exp[(\beta - \beta')E]}{\sum_E P_\beta(E) \exp[(\beta - \beta')E]} . \tag{3}$$

In other words, by using (3) one can extrapolate data generated at temperature T to nearby temperature T'.

The same idea can be used to interpolate data generated from multiple simulations [3]. Consider a series of canonical ensemble Monte-Carlo simulations conducted at r different temperatures. The nth simulation is performed at β_n, and the resulting data are stored and sorted in $N_n(E)$ histograms, where the total number of entries is n_n. The probability distribution corresponding to an arbitrary temperature β is given by

$$P_\beta(E) = \frac{\sum_{n=1}^{r} N_n(E) \exp(-\beta E)}{\sum_{n=1}^{r} n_n \exp(-\beta_n E - f_n)} \tag{4}$$

where

$$\exp[f_n] = \sum_E P_{\beta_n}(E) , \tag{5}$$

and where f_n is a measure of the free energy of the system at temperature β_n. The value of f_n can be found self-consistently by iterating (4) and (5). Multiple histogram reweighing can therefore yield probability distributions over a broad range of temperatures, along with the corresponding relative free energies corresponding to that range. Equation (4) was derived by minimizing the error that arises when all histograms are recombined to provide thermodynamic information over the entire range of conditions spanned by distinct simulations. Note that histograms corresponding to neighboring conditions must necessarily overlap in order for (4) to be applicable.

2.2 Multicanonical Simulations

The range of applicability of histogram extrapolation is large if the fluctuations that arise during the course of the simulation sample states that are also representative of neighboring thermodynamic conditions. This is often the case for small systems, or in the vicinity of a critical point. For large systems, or remote from a critical point, fluctuations are smaller and it is advantageous to sample configurations according to a non-Boltzmann statistical weight, thereby coercing the system to sample configurations that are representative of a wide interval of thermodynamic conditions. Thermal averages, such as the internal energy, E, can then be obtained via a time average over the weighted sequence of visited states. Note, however, that in such a sampling scheme the entropy, S, or free energy, F, cannot be calculated in a direct manner because those quantities cannot be expressed as a function of the particle coordinates; special simulation techniques are required to estimate S and F.

We now discuss several techniques that generate configurations according to weights that are generally constructed in such a way as to provide a more uniform sampling of phase space than the Boltzmann weight. It is instructive to note that a number of simulation methods e.g., multicanonical [4, 6, 21] or Wang-Landau techniques [7,8], have been originally formulated in terms of the pair of thermodynamically conjugated variables consisting of the temperature and the energy. Such methods, however, can be carried over to arbitrary pairs of order parameter or reaction coordinate and conjugated field, e.g., (magnetization and magnetic field), (composition and exchange potential), or (number of particles and chemical potential), by simply replacing the energy by the order parameter and the temperature by the thermodynamically conjugated field.

The free energy, F, can be obtained by thermodynamic integration of the specific heat, c_V, from a reference state interval along a thermodynamically reversible path:

$$\beta F = \beta E - \frac{S}{k_B} = \beta E - \left(\frac{S_0}{k_B} + \int_{T_0}^{T} dT' \, \frac{c_V(T')}{k_B T'} \right). \tag{6}$$

The specific heat can in turn be obtained in canonical-ensemble simulations from the fluctuations of the internal energy,

$$c_V \equiv \frac{\mathrm{d}E}{\mathrm{d}T} = \frac{\langle E^2 \rangle - \langle E \rangle^2}{k_B T^2}. \tag{7}$$

In order to estimate the free energy many canonical simulations at different temperatures are necessary; furthermore, it is often difficult to define a suitable reference state with a known entropy S_0. Two alternatives can be followed to overcome these difficulties: (i) expanded ensemble methods and (ii) multicanonical methods.

In the expanded ensemble method [9], one considers the conjugated field as a Monte Carlo variable and one introduces moves that allow for altering the conjugated field. Each thermodynamic integration can be cast into this form. The conjugated field over which one integrates (e.g., temperature) can adopt discrete values $T_1 \leq T_i \leq T_2$ in the interval of interest, including the boundary values. The values have to be chosen such that the order parameter distributions in the canonical ensemble overlap for neighboring values of the conjugated fields, T_i and T_{i+1}. States of the expanded ensemble are characterized by the particle coordinates, $\{\mathbf{r}_i\}$, and the conjugated field, T_i. The partition function of the expanded ensemble takes the form

$$\mathcal{Z}_{\mathrm{ex}} = \sum_{T_1 \leq T_i \leq T_2} e^{-w(T_i)} Z(n, V, T_i) = \sum_{T_1 \leq T_i \leq T_2} e^{-w(T_i)} \int \mathcal{D}[\{\mathbf{r}_i\}] \exp(-\beta_i E[\{\mathbf{r}_i\}]) \tag{8}$$

where $Z(n, V, T_i) = e^{-\beta_i F}$ denotes the canonical-ensemble partition function. Here we have introduced weight factors, $w(T_i)$, that depend only on the value of the conjugated field but not on the microscopic particle configuration, $\{\mathbf{r}_i\}$. They have to be chosen in such a way as to generate an approximately uniform sampling of the different conjugated fields, T_i, throughout the course of the simulation. The probability of finding the system in state T_i is given by

$$P(T_i) = \frac{e^{-w(T_i)} Z(n, V, T_i)}{\mathcal{Z}_{\mathrm{ex}}}. \tag{9}$$

To achieve uniform sampling of the different canonical ensembles within an expanded ensemble simulation, the weights should ideally obey

$$P(T_i) \approx \mathrm{const} \quad \Rightarrow \quad w(T_i) \approx \ln Z(n, V, T_i) + \mathrm{const}. \tag{10}$$

The free energy is then given by:

$$\beta F = -w(T) - \ln P(T) - \ln \mathcal{Z}_{\mathrm{ex}}. \tag{11}$$

Formally, these equations are valid for an arbitrary choice of the weight factors; a "bad" choice, however, does not allow for sufficient sampling of all values of the conjugated field and dramatically reduces the efficiency of the method. The problem of calculating the free energy is therefore shifted to that of obtaining appropriate weights. An optimal choice of the weights, however, requires a working estimate of the free energy difference between the states.

The second scheme – multicanonical simulations [4, 6] – generates configurations according to a non-Boltzmann distribution designed to sample all values of the order parameter (energy) that are pertinent to the interval of the thermodynamic field (temperature) of interest. Multicanonical methods are advantageous for the study of first order phase transitions, where the order parameter connects the two coexisting phases via a reversible path (see Sect. 2.5), or in circumstances where the conjugated field does not have a simple physical interpretation (e.g., crystalline order in Sect. 3.4). Typically, a canonical simulation can only be reweighted by histogram extrapolation techniques [2] in the vicinity of the temperature at which the canonical distribution was sampled. In contrast, a single multicanonical simulation permits calculation of canonical averages over a range of temperatures which would require many canonical simulations. This feature led to the name "multicanonical" [10].

A multicanonical simulation generates configurations according to the partition function

$$
\mathcal{Z}_{\text{muca}} = \int \mathcal{D}[\{\mathbf{r}_i\}] e^{-w(E[\{\mathbf{r}_i\}])} \tag{12}
$$

where $w(E)$ is a weight function that only depends on the internal energy but not explicitly on the particle coordinates, $\{\mathbf{r}_i\}$. If $\Omega(E)$ denotes the density of states that corresponds to a given value of the order parameter (i.e., the microcanonical partition function, if we utilize the energy), then the probability of sampling a configuration with order parameter E is given by:

$$
P_{\text{muca}}(E) \sim \Omega(E) e^{-w(E)} \tag{13}
$$

and the choice $w(E) = \ln \Omega(E) + C$, where C is a constant, leads to uniform sampling of the energy range of interest. Boltzmann averages of an observable $\mathcal{O}$ at temperature T can be obtained by monitoring the average of this observable, $\mathcal{O}(E)$, at a given energy, E, using

$$
\langle \mathcal{O} \rangle_{nVT} \equiv \frac{\int dE\, \Omega(E) e^{-\beta E} \mathcal{O}(E)}{\int dE\, \Omega(E) e^{-\beta E}} = \frac{\int dE\, P_{\text{muca}}(E) e^{w(E) - \beta E} \mathcal{O}(E)}{\int dE\, P_{\text{muca}}(E) e^{w(E) - \beta E}} \tag{14}
$$

Note that for the specific choice, $w(E) = E/k_B T_0$, a multicanonical simulation corresponds to a canonical simulation at temperature T_0, and the equation above corresponds to the Ferrenberg-Swendsen weighted histogram analysis [2, 3] (cf. (4)). The goal of multicanonical sampling is to explore a much wider range of energies in the course of the simulation, thereby enlarging the extrapolation range. The free energy can be obtained from

$$
\beta F \equiv -\ln \int dE\, \Omega(E) e^{-\beta E} = -\ln \int dE\, P_{\text{muca}}(E) e^{w(E) - \beta E} + C . \tag{15}
$$

Formally, these equations are valid for an arbitrary choice of the weight function, $w(E)$, but a failure to sample all configurations that significantly contribute to canonical averages for a temperature in the interval $T_1 < T < T_2$

introduces substantial sampling errors. A necessary condition is that the simulation samples all energies in the interval $[\langle E \rangle_{nVT_1} : \langle E \rangle_{nVT_2}]$ with roughly equal probability. Again, the problem of efficient multicanonical sampling is shifted to that of obtaining the appropriate weights. The optimal choice of the weights, however, requires a working estimate of the density of states with a given order parameter. Several methods to obtain these weights are discussed in the following section.

2.3 Wang-Landau Sampling and DOS Simulations

As described above, traditional multicanonical algorithms [5] provide an estimate of the density of states through "weights", $w(E)$, constructed to facilitate or inhibit visits to particular state points according to the frequency with which they are visited. The calculation of the weights is necessarily iterative and, depending on the nature of the problem (e.g., the "roughness" of the underlying free energy profile), can require considerable oversight.

In recent years, a different class of algorithms has emerged for direct calculation of the density of states from Monte Carlo simulations [7,8,11,12]. We refer to these algorithms as "density-of-states" (DOS) based techniques. In a recent, powerful implementation of such algorithms by Wang and Landau, a random walk in energy space has been used to visit distinct energy levels [7]; the density of states corresponding to a particular energy level is modified by an arbitrary factor when that level is visited. By controlling and gradually reducing that factor in a systematic manner, an estimate of the density of states is generated in a self-consistent and "self-monitoring" manner. The algorithm relies on a histogram of energies to dictate the rate of convergence of a simulation [7]. A random walk is generated by proposing random trial moves, which are accepted with probability $p = \min[1, \Omega(E_1)/\Omega(E_2)]$, where E_1 and E_2 denote the potential energy of the system before and after the move, respectively. Every time that an energy state E is visited, a running estimate of the density of states is updated according to $\Omega(E) = f\Omega(E)$, where f is an arbitrary convergence factor. The energy histogram is also updated; once it becomes sufficiently flat, a simulation "stage" is assumed to be complete, the energy histogram is reset to zero, and the convergence factor f is decreased in some prescribed manner (e.g., $f = \sqrt{f}$). The entire process is repeated until f is very close to 1 (the original literature recommends that $\ln(f)$ attain a value of 10^{-9}). Wang and Landau's original algorithm focused on an Ising system in the canonical ensemble [7]. It was later extended to systems in a continuum and to other ensembles [8,13–16] and found to be highly effective.

Because the running estimate of the density of states changes at every step of the simulation, detailed balance is never satisfied. In practice, however, the convergence factor decreases exponentially, and its final value can become so small as to render the violation of detailed balance essentially non-existent. Given that the convergence factor decreases as the simulation proceeds, configurations generated at different stages of the simulation do not contribute equally to the estimated density of states. In the final stages of the simulation the convergence factor is so small that the corresponding configurations make a negligible contribution to the density of states. In other words, many of the configurations generated by the simulation are not used efficiently. As described below, one can alleviate this problem by integrating the temperature [17, 18].

The internal energy of a system is related to entropy S and volume V through

$$dE = T\, dS - p\, dV, \tag{16}$$

where p is the pressure. The temperature of the system is related to the density of states $\Omega(n, V, E)$ by Boltzmann's equation [19] according to

$$\frac{1}{T} = \left(\frac{\partial S}{\partial E}\right)_V = k_B \left(\frac{\partial \ln \Omega(n, V, E)}{\partial E}\right)_V. \tag{17}$$

Equation (17) can be integrated to determine the density of states from knowledge of the temperature:

$$\ln \Omega(n, V, E) = \int_{E_0}^{E} \frac{1}{k_B T}\, dE. \tag{18}$$

Equation (18) requires that the temperature be known as a function of energy. Evans et al. [20] have shown that an intrinsic temperature can be assigned to an arbitrary configuration of a system. This so-called "configurational temperature" is based entirely on configurational information and is given by

$$\frac{1}{k_B T_{\text{config}}} = \frac{\left\langle -\sum_i \nabla_i \cdot \mathbf{F}_i \right\rangle}{\left\langle \sum_i |\mathbf{F}_i|^2 \right\rangle}, \tag{19}$$

where subscript i denotes a particle, and $\mathbf{F}_i$ represents the force acting on that particle. This configurational temperature can be particularly useful in Monte Carlo simulations, where kinetic energy is not explicitly involved. In the past it has been proposed as a useful tool to diagnose programming errors [21].

The estimator for the configurational entropy can be exploited in the context of DOS simulations in the following way: Four histograms are collected during a simulation; one for the density of states, one for the potential energy, one for the numerator of (19), and one for its denominator [17]. Note that two independent sets of density of states are available at the end of each stage: one from the original density-of-states histogram, and one from the configurational temperature, which can be integrated to provide Ω. In principle, either set can be used as a starting point for the next stage of a simulation. In practice, however, using a combination of these is advantageous.

In the early stages of the simulation, detailed balance is grossly violated as a result of the large value of f. The resulting estimates of configurational temperature are therefore incorrect. In order to avoid contamination of late stages by the transfer of incorrect information, the temperature accumulators can be reset at the end of the early stages, once the density of states is calculated from the temperature. For small enough convergence factors (e.g., $\ln f < 10^{-5}$), the slight violations of detailed balance incurred by the method become negligible, and the temperature accumulators need no longer be reset at the end of each stage.

Clearly, in the proposed "configurational temperature" algorithm, the dynamically modified density of states only serves to guide a walker through configuration space. The "true", thermodynamic density of states is calculated from the configurational temperatures accumulated in the simulation. Because all configurations generated in the simulation contribute equally to the estimated temperature, configurations generated at various stages of the simulation contribute equally to Ω.

For lattice systems, or for systems interacting through discontinuous potential energy functions (e.g., hard-spheres), (19) cannot be used. In such cases, one can introduce a temperature by resorting to a microcanonical ensemble formulation [22, 23] of the density-of-states algorithm [18].

Figure 1 shows results for a truncated-and-shifted Lennard-Jones fluid (the potential energy is truncated at $r_c = 2.5\sigma$); σ and ϵ denote the length and energy scale of the potential. Results are shown for a system of $n = 400$ particles at a density of $\rho\sigma^3 = 0.78125$, well within the liquid regime [17]. For the random walk algorithm with configurational temperature, the energy window is set to $-1930 \leq E/\epsilon < -1580$; for the multi-microcanonical ensemble simulation, the energy window is $-1500 \leq E/\epsilon < -500$. In both cases, the energy window corresponds roughly to temperatures in the range $0.85 < T^* \equiv k_B T/\epsilon < 1.5$ (i.e. above and below the critical point).

In random walk simulations with configurational temperature, the calculations are started with a convergence factor $f = \exp(0.1)$. When $f > \exp(10^{-5})$, the density of states calculated from the temperature is used as the initial density of states for the next stage, the convergence factor is reduced by $f_{k+1} = f_k^{1/10}$, and the temperature accumulators are reset to zero at the end of each stage. For later stages, e.g., $f < \exp(10^{-5})$, the density of states generated by the random walker is carried over to the next stage and the convergence factor is reduced according to $f_{k+1} = \sqrt{f_k}$. The simulation is terminated when the convergence factor satisfies $f < \exp(10^{-10})$.

To estimate the statistical errors in the estimated density of states, 7 independent simulations are conducted, with exactly the same code, but with different random-number generator seeds. The calculated density of states always contains an arbitrary multiplier; the estimated densities of states from these 7 runs are matched by shifting each $\ln \Omega(E)$ in such a way as to minimize the total variance.

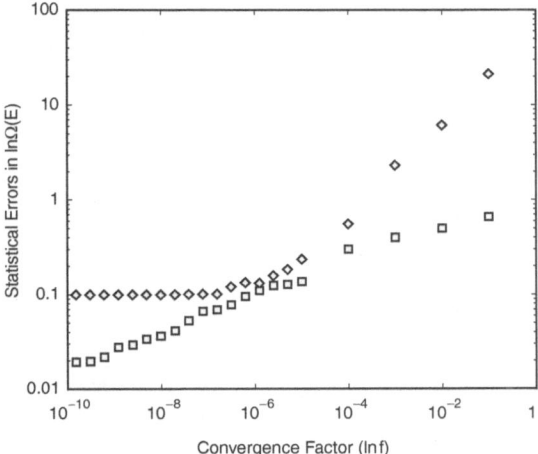

Fig. 1. Statistical errors in the density of states as a function of convergence factor [17]

The diamonds in Fig. 1 show results from the Wang and Landau original algorithm. These errors exhibit two distinct regimes. For large f (e.g., $f > 10^{-4}$), the statistical error is proportional to $\sqrt{f}$. In the small f regime ($f < 10^{-6}$), the error curve levels off, and asymptotically converges to a limiting value of approximately 0.1. The main reason for this is that in the Wang-Landau algorithm, configurations generated at different stages of the simulation do not contribute equally to the histograms. The Wang-Landau algorithm leaves the density of states essentially unchanged once f is reduced to less than 10^{-6}; additional simulations with smaller f only "polish" the results locally, but do little to decrease the overall quality of the data. If phase space has not been ergodically sampled by the time f reaches about 10^{-6}, the final result of the simulation is likely to be inaccurate. Using a more stringent criterion for "flatness" only alleviates the problem partially, because the computational demands required to complete a stage increase dramatically.

The squares in Fig. 1 show the statistical errors in the density of states calculated from configurational temperature. In contrast to the curve generated through a conventional random walk, the errors steadily decrease as the simulation proceeds. The figure also shows that the statistical errors from the configurational temperature method proposed here are always considerably smaller than those from the Wang-Landau technique. At the end of a simulation, the statistical error from a Wang-Landau calculation is approximately 5 times larger than that from configurational temperature. In other words, thermodynamic-property calculations of comparable accuracy would be 25 times faster in the proposed configurational algorithm than in existing random-walk algorithms.

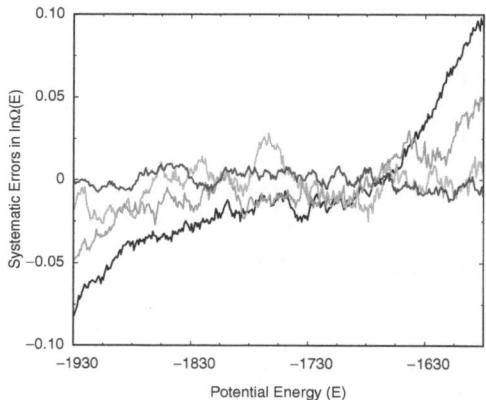

Fig. 2. Systematic errors in configurational temperatures in the first 4 stages of the simulation. With decreasing magnitude of the errors, the corresponding modification factors are $\ln f = 0.1, 0.01, 0.001$, and 0.0001, respectively

Figure 2 shows the difference between the configurational temperature corresponding to the first 4 stages and the final configurational temperature. The figure shows that in the first few stages, the calculated configurational temperature differs from the "true" value in a systematic manner. The reason for such deviations is that in the first few stages the detailed balance condition is grossly violated. For a large convergence factor, the system is expelled into neighboring energy levels, even if the trial configuration does not conform to that energy level. The figure also shows that at the fourth stage ($\ln f = 10^{-4}$), such systematic deviations have already become negligible and are smaller than the random errors; the configurational temperature accumulators need not be reset at the end of each stage once $\ln f \leq 10^{-5}$.

Figure 3 compares the statistical errors in the heat capacity calculated from a Wang-Landau algorithm, from multi-microcanonical ensemble simulations, and from configurational temperature, as a function of actual CPU time. The error in the original Wang-Landau algorithm becomes almost constant once the simulation is longer than about 8 hours [17]. The error from multi-microcanonical ensemble and configurational-temperature simulations are smaller, and continue decreasing as the simulation proceeds.

Recently, it has been proposed that a more efficient version of the original Wang-Landau algorithm can be devised by removing the flatness constraint from the energy histograms [24]. The underlying premise behind non-flat histograms is that the algorithm should maximize the number of round trips of a random walk within an energy range. While that idea has been shown to be effective in the context of spin systems, it is of limited use in the context of complex fluids. The method requires a good initial guess of the density of

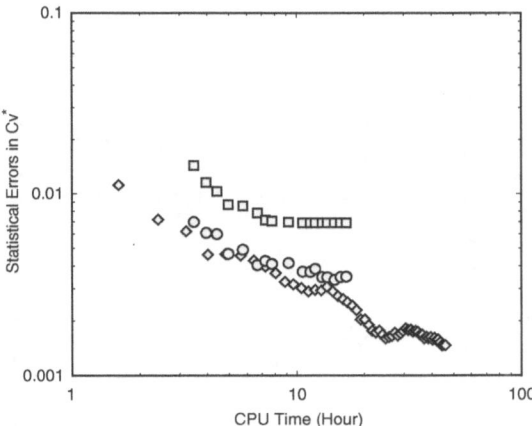

Fig. 3. Statistical errors in heat capacity as a function of CPU time. The *squares* are the results from the Wang-Landau algorithm; the *circles* are the results from configurational temperature calculations, and the *diamonds* are the results from multi-microcanonical ensemble simulations [17]

states, which can only be obtained after considerable effort (or multiple itera-tions of the Wang-Landau algorithm). Our experience suggests that it is more effective to resort to a configurational temperature approach, with multiple overlapping windows [17, 18] whose width and number is adjusted to sample "difficult" regions of phase space more or less exhaustively.

The implementation of configurational temperature density-of-states sim-ulations is illustrated here in the context of a protein. The particular protein discussed in this case is the C-terminal fragment of protein G, which has of-ten been used as a benchmark for novel protein-folding algorithms [14]. This 16-residue peptide is believed to assume a β-hairpin configuration in solu-tion. In this example it is modelled using an atomistic representation and the CHARMM19 force field with an implicit solvent [14, 25, 26]. The total energy range of interest is broken up into smaller but slightly overlapping energy windows $[E_k^- : E_k^+]$ (see Fig. 4). In the spirit of parallel tempering [27], neigh-boring windows are allowed to exchange configurations. The width of individ-ual windows can be adjusted in such a way as to have more configurations in regions where sampling is more demanding. As shown in Fig. 5, high-energy windows include unfolded configurations, whereas low-energy windows are populated by folded configurations. This density of states can subsequently be used to determine any thermodynamic property of interest over the energy range encompassed by the simulation. The heat capacity, for example, which can be determined from fluctuations of the energy according to (7), can be determined to high accuracy and used to extract a precise temperature for the folding transition of this polypeptide. Furthermore, these high accuracy

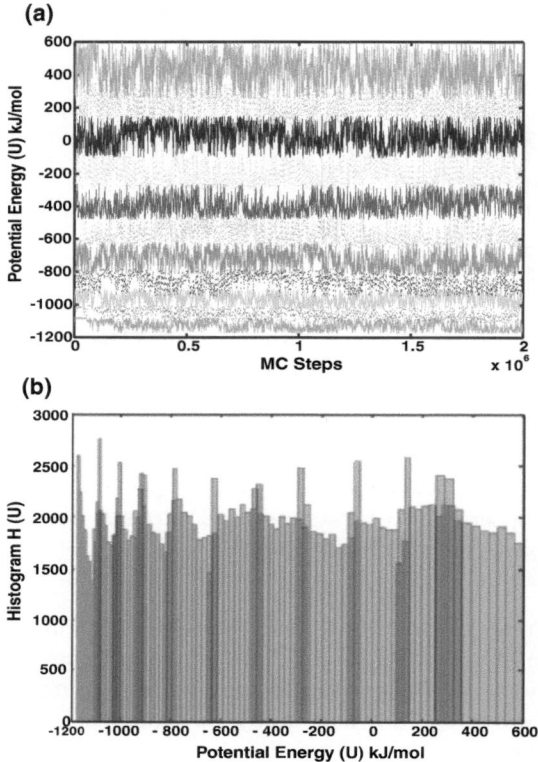

Fig. 4. (a) Evolution of energy in different energy windows during a configurational-temperature DOS simulation of the C-terminal fragment of Protein G. (b) Representative histogram of visited energy states in overlapping windows [18]

estimates of c_v over a relatively wide energy range can be generated on the order of 100 hours of CPU time on a Pentium architecture.

Our description thus far of DOS-based methods has centered on calculation of the density of states. A particularly fruitful extension of such methods involves the calculation of a potential of mean force (Φ), or PMF, associated with a specified generalized reaction coordinate, $\xi(r)$ [28,29]. A PMF measures the free energy change as a function of $\xi(r)$ (where r represents a set of Cartesian coordinates). This potential is related to the probability density of finding the system at a specific value ξ of the reaction coordinate $\xi(r)$:

$$P(\xi(r) = \xi) \equiv Ce^{-\beta\Phi(\xi)} , \tag{20}$$

where C is a normalization constant. As we shall later see, the methods of umbrella sampling rely on (20) for estimating free energy changes by altering the potential function with a biasing function designed to sample phase space more efficiently. This bias is later removed and the simulation data are

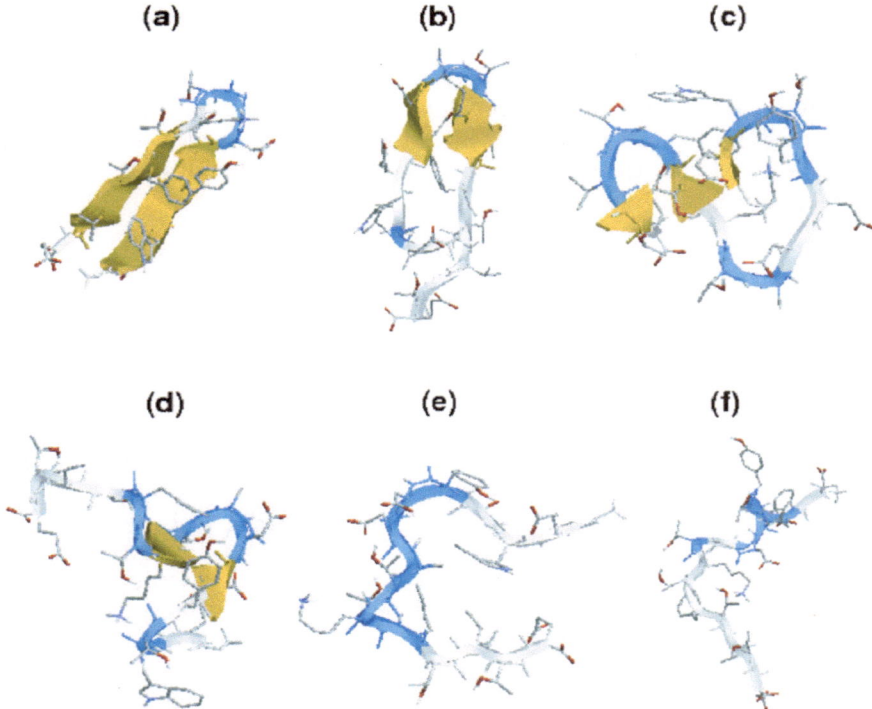

(a)   **(b)**   **(c)**

(d)   **(e)**   **(f)**

Fig. 5. Representative configurations of the C-terminal fragment of protein G corresponding to different energy ranges

reweighted to arrive at the correct probability distribution. The potential of mean force is then calculated from:

$$\Phi(\xi) = -k_B T \ln P(\xi) + C \ .$$

(21)

Other methods arrive at the potential of mean force by calculating the derivative of the free energy with respect to the constrained generalized coordinate ξ in a series of simulation runs. A mean force, $\langle F \rangle_\xi = -\frac{\partial(\Phi(\xi))}{\partial(\xi)}$, can be integrated numerically to yield the corresponding PMF. These simulations involve constraints, and an appropriate correction term must therefore be accounted for in the calculation of the PMF. For a system with no constraints and consisting of N particles with Cartesian coordinates $\boldsymbol{r} = (\boldsymbol{r}_1, \boldsymbol{r}_2, \ldots, \boldsymbol{r}_N)$, the mean force can be written as:

$$F_\xi = \frac{\left\langle \frac{\partial \boldsymbol{r}}{\partial \xi} \left[-\frac{\partial E}{\partial \boldsymbol{r}} + k_B T \frac{\partial \ln |\boldsymbol{J}|}{\partial \boldsymbol{r}} \right] \delta(\xi(\boldsymbol{r}) - \xi) \right\rangle}{\left\langle \delta(\hat{\xi}(\boldsymbol{r}) - \xi) \right\rangle} \ ,$$

(22)

where E is the potential energy function and $\boldsymbol{J}$ is the Jacobian associated with the transformation from Cartesian to generalized coordinates. To compute the mean force acting on the end-to-end distance of a molecule, for example, a suitable reaction coordinate is provided by $\xi = r_{ij} = |\boldsymbol{r}_i - \boldsymbol{r}_j|$, where r_{ij} represents the distance between the two terminal sites i and j. The Jacobian, $|\boldsymbol{J}|$, is a function of the separation r_{ij} and can be taken out of the ensemble average to yield [29]:

$$\langle F \rangle_\xi = \left\langle -\frac{\partial E}{\partial \xi} \right\rangle_\xi + \frac{2k_B T}{\xi} . \tag{23}$$

The mean force therefore includes a contribution from the average mechanical force and another contribution arising from the variations of the volume element associated with the reaction coordinate ξ. The free energy change between two states ξ_1 and ξ_2 can be obtained by integrating (23) according to

$$\Phi(\xi_2) - \Phi(\xi_1) = \int_{\xi_1}^{\xi_2} \left\langle \frac{\partial E}{\partial \xi} \right\rangle d\xi - 2k_B T \ln\left(\frac{\xi_2}{\xi_1}\right) . \tag{24}$$

Constrained simulations rely on the calculation of the first term of (23) from a series of simulations conducted at different values of ξ. This average force is then corrected by adding the second term of (23), and then numerically integrated to give a potential of mean force for the desired range of ξ.

As noted above, the density-of-states method described earlier can be extended to yield accurate estimates of the potential of mean force. The weight factors that dictate the walk in the ξ space can be computed "on the fly" during a simulation in a self-consistent manner. The simulation can be performed without any constraints, which means that the resulting weights can be used directly as in (21) to give the potential of mean force. One can also accumulate the forces acting on the particles that define the reaction coordinate and then use (24) to get the PMF. The computed PMF is therefore available as a continuous function of ξ.

In recent work [28,29] we have explored the use of DOS methods in the context of expanded ensembles, where intermediate states are introduced to facilitate the transition between configurations separated by large energy barriers. We refer to the resulting methods through the acronym EXEDOS (for Expanded Ensemble Density of States). The expanded states are usually defined by some reaction coordinate, ξ, and the sampling in ξ space is governed by unknown weights. This so-called expanded-ensemble density of states method has been employed for studies of suspensions of colloidal particles in liquid crystals [28], colloidal suspensions in polymer solutions [30], and proteins on a surface [31]. In one of the examples at the end of this chapter we discuss it in the context of reversible, mechanical stretching of proteins [32]. In that example, the reaction coordinate, ξ, is chosen to be the end-to-end distance between the N and C terminus of the molecule being stretched. In a different example we discuss it in the context of crystallization [33], where a more elaborate order parameter is necessary.

The goal of the method is to perform a random walk in ξ space. Consider a system consisting of N particles interconnected to form a molecule, and having volume V and temperature T. The end-to-end distance of the molecule (ξ) can be discretized into distinct states; each state is characterized by its end-to-end distance, ξ, in some specified range of interest $[\xi_-,\xi_+]$; ξ_- and ξ_+ represent a lower and an upper bound, respectively. The partition function Ω of this expanded ensemble is given by

$$\Omega = \int Q(N,V,T,\xi)g(\xi)d\xi = \int Q_\xi g(\xi)d\xi \ , \tag{25}$$

where Q_ξ is the canonical partition function for that particular state ξ, and $g(\xi)$ is the corresponding weight factor. The probability of visiting a state having extension ξ can be written as

$$P(\xi) = \frac{Q_\xi g(\xi)}{\Omega} \ . \tag{26}$$

The free energy difference between any two states can therefore be calculated from the weight factors, and the population density can be determined from

$$\Phi(\xi_2) - \Phi(\xi_1) - k_B T \ln \frac{Q_{\xi_2}}{Q_{\xi_1}} = -k_B T \left[\ln \frac{g(\xi_1)}{g(\xi_2)} + \ln \frac{P(\xi_2)}{P(\xi_1)} \right] \ . \tag{27}$$

If each state is visited with equal probability, the second term on the right of (27) disappears and the PMF can be computed from

$$\Phi(\xi) = -k_B T \ln g(\xi) + C \ . \tag{28}$$

In the method discussed here, a running estimate of the weight factors can be computed and refined in a self-consistent and self-monitoring manner. At the beginning of a simulation, $g(\xi)$ is assumed to be unity for all states. Trial Monte Carlo moves are accepted with probability

$$P_{\mathrm{acc}}(\xi_1 \to \xi_2) = \min \left[1, \frac{g(\xi_1)}{g(\xi_2)} \exp(-\beta \Delta E) \right] \ , \tag{29}$$

where ξ_1 and ξ_2 are the end-to-end distances of the system before and after a trial move. After each trial move, the corresponding weight factor is updated by multiplying the current, existing value by a convergence factor, f, that is greater than unity $(f > 1)$, i.e., $g(\xi) \to g(\xi) f$. Every time that $g(\xi)$ is modified, a histogram $H(\xi)$ is also updated. As before, this $g(\xi)$ refinement process continues until $H(\xi)$ becomes sufficiently flat. Once this condition is satisfied, the convergence factor is reduced by an arbitrary amount. We use again $f_{\mathrm{new}} = \sqrt{f_{\mathrm{old}}}$. The histogram is then reset to zero $(H(\xi) = 0)$, and a new simulation stage is started. The process is repeated until f is sufficiently small.

In addition to computing these weight factors from the histograms of visited states, one can obtain a second estimate from the integration of the mean force, as given by (24). The first term on the right hand side of (23) can be estimated in a density of states simulation. The component of the total force acting on the two sites that define the reaction coordinate along the end-to-end vector, $\hat{\xi}$, is accumulated as a function of ξ. At the end of the simulation this mean force is corrected by adding the corresponding second term of (23), and then integrated to yield the PMF.

As mentioned above, in the earlier stages of the simulation ($\ln f > 10^{-5}$), when the convergence factor is large, detailed balance is severely violated. As a result, thermodynamic quantities computed during this time (including average forces) are incorrect. To avoid carrying this error into later stages, the accumulators for average forces are reset at the end of early stages. As the convergence factor decreases (e.g., $\ln f < 10^{-5}$), the violation of detailed balance has a smaller effect, and the accumulators need not be reset anymore.

2.4 Successive Umbrella Sampling

A complementary and alternative technique to reweighting methods are umbrella sampling strategies. The guiding idea of umbrella sampling [34] is to divide the pertinent range of the order parameter, E, into smaller windows and to investigate one small window after the other. If the windows are sufficiently narrow, the free energy profile will not substantially vary within a window, and a crude estimate of the "weights" is often sufficient to ensure uniform sampling [35]. In the limiting case, the windows just consist of two neighboring values of the order parameter.

A histogram $H_k(E)$ monitors how often each state is visited in the kth window $[E_k^- : E_k^+]$. Care must be exercised at the boundaries of a window to fulfill detailed balance [36]: If a move attempts to leave the window, it is rejected and $H(\text{window edge})$ is incremented by unity. Another question which may arise from the discussion of the boundary is the optimum amount of overlap to minimize the uncertainty of the overall ratio. Here we choose the minimal overlap of one state at the interval boundaries, i.e., $E_k^+ = E_{k-1}^-$. This is simple to implement and sufficient to match the probability distributions at their boundaries. A larger overlap may reduce the uncertainty but requires a higher computational effort.

Let $H_{k-} \equiv H_k(E_k^-)$ and $H_{k+} \equiv H_k(E_k^+)$ denote the values of the kth histogram at its left and right boundary, and $R_k \equiv H_{k+}/H_{k-}$ characterize their ratio. After a predetermined number of Monte Carlo steps per window, the (unnormalized) probability distribution can recursively be estimated according to:

$$\frac{P(E)}{P(E_0^-)} = \frac{H_{0+}}{H_{0-}} \cdot \frac{H_{1+}}{H_{1-}} \cdots \frac{H_k(E)}{H_{k-}} = \Pi_{i=1}^{k-1} R_i \cdot \frac{H_k(E)}{H_{k-}} \qquad \text{with} \quad R_i \equiv \frac{H_{i+}}{H_{i-}},$$

(30)

when $E \in [E_k^- : E_k^+]$. The ratios in (30) correspond to the Boltzmann factor associated with the free energy difference across the order parameter interval i.

We now consider how the overall error depends on the choice of the window size [35]: For clarity, we assume that no sampling difficulties are encountered in our system, i.e., the order parameter in which we reweight (e.g., the energy, E, or the particle number, n) is suitable to flatten and overcome all barriers in the (multidimensional) free energy landscape and the restriction of fluctuations of the average order parameter by the window size does not impart sampling errors onto the simulation [37]. Under these circumstances, it has been suggested [38, 39] that small windows reduce computational effort by a factor of N_k, where N_k denotes the total number of windows into which the sampling range is subdivided: The time τ to obtain a predetermined standard deviation δ of the ratio R_k in a single window is proportional to the square of the window size, $\Delta_k = E_k^+ - E_k^-$. With $\tau \sim \Delta_k^2$, we get a total computation time (for all windows) of $t_{\text{cpu}} \sim N_k \Delta_k^2$, implying that the overall error of the simulation is also δ. This contrasts the behavior of a single large window; $N_m' = 1$ and $\Delta_m' = N_k \Delta_k$ yield $t_{\text{cpu}}' = (N_k \Delta_k)^2 = N_k t_{\text{cpu}}$, which suggests that a window size as small as possible should be chosen [38, 39]. Being interested in localizing phase coexistence, however, the pertinent error is related to the free energy difference of the two end points of the order parameter interval that corresponds to the distinct phases. In this case, we have to account for error propagation in (30) and we obtain for the error, Δ, of the ratio $P(E_{N_k}^+)/P(E_1^-)$

$$\Delta \equiv \delta \left(\frac{P(E_{N_k}^+)}{P(E_1^-)} \right) = \sqrt{\sum_{k=1}^{N_k} \delta R_k^2} \sim \mathcal{O}(\delta \sqrt{N_k}) \,.$$

(31)

Due to error propagation across the windows, the error of each individual subinterval has to be smaller than the total error, Δ, by a factor of $\sqrt{N_k}$. Thus, the time that must be spent in each window to achieve an error of $\delta = \Delta/\sqrt{N_k}$ is $N_k \tau$ and the total simulational effort is $N_k^2 \tau \sim (N_k \Delta_k)^2 \sim N_k t_{\text{cpu}}$, which is identical to the time required for a single large window. This argument implies that the statistical error for a given computational effort is independent from the window size, i.e., the number of intervals the range of order parameter is divided into. A more detailed analysis including possible systematic errors due to (i) estimating probability ratios and (ii) correlations between successive windows can be found in [35].

The computational advantage from subdividing the range of order parameter into windows does not stem from an increased statistical accuracy due to the small correlation times within a window, but from the fact that successive umbrella sampling does not require the independent and computationally costly generation of high-quality "weights". Making the window size small we reduce the variation of the free energy across a single interval and achieve sufficiently uniform sampling without or with a very crude approximation of the "weights". In this sense, successive umbrella sampling is as efficient as a multicanonical simulation with very good "weights", except that now the "weights" need not be known beforehand. Additionally, the scheme is easy to implement on parallel computers and the range of order parameter can be easily enlarged by simply adding more windows. This computational scheme has successfully been applied to study phase equilibria and interface properties in polymer-solvent mixtures [40,41], polymer-colloid mixtures [42–45] and liquid crystals [46,47].

As we are sampling one window after the other, efficiency can be increased by combining the scheme with the multicanonical concept. In a multicanonical simulation we replace $H(E)$ in (30) by $H(E) \exp[w(E)]$. In principle, one could use Wang-Landau sampling to estimate the "weights" on the fly. Given that the free energy differences are small a rather crude method often suffices: After $P(E)$ is determined according to (30), $w(E) = \ln[P(E)]$ is extrapolated into the next window. The first window is usually unweighted. If, in this case, states are not accessible, $w(E)$ can be altered by a constant amount of $k_B T$ in each iteration step. In the limit that each window only contains two states we use linear extrapolation for the second and quadratic extrapolations for all subsequent windows. Then, extrapolated and true values for $P(E)$ only differ by a few percent. The basic idea behind this kind of extrapolation is to flatten the free energy landscape even in the small windows. Depending on the steepness of the considered landscape, this may lead to a significant reduction of the error vis-à-vis unweighted umbrella sampling.

Additionally, one should keep in mind that multicanonical simulations in a single, one-dimensional order parameter are not always sufficient when free energy landscapes become more complex, and that the restriction of fluctuations of the average order parameter in a window can impart sampling problems. The problem of simulating too small windows with umbrella sampling is well known, but difficult to quantify [34]. Hence, in practice, the optimum choice of the window size is a compromise between a small value, which allows for a efficient sampling even in the absence of accurate "weights", and a value large enough to avoid getting trapped kinetically. This compromise depends on the specific system of interest and its size. A particular aspect of these general issues is discussed further in the next section.

2.5 Configurations Inside the Miscibility Gap and Shape Transitions

In the previous section we have outlined computational methods to accurately obtain free energy profiles as a function of an order parameter, E. On the one hand, these one-dimensional free energy profiles contain a wealth of information. On the other hand, the configuration space is of very high dimension and a one-dimensional order parameter might not be sufficient to construct a thermodynamically reversible path, i.e., even if the histogram of the order parameter is flat the system might still be forced to overcome barriers in order to sample the entire interval of order parameters. This is readily observable in the time sequence of the order parameter: For a "good" choice of the order parameter, E, the system performs a random walk in E. For a "bad" choice of the order parameter the histogram of E is flat, but the range of E can be divided into subintervals. Within a subinterval the different values of the order parameter are sampled in a random walk-like fashion but there is a barrier associated with crossing from one subinterval to another. The seldom crossing of the barriers between subintervals can slow down the simulation considerably even if the "weights" are optimal. The success of the multicanonical method depends on an appropriate identification of the order parameter (or reaction coordinate) that "resolves" all pertinent free energy barriers.

We illustrate the behavior for a first order transition between a vapor and a dense liquid in the framework of a simple Lennard-Jones model. The condensation of a vapor into a dense liquid upon cooling is a prototype of a phase transition that is characterized by a single scalar order parameter – the density, ρ. The thermodynamically conjugated field is the chemical potential, μ. The qualitative features, however, are general and carry over to other types of phase coexistence, e.g., Sect. 3.4.

At a given temperature, T, and chemical potential, $\mu_{\mathrm{coex}}(T)$, a liquid and a vapor coexist if they have the same pressure, p_{coex}:

$$p_{\mathrm{liq}}(T, \mu_{\mathrm{coex}}) = p_{\mathrm{vap}}(T, \mu_{\mathrm{coex}}) = p_{\mathrm{coex}} \tag{32}$$

The two coexisting phases are characterized by the two values of their order parameter, ρ_{liq} and ρ_{vap}. If one prepares a macroscopic system of volume V, at a fixed density, $\rho_{\mathrm{vap}} < \rho < \rho_{\mathrm{liq}}$, inside the miscibility gap, it will phase separate into two macroscopic domains in which the densities attain their coexistence values. The volumes of the domains, V_{liq} and V_{vap}, are dictated by the lever rule:

$$\rho_{\mathrm{liq}} \frac{V_{\mathrm{liq}}}{V} + \rho_{\mathrm{vap}} \frac{V_{\mathrm{vap}}}{V} = \rho . \tag{33}$$

The pressure of the system inside the miscibility gap is p_{coex} independent from density. This macroscopic behavior is in marked contrast to the van-der-Waals loop of the pressure, p, which is predicted by analytic theories that consider the behavior of a hypothetical, spatially homogeneous system inside the miscibility gap.

Using multicanonical simulation techniques we can sample all configurations in the pertinent interval of density (order parameter) and determine their free energy. It is important to note that the simulation samples all states at a fixed order parameter with the Boltzmann weight of the canonical ensemble. What are the typical configurations that a finite system of volume V adopts inside the miscibility gap in the canonical ensemble [48–55]?

Let us consider the condensation of the vapor phase as we increase the density at coexistence. If the excess number of particles, $\Delta n = (\rho - \rho_{\mathrm{vap}})V$, is small, they will homogeneously distribute throughout the simulation cell and form a supersaturated vapor. In this case the excess free energy defined by

$$\Delta F(\rho) \equiv F(\rho) - F_{\text{vap}} - \frac{F_{\text{liq}} - F_{\text{vap}}}{\rho_{\text{liq}} - \rho_{\text{vap}}} (\rho - \rho_{\text{vap}}) \tag{34}$$

takes the form:

$$\Delta F_{\text{sv}}(\rho) = \frac{V}{2\kappa} \Delta\rho^2. \tag{35}$$

where κ denotes the compressibility of the vapor and $\Delta\rho = \frac{\rho - \rho_{\text{vap}}}{\rho_{\text{liq}} - \rho_{\text{vap}}}$ is the normalized distance across the miscibility gap. Increasing the excess number of particles (or $\Delta\rho$) we increase the excess free energy quadratically and the curvature is proportional to the compressibility.

In a macroscopic system a supersaturated vapor, $\Delta\mu \equiv \mu - \mu_{\text{coex}} > 0$, is metastable and the excess number of particles will condense into a drop of radius R that consists of the thermodynamically stable liquid. In the framework of classical nucleation theory, the excess free energy of such a spatially inhomogeneous system can be decomposed into a surface and a volume contribution:

$$F_{\text{drop}}(R) \approx F_{\text{vap}} + 4\pi\gamma R^2 - \frac{4\pi}{3} R^3 (\rho_{\text{liq}} - \rho_{\text{vap}}) \Delta\mu. \tag{36}$$

The first term describes the increase of the free energy due to the formation of a liquid-vapor interface and γ denotes the interface free energy per unit area (interface tension). The second term describes the free energy reduction by the formation of the thermodynamically stable phase. In the simplest approximation we assume that (i) all excess particles condense into a single drop and (ii) the density of the drop's interior corresponds to the liquid, ρ_{liq}, and its interface tension is given by the macroscopic value γ. These assumptions are reasonable for large drops. Then the size of the drop is given by the lever rule:

$$\frac{4\pi}{3} R^3 (\rho_{\text{liq}} - \rho_{\text{vap}}) = (\rho - \rho_{\text{vap}}) V. \tag{37}$$

Since $F_{\text{liq}} - F_{\text{vap}} = -\Delta\mu V(\rho_{\text{liq}} - \rho_{\text{vap}})$ the excess free energy of a droplet is given by the surface contribution

$$\Delta F_{\text{drop}}(R) \approx 4\pi\gamma R^2. \tag{38}$$

Note that the conjugated field, $\Delta\mu$, does not have a direct interpretation in a multicanonical simulation.

The two equations, (37) and (38), allow us to calculate the excess free energy as a function of the excess density. To first order we obtain, $R \sim \Delta\rho^{1/3}$ and $\Delta F_{\text{drop}} = g(V\Delta\rho)^{2/3}$ with $g = (4\pi/9)^{1/3}\gamma$. A more accurate expression can be obtained by not condensing all excess particles into the drop but allowing the density, ρ', of the surrounding vapor to increase [56]. This increases the free energy of the vapor but it decreases the drop's radius and the associated interface free energy. Then one obtains the excess free energy by minimizing with respect to the drop's radius, R and the vapor density, ρ'

$$\Delta F_{\text{drop}}(\Delta \rho) = \min_{R, \rho'} \left[4\pi \gamma R^2 + \frac{V - \frac{4\pi}{3} R^3}{2\kappa} (\rho' - \rho_{\text{vap}})^2 \right] \quad (39)$$

under the constraint

$$\rho' V + \frac{4\pi}{3} R^3 (\rho_{\text{liq}} - \rho') = V\rho \,. \quad (40)$$

This refinement qualitatively captures that the chemical potential for a drop configuration is not $\Delta\mu = \mu - \mu_{\text{coex}} = 0$ but rather $\Delta\mu = \partial \Delta F / \partial n$. The shift of the chemical potential increases with the interface tension and decreases as the size becomes larger (Kelvin's equation).

If $\Delta\rho$ increases further, the drop grows until its size becomes comparable to the linear dimension $V^{1/3}$ of the simulation cell. At that point it becomes favorable to form a liquid slab which is separated from the vapor by two interfaces of area $V^{2/3}$. In this case the excess free energy

$$\Delta F_{\text{slab}} = 2\gamma V^{2/3} \quad (41)$$

is independent from the excess density $\Delta\rho$. As both, the drop and the slab excess free energies scale like $\gamma V^{2/3}$, the transition from a drop to a slab occurs at a fixed $\Delta\rho$ which depends on the aspect ratio of the simulation cell, but which is independent from its volume or the interface tension. Hence, the largest drops that are observable have a radius $R_{\text{max}} \sim V^{1/3}$.

Increasing the excess density further one observes the reverse set of configurations: From a slab-like configurations one goes to a bubble of vapor surrounded by liquid, and finally to an undersaturated but spatially homogeneous liquid.

The free energy profile obtained from Monte Carlo simulations of a small Lennard-Jones monomer system and the above approximations utilizing values of the interface tension and compressibility extracted from independent simulations are shown in Fig. 6. Of course, the simple expressions overestimate the value of the excess free energy, but the qualitative shape of the free energy profile is predicted well. Snapshots of the simulations are presented in Fig. 7. These corroborate the correct identification of the dominant system configurations.

From the dependence of the free energy profile on the excess density it is apparent that the free energies of the supersaturated vapor and the drop will exhibit an intersection point. For small excess densities the homogeneous supersaturated vapor, $\Delta F_{\text{sv}} \sim \Delta\rho^2$, has the lower free energy while for larger excess densities the drop configuration, $\Delta F_{\text{drop}} \sim \Delta\rho^{2/3}$, will be more stable. The "transition" between the supersaturated vapor and the configuration containing a drop is called droplet evaporation/condensation [50,54,55]. From the two simple expressions we can readily read off that droplet condensation occurs at

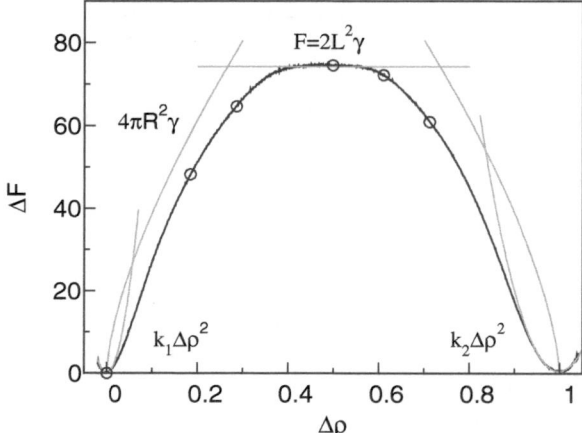

Fig. 6. Grand-canonical simulation of a Lennard-Jones monomer system. Particles interact via a shifted and truncated Lennard-Jones potential of the form: $E_{\mathrm{LJ}} = 4\epsilon\left[\left(\frac{\sigma}{r}\right)^{12} - \left(\frac{\sigma}{r}\right)^6 + \frac{127}{16384}\right]$ for distances, $r \leq 2\sqrt[6]{2}\sigma$, and $E = 0$ for larger particle separations. The free energy profile ΔF (in units of ϵ) has been obtained from the probability distribution of the order parameter, ρ, in grand-canonical simulations for a cubic simulation cell of linear dimension $L = 11.3\ \sigma$ and $k_B T/\epsilon = 0.78$. The phenomenological expressions (35), (38), and (41) for the free energy are also indicated using $\gamma = 0.291\ \frac{\epsilon}{\sigma^2}$. Circles mark densities at which typical configurations are visualized in Fig. 7

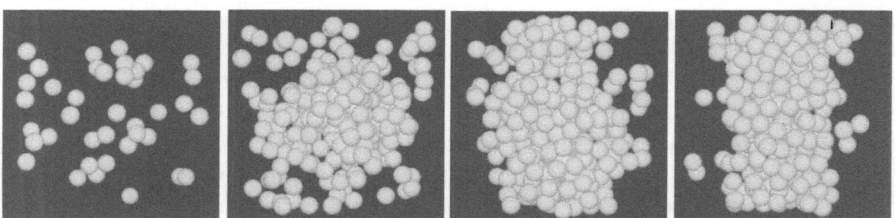

Fig. 7. Typical system configurations for the same parameters as in Fig. 6. The density is $\Delta\rho = 0, 0.184, 0.286, 0.498$. Only a thin slice of the simulation box is shown in the right-most image. Each monomer is represented by a sphere of diameter $1.12\ \sigma$, which corresponds to the minimum of the Lennard-Jones potential

$$\Delta\rho_{\mathrm{dc}} = (2\kappa g)^{3/4}V^{-1/4} \tag{42}$$

where g is defined on p. 88. Of course, for any finite simulation cell going from a supersaturated vapor to a drop is not a sharp transition in a thermodynamic sense because the free energy difference between the two states remains finite. Thus the "transition" will be rounded over a range of densities $\delta\rho = \Delta\rho - \Delta\rho_{\mathrm{dc}}$ where the free energy difference between the two states – supersaturated vapor and drop – is comparable to the thermal energy scale, $\delta F = |\Delta F_{\mathrm{drop}} -$

$\Delta F_{\rm sv}| \sim \mathcal{O}(k_B T)$. Using the above expressions one can expand the free energy difference between the drop and the supersaturated vapor as a function of the distance from the droplet evaporation/condensation:

$$\delta F \approx -\frac{3}{4} \left([2\kappa]^{-1/3} g V \right)^{3/4} \left(\Delta\rho - (2\kappa g)^{3/4} V^{-1/4} \right), \qquad (43)$$

This estimate yields $\delta\rho \sim \left(\kappa^{-1/3} g V \right)^{-3/4}$ for the width of the transition region. As we increase the system size, the excess density at which the supersaturated homogeneous vapor is stable decreases like $V^{-1/4}$ but the smallest drops that are observable at that density are of radius $R_{\rm min} \sim V^{1/4}$, i.e., they increase with system size.

In the following we illustrate in somewhat more detail the droplet evaporation/condensation in a finite-sized system of Lennard-Jones monomers. In order to accurately locate the droplet condensation, we regard the derivative of the free energy $\Delta\mu \equiv \partial\Delta F/\partial n$. The results for such a system are shown in Fig. 8. Inside the miscibility gap and for finite V, $\Delta\mu$ does not remain constant inside the miscibility gap as suggested by macroscopic arguments. First the chemical potential, $\Delta\mu$, linearly increases with the excess density, $\Delta\rho$. This behavior characterizes the homogeneous, supersaturated vapor. Further inside the miscibility gap, $\Delta\mu$ exhibits an s-shaped variation as a function of $\Delta\rho$.

At the densities marked by the symbol ∘ we store configurations for further analysis. After the simulation, the distribution, $P(N_c)$, of cluster sizes was determined. Any ensemble of particles whose distance is smaller than 1.5σ is assumed to belong to the same cluster (Stillinger criterion) [57]. For

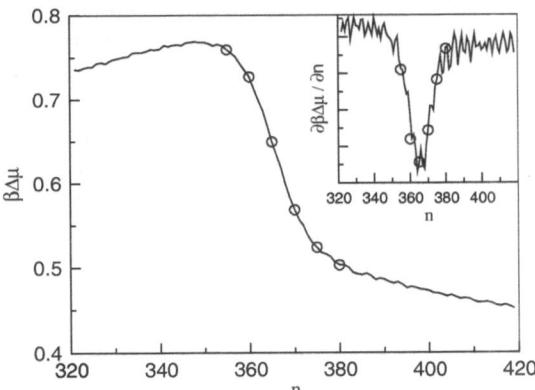

Fig. 8. Plot of $\Delta\mu/k_B T$ vs. the number of particles n, for a Lennard-Jones fluid in a cubic simulation box of size $L = 22.5\sigma$ at $k_B T/\epsilon = 0.68$. $\Delta\mu = \mu - \mu_{\rm coex}$ is the distance from bulk coexistence. ∘ denote states at which configurations have been stored for further analysis (Figs. 9 and 10). The inset shows the derivative of $\Delta\mu$ w.r.t. the number of particles. From MacDowell et al. [55]

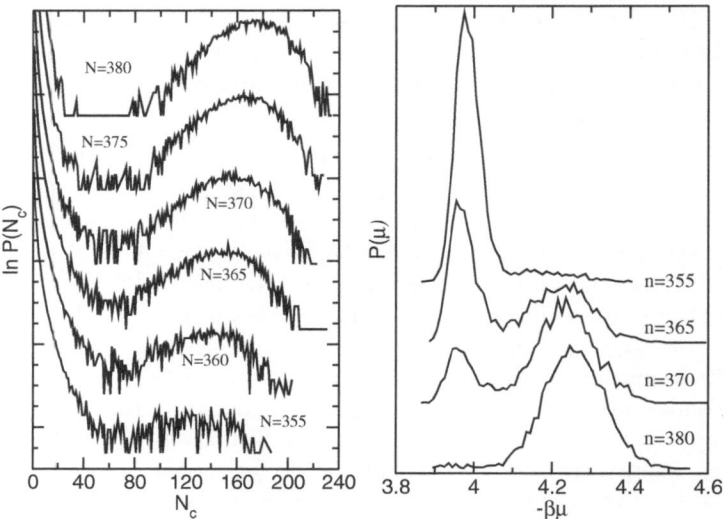

Fig. 9. Distribution $P(N_c)$ of the cluster size N_c for several choices of particle number, n, (*left*) and the corresponding distribution of the chemical potential of the supersaturated gas $P(\Delta\mu)$ obtained by Widom's particle insertion attempts (*right*). System parameters are the same as in Fig. 8. From MacDowell et al. [55]

$n \leq 350$, i.e., on the ascending branch of the $\Delta\mu$ vs. n curve in Fig. 8, $P(N_c)$ is monotonically decreasing with cluster size, N_c (not shown). This distribution characterizes the homogeneous supersaturated vapor. For larger supersaturation, $n \approx 355$, a peak around $N_c \approx 120$ appears. This maximum becomes more pronounced and moves to larger sizes, N_c, as n increases. It corresponds to a single large liquid drop. Such a liquid drop, however, cannot be found in all sampled configurations; but rather some configurations contain a drop while others correspond to a supersaturated vapor. This is most clearly observed when we investigate the probability distribution of the chemical potential, $\Delta\mu$. This quantity has been calculated by performing Widom's particle insertion attempts into the stored configurations. For small particle numbers, n, the distribution of chemical potentials consists of a single peak that characterizes the supersaturated vapor. Upon increasing the particle number a second peak occurs at more negative values of μ. These lower values correspond to configurations with a drop: Some of the excess particles have condensed into the drop and the concomitant reduction of the density of the surrounding vapor gives rise to a lower chemical potential. Therefore the decrease of the chemical potential upon increasing of the particle number in Fig. 8 indicates the droplet evaporation/condensation and gives rise to the s-shaped behavior of the $\Delta\mu$ vs. $\Delta\rho$ curve. From the broad distributions and the fact that supersaturated vapor and drop configurations can be observed over an extended interval of particle numbers (or temperature) it is apparent that in

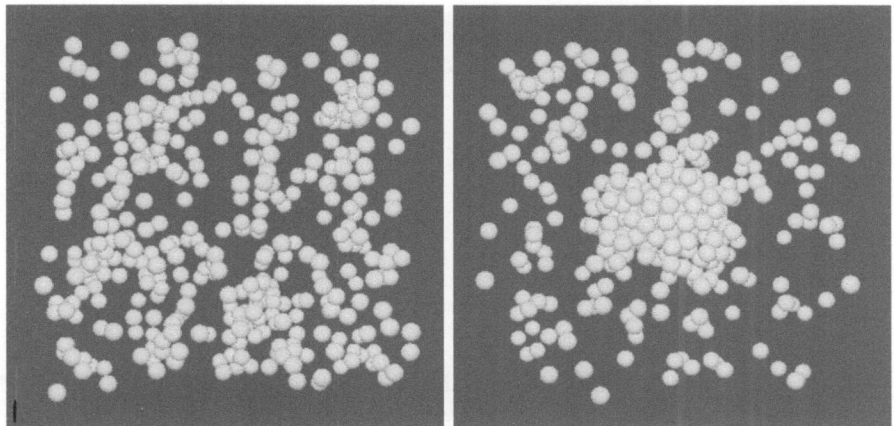

Fig. 10. Two snapshots of configurations at the transition point ($n = 365$ particles): homogeneous, supersaturated vapor (*left*) and drop (*right*). System parameters are the same as in Fig. 8. From MacDowell et al. [55]

a finite-sized system the "transition" is not a sharp one but rather a gradual crossover from configurations dominated by a homogeneous distribution of particles to configurations that contain a drop occurs. Two typical snapshots of these configurations at the same number of particles are presented in Fig. 10. In the course of the simulations the system switches from one to the other conformation forth and back.

The dependence on the system size is explored in Fig. 11: In the left panel we plot the chemical potential vs. density for system sizes ranging from $L = 11.3\sigma$ to 22.5σ. The turning point of the curves shifts closer to the coexistence density of the vapor, $\Delta\rho \to 0$, as we increase the system size. Also the maximum slope increases with increasing L, indicating that for $L \to \infty$ a sharp transition occurs. The right panel of Fig. 11 presents the probability distribution of the energy, U, at the droplet evaporation/condensation for different system sizes. In qualitative agreement with the expectations the droplet evaporation/condensation becomes sharper and both states (the supersaturated vapor and drop) become more separated as we increase the system size.

From the turning points of the $\Delta\mu$ vs. $\Delta\rho$ curves we estimate the location of the droplet evaporation/condensation. The inset of Fig. 11 (right) shows the dependence of $\Delta\rho_{\mathrm{dc}}$ on the system size. The data are compatible with an effective power law $\Delta\rho_{\mathrm{dc}} \sim V^{-0.35}$, while the phenomenological consideration (see (42)) yields an exponent $-1/4$. The deviations can be traced back to the small system sizes: (i) the drop evaporation/condensation still occurs far inside the miscibility gap such that the simple compressibility approximation in (35) breaks down and (ii) the drop is not large enough to neglect additional contributions to the excess free energy, e.g., due to curvature effects. Similarly, only the gross qualitative features of $\Delta F(\Delta\rho)$ in Fig. 6 are described by equations

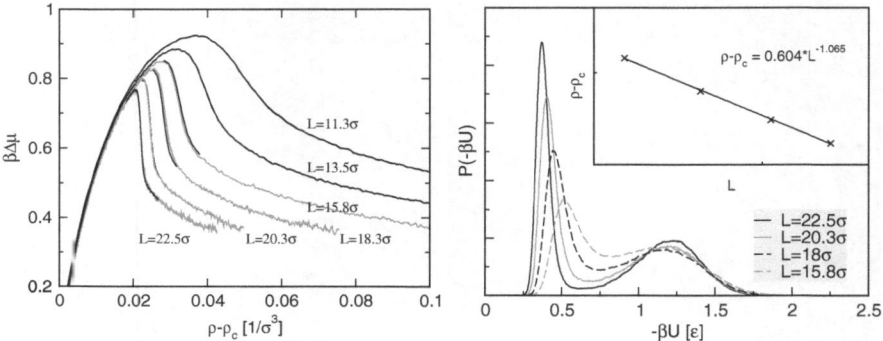

Fig. 11. *Left panel:* Chemical potential vs density loops for $T/k_B T = 0.68$ and several system sizes as indicated in the key. *Right panel:* Distribution of the energy per particle for different system sizes at the droplet condensation. Inset: $\rho - \rho_{\text{vap}}$ as a function $L = V^{1/3}$. ($\rho - \rho_{\text{vap}}(L = 15.8 \, \sigma) = 0.0320 \, \frac{1}{\sigma^3}, \rho - \rho_{\text{vap}}(L = 18 \, \sigma) = 0.0277 \, \frac{1}{\sigma^3}, \rho - \rho_{\text{vap}}(L = 20.3 \, \sigma) = 0.0244 \, \frac{1}{\sigma^3}, \rho - \rho_{\text{vap}}(L = 22.5 \, \sigma) = 0.0220 \, \frac{1}{\sigma^3}$). From MacDowell et al. [55]

(35,38,41), but there are quantitative differences. If one insists on identifying the drop's radius according to (38) (and thereby lumps all approximations into the estimate of the interface tension), these deviations correspond to a reduction of the effective interface tension of small drops of the order 20%.

This simulation study illustrates what kind of qualitatively different conformations are sampled inside the miscibility gap and what macroscopic quantities, γ and κ, can be obtained from the free energy profile. It is important to realize that the interval of excess densities in which supersaturated vapor or drops can be observed depends on the system size. In the limit of large system size, the droplet evaporation/condensation shifts towards the coexistence curve, $\Delta\rho_{\text{dc}} \to 0$. This corresponds to the macroscopic phase coexistence where the excess free energy ΔF vanishes across the miscibility gap. For a finite system, however, the interface contribution will be important – the system chooses a balance between an inhomogeneous density distribution with the corresponding free energy costs associated with the interfaces and the free energy cost of increasing the bulk density homogeneously. The free energy will vary with density and its scale can be estimated by the excess free energy at the droplet evaporation/condensation, $\frac{\Delta F^*}{V} \sim \frac{1}{\sqrt{V}}$.

There are two important consequences for simulations:

- In a finite size simulation box of volume, V, drops of linear dimension R can only be observed in a very limited range of excess densities. The size is limited by $V^{1/4} \sim R_{\min} < R < R_{\max} \sim V^{1/3}$. Consequentially, to study the dependence of the free energy on the drop's radius one also has to vary the system size. Note that the equilibrium drops observed in the canonical ensemble correspond to critical drops in the nucleation theory at a supersaturation $\Delta\mu = \partial\Delta F/\partial n$. At fixed volume, the density of the surrounding vapor (mother phase) does depend on the drop's size. Hence, one cannot study the growth of a drop at fixed supersaturation – the situation assumed in nucleation theory – without a systematic study of finite size effects.

• The "transitions" from the supersaturated vapor to the drop and from the drop to the slab configurations represent barriers in the configuration space which are not removed by the multicanonical reweighting scheme. Hence, these shape "transitions" limit the applicability of reweighting methods in the study of phase equilibria.

The last issue can be clearly observed in the time sequence of the order parameter in the simulation. With a very good reweighting function the probability distribution sampled in the simulation is flat but the system does not perform a random walk in the order parameter. Rather one finds a banded structure of the time evolution where the system explores the configurations within a typical shape – homogenous vapor, drop, slap, bubble, or homogeneous liquid – in a random walk-like manner but only occasionally jumps between different shapes [58]. Typically the free energy barrier associated with "transitions" from one shape to another is a small fraction of the free energy barrier associated with the formation of the interfaces in the slab-like configurations. Neuhaus and Hager [58] pointed out the slowing down of simulations due to these shape "transitions". For a two-dimensional simulation cell of square geometry, $L \times L$, they investigated the barrier for the "transition" between a drop (spot) and a slab (stripe) and compared simulation results of the Ising model with a transition state that had earlier been suggested by Leung and Zia [59]. It consists of a lens-shaped domain whose two arc ends touch each other via the periodic boundaries. The opening angle of the arc is chosen such that the enclosed area of the lens equals the area, L^2/π, at the drop–to–slab transition. Simple geometric considerations yield the arc length $1.135L$ which is larger than the length of the liquid-vapor interface (line), L. Thus the free energy barrier associated with the shape "transition" is 13.5% of the free energy barrier due to the interfaces. Given that this free energy barrier should not exceed the thermal free energy scale $k_B T$ by an order of magnitude total free energy barriers, i.e., $2\gamma L^2$ on the order $10^2 k_B T$ can be efficiently sampled.

3 Applications

3.1 Liquid-Vapor Coexistence

In the first application will will exemplify the accurate localization of the condensation transition for a coarse-grained bead-spring polymer model. As discussed above, the order parameter of the liquid-vapor transition is the monomer number density, ρ. The equation of state of the polymer solution is

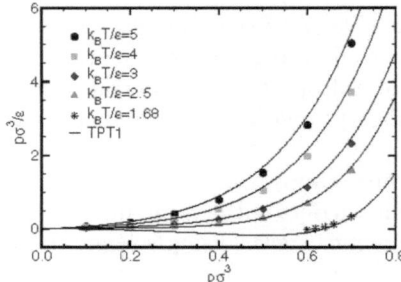 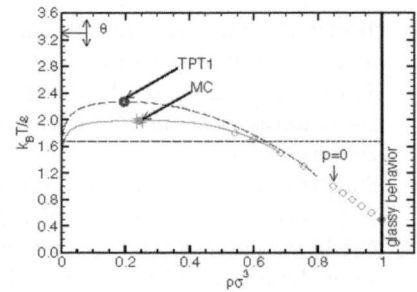

Fig. 12. Equation of state (**a**) and phase diagram (**b**) of a bead-spring polymer model. Monomers interact via a truncated and shifted Lennard-Jones potential as in Fig. 6 and neighboring monomers along a molecule are bonded together via a finitely extensible non-linear elastic potential of the form: $E_{\text{FENE}}(r) = -15\epsilon(R_0/\sigma)^2 \ln\left(1 - \frac{r^2}{R_0^2}\right)$ with $R_0 = 1.5\sigma$. Each chain is comprised of $N = 10$ monomeric units. The pressure of the bulk system in panel (a) has been extracted from canonical Monte Carlo simulations using the virial expression (symbols). Lines show the result of Wertheim's perturbation theory (TPT1). The phase diagram (b) has been obtained from grand-canonical Monte Carlo simulations. Binodal compositions are shown as full lines, while the corresponding results from Wertheim's perturbation theory are presented by dashed lines. Filled and open circles mark the location of the critical point in the simulations and the TPT1 calculations, respectively. The Θ-temperature is indicated by an arrow on the left hand side. The diamonds present the results of constant pressure simulations indicating the densities that correspond to vanishing pressure, which are a good approximation for the coexistence curve at low temperatures. Adapted from Müller and Mac Dowell [60]

presented in Fig. 12 (a) and the Monte Carlo results (symbols) are compared to Wertheim's thermodynamic perturbation theory (TPT1) represented by lines without adjustable parameter [60]. As we reduce the temperature, the pressure decreases and the analytic theory predicts a van-der-Waals loop with negative pressures for a hypothetical homogeneous system. In the simulation, however, the system spatially separates into two coexisting phases – a low-density vapor and a dense liquid. The monomer number density, $\rho = nN/V$ (n being the number of chains and N the number of monomers per chain), distinguishes the two phases. The sequence of typical configurations inside the miscibility gap has been discussed in Sect. 2.5. Upon increasing the system size the equation of state does not exhibit a loop but develops a plateau, p_{coex}, in the miscibility gap.

The grand-canonical ensemble is particularly well suited for studies of liquid-vapor phase coexistence: (i) Fluctuations of the order parameter, i.e., the density, are efficiently relaxed. Since the density is not conserved, spatial fluctuations do not decay via slow diffusion of polymers but relax much faster through insertion/deletion moves. In the grand-canonical ensemble one controls the temperature, T, the volume, V, and the chemical potential, μ, in the simulations and the number of particles, n, fluctuates. To realize this ensemble in Monte Carlo simulations, insertion/deletion moves are utilized to supplement canonical moves that displace particles and alter the conformations of the polymers. Note that insertion of extended particles into a dense liquid is difficult, and several methods have been devised to overcome the concomitant sampling difficulties. In our application we utilize configurational bias Monte Carlo [61] to grow the polymer chains into the system.

(ii) The key quantity to monitor in the simulation is the probability distribution of the number of particles in the simulation cell. It contains much information about the coexisting phases and the interface that separates them (c.f. Sect. 2.5). Additionally, there exist sophisticated methods to control finite-size effects and to accurately locate critical points.

In the vicinity of the phase coexistence, $\mu = \mu_{\mathrm{coex}}(T)$, the distribution $P(\rho)$ exhibits two pronounced peaks that are separated by a deep "valley". The corresponding free energy profile, $F(\rho) = -k_B T \ln P(\rho)$, for polymers that are comprised of $N = 10$ effective segments and temperature $k_B T/\epsilon = 1.68$, is presented in Fig. 13.

The location of phase coexistence can be accurately estimated by the equal-weight rule [63]: Two phases coexist if the weights of the two corresponding peaks in the distribution function of the order parameter are equal. It is important for the system to "tunnel" often between the two phases in

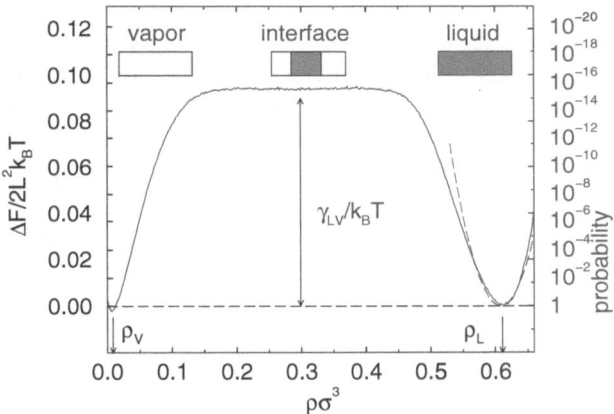

Fig. 13. Illustration of the grand-canonical simulation technique for temperature $k_B T/\epsilon = 1.68$ and $\mu = \mu_{\mathrm{coex}}$. A cuboidal system geometry $13.8\sigma \times 13.8\sigma \times 27.6\sigma$ is used with periodic boundary conditions in all three directions. The *solid line* corresponds to the negative logarithm of the probability distribution, $P(\rho)$, in the grand canonical ensemble. The two minima correspond to the coexisting phases and the arrows on the ρ axis mark their densities. The height of the plateau yields an accurate estimate for the interfacial tension, γ_{LV}. The *dashed line* is a parabolic fit in the vicinity of the liquid phase employed to determine the compressibility. Representative system configurations are sketched schematically. From [62]

the course of the simulation in order to equilibrate the weight between the two phases. Without multicanonical techniques, the large free energy barrier between the two phases would render the simultaneous sampling of both phases in a single simulation infeasible. If we assume that in a multicanonical simulation the system performs a random walk along the order parameter then the statistical uncertainty of the peak weights will scale as $1/\sqrt{\#}$, where $\#$ is the number of transitions (or tunneling events) from one phase to the other. The concomitant statistical uncertainty in the free energy difference of the two coexisting phases is on the order $\mathcal{O}(k_B T/\sqrt{\#})$.

The rule for phase coexistence can be rationalized as follows: Using the relation between the the probability distribution, $P(n)$, and the grandcanonical partition function, $\mathcal{Z}_{\mathrm{gc}}$, we obtain for the ratio of the partition functions of the two phases:

$$\frac{\mathcal{Z}_{\mathrm{gc}}^{\mathrm{liq}}}{\mathcal{Z}_{\mathrm{gc}}^{\mathrm{vap}}} = \frac{\sum_{n\in\mathrm{liq}} \frac{\exp(\beta\mu n)}{n!} \int \mathcal{D}_n[\{r\}] \exp(-\beta E[\{r\}])}{\sum_{n\in\mathrm{vap}} \frac{\exp(\beta\mu n)}{n!} \int \mathcal{D}_n[\{r\}] \exp(-\beta E[\{r\}])} \tag{44}$$

where $\mathcal{D}_n[\{r\}]$ sums over all configurations of the system with n polymers, and $E[\{r\}]$ denotes the energy of this microscopic configuration, $\{r\}$. The condition $n \in \mathrm{liq}$, characterizes all particle numbers that correspond to the liquid phase; if $n \in \mathrm{vap}$, the configuration belongs to the vapor phase. The detailed way in which the interval of particle numbers is divided into vapor and liquid phases affects the estimate of the ratio only by an very small amount. Furthermore, that amount decreases exponentially with the system size because the configurations between the peaks are strongly suppressed due to the free energy cost of the interface between the coexisting phases (cf. Sect. 3.5). The pressure, p, is related to the grand-canonical partition function via $pV = k_B T \ln \mathcal{Z}_{\mathrm{gc}}$. Therefore one obtains:

$$\frac{\mathcal{Z}_{\mathrm{gc}}^{\mathrm{liq}}}{\mathcal{Z}_{\mathrm{gc}}^{\mathrm{vap}}} = 1 \quad \Rightarrow \quad p_{\mathrm{liq}} = p_{\mathrm{vap}} \qquad \text{(equal weight rule)} \tag{45}$$

Thus the equal weight-rule rephrases the macroscopic definition of phase coexistence: Two phases coexists if they have equal pressure at equal chemical potential.

Using histogram extrapolation [2] we determine the coexistence value of the chemical potential. The properties of the individual phases can be extracted by taking averages over the corresponding regions of the order parameter. The densities of the two coexisting phases – the binodal densities – are presented in Fig. 12(b). From the curvature of the free energy profile at the minima we can also extract the compressibility of the liquid and the vapor. Additional information about interfaces between the coexisting phases can also be obtained as discussed in Sect. 3.5.

3.2 Demixing of Binary Blends

Another example of phase coexistence that is described by a single scalar one-component order parameter is provided by incompressible binary mixtures. One considers a dense liquid of two components, A and B; the composition of the mixture, $\phi = \frac{n_A}{n_A + n_B}$, distinguishes the two coexisting phases. The total number of particles, $n_A + n_B$, is assumed to be independent of composition. The more general case of compressible mixtures will be discussed in the following section.

Sariban and Binder [64] employed simulations in the semi-grand canonical ensemble for investigating the phase behavior of an incompressible binary polymer blend at constant volume. In this ensemble, the total monomer density is fixed, the composition of the blend fluctuates, and the chemical potential difference, $\Delta\mu$, between the species is controlled. The Monte Carlo scheme comprises two types of moves. Canonical updates relax the conformation of the macromolecules, whereas semi-grand canonical moves convert A polymers into B polymers and vice-versa. This ensemble is particularly well suited for the study of strictly symmetric mixtures where the two components have the same chain architecture and intramolecular potentials, but different species repel each other. In this limit, semi-grand canonical moves consist of a mere identity exchange (a switch of labels). Note that the algorithm implies that both phases have identical densities but it does not automatically ensure that they have identical pressure. The algorithm can be extended to some degree of structural asymmetry (e.g., different chain lengths between the species [65]). Overall speaking, it is reasonably efficient for a modest degree of structural asymmetry between the different constituents, but the extension to pronounced structural asymmetries is a challenging task. Improvement might be achieved via gradually "mutating" one species into another [66].

Figure 14 exemplifies two computational methods to determine the probability distribution of composition for binary polymer blends described by the bond fluctuation model [67]. Phase coexistence can be extracted from these data via the equal-weight rule. For the specific example of a symmetric blend, the coexistence value of the exchange chemical potential, $\Delta\mu$, is dictated by the symmetry. One can simply simulate at $\Delta\mu_{\text{coex}} \equiv 0$ and monitor the composition. Nevertheless, the probability distribution contains additional information, as discussed in Sect. 3.5.

Panel (a) illustrates the application of the Wang-Landau algorithm [7]. The different probability distributions obtained at successive steps of the modification parameter, f are shown. As f decreases the distribution gradually converges towards the Boltzmann distribution. Note that the starting value of the modification factor, $f = 1.02$, is chosen much smaller than in the original application to a spin system in order to match the relaxation time of the system (chain conformations) with the time scale for exploring the entire composition range.

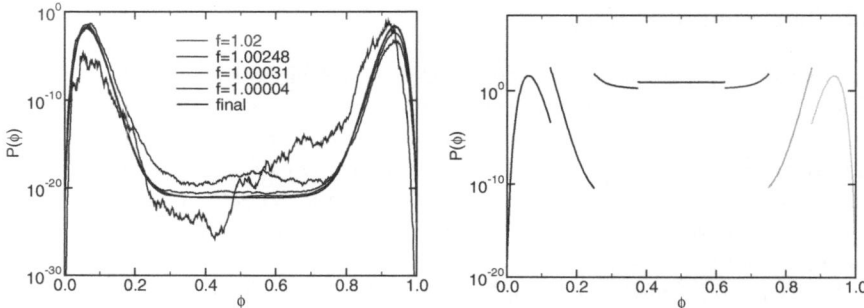

Fig. 14. (a) Probability distribution of the composition, ϕ, of a symmetric binary polymer blend within the framework of the bond fluctuation model. In this coarse-grained lattice model monomeric units are represented by unit cubes on a three-dimensional cubic lattice. Each monomer blocks the 8 sites of a cube from double occupancy. This mimics excluded volume interactions. Monomers along a polymer chain are connected via one of 108 bonding vectors that can adopt lengths, $b = 2, \sqrt{5}, \sqrt{6}, 3$, and $\sqrt{10}$ in units of the lattice spacing. This represents the connectivity along the polymer backbone. Monomers of the same type attract each other via a square-well potential of depth $-\epsilon$ that extends over the nearest 54 lattice sites. "Unlike" monomers repel each other via a potential of opposite sign. The simulation data are obtained at the different stages of the Wang-Landau sampling for chain length $N = 32$ and $\epsilon/k_B T = 0.02$ in a simulation cell of geometry $64 \times 64 \times 128$ in units of the lattice spacing ($R_e = 17$). The convergence factors, f, are indicated in the key. A flatness of 50% was required for the histogram of visited to particle numbers to reduce the convergence factor from $f \rightarrow f' = \sqrt{f}$. (b) Piecewise probability distributions obtained by successive umbrella sampling. By patching the distributions together in regions of mutual overlap it is possible to construct the probability distribution over the complete composition range

Panel (b) illustrates the application of successive umbrella sampling [35], where each piece of the curve corresponds to several intervals used in the successive sampling. The intervals overlap just by one particle number, which is sufficient if one takes due account of detailed balance at the interval boundaries [36]. Matching the pieces at their boundaries one obtains the distribution across the entire composition range. Of course, both methods yield identical results for the final probability distribution and, provided that one starts with an appropriate modification factor, f, they require comparable amounts of CPU time.

One advantage of these Monte Carlo simulations is that both the thermodynamic and structural properties of the binary polymer mixture are simultaneously accessible, and can quantitatively be compared to analytic approaches. The mean field theory of polymers makes detailed predictions for the bulk [68–70] and interface properties [71–77] as a function of the incompatibility, $\tilde{\chi}N$, the spatial extension of the molecules R_e as measured by the root mean squared end-to-end distance of the Gaussian coils, and the invariant degree of polymerization $\bar{\mathcal{N}} \equiv \left(\rho R^3/N\right)^2$. The last parameter describes the strength of fluctuation effects that are neglected by mean field theory; the analytic approach is thought to describe the limit $\bar{\mathcal{N}} \to \infty$.

One question that these simulations can address is the relation between different estimates of the incompatibility between polymer species, i.e.: How can one identify the mutual repulsion between polymer species parameterized by the product of the Flory-Huggins parameter and chain length, $\tilde{\chi}N$, for a specific microscopic polymer model [65, 78]?

In simulations, as well as in experiments, it is common practice to measure the Flory-Huggins parameter by comparing the results of simulations to the predictions of the mean field theory. Since the predictions are affected by fluctuation effects to different extents, not all quantities yield mutually compatible estimates of the Flory-Huggins parameter, $\tilde{\chi}$.

Within mean field theory, for a symmetric blend the excess free energy of mixing per volume is given by the Flory-Huggins expression:

$$\frac{\Delta F_{\mathrm{mix}}}{k_B T(V/R_e^3)} = \sqrt{\bar{\mathcal{N}}}\left[\phi \ln \phi + (1 - \phi)\ln(1 - \phi) + \tilde{\chi}N\phi(1 - \phi)\right] \qquad (46)$$

From this expressions one can calculate the binodal curves (composition of the two coexisting phases)

$$\ln \frac{\phi}{1 - \phi} + \tilde{\chi}N(1 - 2\phi) = \frac{\Delta \mu_{\mathrm{coex}}}{k_B T} = 0 , \qquad (47)$$

the location of the critical point, $\tilde{\chi}_c N = 2$ and $\phi_c = 1/2$, which marks the onset of phase separation, and the composition fluctuations in the one phase region

$$\frac{N}{S(k \to 0)} = \frac{1}{\phi} + \frac{1}{1 - \phi} - 2\tilde{\chi}N \qquad \text{for} \qquad \tilde{\chi}N < 2 \qquad (48)$$

as measured by the structure factor, $S(k)$, of composition fluctuations in the limit where the wavevector k vanishes. The latter quantity is often used in experiments to determine the Flory-Huggins parameter for miscible blends.

A quantitative comparison between the mean field prediction and the Monte Carlo results is presented in Fig. 15. The main panel plots the inverse scattering intensity vs. $\tilde{\chi}N$. At small incompatibility, the simulation data are compatible with a linear prediction (cf. (48)). From the slope, it is possible to estimate the relation between the Flory-Huggins parameter, $\tilde{\chi}$, and the depth of the square well potential, ϵ, in the simulations of the bond fluctuation model. As one approaches the critical point of the mixture, deviations between the predictions of the mean field theory and the simulations become apparent; the theory cannot capture the strong universal (3D Ising-like) composition fluctuations at the critical point [64, 79, 80] and it underestimates the incompatibility necessary to bring about phase separation. If we fitted the behavior of composition fluctuations at criticality to the mean field prediction, we would obtain a quite different estimate for the Flory-Huggins parameter.

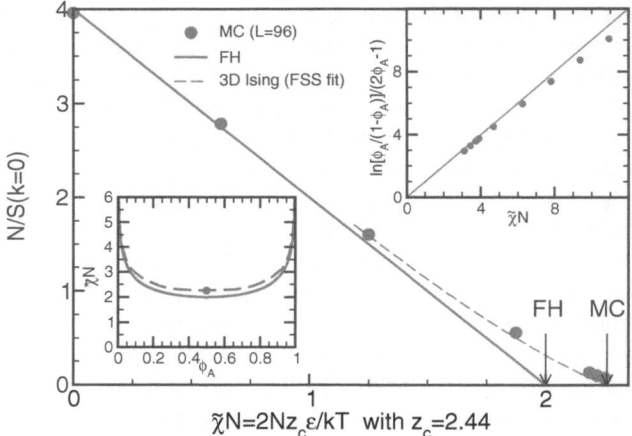

Fig. 15. Inverse maximum of the collective structure factor of composition fluctuations, $N/S(k \to 0)$, as a function of the incompatibility, $\tilde{\chi}N$. Symbols correspond to Monte Carlo simulations of the bond fluctuation model, the *dashed curve* presents the results of a finite-size scaling analysis of simulation data in the vicinity of the critical point, and the straight, *solid line* indicates the prediction of the Flory-Huggins theory. The critical incompatibility, $\tilde{\chi}_c N = 2$ predicted by the Flory-Huggins theory and that obtained from Monte Carlo simulations of the bond fluctuation model ($\mathcal{N} \approx 240$, $N = 64$, $\rho = 1/16$ and $R_e = 25.12$) are indicated by arrows. The left inset compares the phase diagram obtained from simulations with the prediction of the Flory-Huggins theory (c.f. (47)). The right inset depicts the compositions at coexistence such that the mean field theory predicts them to fall onto a straight line. From Müller [78]

This estimate, however, would not characterize the incompatibility between the polymer species too well, but rather quantify the inability of the mean field theory to cope with Ising-like order parameter fluctuations.

In the insets of Fig. 15 we show binodal curves for the symmetric blend. Again, we find deviations in the immediate vicinity of the critical point but for larger incompatibilities, $\tilde{\chi}N \gg 2$, the mean field predictions provide an adequate description of the phase boundary utilizing the Flory-Huggins parameter extracted from the composition fluctuations in the one-phase region, $\tilde{\chi}N < 2$.

3.3 Compressible Blends

The phase behavior of binary blends becomes much more complex if one relaxes the assumption of incompressibility. Then both liquid-liquid demixing and liquid-vapor phase separation are possible and the system is described by two scalar order parameters – density and composition (ρ, ϕ) or the two densities of the species (ρ_{CO_2}, ρ_{HD}). The interplay between the two types of phase

Table 1. Critical points of the pure components from MC simulations and experiments. By comparing the critical temperatures from simulation and experiment, we identify $\epsilon_{CO_2} = 4.201 \cdot 10^{-21} J$, $\epsilon_{HD} = 5.787 \cdot 10^{-21} J$ and $\epsilon_{CO_2}/\epsilon_{HD} = 0.726$. From the critical densities we derive $\sigma_{CO_2} = 3.693 \cdot 10^{-10}$ m, $\sigma_{HD} = 4.523 \cdot 10^{-10}$ m and $\sigma_{CO_2}/\sigma_{HD} = 0.816$. ρ refers to monomer number densities

	T_{crit}	ρ_{crit}	p_{crit}
MC CO$_2$	$0.999 \frac{\epsilon}{k}$	$0.32 \frac{1}{\sigma^3}$	$0.088 \frac{\epsilon}{\sigma^3}$
EXP CO$_2$	$304\ K$	$0.464 \frac{g}{cm^3}$	73.87 bar
MC HD	$1.725 \frac{\epsilon}{k}$	$0.27 \frac{1}{\sigma^3}$	$0.022 \frac{\epsilon}{\sigma^3}$
EXP HD	$723\ K$	$0.219 \frac{g}{cm^3}$	13.98 bar

separation gives rise to 6 qualitatively different types of phase behavior in the classification scheme of Konynenburg and Scott [81]. An example of a phase diagram of type I and type III (in this classification) is presented in Fig. 16. The simulations study the phase behavior of a short polymer (hexadecane, HD) in a supercritical solvent (carbon dioxide, CO$_2$) within a coarse-grain model. The polymer is represented by a Lennard-Jones bead spring model such as that discussed in Sect. 3.1. We utilize a chain length of $N = 5$, which roughly mimics the geometrical shape of the monomer. Comparing the two parameters of the Lennard-Jones potential, ϵ and σ, with experimental results for the critical temperature and pressure, we identify energy and length scales [40, 41].

Carbon dioxide (CO$_2$) is represented by a single Lennard-Jones bead; the interactions between CO$_2$-beads are adjusted to reproduce the experimental critical point. A modified Lorentz-Berthelot mixing rule is used to describe the interactions between hexadecane and CO$_2$ beads [82, 83]:

$$\sigma_{HD-CO_2} = \frac{\sigma_{HD} + \sigma_{CO_2}}{2} \quad \text{and} \quad \epsilon_{HD-CO_2} = \xi\sqrt{\epsilon_{HD}\epsilon_{CO_2}} \qquad (49)$$

where ξ describes the deviations from the standard mixing rule. Figure 16 shows projections of the phase diagram in the plane of temperature, T, and pressure, p, for two choices of ξ. The value $\xi = 1$ corresponds to the standard Berthelot rule which describes the mixing of van-der-Waals interactions, while the value $\xi = 0.886$ matches the experimental behavior [84, 85] most closely.

Solid lines represent the liquid-vapor phase equilibria of the two pure components that end in critical points marked by arrows. When a small amount of solvent is added to the pure polymer, the liquid-vapor coexistence shifts and so does the critical point. The loci of critical points for the binary system form a critical line that is shown by the dashed line with squares for $\xi = 1$ and triangles for $\xi = 0.886$. In the former case – phase behavior of type I – the critical line connects the critical points of the two pure components and the two coexisting phases gradually change from vapor and solvent-rich liquid

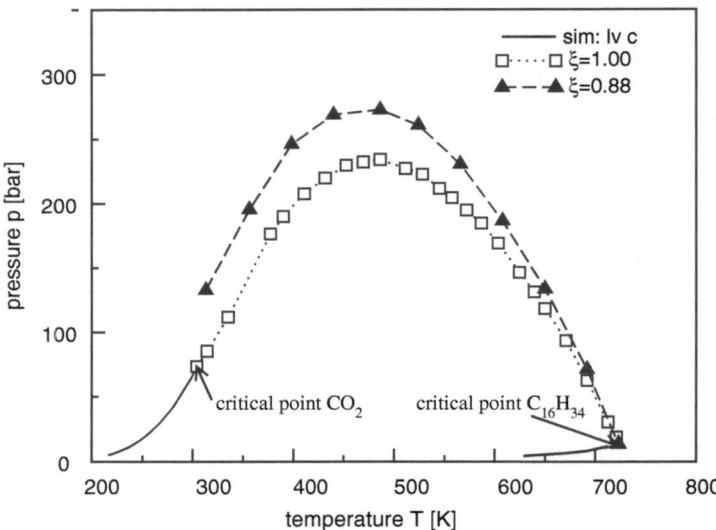

Fig. 16. Projection of the global phase diagram for a compressible mixture of hexadecane, $C_{16}H_{34}$, and carbon dioxide, CO_2, into the temperature-pressure plane for two values of the mixing parameter, $\xi = 1$ (*square*) and $\xi = 0.886$ (*triangle*). Simulation results for the liquid-vapor coexistence of the pure components are shown by solid lines and end in critical points that are indicated by arrows. The line of critical points that emerges from the critical point of the less volatile polymer component is indicated by symbols. From Virnau et al. [40]

to vapor and polymer-rich liquid. This is not, however, what is observed in experiments where a phase diagram of type III is encountered. In a type III diagram, the critical line that emerges from the liquid-vapor critical point of the polymer is not connected to the critical point of the pure solvent but gradually changes its character from a liquid-vapor coexistence to a liquid-liquid coexistence between a dense solvent-liquid and a polymer-liquid. Just a slight change of the interaction between the different components from $\xi = 1$ to $\xi = 0.886$ brings about a qualitative change of the phase diagram. The reduction of the attractions between CO_2 and hexadecane can be traced back to quadrupolar interactions between CO_2–molecules [86]. The well depth ϵ_{CO_2} of the pure solvent parameterized the attraction between carbon dioxide molecules that is caused by the joined effect of both, van-der-Waals and quadrupole interactions. The Berthelot mixing rule, however, implicitly assumes that all attractions result from van-der-Waals interactions and utilizes the full ϵ_{CO_2}, which only partially stems from dipolar interactions.

The techniques (e.g., weighted histogram analysis, equal weight rule, etc.) utilized for one-component systems or incompressible binary mixtures can be readily carried over to compressible systems. Since the system is described by two order parameters one monitors the joint probability distribution of

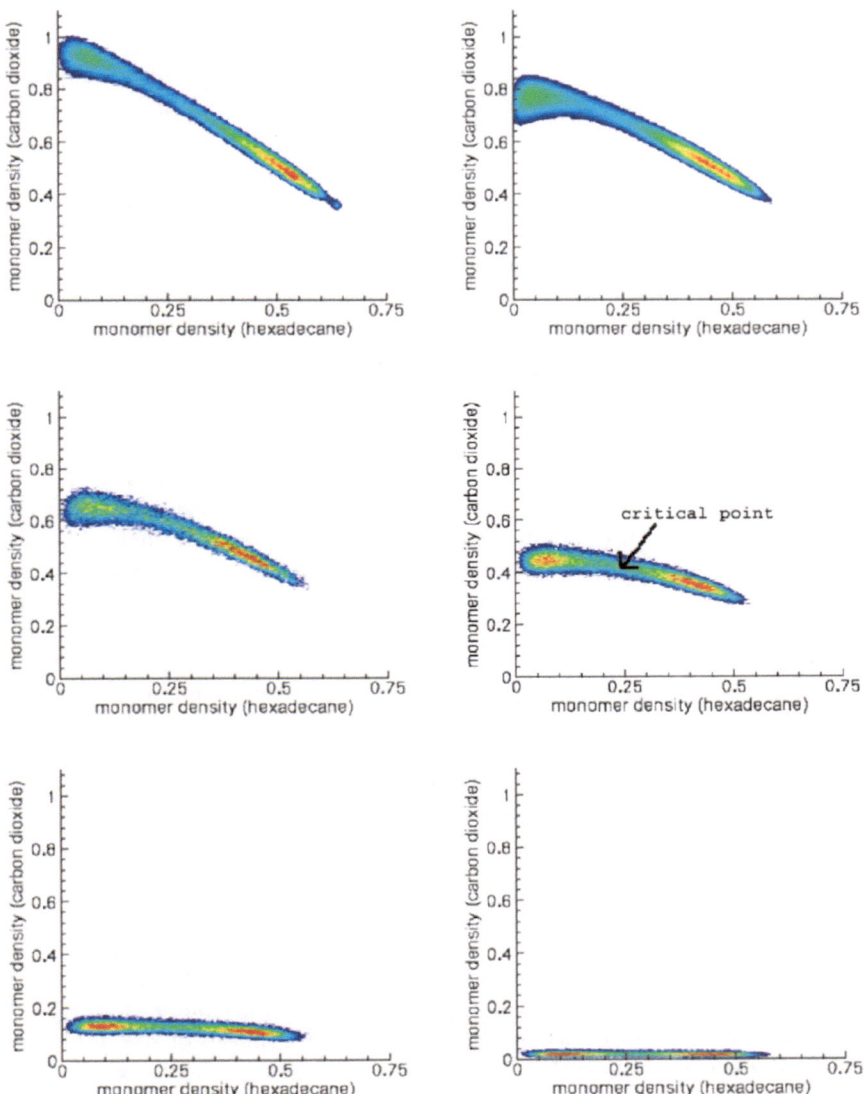

Fig. 17. Joint probability distribution of the solvent and the polymer density along the critical line in a finite-size simulation cell for $\xi = 0.886$. The distributions correspond to temperatures $T = 314$ K, 356 K, 398 K, 486 K, 650 K and 713 K from up/left to down/bottom. For temperatures $T = 314$ K and 356 K the box size is $L = 6.74\ \sigma_{HD}$, while $L = 9\ \sigma_{HD}$ was employed for other temperatures. The arrow in the middle right panel indicates our finite-size estimate for the location of a critical point as an average over both peaks. From Virnau et al. [40]

ρ_{HD} and ρ_{CO_2}. These two-dimensional probability distributions are shown in Fig. 17 for various temperatures. For a finite-size system, the probability distribution at the critical point is bimodal. The two peaks indicate the densities of the two coexisting phases slightly below the critical temperature. Unlike a simulation at (a first-order) phase coexistence, however, the positions of the peaks depend on the system size, and they collapse onto a single density in the thermodynamic limit. The average over the composition of the bimodal distribution of the finite-size system yields an estimate for the critical density. The two-dimensional order parameter distribution [40] at the estimate of the critical point clearly reveals the gradual crossover from a phase coexistence between a polymer-liquid and a vapor characterized by a horizontal ridge in the distribution to a liquid-liquid coexistence marked by a diagonal ridge. If the phase diagram were of type I, however, the distribution would rotate counterclockwise and end up vertically at low temperatures. In order to locate phase coexistence we utilize the equal-weight rule, i.e., we divide the plane of order parameters into regions which correspond to the coexisting phases and tune both chemical potentials as to fulfill the equal weight condition. Looking for phase coexistence in this larger parameter space is hardly feasible without the help of weighted histogram analysis. In practice, one often just chooses a suitable linear combination of the two order parameters that clearly distinguishes between the two coexisting phases, i.e., one divides the order parameter plane by a line. At low pressure the total density will be a suitable order parameter; at higher pressure one should use the composition. The equal weight condition will provide us with $\mu_{HDcoex}(T, \mu_{CO_2})$, i.e., at constant temperature there is a set of phase coexistences characterized by the chemical potential μ_{CO_2} or the pressure, p. If we sampled the entire interval of both order parameters we could not only locate the critical point but we could construct an entire slice of the phase diagram as a function of pressure and composition at constant temperature from a single simulation.

3.4 Crystallization

The simulation techniques presented above can be applied to all first order phase transitions provided that an appropriate order parameter is identified. For vapor-liquid equilibria, where the two coexisting phases of the fluid have the a similar structure, the density (a thermodynamic property) was an appropriate order parameter. More generally, the order parameter must clearly distinguish any coexisting phases from each other. Examples of suitable order parameters include the scalar order parameter for study of nematic-isotropic transitions in liquid crystals [87], a density-based order parameter ψ for block copolymer systems [88], or a bond order parameter for study of crystallization [89]. Having specified a suitable order parameter, Φ, we now show how the EXEDOS technique introduced earlier can be used to obtain $P_{\beta,p}(\Phi(q^n))$ in a particularly effective manner for simulations of crystallization [33]. The Landau free energy of the system $\Lambda(\Phi)$ can then be related to $P_{\beta,p}(\Phi(q^n))$

by (50) [90]:

$$\Lambda(\Phi(q^n)) = \text{constant} - k_B T \ln\{P(\Phi(q^n))\} . \tag{50}$$

Depending on whether $P(\Phi)$ is obtained from EXEDOS in a constant (n, V, T) or constant (n, p, T) ensemble, Λ will correspond to the Helmholtz free energy or the Gibbs free energy of the system, respectively.

Away from a phase transition, Λ is expected to exhibit a single minimum as a function of Φ. Close to coexistence, however, Λ develops two minima; the barrier between the minima corresponds to the free energy associated with the formation of an interface [91] (cf. Sect. 3.5). The method described above is capable of providing a free energy profile (including free energy barriers) at any given temperature and pressure. In general, however, the coexistence temperature (i.e., melting temperature) or pressure are not known. In order to obtain the precise conditions under which two phases coexist it is therefore necessary to conduct a series of exploratory simulations, until a set of conditions is found for which the free energy minima corresponding to the two equilibrium states became identical. This process can be facilitated by resorting to the following weighted histogram analysis [2,33]. Energy and volume data from EXEDOS simulations must be saved and sorted in histograms according to the order parameter. The probability of a system to be in bin m, denoted $P_{\beta_2,p}(\Phi_m, \Phi_{m+1})$, at temperature β_2 can be obtained from a simulation at a different temperature β_1 via weighted histogram analysis (cf. (4))

$$P_{\beta_2,p}(\phi_m, \phi_{m+1}) = \frac{P_{\beta_1,p}(\phi_m, \phi_{m+1}) \sum_{i=1}^{N_m} \exp(-(\beta_1 - \beta_2)(E_{i,m} + pV_{i,m}))}{\sum_{m=1}^{N_b} P_{\beta_1,p}(\phi_m, \phi_{m+1}) \sum_{i=1}^{N_m} \exp(-(\beta_1 - \beta_2)(E_{i,m} + pV_{i,m}))} \tag{51}$$

Where $E_{m,i}$ and $V_{m,i}$ are the energy and volume entries in the mth order parameter histogram bin, N_b is the total number of histogram bins, and N_m is the number of entries in bin m. Equation (51) can be used to determine the precise values of coexistence temperature T^* and pressure p^* for which the free energy (50) exhibits two minima of equal weight [63]. This weighted histogram analysis works particularly well in conjunction with multicanonical methods because we do not only sample configurations which have high Boltzmann weights but configurations which might have a low Boltzmann weight but are important for free energy estimations.

Having determined the free energy Λ as a function of $\Phi(q^n)$, configurations corresponding to the top of the free energy barrier can be used to explore the transition pathways from one phase to another, as discussed later in this work in the context of nucleation mechanisms [92].

Here we consider a system of particles interacting through a repulsive Lennard-Jones (LJ) potential $E(r) = \epsilon(\frac{\sigma}{r})^{12}$ truncated at a distance of $r_c = 2\sigma$. We measure temperature and pressure in Lennard-Jones units, $T^* \equiv k_B T/\epsilon$ and $p^* = p\sigma^3/\epsilon$, respectively. The long-range contribution to the potential was calculated under the assumption of constant density. For this system, the stable nuclei at melting are believed to be face-centered-cubic (fcc) [93–95]. Previous simulations by ten Wolde et al. [92] used umbrella sampling to generate free energy curves and to examine the structure of the critical nuclei that form in a deeply supercooled liquid.

The order parameter used is the bond orientational order parameter, Q_6, originally introduced by Steinhardt, Nelson and Ronchetti [89] and later used by van Duijneveldt and Frenkel [96] to simulate crystallization. It is sensitive to the overall degree of crystallinity in the system, irrespective of the crystal structure, i.e., it distinguishes the liquid from the crystal. The values of Q_6 for pure fcc, body-centered-cubic (bcc), and for the liquid are 0.57452, 0.51069, and 0.0, respectively. Previous work [96] has shown that a defective fcc crystal, which crystallizes from the liquid in simulations, has an order parameter below 0.5 and that finite-size effects lead to small positive values for the order parameter in the liquid phase.

Our simulations were performed on systems of various sizes, ranging from 108 and 1364 particles. The order parameter range explored here goes from 0.05 to 0.5 for the small system and from 0.02 to 0.45 for the larger systems. In all simulations the order parameter was calculated at every step and was used in the Monte Carlo acceptance criteria. Following our previous work, the range of order parameter was split into multiple overlapping windows (intervals of order parameter) [32, 97]. 10 overlapping windows with a 50% of overlap between adjacent windows were chosen in all our simulations. Configuration swaps were implemented to facilitate convergence. During the simulation we observed the tunneling of individual replicas between the liquid and the crystalline state. The total number of tunneling events for each system size exceeds 20000. Thus the statistical uncertainty of the free energies we calculate is on the order $\Delta A \sim \mathcal{O}\left(\frac{k_B T}{\sqrt{\#}}\right) \approx 0.01 k_B T$. The corresponding errors ΔT and Δp are on the order of ≈ 0.001 and ≈ 0.003, respectively. The accuracy of the coexistence temperature and pressure in EXEDOS is comparable to that achieved in techniques such as phase switch Monte-Carlo [98], self-referential Monte-Carlo [99], and techniques using thermodynamic integration and Gibbs-Duhem integration [100]. The added advantage with EXEDOS is that, in addition of obtaining accurate coexistence conditions, it also provides an accurate free energy curve as a function of the chosen reaction coordinate. An alternative technique, transition path sampling, can also provide both coexistence conditions and a free energy curve. In a recent study [101] on crystallization, a free energy barrier and a transition path were also obtained for the Lennard-Jones system; transition path sampling, however, requires that a large number of trajectories be run, and is therefore highly computationally demanding.

For a system of 864 particles, the starting guess for coexistence temperature and pressure was $T^* = 1.14$ and $p^* = 24.21$. The order parameter range was set between 0.02 and 0.45. EXEDOS simulations were performed in multiple windows. The free energy curves obtained at this temperature and pressure (see Fig. 18) show that, as a result of finite-size effects, $T^* = 1.14$

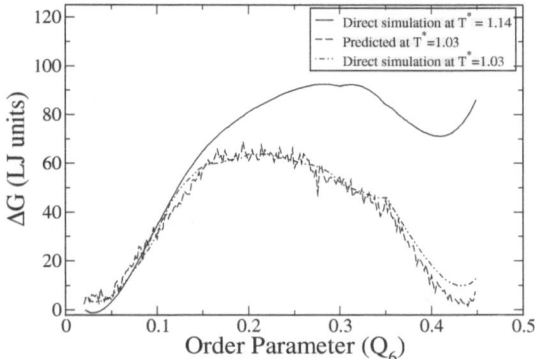

Fig. 18. Free energy curve for crystallization of an 864-particle repulsive LJ system. EXEDOS simulations at $T^* = 1.14$ *solid line*, predicted free energy profile by reweighing $T^* = 1.14$ data to $T^* = 1.03$:*dashed line*, and EXEDOS simulations at $T^* = 1.03$ *dot-dot-dashed line*. From Chopra et al. [33]

is not a good estimate of the phase transition temperature for the larger system. To obtain the coexistence point the free energy curves were reweighted to arrive at a melting temperature of $T_\sigma^* = 1.03$. The histogram-predicted free energy profile was confirmed by a subsequent EXEDOS simulation at the same temperature. The above result is indicative of strong finite-size effects. To verify the finite-size scaling and to predict the coexistence temperature for an infinite system we also performed calculations on systems of 256 and 500 particles. Figure 19 shows the variation of T_σ^* with the volumetric length scale; a clear finite-size scaling can be seen in the behavior of T_σ^* with volume. The extrapolation to infinite size predicts a coexistence temperature of $T_{\sigma,\infty}^* = 0.99(1)$. Note that in the EXEDOS algorithm the nucleation barrier necessary to expand a solid nucleus is removed but the barrier associated with transforming an ordered but homogeneous liquid into an inhomogeneous liquid with an initial solid nucleus at the same value of the order parameter, Q_6, remains [55, 58]. This formation of solid bodies of high Q_6 allows the larger system to move more easily between high and low order-parameter, thereby accelerating the convergence of the simulation.

A subsequent structural examination of the trajectories can be performed using Voronoi analysis and using the analysis outlined by ten Wolde et al. [92]. In this procedure, a combined distribution of local invariants q_4, q_6, w_4, and w_6 was generated for a given configuration. A detailed explanation of local invariants q_4, q_6, w_4, and w_6 can be found in the literature [102]. The distribution of local invariants was then decomposed into contributions for thermally equilibrated bcc, fcc and liquid configurations according to (52):

$$\Delta^2 = [\tilde{v}_{cl} - (f_{\text{liq}}\tilde{v}_{\text{liq}} + f_{\text{bcc}}\tilde{v}_{\text{bcc}} + f_{\text{fcc}}\tilde{v}_{\text{fcc}})]^2, \tag{52}$$

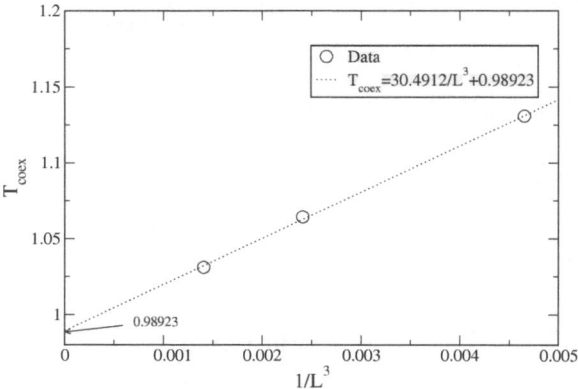

Fig. 19. Melting temperature T^* as a function of system size. The arrow points to the limiting (infinite-size) value. From Chopra et al. [33]

where $\tilde{v}_{cl}$, $\tilde{v}_{\text{liq}}$, $\tilde{v}_{\text{fcc}}$, and $\tilde{v}_{\text{bcc}}$ are combined distributions for the cluster to be analyzed, thermally equilibrated liquid, fcc, and bcc configurations respectively. In (52) f_{liq}, f_{bcc}, and f_{fcc} are weight coefficients for liquid, bcc and fcc configurations. The analysis confirmed that the first minimum in the free energy profile corresponds to a liquid phase, and the second minimum corresponds to a thermally equilibrated fcc phase.

The analysis also provides an explanation for the asymmetric shape and the kinks that appear in the free energy curve of Figure 18. These features were not observed in previous simulations of the same system, which used a different simulation approach, and it is therefore of interest to discuss their origin. This can be achieved by close inspection of the configurations that the system adopts for specific values of the order parameter [48,49,51,52,55,58]. For the order parameter range of 0.10-0.15, we observe the occurrence of spherical nuclei in the configurations. In the range from 0.15 to 0.25 we observe the formation of solid-fluid interfaces. An interesting point to note here is that the interface has a predominately bcc character, hinting at the formation of thin interfacial layers which are structurally different from the bulk solid. The fact that we observe a bcc interface coincides with the fact that the liquid-bcc tension is smaller than the liquid-fcc tension [103]. Moving further along the free energy profile to the range 0.25-0.35 we observe the growth of an interface that spans the entire simulation box, with a defective bcc-like solid. Upon further increasing the order parameter we observe that, beyond 0.35, this defective bcc-like solid rearranges into a defective fcc-like solid. Figure 20 shows snapshots of various representative configurations extracted from our simulations. The asymmetric shape of the free energy profile can be attributed to this two-step transition from an interface between a liquid and defective bcc-like structures, and then from defective bcc-like structures to defective fcc-like structures.

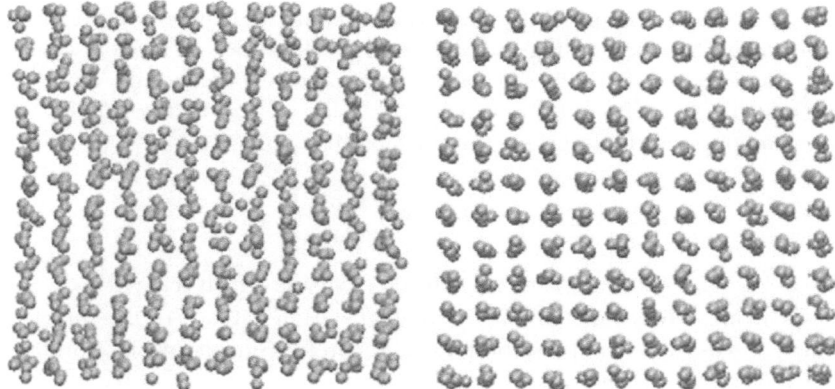

Fig. 20. Snapshots of various representative configurations observed during a typical simulation trajectory. Deffective bcc configuration (*left*) and fcc configuration (*right*)

3.5 Interface Free Energies

The probability distribution of the order parameter does not only characterize the coexisting phases in the bulk (i.e., their density, compressibility, and coexistence pressure) but additional information about spatially inhomogeneous system configurations can be extracted. Let us first discuss the case of liquid-vapor coexistence (c.f. Sect. 3.1) where the total number of polymers is used as order parameter along which a free energy profile is obtained. Recall that at each fixed value of the number of polymers the simulation samples configurations according to the Boltzmann weight independent of the reweighting function. As discussed in Sect. 2.5 the typical configurations consist of (1) supersaturated homogeneous vapor, (2) liquid drop, (3) slab of liquid separated by two planar interfaces from the coexisting vapor, (4) bubble of vapor in liquid and (5) undersaturated spatially homogeneous liquid as one increases the density.

The excess free energy of the configuration (3) that contains a slab of liquid is dominated by the interface free energy [91], $\Delta F = 2L^2\gamma$ provided the system is large enough for the two interfaces not to interact. Here, L^2 is the area of the interface, γ is the interface tension and the factor 2 arises because in a system with periodic boundary conditions two liquid-vapor interfaces are present.

The independence of the two interfaces can be verified by the probability distribution itself: If the interfaces do not interact, one can change the distance between them by varying the amount of liquid at no free energy costs. Therefore one will observe a plateau in the free energy profile [104]. No such plateau is observed in small, cubic systems (cf. Fig. 6) but a clear plateau is observed in Figs. 12 or 14. Often it is useful to utilize a cuboidal simulation box where one dimension, L_z is larger than the two lateral extensions, L [105,106]. This geometry displaces the two interfaces farther apart and thereby reduces their interaction without increasing their free energy costs and the concomitant free energy barrier. If one observes a plateau in the free energy profile and verifies that the typical configurations consist of a slab of the two coexisting bulk phases one obtains the interface tension via [91]:

$$\frac{\gamma L^2}{k_B T} = \frac{1}{2} \ln \frac{P_{\text{bulk}}}{P_{\text{slab}}} \tag{53}$$

There are several corrections to the interface tension which scale like $\Delta\gamma \sim \frac{1}{L^2}$ or $\sim \frac{\ln L}{L^2}$ which stem *inter alia* from the translation entropy of the interfaces, capillary waves, bulk compressibility, etc. A careful consideration of finite size effects is necessary to obtain accurate estimates.

The application of this technique to symmetric binary blends is discussed in Fig. 21, where we plot the results of simulations of the bond fluctuation model. The interface tension has been obtained for different chain lengths and the results are compared to mean field theory [104, 107, 108]. This comparison utilizes the identification of the incompatibility $\tilde{\chi}N$ obtained from the simulation of the spatially homogeneous system [65] in Sect. 3.2.

The mean field theory suggests a simple picture of the structure of the interface between two immiscible polymers which is illustrated in Fig. 22. The cost of each loop into the "hostile" phase is comparable to the thermal energy

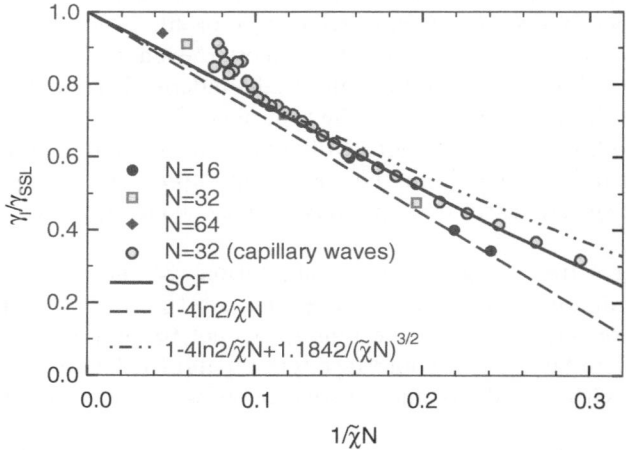

Fig. 21. Ratio between the interface tension γ and the simple expression for the strong segregation limit γ_{SSL} in (54) as a function of inverse incompatibility. Symbols correspond to Monte Carlo results for the bond fluctuation model, the *solid line* shows the result of the SCF theory, and the *dashed line* presents first corrections to (54) calculated by Semenov. Also an estimate of the interface tension from the spectrum of capillary waves is shown to agree well with the results of the reweighting method. Adapted from Schmid and Müller [107]

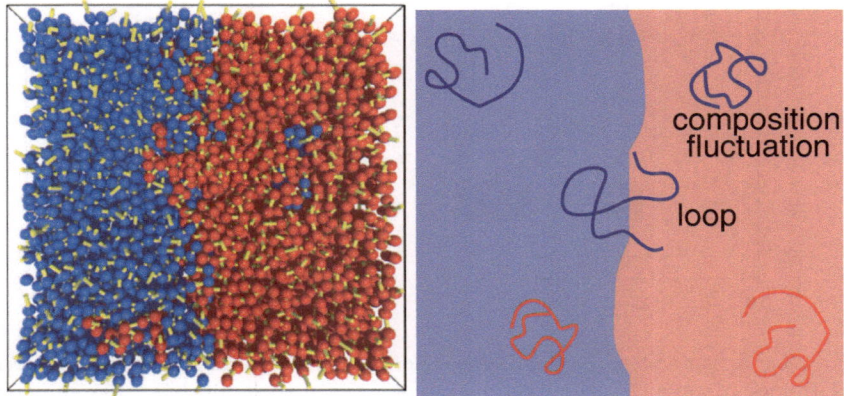

Fig. 22. (a) Snapshot of an interface between two coexisting phases in a binary polymer blend in the bond fluctuation model (invariant polymerization index $\bar{\mathcal{N}} = 91$, incompatibility $\chi N \approx 17$, linear box dimension $L \approx 7.5 R_e$, or number of effective segments $N = 32$, interaction $\epsilon/k_B T = 0.1$, monomer number density $\rho_0 = 1/16.0$). (b) Cartoon of the configuration illustrating loops of a chain into the domain of opposite type, fluctuations of the local interface position (capillary waves) and composition fluctuations in the bulk and the shrinking of the chains in the minority phase. From Müller [109]

scale, $k_B T$. Each monomer along the loop contributes to this cost an amount $\tilde{\chi} k_B T$. Thus the typical number of monomers of a loop is $1/\tilde{\chi}$. If one assumes that loops contain many monomers and that the spatial statistics of loops corresponds to the Gaussian behavior of the entire coil the spatial extent of a loop is given by $w/R_e \sim (1/\tilde{\chi})/N$ (R_e being the end-to-end distance). This characterizes the (intrinsic) width of the interface. Each monomer within this interfacial zone contributes to the free energy cost of the interface an amount $\tilde{\chi}$. The free energy per unit area can be calculated to $\gamma \sim \rho w \tilde{\chi}$. The mean field theory corroborates this simple picture of the interface at strong segregation $(1 \ll \tilde{\chi} N \ll N)$ and yields [71]

$$\gamma_{\text{SSL}} = \rho R_e \sqrt{\tilde{\chi}/6N} = \frac{\sqrt{\bar{\mathcal{N}}}}{R_e^2} \sqrt{\tilde{\chi}N/6} \tag{54}$$

This estimate has been used to normalize the interface tension in Fig. 21. The collapse of the data for different chain lengths onto a common curve shows that the interface tension indeed only depends on the combination $\tilde{\chi}N$ and the data are well described by numerical mean field calculations [107] and analytic predictions by Semenov [110].

Monitoring a joint histogram of the order parameter and other quantities of interest (e.g., energy or surfactant concentration) one can calculate excess properties of the interface. An example is shown in Fig. 23 where we consider a

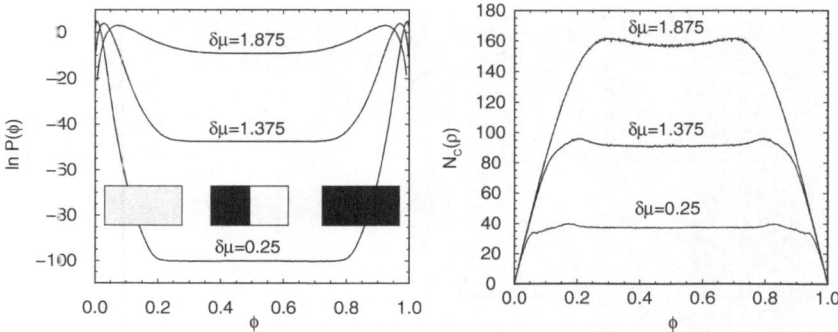

Fig. 23. Ternary blend containing two homopolymers A and B and a symmetric AB diblock copolymer within the bond fluctuation model. All chains have identical length, $N = 32$. (**a**) Probability distribution at $\epsilon = 0.054$ and system size $48 \times 48 \times 96$ in units of the lattice spacing ($R_e = 17$). Upon increasing the chemical potential $\delta\mu$ of the copolymers the "valley" becomes shallower, indicating that the copolymers decrease the interfacial tension. One clearly observes a plateau around $\phi = 1/2$. This assures, that our system size is large enough to neglect interfacial interactions in the measurement of the interfacial tension. (**b**) Average number of copolymers as a function of the composition. The copolymer number is enhanced in the configuration containing two interfaces. From Müller and Schick [105]

mixture of two homopolymers A and B and a symmetric AB diblock copolymer of identical length, $N = 32$, within the bond fluctuation model [105]. The diblock copolymer consists of two short homopolymers, A and B, each of length $N/2$ joined together at one end. It acts like a surfactant and segregates to the interface as to extends its two blocks into the respective bulk phases. In panel (a) we show the probability distribution of the composition for various values of the exchange chemical potential $\delta\mu$ between the homopolymers and diblock copolymers. As we increase the chemical potential of the diblock copolymers the "valley" between the two peaks becomes more shallow indicating that the diblock copolymers reduce the interface tension. In the same simulation we can also monitor the number of copolymers as a function of the composition. Close to the composition of the coexisting phases the number of copolymers is small in the simulation box. From the value we can determine the small solubility of the diblock in the bulk phases. For intermediate compositions the simulation cell contains interfaces and the number of copolymers is increased compared to the bulk phases. The additional copolymers segregate to the interface and we can readily determine the interfacial excess of copolymers. Monitoring the energy as a function of composition we can also estimate the excess energy of the interface.

The adsorption of diblock copolymers and the reduction of the interface tension are presented in Fig. 24 and compared to the prediction of the mean

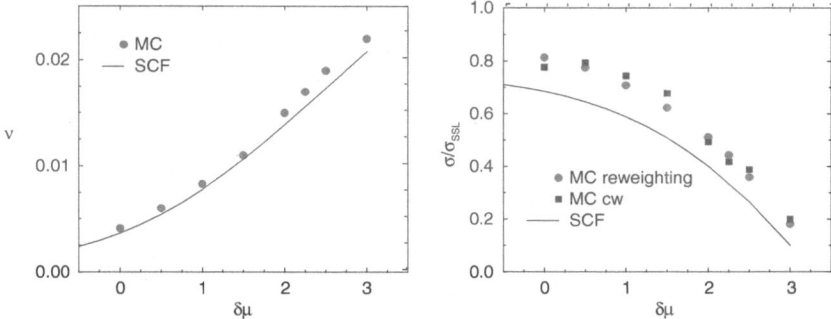

Fig. 24. (a) Adsorption of diblock copolymers at a homopolymer/homopolymer interface as a function of the chemical potential of the copolymer at $\tilde{\chi}N \approx 17$ and chain length $N = 32$. (b) Reduction of the interfacial tension upon adding copolymers. Symbols mark the results of simulations within the bond fluctuation model while lines represent SCF calculations. From Werner et al. [111]

field theory without any adjustable parameter. Good quantitative agreement is found.

3.6 Protein Simulations

We end this chapter with a final example that illustrates the performance of novel Monte Carlo methods vis-à-vis molecular dynamics calculations. More specifically, we briefly consider the reversible mechanical deformation (unfolding) of a model protein. This example is of interest in several respects. In recent years, single-molecule "pulling" experiments have been shown to provide valuable insights into the structure and folding mechanisms of large biomolecules. These experiments have often been interpreted by resorting to molecular dynamics simulations that mimic the experimental pulling action of an atomic force microscope. Unfortunately, the calculations must be performed at pulling rates that are much higher than those employed in experiments. The question that arises is whether fast-pulling calculations provide a realistic picture of the laboratory experiment.

As discussed earlier, EXEDOS simulations can be used to determine free energy differences (or PMF's) with remarkable accuracy. If the reaction coordinate, ξ, is chosen to be the end-to-end distance between the N and C termini of the protein molecule being stretched, then the resulting PMF and its derivative should correspond to the actual force measured in the laboratory, provided the molecule is pulled slowly (i.e. reversibly) [32]. In this example, Monte Carlo simulations therefore provide an ideal bound against which results of molecular dynamics can be compared. We present results for a 15-segment polyalanine molecule, which adopts a stable α-helical conformation in an implicit solvent [26]. By applying an external stretching force, the molecule can be forced to undergo a helix–coil transition. As in the previous example with Protein G, for these simulations we use the CHARMM19 force field with a united atom representation. We use the EEF1 model parameters [25], where the partial charges on the amino acids are modified to neutralize the side chains and the patched molecular termini. The peptide is stretched at $T = 300\,\mathrm{K}$ to a length much larger than that sampled in conventional canonical simulations (molecular dynamics or Monte Carlo) at this temperature. To facilitate convergence and sampling, the ξ space is subdivided into smaller overlapping windows, each box representing a specific range of ξ that overlaps with that of adjacent boxes. Sampling is again enhanced by using swap moves, in addition to regular MC moves. Figure 25 shows the PMF obtained for 15-mer alanine; the free energy minimum is arbitrarily set to zero. The ξ space explored in the simulation spans both the compression and tensile regimes. The forces therefore are negative when the helix is compressed to lengths smaller than the equilibrium end-to-end distance at $300\,\mathrm{K}$. As expected, the PMF computed from the weights is in agreement with that computed by integrating the average force. Figure 26 shows a comparison of free energy profiles obtained from steered molecular dynamics simulations (SMD) [112], SMD simulations in which the PMF is extracted directly from a Nosé-Hoover thermostat [32] (SMD-NH), and density of states (EXEDOS) calculations. Steered MD simulations mimic a single molecule AFM experiment, where one end of the molecule is fixed and the other end is pulled using a cantilever spring. A time dependent external force is applied by moving the other end of the cantilever spring at a constant velocity in the desired direction, which is usually taken as the vector joining the fixed atom to the pulled atom. Assuming that the pulling speed is sufficiently slow, multiple

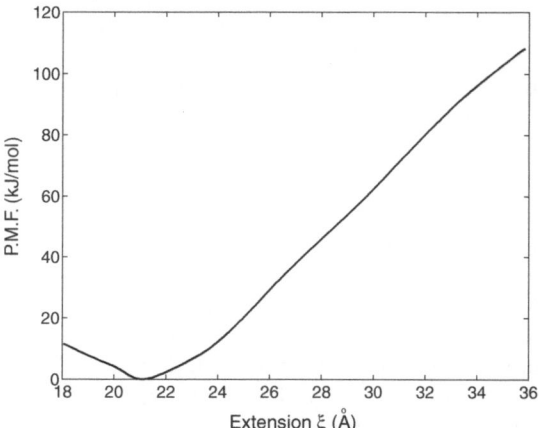

Fig. 25. Potential of mean force for 15-mer alanine as obtained from density-of-states simulations. The estimates computed directly from weights and by integrating the average force were in agreement with each other

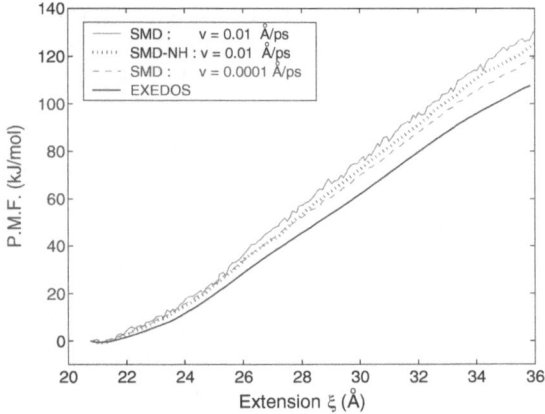

Fig. 26. Potentials of mean force as computed with SMD, SMD-NH and EXEDOS simulations. As the pulling velocity is reduced, the SMD estimates converges towards the EXEDOS results (from Rathore et al. [32])

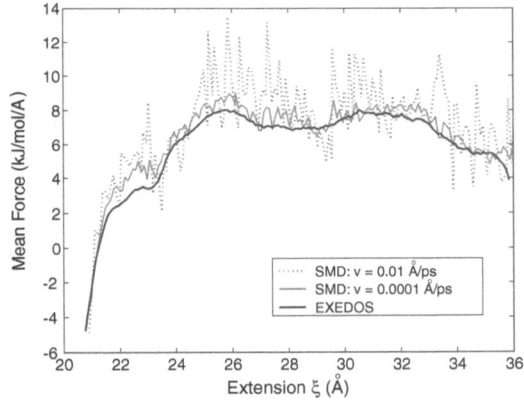

Fig. 27. Mean force-extension profiles for 15-mer alanine obtained using SMD and EXEDOS methods (from Rathore et al. [32])

SMD trajectories can in principle be averaged using Jarzynski's equality [113] to obtain the potential of mean force for the pulling process.

The corresponding force-extension profiles for these calculations are shown in Fig. 27. The data presented for SMD and SMD-NH correspond to an individual trajectory. In order to arrive at a meaningful basis for comparisons between different schemes, the PMF for SMD and SMD-NH simulations is time averaged over 0.1 Å(the bin width in EXEDOS), thereby reducing some of the statistical noise. For higher pulling rates (rates comparable to those employed in the literature), the forces and the PMF obtained from steered MD without a cantilever (SMD-NH) exhibit less noise than those obtained from conventional SMD. As the pulling velocity is reduced, the results of SMD and SMD-NH approach the results of equilibrium, EXEDOS simulations.

In EXEDOS runs, once the simulation has converged and the correct weight factors have been estimated, one can run a production run starting with these converged weights. The system then visits each state with uniform probability and relevant properties can be sampled efficiently. The main advantage of the simulations proposed here is that by resorting to Monte Carlo, density of states sampling, one can generate the true, reversible potential mean force, thereby avoiding any artifacts that might be introduced by the high pulling rates that are generally used in molecular dynamics simulations. Monte Carlo simulations can also offer advantages in that it is possible to use advanced trial moves to further enhance sampling. The cost associated with the density of states simulations as implemented in the example above is comparable to that of SMD, and the accuracy is better.

4 Conclusions and Outlook

Molecular simulation methods have experienced remarkable advances over the past decade. The examples presented in this chapter serve to illustrate that Monte Carlo simulations can now be used to address a wide variety of problems, including phase separation in complex fluids, nucleation and growth, crystallization, or protein folding. Whether one uses a multicanonical method, a density of states based technique, or umbrella sampling, the guiding principle behind most advanced Monte Carlo techniques is to achieve a uniform sampling of phase space. Infrequent events and states can then be efficiently examined and a wealth of structural and thermodynamic information is simultaneously accessible. Distinct methods differ only in the approach that is followed to achieve uniform sampling of the pertinent states either by iterative calculation of weights or umbrella sampling.

In spite of these advances, however, important challenges remain: Advanced Monte Carlo moves must still be devised to simulate complex molecules. Without such moves, a system cannot explore configuration space efficiently. Monte Carlo methods that mimic the steady state response of a system under an external field (e.g., shear flow) would also find widespread use. Flowing systems are currently simulated by resorting to non-equilibrium molecular dynamics techniques, which do not take advantage of the benefits of a Monte Carlo approach. Recent developments in the area of "beyond-equilibrium" thermodynamics [114] offer considerable promise in that respect. One more area where additional advances are needed is in the development of coarse grain models [115–118] and methods for coarse graining [119–121]. Methods in which one can jump back and forth between different levels of description in a systematic manner would expand considerably the range of problems amenable to direct simulation, and increase our understanding of the structure and dynamics of complex fluids considerably.

Acknowledgments

We have benefited from fruitful and enjoyable collaborations with K. Binder, M. Chopra, R. Faller, T.F. Knotts, L.G. MacDowell, N. Rathore, M. Schick, F. Schmid, P. Virnau, A. Werner, and Q. Yan. Financial support from the DFG under grant Mu1674/3, the Volkswagen foundation, the Department of Energy and the National Science Foundation are acknowledged.

References

1. A. Z. Panagiotopoulos (1992) Direct determination of fluid-phase equilibria by simulation in the gibbs ensemble – A review. *Molecular Simulation.* **9**, pp. 1–23
2. A. M. Ferrenberg and R. H. Swendsen (1988) New Monte-Carlo technique for studying phase-transitions. *Phys. Rev. Lett.* **61**, pp. 2635–2638
3. A. M. Ferrenberg and R. H. Swendsen (1989) Optimized Monte-Carlo data-analysis. *Phys. Rev. Lett.* **63**, pp. 1195–1198
4. B. A. Berg and T. Neuhaus (1992) Multicanonical ensemble – a new approach to simulate 1st-order phase-transitions. *Phys. Rev. Lett.* **68**, pp. 9–12
5. B. A. Berg and T. Neuhaus (1991) Multicanonical algorithms for 1st order phase-transitions. *Phys. Lett. B* **267**, pp. 249–253
6. J. Lee (1993) New Monte-Carlo algorithm – entropic sampling. *Phys. Rev. Lett.* **71**, pp. 211–214
7. F. G. Wang and D. P. Landau (2001) Efficient, multiple-range random walk algorithm to calculate the density of states. *Phys. Rev. Lett.* **86**, pp. 2050–2053
8. Q. L. Yan, R. Faller, and J. J. de Pablo (2002) Density-of-states Monte Carlo method for simulation of fluids. *J. Chem. Phys.* **116**, pp. 8745–8749
9. A. P. Lyubartsev, A. A. Martsinovski, S. V. Shevkunov, and P. N. Vorontsov-velyaminov (1992) New approach to Monte-Carlo calculation of the free-energy – method of expanded ensembles. *J. Chem. Phys.* **96**, pp. 1776–1783
10. B. A. Berg (2003) Multicanonical simulations step by step. *Comp. Phys. Comm.* **153**, pp. 397–406; *ibid.* (2004) Markov Chain Monte Carlo Simulations and their Statistical Analysis. World Scientific P. 380
11. P. M. C. de Oliveira, T. J. P. Penna, and H. J. Herrmann (1996) Broad His-togram Method. *Braz. J. Phys.* **26**, pp. 677–683
12. O. Engkvist and G. Karlstrom (1996) A method to calculate the probability distribution for systems with large energy barriers. *Chem. Phys.* **213**, pp. 63–76
13. N. Rathore and J. J. de Pablo (2002) Monte Carlo simulation of proteins through a random walk in energy space. *J. Chem. Phys.* **116**, pp. 7225–7230
14. N. Rathore, T. A. Knotts, and J. J. de Pablo (2003) Configurational temper-ature density of states simulations of proteins. *Biophys. J.* **85**, pp. 3963–3968
15. T. S. Jain and J. J. de Pablo (2002) A biased Monte Carlo technique for calculation of the density of states of polymer films. *J. Chem. Phys.* **116**, pp. 7238–7243
16. T. S. Jain and J. J. de Pablo (2003) Calculation of interfacial tension from density of states. *J. Chem. Phys.* **118**, pp. 4226–4229
17. Q. Yan and J. J. de Pablo (2003) Fast calculation of the density of states of a fluid by Monte Carlo simulations. *Phys. Rev. Lett.* **90**, 035701

18. N. Rathore, T. A. Knotts, and J. J. de Pablo (2003) Configurational temperature density of states simulations of proteins. *Biophys. J.* **85**, pp. 3963–3968
19. D. A. McQuarrie (1976) *Statistical Mechanics*. HarperCollins Publishers Inc., New York
20. O. G. Jepps, O. Ayton, and D. J. Evans (2000) Microscopic expressions for the thermodynamic temperature. *Phys. Rev. E.* **62**, pp. 4757–4763
21. B. D. Butler, G. Ayton, O. G. Jepps, and D. J. Evans (1998) Configurational temperature: Verification of Monte Carlo simulations. *J. Chem. Phys.* **109**, pp. 6519–6522
22. J. R. Ray (1991) Microcanonical ensemble Monte-Carlo method. *Phys. Rev. A* **44**, pp. 4061–4064
23. R. Lustig (1998) Microcanonical Monte Carlo simulation of thermodynamic properties. *J. Chem. Phys.* **109**, pp. 8816–8828
24. P. Dayal, S. Trebst, S. Wessel, D. Wurtz, M. Troyer, S. Sabhapandit, and S. N. Coppersmith (2004) Performance limitations of flat-histogram methods. *Phys. Rev. Lett.* **92**, 097201
25. T. Lazaridis and M. Karplus (1999) Effective energy function for proteins in solution. *Proteins* **35**, pp. 133–152
26. P. Ferrara, J. Apostolakiz, and A. Caflisch (2002) Evaluation of a fast implicit solvent model for molecular dynamics simulations. *Proteins* **46**, pp. 24–33
27. Q. L. Yan and J. J. de Pablo (1999) Hyper-parallel tempering Monte Carlo: Application to the Lennard-Jones fluid and the restricted primitive model. *J. Chem. Phys.* **111**, pp. 9509–9515
28. E. B. Kim, R. Faller, Q. Yan, N. L. Abbott, and J. J. de Pablo (2002) Potential of mean force between a spherical particle suspended in a nematic liquid crystal and a substrate. *J. Chem. Phys.* **117**, pp. 7781–7787
29. N. Rathore, Q. L. Yan and J. J. de Pablo (2004) Molecular simulation of the reversible mechanical unfolding of proteins. *J. Chem. Phys.* **120**, pp. 5781–5788
30. M. Doxastakis, Y. L. Chen, and J. J. de Pablo (2005) Potential of mean force between two nanometer-scale particles in a polymer solution. *J. Chem. Phys.* **123**, 034901
31. T. A. Knotts, N. Rathore, and J. J. de Pablo (2005) Structure and stability of a model three-helix-bundle protein on tailored surfaces. *Proteins-Structure Function and bioinformatics* **61**, pp. 385–397
32. N. Rathore, Q. L. Yan, and J. J. de Pablo (2004) Molecular simulation of the reversible mechanical unfolding of proteins. *J. Chem. Phys.* **120**, pp. 5781–5788
33. M. Chopra, M. Müller, and J. J. de Pablo (2006) Order-parameter-based Monte Carlo simulation of crystallization. *J. Chem. Phys.* **124** p. 134102
34. J. Valleau (1999) Thermodynamic-scaling methods in Monte Carlo and their application to phase equilibria. *Adv. Chem. Phys.* **105**, pp. 369–404
35. P. Virnau and M. Müller (2004) Calculation of free energy through successive umbrella sampling. *J. Chem. Phys.* **120**, pp. 10925–10930
36. B. J. Schulz, K. Binder, M. Müller, and D. P. Landau (2003) Avoiding boundary effects in Wang-Landau sampling. *Phys. Rev. E* **67**, 067102
37. Certainly, restricting the window size limits order parameter fluctuations to far less than those explored in a grandcanonical simulation and each subsimulations resembles more closely a simulation in the canonical ensemble than in the grandcanonical ensemble. We emphasize, however, that local density (order parameter) fluctuations are not restricted and that, ideally, configurations

with a fixed order parameter have identical statistical weight in the canonical ensemble, in the ensemble used in our simulation and in the grandcanonical ensemble.

38. D. Chandler (1987) *Introduction to Modern Statistical Mechanics*. Oxford University Press, New York
39. D. Frenkel and B. Smith (1996) *Understanding Molecular Simulation*. Academic, Boston
40. P. Virnau, M. Müller, L. G. MacDowell, and K. Binder (2004) Phase behavior of n-alkanes in supercritical solution: A Monte Carlo study. *J. Chem. Phys.* **121**, pp. 2169–2179
41. K. Binder, M. Müller, P. Virnau, and L. G. MacDowell (2005) Polymer plus solvent systems: Phase diagrams, interface free energies, and nucleation. *Adv. Polym. Sci.* **173**, pp. 1–104
42. R. L. C. Vink and J. Horbach (2004) Grand canonical Monte Carlo simulation of a model colloid-polymer mixture: Coexistence line, critical behavior, and interfacial tension. *J. Chem. Phys.* **121**, pp. 3253–3258
43. R. L. C. Vink, J. Horbach, and K. Binder (2005) Capillary waves in a colloid-polymer interface. *J. Chem. Phys.* **122**, p. 134905
44. R. L. C. Vink, J. Horbach, and K. Binder (2005) Critical phenomena in colloid-polymer mixtures: Interfacial tension, order parameter, susceptibility, and coexistence diameter. *Phys. Rev. E* **71**, 011401
45. R. L. C. Vink, M. Schmidt (2005) Simulation and theory of fluid demixing and interfacial tension of mixtures of colloids and nonideal polymers. *Phys. Rev. E* **71**, 051406
46. R. L. C. Vink, and T. Schilling (2005) Interfacial tension of the isotropic-nematic interface in suspensions of soft spherocylinders. *Phys. Rev. E* **71**, 051716
47. R. L. C. Vink, S. Wolfsheimer, and T. Schilling (2005) Isotropic-nematic interfacial tension of hard and soft rods: Application of advanced grand canonical biased-sampling techniques. *J. Chem. Phys.* **123**, 074901
48. J. E. Mayer and W. W. Wood (1965) Interfacial Tension Effects in Finite, Periodic, Two-Dimensional Systems. *J. Chem. Phys.* **42**, pp. 4268–4274
49. K. Binder and M. H. Kalos (1980) Critical clusters in a supersaturated vapor - theory and Monte-Carlo simulation. *J. Stat. Phys.* **22**, pp. 363–396
50. H. Furukawa and K. Binder (1982) 2-phase equilibria and nucleation barriers near a critical-point. *Phys. Rev. A* **26**, pp. 556–566
51. B. A. Berg, U. Hansmann, and T. Neuhaus (1993) Properties of interfaces in the 2 and 3-dimensional ising-model. *Z. Phys. B* **90**, pp. 229–239
52. J. E. Hunter and W. P. Reinhardt (1995) Finite-size-scaling behavior of the free-energy barrier between coexisting phases – determination of the critical-temperature and interfacial-tension of the Lennard-Jones fluid. *J. Chem. Phys.* **103**, pp. 8627–8637
53. M. Biskup, L. Chyes, and R. Kotecky (2002) On the formation/dissolution of equilibrium droplets. *Europhys. Lett.* **60**, pp. 21–27
54. K. Binder (2003) Theory of the evaporation /condensation transition of equilibrium droplets in finite volumes. *Physica A* **319**, pp. 99–114
55. L. G. MacDowell, P. Virnau, M. Müller, and K. Binder (2004) The evaporation/condensation transition of liquid droplets. *J. Chem. Phys.* **120**, pp. 5293–5308

56. Generally, the density of the liquid inside the drop will also deviate from the coexistence density of the liquid. Since the compressibility of the liquid phase, however, is much smaller than that of the vapor the deviation of the density inside the drop from the coexistence value will be much smaller than the deviation in the vapor phase.

57. F. H. Stillinger Jr. (1963) Rigorous Basis of the Frenkel-Band Theory of Association Equilibrium. *J. Chem. Phys.* **38**, pp. 1486–1494

58. T. Neuhaus and J. S. Hager (2003) 2D crystal shapes, droplet condensation, and exponential slowing down in simulations of first-order phase transitions. *J. Stat. Phys.* **113**, pp. 47–83

59. K. Leung and R. K. P. Zia (1990) Geometrically induced transitions between equilibrium crystal shapes. *J. Phys.* A **23**, pp. 4593–4602

60. L. G. MacDowell, M. Müller, C. Vega, and K. Binder (2000) Equation of state and critical behavior of polymer models: A quantitative comparison between Wertheim's thermodynamic perturbation theory and computer simulations. *J. Chem. Phys.* **113**, pp. 419–433

61. J. I. Siepmann (1990) A method for the direct calculation of chemical-potentials for dense chain systems. *Mol. Phys.*, **70**, pp. 1145–1158; D. Frenkel, G. C. A. M. Mooij, and B. Smit (1992) Novel scheme to study structural and thermal-properties of continuously deformable molecules. *J. Phys. Condens. Matter* **4**, pp. 3053–3076; M. Laso, J. J. dePablo, U. W. Suter (1992) Simulation of phase-equilibria for chain molecules. *J. Chem. Phys.* **97**, pp. 2817–2819

62. M. Müller and L. G. MacDowell (2000) Interface and surface properties of short polymers in solution: Monte Carlo simulations and self-consistent field theory. *Macromolecules* **33**, pp. 3902–3923

63. C. Borgs and R. Kotecky (1990) A rigorous theory of finite-size scaling at 1st-order phase-transitions. *J. Stat. Phys.* **61**, pp. 79–119; *ibid.* (1992) Finite-size effects at asymmetric 1st-order phase-transitions. *Phys. Rev. Lett.* **68**, pp. 1734–1737

64. A. Sariban and K. Binder (1988) Phase-Separation of polymer mixtures in the presence of solvent. *Macromolecules* **21**, pp. 711–726; *ibid.* (1991) Spinodal decomposition of polymer mixtures – a Monte-Carlo simulation. **24**, pp. 578–592; *ibid.* (1987) Critical properties of the Flory-Huggins lattice model of polymer mixtures. *J. Chem. Phys.* **86**, pp. 5859–5873; *ibid.* (1988) Interaction effects on linear dimensions of polymer-chains in polymer mixtures. *Makromol. Chem.* **189**, pp. 2357–2365

65. M. Müller (1999) Miscibility behavior and single chain properties in polymer blends: a bond fluctuation model study. *Macromol. Theory Simul.* **8**, pp. 343–374; M. Müller and K. Binder (1995) Computer-simulation of asymmetric polymer mixtures. *Macromolecules* **28**, pp. 1825–1834; *ibid.* (1994) An algorithm for the semi-grand-canonical simulation of asymmetric polymer mixtures. *Computer Phys. Comm.* **84**, pp. 173–185

66. R. D. Kaminski (1994) Monte-Carlo evaluation of ensemble averages involving particle number variations in dense fluid systems. *J. Chem. Phys.* **101**, pp. 4986–4994

67. I. Carmesin and K. Kremer (1988) The bond fluctuation method – a new effective algorithm for the dynamics of polymers in all spatial dimensions. *Macromolecules* **21**, pp. 2819–2823; H.-P. Deutsch and K. Binder (1991) Interdiffusion and self-diffusion in polymer mixtures – a monte-carlo study. *J. Chem. Phys.* **94**, pp. 2294–2304

68. M. L. Huggins (1941) Solutions of Long Chain Compounds. *J. Chem. Phys.*
    9, p. 440; P. J. Flory (1941) Thermodynamics of High Polymer Solutions. *J.
    Chem. Phys.* **9**, pp. 660–661

69. K. S. Schweizer and J. G. Curro (1997) Integral equation theories of the struc-
    ture, thermodynamics, and phase transitions of polymer fluids. *Adv. Chem.
    Phys.* **98**, pp. 1–142.

70. K. W. Foreman and K. F. Freed (1998) Lattice cluster theory of multicompo-
    nent polymer systems: Chain semiflexibility and specific interactions. *Advances
    in Chemical Physics* **103**, pp. 335–390; K. F. Freed and J. Dudowicz (1998)
    Lattice cluster theory for pedestrians: The incompressible limit and the misci-
    bility of polyolefin blends. *Macromolecules* **31**, pp. 6681–6690

71. E. Helfand and Y. Tagami (1972) Theory of interface between immiscible poly-
    mers .2. *J. Chem. Phys.* **56**, p. 3592; E. Helfand (1975) Theory of inhomoge-
    neous polymers – fundamentals of Gaussian random-walk model. *J. Chem.
    Phys.* **62**, pp. 999–1005

72. K. M. Hong and J. Noolandi (1981) Theory of inhomogeneous multicompo-
    nent polymer systems. *Macromolecules* **14**, pp. 727–736; *ibid.*, (1982) Inter-
    facial properties of immiscible homopolymer blends in the presence of block
    copolymers. **15**, pp. 482–492

73. K. R. Shull (1993) Interfacial phase-transitions in block copolymer homopoly-
    mer blends. *Macromolecules* **26**, pp. 2346–2360

74. J. M. H. M. Scheutjens and G. J. Fleer (1979) Statistical-theory of the adsorp-
    tion of interacting chain molecules .1. Partition-function, segment density dis-
    tribution, and adsorption-isotherms. *J. Phys. Chem.* **83**, pp. 1619–1635; *ibid.*
    (1980) Statistical-theory of the adsorption of interacting chain molecules .2.
    Train, loop, and tail size distribution. **84**, pp. 178–190; *ibid.* (1985) Interaction
    between 2 adsorbed polymer layers. *Macromolecules* **18**, pp. 1882–1900

75. M. W. Matsen (1995) Stabilizing new morphologies by blending homopolymer
    with block-copolymer. *Phys. Rev. Lett.* **74**, pp. 4225–4228

76. G. H. Fredrickson, V. Ganesan, and F. Drolet (2002) Field-theoretic com-
    puter simulation methods for polymers and complex fluids. *Macromolecules*
    35, pp. 16–39

77. M. Müller and F. Schmid (2005) Incorporating fluctuations and dynamics in
    self-consistent field theories for polymer blends. *Adv. Polym. Sci.* **185**, pp. 1–58

78. M. Müller (2005) Monte Carlo Simulations of Binary Polymer Liquids. In *Mole-
    cular Simulation Methods for Predicting Polymer Properties*, V. Galiatsatos
    (ed), pp. 95–152, Wiley Hoboken, NJ.

79. F. S. Bates, M. F. Schultz, J. H. Rosedale, and K. Almdal (1992) Order and
    Disorder in symmetrical diblock copolymer melts. *Macromolecules* **25**, p. 5547;
    M. D. Gehlsen and F. S. Bates (1994) *Macromolecules* **27**, p. 3611; F. S. Bates
    and G. H. Fredrickson (1994) *Macromolecules* **27**, p. 1065

80. D. Schwahn, G. Meier, K. Mortensen, and S. Janssen (1994) On the N-scaling of
    the ginzburg number and the critical amplitudes in various compatible polymer
    blends. *J. Phys. II* (France) **4**, pp. 837–848; H. Frielinghaus, D. Schwahn, L.
    Willner, and T. Springer (1997) Thermal composition fluctuations in binary
    homopolymer mixtures as a function of pressure and temperature. *Physica B*
    241, pp. 1022–1024

81. P. Van Konynenburg and R. L. Scott (1980) Critical lines and phase-equilibria
    in binary vanderwaals mixtures. *Philos. Trans. Soc. London Series A* **298**,
    pp. 495–540

82. H. A. Lorentz (1881) *Annalen Phys.* **12**, p. 127
83. D. C. Berthelot (1898) *r. hebd. Seanc. Acad Sci.* Paris **126**, p. 1703
84. G. Schneider, Z. Alwani, W. Heim, E. Horvath, and E. U. Franck (1967) Phase equilibria and critical phenomena in binary mixtures (CO_2 with N-octane N-undecane N-tridecane and N-hexadecane up to 1500 bar). *Chem. Ing. Techn.* **39**, p. 649
85. T. Charoensombut-Amon, R. J. Martin, and R. Kobayashi (1986) Application of a generalized multiproperty apparatus to measure phase-equilibrium and vapor-phase densities of supercritical carbon-dioxide in normal-hexadecane systems up to 26 mpa. *Fluid Phase Equilibria* **31**, pp. 89–104
86. C. Menduina, C. McBride, and C. Vega (2001) Correctly averaged Non-Gaussian theory of rubber-like elasticity – application to the description of the behavior of poly(dimethylsiloxane) bimodal networks. *Phys. Chem. Chem. Phys.* **3**, p. 1289
87. P. G. de Gennes and J. Prost (1993) *The Physics of Liquid Crystals*. Clarendon Press, Oxford
88. L. Leibler (1980) Theory of microphase separation in block co-polymers. *Macromolecules* **13**, pp. 1602–1617
89. P. J. Steinhardt, D. R. Nelson, and M. Ronchetti (1983) Bond-orientational order in liquids and glasses. *Phys. Rev. B* **28**, pp. 784–805
90. L. D. Landau and E. M. Lifshitz (1980) *Statistical Physics*, 3rd, Pergamon, London
91. K. Binder (1982) Monte-Carlo calculation of the surface-tension for two-dimensional and 3-dimensional lattice-gas models. *Phys. Rev. A* **25**, pp. 1699–1709
92. P. R. Ten Wolde, M. J. Ruiz-Montero, and D. Frenkel (1995) Numerical evidence for BCC ordering at the surface of a critical fcc nucleus. *Phys. Rev. Lett.* **75**, pp. 2714–2717
93. M J. Mandell, J. P. McTaque, and A. Rahman (1976) Crystal nucleation in a 3-dimensional lennard-jones system – molecular-dynamics study. *J. Chem. Phys.* **64**, pp. 3699–3702
94. C. S. Hsu and A. Rahman (1979) Crystal nucleation and growth in liquid rubidium. *J. Chem. Phys.* **71**, p. 4974
95. W C. Swope and H. C. Andersen (1990) 10(6)-Particle molecular-dynamics study of homogeneous nucleation of crystals in a supercooled atomic liquid. *Phys. Rev. B* **41**, pp. 7042–7054
96. J. S. van Duijneveldt and D. Frenkel (1992) Computer-simulation study of free-energy barriers in crystal nucleation. *J. Chem. Phys.* **96**, pp. 4655–4668
97. E. B. Kim, R. Faller, Q. Yan, N. L. Abbott, and J. J. de Pablo (2002) Potential of mean force between a spherical particle suspended in a nematic liquid crystal and a substrate. *J. Chem. Phys.* **117**, pp. 7781–7787
98. N. B. Wilding and A. D. Bruce (2000) Freezing by Monte Carlo phase switch. *Phys. Rev. Lett.* **85**, pp. 5138–5141
99. M. B. Sweatman (2005) Self-referential Monte Carlo method for calculating the free energy of crystalline solids. *Phys. Rev. E* **72**, 016711
100. D. M. Eike, J. F. Brennecke, and E. J. Maginn (2005) Toward a robust and general molecular simulation method for computing solid-liquid coexistence. *J. Chem. Phys.* **122**, 014115
101. D. Moroni, P. Rein ten Wolde, and P. G. Bolhuis (2005) Interplay between structure and size in a critical crystal nucleus. *Phys. Rev. Lett.* **94**, p. 235703

102. P. R. ten Wolde, M. J. Ruiz-Montero, and D. Frenkel (1996) Numerical calcu-
     lation of the rate of crystal nucleation in a Lennard-Jones system at moderate
     undercooling. *J. Chem. Phys.* **104**, pp. 9932–9947
103. B. B. Laird and R. L. Davidchack (2005) Direct calculation of the crystal-melt
     interfacial free energy via molecular dynamics computer simulation. *J. Phys.
     Chem. B* **109**, pp. 17802–17812
104. M. Müller, K. Binder, and W. Oed (1995) Structural and thermodynamic prop-
     erties of interfaces between coexisting phases in polymer blends – a monte-carlo
     simulation. *J. Chem. Soc. Faraday Trans.* **91**, pp. 2369–2379
105. M. Müller and M. Schick (1996) Bulk and interfacial thermodynamics of a
     symmetric, ternary homopolymer-copolymer mixture: A Monte Carlo study.
     J. Chem. Phys. **105**, pp. 8885–8901
106. B. Grossmann and M. L. Laursen (1993) The confined deconfined interface ten-
     sion in quenched qcd using the histogram method. *Nuc. Phys. B* **408**, pp. 637–
     656
107. F. Schmid and M. Müller (1995) Quantitative comparison of self-consistent-
     field theories for polymers near interfaces with monte-carlo simulations. *Macro-
     molecules* **28**, pp. 8639–8645
108. A. Werner, F. Schmid, M. Müller, and K. Binder (1999) "Intrinsic" profiles
     and capillary waves at homopolymer interfaces: A Monte Carlo study. *Phys.
     Rev. E* **59**, pp. 728–738
109. M. Müller (2006) *Soft Matter* vol. 1, Chap. 3, pp. 179–283 edited by G. Gomp-
     per and M. Schick, Wiley-VCH, Weinheim
110. A. N. Semenov (1996) Theory of long-range interactions in polymer systems.
     J. Phys. (France) II, **6**, pp. 1759–1780
111. A. Werner, F. Schmid, and M. Müller (1999) Monte Carlo simulations of
     copolymers at homopolymer interfaces: Interfacial structure as a function of
     the copolymer density. *J. Chem. Phys.* **110**, pp. 5370–5379
112. H. Lu, B. Isralewitz, A. Krammer, V. Vogel, and K. Schulten (1998) Unfolding
     of titin immunoglobulin domains by steered molecular dynamics simulation.
     Biophys. J. **75**, pp. 662–671
113. C. Jarzynski (2001) How does a system respond when driven away from thermal
     equilibrium? *Proc. Nat. Acad. Sci.* **98**, pp. 3636–3638
114. H.C. Öttinger (2005) *Beyond Equilibrium Thermodynamics.* Wiley Inter-
     science, New Jersey
115. J. Baschnagel, K. Binder, P. Doruker, A. A. Gusev, O. Hahn, K. Kremer, W. L.
     Mattice, F. Muller-Plathe, M. Murat, W. Paul, S. Santos, U. W. Suter, and V.
     Tries (2000) Bridging the gap between atomistic and coarse-grained models of
     polymers: Status and perspectives. *Adv. Polym. Sci.* **152**, pp. 41–156
116. J. C. Shelley, M. Y. Shelley, R. C. Reeder, S. Bandyopadhyay, P. B. Moore, and
     M. L. Klein (2001) Simulations of phospholipids using a coarse grain model. *J.
     Phys. Chem. B* **105**, pp. 9785–9752
117. M. Müller, K. Katsov, and M. Schick (2003) Coarse-grained models and col-
     lective phenomena in membranes: Computer simulation of membrane fusion.
     J. Polym. Sci. B **41**, pp. 1441–1450
118. S. O. Nielsen, C. F. Lopez, G. Srinivas, and M. L. Klein (2004) Coarse grain
     models and the computer simulation of soft materials. *J. Phys.: Condens. Mat-
     ter* **16**, pp. R481–R512
119. F. Müller-Plathe (2002) *Chem. Phys. Chem.* **3**, p. 754

120. R. Faller, H. Schmitz, O. Biermann and F. Müller-Plathe (1999) Molecular mobility in cyclic hydrocarbons: A simulation study. *J. Comput. Chem.* **20**, p 1009; *ibid.* (2004) *Polymer* **45**, p. 3869

121. L. Delle Site, C. F. Abrams, A. Alavi, and K. Kremer (2002) Polymers near metal surfaces: Selective adsorption and global conformations. *Phys. Rev. Lett.* **89**, p. 156103; M. Praprotnik, L. Delle Site, and K. Kremer (2005) *J. Chem. Phys.* **123**, p. 224106

Waste-Recycling Monte Carlo

D. Frenkel

FOM Institute for Atomic and Molecular Physics (AMOLF), Kruislaan 407, 1098 SJ Amsterdam, The Netherlands
frenkelamolf.nl

Daan Frenkel

D. Frenkel: *Waste-Recycling Monte Carlo*, Lect. Notes Phys. **703**, 127–137 (2006)
DOI 10.1007/3-540-35273-2_4

The Metropolis (Markov Chain) Monte Carlo method is simple and powerful. Since 1953, many extensions of the original Markov Chain Monte Carlo method have been proposed, but they are all based on the original Metropolis prescription that only states belonging to the Markov Chain should be sampled. In particular, if trial moves to a potential target state are rejected, that state is not included in the sampling. I will argue that the efficiency of effectively all Markov Chain MC schemes can be improved by including the rejected states in the sampling procedure. Such an approach requires only a trivial (and cheap) extension of existing programs. I will demonstrate that the approach leads to improved estimates of the energy of a system and that it leads to better estimates of free-energy landscapes.

1 Introduction

Monte Carlo simulations are used to estimate the average properties of systems that have a very large number of states L. Rather than generating *all* states ($i = 1, L$) and including them in the average with their appropriate weight ρ_i (e.g. their Boltzmann weight), Monte Carlo simulations approximate the average by generating a small but representative subset of all possible states. The key problem is to generate a representative subset with the least expenditure of computing time.

The attempt to add a state to this representative subset is called a "trial move". The possibility that trial moves can be rejected constitutes the essential difference between the Markov-chain Monte Carlo (MCMC) method of Metropolis et al. [1] and the earlier "random sampling" Monte Carlo of Fermi and Ulam. Random-sampling Monte Carlo had been used (but never published) by Fermi before the advent of electronic computers. Its use in electronic computers (to calculate neutron transport) had been suggested by Ulam. With the support of von Neumann and Metropolis, it had been implemented in 1947 and a description of the method was published in 1949 [2]. Interestingly, Ulam already refers to this Monte Carlo approach as a computer "experiment". While outlining a simulation of the sequence of events initiated by an energetic cosmic ray entering the atmosphere, the 1949 paper says:... *one might try to perform a series of "experiments" and obtain a class or sample of possible of genealogies. These experiments will of course be performed not with any physical apparatus, but theoretically.* Although the Metropolis Monte Carlo method shares its name with the scheme of Ulam, it is conceptually very different. One key difference is that it is a dynamic scheme: in the ("static") random MC scheme, every "move" (an attempt to add a representative state to our sample) starts from scratch, but in the dynamical Metropolis scheme, every move starts from a state (configuration) that has already been reached in an earlier move.

A naive dynamic scheme would simply perform a random walk through the space of all states (configurations, in the case of many-body systems). This approach is fine as long as a reasonably large fraction of all possible states of the system contribute to the desired average. However, for a typical dense liquid, the fraction of states with a non-vanishing Boltzmann weight is very small (less than 1 in 10^{200} for a liquid of 100 hard-spheres near the freezing density – and even much less for larger systems).

The Monte Carlo scheme of Metropolis et al. [1] was designed to overcome this problem by limiting the sampling to states with a non-negligible Boltzmann weight. This is achieved by combining the idea of a random walk through the space of all states with a rule that ensures that attempted steps to states with a low Boltzmann weight are likely to be rejected. In practice, Metropolis et al. used the following procedure (that is still the core of any Monte Carlo program) to estimate the average of an observable A:

1. From the present state of the system (denoted by the symbol o, a trial move is attempted to a trial state n. In the Metropolis scheme, the (stochastic) rule for the generation of these trial moves is such that the probability α_{on} to attempt a trial move to n, given that the system is initially in o, is equal to the probability α_{no} to generate a trial move to o, given that the system is initially in n.

2. The trial move from o to n is then accepted with a probability equal to $\mathrm{Min}\{1, \rho_n/\rho_o\}$, where ρ_n (ρ_o) denote the (Boltzmann) weights of, respectively, the trial state and the original state.

3. Compute the quantity A for the current state of the system. Depending on the outcome of the previous step, the system may either be in state n or state o.

The Boltzmann-weighted average of A is approximated by an unweighted average of all values of A computed in step 3. The key feature of the Metropolis scheme is that (on average) every state of the system is visited with a frequency that is proportional to its Boltzmann weight. That is why the weighted average over all states i,

$$\langle A \rangle = \frac{\sum_i \rho_i a_i}{\sum_i \rho_i} \tag{1}$$

can be estimated by computing the unweighted average over all states visited:

$$\frac{\sum_n a_n}{\sum_n} \tag{2}$$

where a state n is included in the sum, every time it is visited. There exist many variations of the Metropolis scheme that do not make the assumption of step 1 that the *a priori* probabilities to attempt a trial move from o to n and from n to o are equal. Such schemes have different acceptance rules than the one given in step 2. However, *all* Metropolis-style MC schemes share the feature that trial moves may be rejected. When that happens, all information about the trial state n is discarded.

It would seem that this is wasteful: once the computational effort has been been invested to generate a trial state, it does not make sense to waste all information thus generated. Yet, the Metropolis rejection rule is necessary to guarantee that, during the random walk through state space, all states are visited with a frequency proportional to their Boltzmann weights. Fortunately, we have some freedom in the choice of the quantity that is sampled. It need not be a itself – as long as its average is equal to the desired average $\langle A \rangle$. Below, I will show that it is possible to construct other observables that yield $\langle A \rangle$ as average, but *do* retain information about the rejected states.

2 Waste Recycling

In order to discuss schemes that allow us to include information about "rejected" states in our Monte Carlo sampling, it is useful to review the basic equations that underlie Metropolis importance sampling. Our aim is to sample a distribution ρ. The weight of state i is denoted by $\rho(i)$. The probability that the system that is initially in state m will undergo a transition to state n is denoted by π_{mn}. This probability is normalized as the system must end in *some* state (possibly the original state o). The normalization condition

$$\sum_n \pi_{mn} = 1 \tag{3}$$

will be used below. Any valid Monte Carlo algorithm should maintain the equilibrium distribution [3]:

$$\sum_m \rho(m)\pi_{mn} = \rho(n) \tag{4}$$

If we insert the above equation in expression 1, we get

$$\langle A \rangle = \frac{\sum_i \rho(i)a_i}{\sum_i \rho(i)} = \frac{\sum_i \sum_j \rho(j)\pi_{ji}a_i}{\sum_i \sum_j \rho(j)\pi_{ji}} \tag{5}$$

Now we can exchange the order of summation over i and j. Using (3), we can write

$$\sum_j \sum_i \rho(j)\pi_{ji} = \sum_j \rho(j) \tag{6}$$

and hence,

$$\langle A \rangle = \frac{\sum_j \sum_i \rho(j)\pi_{ji}a_i}{\sum_j \sum_i \rho(j)\pi_{ji}} = \frac{\sum_j \rho(j)\sum_i \pi_{ji}a_i}{\sum_j \rho(j)} \tag{7}$$

Hence, $\langle A \rangle$ can be rewritten as a Boltzmann average over all states j of the quantity $\sum_i \pi_{ji}a_i$. As before we can use normal (e.g. Metropolis) MCMC sampling to estimate this average

$$\frac{\sum_n \left[\sum_m \pi_{nm}a_m \right]}{\sum_n} \to \langle A \rangle \tag{8}$$

where the sum runs over the visited states n. We now use the fact that, in a Monte Carlo algorithm, the transition probability π_{nm} is the product of two factors: the a priori probability α_{nm} to attempt a trial move to m, given that the system is initially in state n and the probability $P_{acc}(nm)$ to accept this trial move.

$$\pi_{nm} = \alpha_{nm} \times P_{acc}(nm) \tag{9}$$

With this definition, our estimate for $\langle A \rangle$ becomes

$$\langle A \rangle \leftarrow \frac{\sum_n [\sum_m \alpha_{nm} \times P_{acc}(nm)a_m]}{\sum_n} \tag{10}$$

This expression is still not very useful, because it would require us to compute a_m for *all* all states m that can be reached in a single trial move from n. In what follows, we use the following shorthand notation:

$$\langle a \rangle_n \equiv \sum_m \alpha_{nm} P_{acc}(nm)a_m \tag{11}$$

In simulations, we do not compute $\langle a \rangle_n$ explicitly, but use the fact that α_{nm} is a normalized probability. We can therefore estimate $\langle a \rangle_n$ by drawing, with a probability α_{nm}, trial moves from the total set of all moves starting at n:

$$\langle a \rangle_n = \langle P_{acc}(nm)a_m \rangle_{\alpha_{nm}} \tag{12}$$

Our scheme to estimate $\langle A \rangle$ is then to perform the following sampling

$$\langle A \rangle \leftarrow \frac{\sum_n [\sum_{m'} P_{acc}(nm')a'_m]}{\sum_n} \tag{13}$$

where m' is the set of states (possibly, a single state) generated with the probability α_{nm}. If we use a MC algorithm for which a trial move generates only a single state m, then the expression for $\langle A \rangle$ becomes

$$\langle A \rangle \leftarrow \frac{\sum_n [(1 - P_{acc}(nm'))a_n + P_{acc}(nm')a'_m]}{\sum_n} \tag{14}$$

The crucial point to note is that the above average combines information about both the "accepted" and the rejected state of a trial move. Note that the Monte Carlo algorithm used to generate the random walk among the states n need not be the same as the one corresponding to π_{nm}. For instance, we could use standard Metropolis to generate the random walk, and use the symmetric rule [4]

$$\pi_{mn} = \alpha_{mn} \frac{\rho(n)}{\rho(n) + \rho(m)} \tag{15}$$

to sample the a_m's. However, one may also to use the same algorithm to sample the microscopic observables and to carry out the subsequent MC move.

2.1 Statistical Errors

As we shall briefly discuss below, the Waste-Recycling Monte Carlo (WRMC) scheme is most useful for Monte Carlo trial moves that generate not one, but a large number of trial states. However, it pays to incorporate WRMC even in the simplest form of Metropolis MC sampling. This can be done at no added cost, as all the computational work is involved in generating the trial states. To use WRMC sampling rather than normal MCMC sampling only requires the addition of a a few lines of code in the part of the program that accumulates averages. I will give two examples where it can easily be shown that the WRMC approach reduces statistical errors with respect to standard MC sampling. The first example concerns the estimate of the energy of a system. In what follows, I will assume that a standard MC algorithm is used to construct the random walk through state space, such that every state i is visited with a frequency proportional to its Boltzmann weight $\exp(-\beta E_i)$, where $\beta = 1/(k_B T)$.

Energy

The average energy of the system is

$$\langle E \rangle = \frac{\sum_i E_i \exp(-\beta E_i)}{\sum_i \exp(-\beta E_i)} \tag{16}$$

The variance in the energy is equal to $\sigma_E^2 = \langle E^2 \rangle - \langle E \rangle^2$. In normal MCMC simulations, the estimated error in the sampled energy is proportional to $\sqrt{\langle E^2 \rangle - \langle E \rangle^2}$ (the constant of proportionality depends on the number of "independent" samples that is obtained in an MC run (see e.g. [5]). As the prefactor depends on the details of the MC algorithm that generates the random walk through state space, we can use the variance itself as a measure for the statistical error. If we now use WRMC to sample the energy of the same system, the statistical error is not determined by the variance in the energy, but by the variance in the individual estimates of the energy:

$$E_{est}(i) = (1 - p_{ij})E_i + p_{ij}E_j$$

where i denotes the current state of the system, j denotes the trial state (that may, or may not, be accepted) and p_{ij} is a valid MC acceptance criterion. In what follows, I will assume the "Barker" rule

$$p_{ij} = \frac{\exp(-\beta E_j)}{\exp(-\beta E_i) + \exp(-\beta E_j)}$$

It should be stressed that the the random walk in state space need not be generated using this rule – in fact, for the random walk, the original Metropolis rule is usually better suited. The difference in the variance of the energy of WRMC and standard MC is:

$$\Delta\sigma_E^2 = \frac{\sum_i \exp(-\beta E_i) \sum_j \alpha_{ij} \left((1-p_{ij})E_i + p_{ij}E_j\right)^2}{\sum_i \exp(-\beta E_i)} - \frac{\sum_i \exp(-\beta E_i)E_i^2}{\sum_i \exp(-\beta E_i)}$$

$$(17)$$

where we have used the fact that $\langle E \rangle$, and therefore also $\langle E \rangle^2$ is the same for any valid MC scheme. We aim to show that the above expression is always ≤ 0, i.e. that the WRMC variance is less than the one obtained in normal MCMC. In what follows, we use the notation

$$Z \equiv \sum_i \exp(-\beta E_i)$$

As i and j are dummy indices, we can rewrite (17) with i and j permuted. If we add the original and the permuted expression and multiply the sum by Z, we get

$$2Z\Delta\sigma_E^2 = \sum_i \exp(-\beta E_i) \sum_j \alpha_{ij} \left((1-p_{ij})E_i + p_{ij}E_j\right)^2$$

$$+ \sum_j \exp(-\beta E_j) \sum_i \alpha_{ji} \left((1-p_{ji})E_j + p_{ji}E_i\right)^2$$

$$- \left(\sum_i \sum_j \alpha_{ij} \exp(-\beta E_i)E_i^2 + \sum_j \sum_i \alpha_{ji} \exp(-\beta E_j)E_j^2 \right) \quad (18)$$

where we have used the fact that $\sum_j \alpha_{ij} = 1$. As $Z > 0$, the sign of the above expression is determined by the sign of $\Delta\sigma_E^2$. We now use the fact that in normal MCMC sampling $\alpha_{ij} = \alpha_{ji}$. Then we can write

$$2Z\Delta\sigma_E^2 = \sum_i \sum_j \alpha_{ij} \left[\exp(-\beta E_i) \left((1-p_{ij})E_i + p_{ij}E_j\right)^2 \right.$$

$$+ \exp(-\beta E_j) \left((1-p_{ji})E_j + p_{ji}E_i\right)^2$$

$$\left. - \left(\exp(-\beta E_i)E_i^2 + \exp(-\beta E_j)E_j^2\right) \right] \quad (19)$$

As $\alpha_{ij} \geq 0$, we only have to show that all terms between square brackets are less than or equal to zero. To see this, consider an arbitrary term i, j in this sum. First of all, we note that

$$p_{ij} = (1 - p_{ji}) = \frac{\exp(-\beta E_j)}{\exp(-\beta E_i) + \exp(-\beta E_j)}$$

and

$$p_{ji} = \frac{\exp(-\beta E_i)}{\exp(-\beta E_i) + \exp(-\beta E_j)}$$

Below, we use the notation

$$z_{ij} \equiv \exp(-\beta E_i) + \exp(-\beta E_j)$$

We can then write

$$\exp(-\beta E_i)\left((1-p_{ij})E_i + p_{ij}E_j\right)^2 + \exp(-\beta E_j)\left((1-p_{ji})E_j + p_{ji}E_i\right)^2$$
$$= (\exp(-\beta E_i) + \exp(-\beta E_j)) \times \left(\frac{\exp(-\beta E_i)E_i + \exp(-\beta E_j)E_j}{z_{ij}}\right)^2$$
$$= \frac{1}{z_{ij}}\left(\frac{-\partial z_{ij}}{\partial \beta}\right)^2 \tag{20}$$

The second term in (19) can be written as

$$-\left(\exp(-\beta E_i)E_i^2 + \exp(-\beta E_j)E_j^2\right) = -\left(\frac{\partial^2 z_{ij}}{\partial \beta^2}\right) \tag{21}$$

We can then write ij-th term in (19) as

$$z_{ij}\left[\frac{1}{z_{ij}^2}\left(\frac{\partial z_{ij}}{\partial \beta}\right)^2 - \frac{1}{z_{ij}}\left(\frac{\partial^2 z_{ij}}{\partial \beta^2}\right)\right] = -z_{ij}\frac{\partial^2 \ln(z_{ij})}{\partial \beta^2} \tag{22}$$

As the right-hand side of (22) is ≤ 0, *it follows that the variance in the energy in WRMC can never exceed the Metropolis value.*

In conventional MC simulations, one can estimate the heat capacity of the system from the variance in the energy (see e.g. [5]). However, as the variance in the estimated energy in the WRMC scheme is always less than $\langle E^2 \rangle - \langle E \rangle^2$, the latter quantity should be obtained directly by computing $\langle E^2 \rangle$ and $\langle E \rangle$, where

$$\langle E^2 \rangle = \frac{\sum_{i,j}\exp(-\beta E_i)\alpha_{ij}\left((1-p_{ij})E_i^2 + p_{ij}E_j^2\right)}{\sum_i \exp(-\beta E_i)} \tag{23}$$

The gain in sampling efficiency that can be obtained with WRMC depends on the nature of the system under consideration. For instance, in the trivial case of a two level system, one single trial move is enough to sample all properties of the system exactly (provided that we use the "symmetric" rule for p_{ij}). In more realistic cases, the gain will be substantially less. But as there is always a gain at no added cost, it pays to exploit this advantage. There may not be such a thing as a free lunch, but you do not always have to pay for an extra cup of coffee...

Free-Energy Landscapes

For many applications, we need to compute the free energy of a system as a function of some order parameter Q. In practice, this is achieved using the relation

$$F(Q) = -k_B T \ln P(Q) + \text{constant} \tag{24}$$

where $P(Q)$ is the probability (density) to observe a spontaneous fluctuation that yields an order-parameter value Q. In simulations, $P(Q)$ is usually sampled by accumulating a histogram of sampled Q values. Suppose that we observe that for $N(Q)$ out of N observations the order parameter is between Q and $Q + \Delta Q$ (where ΔQ is the width of the bin), then $P(Q) \sim N(Q)$. The histogram approach works well if the probability to observe this value of Q is not too low (more precisely, if $N(Q) \gg 1$). If this is not the case, biased sampling techniques may be used to improve the statistics. WRMC does not obviate the need for such tricks. However, it allows us to make better use of the data generated in every individual run, by including the information about rejected trial moves in the histogram construction. To illustrate this, we consider a simple example: the computation of $P(E)$, the probability to find the system in a state with total energy E. In normal MCMC, we estimate $P(E)$ by sampling $N(E)$, the number of times that the states of the Markov chain have an energy between E and $E + \Delta E$. The variance in $N(E)$ is given by $\langle N^2(E) \rangle - \langle N(E) \rangle^2 = \langle N(E) \rangle$, where we have made use of the fact that $N(E)$ obeys Poisson statistics. Now consider the WRMC computation of $P(E_j)$. As before, we first focus on two "connected" states i and j (i.e. states for which $\alpha_{ij} \neq 0$). For the present pair of states, we get two contributions: one, with a value $(1 - p_{ji})$, when sampling state j, the other, with a value p_{ij} when sampling state i. If we assume the symmetric rule for p_{ij} (but not necessarily for the underlying Markov chain), we can see that the contribution to the histogram from this pair of states is:

$$\alpha_{ij} \left[\langle N_j \rangle \frac{\exp(-\beta E_j)}{z} + \langle N_i \rangle \frac{\exp(-\beta E_j)}{z} \right] = \alpha_{ij} \langle N_j \rangle \qquad (25)$$

where, as before, $z \equiv \exp(-\beta E_i) + \exp(-\beta E_j)$. If we sum over all states i, we see that WRMC does indeed yield the correct estimate for $\langle N_j \rangle$. Next consider the variance in the estimate of $P(E_j)$. Again, it is best first to consider a single pair of states. We use the fact that the variance in the occupation of states i and j is given by $\langle N_i \rangle$ and $\langle N_j \rangle$ respectively (Poisson statistics). The contribution to the variance in the WRMC estimate of $\langle N_j \rangle$ is;

$$\alpha_{ij} [\langle N_j \rangle + \langle N_i \rangle] \left(\frac{\exp(-\beta E_j)}{z} \right)^2 = \alpha_{ij} \frac{\langle N_j \rangle^2}{\langle N_i \rangle + \langle N_j \rangle} \qquad (26)$$

The total variance is

$$\sigma^2(N_j) = \langle N_j \rangle \sum_i \frac{\alpha_{ij} \langle N_j \rangle}{\langle N_i \rangle + \langle N_j \rangle} \qquad (27)$$

As the sum on the right-hand side is always less than one, we see that the WRMC variance is always less than the corresponding MCMC quantity $(\langle N_j \rangle)$.

3 Conclusions

Thus far, I have only described WRMC as an "add-on" to regular MCMC sampling. It improves the statistics only moderately – but it does so at no added cost. However, much greater gains can be made when incorporating WRMC in algorithms that generate many trial states simultaneously. One example is the Swendsen-Wang cluster algorithm (that generates 2^n trial states if the system can be decomposed in n clusters [7]). Another, more important example is the combination of WRMC with parallel tempering [8]. In this case, n simulations are run in parallel and any swap between a pair of states constitutes a potential trial state. At this stage, the WRMC algorithm has only been applied in a small number of cases (see e.g. [9]). However, in view of the extensive use of MCMC algorithms, it seems worthwhile to expand the use of the WRMC time-saving device.

Acknowledgments

The work of the FOM Institute AMOLF is part of the research program of the "Stichting voor Fundamenteel Onderzoek der Materie (FOM)", which is financially supported by the "Nederlandse Organisatie voor Wetenschappelijk Onderzoek (NWO)". I gratefully acknowledge the important input of Georgios Boulougouris and Ivan Coluzza.

References

1. N. Metropolis et al. (1953) Equation of State Calculations by Fast Computing Machines. *J. Chem. Phys.*, **21**, p. 1087
2. S. Ulam and N. Metropolis (1949) The Monte Carlo Method. *J. Am. Stat. Assoc.*, **44**, p. 335
3. V. I. Manousiouthakis and M. W. Deem (1999) Strict detailed balance is unnecessary in Monte Carlo simulation. *J. Chem. Phys.*, **110**, p. 2753
4. A. A. Barker (1965) Monte Carlo calculations of the radial distribution functions for a proton-electron plasma. *Aust. J. Phys.*, **18**, p. 119
5. *Understanding Molecular Simulations: from Algorithms to Applications* (2nd Edition). (Academic Press, San Diego, 2002)
6. R. H. Swendsen and J. S. Wang (1987) Nonuniversal critical dynamics in Monte Carlo simulations. *Phys. Rev. Lett.*, **58**, p. 86
7. D. Frenkel (2004) Speed-up of Monte Carlo simulations by sampling of rejected states. *Proc. Nat. Acad. Sci.*, **101**, p. 17571
8. I. Coluzza and D. Frenkel (2005) Virtual-move parallel tempering. *Chem. Phys. Chem.*, **6**(9), p. 1779
9. G. Boulougouris and D. Frenkel (2005) Monte Carlo sampling of a Markov web. *J. Chem. Theory Comput.*, **1**, p. 389

Equilibrium Statistical Mechanics, Non-Hamiltonian Molecular Dynamics, and Novel Applications from Resonance-Free Timesteps to Adiabatic Free Energy Dynamics

J.B. Abrams[1], M.E. Tuckerman[1,2], and G.J. Martyna[3]

[1] Dept. of Chemistry, New York University, New York, NY 10003, USA
[2] Courant Institute of Mathematical Sciences, New York University, New York, NY 10003, USA
mark.tuckerman@nyu.edu
[3] T.J. Watson Research Center, International Business Corporation, P.O. Box 218, Yorktown Heights, NY 10598, USA

Glenn J. Martyna and Mark E. Tuckerman

J.B. Abrams et al.: *Equilibrium Statistical Mechanics, Non-Hamiltonian Molecular Dynamics, and Novel Applications from Resonance-Free Timesteps to Adiabatic Free Energy Dynamics*, Lect. Notes Phys. **703**, 139–192 (2006)
DOI 10.1007/3-540-35273-2_5

1 Introduction

Levinthal's paradox [1,2], first introduced in the 1960's (early in the childhood of simulations in Chemistry), serves as a good illustration of the limitations we still face in the application of molecular dynamics (MD). Levinthal reasoned that if we were to assume that every residue in a polypeptide has a least two stable conformations, then a small 100 residue polypeptide would have 2^{100} possible states. If we were to study such a protein using traditional, state of the art, MD techniques, the native state would only be deduced after a little more than a billion years.

This problem is not simply limited to the computation study of polypeptides, but is endemnic of the greater problem that still plagues the simulation of chemically relevant systems [3], whose timescales exceed nanoseconds and even some that would require simulations longer than micro- or milliseconds: current methods and computers are simply unable to reach these timescales in a reasonable amount of computational time.

The problem can further be reduced to two aspects: the frequency and computational cost of force calculations. Force calculations, the centres of every MD timestep, range from being very costly to prohibitively costly. The approaches that are currently being pursued in order to solve this problem are quite numerous and include techniques such as the use of multiple timestep methods(r-RESPA) [4], resonance free multiple timestep methods(iso-NHC-RESPA) [5,6], using coarse grained models [3], and other novel approaches.

The organization of this document is as follows: The basis of all of these methods, equilibrium statistical mechanics, will be reviewed in Sect. 2. Section 3 will discuss the use of non-Hamiltonian systems to generate important ensembles. Novel non-Hamiltonian method, such as variable transformation techniques and adiabatic free energy dynamics will be discussed in Sect. 4. Finally, some conclusions and remarks will be provided in Sect. 5.

2 Thermodynamics and Equilibrium Classical Statistical Mechanics

The goal of most simulation studies presented in the literature is to sample states from the equilibrium distribution and to predict structural and thermodynamic properties. Consider for example the protein folding problem. Although it may be very interesting to study the pathways that lead from the unfolded state to the folded state, the quantity of interest is ultimately the structure of the folded state. Knowledge of the folded state allows biologist to use structure-function relations to predict function and to develop drugs to block function as part of a rational drug design process. Thus, the importance of a basic understanding of equilibrium properties cannot be underestimated.

In order to understand the underpinning of equilibrium molecular simulation, it is therefore necessary to review thermodynamics and equilibrium classical statistical mechanics. Here, thermodynamics is presented in a concise fashion. Gibb's ensemble postulates are then introduced and the microcanonical derived from Newtonian mechanics. The canonical ensemble and the isothermal isobaric ensemble are constructed next. It should be noted that the theory of dynamical processes is built upon the foundation presented here but is beyond the scope of the present discussion. The presentation is basic and borrows from many sources including those of McQuarrie, Fermi, and Castellan.

2.1 Thermodynamics

Thermodynamics is a beautiful and elegent theory. It does not depend on the existence of atoms and molecules. Rather, it holds for any system which satifies the postulates or laws which are decribed below. Thus, thermodynamics was developed well before quantum theory but remains valid.

The first law, Conservation of Energy, states that Heat (q) plus Work (w) equals the change in internal Energy (ΔE).

$$\Delta E = q + w \tag{1}$$

This law is movitated by the fact that perpetual motion of the first kind is not possible, i.e. no work without consuming fuel. The first law can be written in the differential form

$$dE = dq + dw$$
$$\oint dE = 0 \tag{2}$$

which indicates the energy is independent of path and Energy is thereby a state function (only depends on the conditions not on history).

The Second Law states that Entropy is a state function. The purpose of the second law is to obviate perpetual motion of the 2nd kind i.e. heat cannot be converted to work with 100% efficiency. The differential form of the Second Law is

$$dS = \frac{dq_{rev}}{T}$$
$$\oint dS = 0 \tag{3}$$

where T is the temperature and q_{rev} is the reversible heat (the heat produced by performing the process reversibly or infinitely slowly).

Combining the First and Second Law yields

$$dE = dq + dw = dq_{rev} + dw_{rev} \quad \text{(E is a state function)}$$
$$dE = TdS - pdV + \mu dN \tag{4}$$

where p is the pressure, V is the volume, μ is the chemical potential and N is the particle number. Since Energy is a state function, dE is an exact differential and S, V and μ must also be variables of state. It is difficult to measure Entropy and it is, thus, more useful to write Entropy as the independent variable

$$dS = \frac{1}{T}dE + \frac{p}{T}dV - \frac{\mu}{T}dN$$

which implies S(NVE) can be used to determine

$$\frac{1}{T} = \left.\frac{\partial S}{\partial E}\right|_{V,N}$$

$$\frac{p}{T} = \left.\frac{\partial S}{\partial V}\right|_{E,N}$$

$$\frac{\mu}{T} = -\left.\frac{\partial S}{\partial N}\right|_{E,V} \tag{5}$$

The above form shall be used below to develop a connection to molecular systems.

In the construction of the mathematical formalism underlying thermodynamics, it has implicitly been assumed that the integrating factor is trivial. That is, as the temperature goes to zero, the entropy must evolve to trivial constant. Generally, that constant is taken to be zero, and the limit of zero entropy at $T = 0$ is referred to as the Third Law. Note, thermodynamics explictly considers macroscopic amounts of material. Therefore, systems with degenerate ground states still yield essentially zero entropy (the number of degenerate states will not be macroscopic).

Let us return for a moment to the original differential form

$$dE = TdS - pdV + \mu dN \tag{6}$$

which yields the relationships

$$T = \left(\frac{\partial E}{\partial S}\right)_{VN}$$

$$p = -\left(\frac{\partial E}{\partial V}\right)_{SN}$$

$$\mu = \left(\frac{\partial E}{\partial N}\right)_{SV} \tag{7}$$

Derivatives of the extensive variable, E, with respect to three independent, extensive, variables (S,V,N), yield three corresponding dependent, intensive, thermodynamic variables T, p, μ, temperature, pressure and chemical potential. It would, therefore, be nice to have a formalism that allows thermodynamic functions to be defined in which the independent variables S or V or N could be interchanged with their corresponding intensive counterpart T or p or μ, respectively. Indeed, many experiments are performed under conditions such as NVT, NPT or μVT. The fact that the variables in question are related by first derivatives suggest that an elegent extension of thermodynamics should be possible or our conceptions about physical conditions are simply structured in accord with the mathematical framework of the theory!

In order to develop the required formalism, the Legendre transform is introduced. Any single valued function, $f(x)$, can be represented at any point, x, by its derivative, and y intercept, $b(x)$,

$$f(x) = x\frac{df(x)}{dx} + b(x) \tag{8}$$

It is therefore correct to write

$$b(x) = f(x) - x\frac{df(x)}{dx} \tag{9}$$

because the intercept is determined by the function and its slope for all x. Therefore,

$$b(y) = f(x(y)) - x(y)y$$
$$b(y) = f(y) - xy \tag{10}$$

where $y = df(x)/dx$ is now the independent variable.

The thermodynamic function with NVT as independent variables, the Helmholtz Free Energy, A(NVT), is written as

$$A = E - TS$$

Since E, T and S are state functions, A must be a state function. Taking the total derivative of both sides yields

$$dA = dE - SdT - TdS$$
$$dA = TdS - pdV + \mu dN - SdT - TdS$$
$$dA = -SdT - pdV + \mu dN \tag{11}$$

which implies

$$S = -\left(\frac{\partial A}{\partial T}\right)_{VN}$$

$$p = -\left(\frac{\partial A}{\partial V}\right)_{TN}$$

$$\mu = \left(\frac{\partial A}{\partial N}\right)_{TV} \tag{12}$$

The Gibbs free energy, G(NpT), which describes systems under the conditions of constant particle number, pressure and temperature is given by

$$G = A + PV$$

Following the above treatment yields

$$dG = dA + p\,dV + V\,dp$$
$$dG = -S\,dT - p\,dV + \mu\,dN + p\,dV + V\,dp$$
$$dG = -S\,dT + V\,dp + \mu\,dN \tag{13}$$

and

$$S = -\left(\frac{\partial G}{\partial T}\right)_{pN}$$

$$V = \left(\frac{\partial G}{\partial p}\right)_{TN}$$

$$\mu = \left(\frac{\partial G}{\partial N}\right)_{TV}$$

$$G = \mu N \tag{14}$$

Where did the last equation come from? It has simply to do with the fact that we are dealing with a mixture of extensive and intensive variables which has important consequences.

In order to understand the relationship $G = \mu N$, let us investigate the theory of homogeneous functions. If a function, $f(x_1 \ldots x_M)$ is homogeneous function of degree one in the variables $x_1 \ldots x_M$ then scaling the x_i by a constant will result in the scaling of f by the same constant

$$f(\lambda x_1 \ldots \lambda x_M) = \lambda f(x_1 \ldots x_M) \tag{15}$$

Taking the λ derivative of both sides

$$\lambda f(x_1 \ldots x_M) = f(\lambda x_1 \ldots \lambda x_M)$$

$$f(x_1 \ldots x_M) = \sum_i \frac{\partial(\lambda x_i)}{\partial \lambda} \frac{\partial f(\lambda x_1 \lambda x_M)}{\partial \lambda x_i}$$

$$f(x_1 \ldots x_M) = \sum_i x_i \frac{\partial f(\lambda x_1 \ldots \lambda x_M)}{\partial \lambda x_i}$$

$$f(x_1 \ldots x_M) = \sum_i x_i \frac{\partial f(x_1 \ldots x_M)}{\partial x_i}$$

where the value $\lambda = 1$ was inserted in the last step. Thus, there are important relationships among the thermodynamic quantities that must be investigated.

Now, $G(T,p,N)$ is extensive function. However, it depends on only **one** extensive variable, N. Therefore, G must be a homogeneous function of degree 1 in N. That is decreasing/increasing the amount of material in a system by a scaling factor must increase the free energy by the same scaling factor (doubling the ingredients in a cookie recipe doubles the number of cookies). Thus,

$$G = N \left(\frac{\partial G}{\partial N} \right)_{T,p} = \mu N$$

for a one component system. Similar alternative definitions exist for other thermodyamic quantities as well.

Continuing the tour of thermodynamic functions, let us define, ϕ,

$$\phi = A - \mu N$$

which depends on $\mu V T$

$$d\phi = dA - \mu dN - N d\mu$$
$$d\phi = -SdT - pdV + \mu dN - \mu dN - N d\mu$$
$$d\phi = -SdT - pdV - N d\mu$$

The corresponding dependent variables are defined by

$$S = -\left(\frac{\partial \phi}{\partial T} \right)_{V\mu}$$

$$p = -\left(\frac{\partial \phi}{\partial V} \right)_{T\mu}$$

$$N = \left(\frac{\partial \phi}{\partial \mu} \right)_{TV}$$

$$\phi = -pV \tag{16}$$

and the alternative definition is given as above. Indeed, the variable ϕ is never used. The free energy is always written explicitly as "$-pV$".

Finally, it is well known that systems cannot exist under conditions of μpT. Free energies must be extensive functions (or they cannot describe cookies, an important thermodynamic system, indeed!). None of the variables, μpT, are extensive. Therefore, any function which depends on μpT cannot, itself, be extensive. Mathematically, we cannot form a thermodynamic function of μpT by Legendre transform

$$\psi = \phi + pV = -pV + pV = 0$$
$$\psi = G - \mu N = \mu N - \mu N = 0. \tag{17}$$

Physically, a system separated from a particle reservoir by a semipermiable membrane, connected to a heat bath with the walls of the membrane moving in response to pressure variations of the system is **not** at equilibrium. Any fluctuation can give rise to explosive growth or decay in system size.

2.2 Equilibrium Classical Statistical Mechanics

How does thermodynamics arise from the complex dynamical motions of atoms/molecules around us? Let us start the analysis by examining the dynamical equations that govern classical mechanics. Newton stated that for every action there is an equal and opposite reaction. For the most basic case, we write, the mass of a particle times its acceleration equals the force

$$ma = F$$
$$m\ddot{q} = F \tag{18}$$

which can be written in the form

$$\dot{q} = \frac{p}{m}$$
$$\dot{p} = F \tag{19}$$

where q is the particle position and p its momentum.

Hamilton's equation are a generalization/reformulation of Newton's equations. Briefly, the Hamiltonian, $H(p, q)$ is defined such that

$$\dot{q} = \frac{\partial H}{\partial p}$$
$$\dot{p} = -\frac{\partial H}{\partial q} \tag{20}$$

which yields

$$\frac{dH}{dt} = \dot{q}\frac{\partial H}{\partial q} + \dot{p}\frac{\partial H}{\partial p} = 0$$
$$= \frac{\partial H}{\partial p}\frac{\partial H}{\partial q} - \frac{\partial H}{\partial q}\frac{\partial H}{\partial p} = 0 \tag{21}$$

For a function A depending both on postion and momenta, the Poisson Bracket is given by

$$\{A, H\} = \frac{dA}{dt} = \frac{\partial H}{\partial p}\frac{\partial A}{\partial q} - \frac{\partial H}{\partial q}\frac{\partial A}{\partial p} \tag{22}$$

For the most basic case,

$$H = \frac{p^2}{2m} + \phi(q)$$
$$\dot{q} = \frac{p}{m}$$
$$\dot{p} = -\frac{\partial \phi(q)}{\partial q} = F \tag{23}$$

where $p^2/2m$ is the kinetic energy and $\phi(q)$ is the potential energy.

In general, Hamilton's equations can be written in the form

$$\dot{\Gamma} = \xi(\Gamma, t) \tag{24}$$

where Γ is an $2n$-dimensional vector, $\Gamma = (q^1, \ldots, p^n)$, and $\xi(\Gamma, t)$ is an $2n$-dimensional vector function of Γ and t. The formal solution can be written in terms of the initial conditions

$$\Gamma_t^i = \Gamma_t^i(t; \Gamma_0^1, \ldots, \Gamma_0^{2n}) \tag{25}$$

where Γ_t^i is the coordinate i of the vector Γ at time t.

We can consider the solution to Hamilton's equations of motion as a transformation from the initial coordinates at time t_0, to the coordinates at time t and determine the evolution of the volume element

$$d\Gamma_t = J(\Gamma_t; \Gamma_0) d\Gamma_0 \tag{26}$$

where

$$J(\Gamma_t; \Gamma_0) = \frac{\partial(\Gamma_t^1 \cdots \Gamma_t^{2n})}{\partial(\Gamma_0^1 \cdots \Gamma_0^{2n})} \tag{27}$$

The Jacobian is the determinant of a matrix $\mathbf{M}$

$$J(\Gamma_t; \Gamma_0) = \det(\mathbf{M}) = e^{\text{Tr}(\ln \mathbf{M})} \tag{28}$$

An equation of motion for $J(\Gamma_t; \Gamma_0)$ can be derived by

$$\frac{dJ}{dt} = J\text{Tr}\left(\mathbf{M}^{-1}\frac{d\mathbf{M}}{dt}\right) = J\sum_{i,j} M_{ij}^{-1}\frac{dM_{ji}}{dt} \tag{29}$$

subject to the obvious initial condition $J(0) = 1$. Here, the matrix elements of $\mathbf{M}^{-1}$ and $d\mathbf{M}/dt$ are:

$$M_{ij}^{-1} = \frac{\partial \Gamma_0^i}{\partial \Gamma_t^j}$$

$$\frac{dM_{ji}}{dt} = \frac{\partial \dot{\Gamma}_t^j}{\partial \Gamma_0^i} \tag{30}$$

The equation of motion for J reduce to

$$\frac{dJ}{dt} = J \sum_{i,j} \frac{\partial \Gamma_0^i}{\partial \Gamma_t^j} \frac{\partial \dot{\Gamma}_t^j}{\partial \Gamma_0^i}$$

$$= J \sum_{i,j,k} \frac{\partial \Gamma_0^i}{\partial \Gamma_t^j} \frac{\partial \dot{\Gamma}_t^j}{\partial \Gamma_t^k} \frac{\partial \Gamma_t^k}{\partial \Gamma_0^i}$$

$$= J \sum_{j,k} \delta_{jk} \frac{\partial \dot{\Gamma}_t^j}{\partial \Gamma_t^k}$$

$$= J \sum_i \frac{\partial \dot{\Gamma}_t^i}{\partial \Gamma_t^i}$$

$$= J \kappa(\Gamma_t) \tag{31}$$

The quantity

$$\kappa(\Gamma, t) = \nabla_\Gamma \cdot \dot{\Gamma} = \nabla_\Gamma \cdot \xi(\Gamma, t) \tag{32}$$

is known as the *compressibility* of the dynamical system. Using the definition of $\kappa(\Gamma, t)$ and Equation (32)

$$\kappa(\Gamma, t) = -\sum_i \frac{\partial^2 H}{\partial p_i \partial q_i} + \sum_i \frac{\partial^2 H}{\partial q_i \partial p_i} = 0 \tag{33}$$

and $J(t) \equiv 1$. Thus, Hamiltonian systems preserve the phase space volume,

$$d\Gamma_0 = d\Gamma_t \tag{34}$$

a result referred to as Liouville's theorem.

It is now possible to consider a normalized distribution function, $f(\Gamma, t)$, describing an ensemble of systems evolving according to Hamilton's equations. An equation of motion for the distribution function can be derived by balancing the rate of change of the number of ensemble members inside a phase space volume v by the flux through the boundary surface

$$\frac{d}{dt} \int_{D(v)} d\Gamma f(\Gamma, t) = -\int_{\partial V} dS_\Gamma [\hat{n} \cdot \xi(\Gamma, t)] f(\Gamma, t)$$

$$\int_{D(v)} d\Gamma \frac{\partial f(\Gamma, t)}{\partial t} = -\int_{D(v)} \sum_i \nabla_i [\xi_i(\Gamma) f(\Gamma, t)]$$

$$= -\int_{D(v)} \sum_i \xi_i(\Gamma, t) \nabla_i f(\Gamma, t) \tag{35}$$

where phase space is assumed to be Euclidean or flat (it is). In order for the result to hold for all possible volumes, the local results holds

$$\frac{\partial f(\Gamma, t)}{\partial t} + \sum_i \xi_i(\Gamma, t) \nabla_i f(\Gamma, t) = 0$$

$$\frac{df(\Gamma, t)}{dt} = 0 \tag{36}$$

We wish to examine equilibrium solutions, $f(\Gamma)$ or $\partial f/\partial t = 0$. Clearly, any function consisting of conserved quantities of the dynamics, $f(C_1...C_M)$ will satisfy the Liouville Equation where $dC_k/dt = 0$. We wish to consider a distribution function that will allow us to visit **all** points in phase space with equal a priori probability subject to the constraints embodied by the conserved quantity, H. Clearly,

$$f(H(\Gamma)) = K\delta(H(\Gamma) - E) \tag{37}$$

where K is a constant. Therefore, we can define the phase space volume

$$\Omega(NVE) = \frac{1}{N!h^N} \int d\Gamma\delta(H(\Gamma) - E)$$

and static or phase space averages

$$\langle O\rangle = \frac{1}{N!h^N\Omega(NVE)} \int d\Gamma\delta(H(\Gamma) - E)O(\Gamma) \tag{38}$$

where h is Plank's constant, when only the Hamiltonian is assumed to be conserved. The time average of a single trajectory will produce the phase space average

$$\langle O\rangle = \lim_{T\to\infty} \frac{1}{T} \int_0^T dtO(\Gamma_t) \tag{39}$$

if and only if the system is **ergodic. We are now ready to connect to thermodynamics!!**

As thermodynamics required postulates or laws, so does statistical mechanics. Gibbs postulates which define statistical mechanics are (1) Thermodynamic quantities can be mapped onto averages over all possible microstates consistent with the few macrosopic parameters required to specify the state of the system (here, NVE). (2) We construct the averages using an "ensemble". An ensemble is a collection of systems identical on the macroscopic level but different on the microscopic level. (3) The ensemble members obey the principle of "equal a priori probability". That is, no one ensemble member is more important or probable than another.

It is clear from the derivation presented above that the phase space average, Equation (38), is exactly equal to the desired ensemble average. That is, all phase points with energy E are included with equal probability. Consider the phase space volume, $\Omega(NVE)$, the "number of states" with energy E given physical volume, V, and N particles. As the phase space volume increases, obviously, the number of microstates increases and the entropy should increase. This suggest that we postulate that $S(NVE) = F(\Omega(NVE))$ where $\Omega(NVE)$ is now referred to as the microcanonical partion function and F must be a monotonically increasing, function to be determined.

From thermodynamics, we know that

$$\frac{1}{T} = \left.\frac{\partial S}{\partial E}\right|_{V,N}$$

$$\frac{p}{T} = \left.\frac{\partial S}{\partial V}\right|_{E,N}$$

$$\frac{\mu}{T} = -\left.\frac{\partial S}{\partial N}\right|_{E,V} \tag{40}$$

From our postulates these quantities MUST arise from phase space averages. Therefore, F must be a constant times the logarithm ($k \log$) so that, for example,

$$\frac{1}{T} = \frac{\partial S}{\partial E}$$

$$\frac{1}{T} = \frac{1}{N! h^N \Omega(NVE)} \int d\Gamma \delta(H - E) \frac{\partial k \log \delta(H - E)}{\partial E}$$

$$= \langle \hat{T}^{-1} \rangle \tag{41}$$

where k is Bolzmann's constant and $\hat{T}^{-1}$ is the inverse temperature "estimator".

We next consider the ideal gas in three spatial dimensions, in the large N limit to test the theory

$$\Omega(NVE) = \frac{1}{N! h^{3N}} \int dp^N dq^N \delta \left(\sum_k \mathbf{p}_k^2/2m - E \right) \tag{42}$$

The above integral can be performed analytically [7] (Ref. [7] p. 28) as the phase space maps onto a 3N-dimensional sphere with radius, E, to yield

$$\Omega(NVE) = \left[\frac{V}{h^3} (2\pi m E)^{3/2} \right]^N \left[\frac{1}{EN! \Gamma \left(\frac{3N}{2} \right)} \right] \tag{43}$$

Using the relation $S = k \log \Omega$ and Sterling's approximation [7] to simplify $N! = \Gamma(N + 1)$ and $\Gamma(\frac{3N}{2})$ in the large N limit, yields the desired expression for the entropy

$$S(NVE) = Nk \log \left[\frac{V}{Nh^3} \left(\frac{4\pi m E}{3N} \right)^{3/2} \right] + \frac{5}{2}Nk + \mathcal{O}(N^0)$$

The temperature and the pressure can now be determined and

$$\frac{1}{T} = \frac{3k}{2E}$$

$$\frac{p}{T} = \frac{Nk}{V}$$

the standard results for the free particle are generated as expected!!

Statistical mechanics is defined at finite particle number, N. How then does it produce the thermodynamics limit $N \to \infty$? In thinking about the ensemble hypothesis consider that ensemble members can be thought of as very small chunks of a giant system. Thus, by averaging quantities over many small chunks, it should be possible to produce thermodynamic quantities as ensemble averages with small N. That is, we just need each chunk of the giant system to be large enough to be statistically independent, In addition, it can be shown that as $N \to \infty$ statistical mechanics predicts (properly) that quantities such as pressure approach their average, $\langle p \rangle = p_{ext}$, and all thermodynamic quantities are the same in all ensembles.

Having generated the connection between microcanonical ensemble (NVE), its partition function $\Omega(NVE)$, and thermodynamics, it is natural to consider the connection between the canonical ensemble (NVT), its partition function traditionally given the label $Q(NVT)$ and thermodynamics. That is, we seek to construct new phase space volume, $Q(NVT)$, to replace $\Omega(NVE)$, from which equilibrium averages in the canonical ensemble can be determined. In the new ensemble, temperature, T is constant and not energy. The postulates of statisical mechanics require that all states (points in phase space) with energy, E occur with probability $P(E; \beta(T))$ where $\beta(T)$ is an as yet unspecified function of temperature. Using the microcanonical parition function which tells us number of states with energy E possesed by a system and the probablity, $P(E; \beta(T))$ the new phase space volume, $Q(NV\beta(T))$ can be defined, naturally,

$$Q(NV\beta(T)) = \int dE P(E; \beta(T)) \Omega(NVE) \qquad (44)$$

It is now necessary to derive expressions for $P(E; \beta(T))$ and $\beta(T)$ using only the postulates of statistical mechanics.

1. Imagine a system surrounded by a thermal reservoir that consists of two independent subsystems I and II which implies $(H = H_I + H_{II})$, $\{H_I, H_{II}\} = 0$. The thermal reservoir will exchange energy with each system independently and the total energy, $E = E_I + E_{II}$, E_I, and E_{II} will all fluctuate.

2. The probability of having energy E_I in subsystem I **and** E_{II} in subsystem II must be $P_I P_{II} = P_{I,II}$ as the two subsystems are independent. For the separable system, the canonical parition function takes the form

$$Q_{I,II}(NV\beta) = \int dE_I \int dE_{II} P(E_I; \beta) \Omega_I(NVE_I) P(E_{II}; \beta) \Omega_{II}(NVE_{II})$$

$$= \int dE \int_{-2E}^{2E} d\epsilon P \left(E + \frac{\epsilon}{2}; \beta \right) P \left(E - \frac{\epsilon}{2}; \beta \right)$$

$$\cdot \, \Omega_I \left(NV \left(E + \frac{\epsilon}{2} \right) \right) \Omega_{II} \left(NV \left(E - \frac{\epsilon}{2} \right) \right) \qquad (45)$$

Since postulates of statistical mechanics require

$$Q_{I,II}(NV\beta) = \int dE P(E;\beta)\Omega_{I,II}(NVE) \tag{46}$$

the probablity must obey the relation

$$P(E;\beta) = P\left(E + \frac{\epsilon}{2};\beta\right) P\left(E - \frac{\epsilon}{2};\beta\right) \tag{47}$$

which implies

$$P(E;\beta) = \exp(-\beta E) \tag{48}$$

and for completeness,

$$\Omega_{I,II}(NVE) \equiv \int_{-2E}^{2E} d\epsilon \,\Omega_I\left(NV\left(E + \frac{\epsilon}{2}\right)\right) \Omega_{II}\left(NV\left(E - \frac{\epsilon}{2}\right)\right) \tag{49}$$

3. Using the definition of $P(E;\beta)$ and $\Omega(NVE)$

$$Q(NV\beta) = \int dE \exp(-\beta E)\Omega(NVE)$$

$$= \frac{C_N}{h^{3n}} \int d\Gamma \exp[-\beta H(\Gamma)] \tag{50}$$

where C_N is the combinatorial factor that accounts for identical particles (e.g. $N!$ if all particles are the same). It remains to determine $\beta(T)$:

1. Define the average energy:

$$-\frac{\partial \log[Q(NV\beta)]}{\partial \beta} = \int d\Gamma H(V,\Gamma) \left[\frac{K \exp[-\beta H(V,\Gamma)]}{Q(NV\beta)}\right] = \langle E \rangle \tag{51}$$

where $K = \frac{C_N}{h^{3n}}$.

2. Define the average Pressure:

$$\frac{\partial \log[Q(NV\beta)]}{\partial V} = \beta \int d\Gamma p(V,\Gamma) \left[\frac{K \exp[-\beta H(V,\Gamma)]}{Q(NV\beta)}\right] = \beta\langle P \rangle \tag{52}$$

where $p(V,\Gamma) = -\partial H(V,\Gamma)/\partial V$.

3. Perform a Legendre Transformation: Assume averages are equivalent to thermodynamic quantities. It is therefore possible to perform a Legendre on $\log[Q(V\beta)]$ to generate the thermodynamic function that depends on V and E that must be related to the entropy. Now,

$$d(\beta E + \log[Q(V\beta)]) = \beta dE + \beta PdV \tag{53}$$

can be compared to the results of thermodynamic

$$dS = \frac{1}{T}dE + \frac{p}{T}dV \tag{54}$$
$$S = E/T - A/T$$

to yield

$$\beta = \frac{1}{kT}$$
$$A = -kT\log[Q(NVT)] \tag{55}$$

4. Here, the universal constant k is the familiar Boltzmann's constant.

Using the similar arguments to those used to construct the canonical and microcanonical partition functions, it can be shown that

$$
\begin{aligned}
\Delta(NPT) &= \frac{C_N}{h^{dN}} \int_0^\infty dV\, e^{-\beta P_{\text{ext}}V} \int d^N\mathbf{p} \int_{\mathbf{D(V)}} \mathbf{d^N r}\, e^{-\beta \mathbf{H(p,r)}} \\
&= \frac{C_N}{h^{dN}} \int_0^\infty dV\, e^{-\beta P_{\text{ext}}V} \int dE e^{-\beta E} \\
&\quad \times \int d^N\mathbf{p} \int_{\mathbf{D(V)}} \mathbf{d^N r}\, \delta(\mathbf{H(p,r)} - \mathbf{E}) \\
&= \int_0^\infty dV\, e^{-\beta P_{\text{ext}}V} \int dE e^{-\beta E} \Omega(NVE) \\
&= \int_0^\infty dV\, e^{-\beta P_{\text{ext}}V} Q(NVT)
\end{aligned} \tag{56}
$$

is the isothermal-isobaric partition function where C_N a combinatorial factor, $Q(NVT)$ is the canonical partition function and the $\Omega(NVE)$ is the microcanonical partition function. The connection to thermodynamics is, also, standard,

$$G = -kT\log\Delta(NPT) \tag{57}$$

Finally, we note, that while in thermodynamics independent variables are changed via Legendre Transform, partition functions are changed via Laplace Transforms. In the thermodynamics limit, only the maximum term contributes to the integrals, and the two transformations become manifestly identical $E \to E_{thermo}$, $V \to V_{thermo}$ and $A \to A_{thermo}$. Thus, in the thermodynamic limit,

$$\Delta(NPT) = e^{-\beta P_{\text{ext}}V} e^{-\beta E} \Omega(NVE)$$
$$G = PV + E - TS = H - TS$$

and

$$\Delta(NPT) = e^{-\beta P_{\text{ext}}V} Q(NVT)$$
$$G = PV + A$$

consistent with thermodynamics.

3 Classical non-Hamiltonian Statistical Mechanics

The use of non-Hamiltonian dynamical systems has a long history in mechanics [8] and they have recently been used to study a wide variety of problems in molecular dynamics (MD). In equilibrium molecular dynamics we can exploit non-Hamiltonian systems in order to generate statistical ensembles other than the standard microcanonical ensemble (NVE) that is generated by traditional Hamiltonian dynamics. These ensembles, such as the canonical (NVT) and isothermal-isobaric (NPT) ensembles, are much better than the microcanonical ensemble for representing the actual conditions under which experiments are carried out.

Additionally, non-Hamiltonian dynamics can be used in applications/ methodologies such as Path-integral MD, replica-exchange methods, variable transformation techniques, free energy dynamics methods, and other new applications. Generating these alternative statistical ensembles from simulation requires the use of "extended systems" or "extended phase space" [9]. In these systems, the simulations do not only include the N coordinate and momentum vectors that are needed to describe a classical N-particle system, but they also include a set of additional "control" or "extended" variables that are used to drive the fluctuations required by the ensemble of interest.

3.1 Physical Basis for Generalizing Hamiltonian Principles to Non-Hamiltonian Systems

We begin our analysis of non-Hamiltonian systems by considering a general dynamical system of the form

$$\dot{x} = \xi(x, t) \tag{58}$$

where x is the n-dimensional phase-space vector, $x = (x^1, \ldots, x^n)$ (note the change in notation for the phase-space vector) and $\xi(x, t)$ is an n-dimensional vector function of x and t. The components of x are the physical coordinates and momenta as well as any additional control variables included to represent the coupling to the external surroundings. Equation (58) are solved subject to a set of initial conditions $x_0 = (x_0^1, \ldots, x_0^n)$. The solution leads to a set of n vector functions that depend on time and on the initial phase space coordinates:

$$x_t^i = x_t^i(t; x_0^1, \ldots, x_0^n) . \tag{59}$$

Since the equations of motion constitute a set of coupled ordinary differential equations, the solution in (59) is a *unique* function of the initial conditions. Thus, (59) can be viewed as an invertible coordinate transformation from a set of initial system coordinates x_0 at time, t_0, to the coordinates x_t at time, t. As a result, using (58) and (59), the Jacobian of the transformation will tell us how the phase space volume element dx_0 transforms under (59). We can define the phase space element at time t, dx_t, as a transformation from the initial system, dx_0, as

$$dx_t = J(x_t; x_0)dx_0 , \qquad (60)$$

where $J(x_t; x_0)$ is the Jacobian of the system transformation from time t_0 to time t. Previously, we have shown that the Jacobian obeys an equation of motion of the form (31). A general solution is found using the method of characteristics, which yields

$$J(x_t; x_0) = \exp\left(\int_0^t \kappa(x_s, s)ds\right). \qquad (61)$$

Since $\kappa = d\ln J/dt$, (61) can be rewritten as [10]

$$J(x_t; x_0) = e^{w(x_t,t)-w(x_0,0)} \qquad (62)$$

where $w(x,t)$ is the indefinite time integral of $\kappa(x,t)$. Substituting (62) into (60) leads to [11]:

$$e^{-w(x_t,t)}dx_t = e^{-w(x_0,0)}dx_0 \qquad (63)$$

It is clear from (63) that it is not dx_t that is conserved by the dynamics, but rather it is $\exp[-w(x_t,t)]dx_t$ that is conserved. The factor $\exp[-w(x_t,t)]$ can be viewed as a metric determinant factor, $\sqrt{g(x_t,t)}$, for the coordinate transformation. A detailed discussion of the geometry of the space and the metric tensor can be found in [10,12].

Let's now consider the impact of these results on the evolution of an ensemble of systems that evolve according to the equations of motion, (58), that is described by a phase space distribution function $f(x,t)$. Recall that, for a Hamiltonian system, the distribution function $f(x,t)$ will satisfy the Liouville equation:

$$\frac{df}{dt} = \frac{\partial f}{\partial t} + \dot{x} \cdot \nabla f = 0 \qquad (64)$$

which illustrates the fact that $f(x,t)$ is a conserved distribution function. In the case of the non-Hamiltonian systems, however, this equation is no longer valid.

A generalization of Liouville's equation was presented in [10]. Applying modern techniques in differential geometry [13], as they are applied to dynamical systems [14–16], gives rise to the generalized Liouville equation of the form [17]:

$$\frac{\partial(f\sqrt{g})}{\partial t} + \nabla \cdot (f\sqrt{g}\dot{x}) = 0 . \qquad (65)$$

It is clear that for Hamiltonian systems, where $\sqrt{g(x,t)} = 1$ and $\nabla_x \cdot \dot{x} = 0$, that (65) simply reduces to (64). Together, (65) and (31) imply that $f(x,t)$, specifically $df/dt = 0$, is also conserved for non-Hamiltonian systems. Consequently, the ensemble average of any observable property, $A(x)$, can be determined from the invariant measure and the ensemble distribution function, $f(x,t)$,

$$\langle A \rangle_t = \frac{\int dx \sqrt{g(x,t)} A(x) f(x,t)}{\int dx \sqrt{g(x,t)} f(x,t)} \tag{66}$$

Continuing with this analysis, suppose that the equations of motion given in (58) possess a set of n_c conservation laws, $\Lambda_k(x)$, $k = 1, \ldots, n_c$, which satisfy the relation:

$$\frac{d\Lambda_k}{dt} = 0. \tag{67}$$

As a result, a trajectory generated by the dynamics of (58) will not sample the entire phase space, but instead will sample a subspace of the entire phase space surface determined by the intersection of the hypersurfaces $\{\Lambda_k(x) = C_k\}$, where C_k is a set of constants. The microcanonical distribution function, that is generated by these systems, can be constructed from a product of δ-functions that represent these conservation laws:

$$f(x) = \prod_{k=1}^{n_c} \delta(\Lambda_k(x) - C_k) \tag{68}$$

It can be demonstrated by substitution, that (68) satisfies the generalized form of the Liouville equation, (65). In fact, (68) constitutes the complete solution for the microcanonical ensemble, since all the configurations that satisfy the conservation laws have equal probability of being visited by a trajectory.

It is important to note that it is possible to construct a distribution function that satisfies the general Liouville equation, (65), while not generating the particular phase space distribution function corresponding to the given dynamical system. To illustrate, consider a distribution function that is constructed from a product of an arbitrary *subset* of the δ-functions of the conservation laws. This distribution function would also satisfy the generalized Liouville equation and be of the form:

$$f_{\text{reduced}}(x) = \prod_{k=1}^{n_c'} \delta(\Lambda_k(x) - C_k), \tag{69}$$

where $n_c' \leq n_c$ represents a subset of the n_c conservation laws. If, $n_c' < n_c$, the solution in (69) will not correctly describe the phase space distribution of a system with n_c conservation laws. Therefore, it is important to realize that satisfying (65), although a necessary criterion, is **not a sufficient** condition to guarantee that a particular phase space distribution function is generated by a given dynamical system. The true distribution must be consistent with *all* of the conservation laws present.

Finally, the formalism presented above can be used to create a procedure for constructing the equilibrium ensemble partition function that is generated by a non-Hamiltonian dynamical system. First, determine *all* of the conservation laws that are satisfied by the equations of motion. The distribution function, $f(x)$, will then be written in the form of (68). Second, eliminate all variables/solutions that are linearly dependent, driven, trivial, or uncoupled [12]. Next, compute the phase space compressibility, $\nabla_{x'} \cdot \dot{x}'$, of the remaining dynamical system (having already removed all of the trivial and linearly dependent solutions). The compressibility is used to determine the phase space metric, $\sqrt{g(x')}$, and to generate the invariant volume element, $\sqrt{g(x')}dx'$. Finally, the microcanonical partition function is constructed as

$$\Omega(N, V, C_1, \ldots, C_{n_c}) = \int dx' \sqrt{g(x')} \prod_{k=1}^{n_c} \delta(\Delta_k(x') - C_k). \tag{70}$$

3.2 Analysis of a Simple Example – Free Nosé-Hoover Particle

The theory that was developed in the previous section can be applied to the dynamics of a free Nosé-Hoover particle with the associated equations of motion

$$\dot{q} = \frac{p}{m}$$

$$\dot{p} = -\frac{p_\eta}{Q} p$$

$$\dot{\eta} = \frac{p_\eta}{Q}$$

$$\dot{p}_\eta = \frac{p^2}{m} - kT \tag{71}$$

with associated phase space compressibility

$$\kappa = \frac{\partial \dot{q}}{\partial q} + \frac{\partial \dot{p}}{\partial p} + \frac{\partial \dot{\eta}}{\partial \eta} + \frac{\partial \dot{p}_\eta}{\partial p_\eta} = -\frac{p_\eta}{Q} = -\dot{\eta}. \tag{72}$$

where we couple the physical system to a thermostat with position and conjugate momentum, η and p_η, having mass Q.

The analysis from the previous section is used to write the metric tensor for this system, from the phase space compressibility, as

$$\sqrt{g(q, p, t)} = e^{-w} = e^{\eta} \tag{73}$$

where $w = -\eta$ is the indefinite integral over the compressibility, κ. Following the general procedure outlined in [12], the two conservation laws that are associated with this non-Hamiltonian system can be written. In the case of the free Nosé-Hoover particle there exist two relevant conservation laws:

$$H' = \frac{p^2}{2m} + \frac{p_\eta^2}{2Q} + kT\eta$$

$$= H(p, q) + \frac{p_\eta^2}{2Q} + kT\eta = C_1$$

$$K = pe^{\eta} = C_2 \tag{74}$$

where H' is the associated total energy quantity and K is an arbitrary constant related to the center of mass motion of the system. These conservation laws, and the metric, can now be used to write the microcanonical partition function for this system.

$$\Omega(N, V, C_1, C_2) = \int dp\,dq\,dp_\eta\,d\eta\,e^\eta \delta\left(\frac{p^2}{2m} + \frac{p_\eta^2}{2Q} + kT\eta - C_1\right)\delta\left(pe^\eta - C_2\right) \quad (75)$$

Integrating over η and p_η will generate the distribution function in the physical subspace. The action of the second δ-function requires that $pe^\eta - C_2 = 0$ and therefore that $\eta = \ln(C_2/p)$. Applying this delta function we arrive at

$$\Omega(N, V, C_1, C_2) = \frac{1}{C_2} \int dp\,dq\,dp_\eta \frac{C_2}{p}$$
$$\delta\left(\frac{p^2}{2m} + \frac{p_\eta^2}{2Q} + kT\ln\frac{C_2}{p} - C_1\right) \quad (76)$$

The second δ-function further requires that the values of p_η be $\pm(2Q(C_1 - p^2/2m - kT\ln(C_2/p)))^{-1/2}$. Imposing this condition yields the partition function

$$\Omega(N, V, C_1, C_2) = \frac{\sqrt{2Q}}{C_2} \int dp\,dq\left[p^2\left(C_1 - \frac{p^2}{2m} - kT\ln\frac{C_2}{p}\right)\right]^{-1/2} \quad (77)$$

Equation (77) is precisely the momentum distribution that is obtained when (71) are integrated numerically. This result is shown in Fig. 1. It is clear that (77) does not describe a canonical ensemble (as shown by Cho and Joannapolous [18] and by Martyna [19]). Furthermore, careful inspection of (77) shows it to be a distribution that may contain forbidden regions of the phase space corresponding to zero or negative valued arguments of the square root function. It is important to realize, however, that this result does in fact satisfy the generalized Liouville equation despite not generating a canonical distribution function.

The inability of the Nosé-Hoover algorithm to generate canonical distributions is not unique to the case of the free Nosé-Hoover particle. In fact, the Nosé-Hoover method is pathological for any system for which $\sum_i F_i = 0$ and will not generate canonical distributions.

3.3 Solving the Problem – Canonical Ensemble via Nosé-Hoover Chains

The problems endemic of the Nosé-Hoover algorithm can be solved by using the Nosé-Hoover chains (NHC) method of Tuckerman and Martyna [19, 20]. In this method, the fluctuations in the kinetic energy of the thermostat are controlled by coupling it to a second thermostat, which is in turn coupled to a third and a fourth and so on, to form a "chain" of M thermostats.

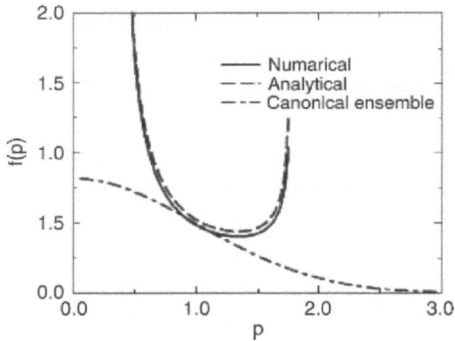

Fig. 1. The momentum distribution function for a one-dimension free particle coupled to a Nosé-Hoover thermostat (*solid line*) $[m = 1, kT = 1, p(0) = 1, q(0) = 0,$ $Q = 1, p_\eta(0) = 1$, time step $\Delta t = 0.05$] compared to the analytical result (*dashed line*) and the canonical ensemble distribution (*dot dashed line*)

The Nosé-Hoover chains method is expressed in terms of the non-Hamiltonian dynamical system with the following equations of motion:

$$\dot{\mathbf{r}}_i = \frac{\mathbf{p}_i}{m_i}$$

$$\dot{\mathbf{p}}_i = \mathbf{F}_i - \frac{p_{\eta_1}}{Q_1}\mathbf{p}_i$$

$$\dot{\eta}_k = \frac{p_{\eta_k}}{Q_k} \qquad k = 1, \ldots, M$$

$$\dot{p}_{\eta_1} = \left[\sum_{i=1}^{N}\frac{\mathbf{p}_i^2}{m_i} - LkT\right] - \frac{p_{\eta_2}}{Q_2}p_{\eta_1}$$

$$\dot{p}_{\eta_k} = \left[\frac{p_{\eta_{k-1}}^2}{Q_{k-1}} - kT\right] - \frac{p_{\eta_{k+1}}}{Q_{k+1}}p_{\eta_k} \qquad k = 2, \ldots, M-1$$

$$\dot{p}_{\eta_M} = \left[\frac{p_{\eta_{M-1}}^2}{Q_{M-1}} - kT\right] \tag{78}$$

where $\eta_1, \ldots, \eta_M$ are the M thermostat coordinates with conjugate momenta, $p_{\eta_1}, \ldots, p_{\eta_M}$, and masses, $Q_1, \ldots, Q_M$. The optimal choice for these thermostat masses was shown in [21] to be

$$Q_i = \begin{cases} LkT\tau^2 & i = 1 \\ kT\tau^2 & i \neq 0 \end{cases} \tag{79}$$

The system defined by (78) has the conserved "energy"

$$H' = H(\mathbf{p}, \mathbf{r}) + \sum_{k=1}^{M} \frac{p_{\eta_k}^2}{2Q_k} + LkT\eta_1 + kT\sum_{k=2}^{M} \eta_k \qquad (80)$$

Assuming that this is the only conservation law for this system, then there exist no linear dependencies. Only η_1 and the thermostat center $\eta_c = \sum_{k=2}^{M} \eta_k$, however, are independently coupled to the dynamics. All other combinations of the thermostat variables $k > 2$ are trivial. As such, the phase space compressibility of the system can be written as

$$\nabla_{\mathbf{x}} \cdot \dot{\mathbf{x}} = -dN\frac{p_{\eta_1}}{Q_1} - \sum_{k=2}^{M} \frac{p_{\eta_k}}{Q_k} = -dN\dot{\eta}_1 - \dot{\eta}_c \qquad (81)$$

and we can write the phase space metric as $\sqrt{g} = \exp[dN\eta_1 + \eta_c]$.

The general proof that these equations of motion, with conjugate conservation laws, generate the correct canonical distributions for the physical subsystem is provided in [19–21].

Let us now examine the case of a one-dimensional free particle coupled to a Nosé-Hoover chain of length $M = 2$ (which is analogous to the system that was already examined using the Nosé-Hoover method). This system is defined in terms of the following equations of motion:

$$\begin{aligned}
\dot{q} &= \frac{p}{m} & \dot{p} &= -\frac{p_{\eta_1}}{Q}p \\
\dot{\eta}_k &= \frac{p_{\eta_k}}{Q_k} & \dot{p}_{\eta_1} &= \frac{p^2}{m} - kT - \frac{p_{\eta_2}}{Q}p_{\eta_1} & \dot{p}_{\eta_2} &= \frac{p_{\eta_1}^2}{Q} - kT
\end{aligned} \qquad (82)$$

Following the same procedure that was outlined above, the phase space compressibility for this system is written as:

$$\kappa = \frac{\partial \dot{q}}{\partial q} + \frac{\partial \dot{p}}{\partial p} + \sum_{k=1}^{2} \left[\frac{\partial \dot{\eta}_k}{\partial \eta_k} + \frac{\partial \dot{p}_{\eta_k}}{\partial p_{\eta_k}} \right] = -\frac{p_{\eta_1}}{Q} - \frac{p_{\eta_2}}{Q} = -\dot{\eta}_1 - \dot{\eta}_2 \qquad (83)$$

Taking the indefinite time integral of κ yields $w = -\eta_1 - \eta_2$. Continuing, the phase space metric is defined, $\sqrt{g} = e^{\eta_1 + \eta_2}$, and the two conservation laws are written as:

$$H' = \frac{p^2}{2m} + \sum_{k=1}^{2} \left[\frac{p_{\eta_k}^2}{2Q} + kT\eta_k \right] = C_1$$
$$K = pe^{\eta_1} = C_2 \qquad (84)$$

where K is an arbitrary constant related to the motion of the center of mass of the system and H' is the conserved quantity. Using these laws, the partition function can be constructed:

$$\Omega(N, V, C_1, C_2) = \int dp\,dq\,dp_{\eta_1}\,dp_{\eta_2}\,d\eta_1\,d\eta_2\, e^{\eta_1 + \eta_2} \delta\left(pe^{\eta_1} - C_2\right)$$

$$\times \delta\left(\frac{p^2}{2m} + \frac{p_{\eta_1}^2}{2Q} + \frac{p_{\eta_2}^2}{2Q} + kT(\eta_1 + \eta_2) - C_1\right) \quad (85)$$

The action of the second δ-function requires that $pe^{\eta_1} - C_2 = 0$, or that $\eta_1 = \ln\frac{C_2}{p}$. Applying this result, the partition function reduces to:

$$\Omega(N, V, C_1, C_2) = \frac{1}{C_2} \int dp\,dq\,dp_{\eta_1}\,dp_{\eta_2}\,d\eta_2\, \frac{C_2}{p} e^{\eta_2}$$

$$\times \delta\left(\frac{p^2}{2m} + \frac{p_{\eta_1}^2}{2Q} + \frac{p_{\eta_2}^2}{2Q} + kT(\ln\frac{C_2}{P} + \eta_2) - C_1\right) \quad (86)$$

The remaining δ-function now requires that

$$\eta_2 = \frac{1}{kT}\left[C_1 - \frac{p^2}{2m} - \frac{p_{\eta_1}^2}{2Q} - \frac{p_{\eta_2}^2}{2Q} - kT\ln\frac{C_2}{p}\right] \quad (87)$$

leading to the final form of the partition function

$$\Omega(N, V, C_1, C_2) \propto e^{C_1/kT} \int dp_{\eta_1}\,dp_{\eta_2}\, e^{-(p_{\eta_1}^2 + p_{\eta_2}^2)/2QkT}$$

$$\times \int dp\,dq\, e^{-p^2/2mkT} \quad (88)$$

It is clear from its form that this partition function will generate a correct canonical distribution for the free one-dimensional particle. The Nosé-Hoover chains have successfully solved the pathology that had existed related to the condition $\sum_i \mathbf{F}_i = 0$. Let's investigate the application of the Nosé-Hoover chains to a slightly more complex problem: a one-dimensional harmonic oscillator with Hamiltonian,

$$H = \frac{p^2}{2m} + \frac{1}{2}m\omega^2 x^2 \quad (89)$$

Figure 2 shows the results for simulations of the same one-dimensional harmonic oscillator with Nosé-Hoover chains of varying lengths. The first column shows the results for a chain of length $M = 1$, which is equivalent to the Nosé-Hoover algorithm. Notice that neither the momentum (b) nor position (c) distributions are canonical in nature. This is further confirmed by the presence of a "Hoover hole" in the Poincaré section depicted in (a).

Conversely, the oscillators that have been coupled to Nosé-Hoover chain thermostats with length $M = 3, 4$ result in momentum and position distributions that match the canonical ensemble distributions. Additionally, the "Hoover hole" has been eliminated from the Poincaré sections for these cases.

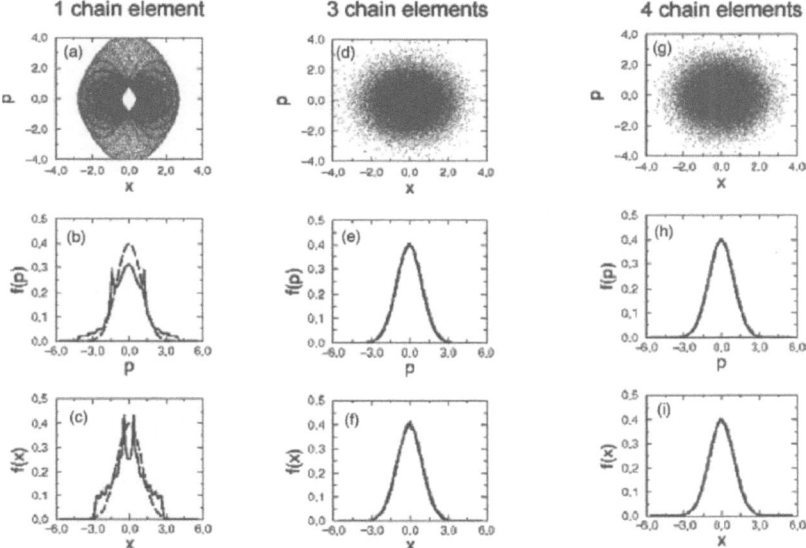

Fig. 2. Simulations of a one-dimensional harmonic oscillator coupled to Nosé-Hoover chains of length $M = 1$ (**a-c**), $M = 3$ (**d-f**), and $M = 4$ (**g-i**). (a),(d),(g) The Poincaré sections for these oscillators. (b),(e),(h) The momentum distribution functions. (c),(f),(i) The position distribution functions

4 Novel Applications of Non-Hamiltonian Molecular Dynamics

4.1 Multiple Time Scales in Molecular Dynamics

The use of reference system algorithms, for integrating problems with separation in time scales and short/long range forces, was first introduced in the early 1990's [22–24]. These methods are extraordinarily useful in obtaining significant speedups for problems, where the force components can be split into "fast" and "slow" components, for systems of different time scales, and into "short" and "long" range components, for systems with disparate distances in the force calculations.

The utility of these methods can be illustrated by investigating a generic molecular mechanics force field. The potential energy function for such a force field is usually divided into 4 parts: bonds, bends, torsions, and non-bonded (electronic and van der waals) interactions.

$$U_{bond}(\overrightarrow{\mathbf{r}}) = \frac{1}{2} \sum_{i=1}^{nbond} k_i \left(d - d_i\right)^2$$

$$U_{bend}(\overrightarrow{\mathbf{r}}) = \frac{1}{2} \sum_{i=1}^{nbend} k_i \left(\theta - \theta_i\right)^2$$

$$U_{tors}(\overrightarrow{\mathbf{r}}) = \sum_{i=1}^{ntors} \sum_{n=0}^{npow} a_n \cos(n\varphi + \delta_n)$$

$$U_{elec}(\overrightarrow{\mathbf{r}}) = \sum_{i=1}^{N-1} \sum_{j=i+1}^{N} \frac{q_i q_j}{r_{ij}}$$

$$U_{vdw}(\overrightarrow{\mathbf{r}}) = \sum_{i=1}^{N-1} \sum_{j=i+1}^{N} \left[\left(\frac{\sigma}{r_{ij}}\right)^{12} - \left(\frac{\sigma}{r_{ij}}\right)^{6} \right] \tag{90}$$

where $\overrightarrow{\mathbf{r}}$ is a vector of position coordinates for a given system and *nbond*, *nbend*, *ntors*, are the number of bonds, bends, and torsions, respectively. In order to achieve adequate sampling of these interactions, the largest time step of the system must be significantly smaller than the fundamental frequency of the fastest interaction. In most cases, the interactions with the fastest motion would be the intramolecular interactions (bonds, bends, torsions) while the non-bonded interactions are significantly slower. Not only slower, these non-bonded interactions scale on the order of N^2 and therefore require significantly more computational effort. This is the prototypical scenario for the use of multiple time scale methods. Splitting the interactions into "fast" (bonded interactions) and "slow" (non-bonded interactions), where the "fast" interactions are much cheaper to calculate, will definitely afford a significant speedup. Additionally, it is possible to further decompose the non-bonded interactions into "short" and "long" range components to the force. Therefore, we can see that the optimal decomposition of a force field integrator will include both separation of time scales and separation into short and long components of the non-bonded interactions.

Reversible Reference System Propagator Algorithm (r-RESPA)

The original RESPA method, introduced by Tuckerman and Berne in 1991 [25], was designed to accelerate the integration of the equations of motion of system whose forces could be decomposed into short and long range components:

$$F(\mathrm{x}) = F_{short}(\mathrm{x}) + F_{long}(\mathrm{x}) \tag{91}$$

The time step, δt, is determined by the short range component of the force and its contribution to the force is calculated at every time step. The contribution from the long range forces, unlike in traditional integration schemes, is only computed every n time steps, thereby significantly reducing the number of forces calculated for every large time step, $\Delta t = n\delta t$, and the CPU time. The original formulation employed a switching function to accomplish the decomposition of the forces. Although this formalism did succeed in reducing computational effort for integrating these equations of motion, the algorithm was not reversible in time and as such remained vulnerable to numerical instabilities after very long times. Moreover, this method could not be combined with any hybrid Monte Carlo methods because these methods require that the molecular dynamics be strictly reversible in order to satisfy detailed balance [26].

The new **reversible-RESPA** [25] exploits the Trotter factorization [27] to make the algorithm reversible. Re-writing the equations of motion, for a system of N particles having coordinates $q_1 \ldots q_N$ and momenta $p_1 \ldots p_N$, to incorporate both the "fast" and "slow" components of the force yields [28]:

$$\dot{q}_i = \frac{p_i}{m_i}$$
$$\dot{p}_i = F_{i,fast} + F_{i,slow} \,. \tag{92}$$

Equation (92) demonstrate that there are now two force dependencies in the equation of motion of p. The Liouville operator, L, for this system can now be written as

$$iL = \{\cdots, H\} = \sum_{j=1}^{N} \left[\dot{q}_j \frac{\partial}{\partial q_j} + (F_{j,fast} + F_{j,slow}) \frac{\partial}{\partial p_j} \right] \tag{93}$$

where $\{\cdots\}$ is the Poisson bracket of the system and $(F_{j,fast} + F_{j,slow})$ is the force, F_j, on the jth degree of freedom. The Hermitian Liouville operator L from (93) can then be separated into two parts: a reference system and a slow part:

$$iL = \left(\frac{p}{m} \frac{\partial}{\partial q} + F_{fast} \frac{\partial}{\partial p} \right) + F_{slow} \frac{\partial}{\partial p} = iL_{ref} + iL_{slow} \tag{94}$$

The propagator, $e^{iL \Delta t}$, can now be factorized symmetrically as,

$$\exp(iL\Delta t) = \exp(iL_{slow}\Delta t/2) \exp(iL_{ref}\Delta t) \exp(iL_{slow}\Delta t/2) + \mathcal{O}(\Delta t^3)$$
$$= \exp(F_{slow} \frac{\partial}{\partial p} \Delta t/2) \exp(iL_{ref}\Delta t) \exp(F_{slow} \frac{\partial}{\partial p} \Delta t/2) \tag{95}$$

where the middle propagator, reflecting the motion of the reference system, can be further factorized by the Trotter theorem as,

$$\exp(iL_{ref}\Delta t) = \left[\exp\left(F_{ref} \frac{\partial}{\partial p} \delta t/2 \right) \exp\left(\dot{q} \frac{\partial}{\partial q} \delta t \right) \exp\left(F_{ref} \frac{\partial}{\partial p} \delta t/2 \right) \right]^n \tag{96}$$

where $\delta t = \frac{\Delta t}{n}$ and n is chosen sufficiently large to generate stable trajectories over long times. This factorization with a small time step, δt, will give rise to a good approximation to the reference system propagator, while the propagators on either side, evolving with time step Δt, will provide good approximations to the slowly evolving forces. Taken together, (95) and (96) give rise to the overall propagator

$$\exp(iL\Delta t) = \exp(iL_{slow}\Delta t/2) \left[\exp(iL_{ref}\delta t)\right]^n \exp(iL_{slow}\Delta t/2) + \mathcal{O}(\Delta t^3)$$

$$= \exp\left[\frac{\Delta t}{2}F_{slow}\frac{\partial}{\partial p}\right]$$

$$\times \left\{\exp\left[\frac{\delta t}{2}F_{fast}\frac{\partial}{\partial p}\right]\exp\left[\delta t\frac{p}{m}\frac{\partial}{\partial q}\right]\exp\left[\frac{\delta t}{2}F_{fast}\frac{\partial}{\partial p}\right]\right\}^n$$

$$\times \exp\left[\frac{\Delta t}{2}F_{slow}\frac{\partial}{\partial p}\right] \tag{97}$$

This algorithm can be implemented in a typical program using the identity $\exp[a(\partial/\partial g(x))]x = g^{-1}[g(x) + a]$ and the "direct translation technique" [29]. Here, the product of operators from (97) are translated into a set of instructions which alleviate the need to calculate analytically the phase space vector $\Gamma(\Delta t)$ in terms of the initial conditions $\Gamma(0)$. A pseudocode for performing these operations would appear as:

$$\mathbf{v}_i \leftarrow \mathbf{v}_i + \frac{\Delta t}{2m_i}\mathbf{F}_i$$

for iref $= 1$ to n

$$\mathbf{v}_i \leftarrow \mathbf{v}_i + \frac{\delta t}{2m_i}\mathbf{F}_i$$

$$\mathbf{r}_i \leftarrow \mathbf{r}_i + \delta t\mathbf{v}_i$$

Get new "Fast" Force

$$\mathbf{v}_i \leftarrow \mathbf{v}_i + \frac{\delta t}{2m_i}\mathbf{F}_i$$

$endfor$

Get new "Slow" Force

$$\mathbf{v}_i \leftarrow \mathbf{v}_i + \frac{\Delta t}{2m_i}\mathbf{F}_i \tag{98}$$

Significant Speedups in Simulations of Large Systems via r-RESPA

Since its inception, r-RESPA has proven to be indispensable for molecular dynamics simulations of large systems. Using r-RESPA to break down the force calculations into different time scales has given rise to the ability to take significantly larger fundamental time steps without suffering a decrease in the energy conservation of the simulation.

An excellent illustration of the power of the reversible RESPA algorithm is the study of HIV-I-protease that is described in [30]. Here, the authors studied the HIV-I protease complexed to an inhibitor, Saquinovir, in explicit water (5041 water molecules). The simulation box had a volume of 28125 Å^3 and contained 18354 atoms.

Table 1. Energy Conservation and CPU times for six (6) simulations of HIV-I-Protease with Saquinivir in solution with different levels of r-RESPA being implemented

$\Delta t(fs)$	n_{srf}	n_{intra}	ΔE ($\times 10^4$)	CPU	Rel. fs[†]
1.0	1	1	1.681	10.4	1.00
1.0	1	2	0.809	10.6	0.98
2.0	1	4	1.043	10.9	1.91
4.0	2	4	1.101	11.9	3.50
6.0	3	4	1.114	12.8	4.88
8.0	4	4	1.386	13.5	6.16

[†] Actual speedup compared to a 1 fs time step without the r-RESPA algorithm

It was found that a significantly larger fundamental time step could be taken when using the r-RESPA algorithm to break down the force calculations into three parts: intramolecular interactions, short-range intermolecular interactions, and long-range intermolecular interactions. For each time step, Δt the long-range interaction force contributions are only computed once. The time step is factorized n_{srf} times, to $\delta t_1 = \Delta t/n_{srf}$, for computing the short-range interactions. Each of these steps is further broken-down an additional n_{intra} times for the intramolecular interactions to steps of size $\delta t_2 = \Delta t/(n_{srf}n_{intra})$.

Table 1 shows the resulting energy conservations (ΔE) and CPU times for different simulations using the r-RESPA algorithm to integrate the equations of motion using different time step factorizations. The energy conservation is defined as,

$$\Delta E = \frac{1}{N_{steps}} \sum_{i=1}^{N_{steps}} \left| \frac{E_i - E_0}{E_0} \right| \tag{99}$$

The first row of the table presents the results for a simulation using a time step of 1 fs, taken without r-RESPA. Time steps in this simulation had an associated CPU time of 10.4 and had an energy conservation of 1.681×10^{-4}. Close examination of the data in rows 2–6 of the table shows that r-RESPA allows for significantly larger fundamental time steps to be taken with energy conservation better or equivalent to simulations not using r-RESPA. Notice, however, that there is a slight overhead for using the r-RESPA algorithm. In the case of the last of the simulations presented above, we see that the simulation with fundamental time step of $\Delta t = 8$ fs is only a little more than 6 times faster than the simulation without any r-RESPA, still significantly faster.

4.2 Resonance Free Multiple Time Scale Molecular Dynamics

The reversible RESPA algorithm, and other multiple time step (MTS) integrators, was a first giant step towards alleviating the time scale separation problem, and it does so in a reversible and straightforward manner that is easy to implement. The fundamental problem that plagues these MTS integrators, even the ones that possess the symplectic property of Hamilton's equations, is that they are derived from perturbation theories and, as such, are plagued by the resonance problems that are endemic of all perturbative techniques. The consequence is that the gains in efficiency that can be achieved by r-RESPA, and other similar techniques, are fundamentally limited by the presence of these resonance phenomena which, in the case of biological systems, restrict the time steps used by these methods to $\Delta t < 8$ fs [31, 32].

A speedup to 8 fs, although significant, is nowhere near sufficient to bridge the time scales that are of interest, which are on the order of microseconds (large scale domain motion of complex proteins). A number of methods have been introduced that are capable of overcoming this resonance problem [31, 32]. Unfortunately, these approaches involve both non-reversible solvers and the introduction of stochastic baths. The first problem with these methods is that adding a stochastic bath overdamps the motions that are of biological importance [33–35]. Additionally, these baths obviate the symplectic property and destroy the time reversal symmetry, which as was discussed in the previous section excludes the possibility of using these methods as part of hybrid Monte Carlo (HMC) schemes [36].

In this section, the resonance phenomenon that is so troublesome in these problems will be investigated. Additionally, a novel set of resonance-free equations of motion based on the isokinetic ensemble that can be used in MD simulations to avoid the resonance problem will be discussed [37].

Illustration of the Resonance Problem

To illustrate the resonance phenomenon, discussed above, consider a single coordinate q with momentum p with associated equations of motion [33]:

$$\dot{q} = v$$
$$\dot{p} = -(\omega^2 + \Omega^2)q \tag{100}$$

where ω and Ω are two components of the force. This system is a harmonic oscillator driven by $(\omega^2 + \Omega^2)$. If $\omega \gg \Omega$ then the system will contain two time scales of the motion and the force terms can be split into slow and fast components as,

$$F_{fast} = -\omega^2 q \qquad F_{slow} = -\Omega^2 q \tag{101}$$

Using a RESPA factorization in which the reference system is solved analytically, the system in (100) and (101) evolves to give the following finite difference equations for the system at time Δt:

$$p'(0) = p(0) - \frac{\Delta t}{2} \Omega^2 q(0)$$

$$q(\Delta t) = q(0) \cos(\omega \Delta t) + \frac{p'}{\omega} \sin(\omega \Delta t)$$

$$p''(0) = p(0) \cos(\omega \Delta t) - \omega q(0) \sin(\omega \Delta t)$$

$$p(\Delta t) = p'' - \frac{\Delta t}{2} \Omega^2 q(\Delta t) \tag{102}$$

It can be shown that the correct propagation from the phase space point $(q(0), p(0))$ at time $t = 0$ to $(q(\Delta t), p(\Delta t))$ at time $t = \Delta t$ can be accomplished as follows

$$\begin{pmatrix} q(\Delta t) \\ p(\Delta t) \end{pmatrix} = A(\omega, \Omega, \Delta t) \begin{pmatrix} q(0) \\ p(0) \end{pmatrix} \tag{103}$$

where $A(\omega, \Omega, \Delta t)$ is the propagation matrix for this system. It is found that this be a symplectic matrix $(det(A) = 1)$ and it is defined as

$$A(\omega, \Omega, \Delta t) = \begin{pmatrix} a_{11} & a_{12} \\ a_{21} & a_{22} \end{pmatrix} \tag{104}$$

with matrix elements

$$a_{11} = a_{22} = \cos(\omega \Delta t) - \frac{\Delta t \Omega^2}{2\omega} \sin(\omega \Delta t)$$

$$a_{12} = \frac{1}{\omega} \sin(\omega \Delta t)$$

$$a_{21} = \left(\frac{\Delta t^2 \Omega^4}{4\omega} - \omega \right) \sin(\omega \Delta t) \tag{105}$$

Depending on the choice of Δt, the eigenvalues of A in (104) are either complex conjugate pairs

$$-2 < \mathrm{Tr}(A) < 2 \tag{106}$$

or the eigenvalues are both real numbers

$$|\mathrm{Tr}(A)| \geq 2. \tag{107}$$

The result is that "resonances" (a change from oscillatory to hyperbolic evolution) of $(|\mathrm{Tr}(A)| \to 2)$ are observed for the dynamics with time steps at $\Delta t = n\pi/\omega$.

Isokinetic Dynamics

The existence of these resonance phenomena has one severe consequence: there are limits to the size of the time step that can be taken and, in most cases, the limit is less than 10 fs. It is possible, however, to prevent these modes from becoming resonant by imposing a constraint on the kinetic energy [5,6]. Doing so will prevent the build up of energy in the resonant mode of the molecule. This constraint on the kinetic energy of the system can be written as,

$$\sum_{i}^{N} \frac{\mathbf{p}_i^2}{2m_i} = \frac{3N-1}{2}kT \qquad (108)$$

for a system of N particles with temperature kT. This constraint is introduced via a Lagrange multiplier to give equations of motion of the form:

$$\dot{\mathbf{r}}_i = \frac{\mathbf{p}_i}{m_i} \qquad \dot{\mathbf{p}}_i = \mathbf{F}_i - \alpha\mathbf{p}_i \qquad (109)$$

where α is the Lagrange multiplier. Taking the derivative of the constraint yields the multiplier α by:

$$\sum_{i} \frac{\mathbf{p}_i}{m_i} \cdot \dot{\mathbf{p}}_i = \sum_{i} \frac{\mathbf{p}_i}{m_i} \cdot [\mathbf{F}_i - \alpha\mathbf{p}_i] = 0$$

$$\Downarrow$$

$$\alpha = \frac{\sum_i \mathbf{F}_i \cdot \mathbf{p}_i/m_i}{\sum_i \mathbf{p}_i \cdot \mathbf{p}_i/m_i} \qquad (110)$$

The compressibility, κ of this system can now be computed:

$$\kappa = \sum_{i} \left[\frac{\partial}{\partial \mathbf{p}_i} \cdot \dot{\mathbf{p}}_i + \frac{\partial}{\partial \mathbf{r}_i} \cdot \dot{\mathbf{r}}_i \right] = e^{-\beta U(\mathbf{r})} \qquad (111)$$

Finally, the microcanonical partition function, based on the kinetic energy constraint and the compressibility, can be written using the approach outlined in Sect. 3 as:

$$\Omega = \int d^N \mathbf{p}\, \delta \left(\sum_{i} \frac{\mathbf{p}_i^2}{2m_i} - \frac{3N-1}{2}kT \right) \int d^N \mathbf{r}\, e^{-\beta U(\mathbf{r})} \qquad (112)$$

where $\beta = 1/kT$ is the temperature of the system. It is clear from (112) that the partition function generated for the isokinetic ensemble is proportional to the canonical configurational partition function,

$$Z(N,V,T) = \int d^N \mathbf{r}\, \exp\left[-\beta U(\mathbf{r})\right]. \qquad (113)$$

Extended system Isokinetic Dynamics

Consider the application of the formalism in the previous section to a one-particle system evolving in one dimension with coordinate (p, x) and mass m. It is obvious that an isokinetic constraint cannot be imposed on the system because, in this case, (109) simply reduce to the trivial result $\dot{x} = p$, $\dot{p} = 0$, for such a case. A simple remedy is to couple this particle to another one-particle, one-dimensional system, with phase space coordinate (p_η, η) with mass Q. The isokinetic constraint is now imposed on the total system such that

$$\frac{p^2}{2m} + \frac{p_\eta^2}{2Q} = \frac{1}{2}kT \tag{114}$$

Suppose that the potential of the combined system, $\Phi(x, \eta)$, is separable,

$$\Phi(x, \eta) = \phi(x) + \xi(\eta). \tag{115}$$

Gauss' principle can be applied to yield the isokinetic ensemble of the combined system with the following equations of motion:

$$\dot{x} = \frac{p}{m}$$

$$\dot{p} = F(x) - \left[\frac{F(x)p/m + G(\eta)p_\eta/m_\eta}{p^2/m + p_\eta^2/m_\eta} \right] p$$

$$\dot{\eta} = \frac{p_\eta}{m_\eta}$$

$$\dot{p}_\eta = G(\eta) - \left[\frac{F(x)p/m + G(\eta)p_\eta/m_\eta}{p^2/m + p_\eta^2/m_\eta} \right] p_\eta \tag{116}$$

where

$$F(x) = -\frac{d\phi(x)}{dx}$$

$$G(\eta) = -\frac{d\xi(\eta)}{d\eta} \tag{117}$$

Again following the analysis presented in Sect. 3, these equations will generate the partition function

$$\Omega = \int dp dp_\eta \delta \left(\frac{p^2}{2m} + \frac{p_\eta^2}{2m_\eta} - \frac{1}{2}kT \right) \int d\eta \exp[-\beta\xi(\eta)]$$

$$\times \int dx \exp[-\beta\phi(x)]$$

$$\propto Z(L, T) = \int dx \exp[-\beta\phi(x)] \tag{118}$$

Notice that the partition function in (118) is not the isokinetic partition function for the one-particle, one-dimensional system. This partition function is again, nevertheless, an isokinetic partition function whose configuration space distribution is canonically distributed. As such, these equations of motion will lead to equilibrium averages of position-dependent properties that are meaningful.

Long Time Step, Resonance-Free, Dynamics via Iso-NHC-RESPA

The extended isokinetic dynamics method was further improved to give the Isokinetic-NHC-RESPA (INR) method that is capable of efficiently generating resonance-free trajectories using long time steps [6]. The equations of motion for the INR method are of the form:

$$\dot{q}_j = v_j \qquad\qquad\qquad j = 1, \ldots, L$$

$$\dot{v}_j = \frac{F_j}{m} - \alpha v_j, \qquad\qquad j = 1, \ldots, L$$

$$\dot{\eta} = -\sum_{j=1}^{L} \left[\frac{Q v_{\eta_2,j} v_{\eta_1,j}^2}{kT} - \sum_{i=2}^{M} v_{\eta_1,j} \right]$$

$$\dot{v}_{\eta_1,j} = -v_{\eta_1,j} v_{\eta_2,j} - \alpha v_{\eta_1,j}, \qquad j = 1, \ldots, L$$

$$\dot{v}_{\eta_i,j} = \frac{G_{i,j}}{Q} - v_{\eta_i,j} v_{\eta_{i+1},j}, \qquad j = 1, \ldots, L, \quad i = 2, \ldots, M-1$$

$$\dot{v}_{\eta_M,j} = \frac{G_{M,j}}{Q}, \qquad\qquad\qquad j = 1, \ldots, L$$

$$G_{i,j} = Q v_{\eta_{i-1},j}^2 - kT \qquad\qquad\qquad\qquad (119)$$

where $F = -\partial\phi(\mathbf{r})/\partial x$, $Q = kT\tau^2$, and the Lagrange multiplier is now of the form:

$$\alpha = \frac{vF - \frac{L}{L+1} \sum_{j=1}^{L} Q v_{\eta_1,j}^2 v_{\eta_2,j}}{LkT}. \qquad (120)$$

These equations of motion ensure that the following constraint is satisfied:

$$2K(v, v_\eta) = mv^2 + \frac{L}{L+1} \sum_{j=1}^{L} Q v_{\eta_1,j}^2 = LkT. \qquad (121)$$

A detailed analysis of the dynamics resulting for these equations of motion is presented in [6]. This method, similarly to its predecessors, also generates partition functions whose configurational space distributions are canonically distributed.

Applying the INR method to problems has allowed for the use of extraordinarily long fundamental time steps and significant improvements in the speed of simulations. The reason for this is that the isokinetic constraint prevents a build-up of kinetic energy in any particular degree of freedom, thereby circumventing the resonance problem. In addition, the combination of the isokinetic constraint with Nosé-Hoover chain thermostats maintains ergodicity in the system. Figure 3 shows the results of INR applied to a harmonic oscillator with a quartic perturbation. The figure illustrates how the INR method successfully eliminates the resonance problem that exists when only the NHC algorithm is used.

The following two results are even more impressive. Figure 4 depicts the radial distribution functions, and conjugate error functions, for a system of

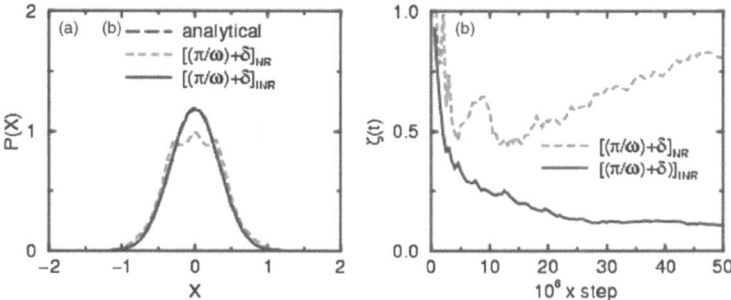

Fig. 3. (a) Distribution function for the harmonic oscillator with a quartic perturbation for the NHC-RESPA and ISO-NHC-RESPA methods using a time step of $\Delta t = \pi/\omega$. (b) The error in the distribution function as a function of time

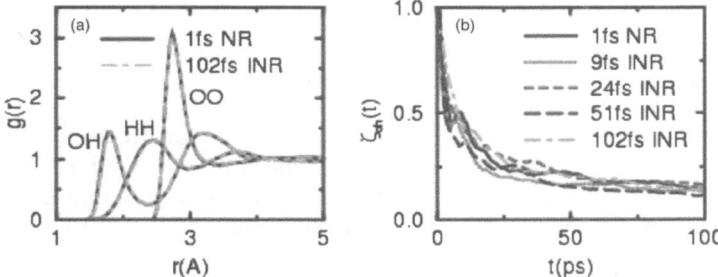

Fig. 4. (a) The radial distribution functions ($g_{OH}(\mathbf{r})$, $g_{OO}(\mathbf{r})$, and $g_{HH}(\mathbf{r})$) of flexible TIP3P liquid water calculated using the NHC-RESPA and ISO-NHC-RESPA methods. (b) The error in the distribution functions as a function of time, where the exact distributions are those that have been generated from a long run with a small time step. The legends report the large time step

TIP3P water molecules calculated using the NHC-RESPA algorithm with a time step of 1 fs and the INR method with time step of 102 fs. It is clear from these results that a time step of up to 102 fs can be used by employing the INR methods while still generating the same distributions as the NHC-RESPA algorithm with small time step.

Finally, the most impressive improvement using the INR method comes from study HIV protease. The distributions depicted in Fig. 5 clearly show that the INR method allows for a time step of up to 102 fs to be taken. It is clear that the INR method can yield efficiency improvements far greater than using NHC-RESPA alone. The single drawback, unfortunately, is that this method can only be used to calculate properties that are position dependent.

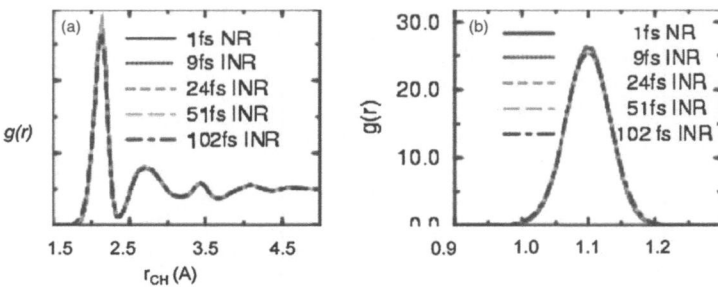

Fig. 5. (a) The radial distribution function, $g_{CH}(\mathbf{r})$, of HIV protease using the NHC-RESPA and the ISO-NHC-RESPA methods. **(b)** The intramolecular portion of the distribution function in part (a). All of the carbon and hydrogen atoms were considered and the legends reports the large time step

4.3 Novel Variable Transformations for Sampling Conformational Equilibria

The human genome which encodes information about all of proteins in the human proteome is, now, known. Reading the code yields information in the form of long strings of letters denoting the sequences of amino acids of each protein. The primary sequence uniquely determines the fold or tertiary structure of a protein. Since it is a difficult and expensive proposition to determine a protein structure experimentally, it is useful to develop computational tools that are capable of structure prediction. There are many methods that attempt to determine structure using databases searches or homology modeling. Others attempt to predict structure using a simplified model potential function that includes only the subset of the degress of freedom present in a coarse grained approach. A third class examines fully atomistic model in approach referred to as "*ab initio*" folding. The latter two approaches require sophisticated sampling schemes to be effective. Any new sampling scheme be it molecular dynamics or Monte Carlo based that improves sampling significantly represents an important contribution to the field.

In this section, a new sampling scheme called REPSWA or reference systems spatial warping algorithm will be developed [38]. It will be contrasted to standard approaches such as umbrella sampling and guiding potentials so that the advantages of the new technique can be underscored.

Conceptual Framework

The canonical partition function takes the form

$$Q = \frac{1}{h} \int dp_x \int dx \, \exp[-\beta H(p_x, x)]$$

$$H(p_x, x) = \frac{p_x^2}{2m} + V(x)$$

as derived above. If there are large barriers separating stable states in $V(x)$ then the dynamics of $H(p_x, x)$ even when coupled to a heat bath so as to sample the canonical distribution, is sluggish. This leads to the question, "Why do I care about preserving dynamics if all I want to do is sample the canonical distribution?" Thus, let us consider making mathematical manipulations that preserve the equilbrium distribution but changes the dynamics. However, even if sampling the equibrium distribution is the goal there is no need to abandon the use of dynamical systems as a general approach as these techniques can be efficiently parallelized. In addition, methods based on dynamical systems can be transformed into Monte Carlo techniques by simply casting them as Hybrid Monte Carlo methods.

In order to proceed, it is necessary to gain insight into a possible solution to the sampling problem. Imagine two groups of people living in two valleys separated by a large mountain. The populations are centered in the valleys. No one spends any time on the mountain top! When they are motivated or activated, people cross the mountain and settle into the other valley. Eventually, the populations in the valleys come to an equilbrium based on the relative desirability of life in each valley. If one is interested in studying equilibrium populations and NOT mountain climbing techniques, then clearly it would be nice to avoid waiting for activated barrier crossing events to equilibrate the system. Let us propose to change the topology or the phase space where the people live. Let us take the two valleys and expand them and the mountainous area and shrink it. In the new topology, the valleys will now be very large and the mountainous region will be very small and shallow. In the new topology, a person simply walks from one valley to the next without need for activation and the dynamics or approach to equilibrium is radically different. However, if the topological change is performed properly, the equilibrium populations will be invariant. The people will spend most of their time in the large valleys and very little time in the mountains. The method will scale as order N since we can alter the topology of any degree of freedom which possesses a barrier without loss of generality and populations come to equilibrium naturally. The challenge is to develop the mathematical formalism within which to implement this idea which we call REPSWA.

Before developing the mathematics required to implement the REPSWA method, lets us consider the conceptual framework behind methods such as umbrella sampling and guiding potentials. Imagine again two populations in two valleys separated by a mountain. Here, we perform a large number of tests in which we PAY people to spend time at various places on plausible routes between the valleys. Populations are now forced to interchange. However, by paying people to spend long periods in inhospitable regions, we have altered the equilibrium populations. The equilibrium population is determined by polling people after all the tests are finished about where they most liked to spend time. This method scales as N^d where d is number of degrees of freedom which possess a barrier as we must poll populations of size N in the d dimensional space to correct for artificially disturbing the equilibrium.

Mathematical Formalism

The mathematical formalism behind various sampling techniques will be discussed next. First, consider the guiding potential method. The method is based on the algebraic identiy

$$1 = \exp[-\beta V_{ref}(x)] \exp[\beta V_{ref}(x)] \tag{122}$$

which may be inserted into the partition function to yield

$$Q = \frac{1}{h} \int dp_x \int dx \exp[-\beta H_{eff}(p_x, x)] \exp[\beta V_{ref}(x)]$$

$$H_{eff}(p_x, x) = \frac{p_x^2}{2m} + V(x) - V_{ref}(x) \tag{123}$$

Note that we have generated a Hamilitonian with a different potential energy function that presumably has smaller barriers. However, sampling from the effective Hamiltonian does not yield the correct probability distribution without correction. That is, any average must be reweighted by the reference potential

$$\langle A(x) \rangle = \frac{\langle A(x) \exp[\beta V_{ref}(x)] \rangle_{(H_{eff})}}{\langle \exp[\beta V_{ref}(x)] \rangle_{(H_{eff})}}$$

since unphysical regions of phase space are visited. As the dimensionality, d, of $V_{ref}(x)$ increases, the amount of unphysical phase space visited increases exponentially. If there are N states in each dimension, the "bad phase" space visited increases as N^d. Hence, this technique is NOT effective for studying systems with important barriers in more than a few degress of freedom.

Next, consider the umbrella sampling method. Here, rather than perform one simulation, a series of canonical distributions are sampled,

$$\tilde{Q}_k = \frac{1}{h} \int dp_x \int dx \exp[-\beta H_{eff}^{(k)}(p_x, x)]]$$

$$H_{eff}^{(k)}(p_x, x) = \frac{p_x^2}{2m} + V(x) + V_{bias}^{(k)}(x)$$

from a modified Hamiltonian. In this way, the system can be dragged through its phase space. However, since $\tilde{Q}_k \neq Q$ a reweighting procedure must be developed. Now,

$$\tilde{P}_k(x) \equiv \frac{\exp[-\beta(V(x) + V_{bias}^{(k)}(x))]}{\tilde{Q}_k}$$

$$P(x) \equiv \frac{\exp[-\beta V(x)]}{Q}$$

$$= \frac{\sum_k \tilde{P}_k(x) \frac{\tilde{Q}_k}{Q}}{\sum_j \frac{\tilde{Q}_j}{Q} \exp[-\beta V_{bias}^{(j)}(x)]}$$

which can be solved self-consistently to yield the required factors, $\tilde{Q}_k/Q$. While this approach is superior to the guiding potential method, it too scales badly with number of degrees of freedom in $V_{bias}^{(k)}(x)$. Each degree of freedom in $V_{bias}^{(k)}(x)$ must be traversed by a series of simulations. Assuming that N simulations are required to traverse each of the d degrees of freedom then the method scales as N^d.

Let us consider, next, a better method, parallel tempering. Here, many simulations at many different inverse temperatures, $\beta_j = 1/[k_B T_j]$ are performed

$$Q_j = \frac{1}{h} \int dp_x \int dx \exp[-\beta_j H(p_x, x)]]$$

$$H(p_x, x) = \frac{p_x^2}{2m} + V(x)$$

Monte Carlo exchange moves (switch inverse temperatures) are attempted at intervals. In this way, configurations from high temperature simulations which have many barrier crossing events trickle down to the temperature of interest. Now, the acceptance of exchange moves depends exponentially on $[\Delta \beta_{jm} \Delta E_{mj}]$. Since ΔE_{mj} grows with N, increasingly fine grained sampling of β is required to accept moves, indicating that the method scales as N^2. However, if the system is near a phase transition, then the method suffers difficulties as exchanges moves around the transition temperature are not accepted (e.g. the character of the configurations changes radically as the system undergoes a phase transition). As protein folding can be thought of a rounded phase transition (e.g. a transition involving a finite system), parallel tempering while a good method is not as effective as one would like for this problem. However, it is possible to combine parallel tempering with umbrella sampling and with the new method, REPSWA to be described next.

Last, the mathemical formalism behind the REPSWA method is discussed. REPSWA is based on the **non-linear transformation**,

$$u = \alpha \left\{ x_0 + \int_{x_0}^{x} d\tilde{x} \exp[-\beta V_{ref}(\tilde{x})] \right\}$$

and the trivial transformation $p_u = p$. Inserting the transformation into the canonical parition function yields

$$Q = \frac{1}{h} \int dp_u \int du \exp[-\beta H_{eff}(p_u, u)]$$

$$H_{eff}(p_u, u) = \frac{p_u^2}{2m} + V(x(u)) - V_{ref}(x(u))$$

The transformation preserves equilibrium distributions BUT changes the dynamics as $H_{eff}(p_u, u) \neq H(p, x)$. However, no reweighting factors are present so that the method will scale as N.

In order to see how REPSWA works consider a double well with the barrier region between $-a < x < a$. Take $V_{ref}(x) = 0$ $x < -a$ and $x > a$ and $V_{ref}(x) = V(x)$ inside the barrier region. Outside the barrier region

$$u = \alpha(x + c)$$

while inside the barrier region

$$u = \alpha \left\{ a + \int_{-a}^{x} d\tilde{x} \exp[-\beta V_{ref}(\tilde{x})] \right\}$$

In this way, u changes slowly when $V_{ref}(x)$ is large. Thus, the u phase space associated with the barrier region is very small as desired. Sampling in u phase space without a barrier is rigorously correct because no time is spent in the barrier region as defined in the x phase space! The authors encourage the reader to carefully consider the different mathematics required by REPSWA and other methods as well as the different conceptual description.

In Fig. 6, a comparison of the results of canonical MD to REPSWA canonical MD is presented for the double potential. Using standard MD, barrier crossing events are rare and equilibrium is not reached in a reasonable amount of cputime (10^6 time steps). RESPWA MD on the other hand crosses the barrier easily and generates the equilibrium populations quickly.

The beauty of the REPSWA approach is that it can easily be employed to remove barriers in many degrees of freedom simultaneously without loss of efficiency. However, the method does require the evaluation of a Jacobian, which is not in general easy. Consider that proteins are polymers. It is, therefore, natural to consider growing in the barriers in the ϕ, ψ angles one at a time. In this way, a lower/upper triangular matrix is formed. Since the Jacobian of an upper/lower triangular matrix is the product of its diagonal elements, REPSWA can be easily implemented.

Treating proteins is beyond the scope of the discussion presented here. Let us consider the case of a united atom n-alkane. Here, the barriers to achieving conformational equilibrium are in the torsional degrees of freedom. If the first three atoms of the molecule are considered to be fixed, then adding 4th atom will generate a barrier in the dihedral angle. In order to transform away this dihedral, r_4 is rotated/translated into a frame in which r_3 is at the origin and $r_3 - r_2$ lies along the z axis to form r_4'. The new vector is then resolved into spherical polar coordinates and REPSWA applied to the azimuthal angle. The transformation can be reversed to generate a u_4' and then a u_4. Of course, there is no need to stop with the 1st dihedral one can simply continue the process up the chain until all dihedral barriers are transformed away.

In Fig. 7, the dynamics of a 400-mer united atom alkane under canonical MD and REPSWA canonical MD is compared. Since REPSWA MD has removed ALL 397 barriers it equilibrates quickly and easily. Standard MD remains stuck and its end to end distance does not reach the equilibrium value in 1ns of simulation. The end to end distance under RESPSWA MD in contrast fluctuates nicely indicating that the equilibrium distribution is being sampled effectively. These simulations were performed without intermolecular interactions as dealing with these forces is beyond the scope of the discussion.

In summary, a promising new method, REPSWA, has been compared and contrasted to existing techniques. Due to its mathematical structure, REP-SWA scales linear with system size and has been shown to perform well in model problems. It can easily be combined with parallel tempering and Hybrid Monte Carlo methods to form interesting and exciting novel sampling schemes. These will be described in future work.

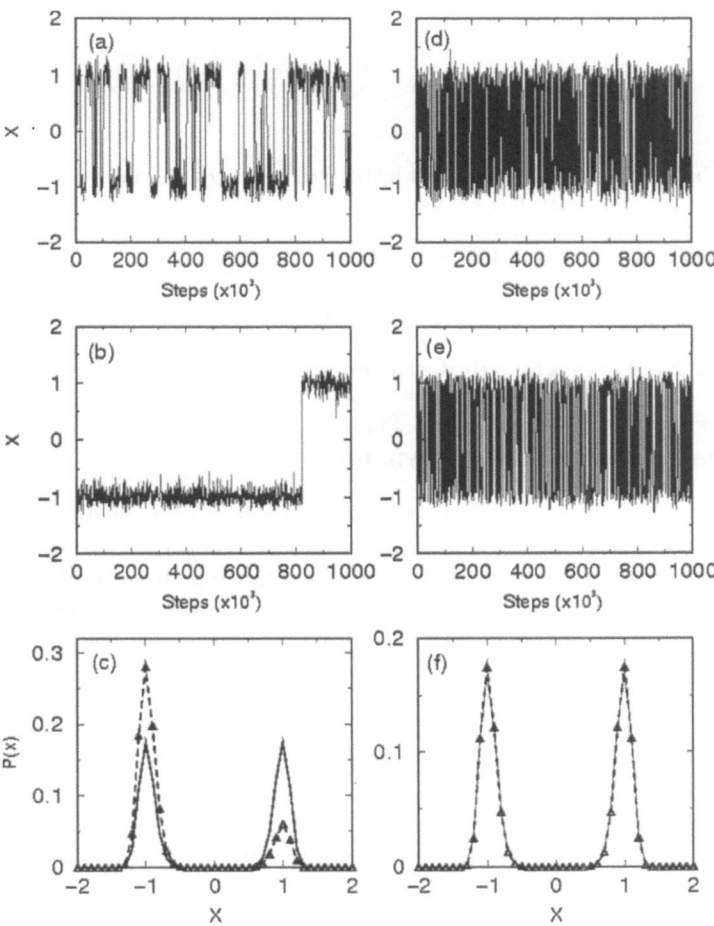

Fig. 6. (**a, d**) Evolution of x in double well potential with a barrier height of $5\,k_\mathrm{B}T$, $k_\mathrm{B}T = 1$ without(a)/with(d) REPSWA. (**b, e**) Same for a barrier height of $10\,k_\mathrm{B}T$ without(b)/with(e) REPSWA. (**c, f**) Probability distribution function of x for a barrier height of $10k_\mathrm{B}T$ without(c)/with(f) REPSWA; the solid line is the analytical result and the dashed line with triangles is the simulation result

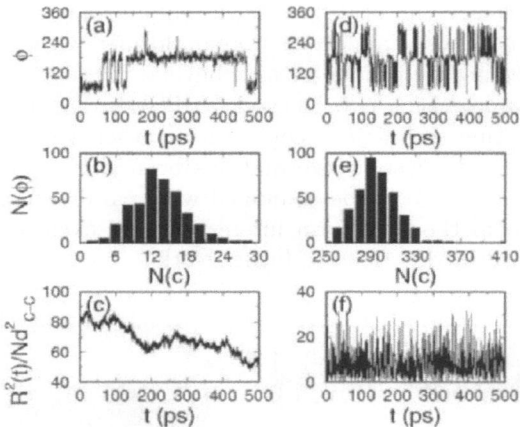

Fig. 7. (a, d) Evolution of the 200th dihedral angle of the united atom 400-mer over a 500 ps thermostatted simulation without(a)/with(d) REPSWA. (b,e) Histogram of the number of dihedral angles crossing potential barriers a given number of times without(b)/with(e) REPSWA. (c), (f) Evolution of the end-to-end distance without(c)/with(f) REPSWA

4.4 Free Energies via Adiabatic Free Energy Dynamics

Generating Multidimensional Free Energy Surfaces via Adiabatic Free Energy Dynamics

Many problems, such as configurational sampling of peptides and protein folding require computing multidimensional free energy surfaces. The state of the art methods that are available for computing these multidimensional surfaces still require multiple biased simulations and a significant amount of post-processing of the simulation data. Adiabatic Free Energy Dynamics (AFED) [39–41] is a very useful method that has been recently introduced by the Tuckerman Group to compute free energy surfaces of systems characterized by rare events, such as barrier crossings. Statistical mechanics provides us with a means to calculate free energy profiles from ensemble averages.

Consider the case of a classical N-particle system at temperature T, the reaction coordinate q can be expressed as a function of the N Cartesian position vectors $\{\mathbf{r}_1, \ldots, \mathbf{r}_N\}$ of the particles, $q = q(\mathbf{r}_1, \ldots, \mathbf{r}_N)$. The free energy profile can now be written as

$$F(q') = -\frac{1}{\beta} \ln P(q') \qquad (124)$$

where

$$P(q') = \langle \delta(q(\mathbf{r}_1, \ldots, \mathbf{r}_N) - q') \rangle \qquad (125)$$

is the probability density function on the reaction coordinate, $q = q'$, and $\beta = \frac{1}{kT}$. This relationship can be used to directly calculate free energy profiles from MD or Monte Carlo simulations. The configurational space of interest in the system can then be characterized by a set of generalized coordinates $q_1(\mathbf{r}), \ldots, q_n(\mathbf{r})$, $n < N$, where the first n of these coordinates will be the coordinates of interest. The following change of variables can then be introduced:

$$
q_1 = q_1(\mathbf{r}_1, \ldots, \mathbf{r}_N)
$$
$$
q_2 = q_2(\mathbf{r}_1, \ldots, \mathbf{r}_N)
$$
$$
\cdots
$$
$$
q_{3N} = q_{3N}(\mathbf{r}_1, \ldots, \mathbf{r}_N) \tag{126}
$$

The transformed partition function for the system can now be written as

$$
Q(N, V, T) = \frac{1}{N! h^{3N}} \int d^N \mathbf{p} \int_{D(V)} d^{3N} q \times
$$
$$
\exp\left\{ -\beta \left[\sum_{i=1}^{N} \frac{\mathbf{p}_i^2}{2m_i} + \tilde{U}(q_1, \ldots, q_{3N}) \right] \right\} \tag{127}
$$

where h is Planck's constant, $D(V)$ is the spatial domain defined by the containing volume, $\tilde{U}(q_1, \ldots, q_{3N}) = U(\mathbf{r}_1(q), \ldots, \mathbf{r}_N(q)) - \frac{1}{\beta} \ln J(q_1, \ldots, q_{3N})$, $J(q) = |\partial \mathbf{r}/\partial q|$ is the Jacobian defined for the transformation from the original Cartesian space into this new coordinate system, and can be generated via thermostated MD using Hamiltonian of the form

$$
H = \sum_{i=1}^{N} \frac{\mathbf{p}_i^2}{2m_i} + \tilde{U}(q_1, \ldots, q_{3N}) \tag{128}
$$

In the AFED method, an adiabatic separation is imposed between the degrees of freedom of interest, $i = 1 \ldots n$, and the rest of the system by choosing a set of masses, m_i, where $m_i > m_j \forall_{j=n \ldots 3N}$. This separation gives rise to an effective AFED Hamiltonian,

$$
H = \sum_{\alpha=1}^{n} \frac{\mathbf{p}_\alpha^2}{2\tilde{m}_\alpha} + \sum_{\alpha=n+1}^{3N} \frac{\mathbf{p}_\alpha^2}{2m_\alpha} + \tilde{U}(q_1, \ldots, q_{3N}) \tag{129}
$$

In addition to the adiabatic separation that is imposed on the coordinates $q_1, \ldots, q_n$, the nature of the events that will be studied must also be considered. The free energy landscape for the rotation about peptide backbone dihedrals is notoriously rough and these coordinates will have to be heated to a temperature higher than the physical system in order to facilitate barrier-crossing events. We can accomplish this separate heating in MD by coupling the "hot" degrees of freedom to thermostats at temperature T_q and the remaining $3N - n$ coordinates to thermostats at temperature T. A detailed

analysis of the dynamics generated at these two temperatures can be found
in [40] and it has been shown that the dynamics, subject to evolution using the
Liouville operator formalism, yields a distribution in terms of the coordinates
$q_1, \ldots, q_n$:

$$P_{adb}(q_1, \ldots, q_n) \propto \left\{ \int dq_{n+1} \cdots dq_{3N} d^{3N} \mathbf{p} \exp[-\beta(K(\mathbf{p}) + \tilde{U}(q_1, \ldots, q_{3N}))] \right\}^{\frac{\beta_q}{\beta}}$$

$$\propto [\hat{Q}(q_1, \ldots, q_n)]^{\frac{\beta_q}{\beta}} \tag{130}$$

where $\beta_q = 1/k_B T_q$, $K(\mathbf{p}) = \sum_{\alpha=1}^{n} \frac{\mathbf{p}_\alpha^2}{2\tilde{m}_\alpha} + \sum_{\alpha=n+1}^{3N} \frac{\mathbf{p}_\alpha^2}{2m_\alpha}$, and $\hat{Q}(q_1, \ldots, q_n)$
is the partition function that is obtained by integrating $\exp(-\beta H)$ over the
phase space variables $q_{n+1}, \ldots, q_{3N}$. By definition, the free energy surface
$F(q_1, \ldots, q_n)$ is given by

$$F(q_1, \ldots, q_n) = -\frac{1}{\beta} \ln \hat{Q}(q_1, \ldots, q_n). \tag{131}$$

It is clear that this same surface can be calculated by substituting the result
from (130) to yield the free energy surface obtained directly from the adiabatic
probability distribution function $P_{adb}(q_1, \ldots, q_n)$

$$F(q_1, \ldots, q_n) = -\frac{1}{\beta_q} \ln P_{adb}(q_1, \ldots, q_n) + const. \tag{132}$$

This result shows that the free energy can be computed directly from the
adiabatic probability distribution function without any unbiasing or any *a
posteriori* processing of the simulation data.

Conformational Sampling of Peptide Backbone Dihedrals via Adiabatic Free Energy Dynamics

The interest in the conformational preferences of small peptides stems, in no
small part, from the wealth of insights that they provide on the secondary
structure motifs that are also present in larger polypeptides. It is the hope
that employing information extracted from the study of small peptides will
provide us with a basis to approach the study of these large proteins. Conformational sampling of proteins has the potential to relate the structure of
certain proteins to their function, to elucidate protein folding events, and
even perhaps the origins of protein misfolding diseases such as Alzheimers,
Creutzfeld-Jacob, and even cataracts. To this end, many new experimental
techniques have emerged to probe key aspects of the conformational preferences and dynamics of proteins, especially small peptides.

In order to exploit the relationship in (132) for studying the conformational
preferences of peptides, the coordinate system must first be transformed from
Cartesian coordinates into a set of generalized coordinates where the dihedral angles of interest will be explicit degrees of freedom of the system. A

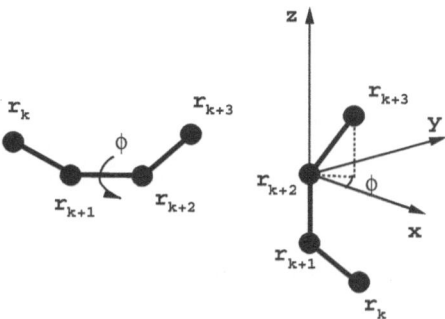

Fig. 8. Illustration of the coordinate transformation used to resolve the backbone dihedrals

formalism for this type of transformation was already presented in Sect. 4.3. Transforming the backbone dihedrals, (ϕ, ψ), into explicit coordinates will allows the Ramachandran surface, $F(\phi, \psi)$, to be mapped for a given pair of dihedrals. Figure 8 illustrates how this transformation can be accomplished. This transformation can be performed for each dihedral angle in succession (excluding the dihedral angles around the peptide bond), leading to a set of generalized coordinates that contain the Ramachandran (ϕ, ψ) angles as explicit coordinates.

This method has already been used to study a number of small (di-, tri-) peptides, both in vacuo and in solution, including alanine di- and tri-peptide, N-acetyl-tryptophan-methylamide (NATMA), and N-acetly-tryptophan-amide (NATA). The results for the studies of alanine di- and tripeptide, in gas phase and solution, are presented in [41]. The conformational sampling studies of the tryptophan dipeptides have not yet been published, but represent a very significant result. Although the Ramachandran surface for the alanine dipeptide is rough, the barriers are not nearly representative of those that might be encountered in the free energy landscape of a protein. Conversely, dipeptides of tryptophan and phenylalanine have significantly more rough free energy landscapes and have been long thought to be able to provide important information about protein folding events.

In general, the initial configurations for these systems are equilibrated for an average of 1 ns at a system temperature of $T = 300$ K in cubic boxes treated with periodic boundary conditions. The constant temperature (NVT) distributions were generated by coupling a thermostat to each degree of freedom in the system at its appropriate temperature. The Generalized Gaussian Moment Thermostats (GGMT) of Liu and Tuckerman [42] were chosen because of their high effectiveness for adiabatic problems. The multiple time scale algorithm, r-RESPA, was employed to exploit the separation of the time scales while integrating the equations of motion. For the simulations performed in the condensed phase, water molecules were modeled using the TIP3P water

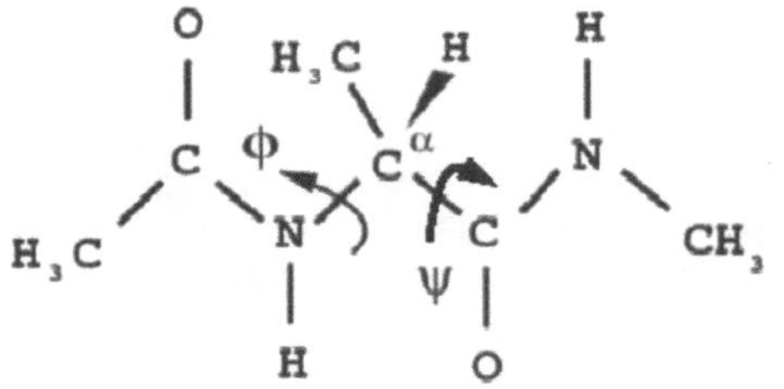

Fig. 9. Schematic drawing of alanine dipeptide with the Ramachandran angles, (ϕ, ψ) depicted

model. All calculations presented are performed with the PINY_MD package [43–45].

In applying the AFED method to the alanine dipeptide (Fig. 9), the central dihedral angles, $(\phi\psi)$, are selected as the coordinates of interest. They were then assigned mass $m_{(\phi,\psi)} = 50m_c$, where m_c is the mass of a carbon atom, and a temperature $T_{(\phi,\psi)} = 1500$ K. The remaining two dihedral angles, (τ, ω), were not treated as adiabatic variables and were allowed to evolve normally. It was found that the free energy surfaces generated by AFED, using the CHARMM22 force field [46], agreed well with the same surfaces generated using the umbrella sampling [47–49] method combined with the weighted histogram approach (WHAM) [50] (Fig. 10). Because of their good agreement, the AFED method can be benchmarked against this well-tested and widely used approach. Generating the free energy surfaces with umbrella sampling required simulations performed in 176 distinct windows, using a force constant of 20 000 K/rad^2, for a total computational cost of 35.2 ns for alanine dipeptide in the gas phase and 50 ns for alanine dipeptide in the condensed phase. Using the AFED method, the same surfaces were generated much faster, in 3.5 ns and 4.7 ns (out of a total of 10 ns run) for the gas phase and condensed phase alanine dipeptide, respectively. This result shows an improvement by up to a factor of 10 when AFED is used to compute these surfaces.

Perhaps almost as impressive as the efficiency increase, using AFED versus umbrella sampling, is the accuracy of the predictions that the CHARMM22 force field, employing the AFED method, makes for the conformational minima. For alanine dipeptide in the gas phase, two equienergetic minima were found at $(\phi, \psi) = \{(-80, 80), (-150, 170)\}$ corresponding to the C7$_{\mathrm{eq}}$ and C5 structures, respectively. This result is in perfect agreement with the DFT calculations performed by Wei and coworkers [51] who found the two minimum energy conformations at $(-80, 80)$ and $(-160, 160)$, which they identified as C7$_{\mathrm{eq}}$ and C5, respectively. The results for alanine dipeptide in solution were encouraging, but not nearly as conclusive. In this case, the two minimum energy conformations observed were the α_R and PP$_{\mathrm{II}}$ conformations located at $(-80, 60)$ and $(-80, 150)$, respectively. Several studies, both theoretical and experimental, have already been performed on the alanine

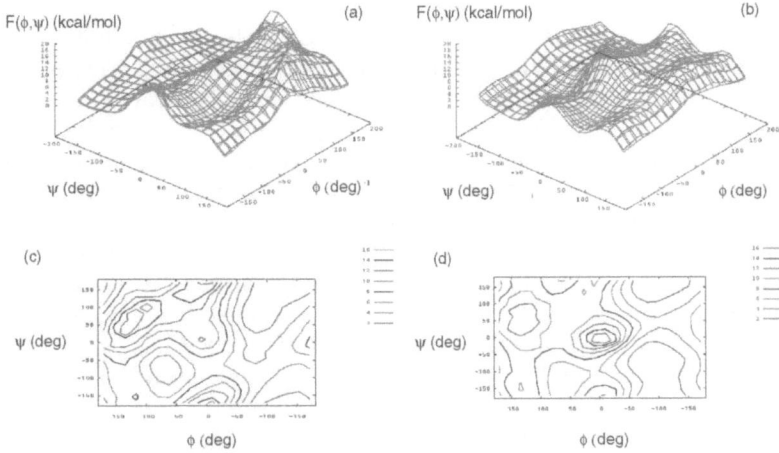

Fig. 10. (a) Free-energy surface $F(\phi, \psi)$ for the alanine dipeptide in the gas phase generated using AFED (*dark gray*) and umbrella sampling (*light gray*). **(b)** Same as for a except in solution. **(c)** Ramachandran contour map corresponding to the free-energy surface in a for the alanine dipeptide in the gas phase. **(d)** Same as in c except corresponding to free-energy surface in b, i.e., solvated alanine dipeptide

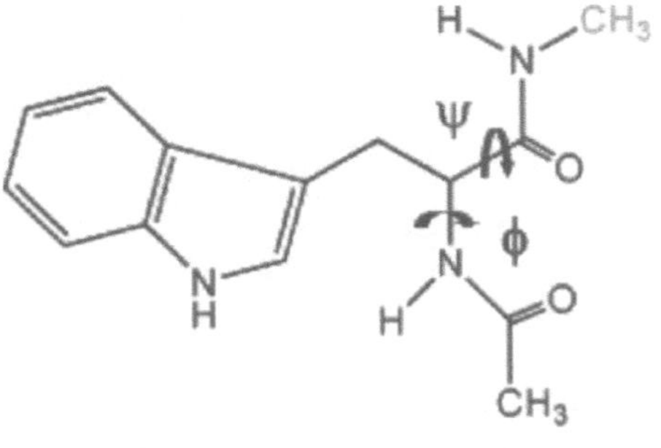

Fig. 11. Schematic drawing of N-acetyl-tryptophan-methyl-amide (NATMA) showing the Ramachandran angles, (ϕ, ψ)

dipeptide in solution [52,53]. All of these studies agree that the major conformation is the PP$_{\mathrm{II}}$ structure, but there is considerable disagreement over the existence of a second minimum energy conformation. NMR, ^{13}CNMR, and CD, studies identify the α_R conformation as a significant conformer while certain DFT and other NMR studies disagree.

In the case of NATMA in vacuo, as discussed, dealing with significantly higher barriers in the free energy surface was anticipated. Consequently, the dihedral angles in NATMA were heated to a much higher temperature, $T_{(\phi, \psi)} = 6000$ K. The higher temperature also required that the dihedral be significantly heavier to ensure adiabatic separation. The masses were therefore increased to $m_{(\phi,\psi)} = 600m_c$. AFED simulations of 10 ns, using the CHARMM22 force field, were performed using these parameters and the free energy surfaces were found to have converged after only 2.5 ns. Although umbrella sampling was not performed on this system, it is safe to say that it would have taken significantly more time to generate these surfaces.

The truly impressive result comes from inspection of the predictions for the free energy minima of NATMA in gas phase, for which the AFED simulations predict that there will be contributions from three distinct conformations. The lowest energy conformer was predicted to be a C5 structure with $(\phi,\psi) = (-160,160)$. The next lowest energy conformation was located at $(-140,140)$ and was also a C5 structure with a relative free energy of $+1.03$ kcal/mol. The final significant conformation $(-100,100)$ had a relative energy of $+2.2$ kcal/mol and was identified as a C7$_{eq}$ structure. These results can be compared with the 2D spectroscopic studies performed by Dian and coworkers [54] who also found three conformational minima for NATMA in vacuo. They observed an absolute minimum C5(AP) conformation, a second C5(AΦ) at $+0.65$ kcal/mol, and a C7$_{eq}$ conformation at $+2.3$ kcal/mol. These encouraging results relating to the abilities of the CHARMM22 force field aside, the true power of the AFED method is evident in the large speedup in generating these multidimensional surfaces versus constrained, and biased, MD. This increase in computational efficiency is expected to be even more dramatic when applied to systems with more than two adiabatic parameters, for which it would be next to impossible to study using a method like umbrella sampling.

Adiabatic Approach to Solvation and Binding Free Energies (λ-AFED)

The ability to compute solvation and binding free energies is a very important one. Drug/Protein binding constants, DNA/protein binding, solvation of biomolecules, are all computational questions of extreme interest. Unfortunately, solvation, binding, and alchemical, free energy problems in general represent a set of important problems that remain difficult to study with traditional MD techniques, which tend to suffer from a distinct lack of ergodicity when studying these problems. Enhanced techniques such as thermodynamic integration [55,56], umbrella sampling [47–49], free energy perturbation [57,58], and metadynamics [59], have been introduced to address these issues. Unfortunately, most of these methods require multiple simulations and often require a significant amount of post-processing of simulation data.

Recent developments have led to an extension of the AFED method for computing binding, solvation, and other alchemical, free energy differences. In general, this method will be used to compute the free energy associated with transforming a chemical system between two thermodynamic states, A and B, each defined by a potential energy function, $V_A(\overrightarrow{\mathbf{r}})$ and $V_B(\overrightarrow{\mathbf{r}})$.

To begin, a new degree of freedom, λ, is introduced and will be used to switch between the different forms of the potential. The "meta-potential" is defined

$$V(\overrightarrow{\mathbf{r}}, \lambda) = f(\lambda)V_A(\overrightarrow{\mathbf{r}}) + g(\lambda)V_B(\overrightarrow{\mathbf{r}}) \tag{133}$$

where $f(\lambda)$ and $g(\lambda)$ are switching functions, and $V(\overrightarrow{\mathbf{r}}, \lambda)$ will be used as the potential energy function that will be used in the MD simulations. A set of two restrictions on the forms of f and g are imposed: it is required that

$$V(\overrightarrow{\mathbf{r}}, 0) = V_A(\overrightarrow{\mathbf{r}})$$
$$V(\overrightarrow{\mathbf{r}}, 1) = V_B(\overrightarrow{\mathbf{r}}) \tag{134}$$

and that the choice of switches leads to a barrier in the free energy profile.

The λ-AFED Hamiltonian is constructed by including λ in the kinetic portion of the Hamiltonian and using the "meta-potential"

$$H = \sum_{i=1}^{N} \frac{\mathbf{P}_i^2}{2m_i} + \frac{p_\lambda^2}{2m_\lambda} + V(\overrightarrow{\mathbf{r}}, \lambda) \tag{135}$$

where $m_\lambda \gg m_i \forall_i$ is the mass of λ and we allow λ to evolve under Hamilton's equations where

$$\dot{\lambda} = \frac{\partial H'}{\partial p_\lambda}$$
$$\dot{p}_\lambda = -\frac{\partial H'}{\partial \lambda} . \tag{136}$$

The same treatment as in the AFED method is applied with λ as the degree of freedom of interest. Following the same approach, the dynamics generated by the action of the propogator results in an adiabatic probability distribution function in λ

$$P_{adb}(\lambda) \propto [\hat{Q}(\lambda; \beta)]^{\beta_\lambda/\beta} \tag{137}$$

where $\beta_\lambda = 1/k_B T_\lambda$ is the temperature of λ sufficient to achieve frequent barrier crossing. Substituting this result into the definition of the free energy profile, gives the central result of this work.

$$F(\lambda) = -\frac{1}{\beta_\lambda} \ln P_{adb}(\lambda) + const. \tag{138}$$

Let us now consider a free energy calculation for the transformation in the thermodynamic states that corresponds directly to an important thermodynamic value: The change in free energy associated with the addition of an $(N+1)^{st}$ particle into an N particle system. (i.e. the chemical potential, μ). Here, an N particle Lennard-Jones (LJ) Argon liquid will be considered and has a potential that takes the form,

$$V_{LJ}(\mathbf{r}_1, \dots, \mathbf{r}_N) = \sum_{i=1}^{N-1} \sum_{j=i+1}^{N} 4\epsilon \left[\left(\frac{\sigma}{\mathbf{r}_{ij}} \right)^{12} - \left(\frac{\sigma}{\mathbf{r}_{ij}} \right)^{6} \right] \tag{139}$$

The free energy associated with the solvation of an additional particle into the liquid (i.e. the chemical potential) is computed using a meta-potential constructed from the N particle potential, $V_{LJ}(\mathbf{r}_1, \dots, \mathbf{r}_N)$, and the $N+1$ particle potential, $V_{LJ}(\mathbf{r}_1, \dots, \mathbf{r}_{N+1})$.

Simulation baths of $N = 108$ Lennard-Jones (LJ) particles with $\epsilon/k_B = 90K$, $\sigma = 3.405A$ and mass $m = 39.94$ amu were used. The density of the systems was $\rho\sigma^3 = 0.844$ and the temperature of the systems was set to $kT = 300$ K by coupling each degree of freedom to a GGMT thermostat. Production runs of 200 ps were performed before and after the addition of the $N + 1^{st}$ particle and this configuration was used as the initial configuration for both λ-AFED simulations and thermodynamic integration [55, 56].

A set of eighth order polynomials where chosen for the switching functions,

$$f(\lambda) = (\lambda^2 - 1)^4$$
$$g(\lambda) = ((\lambda - 1)^2 - 1)^4 \tag{140}$$

so that once again a barrier is achieved in the free energy profile of $F(\lambda)$. The λ-AFED simulations were carried out with a time step of $\Delta t = 6$fs, mass of $m_\lambda = 2000m_i \approx 80 \times 10^3$ amu, and at temperatures $kT_\lambda = 200kT = 6 \times 10^4$ K. Thermodynamic integration simulations were performed using 21 evenly spaced values of λ between 0 and 1. Each of these thermodynamic integration simulations was run for 45 ps. The thermodynamic integration result gives a final free energy change of $\Delta F = \mu = 2.49 \approx 2.5$ kcal/mol. It was found that in order to achieve the same converged result using the λ-AFED method, with the same size time step, a simulation of length 180-200 ps was required. However, since the TI approach requires 21 separate calculations each of length 45 ps, the total run time needed for this method would be 21×45ps $= 0.945$ ns ≈ 1 ns. In contrast, only one λ-AFED calculation produces the same result in under 0.2 ns. In this case, it is observed that λ-AFED has achieved a five (5) fold increase in performance over the thermodynamic integration approach.

5 Conclusions

Clearly, there have been significant advances in the field of Molecular Dynamics in the past few decades. The introduction of new and more complex non-Hamiltonian systems have led to methods capable of sampling interesting and useful ensembles germaine to problems in Chemistry and Biology.

The field has indeed progressed significantly in recent years. Novel methods, algorithms, approaches, and faster computers, have led to the ability to study more and more problems of interest. The methods discussed in this text represent only a fraction of the approaches that are being attempted, but it is fair to say that together they make up some of the most significant strides towards solving complex chemical problems computationally.

References

1. C. Levinthal (1968) Are there pathways for protein folding. *J. Chim. Phys.* **65**, p. 44
2. C. Levinthal, P. I. Debrunner, J. C. M. Tsibris, and E. Munck Eds. (1969) *Proceedings of a Meeting held at Allerton House, Monticello, IL*, University of Illinois Press, Urbana, p. 22
3. S. O. Nielsen, C. F. Lopez, G. Srinivas, and M. L. Klein (2004) Coarse grain models and the computer simulation of soft materials. *J. Phys. Condens. Matter* **16**, p. R481
4. M. E. Tuckerman, G. J. Martyna, and B. J. Berne (1990) Reversible multiple time scale molecular-dynamics. *J. Chem. Phys.* **1992**, p. 97
5. P. Minary, G. J. Martyna, and M. E. Tuckerman (2003) Algorithms and novel applications based on the isokinetic ensemble. I. Biophysical and path integral molecular dynamics. *J. Chem. Phys.* **118**, p. 2510
6. P. Minary, G. J. Martyna, and M. E. Tuckerman (2004) Long time molecular dynamics for enhanced conformational sampling in biomolecular systems. *Phys. Rev. Lett.* **93**, p. 150201
7. D. McQuarrie (1976) *Statistical Mechanics*. Harper and Row, New York
8. K. F. Gauss (1829) Ueber ein neues algemeines Grundgesetz der Mechanik" (don't ask how I was able to find this title! I also found a copy of the paper if anyone wants it). *J. Reine Angew. Math* **IV**, p. 232
9. H. Andersen (1980) Molecular-dynamics simulations at constant pressure and-or temperature. *J. Chem. Phys.* **72**, p. 2384
10. M. Tuckerman, C. Mundy, and G. Martyna (1999) On the classical statistical mechanics of non-Hamiltonian systems. *Europhys. Lett.* **45**, p. 149
11. Again, the more mathematically precise statement would be one relating the phase space volume form at $t = 0$ to that at an arbitrary time t
12. M. E. Tuckerman, Y. Liu, G. Ciccotti, and G. J. Martyna (2001) Non-Hamiltonian molecular dynamics: Generalizing Hamiltonian phase space principles to non-Hamiltonian systems. *J. Chem. Phys.* **115**, p. 1678
13. The covariant form of the conservation law is

$$\left(\frac{\partial}{\partial t} + \mathcal{L}_\xi \right) (f\tilde{\omega}) = 0 \,,$$

   where $\mathcal{L}_\xi$ is the Lie derivative and $\tilde{\omega}$ is the volume n-form
14. B. A. Dubrovin, A. T. Fomenko, and S. P. Novikov (1985) *Modern Geometry – Methods and Applications Part I*. Springer-Verlag: 175 Fifth Ave, New York NY 10010
15. B. A. Dubrovin, A. T. Fomenko, and S. P. Novikov (1985) *Modern Geometry – Methods and Applications Part II*. Springer-Verlag: 175 Fifth Ave, New York NY 10010
16. B. Schutz (1987) *Geometrical methods of mathematical physics*. Cambridge University Press: The Pitt Building, Trumpington Street, Cambridge CB2 1RP

17. We have recently become aware of the fact that a generalization of the Liouville equation similar to the one presented was written down (although without proof) some time ago by Ramshaw [60]
18. K. Cho, J. D. Joannopoulos, and L. Kleinman (1993) Constant-temperature molecular-dynamics with momentum conservation. *Phys. Rev. E* **47**, p. 3145
19. G. Martyna (1994) Remarks on constant-temperature molecular-dynamics with momentum conservation. *Phys. Rev. E* **50**, p. 3234
20. M. Tuckerman, B. Berne, G. Martyna, and M. Klein (1993) Efficient molecular-dynamics and hybrid monte-carlo algorithms for path-integrals. *J. Chem. Phys.* **99**, p. 2796
21. G. Martyna, M. Tuckerman, and M. Klein (1992) Nose–Hoover chains – the canonical ensemble via continuous dynamics. *J. Chem. Phys.* **97**, p. 2635
22. M. E. Tuckerman, G. J. Martyna, and B. J. Berne (1990) Molecular-dynamics algorithm for condensed systems with multiple time scales. *J. Chem. Phys.* **93**, p. 1287
23. M. E. Tuckerman, B. J. Berne, and A. Rossi (1990) Molecular-dynamics algorithm for multiple time scales – systems with disparate masses. *J. Chem. Phys.* **94**, p. 1465
24. M. E. Tuckerman and B. J. Berne (1991) Molecular-dynamics algorithm for multiple time scales – systems with long-range forces. *J. Chem. Phys.* **94**, p. 6811
25. M. E. Tuckerman and B. J. Berne (1991) Molecular-dynamics in systems with multiple time scales – systems with stiff and soft degrees of freedom and with short and long-range forces. *J. Chem. Phys.* **95**, p. 8362
26. S. Duane, A. D. Kennedy, B. J. Pendleton, and D. Roweth (1987) Hybrid Monte Carlo. *Phys. Lett. B* **195**, p. 216
27. H. F. Trotter (1959) On the product of semi-groups of operators. *Proc. Am. Math. Soc.* **10**, p. 545
28. The decomposition of the forces for "fast" and "slow" components is presented. The same analysis can be performed for a system of "short" and "long" range forces, where we decompose the Liouville operator L into a reference system and a long range force contribution:

$$iL = \dot{q}\frac{\partial}{\partial q} + (F_{short} + F_{long})\frac{\partial}{\partial p}$$
$$= \left(\frac{p}{m}\frac{\partial}{\partial q} + F_{short}\frac{\partial}{\partial p}\right) + F_{long}\frac{\partial}{\partial p} \; .$$
$$= iL_{ref} + iL_{long}$$

29. G. Martyna, M. Tuckerman, D. Tobias, and M. Klein (1996) Explicit reversible integrators for extended systems dynamics. *Mol. Phys.* **87**, p. 1117
30. Z. Zhu, D. I. Schuster, and M. E. Tuckerman (2003) Molecular dynamics study of the connection between flap closing and binding of fullerene-based inhibitors of the HIV-1 protease. *Biochemistry* **42**, p. 1326
31. T. Schlick, M. Mandziuk, R. D. Skeel, and K. Srinivas (1998) Nonlinear resonance artifacts in molecular dynamics simulations. *J. Comput. Phys.* **140**, p. 1
32. Q. Ma, J. A. Izaguirre, and R. D. Skeel (2003) Verlet-I/r-RESPA/Impulse is limited by nonlinear instabilities. *SIAM J. Sci. Comput.* **24**, p. 1951
33. E. Barth, and T. Schlick (1998) Overcoming stability limitations in biomolecular dynamics. I. Combining force splitting via extrapolation with Langevin dynamics in LN. *J. Chem. Phys.* **109**, p. 1617

34. A. Sandu and T. Schlick (2003) Masking resonance artifacts in force-splitting methods for biomolecular simulations by extrapolative Langevin dynamics. *J. Comput. Phys.* **151**, p. R45

35. S. Chin (2004) Dynamical multiple-time stepping methods for overcoming resonance instabilities. *J. Chem. Phys.* **120**, p. 8

36. S. Hammes-Schiffer (2002) Impact of enzyme motion on activity. *Biochemistry* **41**, p. 13335

37. S. Duane, A. D. Kennedy, B. J. Pendleton, and D. Roweth (1987) Hybrid Monte Carlo. *Phys. Lett. B* **195**, p. 216

38. Z. Zhu, M. E. Tuckerman, S. O. Samuelson, and G. J. Martyna (2002) Using novel variable transformations to enhance conformational sampling in molecular dynamics. *Phys. Rev. Lett.* **88**, p. 100201

39. L. Rosso and M. E. Tuckerman (2002) An adiabatic molecular dynamics method for the calculation of free energy profiles. *Mol. Simul.* **28**, p. 91

40. L. Rosso, P. Minary, Z. Zhu, and M. E. Tuckerman (2002) On the use of the adiabatic molecular dynamics technique in the calculation of free energy profiles. *J. Chem. Phys.* **116**, p. 4389

41. L. Rosso, J. B. Abrams, and M. E. Tuckerman (2005) Mapping the backbone dihedral free-energy surfaces in small peptides in solution using adiabatic free-energy dynamics. *J. Phys. Chem. B* **109**, p. 4162

42. Y. Liu and M. E. Tuckerman (2000) Generalized Gaussian moment thermostatting: A new continuous dynamical approach to the canonical ensemble. *J. Chem. Phys.* **112**, pp. 1685–1700

43. M. E. Tuckerman, G. J. Martyna, M. L. Klein, and B. J. Berne (1993) Efficient molecular-dynamics and hybrid monte-carlo algorithms for path-integrals. *J. Chem. Phys.* **99**, p. 2796

44. M. E. Tuckerman, D. Marx, M. L. Klein, and M. Parrinello (1996) Efficient and general algorithms for path integral Car-Parrinello molecular dynamics. *J. Chem. Phys.* **104**, p. 5579

45. G. J. Martyna, A. Hughes, and M. E. Tuckerman (1999) Molecular dynamics algorithms for path integrals at constant pressure. *J. Chem. Phys.* **110**, p. 3275

46. A. D. MacKerell, D. Bashford, M. Bellott, J. R. L. Dunbrack, J. D. Evanseck, M. J. Field, S. Fischer, J. Gao, H. Guo, S. Ha, D. Joseph-McCarthy, L. Kuchnir, K. Kuczera, F. T. K. Lau, C. Mattos, S. Michnick, T. Ngo, D. T. Nguyen, B. Prodhom, I. W. E. Reiher, B. Roux, M. Schlenkrich, J. C. Smith, R. Stote, J. Straub, M. Watanabe, J. Wiorkiewicz-Kuczera, D. Yin, and M. Karplus (1998) *J. Phys. Chem. B* **102**, p. 3586

47. G. M. Torrie and J. P. Valleau (1974) Monte-carlo free-energy estimates using non-boltzmann sampling – application to subcritical lennard-jones fluid. *Chem. Phys. Lett.* **28**, p. 578

48. G. M. Torrie and J. P. Valleau (1977) Non-physical sampling distributions in monte-carlo free-energy estimation – umbrella sampling. *J. Comput. Chem.* **23**, p. 187

49. J. Kushick and B. J. Berne (1977) Molecular Dynamics: Continuous Potentials, in Modern Theoretical Chemistry: Statistical Mechanics of Time Dependent Processes, ed. B. J. Berne, Plenum, New York

50. S. Kumar, R. H. Swendsen, P. A. Kollman, and J. M. Rosenberg (1992) The weighted histogram analysis method for free-energy calculations on biomolecules 1 the method. *J. Comput. Chem.* **12**, p. 1011

51. D. Wei, H. Guo, and D. R. Salahub (2001) Conformational dynamics of an alanine dipeptide analog: An ab initio molecular dynamics study. *Phys. Rev. E* **64**, 011907

52. S. Gnanakaran and R. M. Hochstrasser (2001) Conformational preferences and vibrational frequency distributions of short peptides in relation to multidimensional infrared spectroscopy. *J. Am. Chem. Soc.* **123**, p. 12886

53. Y. S. Kim, J. Wang, and R. M. Hochstrasser (2005) Two-dimensional infrared spectroscopy of the alanine dipeptide in aqueous solution. *J. Phys. Chem. B* **109**, p. 7511

54. B. C. Dian, A. Longarte, S. Mercier, D. A. Evans, D. J. Wales, and T. S. Zwier (2002) The infrared and ultraviolet spectra of single conformations of methyl-capped dipeptides: N-acetyl tryptophan amide and N-acetyl tryptophan methyl amide. *J. Chem. Phys.* **117**, p. 10686

55. E. A. Carter, G. Ciccotti, J. T. Hynes, and R. Kapral (1989) Constrained reaction coordinate dynamics for the simulation of rare events. *Chem. Phys. Lett.* **156**, p. 472

56. M. Sprik and G. Ciccotti (1998) Free energy from constrained molecular dynamics. *J. Chem. Phys.* **109**, p. 7737

57. C. Jarzynski (1997) Nonequilibrium equality for free energy differences. *Phys. Rev. Lett.* **78**, p. 2690

58. D. A. Hendrix and C. Jarzynski (2001) A 'fast growth' method of computing free energy differences. *J. Chem. Phys.* **114**, p. 5974

59. A. Laio and M. Parrinello (2000) Escaping free-energy minima. *Proc. Natl. Acad. Sci. USA* **99**, p. 9840

60. J. D. Ramshaw (1986) Remarks on entropy and irreversibility in non-hamiltonian systems. *Phys. Lett. A* **116**, p. 110

Simulating Charged Systems with **ESPResSo**

A. Arnold[1,2], B.A.F. Mann[1], and Christian Holm[1,2]

[1] Max-Planck-Institut für Polymerforschung, Ackermannweg 10, 55128 Mainz,
Germany
`arnolda@mpip-mainz.mpg.de`
[2] Frankfurt Institute for Advanced Studies (FIAS), Johann Wolfgang
Goethe-Universität, Frankfurt/Main, Germany
`c.holm@fias.uni-frankfurt.de`

Christian Holm

A. Arnold et al.: *Simulating Charged Systems with* ESPResSo, Lect. Notes Phys. **703**, 193–221
(2006)
DOI 10.1007/3-540-35273-2_6

194   A. Arnold et al.

We give an introduction into the topic of how to compute efficiently long range interactions. We start with reviewing the traditional Ewald sum for 3D Coulomb systems, discuss then in some detail the P^{3}M method of Hockney and Eastwood. We continue with explaining some strategies to perform the sum under partially periodic boundary conditions, where we present two recently developed methods, namely MMM2D and the ELC method for two dimensionally periodic boundary conditions, and the MMM1D method for systems with only one periodic coordinate. After this, we briefly mention alternative ways of dealing with the Coulomb sum, such as the MEMD method of Maggs. In the second part we present our recently developed MD simulation package, ESPResSo, that includes most of the introduced algorithms. We give a short introduction into the capabilities of ESPResSo and its usage. Finally we present some recent simulation results for polymer networks which were obtained by using ESPResSo.

1 Introduction

Nowadays computer simulations are a well established tool in theoretical physics, and many clever algorithms have been derived to allow to simulate larger and larger systems that can have up to the order of 10^8 particles. Systems with only short range interactions scale linearly with the number of particles. However, once long ranged interactions are involved, the tractable system sizes are considerably smaller, namely in the order of 10^6, and often most of the computation time is spent in the calculation of the long-ranged interactions. Therefore, for systems with long range interactions, there is a need for efficient algorithms to compute those [1].

To simulate bulk systems, one will often apply periodic boundary conditions in all spatial directions. For this kind of boundary conditions, a wide variety of algorithms exists, which we will shortly present here. A special focus will be on the classical Ewald method and its most efficient ancestor, the P^{3}M method of Hockney and Eastwood [2]. We explain how the method works and how one can automatically determine the optimal parameters for this method.

However, not all systems require periodic boundary conditions in all spatial directions.For membranes, for example, only two dimensions are periodic, while the third one is finite. In that case, the Ewald method is computationally highly inefficient and would not allow to treat more than a few hundred charged particles. We present two alternative approaches, the MMM2D and ELC methods, which allow for computational efficiency similar to the bulk case. It is also simple to adapt the MMM2D method for systems with only one periodic dimension.

In the second part of this contribution, we want to give an introduction into our simulation package, ESPResSo [3–5]; ESPResSo stands for *E*xtensible *S*imulation *P*ackage for *Res*earch on *So*ft matter systems. The term soft matter, or complex fluids, as they are called in the American literature, describes a large class of materials, such as polymers, colloids, liquid crystals, glasses, and dipolar fluids; familiar examples of such materials are glues, paints, soaps or baby diapers. Also most biological materials are soft matter – DNA, membranes, filaments and other proteins belong to this class. The research in soft matter science has been increased in the last decade due to its high potential usefulness in technology, biophysics, and nanoscience.

Many properties of soft matter emerge on the molecular rather than the atomistic level: the elasticity of rubber is the result of entropy of the long polymers molecules, and the hygroscopic materials used in modern diapers store water inside a polyelectrolyte network. To reproduce these physical effects in a computer simulation on the atomistic level, one would have to incorporate many millions of atoms, which is only possible on very small time scales even with the most powerful modern computers. However, often a much simpler description of the material is sufficient. Polymers such as polyelectrolytes or rubber often can be modeled by bead-spring models, i.e. (charged) spheres connected by springs, where each of the spheres stands for several atoms, often a complete monomer or even larger compounds. Although this model hides most of the chemical properties, it is quite successful in the description of polymers.

Computer simulations on the bead-spring level still incorporate several thousands of spheres and springs and require an efficient simulation software. Moreover, the modeling of the interactions between the coarse-grained beads requires a high flexibility and extensibility of the simulation code. ESPResSo was specifically designed to meet this requirements. The extensibility expresses mainly in that readability is preferred over code optimizations in ESPResSo, the flexibility in the fact that for many of the basic tasks, such as force calculations, standardized interfaces exist. The lack of optimized code is compensated by the use of state of the art algorithms, such as the electrostatics methods presented in the first part of the chapter. ESPResSo is published under the GNU public license, and is available through our web page [3]. We hope that the ESPResSo software will be useful also to some of the participants of the CSCM summer school.

The last section is devoted to an application of ESPResSo to a typical soft matter problem. While the properties of uncharged polymer networks have been well understood by now, this does not hold for charged networks. Even the coarse-grained simulation of such a polymer network requires a large number of beads as well as long time scales, so that such simulations got feasible only recently. We shortly present the used simulation techniques and some of the results.

2 The Ewald Family of Methods

For many physical investigations one wants to simulate bulk properties and therefore introduces periodic boundary conditions in all spatial directions to avoid boundary effects. For this kind of boundary conditions, the famous Ewald sum [6] does a remarkable job in splitting the very slowly, and only conditionally converging sum over the Coulomb potential into two exponentially converging sums. Using optimized parameter sets, the computation time of the method scales with the number of particles N like $\mathcal{O}(N^{3/2})$ [7].

The Ewald method can also be formulated for partially periodic boundary conditions, i.e. systems where only one or two of the three spatial dimensions are periodic. These geometries are applied for example in simulations of membranes or nano-pores, where some dimensions are supposed to have finite extend. In this case, however, the method scales like $\mathcal{O}(N^2)$, which only allows the treatment of moderately large systems with a few hundred charges.

In 3D periodic boundary conditions, the Ewald method can be further accelerated to a computation time scaling of $\mathcal{O}(N \log N)$ using Fast Fourier Transformations (FFT) by smearing the charges onto a regular mesh. Various variants of these mesh based methods exist, such as P^{3}M, PME or SPME, but P^{3}M is known to be the computationally optimal variant [8]. In addition, there are variants of the P^{3}M method for e.g. orthorhombic boundary conditions or dipolar systems. Here we will concentrate on the classical case of a cubic simulation box, and give a brief introduction into both the Ewald method and P^{3}M.

Consider a system of N particles with charges q_i at positions r_i in an overall neutral and, for simplicity, cubic simulation box of length λ. If periodic boundary conditions are applied, the total electrostatic energy of the box is given by

$$E = \frac{1}{2} \sum_{S=0}^{\infty} \sum_{m^2=S} \sum_{i,j=1}^{N}{}' \frac{q_i q_j}{|r_{ij} + m\lambda|} \,. \tag{1}$$

where $r_{ij} = r_i - r_j$, m counts the periodic images, and the prime denotes that for $m = 0$, the $i = j$ (self-interaction) term has to be omitted. Since this sum is only conditionally convergent, its value is not well-defined unless one specifies the summation order. Usually one chooses an approximately spherical order [9], where the summation is over spherical shells, denoted by $\sum_{m^2=S}$, as used in (1).

The Coulomb potential (i.e. $1/r$) bears *two* intrinsic difficulties: it is slowly decaying at large distances (since it is long ranged), but strongly varying at small distances due to the divergence of $1/r$ at $r = 0$. If only one of them was present, everything would be comparatively easy, since a short-ranged potential could be treated by a simple cutoff, as it is done, e.g., for interactions of the Lennard-Jones type, while the periodic sum over a smooth, non-divergent potential can accurately be represented by the first few terms of its Fourier series. The trick is thus to split the problem into two parts by the trivial identity

$$\frac{1}{r} = \frac{\text{erfc}(r)}{r} + \frac{1 - \text{erfc}(r)}{r} \,, \tag{2}$$

where $\operatorname{erfc}(r) := \frac{2}{\sqrt{\pi}} \int_r^\infty e^{-t^2} dt$ denotes the complementary error function. Since the field equations are linear, we can obtain a solution to the original problem by adding up the contributions for each summand separately. Since the error function is rapidly decaying, the first summand is short-ranged and can be treated directly in real space, while the second summand is now a smooth function everywhere and therefore can be treated in Fourier space. Note that the use of the error function is somewhat arbitrary; see [10–12] for other options. However, this choice allows for a particularly simple solution in Fourier space. We obtain the well known Ewald formula for the electrostatic energy of the primary box:

$$E = \frac{1}{2} \sum_{\boldsymbol{m} \in \mathbb{Z}^3} \sideset{}{'}\sum_{i,j} q_i q_j \frac{\operatorname{erfc}(\alpha|\boldsymbol{r}_{ij} + \boldsymbol{m}\lambda|)}{|\boldsymbol{r}_{ij} + \boldsymbol{m}\lambda|} + \frac{1}{2\lambda^3} \sum_{\boldsymbol{k} \in \mathbb{K}\backslash \boldsymbol{0}} \frac{4\pi}{k^2} e^{-\frac{k^2}{4\alpha^2}} |\tilde{\rho}(\boldsymbol{k})|^2$$

$$- \frac{\alpha}{\sqrt{\pi}} \sum_i q_i^2 + \frac{2\pi}{(1 + 2\epsilon')\lambda^3} \boldsymbol{M}^2, \quad (3)$$

where $\mathbb{K} = \frac{2\pi}{\lambda} \mathbb{Z}^3$, $\boldsymbol{M} = \sum_{i=1}^N q_i \boldsymbol{r}_i$ is the net dipole moment of the primary simulation box, α is the Ewald parameter as explained below, and

$$\tilde{\rho}(\boldsymbol{k}) = \int_V \rho(\boldsymbol{r}) e^{-i\boldsymbol{k}\cdot\boldsymbol{r}} d\boldsymbol{r} = \sum_{j=1}^N q_j e^{-i\boldsymbol{k}\cdot\boldsymbol{r}_j} \quad (4)$$

is the Fourier transformed charge density. Forces can be obtained from (3) by simply taking the derivative. The first sum over $\boldsymbol{m}$ in (3) adds up the short-range part of (2) in real space, while the second sum over $\boldsymbol{k}$ calculates the second (smooth, but long-range) part of (2) in Fourier space. For the Fourier-space sum to be truly periodic, one has to also take into account the $i = j$ term for $\boldsymbol{k} = 0$. The third term in (3) corrects for this, and is therefore called the self energy term.

The last term is called the dipole correction term, which has some interesting properties. The variable ϵ' denotes here the dielectric constant at infinity. The spherical summation order in (1) is equivalent to the limit of a large, spherically bounded regular grid of images of the simulation box embedded in vacuum, basically a crystalline ball. If the surrounding space is filled by a homogeneous medium with a dielectric constant ϵ', the particles of the ball will feel a polarization force. This leads to an additional contribution to the energy, that will not vanish even in the limit of an infinite ball, and that is proportional to the squared dipole moment of the simulation cell. Note that for vacuum $\epsilon' = 1$, so that the term is non-zero in general, but it can be omitted for conducting ("tinfoil") boundary conditions, i.e. $\epsilon' = \infty$ [9]. The dipole term also plays a role when changing the order of summation which will be important in Sect. 3.1.

Let us stress here that the Ewald trick rigorously transformed (1) into (3), and the energy is exact and independant of α. However, the main advantage is that the exponentially converging sums over m and k allow the introduction of comparatively small cutoffs without much loss in accuracy. The Ewald parameter α tunes the relative weights of the real space and the reciprocal space contributions. Optimizing the cutoffs r_{max} and k_{max} with respect to α, one can reduce the required computation time to $\mathcal{O}(N^{3/2})$. For given finite real- and reciprocal space cutoffs there exists an optimal α such that the accuracy of the approximated Ewald sum is the highest possible. This optimal value can be determined easily with the help of the excellent estimates for the cutoff errors derived by Kolafa and Perram [13]. Error estimates are important to avoid simulation artifacts due to numerical errors and to tune the methods to be as fast as possible by maintaining a certain precision. In a coarse-grained simulation, one typically tries to keep numerical errors roughly one order of magnitude smaller than the thermal fluctuations of the system.

2.1 The P^{3}M Method

The calculation of the Fourier space contribution is the most time consuming part of the Ewald sum. The essential idea of P^{3}M is to replace the simple continuous Fourier transformations in (3) by discrete Fast Fourier Transformations, that are numerical faster to calculate. The charges are interpolated onto a regular mesh. Since this introduces additional errors, the simple Coulomb Green function $4\pi/k^2$, as used in the second term in (3), is cleverly adjusted in such a way as to make the result of the mesh calculation most closely resemble the continuum solution [2]. Note, that this simple trick is the reason for the superiority of the P^{3}M algorithms over other mesh variants [8]. Due to the use of the FFT, the computational complexity of the Fourier space part reduces to $\mathcal{O}(N \log N)$. For increasing number of particles the real space cutoff can be varied in such a way, that this scaling applies to the full Ewald sum.

The first step, i.e., generating the mesh based charge density ρ_M (defined at the mesh points r_p), is carried out with the help of a charge assignment function W:

$$\rho_M(r_p) = \frac{1}{h^3} \sum_{i=1}^{N} q_i W(r_p - r_i). \tag{5}$$

Here h is the mesh spacing, and the number of mesh points $N_M = L/h$ along each direction should preferably be a power of two, since in this case the FFT is most efficient. For W a cardinal B-spline of order P is chosen, which is a piecewise polynomial function of weight one, spanning P sections on the mesh. Its Fourier transform is

$$\tilde{W}(\boldsymbol{k}) = h^3 \left(\frac{\sin(\frac{1}{2}k_x h)}{\frac{1}{2}k_x h} \frac{\sin(\frac{1}{2}k_y h)}{\frac{1}{2}k_y h} \frac{\sin(\frac{1}{2}k_z h)}{\frac{1}{2}k_z h} \right)^P \tag{6}$$

In a second step the mesh based electric field $\boldsymbol{E}(r_p)$ is calculated by taking the derivative of the electrostatic potential. In P^{3}M, this is done by simply multiplying the Fourier transformed potential with $i\mathbf{k}$. In this case $\mathbf{E}(\mathbf{r}_p)$ can be written as

$$\mathbf{E}(\mathbf{r}_p) = \overleftarrow{\mathrm{FFT}} \left[-i\mathbf{k} \times \overrightarrow{\mathrm{FFT}} \left[\rho_\mathrm{M} \right] \times \hat{G}_{\mathrm{opt}} \right] (\mathbf{r}_p) \tag{7}$$

In words, $\mathbf{E}(\mathbf{r}_p)$ is the *backward* finite Fourier transform of the product of $-i\mathbf{k}$, the *forward* finite Fourier transform of the mesh based charge density ρ_M and the so-called optimal influence function $\hat{G}_{\mathrm{opt}}$, given by

$$\hat{G}_{\mathrm{opt}}(\mathbf{k}) = \frac{i\mathbf{k} \cdot \sum_{\mathbf{m} \in \mathbb{Z}^3} \tilde{U}^2(\mathbf{k} + \frac{2\pi}{h}\mathbf{m}) \tilde{\mathbf{R}}(\mathbf{k} + \frac{2\pi}{h}\mathbf{m})}{|i\mathbf{k}(\mathbf{k})|^2 \left[\sum_{\mathbf{m} \in \mathbb{Z}^3} \tilde{U}^2(\mathbf{k} + \frac{2\pi}{h}\mathbf{m}) \right]^2} \tag{8}$$

with $\tilde{\mathbf{R}}(\mathbf{k}) := -i\mathbf{k}4\pi e^{-k^2/4\alpha^2}/k^2$ and $\tilde{U}(\mathbf{k}) := \tilde{W}(\mathbf{k})/h^3$. Finally, one arrives at the Fourier part of the force on particle i:

$$\mathbf{F}_i = q_i \sum_{\mathbf{r}_p \in \mathrm{M}} \mathbf{E}(\mathbf{r}_p) W(\mathbf{r}_i - \mathbf{r}_p) \tag{9}$$

Although the presented formulas (5–9) look somewhat complicated, they are rather easy to implement in a step by step procedure.

Similar to the Ewald method, there are also error estimates for P^{3}M [14], allowing to tune the method for optimal computational efficiency. Assuming that the charges are homogeneously distributed in the simulation box and that the errors χ_{ij} originating from the pairwise interactions of the charges i and j are uncorrelated, one obtains for the force error $\Delta\mathbf{F}_i$ of the ith particle

$$\langle (\Delta\mathbf{F}_i)^2 \rangle = q_i^2 \left\langle \left(\sum_{j \neq i} q_j \chi_{ij} \right)^2 \right\rangle = q_i^2 \sum_{j \neq i} \sum_{k \neq i} q_j q_k \langle \chi_{ij} \cdot \chi_{ik} \rangle = q_i^2 \chi^2 \mathcal{Q}^2 \tag{10}$$

where $\mathcal{Q}^2 := \sum_{j=1}^N q_j^2$, and $\chi^2 = \langle \chi_{ij}^2 \rangle$ is the average pairwise error. From this the root mean square force error can be estimated by

$$\Delta F_{rms} = \sqrt{\frac{1}{N} \sum_{i=1}^N (\Delta\mathbf{F}_i)^2} \approx \chi \frac{\mathcal{Q}^2}{\sqrt{N}}. \tag{11}$$

The assumptions necessary for (11) to hold are not specific to P^{3}M, and the details of the method are hidden only in the pairwise error average χ. For P^{3}M, the Fourierspace error is given by $\chi^2 = \chi_r^2 + \chi_k^2$, where the real space error contribution, i.e. the total contribution of the omitted vectors $\mathbf{n}$ in (1), is given by [13]

$$\chi_r = \frac{2}{\sqrt{r_{\max}\lambda^3}} \exp(-\alpha^2 r_{\max}^2). \tag{12}$$

The Fourierspace error contribution of P^{3}M is somewhat more complex, namely

$$\chi_k = \frac{1}{\lambda^3} \sum_{k \in \hat{\mathbb{M}}} \left\{ \sum_{m \in \mathbb{Z}^3} \left| \tilde{R}\left(k + \frac{2\pi}{h} m\right) \right|^2 \right.$$

$$\left. - \frac{\left| i k \cdot \sum_{m \in \mathbb{Z}^3} \tilde{U}^2\left(k + \frac{2\pi}{h} m\right) \tilde{R}^*\left(k + \frac{2\pi}{h} m\right) \right|^2}{|ik^2|^2 \left[\sum_{m \in \mathbb{Z}^3} \tilde{U}^2\left(k + \frac{2\pi}{h} m\right) \right]^2} \right\}. \quad (13)$$

Here, the first sum gives the errors introduced by the finite cutoff in k of (1), while the second sum represents the aliasing errors due to the interpolation of the charges onto a mesh. Admittedly, (13) looks rather complicated. Still, in combination with (11) it gives the rms force error of the Fourier space part of P^{3}M, and therefore allows to tune the P^{3}M algorithm a priori. In addition, the computations of χ_k and $\hat{G}_{\text{opt}}$ are quite similar, and have to be done only once in the code, and play a negligible role in the overall computation time.

The other variants of the mesh-based Ewald method, PME or SPME, differ from the P^{3}M method mainly in the choice of the interpolation scheme, the differentiation procedure, or choice of the lattice Green function [8]. Consequently, all these methods are similar in speed, and a good implementation is faster than the conventional Ewald sum already for a couple of hundred charges, and with today's computers, simulations using P^{3}M with 10^6 charges are feasible. For the PME error estimates exist as well, however the method is 1–2 orders of magnitude more inaccurate than the other two, and should hence not be used at all. SPME is slightly less accurate than P^{3}M but its drawback is that no error estimates are known. This leaves P^{3}M as the clear method to chose among all mesh based algorithms. In Sect. 5 we will explain how the P^{3}M implementation of ESPResSo is used, which features an automatic tuning of the parameters using the above-mentioned error estimates.

3 The MMM Family of Methods

Another approach to tackle the conditionally convergent Coulomb sum is used by Strebel and Sperb [15] and called the MMM method. Instead of defining the summation order, the sum is made convergent by means of a convergence factor:

$$\tilde{E} = \frac{1}{2} \lim_{\beta \to 0} \sum_{m \in \mathbb{Z}^3} \sum_{i,j=1}^{N} {}' \frac{q_i q_j e^{-\beta |r_{ij} + m\lambda|}}{|r_{ij} + m\lambda|}. \quad (14)$$

The limit $\tilde{E}$ exists and has as its value that of the Ewald method minus the dipole term [16]. Starting from this convergence factor approach, Strebel and Sperb constructed a method of computational order $\mathcal{O}(N^{7/5})$ or, with a more clever algorithm, $\mathcal{O}(N \log N)$, MMM [15]. Unlike the particle mesh Ewald methods, no mesh is introduced, so that no interpolation errors occur, and it is comparatively easy to find error estimates for the method. Consequently, this method allows much higher precisions compared to P^{3}M or (S)PME.

Due to the convergence factor, it is possible to calculate the total energy by a pairwise potential ϕ such that

$$\tilde{E} = \frac{1}{2} \lim_{\beta \to 0} \sum_{i,j=1}^{N} q_i q_j \phi_\beta(\boldsymbol{r}_{ij}) = \frac{1}{2} \sum_{i,j=1}^{N} q_i q_j \phi(\boldsymbol{r}_{ij}), \tag{15}$$

where ϕ_β follows from (14). The trick is now to split the calculation of ϕ again. But this time, the split occurs spatially: For particles sufficiently far apart, a Fourier transformation can be applied to the inner sum of (14) to obtain a computationally favorable expression for ϕ, the so called *far formula*. For particles close together this formula shows an extremely slow convergence, so that a different formula, the *near formula* has to be used, which has a less favorable computational complexity, but is stable. Choosing the cutoffs optimally results in the $\mathcal{O}(N^{7/5})$ scaling; the $\mathcal{O}(N \log N)$ scaling is obtained by a clever hierarchic evaluation of the far formula.

The MMM approach can also be adapted to systems with only one or two periodic coordinates. The resulting MMM1D and MMM2D algorithms belong to the fastest methods for partially periodic systems with a few hundred particles, and allow for full error control just as the Ewald type methods. Below we will introduce the MMM2D algorithm. For further details on MMM1D and MMM2D, see [17, 18].

As described above, we now introduce the two formulas used to calculate ϕ as in (15), however only for two periodic coordinates x and y. Moreover a cubic simulation box is often inappropriate for partially periodic systems, therefore we use a general simulation box of dimensions $\lambda_x \times \lambda_y \times \lambda_z$. Here, the convergence factor energy $\tilde{E}$ is equal to the energy obtained from the standard spherical summation, and no dipole term occurs. The MMM2D far formula is

$$\phi(\boldsymbol{r}) = 8\pi u_x u_y \sum_{p,q>0} \frac{e^{-\omega_{pq}|r_z|}}{\omega_{pq}} \cos(\omega_p r_x) \cos(\omega_q r_y)$$

$$+ 4\pi u_x u_y \left(\sum_{q>0} \frac{e^{-\omega_q|r_z|}}{\omega_q} \cos(\omega_q r_y) + \sum_{p>0} \frac{e^{-\omega_p|r_z|}}{\omega_p} \cos(\omega_p r_x) \right) - 2\pi u_x u_y |r_z|,$$

$$\tag{16}$$

where $u_\nu = \lambda_\nu$ for $\nu = x, y, z$, $\omega_{pq} = 2\pi\sqrt{(u_x p)^2 + (u_y q)^2}$, $\omega_p = 2\pi u_x p$ and $\omega_q = 2\pi u_y q$. These notations will be used frequently in the following. This formula is obtained from two Fourier transformations of ϕ_β, which are comparatively simple to perform. The resulting formula is well convergent for $|z| \gg 0$, but does not even exist for $z = 0$.

The near formula, i.e. the replacement for $z \approx 0$, reads

$$\phi(\boldsymbol{r}) = 4u_x \sum_{l,p>0} \left(K_0(\omega_p \rho_l) + K_0(\omega_p \rho_{-l}) \right) \cos(\omega_p r_x)$$

$$- 2u_x \sum_{n \geq 1} \frac{b_{2n}}{2n(2n)!} \Re\left((2\pi u_y (r_z + ir_y))^{2n} \right)$$

$$- u_x \sum_{n \geq 0} \binom{-\frac{1}{2}}{n} \frac{\left(\psi^{(2n)}(N_\psi + u_x r_x) + \psi^{(2n)}(N_\psi - u_x r_x) \right)}{(2n)!} (u_x \rho)^{2n}$$

$$+ \sum_{k=1}^{N_\psi - 1} \left(\frac{1}{r_k} + \frac{1}{r_{-k}} \right) - 2u_x \log\left(4\pi u_y \lambda_x \right) +' \frac{1}{r}, \quad (17)$$

where $r_k = |\boldsymbol{r} + (k\lambda_z, 0, 0)|$ are the distances from the origin to the $(k, 0, 0)$-th periodic image of $\boldsymbol{r}$, $\rho_l = \sqrt{(y + l\lambda_y)^2 + z^2}$ and $\rho = \rho_0$ are the in-plane distances to the $(0, l, 0)$-th image; K_0 denotes the modified Bessel function [19], b_n the n-th Bernoulli number, $\Re(\cdot)$ the real part, and $\psi^{(m)}$ the polygamma function of order m. As usual the prime on the last summation sign denotes that it is left out for the self energy, i.e. $r = 0$. N_ψ denotes a constant which has to satisfy $N_\psi \geq u_x \rho + 1$, but can be chosen freely otherwise. Typically, $N_\psi = 2$ is sufficient, so that the overhead due to the direct summation is small. The formula is a little bit awkward to implement, but it converges well and can be evaluated efficiently on a computer. For details and tips, see [17].

The main advantage of the far formula (16) is that it allows for a product decomposition into the particle components and can be evaluated in linear time with respect to the number of particles with constant cutoffs. To draw advantage from this, some additional bookkeeping is necessary, see [17]. For MMM2D error estimates exist, which allow the tuning of the method; this is important since the evaluation is split up into the near and far formula, leading to a highly non-uniform error distribution (see Fig. 1). With optimized parameters, the method can be tuned to scale like $\mathcal{O}(N^{5/3})$.

3.1 Electrostatic Layer Correction

Similar to the Ewald method in full periodic boundary conditions, MMM2D is only fast enough to be used for a couple of hundred charges. For larger systems, one needs a more efficient method, which in this case can be obtained by a combination of a standard method for full 3d periodicity, e.g. P^{3}M, and the far formula of MMM2D.

So far, larger systems were often treated by using a standard 3d method, but leaving a sufficiently large gap between different images in the nonperiodic coordinate to decouple the interactions in this direction (see Fig. 2). However, this naive approach leads to large systematic errors, which do not vanish even in the limit of infinite gap size. Yeh and Berkowitz [20] found out that these

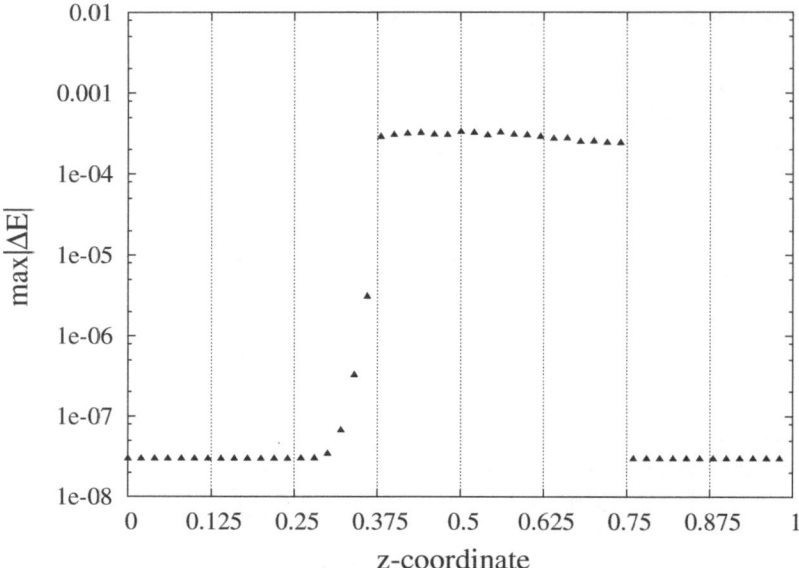

Fig. 1. Absolute MMM2D energy error $|\Delta E|$ depending on the z-position of a randomly placed particle in a $1 \times 1 \times 1$ box; a second particle was fixed at $z = 0.5$. The *dotted lines* the chosen slice boundaries. The adjacent slices, where the near formula used, are located from 0.375 to 0.75 for the fixed particle at $z = 0.5$

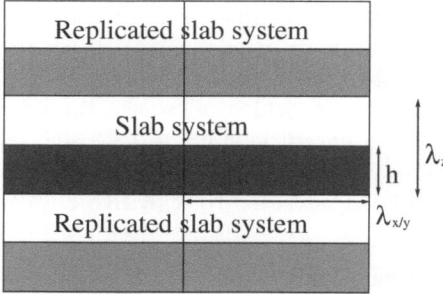

Fig. 2. Schematic representation of a fully periodically replicated slab system

can be be avoided by changing the Ewald spherical summation order to a slabwise summation order. This can be achieved surprisingly easy. If one adds up the particles along z slab-wise, i.e. ordered by increasing z-distance, but radially in x and y, Smith [21] has shown that the dipole term appearing in (3) then only has to be changed to the form $2\pi M_z^2$ instead of using the usual $2\pi M^2/3$. Note that for the slab-wise summation order as well as for partially periodic geometries, the dielectric constant at infinity does not play a role.

This approach converges with $\lambda_z \to \infty$, but no error bounds were known for this approach, so that it had to be checked on a trial and error basis.

Here the MMM2D far formula comes into play: due to the gap, the periodic images are well separated in the z-coordinate, so that the interactions of the images with the primary layer can be calculated by the far formula in linear order. The summation over the image layers can be performed analytically, and the contribution of the image layers can be calculated as

$$
E_{lc} = 8\pi u_x u_y \sum_{i,j=1}^{N} q_i q_j \left[\sum_{p,q>0} \frac{\cosh(\omega_{pq} r_{ij,z})}{\omega_{pq}(e^{\omega_{pq}\lambda_z} - 1)} \cos(\omega_p x) \cos(\omega_q y) \right.
$$
$$
\left. + 4\pi u_x u_y \left(\sum_{p>0} \frac{\cosh(\omega_p z)}{\omega_p(e^{\omega_p \lambda_z} - 1)} \cos(\omega_p x) + \sum_{q>0} \frac{\cosh(\omega_q z)}{\omega_q(e^{\omega_q \lambda_z} - 1)} \cos(\omega_q y) \right) \right].
$$
(18)

This formula is called electrostatic layer correction (ELC) term [22]; forces can be obtained by simple differentiation. Since this yields – up to a desired precision – the contribution of all periodic image layers, they can be subtracted from the result of a standard 3d-method with slabwise summation order to obtain the result for two periodic coordinates. In addition, the error estimates for MMM2D can be adapted for ELC, giving a maximal pairwise error of

$$
|\Delta E_{lc,ij}| \leq \frac{1/2 + (u_x + u_y)/(\pi R)}{e^{2\pi R\lambda_z} - 1} \left(\frac{\exp(2\pi Rh)}{\lambda_z - h} + \frac{\exp(-2\pi Rh)}{\lambda_z + h} \right),
$$
(19)

where R is the cutoff of the sums in the $(u_x p, u_y q)$-space. h denotes the maximal distance between two particles in the primary simulation box, λ_z the artificial periodicity. The estimate gives an upper bound for the error occurring between each particle pair. The reason to use this error estimate is, that the ELC error distribution is highly non-homogeneous (see Fig. 3), so that averaging error estimates underestimates the true error near the surfaces.

For $R = \min(u_x, u_y)$, one can obtain an error estimate also for the Yeh and Berkowitz method, which allows to tune this method. So why should one actually implement the ELC term? First of all because it leads to a faster algorithm, normally by a factor of two or more, since the box size can be made smaller due to the smaller gap size, and therefore one can use a smaller mesh size at constant accuracy. In addition, many implementations of P^{3}M (such as the ESPResSo one) allow only for cubic simulation boxes. But (18) shows that in fact the ratios λ_z/λ_x rsp. λ_z/λ_y dominate the error behaviour, so that one cannot reduce the error in the Yeh and Berkowitz approach with a cubic 3d-method. For such an implementation, ELC is a must.

4 Alternative Methods

There are many other methods for the calculation of electrostatic interactions, of which we want to mention some here. The MMM methods are actually

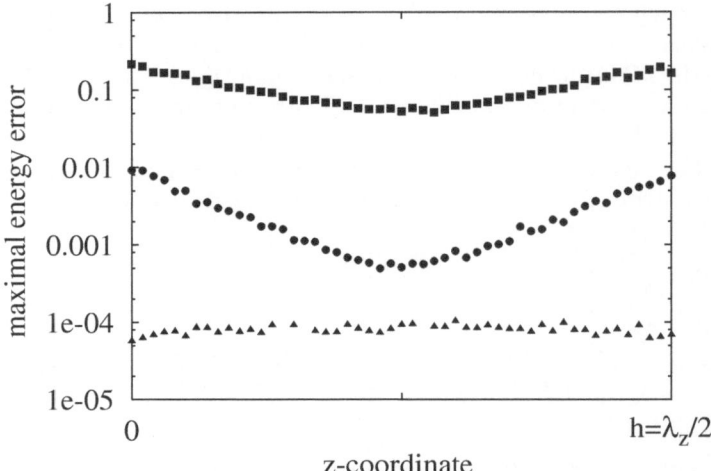

Fig. 3. Maximal ELC force error ΔF_∞ as a function of the particle z-coordinate for a simulation box of size $10 \times 10 \times 5$ and a periodicity of $\lambda_z = 10$. The cutoffs were $R = 0$ (*rectangles*), $R = 0.1$ (*circles*), and $R = 0.3$ (*triangles*). As 3d-method P^3M was used, tuned for an rms force error of 10^{-4}. For $R = 0.3$, the P^3M error dominates

based on the Lekner method. Lekner essentially derived the far formula of MMM1D [23] for the force by direct transformation of the periodic sum. Since the sum is well convergent, it is quite fast for small numbers of particles, and one can also use it for two periodic coordinates by direct summation of the other periodic coordinate [24, 25]. However, one should be aware that the lack of a near formula leads to large numerical artifacts [26], see the section on MMM.

A completely different approach are the so-called multipole methods or tree codes [27], which are optimal for large systems with nonperiodic boundary conditions. The multipole methods are based on a product decomposition in real space by using multipole approximations for distant particle pairs. While the mathematics of the multipole expansions is comparatively simple, the bookkeeping involved in these methods is much more demanding than for the Fourier type algorithms. Their advantage is a computation time scaling which is linear (the fast multipole method) or $\mathcal{O}(N \log N)$ for the tree codes. The fast multipole method however requires enormous amounts of memory, so that tree codes are often faster. Due to a large overhead, the fast multipole method becomes comparable in speed to P^3M only for more than a million particles.

Yet another approach are multigrid methods [28, 29]. These methods are based on solving Poisson's equation numerically on a grid. While periodicity is naturally no problem for these methods, the singularity of the Coulomb potential has to be circumvented, which is done by splitting the equation into a smooth and singular part, of which the latter can be solved analytically. Properly implemented, this methods can achieve a linear computation time scaling, but error analysis is hard for these methods, and there is still a lot of development going on.

4.1 Maggswellian Dynamics

The methods presented so far try to optimize the computation of the classical Ewald sum using mathematical methods. A completely different approach has recently been suggested first by A.C. Maggs' group, and later by B. Dünweg and I. Pasichnyk [30, 31]. Maggs used something similar like the Maxwell equations on a lattice. After all, electrodynamics is a local field theory, so one should be able to implement the equations locally on a lattice, which then results in an algorithm, which must scale linear in N. The algorithm, due to its locality, is especially suited for Monte Carlo algorithms [30]. In Maggs' version, the *pseudo* magnetic field is propagated via a diffusive process through the lattice, and MC moves are constructed that do not violate the Gauss law. Maggs could show that the diffusive dynamics recovers all static properties in thermal equilibrium, i.e. for long times. The whole formalism, where the charges reside on lattice sites, the electric and magnetic field are link variables, and the electromagnetic field tensor analogon is a plaquette variable, is similar in spirit to the one used in lattice gauge theories. A particularly nice feature, which so far has not been exploited, is the possibility to assign a locally varying dielectric constant. This is a problem which puts a formidable constraint on most other methods.

Molecular dynamics variants of this method were presented in [31]. Here the $1/r$ interaction is represented by a local interaction that propagates in waves with a tunable speed of light, c. It is sufficient to make c small enough such that the quasi static approximation is still valid, similar to the Car-Parinello approach, where the electronic degrees of freedom are slowed by an unphysically large mass. Again, the static properties of the system do not depend on the value chosen for c. This, and the lattice spacing, are the main parameters that have to be tuned for speed and accuracy. Preliminary benchmarks presented in [31] show that the so-called MEMD (Maxwell Equation Molecular Dynamics) method is comparable in speed to our P^3M implementation. MEMD is easy to parallelize, and it can be used, in principle, for systems with locally varying dielectric constants, but more results are needed to be able to assess all the advantages and disadvantages compared to the previously presented methods.

5 ESPResSo

In this section, we want to introduce our simulation software ESPResSo, which incorporates some of the presented algorithms [5]. The main goal in the development of ESPResSo was to have an *extensible* simulation package for bead-spring models, that contains many state-of-the-art algorithms. The emphasis was on *extensible*, since many problems in the simulation of coarse-grained polymer models require the implementation of new potentials or forces, or new analysis routines. Moreover, we want to be able to easily incorporate new state-of-the-art algorithms to increase the code performance.

The extensibility paradigm requires that readability of the code is favoured over special code optimization in ESPResSo and that it has standardized interfaces for interaction potentials, analysis, thermostats etc. Even for a beginner it should be possible to implement a new potential in a couple of days.

The wide field of simulation topics investigated by the ESPResSo community requires a high flexibility of the simulation code, which is realized in ESPResSo by the use of the Tcl script language for simulation control. The simulation control script determines all simulation parameters such as the number and type of particles, the type of interactions between these particles, and how the system is propagated; most of the parameters can be changed even during the simulation. This flexibility makes it possible to perform highly complex simulation procedures, such as adapting the interaction parameters to the current configuration during the simulation, cooling down the system in a simulated annealing process, or applying or removing constraints.

ESPResSo is a parallel code, and contains many state-of-the-art algorithms, allowing for simulations of millions of particles on hundreds of CPUs. Since for readability reasons, the code kernel is written in simple ANSI C, it runs on a variety of hardware platforms like PCs (GNU/Linux on IA32 and AMD64 processors), Workstations (MacOS on PowerPC processors) and high performance servers (AIX on Power4 processors or Tru64 on Alpha processors).

ESPResSo is not a self contained code, but relies on other open source packages. Most prominent is the use of the Tcl [32] script language interpreter for the simulation control. For the parallelisation standard MPI routines are used, which on Linux and MacOS are provided e.g. by the LAM/MPI [33] implementation, or MPICH [34]. P^{3}M relies on the FFTW package [35]. Besides these libraries, which are required to be able to have a running version of ESPResSo the development process is supported heavily by the use of the CVS version control system [36], which allows several developers to work simultaneously on the code, and the documentation generation tool Doxygen [37].

ESPResSo is published under the terms of the GPL and can be obtained from the ESPResSo homepage `http://www.espresso.mpg.de`. There you can also find a recent copy of the documentation with further details, and a complete list of the currently implemented features.

5.1 Using ESPResSo

In this section we want to introduce some of the features of ESPResSo by constructing step by step a little simulation script for a simple salt crystal.

We cannot give a full Tcl tutorial here; however, most of the constructs should be self-explanatory. We also assume that the reader is familiar with the basic concepts of a MD simulation here. The code pieces can be copied step by step into a file, which then can be run using `Espresso <file>` from the `Espresso` source directory.

Our script starts with setting up the initial configuration. Most conveniently, one would like to specify the density and the number of particles of the system as parameters:

```
set n_part 200; set density 0.7
set box_l [expr pow($n_part/$density,1./3.)]
```

These variables do not change anything in the simulation engine, but are just standard Tcl variables; they are used to increase the readability and flexibility of the script. The box length is not a parameter of this simulation; it is calculated from the number of particles and the system density. This allows to change the parameters later easily, e.g. to simulate a bigger system.

The parameters of the simulation engine are modified by the `setmd` command. For example

```
setmd box_l $box_l $box_l $box_l
setmd periodic 1 1 1
```

defines a cubic simulation box of size `box_l`, and periodic boundary conditions in all spatial dimensions. We now fill this simulation box with particles

```
set q 1; set type 0
for {set i 0} { $i < $n_part } {incr i} {
   set posx [expr $box_l*[t_random]]
   set posy [expr $box_l*[t_random]]
   set posz [expr $box_l*[t_random]]
   set q [expr -$q]; set type [expr 1-$type]
   part $i pos $posx $posy $posz q $q type $type }
```

This loop adds `n_part` particles at random positions, one by one. In this construct, only two commands are not standard Tcl commands: the random number generator `t_random` and the `part` command, which is used to specify particle properties, here the position, the charge q and the type. In ESPResSo the particle type is just an integer number which allows to group particles; it does not imply any physical parameters. Here we use it to tag the charges: positive charges have type 0, negative charges have type 1.

Now we define the ensemble that we will be simulating. This is done using the `thermostat` command. We also set some integration scheme parameters:

```
setmd time_step 0.01; setmd skin 0.4
set temp 1; set gamma 1
thermostat langevin $temp $gamma
```

This switches on the Langevin thermostat for the NVT ensemble, with temperature **temp** and friction coefficient **gamma**. The skin depth **skin** is a parameter for the link-cell system which tunes its performance, but cannot be discussed here.

Before we can really start the simulation, we have to specify the interactions between our particles. We use a simple, purely repulsive Lennard-Jones interaction to model the hard core repulsion [38], and the charges interact via the Coulomb potential:

```
set sig 1.0; set cut    [expr 1.12246*$sig]
set eps 1.0; set shift [expr 0.25*$eps]
inter 0 0 lennard-jones $eps $sig $cut $shift 0
inter 1 0 lennard-jones $eps $sig $cut $shift 0
inter 1 1 lennard-jones $eps $sig $cut $shift 0
inter coulomb 10.0 p3m tunev2 accuracy 1e-3 mesh 32
```

The first three **inter** commands instruct ESPResSo to use the same purely repulsive Lennard–Jones potential for the interaction between all combinations of the two particle types 0 and 1; by using different parameters for different combinations, one could simulate differently sized particles. The last line sets the Bjerrum length to the value 10, and then instructs ESPResSo to use P^3M for the Coulombic interaction and to try to find suitable parameters for an rms force error below 10^{-4}, with a fixed mesh size of 32. The mesh is fixed here to speed up the tuning; for a real simulation, one will also tune this parameter.

Now we can integrate the system:

```
set integ_steps 200
for {set i 0} { $i < 20 } { incr i} {
  set temp [expr [analyze energy kinetic]/(1.5*$n_part)]
  puts "t=[setmd time] E=[analyze energy total], T=$temp"
  integrate $integ_steps }
```

This code block is the primary simulation loop and runs $20 \times$ **integ_steps** MD steps. Every **integ_steps** time steps, the potential, electrostatic and kinetic energies are printed out (the latter one as temperature). However, the simulation will crash: ESPResSo complains about particle coordinates being out of range. The reason for this is simple: Due to the initial random setup, the overlap energy is around a million kT, which we first have to remove from the system. In ESPResSo, this is can be accelerated by capping the forces, i.e. modifying the Lennard–Jones force such that it is constant below a certain distance. Before the integration loop, we therefore insert this equilibration loop:

```
for {set cap 20} {$cap < 200} {incr cap 20} {
  puts "t=[setmd time] E=[analyze energy total]"
  inter ljforcecap $cap; integrate $integ_steps }
inter ljforcecap 0
```

This loop integrates the system with a force cap of initially 20 and finally 200. The last command switches the force cap off again. With this equilibration, the simulation script runs fine.

However, it takes some time to simulate the system, and one will probably like to write out simulation data to configuration files, for later analysis. For this purpose ESPResSo has commands to write simulation data to a Tcl stream in an easily parsable form. We add the following lines at end of integration loop to write the configuration files "config_0" through "config_19":

```
set f [open "config_$i" "w"]
blockfile $f write tclvariable {box_l density}
blockfile $f write particles {id pos type}
close $f
```

The created files "config_...." are human-readable and look like

```
{tclvariable
        {box_l 10}
        {density 0.7}
}
{particles {id pos type}
        {0 3.51770181433 4.3208975936 5.30529948918 0}
        {1 3.93145531704 6.58506447035 6.95045147034 1}
        . . .
}
```

As you can see, such a *blockfile* consists of several Tcl lists, which are called *blocks*, and can store any data available from the simulation. Reading a configuration is done by the following simple script:

```
set f [open $filename "r"]
while { [blockfile $f read auto] != "eof" } {}
close $f
```

This code will set the Tcl variables box_l and density to the values specified in the file, and the particle positions and types of all 216 particles are restored.

With these configurations, we can now investigate the system. A snapshot of a typical configuration is shown in Fig. 4. As an example, we will create a second script which calculates the averaged radial distribution functions $g_{++}(r)$ and $g_{+-}(r)$. The radial distribution function for a the current configuration can be obtained using the analyze command:

```
set rdf [analyze rdf 0 1 0.9 [expr $box_l/2] 100]
foreach value [lindex $rdf 1] {
   lappend rlist    [lindex $value 0]
   lappend rdflist [lindex $value 1] }
```

The shown analyze rdf command returns the distribution function of particles of type 1 around particles of type 0 (i.e. of opposite charges) for radii between 0.9 and half the box length, subdivided into 100 bins. Changing the

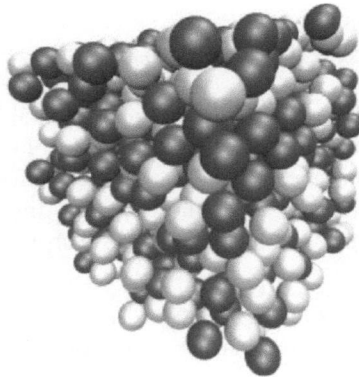

Fig. 4. VMD snapshot of a typical configuration of the salt system

first two parameters to either "0 0" or "1 1" allows to determine the distribution for equal charges. The result is a list of r and $g(r)$ pairs, which the following foreach loop divides up onto two lists `rlist` and `rdflist`.

To average over a set of configurations, we put the two last code snippets into a loop like this:

```
set cnt 0
for {set i 0} {$i < 100} {incr i} { lappend avg_rdf 0}
foreach filename $argv {
    ...read file...
    ...determine rdf...
    set avg_rdf [vec_add $avg_rdf $rdflist]
    incr cnt }
set avg_rdf [vec_scale [expr 1/$cnt] $avg_rdf]
```

Initially, the sum of all $g(r)$, which is stored in `avg_rdf`, is set to 0. Then the loops over all configurations given by `argv`, calculates $g(r)$ for each configuration and adds up all the $g(r)$ in `avg_rdf`. Finally, this sum is normalized by dividing by the number of configurations. Note that `argv` is a predefined variable: it contains all the command line parameters. Therefore this analyzation script should be called like `Espresso <script> <config 1> <config 2>....`

The printing of the calculated radial distribution functions is simple:

```
set plot [open "rdf.data" "w"]
puts $plot "\# r rdf(r)"
foreach r $rlist rdf $avg_rdf { puts $plot "$r $rdf" }
close $plot
```

Figure 5 shows the resulting radial distribution functions, averaged over 100 configurations. In addition, the distribution for a neutral system is given, which can be obtained from our simulation script by simply not turning on P^3M.

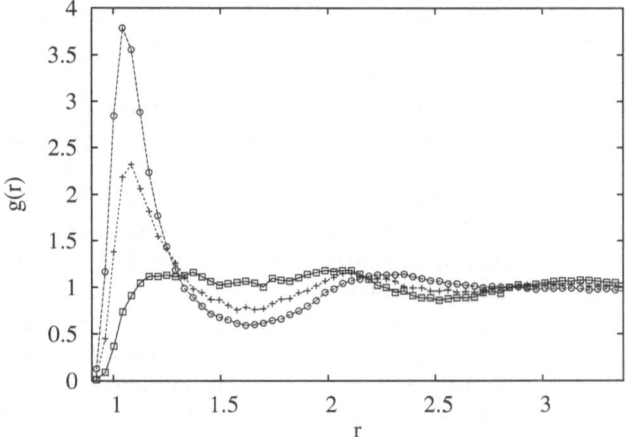

Fig. 5. Radial distribution functions $g_{++}(r)$ between equal charges (*rectangles*) and $g_{+-}(r)$ for opposite charges (*circles*). The plus symbols denote $g(r)$ for an uncharged system

The code example given before is still quite simple, and the reader is encouraged to try to extend the example a little bit, e.g. by using differently sized particle, or changing the interactions. If something does not work, ESPResSo will give comprehensive error messages, which should make it easy to identify mistakes. For real simulations, the simulation scripts can extend over thousands of lines of code and contain automated adaption of parameters or online analysis, up to automatic generation of data plots. Parameters can be changed arbitrarily during the simulation process, as needed for e.g. simulated annealing. The possibility to perform non-standard simulations without the need of modifications to the simulation core was one of the main reasons why we decided to use a script language for controlling the simulation core.

5.2 Features of ESPResSo

In the previous section we have shown how ESPResSo is used to simulate a simple problem. Of course only a small part of ESPResSo's capabilities has been presented, and here we want to give a short overview of other features.

Integrators and thermostats: ESPResSo can currently only perform MD simulations using a Velocity–Verlet integration scheme. Various ensembles can be obtained by different thermostats. For the NVE ensemble, no thermostat is used, for NVT, one can use either a Langevin or DPD thermostat. Constant pressure, i.e. NPT, simulations, can be performed using an algorithm by Dünweg et. al. [39].

Nonbonded potentials: For nonbonded interactions between different particle species, currently the Lennard–Jones, Gay–Berne, and Morse potentials are implemented in ESPResSo. In addition, it is possible to use tabulated potentials. To avoid overlap problems during equilibration, ESPResSo allows to cap the nonbonded potentials.

Bonded potentials: Interactions between two or more specific particles include FENE and harmonic bond potentials, bond angle and dihedral interactions. Again, potentials can also be included as tables.

Long–range potentials: ESPResSo has the P^{3}M, MMM2D, MMM1D and ELC algorithms for the electrostatic interaction included. For all of these, methods, automatic tuning routines are available. In addition, the MEMD scheme (see 4.1) is implemented.

Constraints ESPResSo has the ability to fix some or all coordinates of a particle, or to apply an arbitrary external force on each particle. In addition, spatial constraints such as spherical or cubic compartments, can be used. These constraints interact by any nonbonded interaction with the particles.

Analysis: All ESPResSo analysis routines are available in the simulation engine, allowing for both online analysis (during the simulation) as well as off-line analysis. ESPResSo can of course calculate the energy and (isotropic) pressure, and the forces acting on particles or spatial constraints can be obtained from the simulation engine. There are routines to determine particle distributions and structure factors, and some polymer-specific measures such as the end-to-end distance or the radius of gyration. For visual inspection, ESPResSo has an interface to the VMD visualization software [40].

In addition to these features, ESPResSo is a well scaling parallel code, which can achieve an efficiency of about 70% on 512 Power4+ processors. Since ESPResSo contains some of the fastest currently available simulation algorithms, it also scales well with the number of particles, allowing for the simulation of large systems. Nevertheless, one should be aware that the flexibility of ESPResSo also costs some performance: compared to fast MD programs such as GROMACS, ESPResSo is slower by a factor of 2. However, most of the problems that we use ESPResSo for, cannot be treated with these fast codes at all without massive changes to the simulation engine.

6 A "Real World" Example: Polyectrolyte Networks

In this section, we want to present some recent results which have been obtained by B. Mann using ESPResSo, and explain, how this has been done [41, 43, 47]. The system under investigation is a polyelectrolyte network, i.e. a gel of cross-linked charged polymers. One of the most prominent features of such a network is that it is able to absorb large amounts of the solvent, up to several hundred times its dry mass. Due to its remarkable properties, this hydrogel has many industrial applications, e.g. as superabsorbants in diapers, water treatment or drug delivery. Chemically, polyelectrolytes are polymers which dissociate charges in polar solvents, especially water; a hydrogel is a cross-linked polyelectrolyte. Due to charge-neutrality, the released counterions are confined inside the gel and exert an osmotic pressure which leads to the swelling of the network.

Theoretical approaches are rather simple, studying the swelling of the polyelectrolyte network in pure solvent. Numerical studies of polyelectrolyte networks began to examine this simple model system only recently, usually in an ideal diamond configuration, mainly because the large number of charges involved in these systems requires a large amount of computational power. In this section we can only present some of the results of a large scale survey, which included parameter scans over the strength of the electrostatic interaction, the charge fraction, chain length, and solvent quality. Here, we will only show results for a poor solvent [42], for other cases consult [41, 43, 47].

In these simulations, a perfect and defect-free diamond network of polyelectrolyte chains, connected at their ends to cross-linking sites (nodes), is employed; this network topology is conserved at all times (see Fig. 6). The polymer chains are modeled as bead spring chains of Lennard–Jones particles connected by FENE (spring) bonds, which are standard interactions in ESPResSo. The counterions are modeled as purely repulsive Lennard–Jones particles as well, and have the same size as the polymer beads for simplicity. The solvent properties enter in these simulations only implicitly, as a friction term from the thermostat, and by the presence (poor or Θ-solvent) or absence (good solvent) of an attractive tail in the Lennard–Jones interaction, which models the tendency of the polymer to be exposed to solvent contacts. A fraction of the polymer beads as well as all counterions carry a unit charge, positive for the chain particles, negative for the counterions; in total, the system is always charge neutral. The interactions between these charges are calculated using the P^3M algorithm.

Simulation studies of the swelling behaviour require to use the isothermal-isobaric NPT-ensemble. ESPResSo contains an integration scheme for this ensemble, which was successfully employed for the good solvent regime, but for the studies presented here, the equilibrium is difficult to reach. Therefore, the swelling equilibrium had to be determined from a $p(V)$-diagram, and for each datapoint, a large number of NVT simulations at different volumes had to be performed. Once the equilibrium simulation box size has been determined, the properties of the polyelectrolyte network are obtained from an extended NVT-simulation using this equilibrium box size.

The simulation scripts for these simulations are much more complex than the example of the last section, but still adhere more or less to the same scheme: setting up a random initial configuration, equilibration, integration and analyzation. Of course, there are many additional features, e.g. an outer loop performing the box length changes for the $p(V)$-diagram, and the initialization has to take care to create random polymer chains connecting the nodes of the diamond lattice. But apart from the bonded interactions, all

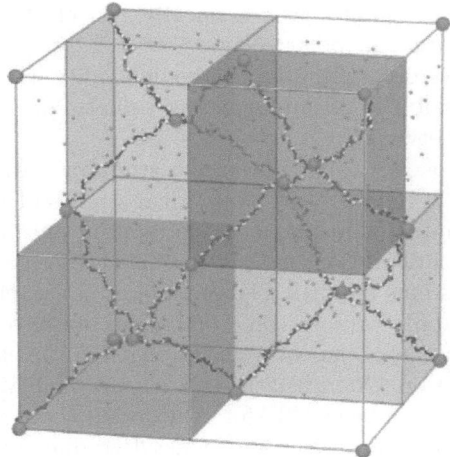

Fig. 6. Snapshot depicting the diamond lattice topology of the polyelectrolyte network. The nodes are shown as oversized spheres, connected by the polymer chains. The counterions are floating freely between the polymers

instructions of ESPResSo necessary to perfom these simulations have been explained in the example script. This shows again the power of the ESPResSo concept of using a script language for the simulation control.

6.1 Conformation in Poor Solvent

A large number of polyelectrolytes are based on a hydrocarbon backbone, which alone is usually non-soluble in water; due to the charged groups, the total polyelectrolyte is however often water-soluble. Under poor solvent conditions, the backbone alone would collapse to minimize the surface exposed to the solvent. However, the repulsion between the like-charged monomers can hinder the collapse. Therefore, whether the system collapses or not depends on the subtle interplay between electrostatic and hydrophobic interactions. For unlinked, single polyelectrolytes, the phase diagram is long known; the conformations range from a collapsed globule to elongated chains [44]. The most remarkable state is the so-called *pearl-necklace* state, when due to a Rayleigh-instability, the polyelectrolyte forms a sequence of small globules, interconnected by elongated chains. These pearl-necklace conformations are predicted both by theory and computer simulations [45], but lack a rigorous experimental confirmation so far.

There is no apparent reason why similar structures observed for the single-chain, unlinked polyelectrolytes at poor solvent conditions should not be present in charged hydrogels as well. The simulations presented in this section are performed at the same conditions as the studies of Limbach et al. [45]. The polyelectrolyte chains connecting the nodes of the diamond lattice consist

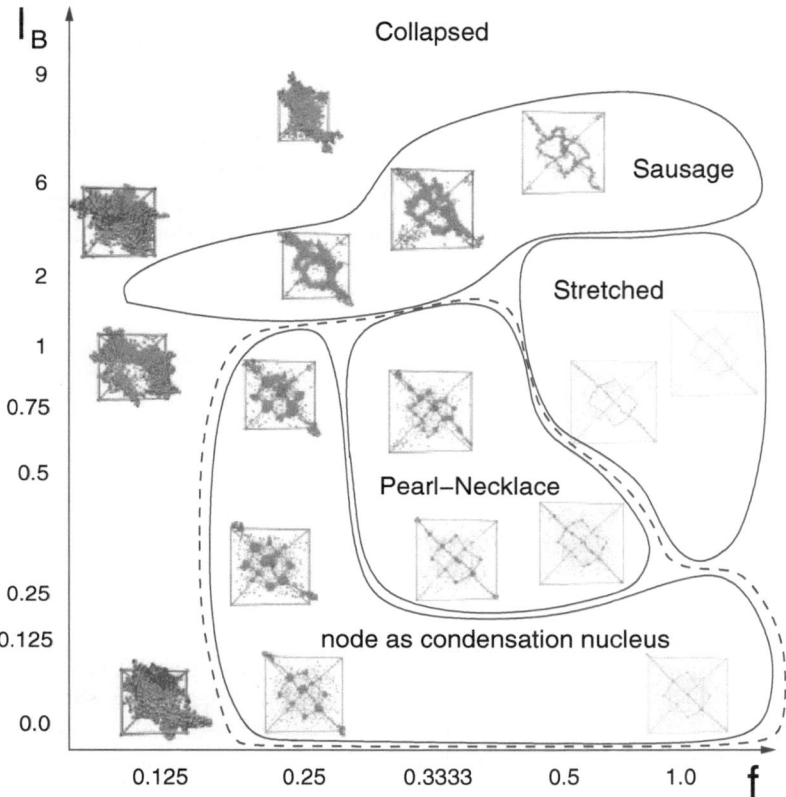

Fig. 7. Structure diagram of the equilibrium swelling conformations charged hydrogels assume in a poor solvent for varying charge fraction f and Bjerrum length ℓ_B

of 199 beads each, of which a fraction f is charged. For $f = 1$, the primary simulation box therefore contains 796 positive and 796 negative charges. The parameters scanned for the phase diagram were the charge fraction f and the Bjerrum length ℓ_B, which determines the strength of the electrostatic interaction compared to thermal fluctuations.

Figure 7 shows the resulting structure diagram, with sketches of the different observed regimes and example configurations. For small charge fractions or very strong electrostatics, one observes collapsed conformations, while pearl-necklaces occur for moderate to high charge fractions and not too strong coulombic coupling. For higher charge fractions and moderate electrostatic interactions, the polyelectrolytes are stretched, which however collapse again into "sausages" for high electrostatic coupling. This behaviour is known from the classical single polyelectrolyte chain, and can be explained by counterion condensation effects. For the polyelectrolyte network, the main difference is that the connecting nodes act as condensation nuclei, so that condensation sets in earlier compared to single polyelectroyte chains [45, 46].

Qualitatively, the phase diagram fits very well to the phase diagram known for single-chain polyelectrolytes, the phase boundaries are only slightly shifted. In principle, the parameter space for polyelectrolytes has far more dimensions, such as the solvent quality parameter, the valency of monomers or counterions, and additional salt concentration in the system. Especially for multivalent counterions, one can expect an even more complex picture, since correlation effects are known to play an important role even for single chains.

7 Concluding Remarks

We have given a short introduction into some methods to compute long range interactions in fully or partially periodic boundary conditions. We tried to give reasons why one should worry about errors, and how error estimates can be used to tune the algorithms to perform optimally for speed and accuracy. We also have tried to convey our understanding to the reader, when which algorithm should be used. A more thorough review on long range interactions has been published by us recently [1].

Most of the algorithms important to the field of charged soft matter have been implemented in the open source program package ESPResSo. We have given a short introduction into the use of ESPResSo, as well as some implementation details. Of course, our hope is that the some of the readers consider ESPResSo as the simulation code base for their next projects. Informations regarding the latest version, or informations on how to participate in the further development of this package can be found on http://www.espresso.mpg.de/.

Acknowledgments

We would like to dedicate this chapter to B.A.F. Mann, who suddenly deceased after completion of this article. In addition we thank M. Deserno, J. Dejoannis, I. Pasichnyk, H.J. Limbach for various contributions to the presented material, T. Schürg for his help in creating the simulation snapshot in Fig. 6, and C. Peter-Tittelbach for a critical reading. Helpful discussions with B. Dünweg and K. Kremer are gratefully acknowledged. We also acknowledge valuable input from the whole ESPResSo team. This work was supported by the BMBF 03N 6500, the DFG within the SFB 625, TR6, HO-1108/8-3, HO-1108/11-2, and a grant from the Volkswagen foundation.

References

1. A. Arnold and C. Holm (2005) Efficient methods to compute long range interactions for soft matter systems. In C. Holm and K. Kremer, editors, *Advanced Computer Simulation Approaches for Soft Matter Sciences II* volume II of, pp. 59–109, *Advances in Polymer Sciences*, Springer Berlin Heidelberg
2. R. W. Hockney and J. W. Eastwood (1988) *Computer Simulation Using Particles* IOP, London
3. ESPResSo (2004) Homepage. http://www.espresso.mpg.de
4. A. Arnold, B. A. Mann, H.-J. Limbach, and C. Holm (2004) ESPResSo – An Extensible Simulation Package for Research on Soft Matter Systems. In Kurt Kremer and Volker Macho, editors, *Forschung und wissenschaftliches Rechnen 2003* volume 63 of *GWDG-Bericht*, pp. 43–59. Gesellschaft für wissenschaftliche Datenverarbeitung mbh, Göttingen, Germany
5. H.-J. Limbach, A. Arnold, B. A. Mann, and C. Holm (2006) Espresso – an extensible simulation package for research on soft matter systems. *Comp. Phys. Comm.* **174**, pp. 704–727
6. P. P. Ewald (1921) Die Berechnung optischer und elektrostatischer Gitterpotentiale. *Ann. Phys.* **64**, pp. 253–287
7. J. Perram, H. G. Petersen, and S. de Leeuw (1988) An algorithm for the simulation of condensed matter which grows as the $3/2$ power of the number of particles. *Mol. Phys.* **65**, p. 875
8. M. Deserno and C. Holm (1998) How to mesh up Ewald sums. I. A theoretical and numerical comparison of various particle mesh routines. *J. Chem. Phys.* **109**, p. 7678
9. S. W. de Leeuw, J. W. Perram, and E. R. Smith (1980) Simulation of electrostatic systems in periodic boundary conditions. i. lattice sums and dielectric constants. *Proc. R. Soc. Lond. A* **373**, pp. 27–56
10. D. M. Heyes (1981) Electrostatic potentials and fields in infinite point charge lattices. *J. Chem. Phys.* **74**(3), pp. 1924–1929
11. H. J. C. Berendsen (1993) In Wilfred F. van Gunsteren, P. K. Weiner, and A. J. Wilkinson, editors, *Computer Simulation of Biomolecular Systems* **2**, pp. 161–81, The Netherlands, ESCOM
12. P. H. Hünenberger (2000) Optimal charge-shaping functions for the particle-particle-particle-mesh (p3m) method for computing electrostatic interactions in molecular simulations. *J. Chem. Phys.* **113**(23), pp. 10464–10476
13. J. Kolafa and J. W. Perram (1992) Cutoff errors in the ewald summation formulae for point charge systems. *Molecular Simulation* **9**(5), pp. 351–68
14. M. Deserno and C. Holm (1998) How to mesh up Ewald sums. II. An accurate error estimate for the particle-particle-particle-mesh algorithm. *J. Chem. Phys.* **109**, p. 7694
15. R. Strebel (1999) Pieces of software for the Coulombic m body problem. Dissertation 13504, ETH Zuerich
16. E. R. Smith (1981) Electrostatic energy in ionic crystals. *Proc. R. Soc. Lond. A* **375**, pp. 475–505
17. A. Arnold and C. Holm (2002) MMM2D: A fast and accurate summation method for electrostatic interactions in 2D slab geometries. *Comp. Phys. Comm.* **148**(3), pp. 327–348
18. A. Arnold and C. Holm (2005) MMM1D: A method for calculating electrostatic interactions in 1D periodic geometries. *J. Chem. Phys.* **123**(14), p. 144103

19. M. Abramowitz and I. Stegun (1970) *Handbook of mathematical functions.* Dover Publications Inc., New York

20. I.-C. Yeh and M. L. Berkowitz (1999) Ewald summation for systems with slab geometry. *J. Chem. Phys.* **111**(7), pp. 3155–3162

21. E. R. Smith (1988) Electrostatic potentials for thin layers. *Mol. Phys.* **65**, pp. 1089–1104

22. A. Arnold, J. de Joannis, and C. Holm (2002) Electrostatics in Periodic Slab Geometries I. *J. Chem. Phys.* **117**, pp. 2496–2502

23. J. Lekner (1989) Summation of dipolar fields in simulated liquid vapor interfaces. *Physica A* **157**, p. 826

24. A. G. Moreira and R. R. Netz (2001) Binding of similarly charged plates with counterions only. *Phys. Rev. Lett.* **87**, p. 078301

25. R. Sperb (1994) Extension and simple proof of lekner's summation formula for coulomb forces. *Molecular Simulation* **13**, pp. 189–193

26. M. Mazars (2001) Lekner summations. *J. Chem. Phys.* **115**(7), p. 2955

27. L. Greengard and V. Rokhlin (1997) A new version of the fast multipole method for the Laplace equation in three dimensions. *Acta Numerica* **6**, pp. 229–269

28. I. Tsukermann (2006) A class of difference schemes with flexible local approximation. *Journal of Computational Physics* **211**, pp. 659–699

29. G. Sutmann and B. Steffen (2005) A particle–particle particle–multigrid method for long–range interactions in molecular simulations. *Comp. Phys. Comm.* **169**, pp. 343–346

30. A. C. Maggs and V. Rosseto (2002) Local simulation algorithms for coulombic interactions. *Phys. Rev. Lett.* **88**, p. 196402

31. I. Pasichnyk and B. Dünweg (2004) Coulomb interactions via local dynamics: A molecular-dynamics algorithm. *Journal of Physics: Cond. Mat.* **16**(38), pp. 3999–4020

32. Tcl/Tk (2003) Tool Command Language / ToolKit – Homepage

33. LAM/MPI (2004) Local Area Multicomputer Message Passing Interface – Homepage

34. MPICH (2004) Message Passing Interface CHameleon – Homepage

35. FFTW (2003) Fastest Fourier Transform in the West – Homepage

36. CVS (2003) Concurrent Versions System – Homepage

37. Doxygen (2005) Doxygen – A documentation generation system

38. G. S. Grest and K. Kremer (1986) Molecular dynamics simulation for polymers in the presence of a heat bath. *Phys. Rev. A* **33**(5), pp. 3628–31

39. A. Kolb and B. Dünweg (1999) Optimized constant pressure stochastic dynamics. *J. Chem. Phys.* **111**(10), pp. 4453–4459

40. W. Humphrey, A. Dalke, and K. Schulten (1996) VMD: Visual molecular dynamics. *Journal of Molecular Graphics* **14**, pp. 33–38

41. B. A. Mann, C. Holm, and K. Kremer (2005) Swelling behaviour of polyelectrolyte networks. *J. Chem. Phys.* **122**(15), p. 154903

42. B. A. Mann, C. Holm, and K. Kremer (2006) Polyelectrolyte networks in poor solvent. *in preparation*

43. B. A. Mann, R. Everaers, C. Holm, and K. Kremer (2004) Scaling in polyelectrolyte networks. *Europhys. Lett.* **67**(5), pp. 786–792

44. H. Schiessel and P. Pincus (1998) Counterion-condensation-induced collapse of highly charged polyelectrolytes. *Macromolecules* **31**, pp. 7953–7959

45. H. J. Limbach and C. Holm (2003) Single-chain properties of polyelectrolytes in poor solvent. *J. Phys. Chem. B* **107**(32), pp. 8041–8055
46. H. J. Limbach, C. Holm, and K. Kremer (2002) Structure of polyelectrolytes in poor solvent. *Europhys. Lett.* **60**(4), pp. 566–572
47. Bernward A. Mann, Swelling Behaviour of Polyelectrolyte Networks, Ph.D. thesis, JoGu Universität Mainz, 2005

Density Functional Theory Based Ab Initio Molecular Dynamics Using the Car-Parrinello Approach

R. Vuilleumier

Laboratoire de Physique Théorique de la Matière Condensée, Tour 24-25, 2ème étage, c.c. 121, Université Pierre et Marie Curie, 4 place Jussieu, F-75005 Paris, France
rodolphe.vuilleumier@lptmc.jussieu.fr

Rodolphe Vuilleumier

R. Vuilleumier: *Density Functional Theory based Ab Initio Molecular Dynamics Using the Car-Parrinello Approach*, Lect. Notes Phys. **703**, 223–285 (2006)
DOI 10.1007/3-540-35273-2_7 © Springer-Verlag Berlin Heidelberg 2006

Ab initio Molecular Dynamics (MD) on the contrary to empirical force field Molecular Dynamics simulations employs an electronic structure calculation at each time-step of the dynamics to determine the forces on the nuclei. This allows for the simulation of materials in a broad range of situations, including during chemical reactions, while chemical bonds are broken or formed. The last few years, use of ab initio MD has spread very rapidly to many fields and is now used by many groups. At the same time many new developments have been pursued including, e.g., the calculation of electronic properties. Ab initio MD is also an integrated part of a new range of techniques to bridge length and time scales: QM/MM, transition path sampling, metadynamics etc. many of whose are discussed in this book.

The purpose of this chapter will be to review the fundamentals of ab initio MD. We will consider here Density Functional Theory based ab initio MD, in particular in its Car-Parrinello version. We will start by introducing the basics of Density Functional Theory and the Kohn-Sham method, as the method chosen to perform electronic structure calculation. This will be followed by a rapid discussion on plane wave basis sets to solve the Kohn-Sham equations, including pseudopotentials for the core electrons. Then we will discuss the critical point of ab initio MD, i.e. coupling the electronic structure calculation to the ionic dynamics, using either the Born-Oppenheimer or the Car-Parrinello schemes. Finally, we will extend this presentation to the calculation of some electronic properties, in particular polarization through the modern theory of polarization in periodic systems.

1 Introduction

Cohesion of matter relies essentially on the properties of the electronic cloud in which nuclei evolve. Computer simulations of materials thus aim at describing nuclei and electrons altogether. However in an approach like the very successful Molecular Dynamics methods based on empirical potentials [1], a system reduction is assumed by an implicit description of electrons through a model interaction between nuclei. The nuclei then evolve according to an empirical force-field designed such that the dynamics and thermodynamics of the reduced systems (atoms or molecules) are as close as possible to the real system properties. This contraction through empirical force-fields allows for the simulation by classical Molecular Dynamics of very large numbers of particles for long simulation times (few nanoseconds).

First-principle Molecular Dynamics on the other hand aims at simulating both nuclei and electrons uniformly [2]. This allows for the simulations of chemical reactions [3–15] where there is large electronic reorganisation following the ionic motion, through bond breaking and bond formation. More generally it allows description of polarization effects or other many-body effects. In particular, within a certain range of physical conditions, first-principle MD ought to be more transferable than empirical force-field MD which may need reparametrization of the analytical force fields. Phase transitions, where also many-body effects may be present, are typical cases of such situations [16–24]. Moderate to strong hydrogen bonds also exhibit many-body and polarization effects [25–30]. Liquid water has indeed attracted a lot of attention [31–39] in first-principle Molecular Dynamics, due also to its importance for chemistry. This is true also for aqueous solutions [40–64], where the solute can be subject to, or induce, polarization effects.

The range of systems studied by first-principle Molecular Dynamics is vast and the number of applications of first-principle MD is growing every day. Apart from the systems previously mentioned, first-principle MD simulations have been for example performed for molecular [3, 48, 65–67] or ionic liquids [68–72], surfaces and interfaces [48, 73–79], solids [27, 80–83] or glasses [84–87] and biological systems [88–96].

Additionally a number of packages implement now first-principle Molecular Dynamics [97–110]; here we will focus more precisely on the widespread plane wave basis set approaches based on density functional theory. The reader can also find many reviews on this subject [51, 98, 111–114].

Section 2 of this chapter notes will be devoted to the framework for separation of the ionic and electronic dynamics through the Born-Oppenheimer approximation. Atomic motion, with forces on the ions at each timestep evaluated through an electronic structure calculation, can then be propagated by Molecular Dynamics simulations, as proposed by first-principle Molecular Dynamics. This allows for a description of the electronic reorganisation following the atomic motion, e.g. bond rearrangements in chemical reactions.

The coupling of electronic states calculations with MD is obviously much more expensive than empirical force fields MD. In the latter, the electronic forces on the atoms are modeled by analytical force fields instead of being the result of an on the fly calculation. The timescales, in the order of a few picoseconds, and length scales achievable by first-principle MD are thus much shorter than those in empirical force-field MD. To reduce this gap and achieve simulations of rather large systems over time scales sufficiently long at the molecular level, an efficient method for on the fly calculation of the electronic state is needed. The method of choice for first-principle MD is Density Functional Theory. Section 3 will introduce the principles of Density Functional Theory and the following Sect. 4 will detail the Kohn-Sham method in the framework of DFT. Section 5 will discuss resolution methods of the Kohn-Sham equations, touching to practical problems like the basis-set, with a large emphasis on plane wave basis sets as the most common basis set until now for first-principle MD, and pseudopotentials.

Section 6 will deal with the core of the problem at hand, coupling of the electronic calculation with Molecular Dynamics for ions. We will describe in particular the Car-Parrinello method introduced in 1985 [2], which is really the starting point for combining on the fly electronic structure calculation to MD and statistical physics. This will be illustrated rapidly by a simulation of 32 water molecules at room temperature.

The analysis of such Car-Parrinello simulation is not limited to analysis of the distribution and dynamics of the ions alone: in Sect. 7 we will discuss the calculations of electronic properties (observables), like polarization, in the frame-work of DFT. We will also introduce the calculation of electronic response properties and excited states energies in DFT.

Finally, in Sect. 8 we will conclude. Emphasis in this chapter are deliberatly put on the electronic calculation that is performed on the fly: we assume that the reader is familiar with MD techniques, whose many recent developments are otherwise discussed in the rest of this book.

2 The Born-Oppenheimer Approximation

We aim at simulating a composite system consisting of ions and N electrons. The instantaneous quantum state of such system is described by the system wavefunction $\Psi_S(\mathbf{R}_I, \mathbf{r}_1, \ldots, \mathbf{r}_N)$ which is an integrable function of the nuclei and electrons coordinates, $\mathbf{R}_I$ and $\mathbf{r}_1, \ldots, \mathbf{r}_N$ respectively. The dynamics of the wavefunction Ψ_S is governed by the Hamiltonian of the system which we write as:

$$\hat{H}_T = \hat{T}_N + \hat{V}_{NN} + \hat{H} \ , \tag{1}$$

where $\hat{T}_N$ is the kinetic energy operator for the nuclei, $\hat{V}_{NN}$ is the potential energy operator acting on nuclei degrees of freedom only and arising from the nuclei-nuclei interaction and $\hat{H}$,

$$\hat{H} = \hat{T} + \hat{V}_{ee} + \hat{V}_{Ne} \ , \tag{2}$$

is the electronic Hamiltonian, including the interaction between the electrons and the nuclei.

Solving for the dynamics of such system is a highly complex problem. The first simplification amounts to assume an adiabatic separations of nuclei and electrons motions. In this approximation a partial factorization of the total system wavefunction is performed and we consider the ions fixed from the point of view of the electrons. This will lead us to the Born-Oppenheimer approximation.

2.1 Adiabatic Separation of Nuclei and Electrons Motions

Considering that the nuclei mass is much larger than the electron mass, we can well approximate in many cases the total system wavefunction by a wavefunction of the form [115]:

$$\Psi_S(\mathbf{R}_I, \mathbf{r}_1, \ldots, \mathbf{r}_N) = \Psi(\mathbf{R}_I)\Psi_0(\mathbf{r}_1, \ldots, \mathbf{r}_N; \mathbf{R}_I) \ , \tag{3}$$

where $\Psi_0(\mathbf{r}_1, \ldots, \mathbf{r}_N; \mathbf{R}_I)$ is the ground state electronic wavefunction of the electronic Hamiltonian $\hat{H}$ at fixed ionic configuration $\{\mathbf{R}_I\}$ and $\Psi(\mathbf{R}_I)$ constitutes the nuclear part of the system wavefunction. We will note $E_0(\mathbf{R}_I)$

$$E_0(\mathbf{R}_I) = \langle \Psi_0 | \hat{H} | \Psi_0 \rangle \tag{4}$$

the (ground-state) energy of the electrons at this ionic configuration. This factorisation of the total system wavefunction assumes that the electrons, which move and relax much faster than the nuclei, due to the large mass ratio, are always found in their ground state and follow adiabatically the nuclei motion. [115] This constitutes the Born-Oppenheimer approximation which greatly simplifies the problem of finding the dynamics of the ions+electrons system. Cases where this approximation fails are thus cases where the electron motion is not fast enough, due to the presence of a low lying excited state, and can not follow adiabatically the nuclei motion, giving rise to spontaneous population of the low lying electronic excited states.

For two wavefunctions of the form (3) $\Psi_S = \Psi \cdot \Psi_0$ and $\Psi'_S = \Psi' \cdot \Psi_0$ the matrix element of $\hat{H}_T$ between these two wavefunctions is equal to

$$\langle \Psi'_S | \hat{H}_T | \Psi_S \rangle = -\sum_I \frac{1}{2M_I} \langle \Psi' | \nabla^2_{\mathbf{R}_I} | \Psi \rangle + \langle \Psi' | \hat{V}_{NN} + E_0(\mathbf{R}_I) | \Psi \rangle \tag{5}$$

$$-\sum_I \frac{1}{M_I} \langle \Psi' | \left(\langle \Psi_0 | \nabla_{\mathbf{R}_I} \Psi_0 \rangle (\{\mathbf{R}\}) \right) | \nabla_{\mathbf{R}_I} \Psi \rangle$$

$$-\sum_I \frac{1}{2M_I} \langle \Psi' | \left(\langle \Psi_0 | \nabla^2_{\mathbf{R}_I} | \Psi_0 \rangle (\{\mathbf{R}\}) \right) | \Psi \rangle$$

The third and fourth term arise from the $\mathbf{R}_I$ dependance of the electronic wavefunction and are the source of the breaking of the Born-Oppenheimer or adiabatic approximation. The third term can couple the ground state to excited states. However if we can limit ourselves to the electronic ground state as assumed here, this term is strictly zero from the normalization condition on Ψ_0: $\langle \Psi_0 | \Psi_0 \rangle = 1$, $\forall \{\mathbf{R}\}$. The fourth term will generally be considered an order of magnitude smaller than the non-adiabatic coupling and often discarded. However we see here that it contributes also when restricting ourselves to the electronic ground states [115].

Thus when these two terms can be neglected, the effective Hamiltonian from (5) governing the evolution of the nuclei wavefunction under the adiabatic approximation (3) is

$$\hat{H}_N = \hat{T}_N + \hat{V}_{NN} + E_0(\hat{\mathbf{R}}_I) \tag{6}$$

from

$$\langle \Psi'_S | \hat{H}_T | \Psi_S \rangle = \langle \Psi'_S | \hat{H}_N | \Psi_S \rangle . \tag{7}$$

The Hamiltonian $\hat{H}_N$ is now a pure nuclear operator dictating, in the Born-Oppenheimer approximation, the evolution of the nuclear wavefunction [115]. The electrons enter $\hat{H}_N$ only through a potential energy term, $E_0(\hat{\mathbf{R}}_I)$, added to the bare nuclei-nuclei interaction $\hat{V}_{NN}$. This potential energy term due to the electrons is the ground state energy of the electronic system at fixed ionic configuration.

2.2 Classical Limit for the Ions

Taking now the classical limit for the nuclei, the potential energy surface for the nuclei classical motion is thus $v_{NN}(\mathbf{R}_I) + E_0(\mathbf{R}_I)$. The forces acting on the ions are the gradient of this electronic energy with respect to ionic positions, $-\frac{\partial E_0}{\partial R_I}$. This term can be evaluated through the Hellmann-Feynman theorem [116] which uses the fact that Ψ_0 is an eigenvector of the electronic Hamiltonian $\hat{H}$:

$$\frac{\partial E_0}{\partial R_I} = \frac{\partial}{\partial R_I} \langle \Psi_0 | \hat{H} | \Psi_0 \rangle \tag{8}$$

$$= \langle \Psi_0 | \frac{\partial}{\partial R_I} \hat{H} | \Psi_0 \rangle + \langle \Psi_0 | \hat{H} | \frac{\partial}{\partial R_I} \Psi_0 \rangle + c.c \tag{9}$$

$$= \langle \Psi_0 | \frac{\partial}{\partial R_I} \hat{H} | \Psi_0 \rangle + E_0 \langle \Psi_0 | \frac{\partial}{\partial R_I} \Psi_0 \rangle + c.c \tag{10}$$

$$= \langle \Psi_0 | \frac{\partial}{\partial R_I} \hat{H} | \Psi_0 \rangle \tag{11}$$

since $\langle \Psi_0 | \frac{\partial}{\partial R_I} \Psi_0 \rangle = 0$. This last equality (11) expresses the Hellmann-Feynman theorem: to calculate the forces on the ions one does not need to know the derivative of the electronic wavefunction with respect to the ionic position but simply need to take the expectation value of the force operator $\frac{\partial}{\partial R_I} \hat{H}$ on the ground state wavefunction. In the electronic Hamiltonian only the term $\hat{V}_{Ne}$ (see (2)) representing the electron-ion interaction, depends on $\mathbf{R}_I$. For a local interaction $\hat{V}_{Ne} \equiv \sum_i v_{Ne}(\mathbf{r}_i; \{\mathbf{R}\})$, $\frac{\partial}{\partial R_I} \hat{V}_{Ne}$ is also local in $\mathbf{r}$, so that the expectation of the force on the ground state wavefunction, equal to the force acting on the ions through the Hellmann-Feynman theorem [116], is

$$\frac{\partial E_0}{\partial R_I} = \langle \Psi_0 | \frac{\partial}{\partial R_I} \hat{V}_{Ne} | \Psi_0 \rangle \tag{12}$$

$$= \int d^3 \mathbf{r} \frac{\partial v_{Ne}(\mathbf{r}, \mathbf{R})}{\partial R_I} n^0(\mathbf{r}), \tag{13}$$

where $n^0(\mathbf{r})$ is the electronic density

$$n^0(\mathbf{r}) = N \int d\mathbf{r}_2 \ldots d\mathbf{r}_N |\psi_0(\mathbf{r}, \ldots, \mathbf{r}_N)|^2. \tag{14}$$

2.3 Born-Oppenheimer Molecular Dynamics

The simulation of a system of classical nuclei and electrons under the Born-Oppenheimer approximation can then be performed within a Molecular Dynamics framework as follows:

1. from the ionic configuration $\mathbf{R}(t)$ at time t compute the total energy $E_0 = E_0[\mathbf{R}(t)]$ given a representation of the electronic state: $\Psi_0(\mathbf{r}_1, \ldots, \mathbf{r}_N; \mathbf{R}_I(t))$.

2. then get the forces from the Hellman-Feynman theorem

$$F_I(t) = -\frac{\partial E_0}{\partial R_I} - \frac{\partial V_{NN}}{\partial R_I} = -\int d^3\mathbf{r} \, \frac{\partial V_{Ne}(\mathbf{r}, \mathbf{R}(t))}{\partial R_I} \, n^0(\mathbf{r}) - \frac{\partial V_{NN}}{\partial R_I}$$

3. advance ionic configuration $\mathbf{R}(t) \to \mathbf{R}(t + \delta t)$ by solving Newton's equations of motion (e.g. using the Verlet algorithm [117])
4. back to step (1)

The total energy $\frac{1}{2} \sum_I M_I \dot{R}_I^2(t) + E_0(t) + V_{NN}(t)$ is in principle a constant of motion of the system.

In empirical force-fields calculations, the information about the electronic system is entirely contracted in the data of the ground state potential energy surface and forces acting on the nuclei. Model potentials and forces are then used to propagate the ionic dynamics, instead of performing an electronic structure calculation. This on the fly quantum calculation is the challenging part of first-principle Molecular Dynamics simulations.

3 Interacting Electrons and Density Functional Theory

3.1 Outline of the Density Functional Theory

The difficulty in solving for the ground state electronic wavefunction at a given ionic configuration arises from the fact that electrons are interacting fermions. The Hamiltonian of an N electrons system in the external field $v_{ext}(\mathbf{r})$ created by the nuclei is:

$$\hat{H} = \hat{T} + \hat{V}_{ee} + \hat{V}_{ext} \tag{15}$$

where $\hat{T}$ is the kinetic energy operator:

$$\hat{T} = -\frac{1}{2} \sum_{i=1}^{N} \hat{\nabla}_i^2 \,, \tag{16}$$

and $\hat{V}_{ee}$ represents the electron-electron repulsion:

$$\hat{V}_{ee} = \frac{1}{2} \sum_{i \neq j} \frac{1}{|\hat{\mathbf{r}}_i - \hat{\mathbf{r}}_j|} \tag{17}$$

($\hat{\mathbf{r}}_i$ is the position operator for electron i).

The electronic ground state $\Psi_0(\mathbf{r}_1, \ldots, \mathbf{r}_N)$ is a function of $3N$ variables which satisfies the fermion antisymmetry property:

$$\Psi_0(\mathbf{r}_1, \ldots, \mathbf{r}_i, \ldots, \mathbf{r}_j, \ldots, \mathbf{r}_N) = -\Psi_0(\mathbf{r}_1, \ldots, \mathbf{r}_j, \ldots, \mathbf{r}_i, \ldots, \mathbf{r}_N). \tag{18}$$

Ψ_0 is the eigenvector of $\hat{H}$ associated with the lowest eigenvalue E_0:

$$\hat{H}\Psi_0\left(\mathbf{r}_1,\ldots,\mathbf{r}_N\right) = E_0\Psi_0\left(\mathbf{r}_1,\ldots,\mathbf{r}_N\right) \tag{19}$$

or, alternatively, Ψ_0 can be obtained from a variational principle:

$$E_0 = \min_{\Psi(\mathbf{r}_1,\ldots,\mathbf{r}_N)} \langle\Psi|\hat{H}|\Psi\rangle, \tag{20}$$

the minimization of $\langle\Psi|\hat{H}|\Psi\rangle$ being realized for $\Psi = \Psi_0$ (non-degenerate ground state).

Since the electrons are interacting the ground state wavefunction Ψ_0 in general can not be factorized in a single Slater determinant:

$$\psi_{Slater} \equiv \frac{1}{\sqrt{N!}} \begin{vmatrix} \phi_1(\mathbf{r}_1) & \phi_2(\mathbf{r}_1) & \ldots & \phi_N(\mathbf{r}_1) \\ \vdots & & & \vdots \\ \phi_1(\mathbf{r}_N) & \ldots\ldots\ldots & & \phi_N(\mathbf{r}_N) \end{vmatrix}, \tag{21}$$

where ϕ_i are one-electron orbitals, which does not exhibit any electron-electron correlation. Thus the electronic wavefunction is a very complicated object that can not even be stored for more than a few electrons. It is for example the purpose of Quantum Monte Carlo techniques to try to perform calculations (calculating the expectation value of the Hamiltonian) using directly a representation of the electronic wavefunction [118,119].

The strength of Density Functional Theory (DFT) [120–122] is based on the fact that the electronic wavefunction $\Psi_0\left(\mathbf{r}_1,\ldots,\mathbf{r}_N\right)$ of the electronic *ground state* of the system can be entirely described only by its electron density $n^0(\mathbf{r})$, as stated by the Hohenberg-Kohn theorem [123]. It is based on a *minimization* principle stating that the ground state electronic density minimizes an *energy functional*. Thus in principle we have to consider and manipulate a much simpler object, the electronic density, which is simply a function of $\mathbb{R}^3$.

3.2 The Hohenberg-Kohn Theorem

The Hohenberg-Kohn theorem [120–123] states that there exists a one to one map between external potentials and the ground state electronic density:

$$n^0(\mathbf{r}) \rightarrow v(\mathbf{r})\,. \tag{22}$$

For an external potential $v(\mathbf{r})$ there is a unique ground-state wavefunction $\Psi_0{}^1$ and as a result it gives rise to a unique ground state density $n^0(\mathbf{r})$. The Hohenberg-Kohn theorem states that for electronic densities $n(\mathbf{r})$ which are v-representable, that is which are the ground state of some external potential, this external potential which gives rise to them is unique. This can be expressed by the following map:

---

[1] For *non-degenerate* ground states, which is the very general case, degenerate systems are to be treated particularly.

$$v(\mathbf{r}) \quad \underset{\longleftarrow}{\longrightarrow} \quad \Psi_0 \quad \longrightarrow \quad n^0(\mathbf{r})$$

$$\text{HK theorem} \tag{23}$$

Knowledge of $n(\mathbf{r})$ gives knowledge of $v(\mathbf{r})$ and of Ψ_0, thus all properties of the electronic system and in particular any expectation value of an observable $\hat{O} = \langle \Psi | \hat{O} | \Psi \rangle$, are functionals of $n(\mathbf{r})$ which will be written $(O[n]$ for an observable $\hat{O})$.

Before proving the Hohenberg-Kohn theorem, we will first show the following lemma [120, 123].

Lemma 1. *Defining the set* $\mathcal{V}$ *of external potentials* $v_{ext}(\mathbf{r})$ *leading to a non-degenerate ground state and* $\{\Psi\}$ *the set of wavefunctions* $\Psi(\mathbf{r}, \ldots, \mathbf{r}_N)$ *which are ground state wavefunctions of a system of N electrons in an external potential, the application:*

$$\mathcal{V} \mapsto \{\Psi\}$$
$$v_{ext} \to \Psi_0 \tag{24}$$

is bijective.

The above application is surjective by definition of the set $\{\Psi\}$ and just means that we are considering v-representable wavefunctions Ψ's. We will now show its injectivity; that is, provided that $v_{ext} - v'_{ext} \neq const.$, v_{ext} and v'_{ext} can not have the same ground state. Indeed if there was such a common ground state Ψ, it would verify:

$$\left(\hat{T} + \hat{V}_{ee} + \hat{V}_{ext} \right) \Psi = E_0 \Psi \tag{25}$$

$$\left(\hat{T} + \hat{V}_{ee} + \hat{V}'_{ext} \right) \Psi = E'_0 \Psi , \tag{26}$$

leading to $(\hat{V}_{ext} - \hat{V}'_{ext})\Psi = (E_0 - E'_0)\Psi$, that is $v_{ext} - v'_{ext} = E_0 - E'_0$ if Ψ is not identically zero, in contradiction with the above assumption that $v_{ext} - v'_{ext}$ was not a constant.

Let's now define the set $\mathcal{N}$ of ground state electron densities.

Lemma 2. *The application*

$$\{\Psi\} \mapsto \mathcal{N}$$
$$\Psi_0 \to n(\mathbf{r}) \tag{27}$$

is bijective

This application is once more surjective by definition of the v-representability of the electronic densities considered. To show its injectivity, let's assume that Ψ and Ψ', ground states of the potentials v and v' $(v - v' \neq const.)$

lead both to the same electron density $n(\mathbf{r})$. Denoting $\hat{H} = \hat{T} + \hat{V}_{ee} + \hat{V}$ and $\hat{H}' = \hat{T} + \hat{V}_{ee} + \hat{V}'$, we have by virtue of the minimization principle, (20):

$$\langle \Psi | \hat{H} | \Psi \rangle < \langle \Psi' | \hat{H} | \Psi' \rangle \tag{28}$$

$$E_0 < \langle \Psi' | \hat{H}' | \Psi' \rangle + \langle \Psi' | (\hat{V} - \hat{V}') | \Psi' \rangle \tag{29}$$

$$E_0 < E_0' + \int d^3\mathbf{r}\, n(\mathbf{r})\, (v(\mathbf{r}) - v'(\mathbf{r})) \tag{30}$$

and similarly

$$E_0' < E_0 + \int d^3\mathbf{r}\, n(\mathbf{r})\, (v'(\mathbf{r}) - v(\mathbf{r})) \tag{31}$$

When the two inequalities are added, one is lead to an obvious contradiction:

$$E_0 + E_0' < E_0 + E_0', \tag{32}$$

which proves Lemma 2.

Putting Lemmas 1 and 2 together proves that the map (23) is a one to one map [120, 123].

3.3 Energy Functional

The total energy of the system is then itself a *functional $E[n]$* of the ground state electronic density [120, 123]. Separating the interaction with the external potential from the rest (kinetic energy and electron-electron interaction) we can write the total energy as

$$E[n] = \int d^3\mathbf{r}\, n(\mathbf{r})v(\mathbf{r}) + F_{HK}[n] \tag{33}$$

which defines the Hohenberg-Kohn functional $F_{HK}[n]$ as

$$F_{HK}[n] = T[n] + V_{ee}[n] \tag{34}$$

$$T[n] = \langle \Psi[n] | \hat{T} | \Psi[n] \rangle \tag{35}$$

$$V_{ee}[n] = \langle \Psi[n] | \hat{V}_{ee} | \Psi[n] \rangle, \tag{36}$$

where $\psi[n]$ is the ground state wavefunction, functional of n. Note that in (33) v is to be understood as a functional $v[n]$ of n through the map (23). We will note that [124] $F_{HK}[n] = \min_{\Psi \to n} \langle \Psi | \hat{T} + \hat{V}_{ee} | \Psi \rangle$, realised for $\Psi = \Psi[n]$, the ground state associated with the electron density n.

In most cases, as in the Born-Oppenheimer Molecular Dynamics scheme discussed earlier (see Sect. 2.3) what is given to us is the external potential $v_0(\mathbf{r})$ instead of the ground state density. In its turn, the ground state density n_0 for the external potential v_0 can be shown to minimize the energy functional [124]:

$$E_{v_0}[n] = \int d^3\mathbf{r}\, n(\mathbf{r})v_0(\mathbf{r}) + F_{HK}[n] \tag{37}$$

in which v_0 is fixed, while $F_{HK}[n]$ is the universal functional defined above (independent on v_0). Indeed

$$E_{v_0}[n] = \langle \Psi[n]|\hat{T} + \hat{V}_{ee} + \hat{V}_0|\Psi[n]\rangle \tag{38}$$

and if $\Psi[n]$ is not ground state of v_0 while $\Psi[n_0]$ is, we have

$$\langle \Psi[n]|\hat{T} + \hat{V}_{ee} + \hat{V}_0|\Psi[n]\rangle > \langle \Psi[n_0]|\hat{T} + \hat{V}_{ee} + \hat{V}_0|\Psi[n_0]\rangle = E[n_0]\,, \tag{39}$$

which proves the above statement.

The Hohenberg-Kohn theorem is a proof of existence of a one to one map between the ground state electronic density and the external potential, for v-representable densities, as well as a proof of the existence of the universal functional $F_{HK}[n]$. As mentioned earlier, this in principle greatly simplifies the interacting electron problem as we have only to consider electronic densities instead of wavefunctions. Unfortunately, this energy functional is unknown and although the Hohenberg-Kohn theorem is exact, we will have to rely on approximate energy functionals. The main difficulty in constructing such a functional turns out to be the kinetic energy term. The Kohn-Sham method is directly aimed at providing a reasonable approximation to it.

4 The Kohn-Sham Method

4.1 Principle

The key point of the Kohn-Sham method is to consider an auxiliary system of N *non-interacting* electrons for estimating the kinetic energy of the real (interacting) system [120–122, 125].

The ground state wavefunction Ψ_s of a system of such an N non-interacting electron system in an external potential $v_s(\mathbf{r})$ is a single slater determinant

$$\Psi_s = \frac{1}{\sqrt{N!}} \det[\phi_1\phi_2 \ldots \phi_N] \tag{40}$$

made of the N orbitals ϕ_i which are the N lowest eigenvectors of the single electron Hamiltonian $\hat{H}_s$:

$$\hat{H}_s = -\frac{1}{2}\nabla^2 + v_s(\mathbf{r})\,. \tag{41}$$

That is the orbitals ϕ_i satisfy

$$\hat{H}_s\phi_i = \epsilon_i\phi_i \text{ with } \langle \phi_i|\phi_j\rangle = \delta_{ij}\,. \tag{42}$$

The ground state density of this non-interacting electron system is

$$n_s(\mathbf{r}) = \sum_{i=1}^{N} |\phi_i(\mathbf{r})|^2 \,. \tag{43}$$

If for a given electronic density $n(\mathbf{r})$ there exists an external potential v_s whose non-interacting electron ground state density is equal to n:

$$n_s(\mathbf{r}) \equiv n(\mathbf{r}) \tag{44}$$

then n is said to be non-interacting v-representable.

The Hohenberg-Kohn theorem is valid also for non-interacting systems such that there exists a one to one map between n and v_s; the map

$$n(\mathbf{r}) \rightleftarrows v_s(\mathbf{r}) \rightleftarrows \{\phi_i(\mathbf{r})\} \tag{45}$$

is unique. The Hohenberg-Kohn functional for the non-interacting electron system is simply the kinetic energy T_s:

$$T_s[n] = \sum_{i=1}^{N} \langle \phi_i[n]| -\frac{1}{2}\nabla^2 |\phi_i[n]\rangle. \tag{46}$$

We will then approximate the kinetic energy functional $T[n]$ for the full system by the kinetic energy functional $T_s[n]$ of this auxiliary non-interacting electron system.

4.2 Energy Decomposition

The total energy of the system can be decomposed into [125]

$$E[n] = T_s[n] + \int d^3\mathbf{r}\, n(\mathbf{r})v(\mathbf{r}) + J[n] + E_{xc}[n] \,, \tag{47}$$

where $J[n]$ is the Hartree term:

$$J[n] = \frac{1}{2} \iint d^3\mathbf{r} d^3\mathbf{r}' \frac{n(\mathbf{r})n(\mathbf{r}')}{|\mathbf{r} - \mathbf{r}'|} \tag{48}$$

and equation (47) defines the exchange-correlation energy $E_{xc}[n]$ as the remainder:

$$E_{xc}[n] = (T[n] - T_s[n]) + (V_{ee}[n] - J[n])\,. \tag{49}$$

The Hartree term $J[n]$ describes the electrostatic energy of the electronic system within a mean field approximation: it is the classical electrostatic energy of a charge density $n(\mathbf{r})$. The exchange correlation energy is mainly the difference between the true electrostatic energy $V_{ee}[n]$ and the mean field term $J[n]$ but also contains a contribution from the kinetic energy, $T[n] - T_s[n]$. Once again, the Hohenberg-Kohn theorem for interacting and non-interacting electrons proves the existence and the uniqueness of the functional $E_{xc}[n]$ but this functional is still unknown to us. It has proved however much easier to find approximations to $E_{xc}[n]$ than to the Hohenberg-Kohn functional $F_{HK}[n]$ (33).

4.3 Self-Consistency Condition

We will now derive a condition for an electronic density $n_0(\mathbf{r})$ to minimize $E_{v_0}[n]$ (37), that is to be the ground state electronic density of the system in the external potential v_0. The variational principle states that if we assume a small variation of n, $\delta n(\mathbf{r})$, which conserves the number of electrons ($\int d^3\mathbf{r}\,\delta n(\mathbf{r}) = 0$),

$$\delta E_v = \delta T_s + \int d^3\mathbf{r} \left(v(\mathbf{r}) + \frac{\partial J}{\partial n(\mathbf{r})} + \frac{\partial E_{xc}}{\partial n(\mathbf{r})} \right) \delta n(\mathbf{r}) = 0 \qquad (50)$$

at the minimum. In the previous equation

$$\frac{\partial J}{\partial n(\mathbf{r})} = v_J(\mathbf{r}) = \int d^3\mathbf{r}'\frac{n(\mathbf{r}')}{|\mathbf{r} - \mathbf{r}'|} \quad \text{is the Hartree potential} \qquad (51)$$

$$\frac{\partial E_{xc}}{\partial n(\mathbf{r})} = v_{xc}(\mathbf{r}) \quad \text{is the exchange-correlation potential} \qquad (52)$$

To derive a condition on n_0 we now need to evaluate δT_s.

From the HK theorem for non-interacting electrons we can associate to n a potential $\bar{v}_s = v_s[n]$, the so-called Kohn-Sham potential, of which n is the ground state density for the non-interacting electron problem. Then n minimizes the energy functional $E_{s,\bar{v}_s}[n]$

$$E_{s,\bar{v}_s}[n] = \sum_i \langle \phi_i | -\frac{1}{2}\nabla^2 + \bar{v}_s | \phi_i \rangle \qquad (53)$$

$$= T_s[n] + \int d^3\mathbf{r}\,\bar{v}_s(\mathbf{r})n(\mathbf{r}) \qquad (54)$$

and for a variation $\delta n(\mathbf{r})$ of n we thus have

$$\delta E_{s,\bar{v}_s} = \delta T_s + \int d^3\mathbf{r}\,\bar{v}_s(\mathbf{r})\delta n(\mathbf{r}) = 0 \quad (n \text{ is the ground state of } \bar{v}_s). \quad (55)$$

Finally

$$\delta T_s = -\int d^3\mathbf{r}\,v_s[n](\mathbf{r})\delta n(\mathbf{r}) \qquad (56)$$

for a variation $\delta n(\mathbf{r})$ of n. In the expression above we make use of the formal functional $v_s[n]$

Substituting δT_s in δE_v, one gets

$$\delta E_v = -\int d^3\mathbf{r}\,v_s[n](\mathbf{r})\delta n(\mathbf{r}) + \int d^3\mathbf{r} \left(v(\mathbf{r}) + \frac{\partial J}{\partial n(\mathbf{r})} + \frac{\partial E_{xc}}{\partial n(\mathbf{r})} \right) \delta n(\mathbf{r}) \quad (57)$$

and the condition for n_0 to be the minimum of E_v is [125]

$$v_s[n_0](\mathbf{r}) = v(\mathbf{r}) + \int d^3\mathbf{r}' \frac{n_0(\mathbf{r}')}{|\mathbf{r} - \mathbf{r}'|} + v_{xc}[n_0](\mathbf{r}) \qquad (58)$$

which is a self-consistent equation that n_0 must fullfill.

Equation (58) means that for an electronic density n_0 to be the ground state density of an external potential v, upon solving the non-interacting electron problem in the external potential constructed by the right hand side expression of (58) the ground state density thus found must be n_0 itself. Indeed, recognising that two potentials are equal if and only if they lead to the same ground state electronic density for the non-interacting electron problem (HK theorem), this self-consistent equation can be formally rewritten:

$$n_0(\mathbf{r}) = n_{NI}\left[v(\mathbf{r}) + \int d^3\mathbf{r}' \frac{n_0(\mathbf{r}')}{|\mathbf{r} - \mathbf{r}'|} + v_{xc}[n_0](\mathbf{r})\right](\mathbf{r}) \qquad (59)$$

where we formally denote by $n_{NI}[u]$ the ground state electron density of a non-interacting electron system subject to an external potential u.

In the following we will discuss in more details methods for solving practically this self-consistent equation in order to find the ground state of a chemical system, however we first need to have tractable approximations for the exchange-correlation functional E_{xc}.

4.4 Exchange-Correlation Energy Functionals

Local Density Approximation

The first approximation to the exchange-correlation energy functional $E_{xc}[n]$ is the Local Density Approximation (LDA) defined as:

$$E_{xc}^{LDA} = \int d^3\mathbf{r}\, \epsilon_{xc}\left(n(\mathbf{r})\right) \cdot n(\mathbf{r}) \qquad (60)$$

where $\epsilon_{xc}(n)$ is the exchange and correlation energy *per electron* of the homogeneous electron gas with density n. It assumes that the electronic density is a smooth function of space. Any region of space can then be locally seen as a homogeneous electron gas of density $n(\mathbf{r})$. The total exchange-correlation energy is then the sum over all electrons in every region of space of the local exchange-correlation energy.

Traditionally one separates the LDA exchange-correlation energy per electron in an exchange part and a correlation part:

$$\epsilon_{xc}(n) = \epsilon_x(n) + \epsilon_c(n), \qquad (61)$$

in which $\epsilon_x(n)$ is the Dirac exchange energy

$$\epsilon_x(n) = C_x\, n^{\frac{1}{3}};\; C_x = -\frac{3}{4}\left(\frac{3}{\pi}\right)^{\frac{1}{3}} \qquad (62)$$

the exchange energy of the homogeneous electron gas. The correlation part can be fitted on the total energy of the homogeneous electron gas. The current reference calculation of the energy of a homogeneous electron gas is the Quantum Monte-Carlo result of Ceperley and Alder [118]. Parametrisations of the LDA correlation energy using Padé approximants based on these results are those of Volsko, Wilk, Nussair [126] Perdew-Wang [127] and Perdew-Zunger [128]. However, it may be more convenient numerically to approximate $\epsilon_{xc}(n) = \epsilon_x(n) + \epsilon_c(n)$ altogether using Padé approximants [129].

Generalized Gradient Approximations

The Generalized Gradient Approximations (GGA) are an extension and an amelioration of LDA to inhomogeneous systems: the local exchange-correlation energy depends not only on the local charge density but also on the local charge density gradient. There are basically three types of GGA's:

ab initio: These functionals are based on exact results, asymptotics etc. The exchange and correlation parts are treated independently. These are typically Perdew's PBE [130] or PW91 [131, 132] etc.

atom based GGA's: They also include some exact results but functional parameters are fitted on closed-shell atoms properties. Exchange and correlation are again treated separately. These are for example Becke's GGA for exchange [133] and Lee Yang and Parr (LYP) correlation [134].

empirical: Exchange and correlation are considered as a whole. Functional parameters are determine by fitting on a set of atomic and molecular properties, including hydrogen bonded complexes. These are Handy's HCTH functionals. [135–137]

The GGA functionals represent a noticeable improvement upon LDA in particular for molecular systems and often work very well for atoms and solids [138]. They can in particular reproduce well hydrogen bonded system like liquid water [139].

Other types of functionals include hybrid functionals [140], incorporating a fraction of Hartree-Fock exchange, orbital dependent functionals, kinetic energy functionals and so on [120].

4.5 Summary

The complexity of the electronic structure problem is in principle dramatically reduced in the framework of Density Functional Theory as the many-body problem amounts to solving a minimization problem of an energy functional $E_{HK}[n]$ or solving the much simpler problem of non-interacting electrons. This is ensured by the HK and KS theorems which state the existence of the functionals $E_{HK}[n]$ and $E_{xc}[n]$ (assuming v-representability). Unfortunately, although we have proved their existence, the $E_{HK}[n]$ and $E_{xc}[n]$ functional are unknown and one has to rely in practice on approximations.

The most noticeable deficiencies of the local density approximations to $E_{xs}[n]$ or GGA's include the lack of long range dispersion forces (van der Waals forces $\propto \frac{1}{R^6}$), because they arise from strongly non-local electron correlation. Although it should be noted that they include short range dispersion (decreasing exponentially instead of algebraically). Another defect is the so-called self-interaction [120]: the Hartree potential includes the interaction of an electron with itself, this should be compensated in principle by the exchange term in the exchange correlation functional (note that Hartree-Fock is self-interaction free) but is only approximately so with LDA and GGA functionals. As a result, the energy of the hydrogen atom for example is not -0.5 Ry because of the inaccuracy of $v_x^{LDA}(\mathbf{r})$. However different Self-Interaction Corrections (SIC) have been proposed [63, 128, 141, 142].

Despite all this, DFT is credited of many successes and can be very useful, being computationally tractable for complex systems. DFT results with GGA functionals are often of much better quality than Hartree-Fock. Although DFT is often seen as a mean field approach (from its structure and since setting $E_{xc} = 0$ amounts to the mean field Hartree solution of the electronic structure problem), approximate functionals like LDA or GGA's incorporate terms that are non mean field contributions. Furthermore they do so while fulfilling some exact relations (sum rules etc.).

5 Solution of the Kohn-Sham Equations

Now that we have discussed the foundations of DFT, we will shortly describe methods for solving the Kohn-Sham problem, particularizing to the case of plane wave basis sets at the end of this section.

There are basically two methods to find the ground-state density within the Kohn-Sham framework for a given external potential, which will be introduced successively. The first one focuses on the self-consistency equation (58) that must be fullfilled by the ground-state density, while the second one aims at directly minimizing the energy functional at fixed external potential.

5.1 Diagonalization Technique

From the analysis described in the previous section, we look for the ground state density as the density $n(\mathbf{r})$ which satisfies the self-consistency condition (58). Starting from a guess density $n(\mathbf{r}) \equiv n^{old}(\mathbf{r})$ we construct the associated Kohn-Sham potential:

$$v_s^{old}(\mathbf{r}) = v(\mathbf{r}) + \int d^3\mathbf{r}' \frac{n^{old}(\mathbf{r}')}{|\mathbf{r} - \mathbf{r}'|} + v_{xc}(\mathbf{r}; n^{old}). \qquad (63)$$

Then, the Schrödinger equation for the non-interacting electrons

$$-\frac{1}{2}\nabla^2 \phi_i^{new}(\mathbf{r}) + v_s^{old}(\mathbf{r})\phi_i^{new}(\mathbf{r}) = \epsilon_i^{new}\phi_i^{new}(\mathbf{r}) \qquad (64)$$

is solved. In order to do so, the orbitals are expanded in a basis set $\{\varphi_\alpha\}$

$$\phi_i(\mathbf{r}) = \sum_\alpha c_i^\alpha \varphi_\alpha(\mathbf{r}). \tag{65}$$

The Kohn-Sham Hamiltonian (41) for the non-interacting electrons is written in this basis set as a matrix $H_{\alpha\beta} = \langle \alpha | H_S | \beta \rangle$. Solving the Schrödinger equation in the basis set $\{\varphi_\alpha\}$ then amounts to diagonalizing the matrix $(H_{\alpha\beta})$:

$$H_{\alpha\beta} c_i^\beta = \epsilon_i c_i^\alpha, \tag{66}$$

where we have assumed for simplicity that the basis set $\{\varphi_\alpha\}$ is orthonormal (which is the case for plane waves, see later).

The diagonalization can be performed by explicit construction of the matrix $(H_{\alpha\beta})$ which is then diagonalized by standard methods when the basis set is not too large. For the case of large systems and/or large basis sets, we will prefer iterative techniques, like the Lanczos method [74, 143–145], which avoid the explicit construction of the Kohn-Sham matrix: it is sufficient in these methods to have a procedure to apply (successively) the Kohn-Sham matrix on vectors $\{c_i^\alpha\}$. Only vectors $\{c_i^\alpha\}$ need then to be stored.

Once we have solved the non-interacting electron problem in the external potential $v_s^{old}(\mathbf{r})$ to obtain one electron orbitals $\{\phi_i^{new}(\mathbf{r})\}$ (expanded on the basis set $\{\varphi_\alpha\}$) we construct the ground state density of the non-interacting electron system simply as

$$n^{new}(\mathbf{r}) = \sum_{i=1}^N |\phi_i^{new}(\mathbf{r})|^2. \tag{67}$$

Self consistency is reached if $n^{new}(\mathbf{r}) = n^{old}(\mathbf{r})$. If this is not the case, the procedure is iterated with a new guess density constructed from $n^{new}(\mathbf{r})$. In the most simple scheme for example, the new guess is $n^{new}(\mathbf{r})$ itself [146]. In practice the iteration is stopped when self consistency is only approximately achieved, when for example n^{new} and n^{old} satisfy $|n^{new}(\mathbf{r}) - n^{old}(\mathbf{r})| < \epsilon$ for all $\mathbf{r}$.

5.2 Constrained Minimization Technique

Here the point of view is somewhat different and instead of looking for solving the self-consistency equation (58) itself, we look for minimizing the Kohn-Sham energy functional. With a slight extension with respect to the previous section, (47), we will in fact look for minimizing the total energy:

$$E\left[\{\phi_i\}\right] = -\sum_i \langle \phi_i | \frac{1}{2} \nabla^2 | \phi_i \rangle + \int d^3\mathbf{r}\, n(\mathbf{r}) v(\mathbf{r}) + J[n] + E_{xc}[n] \tag{68}$$

where $n(\mathbf{r}) = \sum_{i=1}^N |\phi_i(\mathbf{r})|^2$, with respect to variations of $\{\phi_i\}$

$$E_0 = \min_{\{\phi_i\}} E\left[\{\phi_i\}\right] \tag{69}$$

and with orthonormality constraints: $\langle \phi_i | \phi_j \rangle = \delta_{ij}$. Minimizing with respect to variations of $\{\phi_i\}$ instead of minimizing with respect to variations of the electronic density $n(\mathbf{r})$ itself avoids the construction of the implicit functional $T_s[n]$ replacing it by the explicit kinetic energy term $-\sum_i \langle \phi_i | \frac{1}{2} \nabla^2 | \phi_i \rangle$. As we will see, both approaches are consistent and lead to the same minimal density $n^0(\mathbf{r})$ and energy E_0.

In order to perform this minimization under the orthonormality constraints, we use the Lagrange multiplier technique, thus minimizing

$$\tilde{E}\left[\{\phi_i\}\right] = E\left[\{\phi_i\}\right] - \sum_{ij} \Lambda_{ij} \left(\langle \phi_i | \phi_j \rangle - \delta_{ij} \right) \tag{70}$$

where we have introduced the Lagrange multipliers Λ_{ij}, one for each constraint $(\Lambda_{ij} = \Lambda_{ji})$. At the minimum of $\tilde{E}$ we have

$$\frac{\partial E}{\partial \phi_i^*} = \sum_j \Lambda_{ij} \phi_j. \tag{71}$$

Evaluating $\dfrac{\partial E}{\partial \phi_i^*}$:

$$\frac{\partial E}{\partial \phi_i^*(\mathbf{r})} = -\frac{1}{2} \nabla^2 \phi_i(\mathbf{r}) + \int d^3 \mathbf{r}' \left(v(\mathbf{r}') + \frac{\partial J}{\partial n} + \frac{\partial E_{xc}}{\partial n} \right) \frac{\partial n(\mathbf{r}')}{\partial \phi_i^*(\mathbf{r})} \tag{72}$$

which involves the derivative $\frac{\partial n(\mathbf{r}')}{\partial \phi_i^*(\mathbf{r})}$ of the electronic density n with respect to variations of the orbital ϕ_i:

$$\frac{\partial n(\mathbf{r}')}{\partial \phi_i^*(\mathbf{r})} = \delta(\mathbf{r} - \mathbf{r}') \phi_i(\mathbf{r}) \tag{73}$$

we thus get

$$\frac{\partial E}{\partial \phi_i^*(\mathbf{r})} = \left[-\frac{1}{2} \nabla^2 + v_s(\mathbf{r}) \right] \phi_i(\mathbf{r}) \equiv \hat{H}_s \phi_i. \tag{74}$$

In order to minimize $\tilde{E}\left[\{\phi_i\}\right]$, the orbitals are once more expanded on a basis set $\{\varphi_\alpha\}$ and the energy functional is written as a function of the vectors $\{c_i^\alpha\}$. The minimization of $\tilde{E}\left[\{\phi_i\}\right]$ then amounts to minimizing a function in a high dimension space which can be performed by standard iterative methods like the conjugate gradient method [101, 147] or Direct Inversion in Iteractive Space method [148–150] (DIIS). These methods rely on the calculation of the gradient $\frac{\partial E}{\partial \phi_i^*(\mathbf{r})}$; here again one does not need to construct $\hat{H}_s$, as we only need to have a way to apply $\hat{H}_s$ on ϕ_i.

At the minimum of $\tilde{E}$ the set of orbitals $\{\phi_i\}$ then satisfy

$$\hat{H}_s \phi_i = \Lambda_{ij} \phi_j. \tag{75}$$

This appears at first sight different from the condition that the one-electron orbitals of the auxiliary non-interacting electron system diagonalize the Kohn-Sham Hamiltonian:

$$\hat{H}_s \tilde{\phi}_i = \epsilon_i \tilde{\phi}_i \ . \tag{76}$$

However the Λ_{ij} matrix can be diagonalized

$$\left(U^T \cdot \Lambda \cdot U \right)_{ij} = \epsilon_i \delta_{ij} \ , \tag{77}$$

where U_{ij} is a unitary matrix as Λ_{ij} is a symmetric matrix. It is easy to show that the one-electron orbitals

$$\tilde{\phi}_i = U_{ij}^T \phi_j \tag{78}$$

now diagonalize the Kohn-Sham Hamiltonian with orbital energies ϵ_i. This unitary transformation $\phi \rightarrow \tilde{\phi} = U^T \phi$ leaves both the density $n(\mathbf{r})$ and the kinetic energy $\sum_i \langle \phi_i | \frac{1}{2} \nabla^2 | \phi_i \rangle$ invariant, so that the total energy E is also invariant by this unitary transformation.[2] Finding the minimum of $E\left[\{ \phi_i \} \right]$ with respect to variations of $\{ \phi_i \}$ is thus strictly equivalent to solving the Kohn-Sham problem. The one-electron orbitals $\{ \tilde{\phi}_i \}$ which diagonalize the Kohn-Sham Hamiltonian are called the canonical Kohn-Sham orbitals of the system, while any unitary transformation $\{ \phi_i \}$ of these which also minimizes E are called minimal orbitals.

5.3 Two Expressions for the Ground State Energy

We have written the ground state total energy of the system as

$$E_0 = - \sum_i \langle \phi_i | \frac{1}{2} \nabla^2 | \phi_i \rangle + \int d^3 \mathbf{r} \, n(\mathbf{r}) v(\mathbf{r}) + J[n] + E_{xc}[n] \ , \tag{79}$$

however we note that the total energy of the auxiliary system of non-interacting electrons is given by the data of the Kohn-Sham energies ϵ_i:

$$E_0' \left[\{ \phi_i \} \right] = - \sum_i \langle \phi_i | \frac{1}{2} \nabla^2 | \phi_i \rangle + \int d^3 \mathbf{r} \, n(\mathbf{r}) v_s(\mathbf{r}) = \sum_i \epsilon_i. \tag{80}$$

We can thus write the total energy as

$$E_0 = \sum_i \epsilon_i - \int d^3 \mathbf{r} \, n(\mathbf{r}) v_s(\mathbf{r}) + \int d^3 \mathbf{r} \, n(\mathbf{r}) v(\mathbf{r}) + J[n] + E_{xc}[n] \ . \tag{81}$$

Making use of the self-consistency equation (58) we obtain the following alternative form for the ground state energy [121]:

---

[2] In fact, this unitary transformation leaves the whole system wavefunction of the auxiliary set of non-interacting electrons invariant as it is a single Slater determinant.

$$E_0 = \sum_i \epsilon_i - J[n] + E_{xc}[n] - \int d^3\mathbf{r}\, n(\mathbf{r})v_{xc}(\mathbf{r}) \, . \qquad (82)$$

In practice, when convergence is hard to achieve verifying the equality of these two expressions (79) and (82) of the total energy is a good check of self-consistency.

Finally, the alternative expression of the total energy (82) can be made an explicit functional of the density. Defining the energies $\epsilon_i[n]$ as the orbital energies of the one electron Schrödinger equation in the external potential given in an explicit form by the RHS of (58), they can be constructed by solving the non-interacting electron problem in that potential. It can be shown then that this functional is stationary with respect to variations of the electronic density at the ground state density. It is not ensured however that it is a minimum.

5.4 Plane Waves

As we have seen, in order to solve the Kohn-Sham problem, the one-electron orbitals $\{\phi_i\}$ are expanded on a basis set $\{\varphi_\alpha\}$:

$$\phi_i(\mathbf{r}) = \sum_\alpha c_i^\alpha \varphi_\alpha(\mathbf{r}) \, .$$

In this section we will quickly discuss the special case of plane wave basis sets, as they have been often used in first-principle Molecular Dynamics simulations.

In that case the basis set elements are defined as [111, 147]

$$\varphi_\alpha(\mathbf{r}) = \frac{1}{\sqrt{\Omega}} e^{-i\mathbf{G}_\alpha \cdot \mathbf{r}} \qquad (83)$$

where Ω is the volume of the simulation box. For an orthorhombic box with lengths L_x, L_y and L_z, the wavevectors $\mathbf{G}$ are

$$\mathbf{G} = i \cdot \frac{2\pi}{L_x} \cdot \mathbf{x} + j \cdot \frac{2\pi}{L_y} \cdot \mathbf{y} + k \cdot \frac{2\pi}{L_z} \cdot \mathbf{z} \, ; \text{ with } i, j, k \in \mathbb{Z}. \qquad (84)$$

The main advantage of a plane wave basis set, in view of Molecular Dynamics, is the independence of the basis set elements with respect to the ionic positions. [111] As a result, the Hellmann-Feynman theorem can be applied straightforwardly, without additional so-called Pulay terms arising from a basis set that would be dependent on the nuclei positions. The forces on the ions will be calculated at virtually no extra-cost. There is also no Basis Set Superposition Error for the same reasons. Another advantage of plane wave basis sets is that their quality depends only on the number of wave-vectors considered ("cutoff", see later); it is thus easier both to compare results and to make convergence studies with only one number defining the quality of the basis set. Finally, on the computational side, plane wave basis sets have the further advantage of being orthonormal. As a result also, they can not become over-complete (one element being exactly or even approximately a linear combination of other elements, leading to divergences).

However, plane wave basis sets also have disadvantages. The first one is probably the very large number of basis set elements which can range from a few 10000 to a few 10^6. To avoid this number of basis set elements to become even higher so that calculations would become untractable it is absolutely necessary to employ pseudopotentials: Only valence electrons are considered, not core electrons; it results from this that the electron-ion interactions are not simply the fundamental coulomb attraction.

Also, not being atom centered orbitals, plane wave basis sets do not easily lead to chemical insight on the electronic structure of the system studied: it is hard to describe the result of a plane wave calculation in a Linear Combination of Atomic Orbital (LCAO) framework, although it is the basis of many simplified, but qualitative, electronic structure models.[3] We will see later that tools have been designed to construct such chemical insight when using plane wave basis sets.

In such a discussion of pros and cons of plane wave basis sets, some features of them can be considered border-line, and whether they represent an advantage or a disadvantage depends on the application considered. First, plane wave basis set elements fill equally the whole simulation box. Thus, regions of high electronic densities near the atomic cores are described at the same level as regions of lower density like the valence region which leads as already mentioned to the necessary use of pseudo-potentials. When the system is an isolated cluster this turns out to be a further disadvantage as the vaccuum region around the cluster (the tails of the electronic density) is still described at the same level of accuracy. This is not such a problem when condensed matter system are considered, when no region of very low electronic density is present. This property of plane waves may on the other hand be an advantage in situations where the electron density presents some peculiarities: when an electron is localized in a cavity instead of around an atom for example. This is the case for a solvated electron [56]. Another example is given by a solvated silver atom in aqueous solution [57, 151]. By coupling between the highly polarisable atom and the strong fluctuation electric fields in water, a spontaneous dipole moment is created on the silver atom and the $5s$ electron was found not to be centered around the silver nucleus but some 0.5 Å away (see Fig. 1), thus giving to the atoms a dipole of about 2 Debye. Such a situation would be hard to accurately describe with atom centered orbitals unless specific sets of polarization orbitals are added to the basis set.

Another borderline feature of plane waves is that they are naturally periodic. This is obviously a good thing for describing crystals. This is true also for other condensed matter systems like liquids for which such periodic

---

[3] The reason why such a large number of plane wave basis elements is usually needed is that there has no chemical insight been put into them.

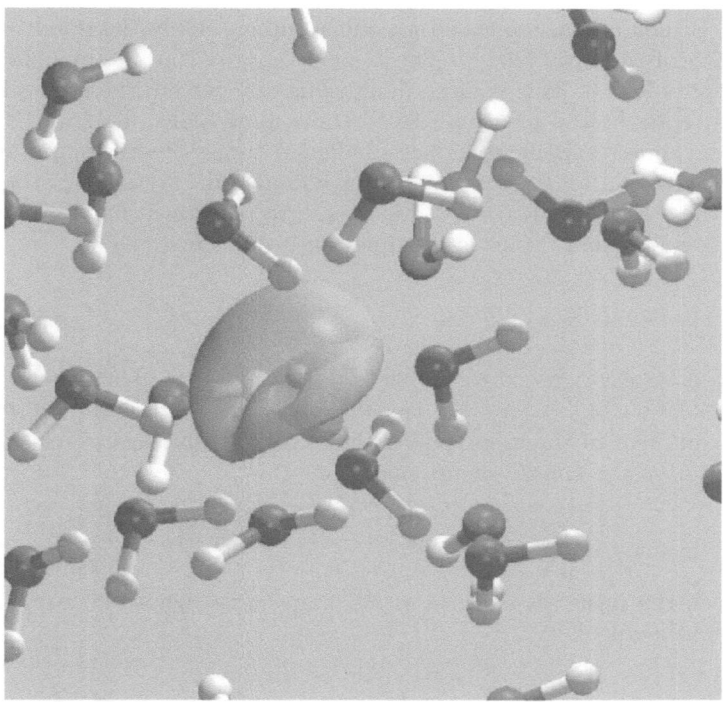

Fig. 1. Instantaneous configuration of a silver atom (*dark gray ball*) in liquid water. *Medium gray balls*: Oxygen atoms; white balls: hydrogen atoms. A isocontour of the $5s$ electron of the silver atom is shown in light gray. The center of that orbital is 0.5 Å away from the silver nucleus (for visualizing purposes this orbital center is displayed as a small light gray ball)

boundary conditions are often assumed in simulations. However, when the periodic boundary conditions do not apply, special care needs to be taken, in particular for the electrostatic interactions [152, 153].

When treating periodic systems, the orbital expansion given so far is incomplete in principle. Although, the charge density is necessarily periodic, there can be for the wavefunction itself a phase factor from one periodic image to the other. This is the essence of the Bloch theorem stipulating that orbitals can be written as

$$\phi_i(\mathbf{r}) = e^{i\mathbf{k}\mathbf{r}} u_i(\mathbf{r}) \tag{85}$$

where $u_i(\mathbf{r})$ is a periodic function and $\mathbf{k}$ is a vector belonging to the first Brillouin zone of the cristal and defines the phase factor between one unit cell and the neighbouring one. The function $u_i(\mathbf{r})$ being periodic can then be expanded on a plane wave basis set as described in the previous paragraphs. The charge density, the kinetic energy etc. are averaged over the $\mathbf{k}$ vectors of the first Brillouin zone, so-called k-point sampling. Only then do we recover extensivity of the total energy. The Γ point approximation amounts to considering only $\mathbf{k} = 0$. This is often done using a supercell: the unit cell of the cristal is reproduced a few times in all three dimensions, making the system bigger and thus the Brillouin zones smaller. Although more computationally expensive this is a sensitive approach for non-metallic systems. In the case of liquids, the system needs to be large enough to reproduce the disorder of the liquid state and the Γ point approximation is often justified.

Size of the Basis Set

The size of the basis set is determined [111] by a cut-off of the kinetic energy associated to $\varphi_{\mathbf{G}}(\mathbf{r}) = \frac{1}{\sqrt{\Omega}} e^{i\mathbf{G}\cdot\mathbf{r}}$ (we drop here the label α for the basis set elements in favor of the more explicit label $\mathbf{G}$). These basis set elements are eigenvectors of the kinetic energy operator:

$$-\frac{1}{2}\nabla^2 \varphi_{\mathbf{G}}(\mathbf{r}) = \frac{1}{2}\|\mathbf{G}\|^2 \varphi_{\mathbf{G}}(\mathbf{r}) . \tag{86}$$

In general, the basis set consists of all plane waves whose kinetic energy is below a given cut-off:

$$E_{kin} = \frac{1}{2}\|\mathbf{G}\|^2 < E_{cut} . \tag{87}$$

This means specifying the highest spatial frequency described by the basis set. The quality of the basis set depends only on this cutoff E_{cut} and, as already discussed, convergence checks are easily performed.

We can evaluate the number of basis set elements considered for a given cutoff. The $\mathbf{G}$-vectors considered for constructing the basis set are points on a regular mesh within a sphere of radius $G_0 = \sqrt{2E_{cut}}$, see Fig. 2. The regular mesh on which $\mathbf{G}$-vectors are situated is given by (84); a volume element for this mesh is thus $dV = \frac{(2\pi)^3}{L_x L_y L_z} = \frac{(2\pi)^3}{\Omega}$. The total number of $\mathbf{G}$-vectors considered can then be well approximated by the ratio of the volume of the sphere of radius G_0 and the volume element dV of the regular mesh of $\mathbf{G}$-vectors

$$\frac{(2\pi)^3}{\Omega} N_{PW} \simeq \frac{4}{3}\pi G_0^3 = \frac{4}{3}\pi 2^{\frac{3}{2}} E_{cut}^{\frac{3}{2}} \tag{88}$$

$$N_{PW} \simeq \frac{1}{2\pi^2} \frac{2^{\frac{3}{2}}}{3} \Omega E_{cut}^{\frac{3}{2}} \approx \frac{1}{2\pi^2} \Omega E_{cut}^{\frac{3}{2}} . \tag{89}$$

Apart from an expected dependence of the number of plane waves N_{PW} on the kinetic energy cutoff, we find that it also depends on the volume of the simulation cell Ω. These are important considerations when assessing the time and memory requirements of a simulation.

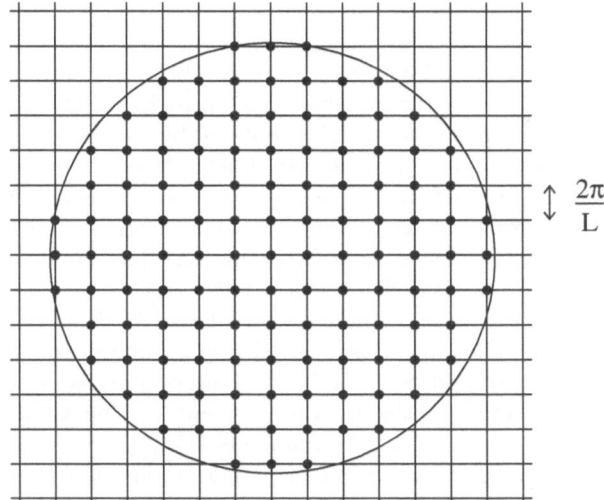

Fig. 2. 2 dimensional representation of the **G**-vectors considered in constructing the plane wave basis-set. The **G**-vectors considered (*black dots*) are vectors on a regular mesh of interspacing $\frac{2\pi}{L}$, within a disk of radius G_0 (see text)

Auxiliary Real Space Grid

The coefficients $c(\mathbf{G})$ of an orbital $\phi(\mathbf{r})$ can be obtained uniquely by discrete Fourier transform from the knowledge of $\phi(\mathbf{R}_i)$ on a regular grid $\{\mathbf{R}_i\}_{i=1\ldots N}$ in real space [111]:

$$c(\mathbf{G}) = \frac{\sqrt{\Omega}}{N} \sum_{i=1}^{N} \phi(\mathbf{R}_i) e^{i\mathbf{G}\cdot\mathbf{R}_i} \tag{90}$$

with $\mathbf{R}_i = (X_i, Y_i, Z_i)$:

$$\begin{aligned}
X_i &= n\,\Delta, \quad n = 0\ldots N_x - 1 \\
Y_i &= m\,\Delta, \quad m = 0\ldots N_y - 1 \\
Z_i &= l\,\Delta, \quad l = 0\ldots N_z - 1
\end{aligned} \tag{91}$$

in which i designates the triplet (n, m, l) and $N = N_x N_y N_z$. It is necessary however that the norm of the largest **G**-vector, G_0, be below the Nyquist critical frequency $f_c = \frac{2\pi}{2\Delta}$ of the real space grid. The real space grid spacing Δ will then depend on the cutoff E_{cut} considered.

If N_x, N_y and N_z are products of small prime numbers, one can use Fast Fourier Transform (FFT) techniques "which are very efficient" to go from the real-space grid to the reciprocal-space grid and back. These algorithms indeed scale as $N \log N$ instead of N^2 and are thus heavily used in plane wave codes [97–103, 105, 106, 108].

Pseudopotentials

As already mentioned, pseudopotentials are needed when using plane wave basis sets in order to limit the number of plane waves. The bare coulomb attraction between electrons and nuclei is replaced by pseudopotentials for two main goals. First, only valence electrons are thus included in most calculations, as they are the ones involved in the chemistry of the system while the charge density of core electrons is very steep and would require a much too large cutoff to describe. The valence orbitals themselves have nodes in the core region to ensure orthogonality with the core electron orbitals; the second role of pseudopotentials is to replace the valence orbitals by smooth functions in the core region, while keeping a very good approximation of them outside the core region. It is required however that pseudopotentials are transferable, that is, that the pseudopotential for one element can be valid in different calculations, in different chemical environments [154].

A major step forward in the theory of pseudopotentials was the introduction of a "norm-conservation" condition with the proposition by Hamann, Schlüter and Chiang [155] of a set of conditions to ensure transferability of pseudopotentials. Different recipes for constructing pseudopotentials satisfying these conditions have been proposed. They can be either analytic [155,156] or numeric [157] and also differ by the way valence wave functions are made smooth in the core region.

For these norm-conserving pseudopotentials, a different potential needs to be applied on each orbital depending on its angular momentum. These pseudopotentials then have a semi-local form:

$$V^{ps}(\mathbf{r}, \mathbf{r}') = \sum_{lm} V_l^{ps}(r)\delta_{r,r'}|Y_{lm}(\mathbf{r})\rangle\langle Y_{lm}(\mathbf{r}')| ,   (92)$$

where $Y_{lm}(\mathbf{r})$ are the spherical hamonics. However, in practical implementations, a fully non-local form is often preferred for efficiency:

$$\hat{V}^{ps} = V_{local}(\hat{r}) + \sum_{kl} \hat{P}_k h_{kl} \hat{P}_l ,   (93)$$

where $V_{local}(\hat{r})$ is an isotropic local potential and $\hat{P}_l$ projectors constructed from atomic orbitals (h_{kl} are parameters, real numbers). There exist methods for going from a semi-local to a fully non-local form [158,159].

Recently, a new kind of analytic pseudo-potentials, directly in a fully non-local form, has been proposed [129,160]. Coefficients are fitted to minimize penalty functions, like atomic properties, to ensure that these properties are well reproduced for the pseudoatom. It has been further generalized for use in the context of QM/MM situations [161] or to include semi-empirically the long-range van der Waals attraction [161,162].

Much lower cutoffs can be achieved with ultrasoft pseudo-potentials, relaxing the norm-conservation condition [163–167]. An augmentation charge to compensate the missing electronic density is then needed. Alternatively, so called augmented-wave methods have been proposed which divide the wavefunction in two parts, an inner core region described by linear superposition of atom-centered orbitals and an outer part described with plane waves [159,168].

6 First-principle Molecular Dynamics

6.1 Born-Oppenheimer Molecular Dynamics

Our primary goal was the simulation of entire atomic systems, thus made of electrons and nuclei. As mentioned earlier (see Sect. 2.3), in a large class of systems (e.g. not too high temperature) one can decouple the motion of nuclei and electrons within the Born-Oppenheimer approximation. The previous section was then devoted to the Density Functional Theory solution of the electronic structure problem at fixed ionic positions. By computing the Hellmann-Feynman forces (11) we can now propagate the dynamics of an ensemble of (classical) nuclei as described in Sect. 2.3, using e.g. the velocity verlet algorithm [117].

Such a simulation should then conserve the total energy

$$E_{tot} = \frac{1}{2} \sum_I M_I \dot{R}_I^2(t) + E_0 \left(\{R_I(t)\} \right) \tag{94}$$

where $E_0 \left(\{R_I(t)\} \right)$ is now the total energy of the system at the minimum of the energy density functional, at fixed ionic positions $\{R_I(t)\}$.

For a large enough system, for which we can assume molecular chaos, the configurations generated by a Born-Oppenheimer Molecular Dynamics should then be representative of the microcanonical distribution:

$$f_{MC} \left(\{P_I\} ; \{R_I\} \right) \propto \delta \left(E - \frac{1}{2} \sum_I \frac{P_I^2}{M_I} - E_0 \left(\{R_I\} \right) \right). \tag{95}$$

Similarly, if thermostating (e.g. using a Nosé-Hoover thermostat [169,170]) is performed while integrating the equations of motions for the nuclei, this will generate the canonical distribution

$$f_C \left(\{P_I\} ; \{R_I\} \right) \propto \exp \left[-\beta \left(\frac{1}{2} \sum_I \frac{P_I^2}{M_I} + E_0 \left(\{R_I\} \right) \right) \right], \tag{96}$$

where $\beta = \frac{1}{k_B T}$, with T the thermostat temperature. The reader should refer to the chapter by J. Abrams et al. in this book for further discussion along this line [171].

However, for the Hellman-Feynman theorem, (11), to be valid, the electronic density should be the one minimizing the total energy density functional (see [121] for a discussion of the Hellmann-Feynamn theorem in the framework of DFT). Numerical methods for minimization, as described in the previous section, only lead us to an approximate minimal Kohn-Sham density (or approximate minimal orbitals), dictated by the convergence criterium or the number of iterations performed during the minimization. Molecular Dynamics simulations using these forces will then only approximately conserve the total energy E_{tot} due to this intrinsic limitations, and very tight convergence criteria have to be used in Born-Oppenheimer MD, which then needs highly performant minimization algorithms [100, 101, 110, 172].

6.2 Car-Parrinello Molecular Dynamics

Principle

Car and Parrinello in their celebrated 1985 paper [2] proposed an alternative route for molecular simulations of electrons and nuclei altogether, in the framework of density functional theory. Their idea was to reintroduce the expansion coefficients $c_i(\mathbf{G})$ of the Kohn-Sham orbitals in the plane wave basis set, with respect to which the Kohn-Sham energy functional should be minimized, as degrees of freedom of the system. They then proposed an extended Car-Parrinello Lagrangian for the system, which has dependance on the fictitious degrees of freedom $c_i(\mathbf{G})$ and their time derivative $\dot{c}_i(\mathbf{G})$:

$$\mathcal{L}^{CP}(\mathbf{c}, \dot{\mathbf{c}}, \mathbf{R}, \dot{\mathbf{R}}) = \mu \sum_{i,\mathbf{G}} \dot{c}_i^*(\mathbf{G})\dot{c}_i(\mathbf{G}) + \frac{1}{2}\sum_I M_I \dot{\mathbf{R}}_i^2 - E(\mathbf{c}, \mathbf{R}) \qquad (97)$$

$$+ \sum_{ij} \Lambda_{ij}\left[\sum_{\mathbf{G}} c_i^*(\mathbf{G})c_j(\mathbf{G}) - \delta_{ij}\right]. \qquad (98)$$

In this last expression, μ is an additional, non-physical parameter, which represents the fictitious mass assigned to the additional degrees of freedom, $c_i(\mathbf{G})$'s, of the system. The potential energy of the system as a whole is $E(\mathbf{c}, \mathbf{R}) = E(\phi_i(\mathbf{c}), \mathbf{R})$, the electron+nuclei total energy functional in the Khon-Sham framework. Finally, Λ_{ij} are Lagrange multipliers introduced to satisfy at all times the orthonormality constraints of the Kohn-Sham orbitals.
This leads to the coupled set of equations of motion:

$$\mu\ddot{c}_i(\mathbf{G}) = -\frac{\partial E(\mathbf{c}, \mathbf{R})}{\partial c_i^*(\mathbf{G})} + \sum_j \Lambda_{ij}c_j(\mathbf{G}) \qquad (99)$$

$$M_I\ddot{\mathbf{R}}_I = -\frac{\partial E(\mathbf{c}, \mathbf{R})}{\partial \mathbf{R}_I} \qquad (100)$$

under the orthonormality constraints.
Furthermore

$$\frac{\partial E(\mathbf{c}, \mathbf{R})}{\partial c_i^*(\mathbf{G})} = \sum_{\mathbf{G}'} \langle \mathbf{G} | \mathbf{H}_{KS} | \mathbf{G}' \rangle c_i(\mathbf{G}') \tag{101}$$

$$\frac{\partial E(\mathbf{c}, \mathbf{R})}{\partial \mathbf{R}_I} = \int d^3 \mathbf{r} \, \frac{\partial v_{ext}(\mathbf{r})}{\partial \mathbf{R}_I} \sum_i |\phi_i(\mathbf{r})|^2 . \tag{102}$$

These equations then conserve the CP Hamiltonian

$$\mathcal{H}^{CP}(\mathbf{c}, \dot{\mathbf{c}}, \mathbf{R}, \dot{\mathbf{R}}) = \mu \sum_{i, \mathbf{G}} \dot{c}_i^*(\mathbf{G}) \dot{c}_i(\mathbf{G}) + \frac{1}{2} \sum_I M_I \dot{\mathbf{R}}_i^2 + E(\mathbf{c}, \mathbf{R}) . \tag{103}$$

To integrate these equations of motion [173, 174] one can use a velocity verlet algorithm [117] propagating similarly the nuclei positions and the electronic degrees of freedom $c_i(\mathbf{G})$. The orthonormality conditions can be taken into account by the SHAKE and RATTLE method [173, 175, 176]. Constant ionic temperature can also be imposed through the use of thermostats [169, 170, 173, 177].

We thus have defined a coupled dynamics to propagate in parallel electrons and nuclei in which electronic orbitals follow a fictitious dynamics. How can such electron-nuclei dynamics reproduce the dynamics of the physical system?

Let's consider once more what happens if the ions are kept fixed. The dynamics of the $c_i(\mathbf{G})$ occurs then on the potential energy surface $E(\mathbf{c}, \mathbf{R})$, that is, the Kohn-Sham energy functional. Under normal conditions, the expansion coefficients will oscillate around the minimum position $\mathbf{c}^0$, the minimal orbitals, corresponding to the Kohn-Sham energy E_0 of the system with ionic positions $\mathbf{R}$. In simulated annealing techniques, the minimum E_0 would simply be found by adding a friction term to the orbital dynamics so that it stops at the bottom of the energy functional. The frequency of the oscillations around $\mathbf{c}^0$ will depend on the fictitious mass μ and will be faster when μ is small. The fictitious electron mass will be choosen such that the dynamics of the orbital expansion coefficients $c_i(\mathbf{G})$ is fast compared to the ion dynamics.

Providing the dynamics of $c_i(\mathbf{G})$'s is then fast enough, the electrons respond nearly adiabatically to the change of the slowly varying potential $E(\mathbf{c}, \mathbf{R})$ acting on them, due to the "slow" ionic motion. By choosing a small enough fictitious mass μ of the electrons, to make their dynamics much faster than that of the nuclei, the orbitals will then be "approximately" minimal and should stay "close" to the Born-Oppenheimer surface, $\mathbf{c}^0(\mathbf{R})$.

These ideas have been given a precise mathematical meaning [178] and it has been shown [179, 180] that indeed as $\mu \to 0$ the nuclei trajectory under the CP equations of motion converges towards the Born-Oppenheimer trajectory and it stays in the vicinity of the latter for finite μ.

In the following we will discuss with some more details the conditions that the fictitious electronic mass should satisfy. Let's first spend some short time illustrating the Car-Parrinello method on a system of 32 water molecules to simulate liquid water at ambient conditions.

Example: 32 Water Molecules

With the development of GGA functionals, description of molecular systems with the Kohn-Sham method reached a precision similar to other quantum theory methods. It was quickly shown that the GGA's could also well reproduce the hydrogen bond properties. Short after, liquid water at ambient condition was first simulated by Car-Parrinello MD, with a sample of 32 water molecules with periodic boundary conditions [31]. Since then, many simulations of liquid water at different temperatures and pressures and of water solutions have been performed [32–39]. Nowadays, Car-Parrinello MD has become a major tool for the study of aqueous solutions [40–64].

Here we present a short simulation of a sample of 32 water molecules with periodic boundary conditions. The cubic box size is 9.865 Å in order to reproduce the experimental density of liquid water. The GGA functionals used were Becke88 and LYP (BLYP) [133, 134] and a plane wave basis set with a cutoff $E_{cutoff} = 70\,\mathrm{Ry}$ was employed to expand the Kohn-Sham orbitals, with Troullier-Martins pseudo potentials [157]. The timestep for integrating the CP equations of motions was $\delta t = 5\,\mathrm{A.U.}$ ($\delta t = 0.12\,\mathrm{fs}$), which was chosen consistently with a fictitious electronic mass $\mu = 500\,\mathrm{A.U.}$ This simulation was made with the CPMD package [97]. The instantaneous temperature of the system T_{ions} (or equivalently the ions kinetic energy K_{ions}), see Fig. 3, fluctuates around a mean temperature close to 280 K, while the instantaneous Kohn-Sham energy, E_{KS}, shows similar deviations from its mean value but opposite. Indeed, see Fig. 3-second to bottom, the classical hamiltonian $E_{class} = E_{KS} + K_{ions}$ exhibits fluctuations from its mean value an order of magnitude smaller than E_{KS}. These fluctuations though are still relatively large as E_{class} is not the conserved quantity of the CP equations of motions. Focusing now on the fictitious electronic degrees of freedom, we can see, Fig. 3-top, that, in that example, the fictitious kinetic energy, K_{el}, oscillates around 0.0165 A.U. (1.3×10^{-4} A.U./e). Only when added to the classical energy, E_{class} do we obtain a well conserved and stable CP hamiltonian, $E_{ham} = E_{class} + K_{el}$, see Fig. 3-bottom, with fluctuations nearly two orders of magnitude smaller.

Some Conditions on μ

We have stated earlier that the fictitious electronic mass μ should be such that the electronic dynamics is fast enough to follow adiabatically the ionic motion. We will try to give here some conditions for choosing μ [111, 178].

First, μ should be small enough so that the fictitious kinetic energy satisfies

$$\left\langle \frac{1}{2}\mu \dot{c}_i^2(\mathbf{G}) \right\rangle \ll \left\langle \frac{1}{2}M_I \dot{\mathbf{R}}_I^2 \right\rangle. \tag{104}$$

Otherwise stated, the fictitious kinetic energy should be a small term in the CP Hamiltonian.

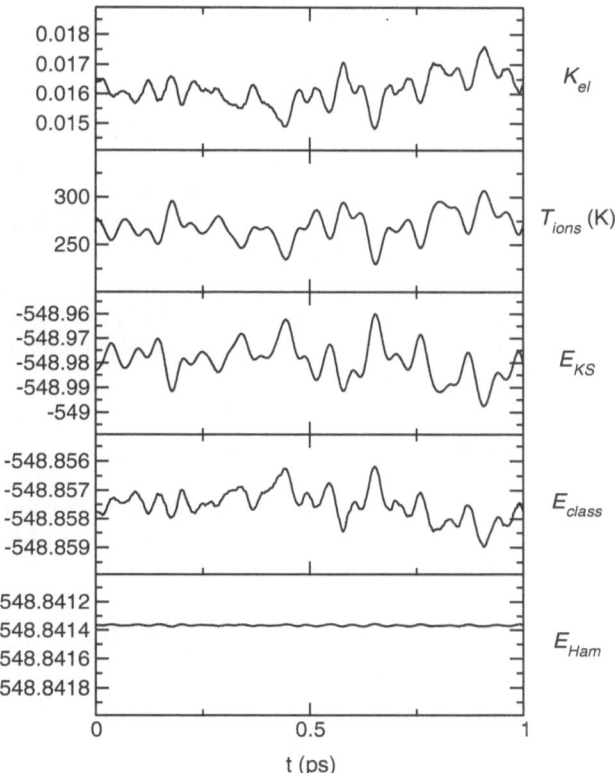

Fig. 3. Total energy and energy components for a system of 32 water molecules (simulations parameters: see text). *Top*: fictitious kinetic energy of the electrons (K_{el}), second from top: instantaneous ionic temperature, T_{ions} (proportional to the ions kinetic energy, K_{ions}), *middle*: instantaneous Kohn-Sham energy E_{KS}, second from bottom: classical hamiltonian $E_{class} = E_{KS} + K_{ions}$, *bottom*: CP hamiltonian, $E_{ham} = E_{class} + K_{el}$. Note the change of scale of the vertical axis from one frame to the other

We can also evaluate the lowest typical frequency for the electronic dynamics from the gap E_{gap}, energy difference between the Lowest Unoccupied Molecular Orbital (LUMO) and the Highest Occupied Molecular Orbital (HOMO), of the Kohn-Sham non-interacting electron system, which determines the lowest curvature of the $E(\mathbf{c})$ Kohn-Sham energy functional:

$$\omega_e^{min} \approx \left(\frac{E_{gap}}{\mu} \right)^{1/2} . \tag{105}$$

μ should then be small enough so that this frequency is much higher than the typical atom dynamics time-scales.

On the contrary, the highest typical frequency for the electronic dynamics is

$$\omega_e^{max} \approx \left(\frac{E_{cut}}{\mu}\right)^{1/2} . \tag{106}$$

This frequency should be the highest in the system dynamics and thus determines the timestep δt that one can use in conjunction with a given fictitious mass μ. As a result, the time step used in CP-MD is smaller than typical timesteps in empirical force fields MD or BO-MD. In the previous example, we have indeed used a timestep $\delta t = 0.12$ fs, while a typical timestep with empirical models of water would be around 0.5-1 fs. However, because of the square root dependence of ω_e^{max} on μ, dividing μ by 2 should lead to a timestep smaller only by a factor $\sqrt{2} \approx 1.4$.

Effect of the Fictitious Electronic Mass

Strictly speaking though, the CP dynamics tends towards the BO dynamics only in the $\mu \to 0$ limit. In order to get an idea of the effect of a finite μ, let's start by recalling the CP extended Lagrangian:

$$\begin{aligned}\mathcal{L}^{CP}(\{\psi\}, \{\dot{\psi}\}, \mathbf{R}, \dot{\mathbf{R}}) =& \mu \sum_i \langle \dot{\psi}_i | \dot{\psi}_i \rangle + \frac{1}{2} \sum_I M_I \dot{\mathbf{R}}_I^2 - E(\{\psi\}, \mathbf{R}) \\ &+ \sum_{ij} \Lambda_{ij} \left[\langle \psi_i | \psi_j \rangle - \delta_{ij} \right] ,\end{aligned} \tag{107}$$

where for commodity reasons, the orbital coefficients $c_i(\mathbf{G})$ have been represented by the orbital $|\psi_i\rangle$ they expand to. If we have chosen the fictitious mass small enough we can approximate the orbitals $|\psi_i\rangle$ by the minimal orbitals $|\psi_i\rangle \approx |\psi_i^0(\mathbf{R})\rangle$ minimizing the Kohn-Sham energy functional, leading also to $E(\psi, \mathbf{R}) \approx E_0(\mathbf{R})$. Then

$$\frac{d}{dt}|\psi_i^0(\mathbf{R})\rangle = \left|\frac{\partial \psi_i^0}{\partial \mathbf{R}_I}\right\rangle \dot{\mathbf{R}}_I \tag{108}$$

thus the CP Lagrangian can be approximated by

$$\mathcal{L}^{CP} \approx \mu \left\langle \frac{\partial \psi_i^0}{\partial \mathbf{R}_I} \middle| \frac{\partial \psi_i^0}{\partial \mathbf{R}_I} \right\rangle \dot{\mathbf{R}}_I^2 + \frac{1}{2} \sum_I M_I \dot{\mathbf{R}}_I^2 - E_0(\mathbf{R}) , \tag{109}$$

in which the electronic degrees of freedom have been reabsorbed. We observe in that expression a renormalization of the atomic masses

$$M_\mu = \left(M + \mu \left\langle \frac{\partial \psi}{\partial \mathbf{R}} \middle| \frac{\partial \psi}{\partial \mathbf{R}} \right\rangle \right) \tag{110}$$

coming from the fictitious electronic kinetic energy. A more rigorous demonstration of this mass renormalization effect, in a controlled situation where the system is composed of ions around which the electronic density moves solidly, can be found in [181].

It is very interesting to note that at first order the effect of the fictitious mass μ is a renormalization of atomic masses. If we consider a canonical ensemble at temperature T, this effect should not alter the static properties of the system which depend only on the potential energy surface $E_0(\mathbf{R})$. However, the temperature of the system is usually defined from the kinetic energy of the ions:

$$K_{ions} = \frac{1}{2} \sum_I M_I \dot{\mathbf{R}}_I^2 \tag{111}$$

$$T_{ions} = \frac{K_{ions}}{\frac{3}{2} N k_B}. \tag{112}$$

The effective kinetic energy of the ions is modified by the fictitious electronic mass μ and this leads to uncertainties in the temperature of the system. This could be circumvented by defining the temperature as the configurational temperature [182] but this is too expensive in ab initio calculations and has not been tried so far. At a more empirical level, appropriate corrections to the atomic masses can be applied [183]. By satisfying the first condition given in the previous paragraph we can make the uncertainties for the temperature determination small.

If we are interested in dynamical quantities on the other side (vibrational density of states, infrared or Raman spectra), these quantities will then depend on the fictitious mass μ through the atomic mass renormalization. This dependence of the typical ionic frequency on μ will be linear at small μ. Thus, for extracting vibrational frequencies from Car-Parrinello MD, the criterium for choosing μ should certainly be tighter than described in the previous paragraph.

Advantages and Disadvantages of CP-MD with Respect to BO-MD

Having presented the Car-Parrinello method and discussed some of its properties, we can now try to assess its advantages and disadvantages with respect to the Born-Oppenheimer MD.

Probably the main drawback of CP-MD is that we are always a little bit off the Born-Oppenheimer surface. We have seen in the previous paragraphs that this can be well controlled. There are some cases however where caution is needed, this is typically the case for electron transfer reactions where there is a crossing between two electronic states, see Fig. 4. When the electronic gap closes in such situations CP-MD may fail to follow the adiabatic state as inertia is favorable for the system to stay in the same diabatic state.

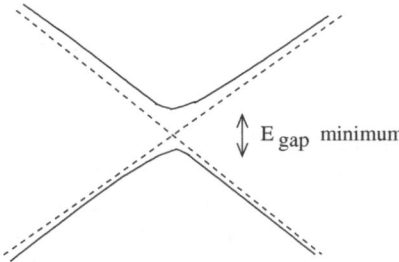

Fig. 4. Schematic representation of a correlation diagram in the case of a level crossing: two diabatic states (*dashed line*) cross for some value of the control parameter and the gap at the crossing point is due to a small coupling between them. The resulting adiabatic states are displayed as solid lines: only a very small gap results from the avoided-crossing

The main advantage of CP-MD however is the exact conservation of the CP Hamiltonian which leads to a stable dynamics. While thermostating is often needed in BO-MD to avoid heating up of the system, micro-canonical simulations over long times can be performed with CP-MD. This stability ensures that the system stays in the same thermodynamics state.

There is finally an aspect of CP-MD which is borderline. There is no self-consistency loop needed to minimize the Kohn-Sham energy functional at each time-step, as the electronic degrees of freedom are propagated alongside the nuclei. On the other hand, the timestep needed for integrating the CP motion is determined by the fastest electronic motion and is a few times shorter in BO-MD than in CP-MD. As this is also guided by the electronic gap of the system, through the choice of μ, CP-MD will be favored for isolating molecular systems, like water, while conducting, metallic systems where the gap closes (see the discussion above) impose the use of Born-Oppenheimer MD. The basis set used is another element for choosing between CP and BO. For plane wave basis set, the parameter space for the minimization is very large and it is certainly advantageous to propagate all parameters at the same time. Furthermore, the calculation of forces in plane wave basis set is very efficient due to the absence of Pulay terms, and calculating them more often (at each CP timestep instead of at the end of a minimization) comes at no extra cost. For atomic basis set, like gaussian basis sets this may not be true. Indeed, new first-principle Molecular Dynamics implementations [38, 109, 110, 172] based on the use of gaussian basis sets propagate the ionic motions through a Born-Oppenheimer Molecular Dynamics, while Car-Parrinello MD should stay the method of choice for plane wave basis set simulations.

Other Applications of the Extended Lagrangian Technique

The extended Lagrangian technique on which the Car-Parrinello method is based can be used also in other contexts. Whenever the forces on some atoms along an MD simulation have to be calculated through a minimization procedure in a large parameter space, the variables in this space can be reintroduced as extended degrees of freedom of the atomic system and propagated alongside the atoms with a fictitious mass. This has been applied in the context of an empirical polarisable model of water by Sprik et al. [184,185], where point dipoles on the water molecules are given by minimization of the electrostatic energy of the system. Recently, Marchi et al. [186] used an extended Lagrangian to implement a Molecular Dynamics simulation of a system surrounded by an implicit solvent, not described at the molecular level but through a statistical version of the Density Functional Theory. It has also been applied by us for studying the quantum ground state of a proton in a very strong hydrogen-bond [187]. A final example of the extended Lagrangian is also given in this book by Abrams et al. in their implementation of a dynamics in parameter space [171].

7 Observables and Properties Calculation

From the electronic structure of the system determined on the fly during the simulation, we have so far extracted only the total energy and the Hellmann-Feynman forces on the nuclei to generate MD trajectories. More information can be gained from the electronic structure of the system, allowing for direct comparison with experimental data or for interpreting the structure in chemical terms. This is an advantage of first-principle MD with respect to empirical force fields MD, in which the model for the forces may not give a proper descriptions of observables and a specific model may be needed for simulating some experimental quantity. This is the case for example when trying to calculate the Raman spectrum of a liquid from empirical force field MD, which in general does not account for polarizability.

However, in Density Functional Theory, instead of manipulating the full electronic ground state wavefunction, we have access only to the electronic ground state density, or the minimal Kohn-Sham orbitals. How can we calculate general observables in this framework?

As we have already underlined the Hohenberg-Kohn theorem means that knowledge of $n(\mathbf{r})$ gives knowledge of $v(\mathbf{r})$ and of the ground state wavefunction Ψ_0. Thus all properties of the electronic system and in particular any expectation value of an observable $O = \langle \Psi | \hat{O} | \Psi \rangle$, are functionals of $n(\mathbf{r})$ ($O[n]$ for the observable $\hat{O}$). This functional is straightforward when $\hat{O}$ is a single particle operator depending only of the position operator $\hat{\mathbf{r}}$: $\hat{O} \equiv O(\hat{\mathbf{r}})$ then

$$O = \langle \Psi | \hat{O} | \Psi \rangle = \int d^3\mathbf{r}\, n(\mathbf{r}) O(\mathbf{r}). \tag{113}$$

This is the case for example for the dipole moment, $\hat{M} \equiv -e \cdot \hat{r}$, the electrostatic potential at a point $\mathbf{R}$, $\hat{V} \equiv \frac{-e}{|\hat{r}-\mathbf{R}|} \cdots$

7.1 Position Operator in Periodic Boundary Conditions

A problem arises already with the position operator $\hat{\mathbf{r}}$ in Periodic Boundary Conditions (PBC). Indeed, the position operator is ill-defined in that case. This can be illustrated as follows. In PBC, all wavefunctions are periodic $\psi(\mathbf{r}) = \psi(\mathbf{r} + \mathbf{L})$ and so should be the result of applying an operator $\hat{\mathcal{O}}$ on ψ

$$\hat{\mathcal{O}}\psi(\mathbf{r}) = \phi(\mathbf{r}) = \phi(\mathbf{r} + \mathbf{L}). \tag{114}$$

This is not true however for the position operator

$$\mathbf{r}\,\psi(\mathbf{r}) \neq (\mathbf{r} + \mathbf{L})\psi(\mathbf{r} + \mathbf{L}). \tag{115}$$

Resta's Solution

This problem was solved by Resta [188, 189] who proposed, using the Berry phase approach [190–192], to define the expectation value of the position operator in PBC as (here for the x axis only):

$$\langle x \rangle = \frac{L}{2\pi}\Im\ln\langle\Psi|e^{i\frac{2\pi}{L}\hat{x}}|\Psi\rangle. \tag{116}$$

Note that $\langle x \rangle$ is then defined modulo L.

$\langle\Psi|e^{i\frac{2\pi}{L}\hat{x}}|\Psi\rangle$ is a many-body quantity as $|\Psi\rangle$ is the all-electron wavefunction and the operator $e^{i\frac{2\pi}{L}x}$ (which is now periodic as we defined it in the last paragraph) is not the sum of one-electron operators, but the product. In the Kohn-Sham method we will replace $|\Psi\rangle$ by the wavefunction of the auxiliary non-interacting electron system $|\Psi_s\rangle$ which is a single Slater determinant constructed from the minimal Kohn-Sham orbitals:

$$\Psi_s = \frac{1}{\sqrt{N!}}\det[\phi_1\phi_2\ldots\phi_N]. \tag{117}$$

In that case the expectation value of the position operator takes the form:

$$\langle x \rangle = \frac{L}{2\pi}\Im\ln\det\langle\phi_i|e^{i\frac{2\pi}{L}\hat{x}}|\phi_j\rangle. \tag{118}$$

It should be noted also that in principle the electronic polarization (dipole per unit length) is defined from the expectation value of x in the limit of a large system:

$$P_{el} = \lim_{L\to\infty}\frac{e}{2\pi}\Im\ln\langle\Psi|e^{i\frac{2\pi}{L}x}|\Psi\rangle. \tag{119}$$

Resta then showed that

$$P_{el}(t) - P_{el}(0) = \int_0^t dt'\,j(t') \tag{120}$$

for a quasi-static transformation, where j is the charge current. This justifies the definition taken for the polarization. A question was raised during the school in Erice on whether it would not be then more practicable to compute the charge current j, well defined in PBC, instead of P_{el}. The charge density however is a sum of one-body operators but is not a function of the position operator $\hat{r}$, it is a function of the momentum operator $\hat{p}$. The evaluation of such an operator in the course of a quasistatic transformation is non trivial and requires about the same effort as a response calculation (see later).

Resta's solution for the expectation value of the position operator in PBC was a big step forward. Not only did it allow for the computation of the polarization of electronic systems in PBC (e.g. ferroelectric crystals) but it opens the road for computation of infrared spectroscopy [193–195], polarizability [196, 197] or the simulation of periodic systems embedded in a permanent electric field [198–200].

We will illustrate briefly the use of (118) for the calculation of the infrared spectrum of liquid water.

Example: Infrared Spectrum of Water

From statistical linear response theory, the infrared spectrum of a sample at temperature T is[4]

$$I(\omega) = \omega \epsilon''(\omega) = \frac{\beta \omega^2}{2} \int_{-\infty}^{\infty} e^{i\omega t} \langle M(0)M(t) \rangle_0 dt \qquad (121)$$

where $\langle M(0)M(t) \rangle_0$ is the autocorrelation function of the total dipole moment of the system at equilibrium. For the system composed of 32 water molecules at 280 K (see earlier in the text) we have computed the dipole moment using Resta's formula [188], assuming that we could consider it to be close to the $L \to \infty$ limit. A typical time evolution of the x component of the dipole moment is shown Fig. 5. The dipole moment displays fluctuations at different time-scales, it is precisely the goal of IR spectroscopy to identify these characteristic times of the dynamics. By calculating the autocorrelation function of the x component of the dipole moment, see Fig. 6, some of these timescales, of the order of 100 fs, appear clearly. However the quantity directly

---

[4] An issue regarding the calculation of infrared spectra from MD simulations are the so-called quantum-classical corrections as nuclei are assumed classical in the simulations but quantum effects can be noticeable at room temperature. The quantum correction implied by (121) has been shown to be the most accurate one [201, 202]. It simply assumes that the quantum response function can be well approximated by the classical one (while this is not true for the correlation functions themselves). There is something interesting in the history of quantum-classical corrections. Being in great part empirical, choice of the best correction is done through comparison of the obtained data with experimental data, in particular on the peak intensities. As empirical force fields simulations give wrong

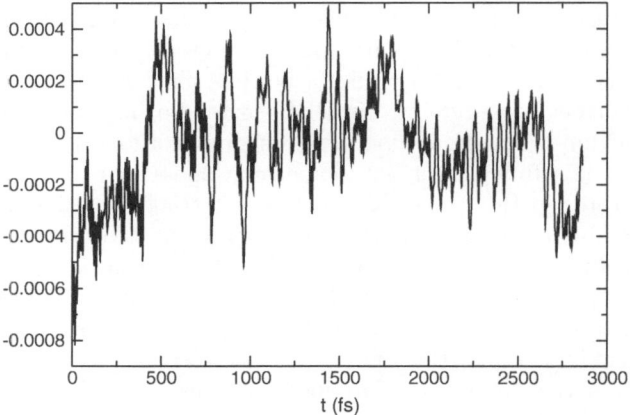

Fig. 5. Typical time evolution of the x component of the dipole moment of a sample of 32 water molecules at ambient temperature

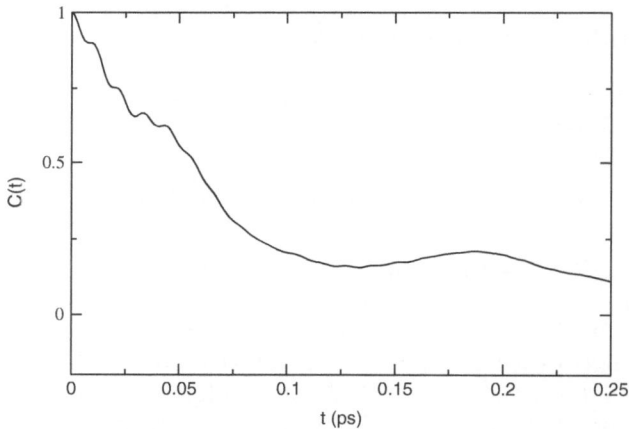

Fig. 6. Autocorrelation function of the x component of the dipole moment of a sample of 32 water molecules at ambient temperature

comparable to experimental data is the Fourier transformed of $\langle M(0)M(t)\rangle_0$ as expressed by (121). The infrared spectrum computed from CP-MD of 32 water molecules, similar to the one of [193], is displayed in Fig. 7. The overall shape of the IR spectrum is well reproduced with respect to experimental data [203]. The too low frequency of the OH stretch peak above 3000 cm^{-1} in the CP-MD data is due about equally to the defects of the density functional and to the fictitious electronic mass (see Sect. 6.2). The main advantage over

---

IR intensities by lack of polarization effects, this has lead to a different correction than the one now accepted for first-principle MD.

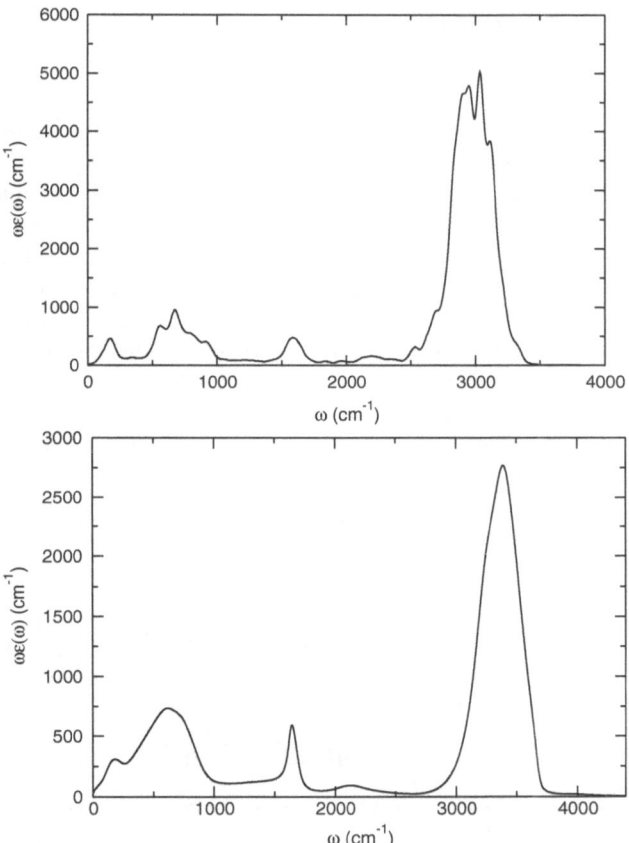

Fig. 7. Infrared spectrum of liquid water from CP-MD simulation (*top*) and from experiment (*bottom*)

empirical force field calculations is that the intensities are much better reproduced in CP-MD: The overall aspect of the spectrum with low intensity below 3000 cm^{-1} and a high intensity above is not reproduced by empirical force fields in which these two regions have about equal intensity. Note also that the observed band at 200 cm^{-1} has no intensity in empirical force fields calculations [29]. Many applications of CP-MD for calculating IR spectra can now be found [194, 195, 202]. For more complex molecules in solution, like NMA in liquid water CP-MD shows a dramatic improvment with respect to empirical force fields [195].

7.2 Response Properties

Experimental Properties as Response Quantities

It may appear as an important limitation of DFT that one class only of observables expectation value (one electron operators depending only on position) can be calculated as explicit functionals of the density. Fortunately, most of the experimentally accessible quantities are not observable expectation values but response properties. For example, we are barely interested in the expectation value of $\hat{M}^2$, $\langle \Psi_0 | \hat{M}^2 | \Psi_0 \rangle$, but rather in the polarizability tensor $\bar{\bar{\alpha}}$.

The polarizability is a response coefficient being the derivative of the dipole with respect to infinitesimal variations of the electric field: $\alpha_{ij} = \frac{\partial M_i}{\partial \mathcal{E}_j}$. Although we did not stress it earlier, the dipole moment M_i can be also viewed as a response quantity: Adapting the Hellmann-Feynman theorem to application of infinitesimal electric field, it can be shown that

$$M = -\frac{\partial E_0}{\partial \mathcal{E}} = \langle \Psi_0 | \hat{M} | \Psi_0 \rangle \tag{122}$$

Thus,

$$\alpha_{ij} = \frac{\partial M_i}{\partial \mathcal{E}_j} = -\frac{\partial^2 E_0}{\partial \mathcal{E}_j \, \partial \mathcal{E}_i} \ . \tag{123}$$

Another response property is for example the Hessian of the system:

$$H_{IJ} = -\frac{\partial^2 E_0}{\partial \mathbf{R}_I \, \partial \mathbf{R}_J} = \frac{\partial \mathbf{F}_J}{\partial \mathbf{R}_I} \tag{124}$$

which can also be expressed alternatively as a second order derivative of the energy or as a first order reponse of the forces. Such quantities can then be calculated either by finite difference or by Density Functional Perturbation Theory [204, 205] (DFPT). In the following we will discuss the latter.

Density Functional Perturbation Theory

Writing the external potential $v(\mathbf{r})$ as a zero'th order term plus a perturbation depending on a parameter λ [204, 205]:

$$v(\mathbf{r}) \to v^{(0)}(\mathbf{r}) + \lambda v^{(1)}(\mathbf{r}) \tag{125}$$

we can expand the charge density $n(\mathbf{r})$ or else the minimal Kohn-Sham orbitals $\{\phi\}$ in a power series of λ:

$$n(\mathbf{r}) = n^{(0)}(\mathbf{r}) + \lambda n^{(1)}(\mathbf{r}) + \lambda^2 n^{(2)}(\mathbf{r}) \ldots \tag{126}$$

$$\phi_i(\mathbf{r}) = \phi_i^{(0)}(\mathbf{r}) + \lambda \phi_i^{(1)}(\mathbf{r}) + \lambda^2 \phi_i^{(2)}(\mathbf{r}) \ldots \tag{127}$$

If $v^{(1)}(\mathbf{r}) = -e\mathcal{E} \cdot \mathbf{r}$ represents the interaction with a constant electric field, the polarizability $\bar{\bar{\alpha}}$ will be calculated from $n^{(1)}(\mathbf{r}) = \frac{\partial n(\mathbf{r})}{\partial \mathcal{E}_j}$ as

$$\alpha_{ij} = \frac{\partial M_i}{\partial \mathcal{E}_j} = \frac{\partial}{\partial \mathcal{E}_j} \int d^3\mathbf{r}\, r_i\, n(\mathbf{r}) = \int d^3\mathbf{r}\, r_i\, n^{(1)}(\mathbf{r}). \qquad (128)$$

A very important aspect of DFPT is the extension to perturbations of the variational principle. In a given external potential $v(\mathbf{r})$ the charge density and Kohn-Sham orbitals are obtained by minimizing the functional (see Sect. 4)

$$F = E - \Lambda_{ij}\left(\langle\phi_i|\phi_j\rangle - \delta_{ij}\right) \qquad (129)$$

$$= -\sum_i \langle\phi_i|\frac{1}{2}\nabla^2|\phi_i\rangle + \int d^3\mathbf{r}\, n(\mathbf{r})v(\mathbf{r}) + J[n] + E_{xc}[n] - \Lambda_{ij}\left(\langle\phi_i|\phi_j\rangle - \delta_{ij}\right).$$

This functional can be expanded in a power series of λ:

$$F = F^{(0)}\left(\left\{\phi_i^{(0)}\right\}\right) + \lambda F^{(1)}\left(\left\{\phi_i^{(0)}\right\}\right) + \lambda^2 F^{(2)}\left(\left\{\phi_i^{(0)}\right\};\left\{\phi_i^{(1)}\right\}\right)\cdots \quad (130)$$

Expanding to first order the orthonormality conditions

$$\langle\phi_i^{(0)}|\phi_i^{(1)}\rangle + \langle\phi_i^{(1)}|\phi_i^{(0)}\rangle = 0 \qquad (131)$$

it can be shown that $F^{(2)}$

$$F^{(2)}\left(\left\{\phi_i^{(0)}\right\};\left\{\phi_i^{(1)}\right\}\right) = E^{(2)}\left(\{\phi_i\}\right) - \left(\Lambda_{ij}\left(\langle\phi_i|\phi_j\rangle - \delta_{ij}\right)\right)^{(2)} \qquad (132)$$

the second order term of F depends only on the zeroth order $\{\phi_i^{(0)}\}$ and first order $\{\phi_i^{(1)}\}$ KS orbitals but not on the second order orbitals $\{\phi_i^{(2)}\}$ [204,205]. Furthermore, the first order KS orbitals minimize $F^{(2)}(\{\phi_i^{(0)}\};\{\phi_i^{(1)}\})$.

More generally, a $2n + 1$ theorem can be shown [204, 205] which states that the $2n$'th and $2n + 1$'th order terms in the expansion of F depend only on term up to order n of the expansion of the orbitals. $\{\phi_i^{(n)}\}$ also minimizes $F^{(2n)}$, given $\{\phi_i^{(0)}\}\ldots\{\phi_i^{(n-1)}\}$.

Returning to the first order response of the density, we take the second order in the perturbation λ and minimize with respect to $\{\phi_i^{(1)}\}$. We obtain that $\{\phi_i^{(1)}\}$ satisfies the Sternheimer equation [196, 204, 205]:

$$\left(H^{(0)}\delta_{ij} - \Lambda_{ij}^{(0)}\right)\phi_j^{(1)} = \left(H^{(1)}\delta_{ij} - \Lambda_{ij}^{(1)}\right)\phi_j^{(0)}. \qquad (133)$$

This is similar to the usual perturbation theory in quantum mechanics, in particular when solving the Sternheimer equation for $\phi_j^{(1)}$ as a sum over states. Note however that it is, in the case of DFPT, a self consistency equation, as $H^{(1)}$ depends on $n^{(1)}(\mathbf{r})$. As with standard DFT, two methods can be used for determining $\{\phi_i^{(1)}\}$: either by direct minimization of $F^{(2)}$ [196] or by successive resolution of the Sternheimer equation until self-consistency is reached.

Many quantities can be calculated using DFPT. Mentioning only applications in the context of first-principle MD, DFPT has been applied to the calculation of polarizability and Raman spectrum of ice [197], NMR chemical shifts in liquid water [206] or other systems [196, 207–209], chemical hardness [47], atomic polar tensors in liquid water [28] and so on. DFPT is a growing field for ab initio simulations.

7.3 Excited States – Time-Dependent Density Functional Theory

Density Functional Theory is a ground state theory in principle; this is due to the fact that the variational principle is an essential element of DFT. Despite this, determination of excited state energies and properties is still possible in the framework of DFT. There are two main approaches to this problem.

For the lowest excited *singlet* state Frank et al. [210] introduced a Restricted Open-shell Kohn-Sham method (ROKS). The wavefunction of non degenerate single excited states of closed-shell systems can not be described with a single determinant but a minimum of two Slater determinant are necessary. Frank et al. constructed a spin symmetry adapted wavefunction *à la Roothman*, based on the triplet and mixed spin first excited states of the non-interacting electron system, combined with Kohn-Sham Density Functional Theory to formulate an expression of the total energy. ROKS was designed from the start for Molecular Dynamics simulations in the S^1 electronic excited state and was used by Frank et al. to simulate photoisomerization [210].

A totally different point of view is proposed by Time-Dependent Density Functional Theory [211–215] (TD-DFT). This important extension of DFT is based on the Runge-Gross theorem [216]. It extends the Hohenberg-Kohn theorem to time-dependent situations and states that there is a one to one map between the time-dependent external potential $v_{ext}(\mathbf{r}, t)$ and the time-dependent charge density $n(\mathbf{r}, t)$ (provided we know the system wavefunction at $t = -\infty$). Although it is linked to a stationary principle for the system action, its demonstration does not rely on any variational principle but on a step by step construction of the charge current.

In particular, the time-dependent KS orbitals satisfy the time-dependent Schrödinger equation for non-interacting electrons [216]

$$i\hbar \frac{\partial \phi_i}{\partial t} = -\frac{1}{2}\nabla^2 \phi + v_s(\mathbf{r}, t)\phi \tag{134}$$

where $v_s(t)$ is the time-dependent KS potential, different from the applied potential:

$$v_s(\mathbf{r}, t) = v_{ext}(\mathbf{r}, t) + v_H(\mathbf{r}, t; [n(\tau)]) + v_{xc}(\mathbf{r}, t; [n(\tau)]). \tag{135}$$

In the definition of the KS potential the Hartree-potential v_H and the exchange-correlation term are functionals of $n(\mathbf{r}, \tau)$, function of time and space. The Hartree potential however is defined as

$$v_H(\mathbf{r}, t; [n(\tau)]) = \int d^3\mathbf{r}' \frac{n(\mathbf{r}', t)}{|\mathbf{r} - \mathbf{r}'|} \tag{136}$$

and turns out to be local in time. On the contrary the exchange correlation time-dependent potential $v_{xc}(\mathbf{r}, t; [n(\tau)])$ is non-local in space *and time*. As for DFT, one has to make approximations to v_{xc} and the most common one is the Adiabatic Local Density Approximation(ALDA). ALDA assumes that

$v_{xc}(\mathbf{r}, t; [n(\tau)])$ is local in time and that we can use the ground-state exchange correlation functional (LDA or GGA) evaluated for $n(\mathbf{r}')$ at time t to determine $v_{xc}(\mathbf{r})$.

TD-DFT extends ground state DFT in three aspects [217]:

- Its first use is perhaps the simulation of electronic dynamics like atoms and molecules in strong electric fields for example [211–214]. This is also an approach to electronic spectroscopy, calculating the electronic response to applied time-dependent potentials. The time-dependant electronic susceptibility is indeed the time-dependent reponse to a Dirac like perturbation.
- Electronic spectroscopy is rather obtained from the linear response to small oscillating electric fields [218–220]. In this framework, the excited state energies are characterized by being the poles of the response function [221, 222]. This is the main approach to compute excited states energies from TD-DFT [59, 168, 222–227].
- TD-DFT can also be used to get better approximations to the *ground-state exchange-correlation energy* based on frequency dependent response functions. For example, the long-range dispersion $\frac{1}{R^6}$ term between two well-separated systems can be obtained from the frequency dependent susceptibility of the two systems [228, 229].

Finally, using response theory it is also possible to determine some excited states properties like the dipole moment, from response of the excited state energy to an applied constant electric field, or forces [222, 230, 231]. It is then possible to perform Molecular Dynamics simulations on electronic excited states surfaces, to describe the dynamics of photo-chemical reactions for example [210, 232].

Work is also done on trying to incorporate non-adiabatic effects in the coupled nuclei+electrons dynamics [233, 234] (using ROKS excited states).

7.4 Localized Orbitals

We have concentrated so far on the determination of experimental quantities within DFT. The calculated electronic structure can also be studied with the goal to get chemical insight from it and answer questions like: what is the bonding pattern? How are bonds formed and broken? Determination of localized orbitals has proved to be a useful tool in this direction.

As noted in Sect. 4, a unitary transformation $\phi \rightarrow \tilde{\phi} = U^T \phi$ leaves both the density $n(\mathbf{r})$ and the total energy invariant. Any unitary transformation of the Kohn-Sham orbitals is thus a valid set of orbitals. Canonical orbitals are a special set of such orbitals which diagonalize the Kohn-Sham Hamiltonian. Localized orbitals on the other hand are obtained by finding the unitary transformation U so as to optimize the expectation value of a *two electrons* operator Ω:

$$\langle \Omega \rangle = \sum_{i=1}^{N} \langle \phi_i(1)\phi_i(2)|\Omega|\phi_i(1)\phi_i(2)\rangle. \tag{137}$$

Different prescriptions for Ω have been given: Boys-Foster: Minimization of the spatial extend [235] of the orbital i

$$\Omega = |\mathbf{r}_1 - \mathbf{r}_2|^2$$

Edminton-Ruedenberg: Maximization of the Coulomb self-repulsion energy [236]

$$\Omega = \frac{1}{|\mathbf{r}_1 - \mathbf{r}_2|}$$

von Nissen: Maximization of the self-charge [237]

$$\Omega = \delta(\mathbf{r}_1 - \mathbf{r}_2).$$

Let's consider for a moment the Boys-Foster localization; we can rewrite $\langle \Omega \rangle$ as

$$\langle \Omega \rangle = \sum_i \langle \phi_i \phi_i | (\mathbf{r}_1 - \mathbf{r}_2)^2 | \phi_i \phi_i \rangle = \sum_i \langle \phi_i \phi_i | \mathbf{r}_1{}^2 + \mathbf{r}_2{}^2 - 2\mathbf{r}_1\mathbf{r}_2 | \phi_i \phi_i \rangle \quad (138)$$

$$= 2 \sum_i \left(\langle \phi_i | \mathbf{r}_1{}^2 | \phi_i \rangle - \langle \phi_i | \mathbf{r}_1 | \phi_i \rangle \langle \phi_i | \mathbf{r}_2 | \phi_i \rangle \right) \quad (139)$$

$$= 2 \sum_i \left(\langle \phi_i | \mathbf{r}_1{}^2 | \phi_i \rangle - \langle \phi_i | \mathbf{r}_1 | \phi_i \rangle^2 \right). \quad (140)$$

It now becomes apparent that this amounts to minimizing the sum of the spread of the orbitals, leading to localized orbitals.

It should be noted also that $\sum_i \langle \phi_i | \mathbf{r}_1{}^2 | \phi_i \rangle = \mathrm{Tr}\left[(\mathbf{r}_1{}^2)_{ij} \right]$ is invariant by unitary transformation and as a results miminizing $\Omega = |\mathbf{r}_1 - \mathbf{r}_2|^2$ amounts to maximizing the sum of squares of the dipole integrals

$$\langle \tilde{\Omega} \rangle = \sum_i \langle \phi_i | \mathbf{r}_1 | \phi_i \rangle^2. \quad (141)$$

In fact, further manipulation of $\langle \tilde{\Omega} \rangle$ leads to

$$\langle \tilde{\Omega} \rangle = \frac{1}{2N} \left[\sum_{i \neq j} \left(\langle \phi_i | \mathbf{r}_1 | \phi_i \rangle - \langle \phi_j | \mathbf{r}_1 | \phi_j \rangle \right)^2 + 2 \left(\sum_i \langle \phi_i | \mathbf{r}_1 | \phi_i \rangle \right)^2 \right] \quad (142)$$

and as $\sum_i \langle \phi_i | \mathbf{r}_1 | \phi_i \rangle$ is also invariant by a unitary transformation, we seek for maximizing the inter-center distances:

$$\langle \tilde{\Omega}' \rangle = \sum_{i \neq j} \left(\langle \phi_i | \mathbf{r}_1 | \phi_i \rangle - \langle \phi_j | \mathbf{r}_1 | \phi_j \rangle \right)^2. \quad (143)$$

In PBC, the spread of an orbital $\Omega = \langle \mathbf{r}^2 \rangle - \langle \mathbf{r} \rangle^2$ (Boys-Foster), whose sum we would like to minimize, is again ill-defined. A generalization of Resta's formula for the position operator expectation value leads to a definition of the spread in PBC as [238–241]

$$\Omega = \frac{-1}{(2\pi)^2} w_\alpha \ln|z_\alpha|^2 \; ; \; z_\alpha = \langle \Psi | e^{i\frac{2\pi}{L} r_\alpha} | \Psi \rangle, \tag{144}$$

where w_α are weights assigned to each axis x, y and z. Maximally localized Wannier orbitals are defined from the choice of the unitary transformation which minimizes the sum of spreads (144) of the orbitals.

Having now to deal with localized orbitals, they should have a well-defined center, referred to as the Wannier center in PBC and defined as [239, 240, 242]

$$r_\alpha^w = \frac{L}{2\pi} \Im \ln \langle \phi^w | e^{i\frac{2\pi}{L} r_\alpha} | \phi^w \rangle, \tag{145}$$

for each Wannier orbital ϕ^W.

Figure 8 displays the maximally localized Wannier orbitals computed for one isolated water molecule. We can straight away identify two OH bond orbitals and two lone pair orbitals. The Wannier centers for these orbitals are shown on Fig. 9; they form approximately a tetrahedron around the oxygen of the water molecules.

By analyzing the Wannier orbitals of a sample of 32 water molecules, Silvestrelli et al. [242] have determined the average molecular dipole moment of water molecules in the liquid phase. They have assigned all Wannier centers

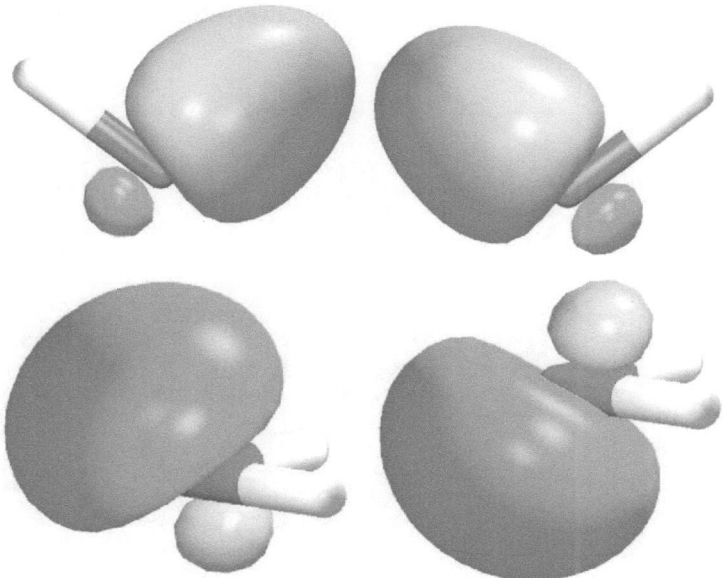

Fig. 8. Maximally localized Wannier orbitals of one water moelcule

Fig. 9. Wannier centers of one water molecule

to individual water molecules, see Fig. 10 for a typical configuration of nuclei and Wannier centers for a sample of 32 water molecules. For each molecule, a dipole moment is calculated which can be done simply by assigning a charge $-2e$ to each Wannier center (each orbital is doubly occupied) and their valence charge to the oxygens and hydrogens. Silvestrelli et al. [242] thus obtained an average molecular dipole moment of 3.0 Debye, compare to 1.8 Debye in the gas phase. This showed for the first time the enhancement of the water dipole due to the hydrogen bond network of liquid water.

We can now come back to the aqueous silver atom example [57, 151] mentioned in Sect. 5.4. The orbital displayed on Fig. 1 is one of the five maximally localized Wannier orbitals assigned to the silver atom and which has been

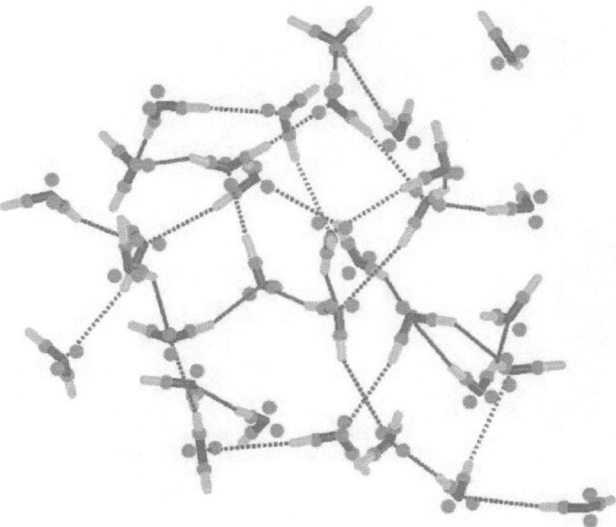

Fig. 10. Wannier centers of a snapshot of a simulation of 32 water molecules

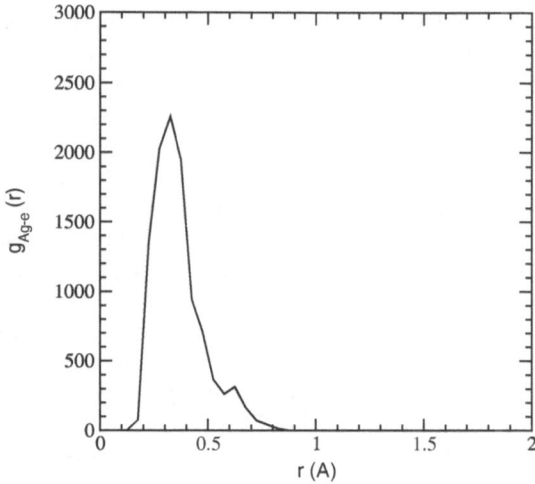

Fig. 11. Radial distribution function between the Wannier center of the $5s$ electron of a silver atom and the silver nucleus.

identified as the $5s$ electron. The light grey sphere on Fig. 1 represents its Wannier center. To show that this Wannier center is detached from the silver nucleus, due to the interaction with the fluctuating electric field of water, we have plotted the silver-$5s$ electron radial distribution function on Fig. 11. It can be seen that it has a maximum at 0.35 Å, with an average silver-$5s$ electron distance equal to 0.5 Å. The aqueous silver atom thus has a spontaneous dipole moment and is not spherical on average.

Localized orbitals have also been used as a tool to extract the infrared spectrum of a solute in solution [194,195,202] or to decompose the IR spetrum in intramolecular and intermolecular contributions [202]. Model electrostatics of solute molecules was also based on localized orbitals [242, 243], not only at the dipolar level [244]. As an extension we also defined molecular states from localized orbitals to study the electronic states of liquid water [245], or of solvated ions [47]. It is also possible to perform CP-MD propagating the Wannier orbitals, by constraining the Kohn-Sham orbitals to stay in a Wannier gauge [246].

Finally, Maximally localized orbitals have recently attracted a lot of attention, as they may be a route towards order N methods for electronic structure calculations.

8 Conclusion

More than an on the fly determination of ionic forces through successive electronic calculation, first-principle Molecular Dynamics aims at simulating the system of electrons and nuclei as a whole. Through the Born-Oppenheimer approximation, physically footed on the large mass difference between ions and electrons, a separation of their dynamics is performed. The ions are then evolved classically on a potential energy surface given by the ground state energy of the electronic system at fixed ion positions.

We have introduced here the Density Functional Theory for electronic structure calculation as it is probably the most used electronic theory in present first-principle Molecular Dynamics simulations. The use of the Hohenberg-Kohn theorem greatly reduces the complexity of electronic structure calculations, making rather large systems amenable to first-principle MD, still retaining good accuracy. Futhermore, using its variational character, Car and Parrinello designed an extended Lagrangian technique to propagate jointly electronic orbitals and ions. This is particularly efficient with plane wave basis sets, for molecular systems like liquid water.

Belonging to the class of MD simulations, any method of Molecular Dynamics can be applied to first-principle MD. This includes thermostating and barostating [173], use of constraints and generalized constraints [6,12,247], accelerating techniques for rare event [248–250], (metadynamics [24,251], path sampling [8,252] and so on), structure and dynamical analysis [1] etc. The specificity of ab initio MD with respect to these comes both from computer time issues and from the type of systems studied. However, it clearly departs from its empirical counterpart in the explicit description of the electronic systems: we have tried here to give some hints on how this can be exploited to extract properties directly comparable to experimental results (polarization and IR spectra, response properties and excited states) or to get a better understanding of the "chemistry" of the system. We did not mention though techniques to *manipulate* the electronic structure to study rare events [253–255], just like the ionic dynamics can be manipulated.

We did not address also other extensions of first-principle MD to free-energy Density Functional Theory for metals [145,256,257], simulations with a variable number of electrons [11,14,67,258–260] (fixed electron chemical potential) or otherwise path-integral Molecular Dynamics in the Car-Parrinello framework [27,43,51,80,81,208,261–268]. We hope however that the present introduction on first-principle MD will give the reader both the background and the enthusiasm to look in this vast literature.

Finally, can we dare to ask what is the future of first-principle MD? It would be hard to be highly predictive. However we would like to quote the following directions of research: QM/MM methods to treat quantum systems in an environment [92–94,225,226,269–272], Gaussian basis sets [23,30,38, 63,110,172] or Gaussian augmented plane waves methods [168] in search for order N methods [273,274] etc. Also, in order to go beyond Density Functional Theory, Quantum-Monte Carlo techniques are very attractive [119]. Some of these topics are already well-advanced and are discussed here in this book.

Acknowledgements

This manuscript owes a lot to tutorials given at CECAM from 2000 to 2005 with A. Alavi, E. J. Meijer, J. Hutter and M. Sprik, to which I am deeply indebted. I would also like to warmly acknowledge B. Guillot, V. Marry, W. Quester and M. Rasander for careful reading of this manuscript.

References

1. D. Frenkel and B. Smit (2001) *Understanding Molecular Simulation.* Academic Press, London
2. R. Car and M. Parrinello (1985) Unified Approach for Molecular Dynamics and Density-Functional Theory. *Phys. Rev. Lett.* **55**, p. 2471
3. A. Curioni, M. Sprik, W. Andreoni, H. Schiffer, J. Hutter, and M. Parrinello (1997) Density functional theory-based molecular dynamics simulation of acid-catalyzed chemical reactions in liquid trioxane. *J. Am. Chem. Soc.* **119**, p. 7218
4. E. J. Meijer and M. Sprik (1998) A Density Functional Study of the Addition of Water to SO_3 in the Gas Phase and in Aqueous Solution. *J. Phys. Chem. A* **102**, p. 2893
5. E. J. Meijer and M. Sprik (1998) Ab Initio Molecular Dynamics Study of the Addition Reaction of Water to Formaldehyde in Sulfuric Acid Solution. *J. Am. Soc. Chem.* **120**, p. 6345
6. M. Sprik (2000) Computation of the pK of liquid water using coordination constraints. *Chem. Phys.* **258**, p. 139
7. K. Doclo and U. Rothlisberger (2000) Conformational Equilibria of Peroxynitrous Acid in Water: A First-Principles Molecular Dynamics Study. *J. Phys. Chem. A* **104**, p. 6464
8. P. Geissler, C. Dellago, D. Chandler, J. Hutter, and M. Parrinello (2001) Autoionization in Liquid Water. *Science* **291**, p. 2121
9. B. Ensing, E. J. Meijer, P. E. Blochl, and E. J. Baerends (2001) Solvation Effects on the SN2 Reaction between CH_3Cl and Cl- in Water. *J. Phys. Chem. A.* **105**, p. 3300
10. N. L. Doltsinis and M. Sprik (2003) Theoretical pKa estimates for solvated $P(OH)_5$ from coordination constrained Car-Parrinello dynamics. *Phys. Chem. Chem. Phys.* **5**, p. 2612
11. J. Blumberger, L. Bernasconi, I. Tavernelli, R. Vuilleumier, and M. Sprik (2004) Electronic structure and solvation of copper and silver ions: A theoretical picture of a model aqueous redox reaction. *J. Am. Chem. Soc.* **126**, p. 3928
12. J. Blumberger and M. Sprik (2004) Free energy of oxidation of metal aqua Ions by an enforced change of Coordination. *J. Phys. Chem. B.* **108**, p. 6529
13. P. R. L. Markwick, N. L. Doltsinis, and D. Marx (2005) Targeted Car-Parrinello molecular dynamics: Elucidating double proton transfer in formic acid dimer. *J. Chem. Phys.* **122**, 054112
14. Y. Tateyama, J. Blumberger, M. Sprik, and I. Tavernelli (2005) Density-functional molecular-dynamics study of the redox reactions of two anionic, aqueous transition-metal complexes. *J. Chem. Phys.* **122**, p. 234505
15. F. Zipoli, M. Bernasconi, and A. Laio (2005) Ab initio simulations of Lewis-acid-catalyzed hydrosilylation of alkynes. *Chem. Phys. Chem.* **6**, p. 1772

16. A. D. Vita, G. Galli, A. Canning, and R. Car (1996) A microscopic model for surface-induced diamond-to-graphite transitions. *Nature* **379**, p. 523
17. C. Cavazzoni, G. L. Chiarotti, S. Scandolo, E. Tosatti, M. Bernasconi, and M. Parrinello (1999) Superionic and metallic states of water and ammonia at giant planet conditions. *Science* **283**, p. 44
18. R. Rousseau, M. Boero, M. Bernasconi, M. Parrinello, and K. Terakura (2000) Ab initio Simulation of Phase Transitions and Dissociation of H_2S at High Pressure. *Phys. Rev. Lett.* **85**, p. 1254
19. D. D. Klug, R. Rousseau, K. Uehara, M. Bernasconi, Y. L. Page, and J. S. Tse (2001) Ab initio molecular dynamics study of the pressure-induced phase transformations in cristobalite. *Phys. Rev. B* **63**, p. 104106
20. C. J. Wu, J. N. Glosli, G. Galli, and F. H. Ree (2002) Liquid-liquid phase transition in elemental carbon: A first-principles investigation. *Phys. Rev. Lett.* **89**, p. 135701
21. L. M. Ghiringhelli and E. J. Meijer (2005) Phosphorus: First principle simulation of a liquid–liquid phase transition. *J. Chem. Phys.* **122**, p. 184510
22. F. Zipoli, M. Bernasconi, and R. Martoňák (2004) Constant pressure reactive molecular dynamics simulations of phase transitions under pressure: The graphite to diamond conversion revisited. *Eur. Phys. J. B* **39**, p. 41
23. M. McGrath, J. Siepmann, I.-F. W. Kuo, C. J. Mundy, J. VandeVondele, M. Sprik, J. Hutter, F. Mohamed, M. Krack, and M. Parrinello (2005) Toward a Monte Carlo program for simulating vapor-liquid phase equilibria from first principles. *Comp. Phys. Comm.* **169**, p. 289
24. R. Martoňák, A. Laio, M. Bernasconi, C. Ceriani, P. Raiteri, F. Zipoli, and M. Parrinello (2005) Simulation of structural phase transitions by metadynamics. *Zeitschrift für Kristallographie* **220**, p. 489
25. T. Ikeda, M. Sprik, K. Terakura, and M. Parrinello (1998) Pressure Effects on Hydrogen Bonding in the Disordered Phase of Solid Hbr. *Phys. Rev. Lett.* **81**, p. 4416
26. T. Ikeda, M. Sprik, K. Terakura, and M. Parrinello (2000) Hydrogen elimination and solid-state reaction in hydrogen-bonded systems under pressure: The case of Hbr. *J. Phys. Chem. B* **104**, p. 11801
27. M. Benoit, A. H. Romero, and D. Marx (2002) Reassigning Hydrogen-Bond Centering in Dense Ice. *Phys. Rev. Lett.* **89**, p. 145501
28. A. Pasquarello and R. Resta (2003) Dynamical monopoles and dipoles in a condensed molecular system: The case of liquid water. *Phys. Rev. B* **68**, p. 174302
29. M. Sharma, R. Resta, and R. Car (2005) Intermolecular Dynamical Charge Fluctuations in Water: A Signature of the H-Bond Network. *Phys. Rev. Lett.* **95**, p. 187401
30. M. McGrath, J. Siepmann, I.-F. W. Kuo, C. J. Mundy, J. VandeVondele, J. Hutter, F. Mohamed, and M. Krack (2005) First Principles: Application to Liquid Water at Ambient Conditions. *Chem. Phys. Chem.* **6**, p. 1894
31. K. Laasonen, M. Sprik, M. Parrinello, and R. Car (1993) Ab initio liquid water. *J. Chem. Phys.* **99**, p. 9080
32. J. C. Grossman, E. Schwegler, E. W. Draeger, F. Gygi, and G. Galli (2004) Towards an assessment of the accuracy of density functional theory for first principles simulations of water. *J. Chem. Phys.* **120**, p. 300
33. E. Schwegler, J. C. Grossman, F. Gygi, and G. Galli (2004) Towards an assessment of the accuracy of density functional theory for first principles simulations of water. II. *J. Chem. Phys.* **121**, p. 5400

34. I. Bakó, J. Hutter, and G. Pálinkás (2002) Car-Parrinello molecular dynamics simulation of the hydrated calcium ion. *J. Chem. Phys.* **117**, p. 9838
35. M. Cavalleri, M. Odelius, A. Nilsson, and L. G. M. Pettersson (2004) X-ray absorption spectra of water within a plane-wave Car-Parrinello molecular dynamics framework. *J. Chem. Phys.* **121**, p. 10065
36. M. Allesch, E. Schwegler, F. Gygi, and G. Galli (2004) A first principles simulation of rigid water. *J. Chem. Phys.* **120**, p. 5192
37. P. H.-L. Sit and N. Marzari (2005) Static and dynamical properties of heavy water at ambient conditions from first-principles molecular dynamics. *J. Chem. Phys.* **122**, p. 204510
38. I.-F. W. Kuo, C. J. Mundy, M. J. McGrath, J. I. Siepmann, J. VandeVondele, M. Sprik, J. Hutter, B. Chen, M. L. Klein, F. Mohamed, M. Krack, and M. Parrinello (2004) Liquid water from first principles, investigation of different sampling approaches. *J. Phys. Chem. B* **108**, p. 12990
39. J. VandeVondele, F. Mohamed, M. Krack, J. Hutter, M. Sprik, and M. Parrinello (2005) The influence of temperature and density functional models in ab initio molecular dynamics simulation of liquid water. *J. Chem. Phys.* **122**, 014515
40. D. Marx, J. Hutter, and M. Parrinello (1995) Density functional study of small aqueous Be^{2+} clusters. *Chem. Phys. Lett.* **241**, p. 457
41. M. E. Tuckerman, K. Laasonen, M. Sprik, and M. Parrinello (1995) Ab initio molecular dynamics simulation of the solvation and transport of H_3O^+ and OH- ions in water. *J. Phys. Chem.* **99**, p. 5749
42. D. Marx, M. Sprik, and M. Parrinello (1997) Ab initio molecular dynamics of ion solvation. The case of Be^{2+} in water. *Chem. Phys. Lett.* **273**, p. 360
43. D. Marx, M. Tuckerman, J. Hutter, and M. Parrinello (1999) The nature of the hydrated excess proton in water. *Nature* **397**, p. 601
44. M. Tuckerman, K. Laasonen, M. Sprik, and M. Parrinello (1995) Ab initio molecular dynamics simulation of the solvation and transport of hydronium and hydroxyl ions in water. *J. Chem. Phys.* **103**, p. 150
45. L. M. Ramaniah, M. Bernasconi, and M. Parrinello (1999) Ab initio molecular-dynamics simulation of K^+ solvation in water. *J. Chem. Phys.* **111**, p. 1587
46. C. J. Mundy, J. Hutter, and M. Parrinello (2000) Microsolvation and chemical reactivity of sodium and water clusters. *J. Am. Chem. Soc.* **122**, p. 4837
47. R. Vuilleumier and M. Sprik (2001) Electronic properties of hard and soft ions in solution: Aqueous Na^+ and Ag^+ compared. *J. Chem. Phys.* **115**, p. 3454
48. Z.-P. Liu, P. Hu, and A. Alavi (2001) Mechanism for the high reactivity of CO oxidation on a ruthenium-oxide. *J. Chem. Phys.* **114**, p. 5956
49. T. S. van Erp and E. J. Meijer (2003) Ab initio molecular dynamics study of aqueous solvation of ethanol and ethylene. *J. Chem. Phys.* **118**, p. 8831
50. J. M. Heuft and E. J. Meijer (2003) Density functional theory based molecular-dynamics study of aqueous chloride solvation. *J. Chem. Phys.* **119**, p. 11788
51. M. E. Tuckerman, D. Marx, and M. Parrinello (2002) The nature and transport mechanism of hydrated hydroxide ions in aqueous solution. *Nature* **417**, p. 925
52. B. Kirchner, J. Stubbs, and D. Marx (2002) Fast Anomalous Diffusion of Small Hydrophobic Species in Water. *Phys. Rev. Lett.* **89**, p. 215901
53. B. Kirchner and J. Hutter (2002) The structure of a DMSO/Water mixture from Car-Parrinello simulations. *Chem. Phys. Lett.* **364**, p. 497
54. T. Ikeda, M. Hirata, and T. Kimura (2005) Hydration of Y^{3+} ion: A Car-Parrinello molecular dynamics study. *J. Chem. Phys.* **122**, 024510

55. S. Izvekov and G. A. Voth (2005) Ab initio molecular-dynamics simulation of aqueous proton solvation and transport revisited. *J. Chem. Phys.* **123**, 044505

56. M. Boero, M. Parrinello, K. Terakura, T. Ikeshoji, and C. C. Liew (2003) First-Principles Molecular-Dynamics Simulations of a Hydrated Electron in Normal and Supercritical Water. *Phys. Rev. Lett.* **90**, p. 226403

57. C. Nicolas, R. Spezia, A. Boutin, and R. Vuilleumier (2003) Molecular Dynamics simulations of a silver atom in water: dipolar excitonic state evidence. *Phys. Rev. Lett.* **91**, p. 208304

58. M.-P. Gaigeot and M. Sprik (2004) Ab initio molecular dynamics study of uracil in aqueous solution. *J. Phys. Chem. B* **108**, p. 7458

59. L. Bernasconi, M. Sprik, and J. Hutter (2004) Hartree-Fock exchange in time dependent density functional theory: Application to charge transfer excitations in solvated molecular systems. *Chem. Phys. Lett.* **394**, p. 141

60. F. C. Lightstone, E. Schwegler, M. Allesch, F. Gygi, and G. Galli (2005) A first-principles molecular dynamics study of calcium in water. *Chem. Phys. Chem* **6**, p. 1745

61. J. M. Heuft and E. J. Meijer (2005) Density functional theory based molecular-dynamics study of aqueous fluoride solvation. *J. Chem. Phys.* **122**, 094501

62. J. M. Heuft and E. J. Meijer (2005) Density functional theory based molecular-dynamics study of aqueous iodide solvation. *J. Chem. Phys.* **123**, 094506

63. J. VandeVondele and M. Sprik (2005) A molecular dynamics study of the hydroxyl radical in solution applying self-interaction corrected density functional methods. *Phys. Chem. Chem. Phys.* **7**, p. 1363

64. P. Hunt and M. Sprik (2005) On the position of the highest molecular orbital in aqueous solutions of simple ions. *Chem. Phys. Chem.* **6**, p. 1805

65. J. A. Morrone and M. E. Tuckerman (2002) Ab initio molecular dynamics study of proton mobility in liquid methanol. *J. Chem. Phys.* **117**, p. 4403

66. J.-W. Handgraaf, E. J. Meijer, and M.-P. Gaigeot (2004) Density-functional theory-based molecular simulation study of liquid methanol. *J. Chem. Phys.* **121**, p. 10111

67. J. VandeVondele, R. Lynden-Bell, E. J. Meijer, and M. Sprik (2006) Density functional theory study of tetrathiafulvalene and thianthrene in acetonitrile: structure, dynamics and redox properties. *J. Phys. Chem. B* **110**, p. 3614

68. A. Pasquarello, K. Laasonen, R. Car, C. Lee, and D. Vanderbilt (1992) Ab initio molecular dynamics for d-electron systems: Liquid copper at 1500 K. *Phys. Rev. Lett.* **69**, p. 1982

69. J. Sarnthein, A. Pasquarello, and R. Car (1995) Structural and Electronic Properties of Liquid and Amorphous SiO_2: An Ab Initio Molecular Dynamics Study. *Phys. Rev. Lett.* **74**, p. 4682

70. P. Silvestrelli, A. Alavi, and M. Parrinello (1997) Electronic conductivity calculation in ab-initio simulations of metals. Application to liquid sodium. *Phys. Rev. B* **55**, p. 15515

71. M. Pohlmann, M. Benoit, and W. Kob (2004) First-principles molecular-dynamics simulations of a hydrous silica melt: Structural properties and hydrogen diffusion mechanism. *Phys. Rev. B* **70**, p. 184209

72. M. Popolo, R. M. Lynden-Bell, and J. Kohanoff (2005) Ab Initio simulation of a room temperature ionic liquid. *J. Phys. Chem. B* **109**, p. 5895

73. F. Haase, J. Sauer, and J. Hutter (1997) Ab Initio Molecular Dynamics Simulation of Methanol Adsorbed in Chabazite. *Chem. Phys. Lett.* **266**, p. 397

74. A. Alavi, P. Hu, T. Deutsch, P. L. Silvestrelli, and J. Hutter (1998) CO Oxidation on Pt(111): An Ab Initio Density Functional Theory Study. *Phys. Rev. Lett.* **80**, p. 3650

75. A. Y. Lozovoi, A. Alavi, and M. Finnis (2000) Surface stoichiometry and the initial oxidation of NiAl(110). *Phys. Rev. Lett.* **85**, p. 610

76. P. Hu and A. Alavi (2001) Insight into electron-mediated reaction mechanisms: Catalytic CO oxidation on a ruthenium surface. *J. Chem. Phys.* **114**, p. 8113

77. E. S. Boek and M. Sprik (2003) Ab initio molecular dynamics study of the hydration of a sodium smectite clay. *J. Phys. Chem. B* **107**, p. 3251

78. X. Wu, A. Selloni, and S. K. Nayak (2004) First principles study of CO oxidation on TiO_2 (110): The role of surface oxygen vacancies. *J. Chem. Phys.* **120**, p. 4512

79. M. W. Finnis, A. Lozovoi, and A. Alavi (2005) Ab initio calculations on the oxidation of NiAl, *Annual Reviews of Materials Research* **35**, pp. 167–207

80. M. Benoit, D. Marx, and M. Parrinello (1998) Tunnelling and zero-point motion in high-pressure ic. *Nature* **392**, p. 258

81. M. Benoit, D. Marx, and M. Parrinello (1999) The role of quantum effects and ionic defects in high-density ice. *Solid State Ionics* **125**, p. 23

82. A. Alavi, R. Lynden-Bell, and R. Brown (1999) Displacement and distortion of the ammonium ion in reorientational transition states of ammonium chloride and ammonium fluoride. *J. Chem. Phys.* **110**, p. 5861

83. A. Alavi, R. Lynden-Bell, and R. Brown (2000) Pathway for reorientation in ammonium fluoride. *Chem. Phys. Lett.* **320**, p. 487

84. S. Ispas, M. Benoit, P. Jund, and R. Jullien (2001) Structural and electronic properties of the sodium tetrasilicate glass $Na_2Si_4O_9$ from classical and ab initio molecular dynamics simulations. *Phys. Rev. B* **64**, p. 214206

85. S. Ispas, M. Benoit, P. Jund, and R. Jullien (2002) Structural properties of glassy and liquid sodium tetrasilicate: comparison between ab initio and classical molecular dynamics simulations. *Journal of Non-Crystalline Solids* **307**, p. 946

86. J. Sarnthein, A. Pasquarello, and R. Car (1995) Model of vitreous SiO_2 generated by an ab initio molecular-dynamics quench from the melt. *Phys. Rev. B* **52**, p. 12690

87. A. Pasquarello and R. Car (1997) Dynamical Charge Tensors and Infrared Spectrum of Amorphous SiO_2. *Phys. Rev. Lett.* **79**, p. 1766

88. P. Carloni and U. Rothlisberger (2001) Simulations of Enzymatic Systems: Perspectives from Car-Parrinello Molecular Dynamics Simulations. In *Theoretical Biochemistry – Processes and Properties of Biological Systems*, ed. L. Eriksson Elsevier Science, pp. 215–251

89. P. Carloni, U. Rothlisberger, and M. Parrinello (2002) The Role and Perspective of Ab-initio Molecular Dynamics in the Study of Biological Systems. *Acc. Chem. Res.* **35**, p. 455

90. M. Sulpizi, G. Folkers, U. Rothlisberger, P. Carloni, and L. Scapozza (2002) Applications of Density Functional Theory-Based Methods in Medicinal Chemistry. *Quant. Struct.-Act. Rel.* **21**, p. 173

91. U. Röhrig, L. Guidoni, and U. Rothlisberger (2002) Early Steps of the Intramolecular Signal Transduction in Rhodopsin Explored by Molecular Dynamics Simulations. *Biochemistry* **41**, p. 10799

92. M. Colombo, L. Guidoni, A. Laio, A. Magistrato, P. Maurer, S. Piana, U. Röhrig, K. Spiegel, M. Sulpizi, J. VandeVondele, M. Zumstein, and U. Rothlisberger (2002) Hybrid QM/MM Car-Parrinello Simulations of Catalytic and Enzymatic Reactions. *CHIMIA* **56**, p. 13

93. A. Magistrato, W. DeGrado, U. Rothlisberger, and M. Klein (2003) Structural and Dynamical Characterization of Dizinc DF1, a Biomimetic Compound of Diiron Proteins via ab initio and Hybrid (QM/MM) Molecular Dynamics. *J. Phys. Chem. B* **107**, p. 4182

94. K. Spiegel, U. Rothlisberger, and P. Carloni (2004) Cisplatin Binding to DNA Oligomers from Hybrid Car-Parrinello Molecular Dynamics Simulations. *J. Phys. Chem. B* **108**, p. 2699

95. L. Guidoni, K. Spiegel, M. Zumstein, and U. Rothlisberger (2004) Green Oxidation Catalysts: Computational Design of High Efficiency Models of Galactose Oxidase. *Angew. Chem.* **116**, p. 3348

96. F. L. Gervasio, A. Laio, M. Parrinello, and M. Boero (2005) Charge Localization in DNA Fibers. *Phys. Rev. Lett.* **94**, p. 158103

97. *CPMD*, Copyright IBM Corp 1990–2001, Copyright MPI für Festkörperforschung Stuttgart (1997–2004)

98. J. Hutter and M. Iannuzzi (2005) CPMD: Car-Parrinello molecular dynamics. *Z. für Kristallographie* **220**, p. 549

99. S. Baroni, A. D. Corso, S. de Gironcoli, P. Giannozzi, C. Cavazzoni, G. Ballabio, S. Scandolo, G. Chiarotti, P. Focher, A. Pasquarello, K. Laasonen, A. Trave, R. Car, N. Marzari, and A. Kokalj, http://www.pwscf.org/

100. G. Kresse and J. Furthmüller (1996) Efficient iterative schemes for ab initio total-energy calculations using a plane-wave basis set. *Phys. Rev. B* **54**, p. 11169

101. M. C. Payne, M. P. Teter, D. C. Allan, T. A. Arias, and J. D. Joannopoulos (1992) Iterative minimization techniques for ab initio total-energy calculations: molecular dynamics and conjugate gradients. *Rev. Mod. Phys.* **64**, p. 1045

102. S. J. Clark, M. D. Segall, C. J. Pickard, P. J. Hasnip, M. I. J. Probert, K. Refson, and M. C. Payne (2005) First principles methods using CASTEP. *Zeitschrift für Kristallographie* **220**, p. 567

103. J. N. Michel Bockstedte, A. Kley, and M. Scheffler (1997) Density-functional theory calculations for poly-atomic systems: electronic structure, static and elastic properties and ab initio molecular dynamics. *Comput. Phys. Comm.* **107**, p. 187

104. *Nwchem*, developed and distributed by Pacic Northwest National Laboratory, USA

105. X. Gonze (2005) A brief introduction to the ABINIT software packag. *Zeitschrift für Kristallographie* **220**, p. 558

106. R. Dovesi, R. Orlando, B. Civalleri, C. Roetti, V. R. Saunders, and C. M. Zicovich-Wilsonan (2005) CRYSTAL: A computational tool for the ab initio study of the electronic properties of crystals. *Zeitschrift für Kristallographie* **220**, p. 571

107. S. Scandolo, P. Giannozzi, C. Cavazzoni, S. de Gironcoli, A. Pasquarello, and S. Baroni (2005) First-principles codes for computational crystallography in the Quantum-ESPRESSO package. *Zeitschrift für Kristallographie* **220**, p. 574

108. M. E. Tuckerman, D. A. Yarne, S. O. Samuelson, A. L. Hughes, and G. J. Martyna (2000) Exploiting multiple levels of parallelism in molecular dynamics

based calculations via modern techniques and software paradigms on distributed memory computers. *Comp. Phys. Comm.* **128**, p. 333

109. *The cp2k developers group*, http://cp2k.berlios.de/ (2004)
110. J. Vandevondele, M. Krack, F. Mohamed, M. Parrinello, T. Chassaing, and J. Hutter (2005) Quickstep: Fast and accurate density functional calculations using a mixed Gaussian and plane waves approach. *Comp. Phys. Comm.* **167**, p. 103
111. D. Marx and J. Hutter (2000) Ab initio molecular dynamics: Theory and implementation; Modern Methods and Algorithms of Quantum Chemistry. In *NIC Series*, ed. J. Grotendorst Forschungszentrum Jülich 1
112. M. Tuckerman and G. Martyna (2000) Understanding modern molecular dynamics methods: Techniques and Applications. *J. Phys. Chem.* **104**, p. 159
113. D. Sebastiani and U. Rothlisberger (2003) Advances in Density Functional Based Modelling Techniques: Recent Extensions of the Car-Parrinello Approach. In *Medicinal Quantum Chemistry*, eds. P. Carloni and F. Alber Wiley-VCH, Weinheim. Methods and Principles in Medicinal Chemistry, pp. 5–40
114. J. Hutter (2004) Large Scale Density Functional Calculations. In *Multiscale Modelling and Simulation*, eds. S. Attinger and P. Koumoutsakos Springer, Heidelberg. Lecture Notes in Computational Science and Engineering, p. 39
115. A. Messiah (2000) *Quantum Mechanics, 2 volumes Bound as One*. Chap. XVIII, Dover, New York
116. R. P. Feynman (1939) Forces in Molecules. *Phys. Rev.* **56**, p. 340
117. L. Verlet (1967) Computer Experiments on Classical Fluids. I. Thermodynamical Properties of Lennard-Jones Molecules. *Phys. Rev.* **159**, p. 98
118. D. M. Ceperley and B. J. Alder (1980) Ground State of the Electron Gas by a Stochastic Method. *Phys. Rev. Lett.* **45**, p. 566
119. C. Pierleoni, D. M. Ceperley, and M. Holzmann (2004) Coupled Electron-Ion Monte Carlo Calculations of Dense Metallic Hydrogen. *Phys. Rev. Lett.* **93**, p. 146402
120. R. M. Dreizler and E. K. U. Gross (1990) *Density Functional Theory*. Springer-Verlag, Berlin Heidelberg
121. R. G. Parr and Y. Weitao (1994) *Density-Functional Theory of Atoms and Molecules*. Oxford University Press
122. R. M. Martin (2004) *Electronic Structure : Basic Theory and Practical Methods*. Cambridge University Press
123. P. Hohenberg and W. Kohn (1964) Inhomogeneous Electron Gas. *Phys. Rev.* **136**, p. B864
124. M. Levy (1979) Universal Variational Functionals of Electron Densities, First-Order Density Matrices, and Natural Spin-Orbitals and Solution of the v-Representability Problem. *Proc. Nat. Acad. Science* **76**, p. 6062
125. W. Kohn and L. J. Sham (1965) Self-Consistent Equations Including Exchange and Correlation Effects. *Phys. Rev.* **140**, p. A1133
126. S. H. Vosko, L. Wilk, and M. Nusair (1980) Accurate spin-dependent electron liquid correlation energies for local spin density calculations: a critical analysis. *Can. J. Phys.* **58**, p. 1200
127. J. P. Perdew and Y. Wang (1992) Accurate and simple analytic representation of the electron-gas correlation energy. *Phys. Rev. B* **45**, p. 13244
128. J. P. Perdew and A. Zunger (1981) Self-interaction correction to density-functional approximations for many-electron systems. *Phys. Rev. B* **23**, p. 5048

129. S. Goedecker, M. Teter, and J. Hutter (1996) Seperable Dual-Space Gaussian Pseudopotentials. *Phys. Rev. B* **54**, p. 1703
130. J. P. Perdew, K. Burke, and M. Ernzerhof (1996) Generalized Gradient Approximation Made Simple. *Phys. Rev. Lett.* **77**, p. 3865
131. J. P. Perdew, J. A. Chevary, S. H. Vosko, Koblar, A. Jackson, M. R. Pederson, D. J. Singh, and C. Fiolhais (1992) Atoms, molecules, solids, and surfaces: Applications of the generalized gradient approximation for exchange and correlation. *Phys. Rev. B* **46**, p. 6671
132. J. P. Perdew (1991) in *Electronic Structure of Solids '91*, eds. P. Ziesche and H. Eschrig Akademie Verlag, Berlin, p. 11
133. A. D. Becke (1988) Density-functional exchange-energy approximation with correct asymptotic behavior. *Phys. Rev. A* **38**, p. 3098
134. C. Lee, W. Yang, and R. G. Parr (1988) Development of the Colle-Salvetti correlation-energy formula into a functional of the electron density. *Phys. Rev. B* **37**, p. 785
135. F. A. Hamprecht, A. J. Cohen, D. J. Tozer, and N. C. Handy (1998) Development and assessment of new exchange-correlation functionals. *J. Chem. Phys.* **109**, p. 6264
136. A. D. Boese and N. C. Handy (2001) A new parametrization of exchange-correlation generalized gradient approximation functionals. *J. Chem. Phys.* **114**, p. 5497
137. A. D. Boese, N. L. Doltsinis, N. C. Handy, and M. Sprik (2000) New generalized gradient approximation functionals. *J. Chem. Phys.* **112**, p. 1670
138. A. D. Corso, A. Pasquarello, A. Baldereschi, and R. Car (1996) Generalized-gradient approximations to density-functional theory: A comparative study for atoms and solids. *Phys. Rev. B* **53**, p. 1180
139. M. Sprik, J. Hutter, and M. Parrinello (1996) Ab initio molecular dynamics simulation of liquid water: Comparison of three gradient-corrected density functionals. *J. Chem. Phys.* **105**, p. 1142
140. A. Becke (1996) Density-functional thermochemistry. IV. A new dynamical correlation functional and implications for exact-exchange mixing. *J. Chem. Phys.* **104**, p. 1040
141. M. d'Avezac, M. Calandra, and F. Mauri (2005) Density functional theory description of hole-trapping in SiO_2 : A self-interaction-corrected approach. *Phys. Rev. B* **71**, p. 205210
142. I. Ciofini, C. Adamo, and H. Chermette (2005) Effect of self-interaction error in the evaluation of the bond length alternation in trans-polyacetylene using density-functional theory. *J. Chem. Phys.* **123**, p. 121102
143. W. Pollard and R. Friesner (1993) Efficient Fock matrix diagonalization by a Krylov-space method. *J. Chem. Phys.* **99**, p. 6742
144. J. Cullum and R. A. Willoughby (1985) *Lanczos algorithms for large symmetric eigenvalue computations, 1: Theory of Progress in Scientific Computing.* Birkhauser, Boston, 3 edn
145. A. Alavi, J. Kohanoff, M. Parrinello, and D. Frenkel (1994) Ab Initio Molecular Dynamics with Excited Electrons. *Phys. Rev. Lett.* **73**, p. 2599
146. D. G. Anderson (1965) Iterative Procedures for Nonlinear Integral Equations. *J. Assoc. Comput. Mach.* **12**, p. 547
147. M. P. Teter, M. C. Payne, and D. C. Allen (1989) Solution of Schrödinger's equation for large systems. *Phys. Rev. B* **40**, p. 12255

148. P. Pulay (1980) Convergence acceleration of iterative sequences the case of scf iteration. *Chem. Phys. Lett.* **73**, p. 393

149. C. Csaszar and P. Pulay (1984) Geometry optimization by direct inversion in the iterative subspace. *J. Mol. Struc.* **114**, p. 31

150. J. Hutter, H. P. Luthi, and M. Parrinello (1994) Electronic Structure Optimisation in Plane-Wave-Based Density Functional Calculations by Direct Inversion in the Iterative Subspace. *Comput. Mat. Sci.* **2**, p. 244

151. R. Spezia, C. Nicolas, F.-X. Coudert, P. Archirel, R. Vuilleumier, and A. Boutin (2004) Reactivity of an excess electron with monovalent cations in bulk water by mixed quantum classical molecular dynamics simulations. *Mol. Sim.* **30**, p. 749

152. R. W. Hockney (1970) The potential calculation and some applications. *Methods Comput. Phys.* **9**, p. 136

153. G. J. Martyna and M. E. Tuckerman (1999) A reciprocal space based method for treating long range interactions in ab initio and force-field-based calculations in clusters. *J. Chem. Phys.* **110**, p. 2810

154. A. Filippetti, D. Vanderbilt, W. Zhong, Y. Cai, and G. B. Bachelet (1995) Chemical hardness, linear response, and pseudopotential transferability. *Phys. Rev. B* **52**, p. 11793

155. D. R. Hamann, M. Schlüter, and C. Chiang (1979) Norm-Conserving Pseudopotentials. *Phys. Rev. Lett.* **43**, p. 1494

156. G. B. Bachelet, D. R. Hamann, and M. Schlüter (1982) Pseudopotentials that work: From H to Pu. *Phys. Rev. B* **26**, p. 4199

157. N. Troullier and J. L. Martins (1991) Efficient pseudopotentials for plane-wave calculations. *Phys. Rev. B* **43**, p. 1993

158. L. Kleinman and D. M. Bylander (1982) Efficacious Form for Model Pseudopotentials. *Phys. Rev. Lett.* **48**, p. 1425

159. P. E. Blöchl (1990) Generalized separable potentials for electronic-structure calculations. *Phys. Rev. B* **41**, p. 5414

160. C. Hartwigsen, S. Goedecker, and J. Hutter (1998) Relativistic separable dual-space Gaussian pseudopotentials from H to Rn. *Phys. Rev. B* **58**, p. 3641

161. O. A. von Lilienfeld, I. Tavernelli, U. Rothlisberger, and D. Sebastiani (2005) Variational optimization of effective atom centered potentials for molecular properties. *J. Chem. Phys.* **122**, 014113

162. O. A. von Lilienfeld, I. Tavernelli, U. Rothlisberger, and D. Sebastiani (2004) Optimization of Effective Atom Centered Potentials for London Dispersion Forces in Density Functional Theory. *Phys. Rev. Lett.* **93**, p. 153004

163. D. Vanderbilt (1985) Optimally smooth norm-conserving pseudopotentials. *Phys. Rev. B* **32**, p. 8412

164. D. Vanderbilt (1990) Soft self-consistent pseudopotentials in a generalized eigenvalue formalism. *Phys. Rev. B* **41**, p. 7892

165. K. Laasonen, R. Car, C. Lee, and D. Vanderbilt (1991) Implementation of ultrasoft pseudopotentials in ab initio molecular dynamics. *Phys. Rev. B* **43**, p. 6796

166. K. Laasonen, A. Pasquarello, R. Car, C. Lee, and D. Vanderbilt (1993) Car-Parrinello molecular dynamics with Vanderbilt ultrasoft pseudopotentials. *Phys. Rev. B* **47**, p. 10142

167. J. Hutter, M. E. Tuckerman, and M. Parrinello (1995) Integrating the Car-Parrinello equations III. *J. Chem. Phys.* **102**, p. 859

168. M. Iannuzzi, T. Chassaing, T. Wallman, and J. Hutter (2005) Ground and Excited State Density Functional Calculations with the Gaussian and Augmented-Plane-Wave Method. *Chimia* **59**, p. 499

169. S. Nosé (1984) A unified formulation of the constant temperature molecular dynamics methods. *J. Chem. Phys* **81**, p. 511

170. W. G. Hoover (1985) Canonical dynamics: Equilibrium phase-space distributions. *Phys. Rev. A* **31**, p. 1695

171. J. B. Abrams, M. E. Tuckerman, and G. J. Martyna, *Equilibrium statistical mechanics, non-hamiltonian molecular dynamics, and novel applications from resonance-free timesteps to adiabatic free energy dynamic.* Lect. Notes in Phys. **703**, pp. 139–192

172. J. VandeVondele and J. Hutter (2003) An efficient orbital transformation method for electronic structure calculations. *J. Chem. Phys.* **118**, p. 4365

173. M. E. Tuckerman, J. Hutter, and M. Parrinello (1994) Integrating the Car-Parrinello equations I. *J. Chem. Phys.* **101**, p. 1302

174. M. E. Tuckerman, J. Hutter, and M. Parrinello (1994) Integrating the Car-Parrinello equations II. *J. Chem. Phys.* **101**, p. 1316

175. J. P. Ryckaert, G. Ciccotti, and H. J. C. Berendsen (1977) Numerical integration of the Cartesian equation of motion of a system with constraints: molecular dynamics of N-alkanes. *J. of Computational Physics* **23**, p. 327

176. H. C. Andersen (1983) Rattle: A velocity version of the shake algorithm for molecular dynamics calculations. *J. Comp. Phys.* **52**, p. 24

177. G. J. Martyna, M. L. Klein, and M. Tuckerman (1992) Nosé-Hoover chains: The canonical ensemble via continuous dynamics. *J. Chem. Phys.* **97**, p. 2635

178. G. Pastore, E. Smargiassi, and F. Buda (1991) Theory of ab initio molecular-dynamics calculations. *Phys. Rev. A* **44**, p. 6334

179. F. A. Bornemann and C. Schutte (1998) A mathematical investigation of the Car-Parrinello method. *Numer. Math.* **78**, p. 359

180. F. A. Bornemann and C. Schutte (1999) Adaptive accuracy control for Car-Parrinello simulations. *Numer. Math.* **83**, p. 179

181. P. Tangney and S. Scandolo (2002) How well do Car-Parrinello simulations reproduce the Born-Oppenheimer surface? Theory and examples. *J. Chem. Phys.* **116**, p. 14

182. O. G. Jepps, G. Ayton, and D. J. Evans (2000) Microscopic expressions for the thermodynamic temperature. *Phys. Rev. E* **62**, p. 4757

183. P. Tangney (2006) On the theory underlying the Car-Parrinello method and the role of the fictitious mass parameter. *J. Chem. Phys.* **124**, 044111

184. M. Sprik (1991) Hydrogen bonding and the static dielectric constant in liquid water. *J. Chem. Phys.* **95**, p. 6762

185. M. Sprik and M. L. Klein (1988) A polarizable model for water using distributed charge sites. *J. Chem. Phys.* **89**, p. 7556

186. M. Marchi, D. Borgis, N. Levy, and P. Ballone (2001) A dielectric continuum molecular dynamics method. *J. Chem. Phys.* **114**, p. 4377

187. R. Vuilleumier and D. Borgis (2000) Wavefunction quantisation of the proton motion in a $H_5O_2^+$ dimer solvated in liquid water. *Journal of Molecular Structure* **552**, p. 117

188. R. Resta (1998) Quantum-Mechanical Position Operator in Extended Systems. *Phys. Rev. Lett.* **80**, p. 1800

189. R. Resta and S. Sorella (1999) Electron Localization in the Insulating State. *Phys. Rev. Lett.* **82**, p. 370

190. R. Resta (1994) Macroscopic polarization in crystalline dielectrics: the geometric phase approach. *Rev. Mod. Phys.* **66**, pp. 899–915
191. D. Vanderbilt and R. D. King-Smith (1993) Electric polarization as a bulk quantity and its relation to surface charge. *Phys. Rev. B* **48**, p. 4442
192. R. D. King-Smith and D. Vanderbilt (1993) Theory of polarization of crystalline solids. *Phys. Rev. B* **47**, p. 1651
193. P. L. Silvestrelli, M. Bernasconi, and M. Parrinello (1997) Ab initio infrared spectrum of liquid water. *Chem. Phys. Lett.* **277**, p. 478
194. M.-P. Gaigeot and M. Sprik (2003) Ab initio molecular dynamics computation of the infrared spectrum of aqueous uracil. *J. Phys. Chem. B* **107**, p. 10344
195. M. Gaigeot, R. Vuilleumier, M. Sprik, and D. Borgis (2005) Infrared spectroscopy of N-methyl-acetamide revisited by ab initio molecular dynamics simulations. *J. Chem. Theory Comput.* **1**, p. 772
196. A. Putrino, D. Sebastiani, and M. Parrinello (2000) Generalized variational density functional perturbation theory. *J. Chem. Phys.* **113**, p. 7102
197. A. Putrino and M. Parrinello (2002) Anharmonic Raman Spectra in High-Pressure Ice from Ab Initio Simulations. *Phys. Rev. Lett.* **88**, p. 176401
198. I. Souza, J. Íñiguez, and D. Vanderbilt (2002) First-Principles Approach to Insulators in Finite Electric Fields. *Phys. Rev. Lett.* **89**, p. 117602
199. P. Umari and A. Pasquarello (2002) Ab initio Molecular Dynamics in a Finite Homogeneous Electric Field. *Phys. Rev. Lett.* **89**, p. 157602
200. V. Dubois, P. Umari, and A. Pasquarello (2004) Dielectric susceptibility of dipolar molecular liquids by ab initio molecular dynamics: application to liquid HCl. *Chem. Phys. Lett.* **390**, p. 193
201. R. Ramirez, T. Lopez-Ciudad, P. Kumar, and D. Marx (2004) Quantum corrections to classical time-correlation functions: Hydrogen bonding and anharmonic floppy modes. *J. Chem. Phys.* **121**, p. 3973
202. R. Iftimie and M. E. Tuckerman (2005) Decomposing total IR spectra of aqueous systems into solute and solvent contributions: A computational approach using maximally localized Wannier orbitals. *J. Chem. Phys.* **122**, p. 214508
203. J. E. Bertie and Z. Lan (1996) Infrared Intensities of Liquids XX: The Intensity of the OH Stretching Band of Liquid Water Revisited, and the Best Current Values of the Optical Constants of $H_2O(l)$ at 25°C. *App. Spec.* **50**, p. 1047
204. X. Gonze (1995) Adiabatic density-functional perturbation theory. *Phys. Rev. A* **52**, p. 1096
205. X. Gonze (1995) Perturbation expansion of variational principles at arbitrary order. *Phys. Rev. A* **52**, p. 1086
206. D. Sebastiani and M. Parrinello (2002) Ab-initio study of NMR chemical shifts of water under normal and supercritical conditions. *Chem. Phys. Chem.* **3**, p. 675
207. J. Schmidt and D. Sebastiani (2005) Anomalous temperature dependence of nuclear quadrupole interactions in strongly hydrogen-bonded systems from first principles. *J. Chem. Phys.* **123**, 074501
208. M. Benoit and D. Marx (2005) The Shapes of Protons in Hydrogen Bonds Depend on the Bond Length. *Chem. Phys. Chem.* **6**, p. 1738
209. M. Profeta, M. Benoit, F. Mauri, and C. J. Pickard (2004) First-Principles Calculation of the NMR Parameters in Ca Oxide and Ca Aluminosilicates: the Partially Covalent Nature of the Ca-O Bond, a Challenge for Density Functional Theory. *J. Am. Chem. Soc.* **126**, p. 12628

210. I. Frank, J. Hutter, D. Marx, and M. Parrinello (1998) Molecular dynamics in low-spin excited states. *J. Chem. Phys.* **108**, p. 4060
211. E. K. U. Gross and W. Kohn (1990) Time-dependent density functional theory. *Adv. Quant. Chem.* **21**, p. 255
212. E. K. U. Gross, J. F. Dobson, and M. Petersilka (1996) Density-functional theory of time-dependent phenomena. In *Topics in Current Chemistry*, Springer Berlin Heidelberg, pp. 81–172
213. K. Burke and E. K. U. Gros (1998) A guided tour of time-dependent density functional theory. In *Density Functionals: Theory and Applications*, ed. D. Joubert, Springer Berlin Heidelberg, pp. 116–146
214. N. T. Maitra, K. Burke, E. K. U. G. H. Appel, and R. van Leeuwen (2002) Ten topical questions in time-dependent density functional theory. In *Reviews in Modern Quantum Chemistry: A Celebration of the Contributions of R.G. Parr*, ed. K. D. Sen, World Scientific, pp. 1186–1225
215. M. A. L. Marques and E. K. U. Gross (2004) Time-Dependent Density-Functional Theory. *Annu. Rev. Phys. Chem.* **55**, p. 427
216. E. Runge and E. K. U. Gross (1984) Density functional theory for time-dependent systems. *Phys. Rev. Lett.* **52**, p. 997
217. K Burke, J. Werschnik, and E. K. U. Gross (2005) Time-dependent density functional theory: Past, present, and future. *J. Chem. Phys.* **123**, 062206
218. M. Petersilka, E. K. U. Gross, and K. Burke (2000) Excitation energies from time-dependent density-functional theory using exact and approximate potentials. *Int. J. Quant. Chem.* **80**, p. 534
219. T. Grabo, M. Petersilka, and E. K. U. Gross (2000) Molecular excitation energies from time-dependent density-functional theory. *Journal of Molecular Structure (Theochem)* **501**, p. 353
220. H. Appel, E. K. U. Gross, and K. Burke (2003) Excitations in Time-Dependent Density-Functional Theory. *Phys. Rev. Lett.* **90**, 043005
221. N. L. Doltsinis and M. Sprik (2000) Electronic excitation spectra from time-dependent density functional response theory using plane wave methods. *Chem. Phys. Lett.* **330**, p. 563
222. J. Hutter (2003) Excited state nuclear forces from the Tamm-Dancoff approximation to time-dependent density functional theory within the plane wave basis set framework. *J. Chem. Phys.* **118**, p. 3928
223. L. Bernasconi, M. Sprik, and J. Hutter (2003) Time dependent density functional theory study of charge-transfer and intramolecular electronic excitations in acetone–water systems. *J. Chem. Phys.* **119**, p. 12417
224. L. Bernasconi, J. Blumberger, M. Sprik, and R. Vuilleumier (2004) Density functional calculation of the electronic absorption spectrum of Cu^+ and Ag^+ aqua ions. *J. Chem. Phys.* **121**, p. 11885
225. M. Sulpizi, P. Carloni, J. Hutter, and U. Röthlisberger (2003) A hybrid TDDFT/MM investigation of the optical properties of aminocoumarins in water and acetonitrile solution. *Phys. Chem. Chem. Phys.* **5**, p. 4798
226. M. Sulpizi, U. F. Röhrig, J. Hutter, and U. Rothlisberger (2005) Optical properties of molecules in solution via hybrid TDDFT/MM simulations. *Int. J. Quant. Chem.* **101**, p. 671
227. L. Bernasconi and M. Sprik (2005) Time-dependent density functional theory description of on-site electron repulsion and ligand field effects in the optical spectrum of hexa-aquoruthenium(II) in solution. *J. Phys. Chem. B* **109**, p. 12222

228. W. Kohn, Y. Meir, and D. E. Makarov (1998) van der Waals Energies in Density Functional Theory. *Phys. Rev. Lett.* **80**, p. 4153

229. M. Lein, J. F. Dobson, and E. K. U. Gross (1999) Towards the description of van der Waals interactions within density-functional theory. *J. Comput. Chem.* **20**, p. 12

230. M. Odelius, D. Laikov, and J. Hutter (2003) Excited state geometries within time-dependent and restricted open-shell density functional theories. *J. Mol. Struc. (Theochem)* **630**, p. 163

231. N. L. Doltsinis and D. S. Kosov (2005) Plane wave/pseudopotential implementation of excited state gradients in density functional linear response theory: A new route via implicit dierentiation. *J. Chem. Phys.* **122**, p. 144101

232. H. Langer and N. L. Doltsinis (2003) Excited state tautomerism of the DNA base guanine: A restricted open-shell Kohn-Sham study. *J. Chem. Phys.* **118**, p. 5400

233. N. L. Doltsinis and D. Marx (2002) Nonadiabatic Car-Parrinello molecular dynamics. *Phys. Rev. Lett.* **88**, p. 166402

234. N. L. Doltsinis (2004) Excited state proton transfer and internal conversion in o-hydroxybenzaldehyde: New insights from nonadiabatic ab initio molecular dynamics. *Mol. Phys.* **102**, p. 499

235. J. M. Foster and S. F. Boys (1960) Canonical Configurational Interaction Procedure. *Rev. Mod. Phys.* **32**, p. 300

236. C. Edmiston and K. Ruedenberg (1963) Localized Atomic and Molecular Orbitals. *Rev. Mod. Phys.* **35**, p. 457

237. W. von Niessen (1972) Density Localization of Atomic and Molecular Orbitals. I. *J. Chem. Phys.* **56**, p. 4290

238. N. Marzari and D. Vanderbilt (1997) Maximally localized generalized Wannier functions for composite energy bands. *Phys. Rev. B* **56**, p. 12847

239. P. L. Silvestrelli (1999) Maximally localized Wannier functions for simulations with supercells of general symmetry. *Phys. Rev. B* **59**, p. 9703

240. G. Berghold, C. J. Mundy, A. H. Romero, J. Hutter, and M. Parrinello (2000) General and Efficient Algorithms for Obtaining Maximally-Localized Wannier Functions. *Phys. Rev. B* **61**, p. 10040

241. I. Souza, N. Marzari, and D. Vanderbilt (2002) Maximally localized Wannier functions for entangled energy bands. *Phys. Rev. B* **65**, 035109

242. P. L. Silvestrelli and M. Parrinello (1999) Water Molecule Dipole in the Gas and in the Liquid Phase. *Phys. Rev. Lett.* **82**, p. 3308

243. M. Boero, K. Terakura, T. Ikeshoji, C. C. Liew, and M. Parrinello (2000) Hydrogen Bonding and Dipole Moment of Water at Supercritical Conditions: A First-Principles Molecular Dynamics Study. *Phys. Rev. Lett.* **85**, p. 3245

244. B. Kirchner and J. Hutter (2004) Solvent effects on electronic properties from Wannier functions in a dimethyl sulfoxide/water mixture. *J. Chem. Phys.* **121**, p. 5133

245. P. Hunt, M. Sprik, and R. Vuilleumier (2003) Thermal versus electronic broadening in the density of states of liquid water. *Chem. Phys. Lett.* **376**, p. 68

246. R. Iftimie, J. W. Thomas, and M. E. Tuckerman (2004) On-the-fly localization of electronic orbitals in Car–Parrinello molecular dynamics. *J. Chem. Phys.* **120**, p. 2169

247. I. Coluzza, M. Sprik, and G. Ciccotti (2003) Constrained reaction coordinate dynamics for systems with constraints. *Mol. Phys.* **101**, p. 2885

248. J. VandeVondele and U. Rothlisberger (2000) Efficient Multidimensional Free Energy Calculations for ab initio Molecular Dynamics Using Classical Bias Potentials. *J. Chem. Phys.* **113**, p. 4863

249. J. VandeVondele and U. Rothlisberger (2002) Canonical Adiabatic Free Energy Sampling (CAFES): A Novel Method for the Exploration of Free Energy Surfaces. *J. Phys. Chem. B* **106**, p. 203

250. L. Rosso, P. Minary, Z. Zhu, and M. E. Tuckerman (2002) On the use of adiabatic molecular dynamics to calculate free energy profiles. *J. Chem. Phys.* **116**, p. 4389

251. A. Laio and M. Parrinello (2002) Escaping free-energy minima. *Proc. Nat. Acad. Sciences* **99**, p. 12562

252. P. L. Geissler, C. Dellago, D. Chandler, J. Hutter, and M. Parrinello (2000) Ab initio analysis of proton transfer dynamics in $(H_2O)3H^+$. *Chem. Phys. Lett.* **321**, p. 225

253. R. Vuilleumier and M. Sprik (2002) Electronic control using density functional perturbation methods. *Chem. Phys. Lett.* **365**, p. 305

254. J. VandeVondele and U. Rothlisberger (2001) Estimating Equilibrium Properties from Non-Hamiltonian Dynamics. *J. Chem. Phys.* **115**, p. 7859

255. J. VandeVondele and U. Rothlisberger (2002) Accelerating Rare Reactive Events by Means of a Finite Electronic Temperature. *J. Am. Chem. Soc.* **124**, p. 8163

256. P. Silvestrelli, A. Alavi, M. Parrinello, and D. Frenkel (1996) Hot electrons and the approach to metallic behaviour in Kx(KCl)1-x. *Euro. Phys. Lett.* **33**, p. 551

257. R. Vuilleumier, M. Sprik, and A. Alavi (2000) Computation of electronic chemical potentials using free energy density functionals. *Journal of Molecular Structure (THEOCHEM)* **506**, p. 343

258. A. Y. Lozovoi, A. Alavi, J. Kohanoff, and R. M. Lynden-Bell (2001) Ab initio simulation of charged slabs at constant chemical potential. *J. Chem. Phys.* **115**, p. 1661

259. I. Tavernelli, R. Vuilleumier, and M. Sprik (2002) Ab initio molecular dymamics for molecules with variable numbers of electrons. *Phys. Rev. Lett.* **88**, p. 213002

260. J. Blumberger, Y. Tateyama, and M. Sprik (2005) Ab initio molecular dynamics simulation of redox reactions in solution. *Comp. Pys. Comm.* **169**, p. 256

261. D. Marx and M. Parrinello (1996) Ab initio path integral molecular dynamics: Basic ideas. *J. Chem. Phys.* **104**, p. 4077

262. M. E. Tuckerman, D. Marx, M. L. Klein, and M. Parrinello (1996) Efficient and general algorithms for path integral Car-Parrinello molecular dynamics. *J. Chem. Phys.* **104**, p. 5579

263. M. E. Tuckerman, D. Marx, M. L. Klein, and M. Parrinello (1997) On the Quantum Nature of the Shared Proton in Hydrogen Bonds. *Science* **275**, p. 817

264. S. Miura, M. Tuckerman, and M. Klein (1998) An ab initio path integral molecular dynamics study of double proton transfer in the formic acid dimer. *J. Chem. Phys.* **109**, p. 5290

265. M. Benoit, D. Marx, and M. Parrinello (1998) Quantum effects on phase transitions in high-pressure ice. *Comp. Mat. Sci.* **10**, p. 88

266. M. Diraison, G. J. Martyna, and M. E. Tuckerman (1999) Simulation studies of liquid ammonia by classical ab initio, classical, and path-integral molecular dynamics. *J. Chem. Phys.* **111**, p. 1096

267. D. Marx, M. E. Tuckerman, and G. J. Martyna (1999) Quantum dynamics via adiabatic ab initio centroid molecular dynamics. *Comput. Phys. Commun.* **118**, p. 166

268. D. Marx, M. E. Tuckerman, and M. Parrinello (2000) Solvated excess protons in water: quantum effects on the hydration structure. *J. Phys.: Condens. Matter A* **12**, p. 153

269. U. Roethlisberger, M. Sprik, and J. Hutter (2002) Time and length scales in ab initio molecular dynamics. In *Bridging time scales: Molecular simulations for the next decade*, eds. P. Niebala, M. Maraschal and G. Ciccotti, Springer Verlag, p. 413

270. L. Guidoni, P. Maurer, S. Piana, and U. Rothlisberger (2002) Hybrid Car-Parrinello/Molecular Mechanics Modelling of Transition Metal Complexes: Structure, Dynamics and Reactivity. *Quant. Struct.-Act. Rel.* **21**, p. 119

271. U. Röhrig, I. Frank, J. Hutter, A. Laio, J. VandeVondele, and U. Röthlisberger (2003) A QM/MM Car-Parrinello Molecular Dynamics Study of the Solvent Effects on the Ground State and on the First Excited Singlet State of Acetone in Water. *Chem. Phys. Chem.* **4**, p. 1177

272. M. Sulpizi, A. Laio, J. VandeVondele, U. Rothlisberger, A. Cattaneo, and P. Carloni (2003) Reaction Mechanism of Caspases: Insights from Mixed QM/MM Car-Parrinello Simulations. *Proteins-Structure, Function and Genetics* **52**, p. 212

273. Y. Liu, D. A. Yarne, and M. E. Tuckerman (2003) Ab initio molecular dynamics calculations with simple, localized, orthonormal real-space basis sets. *Phys. Rev. B* **68**, p. 125110

274. F. R. Krajewski and M. Parrinello (2005) Stochastic linear scaling for metals and nonmetals. *Phys. Rev. B* **71**, p. 233105

Large Scale Condensed Matter Calculations using the Gaussian and Augmented Plane Waves Method

J. VandeVondele[1], M. Iannuzzi[2], and J. Hutter[2]

[1] Department of Chemistry, University of Cambridge, Lensfield Road, Cambridge CB2 1EW, United Kingdom
[2] Physical Chemistry Institute, University of Zurich, Winterthurerstrasse 190, CH-8057 Zurich, Switzerland

Jürg Hutter

J. VandeVondele et al.: *Large Scale Condensed Matter Calculations using the Gaussian and Augmented Plane Waves Method*, Lect. Notes Phys. **703**, 287–314 (2006)
DOI 10.1007/3-540-35273-2_8

1 Introduction

Density functional theory DFT Kohn-Sham [16] is the method of choice for the calculation of electronic properties of large systems. This is due to the combination of accuracy and efficiency that has been achieved for the Kohn–Sham (KS) method in DFT [19]. DFT based electronic structure calculations are nowadays routinely used by chemists and physicists to support their research. Increasingly complex systems can be treated and the inclusion of environmental effects, through implicit or explicit solvent treatments, as well as the effects of different thermodynamic parameters (temperature, pressure) through first–principles molecular dynamics, opens the door for simulations close to experimental conditions. The accuracy of the method is such that many properties of systems of interest to chemistry, physics, material science, and biology can be predicted in a parameter free way. The success of the KS method makes it also the favorite framework for new developments to improve both, accuracy and efficiency. Better accuracy in this context can be achieved along two lines. On one hand the numerical limit of a given model should be reached and on the other hand more accurate models should be developed (i.e. exchange-correlation functional in DFT). The development of new functionals is an art on its own and will not concern us here. However, it is intimately related to the efficiency problem, as only numerically accurate tests on more and more complex systems can give unambiguous information on the performance of new functionals. The goal of improved algorithms is therefore, to provide methods to accurately and efficiently solve the KS equations.

Our recent work focused on the development of methods to perform KS calculations using accurate basis sets on large systems, including condensed matter systems which require periodic boundary conditions (PBC). The method uses an atom-centered Gaussian-type basis to describe the wave functions and an auxiliary plane wave basis to describe the density. It is therefore called Gaussian and plane waves (GPW) method. With a density represented as plane waves or on a regular grid, the efficiency of Fast Fourier Transforms (FFT) can be exploited to solve the Poisson equation and to obtain the Hartree energy in a time that scales linearly with the system size [28]. The use of an auxiliary basis set to represent the density goes back to the seventies [8, 37] and has become increasingly popular as resolution of the identity (RI) method or density fitting method. Contrary to the GPW method, most RI methods expand the density in an auxiliary basis of the same nature as the primary basis.

Periodic boundary conditions follow naturally from the FFT based treatment of the Poisson equation, and the GPW method scales linearly for three dimensional systems with a small prefactor and an early onset. The GPW method seems therefore best suited for the simulation of large and dense systems, such as liquids and solids, and most recent applications of the method fall in this category [17, 21–23, 35].

In order to reduce the size of the PW basis set pseudo potentials (PP) of the dual-space type [12, 13] are used. The latest implementation of the GPW method [34] has been done within the CP2K program and the corresponding module is called Quickstep [32]. In this implementation the linear scaling calculation of the GPW KS matrix elements is combined with an optimizer based on orbital transformations [33]. This optimization algorithm scales linearly in the number of basis functions for a given system size and, in combination with parallel computers, it can be used for systems with several thousands of basis functions [33, 34].

The GPW method is very efficient in comparison with most other methods. However, since we rely on the pseudo potential approximation in order to limit the size of the PW auxiliary basis, GPW is not suited for those applications that require the all electron density. Moreover, even when PP are used, systems containing second-row transition metals require the inclusion of rather localized semi-core states into the valence region. The resulting density can only be properly described with a PW expansion using a very large basis set, therefore reducing the efficiency of the GPW method. These drawbacks of the GPW method have been overcome with the Gaussian and augmented-plane wave (GAPW) method [21, 27]. The GAPW method, also implemented in the CP2K/Quickstep program, uses the PW representation of the density only for the smoothly varying density between atoms, but relies on localized functions for the rapidly varying density close to the nuclei. The basic idea of this separation was taken from the projector augmented-wave (PAW) scheme proposed by Blöchl [5].

The extensive experience with Gaussian-type basis sets shows that basis set sequences that increase rapidly in accuracy can be constructed in a systematic way [9]. At the same time, a compact description of the wave functions is maintained, and this opens the way for efficient methods to solve for the self consistent field (SCF) equations. Furthermore, as Gaussian functions are localized, the representations of the KS, overlap and density matrix in this basis become sparse with increasing system size [11]. This eventually allows for solving the KS equations using computational resources that scale linearly with system size.

2 Kohn–Sham Density Functional Method

The wavefunction of a system of N non-interacting electrons in a local potential $V_s(\boldsymbol{r})$ can be written as an anti-symmetrized product of one-electron functions

$$\Psi(\boldsymbol{r}_1, \boldsymbol{r}_2, \ldots, \boldsymbol{r}_N) = \mathcal{A}\left(\Pi_{i=1}^N \Phi_i(\boldsymbol{r}_i)\right) , \tag{1}$$

where the orbital functions are orthonormal

$$\int \Phi_i^\star(\boldsymbol{r})\Phi_j(\boldsymbol{r}) \, d\boldsymbol{r} = \delta_{ij} . \tag{2}$$

The total electron density is therefore given by

$$n(\boldsymbol{r}) = \sum_i \|\Phi_i(\boldsymbol{r})\|^2 \ , \tag{3}$$

and the orbitals are solutions to the one-particle Schrödinger equations

$$\left(-\frac{1}{2}\nabla^2 + V_s(\boldsymbol{r})\right)\Phi_i(\boldsymbol{r}) = \epsilon_i\Phi_i(\boldsymbol{r}) \ . \tag{4}$$

In the Kohn-Sham (KS) method [19] the local potential $V_s(\boldsymbol{r})$ has to be chosen such, that the resulting electron density reproduces the density of the system of interest with interacting electrons. Under this condition, the application of the Hohenberg-Kohn theorems [16] results in an explicit expression for the local KS potential and the total electronic energy. The KS total energy is

$$E[\{\Phi_i\}_{i=1}^N] = E_s^{\mathrm{T}}[\{\Phi_i\}_{i=1}^N] + E^{\mathrm{ext}}[n] + E^{\mathrm{H}}[n] + E^{\mathrm{XC}}[n] + E^{\mathrm{II}} \ , \tag{5}$$

where $E_s^{\mathrm{T}}[\{\Phi_i\}_{i=1}^N]$ is the electronic kinetic energy of non-interacting electrons, $E^{\mathrm{ext}}[n]$ is the electronic interaction with an external potential provided by the ionic cores, $E^{\mathrm{H}}[n]$ is the electronic Hartree (classical Coulomb) energy and $E^{\mathrm{XC}}[n]$ is the exchange-correlation energy. The interaction energies of the ionic cores with charges Z_A and positions $\boldsymbol{R}_A$ is denoted by E^{II}. The KS potential is defined by

$$V_s[n](\boldsymbol{r}) = V^{\mathrm{ext}}(\boldsymbol{r}) + V^{\mathrm{H}}[n](\boldsymbol{r}) + V^{\mathrm{XC}}[n](\boldsymbol{r}) \tag{6}$$

where the external potential, the Hartree potential, and the exchange-correlation potential

$$V^{\mathrm{ext}}(\boldsymbol{r}) = \sum_A \frac{Z_A}{|\boldsymbol{r} - \boldsymbol{R}_A|} \ , \tag{7}$$

$$V^{\mathrm{H}}[n](\boldsymbol{r}) = \int \frac{n(\boldsymbol{r}')}{|\boldsymbol{r} - \boldsymbol{r}'|} d\boldsymbol{r}' \ , \tag{8}$$

$$V^{\mathrm{XC}}[n](\boldsymbol{r}) = \frac{\delta E^{\mathrm{XC}}[n]}{\delta n(\boldsymbol{r})} \ , \tag{9}$$

have been introduced.

3 Gaussian and Augmented Plane Waves Method

Like many other approaches in quantum chemistry, the GAPW method uses a basis of contracted Gaussian functions to expand the Kohn-Sham orbitals $\Phi_i(\mathbf{r})$

$$\Phi_i(\mathbf{r}) = \sum_\alpha C_{\alpha i}\, \varphi_\alpha(\mathbf{r}) \,, \qquad (10)$$

$$\varphi_\alpha(\mathbf{r}) = \sum_m d_{m\alpha}\, g_m(\mathbf{r}) \,. \qquad (11)$$

The contraction coefficients $d_{m\alpha}$ are held fixed during a calculation and the functions $g_m(\mathbf{r})$ are primitive Gaussian function characterized by the order of their monomial pre-factor and their exponent α_m

$$g_m(\mathbf{r}) = x^{m_x} y^{m_y} z^{m_z} e^{-\alpha_m r^2} \,. \qquad (12)$$

These simple functions are an efficient basis set to describe atomic and molecular orbitals. Furthermore, all the density independent contributions to the KS Hamiltonian, like the kinetic energy and the electronic interaction with the ionic cores, can be calculated analytically using integral recurrence relations [30]. In order to avoid the cumbersome and time consuming four center integrals needed for the Coulomb terms, the GPW method exploits the PW representation of the density. The Coulomb potential is then calculated in reciprocal space using fast Fourier transforms [26]. In order to describe the rapidly varying electronic density in the vicinity of the atoms, the GAPW method employs a partitioning of the electronic density. The usual expansion of the density using the density matrix P

$$n(\mathbf{r}) = \sum_{\alpha\beta} P_{\alpha\beta}\, \varphi_\alpha(\mathbf{r})\, \varphi_\beta^\star(\mathbf{r}) \,, \qquad (13)$$

is replaced in the calculation of the Coulomb and exchange-correlation (xc) energy by

$$n(\mathbf{r}) = \tilde{n}(\mathbf{r}) + \sum_A n_A(\mathbf{r}) - \sum_A \tilde{n}_A(\mathbf{r}) \,. \qquad (14)$$

The partitioning of the density introduced in (14) is borrowed from the projector augmented-wave approach [5]. Its special form separates the smooth contributions $\tilde{n}$, characteristic of the interatomic regions, from the quickly varying terms close to the atoms n_A. $\tilde{n}_A$ is a smooth local term which compensates for the overlap of the soft and the hard densities in the atomic region, so that the integrals can still be expanded over all space. In our current implementation the densities $\tilde{n}(\mathbf{r}), n_A(\mathbf{r})$, and $\tilde{n}_A(\mathbf{r})$ are expanded in plane-waves and products of primitive Gaussians centered on atom A, respectively

$$\tilde{n}(\mathbf{r}) = \frac{1}{\Omega} \sum_{\mathbf{G}} \tilde{n}(\mathbf{G})\, e^{i\mathbf{G}\cdot\mathbf{r}} \,, \qquad (15)$$

$$n_A(\mathbf{r}) = \sum_{mn\in A} P_{mn}^A\, g_m(\mathbf{r})\, g_n^\star(\mathbf{r}) \,, \qquad (16)$$

$$\tilde{n}_A(\mathbf{r}) = \sum_{mn\in A} \tilde{P}_{mn}^A\, g_m(\mathbf{r})\, g_n^\star(\mathbf{r}) \,. \qquad (17)$$

In (15), $\tilde{n}(\mathbf{G})$ are the Fourier coefficients of the soft density, as obtained from (13) by keeping in the expansion of the contracted Gaussians only the primitives with exponents smaller than a given threshold. Ω denotes the volume of the periodic cell and all wave-vectors $\mathbf{G}$ corresponding to a given grid spacing are included in the expansion. The expansion coefficients P_{mn}^A, and $\tilde{P}_{mn}^A$ are also functions of the density matrix $P_{\alpha\beta}$ and can be calculated efficiently as reported in the following Sect. 5 [27]. The sum of the contributions in (14) gives the correct full density if the following conditions are fulfilled

$$n(\mathbf{r}) = n_A(\mathbf{r}) \qquad \tilde{n}(\mathbf{r}) = \tilde{n}_A(\mathbf{r}) \qquad \text{close to atom } A , \qquad (18)$$

$$n(\mathbf{r}) = \tilde{n}(\mathbf{r}) \qquad n_A(\mathbf{r}) = \tilde{n}_A(\mathbf{r}) \qquad \text{far from atom } A . \qquad (19)$$

The first conditions are exactly satisfied only in the limit of a complete basis set. However, the approximation introduced in the construction of the local densities, can be systematically improved by choosing larger basis sets.

For semi-local xc functionals such as the local density approximation, general gradient approximations or meta functionals using the kinetic energy density, the xc energy can, using (18) and (19), be written as

$$E_{\text{XC}}^{\text{GAPW}}[n] = E_{\text{XC}}[\tilde{n}] + \sum_A E_{\text{XC}}[n_A] - \sum_A E_{\text{XC}}[\tilde{n}_A] . \qquad (20)$$

The first term is calculated on the real-space grid defined by the plane wave expansion and the other two are efficiently and accurately calculated using atom centered meshes.

Due to the non-local character of the Coulomb operator, the decomposition for the electrostatic energy is more complex. In order to distinguish between local and global terms, we need to introduce atom-dependent screening densities, n_A^0 (hard) and $\tilde{n}_A^0$ (soft), that generate the same multipole expansion Q_A^{lm} as the local density $n_A - \tilde{n}_A + n_A^Z$, where n_A^Z is the nuclear charge of atom A.

$$n_A^0(\mathbf{r}) = \sum_{lm} Q_A^{lm} g_A^{lm}(\mathbf{r}) , \qquad (21)$$

$$\tilde{n}_A^0(\mathbf{r}) = \sum_{lm} Q_A^{lm} \tilde{g}_A^{lm}(\mathbf{r}) . \qquad (22)$$

The primitive Gaussians $g_A^{lm}(\mathbf{r})$ and $\tilde{g}_A^{lm}(\mathbf{r})$ are defined with large and small exponents, respectively, and normalized. Since the sum of local densities $n_A - \tilde{n}_A + n_A^Z - n_A^0$ has vanishing multiple moments, it does not interact with charges outside the localization region, and the corresponding energy contribution can be calculated by one-center integrals. The final form of the Coulomb energy in the GAPW method [27] is given by

$$E_{\text{C}}^{\text{GAPW}}[n + n^Z] = E_{\text{H}}[\tilde{n} + \tilde{n}^0] + E_{\text{H}}[n^0] - E_{\text{H}}[\tilde{n}^0] +$$

$$\int d\mathbf{r} V_{\text{H}}[n^0 - \tilde{n}^0]\tilde{n} + \sum_A E_{\text{H}}[n_A + n_A^Z] - \sum_A E_{\text{H}}[\tilde{n}_A + n_A^0] , \qquad (23)$$

where quantities $n^0, \tilde{n}^0$ are summed over all atomic contributions, and $E_H[n]$ and $V_H[n]$ denote the Coulomb energy and potential of a charge distribution n. The first term in (23) can be calculated efficiently using fast Fourier transform methods. The next three terms involve Coulomb integrals over two and three Gaussian functions that can be calculated efficiently using analytic integral formulas [30]. The last two terms are one-center terms and are calculated in our current implementation on radial atomic grids.

The total electronic energy within the GAPW method is therefore calculated from

$$E_{el} = \sum_{\alpha\beta} P_{\alpha\beta} h_{\alpha\beta} + E_C^{GAPW}[n + n^Z] + E_{XC}^{GAPW}[n] , \qquad (24)$$

where $h_{\alpha\beta}$ denotes matrix elements of the core Hamiltonian, the sum of kinetic energy and local part of the external potential [21, 27]. This form of the GAPW energy functional involves several approximations in addition to a standard implementation. The number of reciprocal space vectors included into the expansion of the smooth density $\tilde{n}(\mathbf{r})$ controls the accuracy of the corresponding terms in Coulomb and xc energy. The accuracy of the local expansion of the density is controlled by the flexibility of the product basis of primitive Gaussians. Therefore, we have to consider this approximation as inherent to the primary basis used.

4 Pseudo and Core Potentials

For a wide range of chemically interesting events, such as bond breaking and formation, an accurate description is required only for the valence electrons. Such an accurate description can be obtained using a pseudo potential description of the nuclei. This technique is well established in the plane wave community. We take advantage of the experience with this scheme using the pseudo potentials of Goedecker et al. (GTH) [12, 13]. These accurate and transferable pseudo potentials have an analytic form that allows for an efficient treatment of all terms within the GPW method.

The norm-conserving, separable, dual-space GTH pseudo potentials consist of a local part including a long-ranged (LR) and a short-ranged (SR) term

$$V_{loc}^{PP}(r) = V_{loc}^{LR}(r) + V_{loc}^{SR}(r) \qquad (25)$$

$$= -\frac{Z_{ion}}{r} \mathrm{erf}\left(\frac{r}{\sqrt{2}r_{loc}^{PP}}\right) + \sum_{i=1}^{4} C_i^{PP} \left(\frac{r}{r_{loc}^{PP}}\right)^{2i-2} e^{-\left(\frac{r}{\sqrt{2}r_{loc}^{PP}}\right)^2}$$

where r_{loc}^{PP} is the radius of the local pseudo potential and a non-local part

$$V_{nl}^{PP}(\mathbf{r}, \mathbf{r}') = \sum_{lm} \sum_{ij} \langle \mathbf{r} \,|\, p_i^{lm} \rangle \, h_{ij}^l \, \langle p_j^{lm} \,|\, \mathbf{r}' \rangle \qquad (26)$$

with the Gaussian-type projectors

$$\langle \boldsymbol{r} \mid p_i^{lm} \rangle = N_i^l \, Y^{lm}(\hat{r}) \, r^{l+2i-2} \exp \left[-\frac{1}{2} \left(\frac{r}{r_l} \right)^2 \right] ,$$

where N_i^l are normalisation constants and $Y^{lm}(\hat{r})$ spherical harmonics. The small set of GTH pseudo potential parameters ($r_{\text{loc}}^{\text{PP}}$, C_i^{PP}, r_l, and h_{ij}^l) have been optimized with respect to atomic all electron wave functions as obtained from fully relativistic density functional calculations using a numerical atomic program. The optimised pseudo potentials include all scalar relativistic corrections via an averaged potential, [13] and improve therefore the accuracy for applications involving heavier elements. The emphasis in the construction of these pseudo potentials has been on accuracy, and hence these pseudo potentials are computationally more demanding for plane wave methods, as a large plane wave basis typically is required. The GPW and GAPW methods are less sensitive to the hardness of the pseudo potential since the kinetic energy and the short range pseudo potential can be easily computed analytically as two and three centre overlap integrals.

In the case of the GAPW method we also have the possibility to avoid pseudo potentials and use only the bare nuclear core potential. We can cast this potential into a form similar to the above defined pseudo potentials

$$V^{\text{ext}}(r) = V^{\text{LR}}(r) + V^{\text{SR}}(r) = -\frac{Z_{\text{ion}}}{r} \text{erf}\,(\alpha r) - \frac{Z_{\text{ion}}}{r} \text{erfc}\,(\alpha r) , \qquad (27)$$

where erfc denotes the complementary error function and α is an adjustable parameter that, however, can be easily fixed for most calculations. The long range part of the bare nuclear or pseudo potential can be efficiently treated as part of the electrostatic energy. The short range part can be calculated using analytical formulas and efficient screening procedures.

5 Basis Sets

Significant experience exists with Gaussian basis sets and they are available in a number of formats [1,15]. Polarization and diffuse functions can normally be adopted from published basis sets. For all electron GAPW calculations also the core part of the basis can be used un-altered. The use of GTH PP requires adapted basis sets. A systematically improving sequence of basis sets for use with the GTH PP was optimized for all first- and second-row elements, using the procedure detailed below.

Exponents of a set of primitive Gaussian function have been optimized to yield the lowest pseudo atom energies for all first- and second-row elements with an atomic DFT code employing the appropriate GTH potential for each element. A family basis set scheme has been adopted using the same set of exponents for each angular momentum quantum number of the occupied valence states, i.e. s and p orbitals for the elements from H to Ar. A growing number of primitive Gaussian functions, typically four to six, has been included into these sets to provide an increasingly good description of the pseudo atomic wave function. Finally, these primitive Gaussian functions have been contracted using the coefficients of the respective pseudo atomic wave functions. In addition, a split valence scheme has been applied to enhance the flexibility of the valence basis part. The splitting has been increased in line with the number of primitive Gaussian functions employed from double- (DZV) over triple- (TZV) up to quadruple-zeta valence (QZV). For instance, the basis set sequence for oxygen starts with four primitive Gaussian functions on the DZV level, uses five functions for TZV, and finally six on the QZV level. Moreover, these basis sets have been augmented by polarization functions which were taken from the all electron basis sets cc-pVXZ (X = D, T, Q) of Dunning, [9,38] but only the first p or d set of polarization functions has been used depending on the actual element. In that way a new sequence of basis sets has been created with an increasing number of primitive Gaussian functions and polarization functions for each first- and second-row element. The basis sets have been labeled DZVP, TZVP, TZV2P, QZV2P, and QZV3P according to the applied degree of splitting and the increasing number of provided polarization functions.

The one center densities n_A and $\tilde{n}_A$ are constructed as expansions in a basis set centered on atom A. The conditions in (18) can be satisfied by finding for each A two sets of atom centered functions χ_α^A and $\tilde{\chi}_\alpha^A$ such that

$$n_A(\mathbf{r}) = \sum P_{\alpha\beta}\chi_\alpha(\mathbf{r})\chi_\beta(\mathbf{r}) \ , \tag{28}$$

$$\tilde{n}_A(\mathbf{r}) = \sum P_{\alpha\beta}\tilde{\chi}_\alpha(\mathbf{r})\tilde{\chi}_\beta(\mathbf{r}) \ . \tag{29}$$

where the indexes α and β run over the entire basis set used in the expansion of (13), and $P_{\alpha\beta}$ is the same density matrix. To construct n_A, for example, each $\chi_\alpha(\mathbf{r} - \mathbf{R}_A) \approx \varphi_\alpha(\mathbf{r})$ within the cutoff r_A, whether $\varphi_\alpha(\mathbf{r})$ is centered on A or not. If φ_α is centered on A, $\chi_\alpha = \varphi_\alpha$ and the same expansion as in (11) is used. Otherwise its contribution to the density in the region around A is much smaller and comes from the smooth tail of the function that reaches A. The corresponding χ_α can still be represented through an expansion in the primitive Gaussians of A, but using atom dependent contraction coefficients $\chi_\alpha = \sum_m d_{m\alpha}'^A g_m(\mathbf{r})$. In other words, the latter expansion has to provide the projection of φ_α onto the atomic region of A. To this end we take a new set of Gaussians centered on A, which serves as a projector basis $\{p_m\}$. These are in number equal to the set of primitives $\{g_m\}$, and they are generated from the exponents of the geometric progression $\lambda_m = k^m \lambda_{min}$. The minimum exponent λ_{min} is the smallest exponent representing a Gaussian confined within r_A, while the factor k is determined by the number of primitives, to limit the size of the maximum exponent and to avoid numerical instabilities. Projecting the basis functions then yields

$$\langle p_n | \varphi_\alpha \rangle = \sum_m d'^A_{m\alpha} \langle p_n | g_m \rangle \tag{30}$$

and inverting the $\langle p | g \rangle$ matrix the contraction coefficients $\{d'^A_{m\alpha}\}$ can be determined such that

$$n_A(\mathbf{r}) = \sum_{mn} \left[\sum_{\alpha\beta} P_{\alpha\beta} d'^A_{m\alpha} d'^A_{n\beta} \right] g_m(\mathbf{r}) g_n^*(\mathbf{r}) = \sum_{mn} P^A_{mn} g_m(\mathbf{r}) g_n^*(\mathbf{r}) \tag{31}$$

One could use the same procedure to find the contraction coefficients $\tilde{d}'^A_{m\alpha}$ for the expansion of the $\tilde{\chi}_\alpha$ functions. However, since the contribution coming from a φ_α which is not centered on A is smooth by definition, the corresponding expansion coefficients have to be the same, $d'^A_{m\alpha} = \tilde{d}'_{m\alpha}$. If, instead, $\tilde{\chi}_\alpha$ is centered on A we use the same technique as for the interstitial smooth density $\tilde{n}(\mathbf{r})$, i.e. we take the same expansion as in (11) and simply set to zero the contraction coefficients that multiply primitives with exponents larger than a given threshold.

6 Integrals

Analytic and numerical integration is performed using unnormalized Cartesian Gaussian functions. We define such a function as

$$|\mathbf{n}) = \varphi(\mathbf{r}; \eta, \mathbf{n}, \mathbf{R}) = (x - R_x)^{n_x} (y - R_y)^{n_y} (z - R_z)^{n_z} \exp[-\eta(\mathbf{r} - \mathbf{R})^2], \tag{32}$$

where η is the orbital exponent and $\mathbf{n} = (n_x, n_y, n_z)$ denotes a set of non-negative integers. Functions with the same orbital exponent η, the same origin $\mathbf{R}$, and the same angular momentum $l = n_x + n_y + n_z$ constitute a shell. There are $(l+1)(l+2)/2$ functions in each shell and shells with $l = 0, 1, 2, \ldots$ are referred to as $s, p, d, \ldots$, respectively. An s shell consists of a single component with index $\mathbf{0} = (0, 0, 0)$. The angular momentum indices of a p and d shell are $\mathbf{1}_i = (\delta_{ix}, \delta_{iy}, \delta_{iz})$ $(i = x, y, z)$ and $\mathbf{1}_i + \mathbf{1}_j$ $(i, j = x, y, z)$.

A basic property of Cartesian Gaussian functions is the differential relation

$$\frac{\partial}{\partial R_i} |\mathbf{n}) = 2\eta |\mathbf{n} + \mathbf{1}_i) - N_i(\mathbf{n}) |\mathbf{n} - \mathbf{1}_i), \tag{33}$$

where $N_i(\mathbf{n})$ stands for the value of component i of vector $\mathbf{n}$, i.e. n_i in this case. Nuclear coordinates always appear in the form $r_i - R_i$ and therefore differentiation w.r.t. R_i is related to differentiation w.r.t. r_i

$$\frac{\partial}{\partial R_i} |\mathbf{n}) = -\frac{\partial}{\partial r_i} |\mathbf{n}). \tag{34}$$

6.1 Analytic Integrals

The success of Cartesian Gaussian functions in quantum chemistry is based on efficient algorithms for the calculation of analytic integrals. For a review of the many different schemes that have been proposed, see [14]. We follow the simple scheme of Obara and Saika [30] that relies on recursion relations starting from basic integrals over s type functions. To derive the recursion relations we need two basic identities. First, the shift of angular momentum is achieved by

$$(a|\mathcal{O}(r)|b + 1_i) = (a + 1_i|\mathcal{O}(r)|b) + (A_i - B_i)(a|\mathcal{O}(r)|b) , \qquad (35)$$

and second, we need the invariance of integrals w.r.t. differentiation to the electronic position variable

$$\frac{\partial}{\partial r_i}(a|\mathcal{O}(r)|b) = 0 . \qquad (36)$$

Overlap Integrals

The recursion relation for overlap integrals is

$$(a+1_i \mid b) = (P_i - A_i)(a \mid b) + \frac{1}{2\zeta}N_i(a)(a - 1_i \mid b) + \frac{1}{2\zeta}N_i(b)(a \mid b - 1_i) , \qquad (37)$$

where

$$\boldsymbol{P} = \frac{\alpha\boldsymbol{A} + \beta\boldsymbol{B}}{\alpha + \beta} \quad \text{and} \quad \zeta = \alpha + \beta . \qquad (38)$$

The basic integral over s functions is

$$(0_a \mid 0_b) = \left(\frac{\pi}{\zeta}\right)^{3/2} \exp[-\xi(\boldsymbol{A} - \boldsymbol{B})^2] , \qquad (39)$$

where

$$\xi = \frac{\alpha\beta}{\alpha + \beta} . \qquad (40)$$

Kinetic Energy Integrals

The recursion relation for integrals over the kinetic energy operator $\mathcal{T} = -\frac{1}{2}\nabla^2$ is

$$(a + 1_i|\mathcal{T}|b) = (P_i - A_i)(a|\mathcal{T}|b) + \frac{1}{2\zeta}N_i(a)(a - 1_i|\mathcal{T}|b)$$

$$+ \frac{1}{2\zeta}N_i(b)(a|\mathcal{T}|b - 1_i) + 2\xi\left\{(a + 1_i \mid b) - \frac{1}{2\alpha}N_i(a)(a - 1_i \mid b)\right\} , \qquad (41)$$

with the basic integral

$$(0_a|\mathcal{T}|0_b) = \xi\left\{3 - 2\xi(\boldsymbol{A} - \boldsymbol{B})^2\right\}(a \mid b) . \qquad (42)$$

Nuclear Attraction and Pseudo Potential Integrals

The nonlocal part of the pseudo potential requires the calculation of overlap integrals between the basis functions and the projectors. As the projectors can be written themselves as Cartesian Gaussian functions the same formulas as for the basis set overlap integrals can be used. The error function part of the local pseudo potential and the core potential can be written as a special case of a Coulomb integral

$$
\left(a \left| \frac{\mathrm{erf}[\sqrt{\gamma}|\mathbf{r} - \mathbf{C}|]}{|\mathbf{r} - \mathbf{C}|} \right| b \right) = \left(\frac{\gamma}{\pi} \right)^{3/2} \int\int d\mathbf{r} d\mathbf{r}' g_a(\mathbf{r}) g_b(\mathbf{r}) \frac{\exp[-\gamma(\mathbf{r}' - \mathbf{C})^2]}{|\mathbf{r} - \mathbf{r}'|} ,
$$

(43)

treated in the next section. The remaining part of the local pseudo potential reduces to three center overlap integrals, where the central pseudo potential Gaussian function only assumes even angular momentum values. The recursion relation for these integrals is

$$
(\mathbf{a} + 1_i|\mathbf{c}|\mathbf{b}) = H_i(\mathbf{a}|\mathbf{c}|\mathbf{b})
$$
$$
+ \frac{1}{2(\alpha + \beta + \gamma)} N_i(\mathbf{a})(\mathbf{a} - 1_i|\mathbf{c}|\mathbf{b}) + \frac{1}{2(\alpha + \beta + \gamma)} N_i(\mathbf{b})(\mathbf{a}|\mathbf{c}|\mathbf{b} - 1_i)
$$
$$
+ \frac{1}{2(\alpha + \beta + \gamma)} N_i(\mathbf{c}) \left[(\mathbf{a} + 1_i|\mathbf{c} - 2_i|\mathbf{b}) + (A_i - C_i)(\mathbf{a}|\mathbf{c} - 2_i|\mathbf{b}) \right] , \quad (44)
$$

with the basic integral

$$
(\mathbf{0}_a|\mathbf{0}_c|\mathbf{0}_b) = \left(\frac{\alpha + \beta}{\alpha + \beta + \gamma} \right)^{3/2} \exp\left[-\gamma \frac{\alpha + \beta}{\alpha + \beta + \gamma} (\mathbf{P} - \mathbf{C})^2 \right] (\mathbf{a} \mid \mathbf{b}) , \quad (45)
$$

where

$$
\mathbf{H} = \frac{\beta \mathbf{B} + \gamma \mathbf{C} - (\beta + \gamma)\mathbf{A}}{\alpha + \beta + \gamma} .
$$

(46)

Coulomb Integrals

Coulomb integrals appear in various forms in the GAPW method, either as integrals over two or three functions, with one, two, or three different centers. All these integrals can be treated as special cases of a general auxiliary integral with index m

$$
(\mathbf{ab}, \mathbf{cd}) = (\mathbf{ab}, \mathbf{cd})^{(m=0)} = \int\int d\mathbf{r} d\mathbf{r}' \frac{g_a(\mathbf{r}) g_b(\mathbf{r}) g_c(\mathbf{r}') g_d(\mathbf{r}')}{|\mathbf{r} - \mathbf{r}'|} .
$$

(47)

The starting integrals are

$$
(\mathbf{0}_a\mathbf{0}_b, \mathbf{0}_c\mathbf{0}_d)^{(m)} = 2 \left(\frac{\rho}{\pi} \right)^{1/2} (\mathbf{0}_a \mid \mathbf{0}_b)(\mathbf{0}_c \mid \mathbf{0}_d) F_m(T)
$$

(48)

and the recursion formula reads

$$(a + 1_i b, cd)^{(m)} = (P_i - A_i)(ab, cd)^{(m)} + (W_i - P_i)(ab, cd)^{(m+1)}$$
$$+ \frac{1}{2\zeta} N_i(a) \left[(a - 1_i b, cd)^{(m)} - \frac{\rho}{\zeta}(a - 1_i b, cd)^{(m+1)} \right]$$
$$+ \frac{1}{2\zeta} N_i(b) \left[(ab - 1_1, cd)^{(m)} - \frac{\rho}{\zeta}(ab - 1_i, cd)^{(m+1)} \right]$$
$$+ \frac{1}{2(\zeta + \eta)} \left[N_i(c)(ab, c - 1_i d)^{(m+1)} + (ab, cd - 1_i)^{(m+1)} \right] , \quad (49)$$

where $\eta = \gamma + \delta$ and $\rho = \frac{\zeta\eta}{\zeta+\eta}$ and

$$Q = \frac{\gamma C + \delta D}{\gamma + \delta} \quad \text{and} \quad W = \frac{\zeta P + \eta Q}{\zeta + \eta} . \quad (50)$$

The incomplete Gamma function and its argument T are defined by

$$F_m(T) = \int_0^1 dt \, t^{2m} \exp[-Tt^2]; \qquad T = \rho(P - Q)^2 . \quad (51)$$

6.2 Grids

Regular Grids

Central in the GPW and GAPW methods is the expansion of the density (or part thereof) on regular grids (see (15)). This grid based density is used to calculate the coulomb energy, the xc energy and the corresponding potentials. Fast Fourier Transforms enable a straightforward computation of the coulomb energy that effectively scales linear with the system size. Consistent with the definition of the energies, KS matrix elements are computed as integrals of the corresponding potentials on the same grids. The consistency implies that the KS matrix elements are, within machine accuracy, the exact derivatives of the KS energy with respect to the density matrix elements. This property, which remains valid if multigrids and finite support of the Gaussians are introduced, is highly valuable if an accurate wavefunction optimization is required [34].

The mapping of the full density (13) on to the grids (15) and the corresponding integration can be performed particularly efficiently for a primary Gaussian basis. This can mainly be attributed to the following three techniques: 1) The overlap of two Gaussians, and thus their contribution to (13), decays exponentially with the square of their distance. Therefore, within a given threshold, typically $O(10^{-10})$, only a small number of terms remains in the sum. For large systems, this number scales linearly with system size. Note that this property results from the choice of basis set, and is true for all systems, including e.g. metals where the density matrix elements do not decay rapidly. 2) The Gaussian width ($\sigma = 1/\sqrt{2\alpha}$) sets a well defined length scale that can be exploited using multigrid techniques. In particular, grids with varying spacing can be used to map Gaussians with roughly a fixed number of grid points per unit length. This leads to an accuracy and computational effort which is independent of the extend of the basis functions, resulting in large savings for diffuse functions. 3) The value of a product of Cartesian Gaussians can be computed very effectively, exploiting the property that a Cartesian Gaussian can be factorized in x, y and z components. Furthermore, all $O(l^6)$ angular moment components for a given center can be computed simultaneously in a time $O(l)$, retaining good efficiency also for high quality basis sets and force calculations.

The accuracy of the auxiliary basis, i.e. the density grids, depends only on the grid spacing, and can be systematically improved increasing the number of grid points, more commonly described as increasing the plane wave cutoff. The convergence of the total energy with respect to the plane wave cutoff, illustrated in Fig. 1 is rapid. Since only the auxiliary basis is improved, the convergence is towards a total energy determined by the Gaussian basis, and not towards the basis set limit. This is at variance with traditional plane wave calculations where plane waves are the only basis set employed. A further difference between these methods is the rate of convergence with respect to plane wave cutoff (see also Fig. 1). This is a result of describing part of the KS energy terms, in particular the kinetic energy, in the Gaussian basis. We remark that the rate of convergence is different for the Coulomb term and the exchange and correlation term (not shown). Whereas the Coulomb

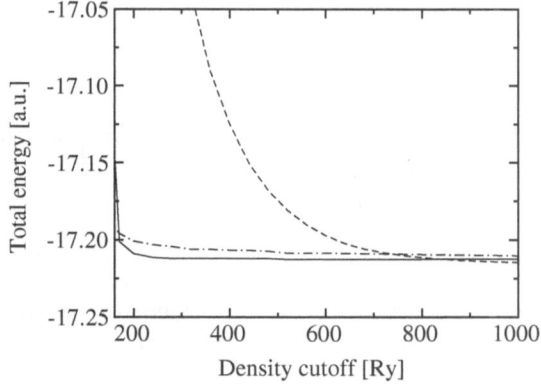

Fig. 1. The convergence of the total energy (BLYP) of a single water molecule with respect to the density cutoff using a traditional plane wave approach (CPMD, *dashed lines*), the GPW method using Fourier space derivatives (*solid lines*) and spline derivatives (*dash-dotted lines*). An aug-TZV2P Gaussian basis has been employed for the GPW calculations. Note that the cutoff reported for CPMD calculations is usually the wavefunction cutoff, which is defined as one fourth of the density cutoff employed here

term converges very smoothly, slower convergence is observed for the xc term, in particular if gradient corrected functionals are employed. As previously discussed, the latter term can cause ripples in the potential energy surface. It was shown in [34] that these ripples can be controlled using a spline based derivative of the density instead of the more intuitive Fourier space derivative.

Atom Centered Grids

The one center terms in (20) and in (23) depend only on the local densities n_A, $\tilde{n}_A$, n_A^0, $\tilde{n}_A^0$, and n_A^Z, which are expanded in terms of the spherical Gaussians centered on the same atom A. By using the spherical form of the Gaussian function, rather than the Cartesian one, we can take advantage of the separation into the radial and the angular contributions. For example, the hard density n_A given in (16) becomes

$$n_A(\mathbf{r}) = \sum_{mn} P_{mn}^A r^{l_m} e^{-\alpha_m r^2} \mathcal{Y}_{l_m m_m}(\theta, \phi)\, r^{l_n} e^{-\alpha_n r^2} \mathcal{Y}_{l_n m_n} = n_A(r, \theta, \phi) \quad (52)$$

The spherical harmonics $\mathcal{Y}_{lm}(\theta, \phi)$ are a complete basis set in the space described by θ and ϕ, i.e. any function defined in this space can be represented as a linear combination of spherical harmonics. In particular the expansion coefficients for the products of two spherical harmonics are tabulated and known as the Clebsch-Gordon coefficients , $C(m, n, LM)$. Here m and n label the angular momenta (l_m, m_m) and (l_n, m_n) of the two harmonics to be multiplied, and LM are the angular momentum and its z component of the harmonics in the expansion. By applying the Clebsch-Gordon expansion the density reads

$$\begin{aligned}
n_A(\mathbf{r}) &= \sum_{mn} P_{mn}^A r^{l_{mn}} e^{-\alpha_{mn} r^2} \sum_{LM} C(m, n, LM) \mathcal{Y}_{LM} \\
&= \sum_{LM} \left[\sum_{mn} n_{mn}^{(r)}(r)\, P_{mn}^A C(m, n, LM) \right] \mathcal{Y}_{LM}(\theta, \phi) \\
&= \sum_{LM} \mathcal{N}_A^{LM}(r) \mathcal{Y}_{LM}(\theta, \phi)
\end{aligned} \quad (53)$$

where $\alpha_{mn} = \alpha_m + \alpha_n$ and $l_{mn} = l_m + l_n$ Similar expansions can be used for the derivatives of spherical harmonics, where the expansion coefficients for a generic function $f(\theta, \phi)$ are simply the scalar products $\langle f | \mathcal{Y}_{LM} \rangle$ in θ and ϕ. As a consequence, beside the density, also the gradient of the density and the XC and Coulomb potential can be expressed as sum of products of a radial part and a spherical harmonic. The one center contributions to the KS matrix elements $\langle m | V_{XC}^A | n \rangle$ and $\langle m | V_C^A | n \rangle$ are integrated numerically on spherical grids that are centered on the same atom A. In our implementation, these local grids are constituted by a logarithmic radial part using a Gauss-Chebyshev quadrature formula [3,20]. Given the number of points N the grid is constructed for $1 \leq i \leq N$ as follows

$$x = \frac{2\,i - N - 1}{N + 1} - \frac{2}{\pi}\left[1 + \frac{2}{3}\left(\sin\left(\frac{\pi\,i}{N+1}\right)\right)^2\right]\cos\left(\frac{\pi\,i}{N+1}\right)\sin\left(\frac{\pi\,i}{N+1}\right)$$

$$r(N + 1 - i) = \frac{1}{\ln 2}\ln\left(\frac{2}{1 - x}\right)$$

$$w(N + 1 - i) = \frac{16}{\ln 2\,(1 - x)}\frac{(r(N+1-i))^2}{3(N+1)}\left(\sin\left(\frac{\pi\,i}{N+1}\right)\right)^4$$

where the points $r \in [0, \infty]$ and w are the weights. The spherical harmonics are instead integrated over a sphere by the Lebedev quadrature, [24] which is based on the octahedral group. Since a Lebedev grid of degree L integrates exactly all the spherical harmonics of degree L or less, and the number of points is approximately $(L + 1)^2/3$, the size of the grid depends on the maximum angular momentum present in the basis set.

The separation of the density in its radial and angular part makes it possible to compute the Coulomb potential generated by the local densities through the multiple expansion of a Gaussian charge distribution

$$V_{mn}^{A}(\mathbf{r}) = P_{mn}^{A}\sum_{LM}C(m,n,LM)\int\frac{r'^{l_{mn}}e^{-\alpha_{mn}r'^2}}{|\mathbf{r}-\mathbf{r}'|}\mathcal{Y}_{LM}(\theta',\phi')r'^2\sin\theta'dr'd\theta'd\phi'$$

$$= P_{mn}^{A}\sum_{\lambda\mu}\frac{4\pi}{2\lambda+1}\int r'^{l_{mn}+2}e^{-\alpha_{mn}r'^2}\frac{r_-^\lambda}{r_+^{\lambda+1}}dr'\ \times$$

$$\times \sum_{LM}C(m,n,LM)\int\mathcal{Y}_{LM}(\theta',\phi')\mathcal{Y}_{\lambda\mu}(\theta,\phi)\mathcal{Y}_{\lambda\mu}(\theta',\phi')\sin\theta'd\theta'd\phi'$$

$$\tag{54}$$

where

$$r_+ = r' \text{ and } r_- = r \qquad\qquad \text{for } r' > r$$
$$r_+ = r \text{ and } r_- = r' \qquad\qquad \text{for } r' < r$$

Since the spherical harmonics are an orthonormal basis set,

$$\int\mathcal{Y}_{\lambda\mu}(\theta',\phi')\mathcal{Y}_{LM}(\theta',\phi')d\theta'd\phi' = \delta_{\lambda L}\delta_{\mu M},\tag{55}$$

the above expression simplifies into

$$V_{mn}^{A}(\mathbf{r}) = P_{mn}^{A}\sum_{LM}C(m,n,LM)\frac{4\pi}{2L+1}\mathcal{Y}_{LM}(\theta,\phi)\int r'^{l_{mn}+2}e^{-\alpha_{mn}r'^2}\frac{r_-^L}{r_+^{L+1}}dr'$$

$$\tag{56}$$

The radial integral can be rewritten for each point r on the radial grid by separating the two contributions

$$\mathcal{I}(r) = \frac{1}{r^{L+1}} \int_0^r r'^{(l_{mn}+L+2)} e^{-\alpha_{mn} r'^2} dr' + r^L \int_r^\infty r'^{(l_{mn}-L+1)} e^{-\alpha_{mn} r'^2} dr'$$

$$(57)$$

Since the coefficients $C(m, n, LM)$ do not vanish only for $l_{mn} + L$ even, the two integrals in (57) can be solved in terms of the Whittaker functions W, [36] given l even and $\alpha > 0$

$$\int_0^r r'^{l+2} e^{-\alpha r'^2} dr' = \frac{r^{l+1} (\alpha r^2)^{(-\frac{(l+1)}{4})} e^{-\frac{\alpha r^2}{2}} \mathrm{W}(\frac{l+1}{4}, \frac{l+3}{4}, \alpha r^2)}{\alpha(l+3)}$$

$$(58)$$

$$\int_r^\infty r'^{l+1} e^{-\alpha r'^2} dr' = \frac{\Gamma(\frac{l+2}{2}, \alpha r^2)}{2\alpha^{(l+2)/2}}$$

$$(59)$$

7 Accuracy

7.1 Small Molecules

As a first accuracy test for the GPW method we optimized the geometry of a set of 39 small molecules consisting of first- and second-row elements using the local density approximation (LDA). Results are compared to the NUMOL results of Dickson and Becke [7]. NUMOL is a purely numerical DFT code and thus considered to be free of basis set effects. The smallest basis set DZVP gives on average slightly too long bond distances, but already the TZVP basis set works fine for most of the molecules. Finally, the TZV2P, QZV2F, and QZV3P show a satisfactory agreement for all bond distances. The agreement for bond angles for the small DZVP and the TZVP basis set is already excellent. Only one data point is off which corresponds to the dihedral angle of H_2O_2. This angle is known to be very sensitive to the number of employed polarization functions. However, for the TZV2P basis set the dihedral angle is already very close to the reference value and for the QZV3P basis set shows more or less a converged result. A summary of the numerical results of the geometry optimizations is provided in Table 1 which shows the

Table 1. Maximum ($\Delta_{\max}$) and root mean square deviation (σ) of bond distances (Å) and bond angles and dihedral angles (°) compared to the NUMOL results for different basis sets

Basis Set	Distances [Å]		Angles [°]	
	$\Delta_{\max}$	σ	$\Delta_{\max}$	σ
DZVP	0.048	0.018	6.4	1.6
TZVP	0.040	0.013	8.5	2.1
TZV2P	0.015	0.006	1.7	0.6
QZV2P	0.012	0.005	2.1	0.6
QZV3P	0.011	0.004	0.7	0.3

Table 2. Maximum average absolute (Δ_{max}) and root mean square standard deviation (σ) of bond distances (Å) and total energies (μHartree) compared to quantum chemistry calculations using a standard approach with the same basis set

Basis Set	Distances [Å]		Total Energy [μHartree]	
	Δ_{max}	σ	Δ_{max}	σ
6-31G*	0.0098	0.0152	466	567
6-311++G(3df,3dp)	0.0068	0.0103	105	165

maximum and the root mean square deviation of all bond distances and angle compared to the NUMOL results based on 52 bond distances and 18 angles and dihedral angles. The errors become smaller for growing basis set size as expected. The TZV2P basis set gives already an excellent overall agreement and for the QZV3P most distances coincide within the expected errors. Note, that a full agreement with the NUMOL values is not possible, since NUMOL uses a slightly different LDA implementation and it employs a frozen core approximation for the elements beyond Beryllium that differs from the GTH pseudo potential.

To assess the accuracy of the GAPW method we compare molecular structures and total energies of a series of small molecules with results obtained by standard quantum chemistry packages for the BLYP density functional [2,25]. The results from the standard calculations can be considered the limiting values of a GAPW calculation, that have to be reached for the case of a complete plane wave and local function basis. The calculations were performed with a 6-31G* and 6-311++G(3df,3dp) Gaussian basis sets. The error is for all cases below 0.01 Å. The test calculations have been performed with a kinetic energy cut off of 250 Rydbergs ensuring almost converged results for the plane wave part. The employed set of molecules and the error in total energy and bond length for each of them are reported in [18]. The expansion of the local atomic densities relies on the primitive Gaussians present in the primary basis set. We therefore can expect that large basis sets will have results closer to the reference value. This is confirmed by the decrease in the maximum difference in the bond length reported for almost all the molecules. This feature is even more evident from the differences in the single point energies and the energies of the optimized structures. While for the smaller basis set the total energy differences are always lower than 1 mHartree, with the 6-311++G(3df,3dp) basis they decrease to values below 0.1 mHartree. This number has to be compared with differences of the same order or larger for auxiliary basis calculations [10].

Calculation methods that involve regular grids or plane wave basis sets for the evaluation of the exchange and correlation integrals all show a periodic dependence of the total energy on the grid spacing. These energy ripples are also known in plane waves calculations and are often very irritating, especially for high precision calculations. In the GPW method the ripples problem is

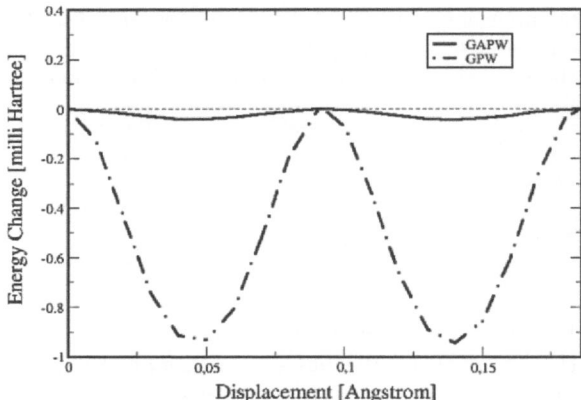

Fig. 2. The total energy (BLYP) as a function of the position of a water molecule with respect to the grid points. A 300 Ry plane wave cut off has been employed for both methods, and the gradient of the density has been computed using the Fourier space derivative in the GPW case

enhanced by a special property of the pseudo potentials employed. The dual-space PP produce atomic densities that approach a zero value at the nuclei and it has been shown that this can lead to numerical problems. However, in the GAPW method this problem is largely avoided by the construction of the smooth density that is used in the plane wave expansion. The effect of this smoothing is easily seen in Fig. 2 where the dependence of the total energy on the grid position is shown for both, a GPW and GAPW calculation of a single water molecule. The maximum change of the total energy is reduced in the GAPW method by more than an order of magnitude using the same plane wave cut off of 300Ry.

7.2 Condensed Phase Systems

Standard benchmarks to judge the quality of an electronic structure method for the liquid phase do not yet exist. Therefore, in order to assess the accuracy in the condensed phase, comparisons with plane wave calculations have been made. Plane waves provide an orthogonal basis that allows to reach conveniently the basis set limit for a given pseudo potential. In principle, identical total energies can be obtained using Gaussian or plane wave basis sets, provided the system is described by the same pseudopotential. Nevertheless, this is a non-trivial and very stringent test for the accuracy of both methods, since the two methodologies and basis sets are very different. In this work the plane wave calculations have been performed using the CPMD code [6]. Figure 3 shows a correlation plot between forces computed with the GPW method and CPMD for the same configuration of a liquid water sample with 32 water molecules in the simulation cell using an LDA density functional. Five different

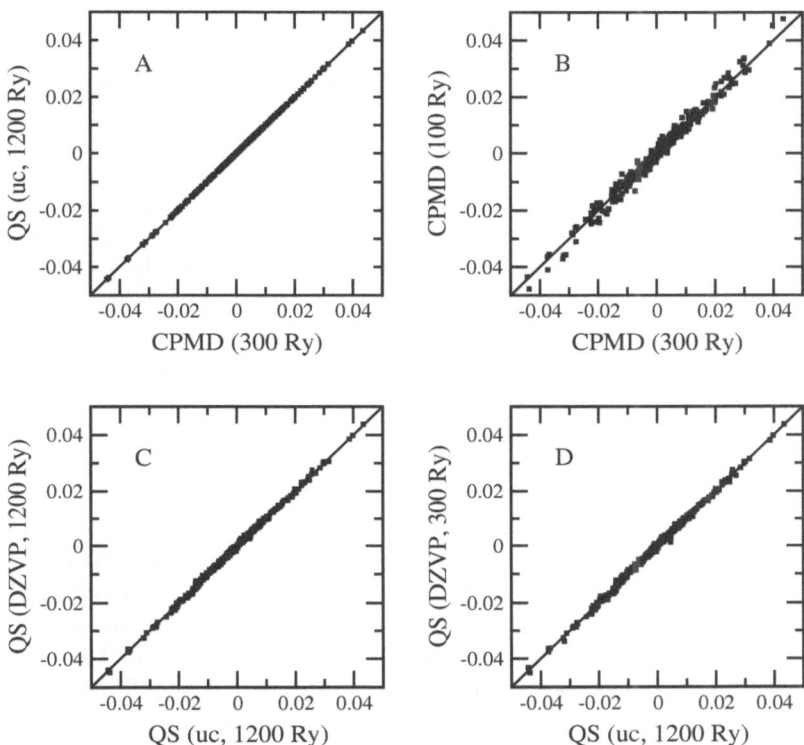

Fig. 3. The four panels show the correlation between all forces F_x, F_y, and F_z calculated for the same configuration of a liquid water sample containing 32 water molecules. Panel A shows that perfect agreement can be obtained between GPW (QS) and CPMD using large basis sets (see text) for both methods; Panel B compares these converged results to CPMD results using slightly more typical settings; Panel C shows the effect of a smaller Gaussian basis set; Panel D additionally shows the effect of a smaller density cut off

methods are being compared. CPMD with a plane wave cut off of 300 Ry and 100 Ry implying a density cut off of 1200 Ry and 400 Ry, respectively. GPW with a large basis (uc, uncontracted with high angular momentum terms, 254 basis functions per water molecule) and a density cut off of 1200 Ry, and a DZVP basis set (23 basis functions per water molecule) and a density cut off of 1200 Ry or 300 Ry. Panel A shows that perfect agreement between CPMD and GPW can be achieved. The root mean square deviation (RMSD) between the CPMD and the GPW forces is below $6 \cdot 10^{-5}$ a.u.. The panels B, C, and D, with RMSDs of $2 \cdot 10^{-3}$, $2 \cdot 10^{-4}$, and $7 \cdot 10^{-4}$ a.u., show that the quality of the forces as obtained with GPW compares favourably with a typical plane waves calculation. Furthermore, we compared the accuracy of the relative energies

Table 3. Root mean square deviation (RMSD) and maximum absolute deviation (MAD) of the total energy in 10^{-3} a.u. (see text for a definition) from a reference BLYP potential energy landscape of 25 configurations of 64 H_2O for a variety of methods

Plane Waves	70 Ry	85 Ry	100 Ry	150 Ry	200 Ry
MAD	15.5	10.8	3.6	1.9	0.0
RMSD	7.7	4.2	1.8	0.7	0.0
GPW (TZV2P)	200 Ry	240 Ry	280 Ry	340 Ry	380 Ry
MAD	16.7	6.5	5.8	5.3	4.9
RMSD	5.9	2.3	2.2	2.0	1.8
GPW (340 Ry)	DZVP	TZVP	TZV2P	QZV2P	QZV3P
MAD	11.1	9.3	5.3	5.2	4.6
RMSD	5.1	4.0	2.0	2.1	1.9

obtained from the different methods by computing a potential energy surface of 25 configurations of 64 water molecules sampled during an MD run, using the BLYP density functional [2, 29]. The reference potential energy surface is computed using CPMD with a plane wave cut off of 200 Ry. Results for several basis sets and several density cut offs are shown in Table 3. The reference potential energy surface itself fluctuates around zero by $9.8 \cdot 10^{-3}$ a.u. RMSD and $18.6 \cdot 10^{-3}$ a.u. MAD. These results show that good agreement can be obtained, but that both a sufficiently good basis set and a sufficiently high density cut off are needed to reach an accuracy that is e.g. similar to a 100 Ry plane wave calculation.

8 Benchmarks

In this section, we show that the accuracy of the GPW and GAPW methods can be achieved with high computational efficiency. We illustrate both the serial performance and the scalability on parallel computers using a high-end supercomputer as well as a modern cluster based on a PC-like architecture.

8.1 X-Ray Structure of a Dinucleoside Monophosphate

In this first section, we illustrate the performance of the GPW method on a desktop computer employing a single Pentium Xeon (3GHz) CPU for the calculations. The system studied is a dinucleoside monophosphate A2'p5'C that contains a 2'-5' link, as opposed to the usual 3'-5'. The system has been crystallised and discussed in detail in [31]. Its structure is available in the nucleic acid database [4] under entry URB001. It crystallises with an orthorhombic unit cell (8.631 Å by 18.099 Å by 16.101 Å) with space group

$P2_1 2_1 2_1$ that contains 280 atoms. The symmetry of the space group has not been exploited in these calculations. Not all hydrogen coordinates are reported in the deposited structure, and as an example application we report here the timings that are relevant in optimising the geometry of the missing hydrogen atoms.

Moderately accurate settings are employed for this task, as our target accuracy for locating the hydrogens is similar to the reported experimental uncertainty of 0.06 Å for hydrogen. We therefore employed a DZVP basis for all atoms (2712 basis functions), a threshold of 10^{-10} for neglecting overlap matrix elements, an auxiliary basis cutoff of 240 Ry ($2.3 \cdot 10^6$ plane waves), and an SCF convergence criterium of $3 \cdot 10^{-6}$, leading to an energy convergence of $\approx 3 \cdot 10^{-8}$ a.u. per atom. Using these settings we obtain single CPU timings of 39 seconds per SCF iteration, and one geometry optimisation step takes on average somewhat less than six minutes. The CPU time per SCF iteration increases to only 85 seconds if the basis is nearly doubled by adopting a TZV2P basis (4652 basis functions). Such a system can thus be studied on a desktop computer.

8.2 Liquid Water

Liquid water is a convenient benchmark system since it can easily be scaled by doubling the number of water molecules in the unit cell, followed by classical equilibration to yield a system without additional symmetries. MD runs for pure liquid water have been conducted within the GPW scheme by using input parameters as appropriate for quality production runs, i.e. GTH PP, TZV2P basis sets for hydrogen and oxygen, a density cut off of 280 Ry for the expansion of the electronic density, a threshold of 10^{-12} a.u. for the overlap integral between two primitive Gaussian functions, and the total energy of the system was converged to 10^{-7} a.u. at every MD time step (0.5 fs).

Table 4 lists the characteristics of the benchmark systems that range in size from 32 to 1024 water molecules, the largest system being several nanometres in all dimensions. The number of contracted Gaussian-type orbital basis functions is growing linearly from 1280 to 40960 functions, and $32.8 \cdot 10^6$ plane waves are required for the auxiliary basis of the largest system. Since matrices like the overlap and Kohn-Sham matrix grow quadratically with system size, it is indispensable to take advantage of the sparsity that is emerging with increasing system size.

As a measure of the performance we have reported the average time needed per MD step for the benchmark systems of Table 4 using the orbital transformation method [33, 34]. These times covers all aspects of the calculation, including the geometry dependent initialisations, SCF iterations and force calculation. Results are reported in Table 5 using an IBM Regatta p690+ system with 32 Power4+ (1.7 GHz) CPUs per node. The orbital transformation method provides significantly improved scaling and has reduced memory requirements for all systems allowing simulation of up to 1024 water molecules.

Table 4. Detailed characteristics of the employed benchmark systems for liquid water at ambient conditions (300 K, 1 bar). The edge length of the cubic simulation cell, the number of atoms, electrons, Gaussian-type orbitals (M), occupied orbitals (N), and plane waves, i.e. grid points, is listed

System	Cell [Å]	Atoms	Electrons	M	N	Grid Points ($\times 10^6$)
32 H_2O	9.9	96	256	1280	128	1.3
64 H_2O	12.4	192	512	2560	256	2.0
128 H_2O	15.6	384	1024	5120	512	4.1
256 H_2O	19.7	768	2048	10240	1024	9.3
512 H_2O	24.9	1536	4096	20480	2048	16.0
1024 H_2O	31.3	3072	8192	40960	4096	32.8

Table 5. CPU time [s] per MD step using the orbital transformation (OT) method for the benchmarks systems of Table 4. The calculations were performed on an IBM Regatta p690+ system with 32 Power4+ (1.7 GHz) per node interconnected by an IBM High Performance Switch (HPS)

# CPUs	Number of Water Molecules					
	32	64	128	256	512	1024
1	151	391	–	–	–	–
2	100	224	684	–	–	–
4	57	124	392	–	–	–
8	28	62	194	835	–	–
16	17	34	98	385	–	–
32	9	19	58	210	912	–
64	9	16	44	139	534	–
128	12	19	47	116	344	1922
256	–	–	–	–	297	1114
512	–	–	–	–	–	966

With the current implementation, at least one or two H_2O molecules (40–80 basis functions) per CPU are needed to maintain efficiency. 10 ps of dynamics of a system containing 64 water molecules can be generated in roughly four days using one Regatta node.

Similar performance tests have been carried out using the GAPW method for both the pseudo potential and the all electrons implementation. For these methods, the TZV2P and the TZVP basis sets have been employed respectively. The smaller basis set used for the all electron calculations is due to memory constraints. We ran single point energy calculations on an IBM Regatta node (32 CPUs) with a plane wave cutoff of 200 Ry and the BLYP functional. As expected, also in this case by increasing the size of the box from 32 up to 1024 (512 for the all electron case) water molecules, the cpu time required for the construction of the KS Hamiltonian and of the density scales linearly, see Table 6. For the GTH calculations we used the orbital transformation method, instead a standard diagonalization scheme has been used in the all electrons case. The use of the orbital transformation scheme in all electron calculations is also possible, provided that the preconditioner takes the broad all electron eigenvalue spectrum into account. The memory and CPU time requirements of the diagonalization step limits the system size to about 512 molecules. While the time required to construct the KS hamiltonian increases moderately with respect to the smaller systems, the optimization procedure slows down dramatically, especially in the all electron case, where a diagonalization scheme is used. This also causes the deviation from linearity in the total time for a SCF iteration.

Table 6. Average CPU time [s] per iteration measured along the SCF cycle in single point energy calculations using the GAPW method. The OT optimizer has been used for the PP calculations while a standard diagonalization scheme has been used to optimize the all electron wavefunctions. The time required for the construction of the KS hamiltonian (KS), the construction of the density n, and for the full SCF step are reported. M indicates the number of basis functions used in the expansion.

System	GTH				AE			
	M	KS	n	SCF	M	KS	n	SCF
32 H_2O	1280	3	0.33	4	896	3	0.23	5
64 H_2O	2560	6	0.61	8	1792	6	0.41	13
128 H_2O	5120	14	1.22	21	3584	12	0.92	38
256 H_2O	10240	31	2.80	57	7168	25	2	145
512 H_2O	20480	63	4.90	172	14336	48	3.6	879
1024 H_2O	40960	152	11.6	929	–	–	–	–

9 Summary

The GPW and GAPW methods allow for efficient and accurate density functional calculations of small and large molecules either as isolated systems or with periodic boundary conditions. It is especially noteworthy that high quality basis sets, which practically reach the DFT basis set limit and guarantee accuracy, can also be employed for large systems. We demonstrated the accuracy of the approach by comparing total energies and structures with results from standard quantum chemistry codes. Efficiency and scaling was illustrated on a series of water systems reaching up to more than 3000 atoms and 40'000 basis functions as well as on the unit cell of a small peptide with more than 1000 atoms. Calculations of this size can be performed on medium sized computers, a single frame of an IBM p690 in our case.

A major advantage of the GAPW method is its capability to compute all electron wavefunctions and thus properties that depend sensitively on the electron density close to the nuclei. Since the GAPW method is well suited and efficient for condensed phase systems, e.g. for molecules in solution, it becomes possible to study systems that have been unaccessible to density functional methods so far. The GAPW method sets therefore a standard for all electron calculations of large systems.

References

1. EMSL Gaussian Basis Set Order Form. http://www.emsl.pnl.gov/forms/basisform.html
2. A. D. Becke (1988) Density-functional exchange-energy approximation with correct asymptotic-behavior. *Phys. Rev. A* **38**(6), pp. 3098–3100
3. A. D. Becke (1988) A multicenter numerical integration scheme for polyatomic molecules. *J. Chem. Phys.* **88**(4), pp. 2547–2553
4. H. M. Berman, W. K. Olson, D. L. Beveridge, J. Westbrook, A. Gelbin, T. Demeny, S.-H. Hsieh, A. R. Srinivasan, and B. Schneider (1992) The nucleic acid database: A comprehensive relational database of three-dimensional structures of nucleic acids. *Biophys. J.* **63**, pp. 751–759
5. P. Blöchl (1994) Projector augmented-wave method. *Phys. Rev. B* **50**(24), pp. 17953–17979
6. CPMD, Version 3.9. copyright IBM Corp. 1990–2004, copyright MPI für Festkörperforschung Stuttgart 1997-2001; http://www.cpmd.org/
7. R. M. Dickson and A. D. Becke (1993) Basis-set-free local density-functional calculations of geometries of polyatomic-molecules. *J. Chem. Phys.* **99**(5), pp. 3898–3905
8. B. I. Dunlap, J. W. D. Connolly, and J. R. Sabin (1979) On first-row diatomic molecules and local density models. *J. Chem. Phys.* **71**(12), pp. 4993–4999
9. T. H. Dunning (1989) Gaussian-basis sets for use in correlated molecular calculations .1. the atoms boron through neon and hydrogen. *J. Chem. Phys.* **90**(2), pp. 1007–1023
10. K. Eichorn, O. Treutler, H. Öhm, M. Häser, and R. Ahlrichs (1995) Auxiliary basis sets to approximate coulomb potentials. *Chem. Phys. Lett.* **240**, pp. 283–290
11. S. Goedecker (1999) Linear scaling electronic structure methods. *Rev. Mod. Phys.* **71**(4), pp. 1085–1123
12. S. Goedecker, M. Teter, and J. Hutter (1996) Separable dual-space Gaussian pseudopotentials. *Phys. Rev. B* **54**(3), pp. 1703–1710
13. C. Hartwigsen, S. Goedecker, and J. Hutter (1998) Relativistic separable dual-space Gaussian pseudopotentials from H to rn. *Phys. Rev. B* **58**(7), pp. 3641–3662
14. T. Helgaker, P. Jørgensen, and J. Olsen (2000) *Molecular Electronic-Structure Theory*. John Wiley & Sons, Ltd, Chichester
15. T. Helgaker and P. R. Taylor (1995) *Modern Electronic Structure Theory, Part II*. World Scientific, Singapore
16. P. Hohenberg and W. Kohn (1964) Inhomogeneous electron gas. *Phys. Rev. B* **136**(3B), pp. B864–B871

17. G. Hura, D. Russo, R. M. Glaeser, T. Head-Gordon, M. Krack, and M. Parrinello (2003) Water structure as a function of temperature from x-ray scattering experiments and ab initio molecular dynamics. *Phys. Chem. Chem. Phys.* **5**, pp. 1981–1991

18. M. Iannuzzi, T. Chassaing, T. Wallman, and J. Hutter (2005) Ground and excited state density functional calculations with Gaussian and augmented method. *Chimia*, **59**, pp. 499–503

19. W. Kohn and L. J. Sham (1965) Self-consistent equations including exchange and correlation effects. *Phys. Rev.* **140**(4A), pp. A1133–A1139

20. M. Krack and A. M. Köster (1998) An adaptive numerical integrator for molecular integrals. *J. Chem. Phys.* **8**(108), pp. 3226–3234

21. M. Krack and M. Parrinello (2000) All-electron ab-initio molecular dynamics. *Phys. Chem. Chem. Phys.* **2**(10), pp. 2105–2112

22. I.-F. W. Kuo and C. J. Mundy (2004) An ab initio molecular dynamics study of the aqueous liquid-vapor interface. *Science* **303**, pp. 658–660

23. I.-F. W. Kuo, C. J. Mundy, M. J. McGrath, J. I. Siepmann, J. VandeVondele, M. Sprik, J. Hutter, B. Chen, M. L. Klein, F. Mohamed, M. Krack, and M. Parrinello (2004) Liquid water from first principles: Investigation of different sampling approaches. *J. Phys. Chem. B* **108**(34), pp. 12990–12998

24. V. I. Lebedev (1977) Spherical quadrature formulas exact to order-25-order-29. *Siberian Mathematical Journal* **18**(1), pp. 99–107

25. C. T. Lee, W. T. Yang, and R. G. Parr (1988) Development of the Colle-Salvetti correlation-energy formula into a functional of the electron-density. *Phys. Rev. B* **37**(2), pp. 785–789

26. G. Lippert, J. Hutter, and M. Parrinello (1997) A hybrid Gaussian and plane wave density functional scheme. *Mol. Phys.* **92**(3), pp. 477–487

27. G. Lippert, J. Hutter, and M. Parrinello (1999) The Gaussian and augmented-plane-wave density functional method for ab initio molecular dynamics simulations. *Theor. Chem. Acc.* **103**(2), pp. 124–140

28. D. Marx and J. Hutter. *ab-initio* Molecular Dynamics: Theory and Implementation. In J. Grotendorst, editor, *Modern Methods and Algorithms of Quantum Chemistry*, volume 1 of *NIC Series*, pages 329–477. FZ Jülich, Germany, 2000. see also http://www.fz-juelich.de/nic-series/Volume1/

29. B. Miehlich, A. Savin, H. Stoll, and H. Preuss (1989) Results obtained with the correlation-energy density functionals of Becke and Lee, Yang and Parr. *Chem. Phys. Lett.* **157**(3), pp. 200–206

30. S. Obara and A. Saika (1986) Efficient recursive computation of molecular integrals over cartesian gaussian functions. *J. Chem. Phys.* **84**(7), pp. 3963–3974

31. R. Parthasarathy, M. Malik, and S. M. Fridey (1982) X–ray structure of a dinucleoside monophosphate a2'p5'c that contains a 2'–5' link found in (2'–5')oligo(a)s induced by interferons: Single-stranded helical conformation of 2'–5'–linked oligonucleotides. *Proc. Natl. Acad. Sci. USA* **79**, pp. 7292–7296

32. The CP2K developers group. http://cp2k.berlios.de/, 2004

33. J. VandeVondele and J. Hutter (2003) An efficient orbital transformation method for electronic structure calculations. *J. Chem. Phys.* **118**(10), pp. 4365–4369

34. J. VandeVondele, M. Krack, F. Mohamed, M. Parrinello, T. Chassaing, and J. Hutter (2005) QUICKSTEP: Fast and accurate density functional calculations

using a mixed gaussian and plane waves approach. *Comp. Phys. Comm.* **167**, pp. 103–128

35. J. VandeVondele, F. Mohamed, M. Krack, J. Hutter, M. Sprik, and M. Parrinello (2005) The influence of temperature and density functional models in ab initio molecular dynamics simulation of liquid water. *J. Chem. Phys.* **122**, p. 014515

36. E. T. Whittaker and G. N. Watson (1990) *A Course in Modern Analysis, 4th ed.* Cambridge University Press

37. J. L. Whitten (1973) Coulombic potential energy integrals and approximations. *J. Chem. Phys.* **58**(10), pp. 4496–4501

38. D. E. Woon and T. H. Dunning (1993) Gaussian-basis sets for use in correlated molecular calculations. III. The atoms aluminum through argon. *J. Chem. Phys.* **98**(2), pp. 1358–1371

Computing Free Energies and Accelerating Rare Events with Metadynamics

A. Laio[1] and M. Parrinello[2]

[1] International School for Advanced Studies – SISSA, Via Beirut 2, 34100, Trieste, Italy
Laio@sissa.it
[2] Computational Science, Dept. of Chemistry and Applied Biosciences, ETH Zurich. c/o USI Campus, Via Buffi 13, 6900 Lugano, Switzerland
parrinello@phys.chem.ethz.ch

Alessandro Laio and Michele Parrinello

A. Laio and M. Parrinello: *Computing Free Energies and Accelerating Rare Events with Metadynamics*, Lect. Notes Phys. **703**, 315–347 (2006)
DOI 10.1007/3-540-35273-2_9

Computer simulations of complex polyatomic systems are nowadays routinely exploited in chemistry, physics and biophysics for extracting informations about equilibrium properties and predictions of long time behavior. As it is well known, a straightforward simulation by molecular dynamics or Monte Carlo can usually provide only limited information, due to the intrinsic complexity of most of the potential energy functions needed to describe real systems. For example, the folding time of most of the proteins is of the order of seconds, while direct simulation of these systems can access at most the microseconds time scale. For this reason the problem of sampling is of great importance in computational physics. One would like to exploit at best the available computer time in order to extract information about events that might happen on a long time scale ("rare events") and to predict the most probable state of a system.

Numerous approaches have been developed to address the sampling problem. Many methods are aimed at reconstructing the probability distribution or the free energy as a function of one or more coordinates [1]: for example thermodynamic integration [2,3], free energy perturbation [4], umbrella sampling [5], weighted histogram techniques [6–8], parallel tempering [9], Jarzynski's identity-based methods [10], adaptive force bias [11, 12] and adiabatic molecular dynamics [13]. Other methods are instead directly aimed at accelerating rare events and constructing reactive trajectories: finite-temperature string method [14], transition path sampling [15,16], milestoning [17], conformational flooding [18] taboo search [19], local elevation [20], multicanonical MD [21], force probe MD [22], nudged elastic band [23], eigenvalue following [24], steered MD [25] and methods based on the minimization of the action [26].

The metadynamics method [27] encompasses several features of other techniques and provides in many cases a unified framework for computing free energies and for accelerating rare events. The algorithm is based on a dimensional reduction, in the spirit of the approach by Kevrekidis [28,29], and requires the preliminary identification of a set of collective variables (CVs) s, which are function of the system coordinates, x, and are able to describe the activated process of interest. The dynamics in the space of the chosen CVs is driven by the free energy of the system and is biased by a history-dependent potential, $F_G(s,t)$, constructed as a sum of Gaussians centered along the trajectory followed by the collective variables up to time t. Similar techniques are also used in other methods aimed at exploring the configurational space, such as local elevation [20] and taboo search [19]. Moreover, as we discussed in [30], our methodology can be viewed as a finite temperature extension of the Wang and Landau algorithm [31] and it is also closely related to the adaptive force bias algorithm [11], in which the derivative of the free energy along a one-dimensional reaction coordinate is reconstructed with a history-dependent bias.

The capability of metadynamics to provide quantitative informations on the equilibrium properties of a system is based on the observation that the time dependent potential $F_G(s, t)$, if properly constructed, provides an unbiased estimate of the free energy in the regions explored by the system up to time t. This property does not follow from any ordinary thermodynamic identity, such as umbrella sampling [5]. It was postulated on a purely heuristic basis in [27], and afterwards verified empirically in several systems of increasing complexity. Successively, we have shown [32] that this property derives from rather general principles, and can be demonstrated rigorously for a system evolving under the action of a Langevin dynamics.

Since the history dependent potential iteratively compensates the underlying free energy, a system evolved with metadynamics will tend to escape from any free energy minimum via the lowest free energy saddle point. This is possibly the most important property of metadynamics. In fact, it makes it a very flexible tool that can be used not only for efficiently computing the free energy, but also for exploring new reaction pathways and accelerating the observation of rare events. When the collective variables are chosen in a sensible manner the system will evolve in a short time through the pathway that it would have taken, in a much longer time, if evolved with ordinary molecular dynamics, namely the lowest free energy saddle point. After crossing this saddle point, the system evolves towards the next minimum which at times is new and unpredicted.

It is obvious that in order not to prejudge the evolution of the system and remain faithful to the unbiased dynamics, the collective variables have to be chosen appropriately. This might sound as a daunting task, since there is no a priori recipe for finding the correct set of collective variables, although one can check *a posteriori* if the description provided by the chosen set is accurate [15, 16]. If an important variable is forgotten the free energy estimate converges very slowly, or does not converge at all, and important histeresis can be observed. Nevertheless, experience has demonstrated that choosing flexible collective variables is indeed possible and, if this is done, metadynamics provides very reasonable transition pathways and discovers new unpredicted stable and metastable states.

For example, for studying crystal structure transformations [33–35] we use the simulation box edges as collective variables. Under the action of metadynamics the box shape changes and the system evolves from the initial structure towards novel and unpredicted structures, which correspond to local or global minima of the Gibbs free energy at the thermodynamics conditions of interest. Similarly, in the study of chemical reactions, generic collective coordinates can be chosen, so as to distinguish between different reaction pathways [36–42]. In the fields of biophysics and ligand docking [43–45] and material science [46–48], system specific collective variables have been developed in order to observe a specific event, for example the translocation of an antibiotic through a membrane pore [43], or the nucleation of the liquid phase in hexagonal ice [48].

In this chapter we provide a general introduction to metadynamics without entering into the details of the choice of the collective variables, which is, as we already underlined, a system-specific issue. In Sect. 1 we describe the original version of the algorithm, introduced in [27] and [30]. This algorithm requires, at every step, the evaluation of the derivative of the free energy with respect to the collective variables, which are then evolved in a discrete fashion in the direction of the maximum gradient. In [36], in order to facilitate the implementation of the algorithm in conjunction with molecular dynamics, we introduced a formulation in which the collective variables are evolved continuously. This significantly broadened the scope of the methodology, so that it can now be applied, in conjunction with Car-Parrinello molecular dynamics [49], to the prediction of reaction pathways for systems described at the density functional theory level [36, 40]. This version of the algorithm is described in Sect. 2. In Sect. 3 we analyze in detail the performance of continuous metadynamics, deriving an analytic expression for the error as a function of the parameters of the method. Moreover, we outline the demonstration that the key assumption of metadynamics, namely that the history dependent potential is an unbiased estimator of the free energy, can be derived rigorously under rather general assumptions. In Sect. 4.1 we show that it is possible to use multiple interacting simulations, *walkers*, for exploring and reconstructing the same free energy surface. Each walker contributes to the history-dependent potential that, in metadynamics, is an estimate of the free energy. We show that the error on the reconstructed free energy does not depend on the number of walkers, leading to a fully linear scaling algorithm even on inexpensive loosely coupled clusters of PCs. Finally, in Sect. 4.2 and 4.3 we describe two different ways of combining metadynamics with weighted histogram analysis [7, 8] and umbrella sampling [5]. These extensions improve significantly the accuracy and the stability of the method.

1 Discrete Metadynamics

Let us consider a system described by a set of coordinates x and a potential $V(x)$ evolving under the action of a dynamics, which could be for instance Newtonian, Langevin, or Monte Carlo, whose equilibrium distribution is canonical at a temperature $1/\beta$. The set of coordinates x may include ordinary atomic positions, but also electronic coordinates, such as in Car-Parrinello molecular dynamics [49], box shape, as in Parrinello-Rahman [50], or any other auxiliary variables. We are interested in exploring the properties of the system as a function of a finite number of collective variables (CVs) s_i, $i = 1, d$ where d is a small number, assuming that they provide a good coarse grained description. The CVs can be any explicit function of x such as, for example, an angle, a distance, a coordination number or the potential energy. The equilibrium behavior of these variables is completely defined by the probability distribution:

$$P(s) = \frac{\exp(-\beta F(s))}{\int ds \exp(-\beta F(s))} \tag{1}$$

where s denotes the d dimensional vector $(s_1, \ldots, s_d)$ and the free energy $F(s)$ given by

$$F(s) = -\frac{1}{\beta} \ln \left(\int dx \exp(-\beta V(x)) \, \delta(s - s(x)) \right). \qquad (2)$$

Consider now a trajectory $x(t)$ at temperature $1/\beta$. If this trajectory could be computed for a very long time, $P(s)$ could be obtained by taking the histogram of the collective variable s along this trajectory, i.e., at time t, $P(s) \sim \frac{1}{t} \int_0^t dt' \delta(s(x(t')) - s)$. If the system displays metastability, the motion of s will be bound in some local minimum of the free energy $F(s)$ (i.e. in a local maximum of $P(s)$) and it will escape from this minimum with a very low probability on the time scale determined by the potential $V(x)$ alone. In thermodynamic integration [2, 3], and weighted histogram analysis method [7, 8], the capability of the system to explore the phase space is increased by adding a constant external potential to the system or a constraint. If, e.g. the dynamics is restrained by an external potential of the form $\frac{k}{2}(s_0 - s(x))^2$ [7], the system will explore preferentially the region around s_0 even if it is not at the bottom of a free energy well. The free energy is then reconstructed using the combined information obtained from simulations restrained at several values of s_0. In this manner it is possible to collect sufficient statistics for every value of s. In these methods the collective variable space is thus explored in a deterministic and systematic manner.

In metadynamics the free energy is instead reconstructed sequentially, beginning from the bottom of the well in a manner that resembles the filling of a basin with water. The discrete version of the algorithm performs this operation in a discrete fashion, constructing a discrete walk in the CV space that iteratively fills the free energy minimum. In order to simplify the notation, we shall assume that the units of measure of the CVs are chosen in such a way that the region in which the free energy is explored is approximately spherical. If the method is used to reconstruct a single free energy minimum, this can be achieved in practice by performing a short Monte Carlo or molecular dynamics run and computing, for each collective variable s_i, the magnitude of its thermal fluctuation $\Delta s_i = \sqrt{\langle (s_i - \langle s_i \rangle)^2 \rangle}$. The units of s_i are then rescaled in such a way that $\Delta s_i = \Delta s_j$ for all i and j. This amounts to an empirical form of preconditioning which makes the free energy minimum nearly spherical in d dimensions and easy to fill with d-dimensional Gaussians.

The exploration of the free energy surface is guided by the forces $f_i = -\partial F / \partial s_i^t$. These forces are estimated in a finite temperature molecular dynamics run, by adding to the normal Lagrangian of the system a term $\sum_{i=1,d} \lambda_i(s_i - s_i^t)$, where λ_i are Lagrange multipliers. Averaging over time, the derivative of the free energy relative to the s_i^t's can be evaluated as $f_i^t = \langle \lambda_i \rangle$ [2, 3]. Kinematic corrections due to the influence on the constraints of inertial terms can also be included [3]. These forces, determined by the microscopic dynamics, are then used to evolve the collective variables according to

$$s_i^{t+1} = s_i^t + \delta s \frac{\widetilde{f}_i^t}{\left|\widetilde{f}^t\right|} \tag{3}$$

where δs is a dimensionless stepping parameter and, for a vector a_i in the CV space, $|a| = \sqrt{\sum_i a_i^2}$. The metadynamics force $\widetilde{f}_i^t$ is the sum of the thermodynamic force f_i^t and a history-dependent force constructed as a sum of Gaussians of height w and width δs centered in all the points explored by the dynamics in the CV space up to time t:

$$\widetilde{f}_i^t = f_i^t - \frac{\partial}{\partial s_i} w \sum_{t' \leq t} \exp\left(-\frac{\left|s - s^{t'}\right|^2}{2\delta s^2}\right) \tag{4}$$

The parameters w and δs determine the accuracy and efficiency of the free energy reconstruction, as we will discuss in Sects. 1.1 and 3. After the collective coordinates are updated using (3), an ensemble with values s_i^{t+1} is prepared, and the new forces f_i^{t+1} are evaluated in another MD or MC run. In this manner the collective variables will perform a "metadynamics" in the CV space, visiting configurations very different from the ones they would have explored under the action of the thermodynamic force alone. In the following we refer to the point that explores the CV space as a *walker*.

The component of the forces coming from the Gaussians discourages the walker from revisiting the same point in configurational space and encourages an efficient exploration of the free energy surface. As the walker diffuses through the collective variable space, the Gaussian potential accumulates and fill the free energy basin, allowing the system to migrate from well to well. After a sufficient time the sum of the Gaussians, F_G, will approximately compensate the underlying free energy, i.e.

$$\lim_{t \to \infty} F_G(s, t) \sim F(s) \tag{5}$$

with

$$F_G(s, t) = -w \sum_{t' \leq t} \exp\left(-\frac{\left|s - s^{t'}\right|^2}{2\delta s^2}\right) \tag{6}$$

Equation 5 is the basic assumption of metadynamics. It states that an equilibrium quantity, namely the free energy, can be estimated by a nonequilibrium dynamics, in which the underlying potential is changed every time a new Gaussian is added. This relation does not derive from any standard identity for the free energy, like the umbrella sampling equality or the perturbation free energy formula. In [27], (5) was postulated in a heuristic manner, observing the behavior of the dynamics 3 on free energy surfaces of known functional form. For example, in Fig. 1 we consider a model free energy with three local minima. For this system we observed that if the dynamics

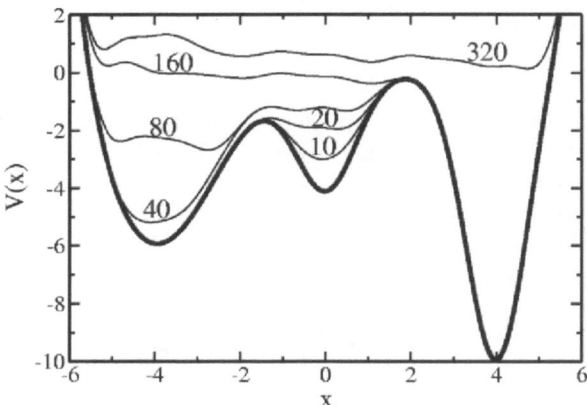

Fig. 1. Time evolution of the sum of a one-dimensional model potential $V(\sigma)$ and the accumulating Gaussian terms of (6). The dynamic evolution (*thin lines*) is labelled by the number of dynamical iterations (3). The starting potential (*thick line*) has three minima and the dynamics is initiated in the second minimum. After [27]

starts from the central local minimum, this is filled by the Gaussians in ~ 20 steps. After that the dynamics escapes over the well from the lowest saddle point, filling the second well in ~ 80 steps. The second highest saddle point is reached in ~ 160 steps, and the full free energy surface is filled in a total of ~ 320 steps. Similar behavior was afterward observed in realistic systems of increasing complexity, such as $NaCl$ in water [27], leading us to assume that (5) holds in general and can be used to compute free energies.

The example of Fig. 1 also provides a demonstration of the two different manners in which metadynamics can be exploited:

- If the method is used for "escaping free energy minima", i.e. for finding the lowest free energy saddle point out of a local minimum, the metadynamics should be stopped as soon as the walker starts exploring a completely new region of space. For the example of Fig. 1, this happens after ~ 20 steps. The crossing of the transition state can be detected in real systems by monitoring the relative orientation of the thermodynamic force f^t, and the forces coming from the Gaussians f_G^t. In fact as the initial well is being filled these two forces approximately balance each other, i.e. $f^t + f_G^t \approx 0$, and the two vectors have roughly opposite directions. After crossing the saddle point and entering the new well the two vectors become instead almost parallel. Therefore, the indicator $f^t \cdot f_G^t / (|f^t||f_G^t|)$ develops at this point a sharp spike which can be used to signal the transition from one basin to the other [33].
- If the aim is to estimate the free energy in a predefined region in the CV space, metadynamics should be stopped when the walker has explored this whole region and its motion becomes diffusive. This can be detected by

checking the *absence* of spikes in the indicator $f^t \cdot f^t_G / (|f^t||f^t_G|)$ in the last part of the dynamics. For the example of Fig. 1, the full free energy profile is filled after ~ 320 steps.

The time required to escape from a local minimum in the free energy surface is determined by the number of Gaussians that are needed to fill the well. This number is proportional to $(1/\delta s)^d$, where d is the number of collective variables used in the system. Hence, the efficiency of the method scales exponentially with the number of dimensions involved. If d is large, the only way to obtain a reasonable efficiency is to use Gaussians with a size comparable to that of the well. On the other hand, a sum of Gaussians can only reproduce features of the FES on a scale larger than $\sim \delta s$.

Already from these simple considerations it is clear that the metadynamics parameters w and δs strongly influence the quality of the reconstructed free energy and that, for a given problem, have to be carefully chosen in order to ensure the right compromise between accuracy and sampling efficiency. Large values for w and δs will allow a fast exploration of the CV space but at the price of a low accuracy.

1.1 Error Control in Discrete Metadynamics

In [30] we addressed explicitly the issue of error control for the discrete algorithm introduced in, [27], analyzing the quality of F_G as a statistical estimator of the free energy F. We showed that the accuracy of the discrete algorithm can be significantly improved introducing three simple modifications:

- In order to reduce the systematicity in the position of the Gaussians that is intrinsic in (3), the stepping parameter should be chosen at random with a uniform distribution between two limiting values. Thus, it is not equal to the Gaussian width δs, as in the implementation of [27].
- If the thermodynamic force is measured in s^t and the Gaussian at time t is also centered in s^t, as it is done in [27], the force in this position will not be changed by the Gaussian. In order to compensate at best the thermodynamic force, it is more convenient to place the center of the Gaussian at a distance δs from s^t in the direction of the force.
- If the metadynamics is terminated at position s^t, the reconstructed free energy will present a bump in a region around s^t whose spread depends on the correlation time of metadynamics. This effect can be lessened if the contribution of the Gaussians placed at the end of the dynamics are weighted less. Therefore, after the metadynamics with constant w is terminated the free energy is reconstructed from $F_R(s) = -\sum_{u \leq t'} w \tanh(\frac{t-t'}{\tau_c}) \exp(-\frac{|s-s^{t'}|^2}{2\delta s^2})$ where τ_c is taken to be larger than the typical time required to sweep the full CV space. This choice reduces significantly the spatial correlations in the reconstructed free energy and, as a consequence, the average error. A similar manner to postprocess the reconstructed free energy has been introduced independently in [42].

These prescriptions can be formalized in the following modified metadynamics equations, which replace (3) and (4):

$$s_i^{t+1} = s_i^t + \Delta s \frac{\widetilde{f}_i^t}{\left|\widetilde{f}^t\right|} \tag{7}$$

$$\widetilde{f}_i^t = f_i^t - \frac{\partial}{\partial s_i} w \sum_{t' \leq t} \exp\left(-\frac{1}{2\delta s^2}\left|s - s^{t'} - \frac{\widetilde{f}^{t'}}{\left|\widetilde{f}^{t'}\right|}\delta s\right|^2\right) \tag{8}$$

where Δs is chosen at random from a distribution with average δs. The reconstructed free energy is now given by

$$F_G(s,t) = -w \sum_{t' \leq t} \tanh\left(\frac{t - t'}{\tau_c}\right) \exp\left(-\frac{1}{2\delta s^2}\left|s - s^{t'} - \frac{\widetilde{f}^{t'}}{\left|\widetilde{f}^{t'}\right|}\delta s\right|^2\right). \tag{9}$$

The capability of this algorithm to reconstruct a pre-assigned $F(s)$ can be directly assessed by computing the average reconstructed free energy $\langle F_G(s,t)\rangle_M$, where with $\langle \cdot \rangle_M$ we indicate the average over several independent metadynamics runs, all of the same time duration. It can be verified that, using the algorithm defined by (8), $\langle F_G(s,t)\rangle_M - F(s)$ is constant in the region explored by metadynamics. Near the boundaries of the explored region $F_G(s,t)$ decays to zero and hence deviates from the true free energy. This is shown, for a model free energy, in Fig. 2. The error in the reconstructed free energy is given by the standard deviation of the difference $F_G(s,t) - F(s)$:

$$\varepsilon(s,t) = \sqrt{\left\langle (F_G(s,t) - F(s))^2 \right\rangle_M - \langle F_G(s,t) - F(s)\rangle_M^2}.$$

Since $\langle F_G(s,t)\rangle_M = F(s)$

$$\varepsilon(s,t) = \sqrt{\langle F_G^2(s,t)\rangle_M - \langle F_G(s,t)\rangle_M^2} \tag{10}$$

This error is also approximately constant in a region $\Omega(s)$ (see Fig. 2) in which the number of accumulated Gaussians, $\approx F_G(s,t)/w$, is significantly larger than a threshold value, conventionally fixed to five. Therefore, we define the error at time t as

$$\bar{\varepsilon}(t) = \sqrt{\frac{\int_{\Omega(s)} ds \varepsilon^2(s,t)}{\int_{\Omega(s)} ds}} \tag{11}$$

The value of $\bar{\varepsilon}$ depends on the Gaussian width δs and on the Gaussian height w. Moreover, as is intuitively obvious, it is influenced by the accuracy of the evaluation of the thermodynamic force used to evolve the system. Since the force is estimated by a molecular dynamics or a Monte Carlo run, its

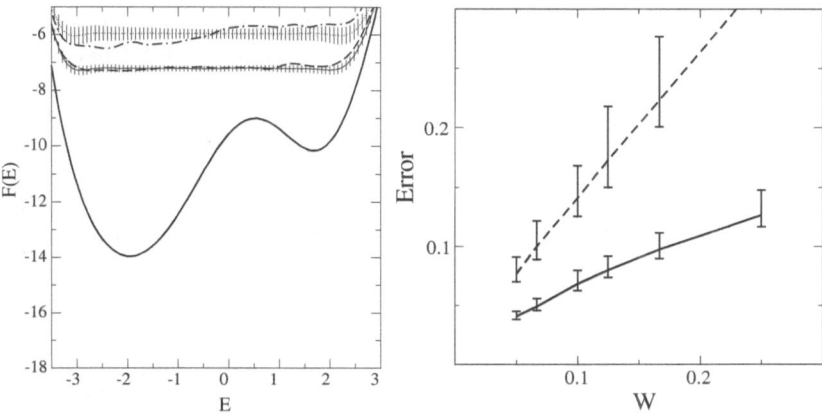

Fig. 2. *Left*: Example of reconstruction of a preassigned (analytic) free-energy profile (*black curve*), $F(s)$, by means of metadynamics runs consisting of 200 Gaussians of spread $\delta\sigma = 0.4$ and height $w = 0.16$. The *dashed* and *dot-dashed* lines denote the filled profile, $\delta F(s) = F(s) - F_G(s)$, obtained respectively with an unsmoothed ($\tau_c = 0$) and smoothed ($\tau_c = 100$) metadynamics according to (9). Notice the shorter correlation lengths and smaller amplitudes of the fluctuations in the second case. The average and dispersion of $\delta F(s)$ (continuous and dotted lines with error bars) are calculated for 1000 independent runs, showing that $F_G(s)$ is an unbiased estimator of $F(s)$ except close to the boundaries. *Right*: average, maximum and minimum values of $\bar{\varepsilon}$ averaged over 1000 runs for an unsmoothed (dashed line) and smoothed (continuous line) metadynamics. The number of Gaussians used in each run is rescaled by $1/w$ in order to work at constant filling volume. After [30]

error will scale with the square root of the total computational time, and this will ultimately determine the cost of the free energy reconstruction. A reasonable choice is to stop the MD or MC sampling when the estimated uncertainty on the force is equal to the maximum force introduced by a single Gaussian, $w \exp(-1/2)/\delta s$. This choice ensures that, for large values of t, the typical force is of the order of $w/\delta s$. If the force were calculated with greater accuracy the repeated superposition of the Gaussians would still lead to the same uncertainty. With this choice the error at fixed δs becomes in practice linear in w, as shown in Fig. 2-b. This allows to control the error performed in a metadynamics reconstruction and to combine different estimates of the free energy obtained in two or more independent metadynamics by the weighted histogram technique that we will describe in Sect. 4.2.

2 Continuous Metadynamics

The discrete version of metadynamics requires, at every "metastep" the evaluation of a thermodynamic average in order to evaluate the forces that are used to evolve the walker. Its position is thus changed in a discrete manner, and the microscopic system jumps discontinuously in the CV space. In [36], in order to facilitate the implementation of the algorithm in conjunction with Car-Parrinello molecular dynamics, we introduced a formulation in which the collective variables are evolved continuously, and the extra forces due to the time dependent potential act directly on the coordinates of the system. As we will show in Sect. 3, using this formulation it is possible to obtain an explicit expression for the error that takes explicitly into account the relaxation properties of the system in the CV space, and to show rigorously under which conditions Eq. (5) is valid.

The continuous metadynamics algorithm can be applied to any system evolving under the action of a dynamics whose equilibrium distribution is canonical at an inverse temperature $1/\beta$. In a molecular dynamics scheme this requires that the evolution is carried out at constant temperature, by using a suitable thermostat [51]. In the continuous version of metadynamics, Gaussians are added at every MD step and act directly on the microscopic variables. This generates at time t extra forces on x that can be written as

$$f^G(t) = -\frac{\partial}{\partial x} \frac{w}{\tau_G} \int_0^t dt' \exp\left(-\frac{(s(x) - s(x(t')))^2}{2\delta s^2}\right) \tag{12}$$

where the parameter w/τ_G controls the weight of the Gaussians that are deposed and $x(t')$ is the trajectory of the system. It should be remarked that since these extra forces act directly on the coordinates x of the system, continuous metadynamics is a non-equilibrium process. This is an important difference with respect to the discrete version, in which the microscopic dynamics of the system is, at every metastep, a normal equilibrium dynamics aimed only at computing the (equilibrium) derivatives of the free energy with respect to the CVs.

The reconstructed free energy is given by

$$F_G(s,t) = -\frac{w}{\tau_G} \int_0^t dt' \exp\left(-\frac{(s - s(x(t')))^2}{2\delta s^2}\right). \tag{13}$$

and, as in (5), is an approximation of $F(s)$ in the region explored by the walker up to time t:

$$\lim_{t\to\infty} F_G(s,t) \sim F(s). \tag{14}$$

This equation is an explicit relation between an equilibrium quantity, the free energy, and an observable computed in a non-equilibrium run, $F_G(s,t)$. As we shall see in Sect. 3 it is possible to prove this equation rigorously if the dynamics in the CV space can be modelled by a Langevin equation. More importantly, it has been verified to hold in practice in many different systems, using the most varied collective variables.

Equation (5) can be qualitatively understood, for the continuous algorithm, in the limit of slow "deposition" (i.e. $w/\tau_G \to 0$). In this limit, the probability distribution is always approximately proportional to $\exp\left[-\beta\left(F(s) - F_G(s, t)\right)\right]$. If the function $F(s) - F_G(s, t)$ has some local minimum, $s\left(x\left(t\right)\right)$ will preferentially be localized in the neighborhood of this minimum and increasing numbers of Gaussians will be deposed there until this minimum is flattened. Let us consider instead the case in which $F(s) \sim F_G(s, t)$ in a region $\Omega\left(s\right)$. The probability distribution will be approximately flat in this region, and the location of the new Gaussians will not be affected by the bias deriving from the difference $F(s) - F_G(s, t)$. Hence, if $w/\tau_G \to 0$, the only corrugations in the free energy that are not flattened by the dynamics will be of the order of the size of the new Gaussians that are deposed.

Using the continuous formulation, metadynamics can be implemented almost straightforwardly in any code that can perform molecular dynamics. In the routine computing the interatomic forces, one should add a routine performing the following operations:

- Compute the value of each of the d collective variables at time t and store it in an array s_i^t, $i = 1, \ldots, d$.
- For every atom i of position r_i add to the ordinary force f_i the metadynamics force:

$$f_i \to f_i - \frac{wdt}{\tau_G} \sum_{t' < t} \sum_{j=1}^{d} \left(\frac{s_j^t - s_j^{t'}}{\delta s^2}\right) \exp\left(-\frac{\left|s^t - s^{t'}\right|^2}{2\delta s^2}\right) \left.\frac{\partial s_j\left(r\right)}{\partial r_i}\right|_{r=r(t)}.$$

(15)

This requires, for every collective variable $s_j(r)$, the implementation of its explicit derivative with respect to the atomic positions.

2.1 Lagrangian Metadynamics

If the metadynamics method is applied to the simulation of chemical reactions in conjunction with Car-Parrinello molecular dynamics [36,40,49], the history dependent potential has to force the system to cross barriers of several tenths of kcal/mol in a very short time, usually a few picoseconds. This implies that a lot of energy is injected in the degrees of freedom associated with the collective variables. This might lead to a significant dishomogeneity in the temperature distribution of the system, and possibly to instabilities in the dynamics.

To address this problem in [36] we introduce, in the spirit of the extended Lagrangian approach [49,50,52], an auxiliary variable $\widetilde{s}$ coupled to the system by an harmonic restraining potential of the form $\frac{1}{2}k(\widetilde{s} - s(x))^2$. We also assign to the auxiliary variables a fictitious kinetic energy $\frac{1}{2}M\dot{\widetilde{s}}^2$. In this way, the dynamics of these extra degrees of freedom can be explicitly controlled by suitable thermostats and the trajectory of the $\widetilde{s}$ can be made so smooth that the stability of the algorithm is kept under control.

The modified potential for the system is

$$\widetilde{V}(x,\widetilde{s}) = V(x) + \frac{1}{2}k\left(\widetilde{s} - s(x)\right)^2 \tag{16}$$

The free energy for this system as a function of the $\widetilde{s}$ variables is given by

$$\widetilde{F}(\widetilde{s}) = -\frac{1}{\beta}\ln\left(\int dx d\widetilde{s}' \exp\left(-\beta\left[V(x) + \frac{1}{2}k\left(\widetilde{s}' - s(x)\right)^2\right]\right)\right)\delta\left(\widetilde{s}' - \widetilde{s}\right)$$

$$= -\frac{1}{\beta}\ln\left(\int dx \exp\left(-\beta\left[V(x) + \frac{1}{2}k\left(\widetilde{s} - s(x)\right)^2\right]\right)\right)$$

Since $\lim_{k\to\infty}\exp(-\beta\frac{1}{2}k(\widetilde{s} - s(x))^2) \propto \delta(\widetilde{s} - s(x))$, we have $\lim_{k\to\infty}\widetilde{F}(\widetilde{s}) = F(\widetilde{s})$, modulus an additive constant. Hence, the free energy of a system of potential $V(x)$ can be obtained by performing metadynamics on the extended system of potential given by (16) if k is large enough. In practice, the value of this parameter is assigned by performing an ordinary MD run on the extended system. k must be chosen in such a way that the typical value of the difference $\widetilde{s} - s(x)$ is smaller than the length on which the free energy varies of approximately $1/\beta$. This leads to the condition

$$\left\langle\left(\widetilde{s} - s(x)\right)^2\right\rangle \sim \frac{1}{\beta k} \ll \langle\widetilde{s}^2\rangle - \langle\widetilde{s}\rangle^2 \tag{17}$$

where the averages are taken at a temperature $1/\beta$ and in the absence of the metadynamics bias.

The value of the mass is a free parameter, that can be tuned in order to obtain a smooth evolution of the $\widetilde{s}$. Within a Car-Parrinello scheme [49] an important requirement is the adiabatic separation with respect to the electronic degrees of freedom. Since the extra term in the Hamiltonian introduces frequencies of the order of $\sqrt{\frac{k}{M}}$, and since k is fixed by (17), this defines a lower bound for M. On the other hand, if M is very large, the collective variables will relax to the equilibrium distribution very slowly, reducing the efficiency of the method. For a very detailed review about the proper way to tune this parameters, see [53].

With the goal of exploiting at best the short computational time available, in [36] we also introduced a variant of the Gaussian potential defined in (13):

$$F_G^{tube}(s,t) = \int_0^t dt' |\dot{\widetilde{s}}(t')| w \exp\left\{-\frac{(\widetilde{s} - \widetilde{s}(t'))^2}{2\delta s^2}\right\} \delta\left(\frac{\dot{\widetilde{s}}(t')}{|\dot{\widetilde{s}}(t')|} \cdot (\widetilde{s} - \widetilde{s}(t'))\right) \tag{18}$$

This describes a d-dimensional Gaussian tube, with axis along the trajectory. $F_G^{tube}(s,t)$ results from the accumulation of tube-slices of infinitesimal thickness $dt'|\dot{\widetilde{s}}(t')|$ in the direction of the trajectory, whereas, in the orthogonal direction, their size is given by δs. This functional form can be used only if the trajectory of the $\widetilde{s}$ is smooth enough, i.e. for a large enough mass M. It is designed so as to deposit small Gaussians if the velocity of the CVs is small, and larger Gaussians if the velocity is large, as, for example, at the bottom of the wells. This functional form for the history dependent potential improves the efficiency of the algorithm in filling the wells, but no systematic comparison with the performance of the normal Gaussian potential has been performed so far.

3 Estimating the Error

In order to allocate in the best manner the computational resources available for a given problem, it is important to know explicitly the performance of metadynamics, i.e. the best possible accuracy that can be obtained on the free energy of a system for a given simulation time. It is obvious that this performance will depend significantly on the system and on the choice of the parameters of the algorithm. In Sect. 1 we already described how accuracy can be kept under control in discrete metadynamics. In the continuous version of the algorithm the free parameters are the Gaussian width δs and the ratio w/τ_G between the Gaussian height and the time between two successive Gaussians. Since metadynamics is usually applied for reconstructing free energies in several dimensions, one can in principle fix independently the Gaussian width in each dimension. Nevertheless, as we already discussed in Sect. 1, it is always convenient to chose the relative units of different collective variables in such a way that the shape of the region that has to be filled with Gaussians is approximately spherical. If the region of interest is a single well, this can be done in a preliminary finite temperature run computing the standard deviations of the CVs. Therefore, we will consider in the following the idealized case of a free energy well with a spherical symmetry.

In order to understand how the algorithm actually works and to construct an explicit expression for the error it is not convenient to work with the metadynamics equations (12) in their full generality. Instead, we notice that the finite temperature dynamics of the collective variables satisfies, under rather general conditions, a stochastic differential equation [54, 55]. Furthermore, in real systems the quantitative behavior of metadynamics is perfectly reproduced by the Langevin equation in its strong friction limit [56]. This is due to the fact that all the relaxation times are usually much smaller than the typical diffusion time in the CV space. Hence, we model the CVs' evolution with a Langevin type dynamics:

$$ds = -\beta D \left(\frac{d}{ds} \left(F(s) + \frac{w}{\tau_G} \int_0^t dt' \exp \left(-\frac{(s - s(t'))^2}{2\delta s^2} \right) \right) \right) dt + \sqrt{2D} dW(t).$$

$$(19)$$

where $dW(t)$ is a Wiener process and D is the diffusion coefficient. The motion of the walker described by (19) is assumed to satisfy reflecting boundary conditions at the boundary of a region Ω.

The free energy $F(s)$ and the diffusion coefficient D are defined in terms of thermodynamic averages. In particular, $F(s)$ is defined by (2) and

$$D = \int_0^\infty dt \left\langle \dot{s}(0)\dot{s}(t) \right\rangle \tag{20}$$

As we checked in practical applications, the evolution of the collective variables can be modelled with a Langevin equation also for the Lagrangian metadynamics introduced in Sect. 2.1. Therefore, the error analysis performed in this section can be applied also for Lagrangian metadynamics.

Equation (19), for $w/\tau_G = 0$ describes the relaxation to the correct probability distribution $P(s) \propto \exp(-\beta F(s))$. The time required to relax to this distribution is determined by the diffusion time of the system through the CV space, which is inversely proportional to D. Other characteristic times or memory can be included by an inertial term or by a time-dependent friction kernel. These generalizations can be rather straightforwardly introduced in the model, but we checked that they do not affect the qualitative picture that we will present below.

Equation (19) describes a non-Markovian process in the CV space. In fact, the forces acting on the CVs depend explicitly on their history. Due to this non-Markovian nature, it is not clear if, and in which sense, the system can reach a stationary state under the action of this dynamics. In [32] we introduced a formalism that allows to map this history-dependent evolution into a Markovian process in the original variable and in an auxiliary field that keeps track of the visited configurations. Defining

$$\varphi(s,t) = \int_0^t dt' \delta(s - s(t')) \tag{21}$$

Equation (19) can in fact be written as

$$d\varphi = \delta(s - s(t))\, dt$$
$$ds = -\beta D\left(\frac{d}{ds}\left(F(s) + \int ds'\varphi(s',t)\, g(s',s(t)) \right) \right) dt + \sqrt{2D}\, dW(t) \tag{22}$$

with $g(s,s') = \frac{w}{\tau_G}\exp(-\frac{(s'-s)^2}{2\delta s^2})$. These equations are fully Markovian, i.e. the state of the system at time $t + dt$, $(s(t+dt), \varphi(s,t+dt))$, depends only on the state of the system at time t, $(s(t), \varphi(s,t))$. Using this property it is possible to analyze in a rigorous manner (19), studying, for instance, its behavior for long t.

The state of the system is described by a probability distribution $P(s,\varphi,t)$, which is a function of the walker position s and a functional of the field φ. $P(s,\varphi,t)$ satisfies a Fokker-Plank equation that can be directly derived from (22) using standard techniques [57]. In [32] we show that, for large t, $P(s,\varphi,t)$ converges to a distribution $P_\infty(\varphi)$ that does not depend on s. This distribution is Gaussian in functional space and centered around a function $\varphi_0(s')$ that is the field corresponding to the free energy of the system $F(s)$:

$$P_\infty(\varphi) \propto \exp\left(\frac{\beta D}{2}\int dsds'\left(\varphi(s') - \varphi_0(s')\right)\partial_s^2 g(s, s')\left(\varphi(s) - \varphi_0(s)\right)\right) \quad (23)$$

$$\varphi_0(s') : \int ds'\varphi_0(s')g(s, s') = F(s) \quad (24)$$

The knowledge of $P(s, \varphi, t)$ allows to estimate the probability to observe, at time t, the reconstructed free energy

$$F_G(s, t) = \int ds'\varphi(s', t)g(s, s') \quad (25)$$

Eq. (24) states that, after a long Metadynamics run, it is in principle possible to observe any reconstructed free energy $F_G(s)$. But the probability to actually observe a given realization, will be given by $P_\infty(\varphi)$ evaluated for the corresponding field, as defined by (25). This probability will be low for a φ very different from φ_0.

Using (24) we can prove that the average value of $F_G(s, t)$ over several independent metadynamics runs is exactly equal to $F(s)$. In fact, denoting by $\langle\cdot\rangle_M$ the average over several metadynamics realizations,

$$\langle F_G(s)\rangle_M = \int ds'g(s, s')\langle\varphi(s')\rangle_M$$

$$= \int ds'g(s, s')\int d\varphi P_\infty(\varphi)\varphi$$

$$= \int ds'g(s, s')\varphi_0(s') = F(s)$$

This provides a rigorous proof of (5).

This property is also illustrated in Fig. 3 in which we show the results obtained integrating numerically (19) for four different profiles $F(s)$ with respectively zero, one, two and three minima. The metadynamics parameters are $\delta s = 0.1$ and $w/\tau_G = 4\times10^{-4}$. We also have $D = 0.0005$ and $\beta = 1$ and the CVs satisfy reflecting boundary conditions in a region of length 8. The average value of $F_G(s, t) - F(s)$ at $t = 5 \times 10^5$ is represented as a continuous line in all the four profiles, and is constant in all the explored region, as predicted by (5).

Equation (24) also allows for an explicit estimate of the metadynamics error as a function of δs, w/τ_G and the system parameters. Using (10) and (11) we have

$$\bar{\varepsilon}^2 = \frac{1}{S^d}\int ds\left(\int d\varphi P_\infty(\varphi)F_G^2(s, t) - \left(\int d\varphi P_\infty(\varphi)F_G(s, t)\right)^2\right) \quad (26)$$

with $F_G(s, t)$ given by (25). The functional integrals over $d\varphi$ are conveniently computed in Fourier space, in which the probability distribution becomes diagonal. After some algebra [32], this leads to,

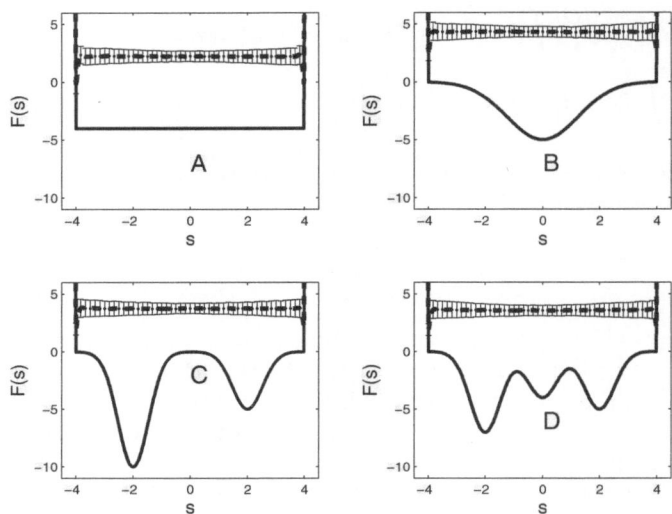

Fig. 3. Metadynamics results for four different free energy profiles. A: $F(s) = -4$; B: $F(s) = -5\exp\left(-\left(\frac{s}{1.75}\right)^2\right)$; C: $F(s) = -5\exp\left(-\left(\frac{s-2}{0.75}\right)^2\right) - 10\exp\left(-\left(\frac{s+2}{0.75}\right)^2\right)$; D: $F(s) = -5\exp\left(-\left(\frac{s-2}{0.75}\right)^2\right) - 4\exp\left(-\left(\frac{s}{0.75}\right)^2\right) - 7\exp\left(-\left(\frac{s+2}{0.75}\right)^2\right)$. The average $\langle F(s) - F_G(s,t)\rangle$ computed over 1000 independent trajectories is represented as a dashed line, with the error bar given by (10). The units of the error is the same as the units of the free energy. After [56]

$$\bar{\varepsilon}^2 = \frac{S^2 w}{\beta D \tau_G}\left(\frac{\delta s}{S}\right)^d (2\pi)^{\frac{d}{2}} \sum_k \frac{1}{\pi^2 k^2} \exp\left(-\frac{k^2\pi^2}{2}\left(\frac{\delta s}{S}\right)^2\right) \qquad (27)$$

where the sum is performed over all the d dimensional vectors of integers of norm different from zero.

In a previous work [56] we deduced an alternative expression for the error (11) by performing extensive numerical simulations of the stochastic differential equation (19). The metadynamics parameters, w/τ_G, $\delta\sigma$, and the system-dependent parameters, β, D and S were systematically varied, and for each choice of the parameters the error (11) was computed by repeating several metadynamics reconstructions. Fitting the results, we obtained that the data were reproduced within an accuracy of $\sim 20\%$ by

$$\bar{\varepsilon}^2_{\text{approx}} = C_d \frac{S^2 w}{\beta D \tau_G}\frac{\delta s}{S} \qquad (28)$$

where C_d is a constant that depends only on the dimensionality. The two expressions for the error depend in the same manner on w/τ_G, β, D and S. The dependence on $\delta s/S$ in (27) is instead much more complicated than it was possible to infer by simple numerical simulations. The ratio between the two expressions is approximately a constant as a function of δs only for $d = 1$ and $d = 2$, while significant deviations are observed in higher dimensionality.

The dependence of the error on the simulation parameters becomes more transparent if $\bar{\varepsilon}$ is expressed as an explicit function of the total simulation time. Consider in fact a free energy profile $F(s)$ that has to be filled with Gaussians up to a given level $F_{\max}$, for example the free energy of the lowest saddle point in $F(s)$. The total computational time needed to fill this profile can be estimated as the ratio between the volume that has to be filled and the volume of one Gaussian times τ_G :

$$t_{sim} \approx \tau_G \frac{F_{\max}}{w} \left(\frac{S}{\delta s}\right)^d \tag{29}$$

Substituting in (27) gives

$$\bar{\varepsilon}^2 \approx \frac{\tau_S}{t_{sim}} \frac{F_{\max}}{\beta} f_d \left(\frac{\delta s}{S}\right) \tag{30}$$

where $\tau_S \doteq \frac{S^2}{D}$ is the average time required for the collective variables to diffuse on a distance S and

$$f_d \left(\frac{\delta s}{S}\right) = (2\pi)^{d/2} \sum_k \frac{1}{k^2 \pi^2} \exp\left(-\frac{k^2 \pi^2}{2} \left(\frac{\delta s}{S}\right)^2\right)$$

is function only of $\frac{\delta s}{S}$ and of the dimensionality. Equation (30) states that the error of a metadynamics reconstruction is inversely proportional to the square root of the total simulation time, measured in units of the diffusion time. The error will be large for slowly diffusing systems, in which the walker takes a long time to explore the collective variables space.

The error at fixed simulation time is determined by the prefactor $f_d(\frac{\delta s}{S})$, plotted in Fig. 4 for $d = 2$ and $d = 3$ as a function of $\frac{\delta s}{S}$. For a comparison, we also plot the function $f_d^{real}\left(\frac{\delta s}{S}\right) = \bar{\varepsilon}_{fixed}^2$, where $\bar{\varepsilon}_{fixed}^2$ is the error measured by (11) on several independent metadynamics reconstructions performed in a parabolic well. All the runs are performed changing $\frac{\delta s}{S}$ and choosing w in such a way that the filling speed is the same, i.e. $w \propto \left(\frac{S}{\delta s}\right)^d$. Moreover, the parameters of the system are chosen in such a way that $\frac{\tau_S}{t_{sim}} \frac{F_{\max}}{\beta} = \frac{S^2 w}{\beta D \tau_G} \left(\frac{\delta s}{S}\right)^d = 1$. The function $f_d^{real}\left(\frac{\delta s}{S}\right)$ measured under these conditions is an estimate of the error at fixed filling time in a free energy well. For a fixed total simulation time, the error always decreases as a function of δs and the higher the dimensionality the more this dependence is significant. For small $\frac{\delta s}{S}$ the error is accurately predicted by (27). Nevertheless, deviations are observed when δs becomes comparable with the dimension of the well. Indeeed, the "filling" behavior of metadynamics depicted in Fig. 1 can be observed only if δs is smaller than the typical length scale of variation of the free energy. Equtaion (27) describes the error only when the walker is freely diffusing in a confined region

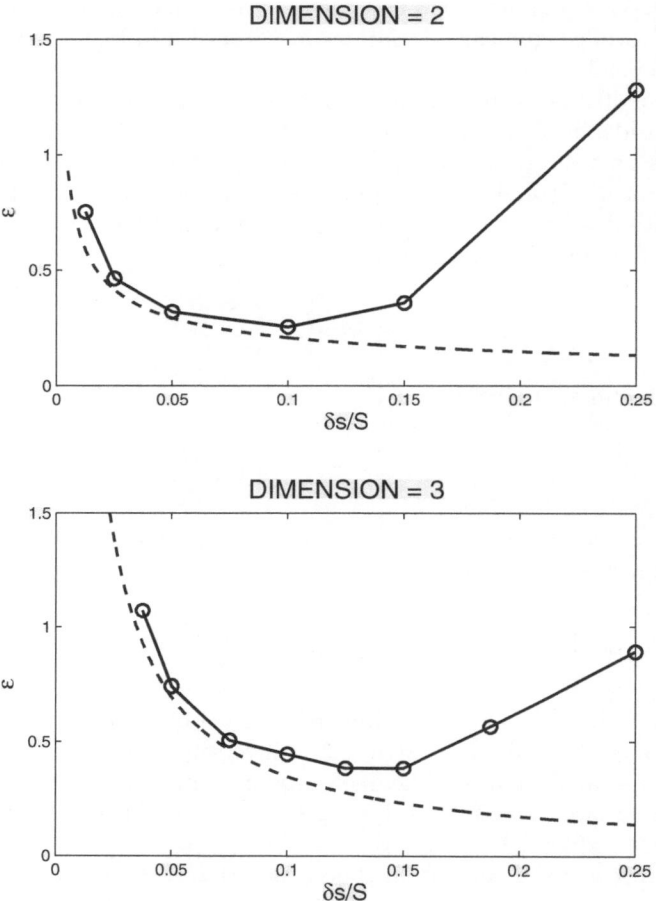

Fig. 4. Effect of the $\delta s/S$ on the error ε computed with (11) for a system in $d = 2$ (*top panel*) and $d = 3$ (*bottom panel*), with $S = 8$, $\beta = 1$ and $D = 0.0005$. The dashed line corresponds to the theoretical error given by (27). w/τ_G is chosen in such a way that $\frac{\tau_S}{t_{sim}} \frac{F_{\max}}{\beta} = \frac{S^2 w}{\beta D \tau_G} \left(\frac{\delta s}{S} \right)^d = 1$

in which reflecting boundary conditions are imposed. For a large δs the error will be large while the walker is filling each well.

In conclusion, δs should be chosen as large as possible, but an upper bound is imposed by the features of the free energy profile that has to be filled.

3.1 Predicting the Error in a Real System

In [56] we estimate the performance of metadynamics for a system consisting of a tetracationic cyclophane (Cyclobis(paraquat-p-phenylene)$_8^{4+}$) and a 1,5-dihydroxy-naphthalene solvated in acetonitrile (see Fig. 5). For this system, which is a simple model for a class of nanomachines [58], we compute the free energy as a function of two collective variables, the distance between the centroids of the cyclophane and the naphthalene and the coordination number of the naphthalene with the atoms of the acetonitrile. This is defined as:

$$s_2 = \sum_{\substack{i \in naphthalene \\ j \in \ solvent}} \frac{1 - (r_{ij}/r_0)^8}{1 - (r_{ij}/r_0)^{14}}$$

where r_{ij} is the distance between the two chosen atom types and $r_0 = 4\mathring{A}$.

We first obtained a free energy surface as a function of these two variables by a bi-dimensional umbrella sampling using the WHAM scheme [7] (see [56] for technical details). This free energy (see Fig. 5) has two minima, one corresponding to the threaded state ($s_1 \sim 0$ and $s_2 \sim 6$) and one corresponding to the unthreaded state. The barrier towards the dissociated state is of ~ 12 kcal/mol and is located at $s_1 \sim 6$ and $s_2 \sim 8.5$.

We then performed several metadynamics runs with different choices of w, δs and τ_G and, since the exact free energy for the system has been computed by umbrella sampling explicitly, we estimate the error with (11). In total, we tested more than fifty different combinations of parameters, for a total of almost 200 ns of molecular dynamics. All the simulations are started from the same initial condition in which the naphthalene is inserted in the catenane, and are stopped when the system escapes this free energy minimum.

In Fig. 6 we compare the error computed with (11) with the one predicted by (28), including a correction factor for a finite τ_G derived in [56]. The size S of the system is here estimated from the WHAM free energy, and is approximately equal to 7. If the free energy is not known a priori, the value of S can be estimated exploiting some knowledge of the physics of the system or a preliminary metadynamics run. This introduces an additional uncertainty in the error estimate. Choices of the parameters for which the error is not approximately constant during the filling procedure are discarded (e.g. $\delta s > 1$). The diffusion tensor is computed with (20) and its three independent components are given by $D_{11} = 15 \ 10^{-5}\mathring{A}^2fs^{-1}$, $D_{22} = 14 \ 10^{-5}fs^{-1}$ and $D_{12} = 10^{-5}\mathring{A}fs^{-1}$. The difference between the predicted and the computed value is on average only 10%.

This important result requires some comment. The expressions for the error (27) and (28) are derived under the assumption that metadynamics can be modelled with a stochastic differential equation of the form (19). In this equation the noise is independent on the position in the CV space, there are no inertial effects, and the relaxation to equilibrium is described by a simple diffusion coefficient. These approximations might sound severe, but the

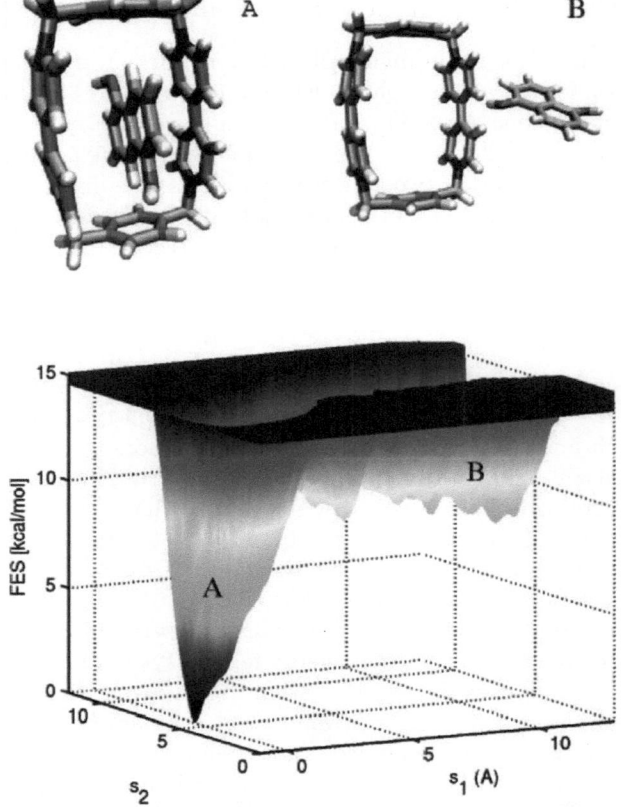

Fig. 5. *Top*: a schematic representation of the threaded and unthreaded configurations are reported. For clarity the solvent molecules are omitted. *Bottom*: the FES of the threading process reconstructed with umbrella sampling and weighted histogram analysis. After [59]

example described in this Section shows that in real systems the *quantitative* behavior of metadynamics is perfectly reproduced by this equation. This is due to the fact that even if (19) does not in general describe the relaxation to equilibrium of all the systems, the only relaxation time that determines the error in metadynamics is the time required to the collective variable to diffuse through the region in which the free energy is reconstructed, S^2/D. The other characteristic times, that are certainly present in real systems, are averaged out during the metadynamics reconstruction.

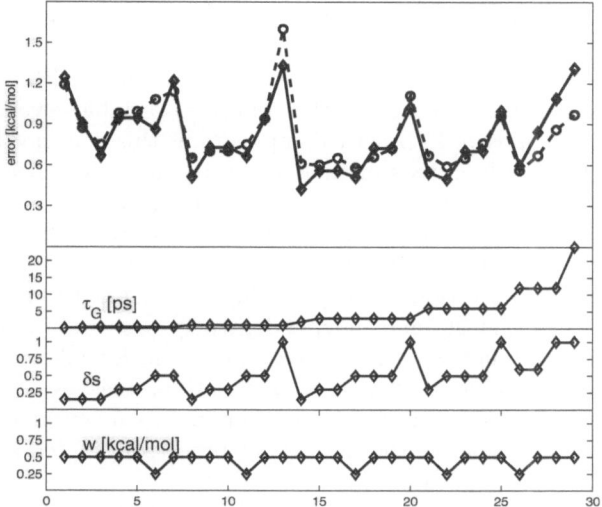

Fig. 6. Metadynamics error on the free energy for the pseudo-rotaxane naphthalene complex as a function of the run number. *Upper panel:* the error computed by (11) (dashed line with circles) and the error predicted by (28), including the correction factor for a finite τ_G derived in [56] (continuous line with diamonds). *Lower panels:* the metadynamics parameters w, δs and τ_G as a function of the run number. After [56]

4 Advanced Topics

4.1 Parallel Metadynamics

In [59] we introduced a parallel version of metadynamics that is intrinsically linear scaling and that preserves the same accuracy of the original algorithm. The working principle is extremely simple: we imagine N_W metadynamics simulations, i.e. N_W walkers, that fill simultaneously the same free energy well. The walkers are completely independent and are free to wander in the whole CV space. The only knowledge they share is the history dependent potential, since the Gaussians that each walker adds are felt by all the walkers. Therefore, the metadynamics forces on the coordinates x_i of the walker i at time t are given by

$$f_i^G(t) = -\frac{\partial}{\partial x_i} \frac{w}{\tau_G} \int_0^t dt' \sum_{j=1}^{N_W} \exp' \left(-\frac{(s(x_i) - s(x_j(t')))^2}{2\delta s^2} \right) \qquad (31)$$

where $x_j(t)$ is the trajectory of the walker j. The generalized form of the filling potential (13) is

$$F_G(s,t) = \frac{w}{\tau_G} \int_0^t dt' \sum_{j=1}^{N_W} \exp \left(-\frac{(s - s(x_j(t')))^2}{2\delta s^2} \right) \qquad (32)$$

i.e. the estimated free energy is given by the sum of the Gaussians laid by all the walkers.

The amount of information exchanged between the walkers is very small, since it includes only the values of the collective variables every time that a new Gaussian is deposed. In practical applications this does not happen every time step, but every hundreds of time steps. In fact, the collective variables are usually much slower than most of the other degrees of freedom. Hence, the multiple walkers algorithm can be straightforwardly implemented using an external file, shared by all the processes. The file is periodically read from each walker and the position of the Gaussian is appended to it. Since the file is accessed asynchronously and independently by each walker, the dynamics can be run on machines having different speed and a walker can be started or stopped without interfering with the simulation. This gives to the method stability with respect to crashes of one or more nodes, an event which becomes more and more probable as the number of processors increases, as it is the present trend in high end computers.

Since each independent walker fills the free energy surface independently, the clock time required to obtain an estimate of the free energy is inversely proportional to N_W. For example, if a profile is reconstructed by a single walker in one nanosecond, only 100 psec will be required using 10 walkers. We have seen in Sect. 3 that the error is inversely proportional to the filling time. Hence, one could anticipate that the error obtained reconstructing the free energy with several walkers will be large. Rather surprisingly, this is not the case, as shown in Fig. 7 for a simplified Langevin model and in Fig. 8 for the system described in Sect. 3.1. The error is instead independent on the number of walkers, even if the filling speed grows linearly with N_W. The intuitive reason of this behavior is that the walkers explore the CV space in an independent manner. This leads to an effective diffusion coefficient for the full system that is larger than the one of the single walker. The two effects (larger filling speed and larger diffusion coefficient) exactly compensate each other.

For this reason, filling a free energy profile with several walkers can improve significantly the efficiency of the reconstruction on a parallel machine. Still, an upper limit to the number of walkers is imposed by the intrinsic diffusivity properties of the system. In fact, if the free energy surface is not known a priori, the walkers have to be initialized, in the worst case scenario, in the same position. A natural choice is to place them at the $F(s)$ minimum. Before the free energy reconstruction converges, the walkers have to loose memory of this initial position. For a given choice of parameters, the free energy reconstruction is meaningful only if the probability distribution of the walkers has lost memory of the initial value. This happens after a relaxation time t_{REL}, that will be a function of the metadynamics parameters and of the number of walkers. In [59] we show that if $\frac{S}{\delta s} \gg 1$ and $\frac{S^2}{\tau_G D} \beta w N_W \left(\frac{\delta s}{S}\right)^d \gg 1$ the relaxation time is given by

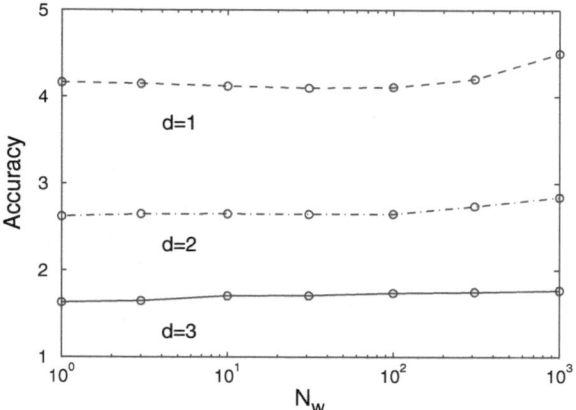

Fig. 7. Error as a function of the number of walkers for a spherical well in $d = 1$, $d = 2$ and $d = 3$. The metadynamics parameters are: $w = 1$, $\delta s = 0.3$, $tau_G = 50$, $S = 4$, $\beta = 1$, $D = 0.0005$. After [59]

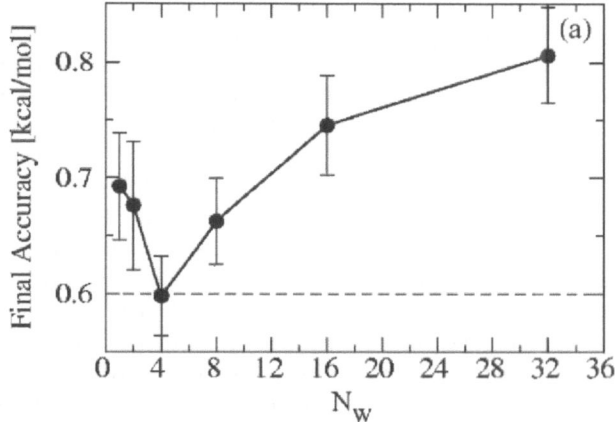

Fig. 8. Accuracy of the reconstructed FES as a function of the number of walkers on the free energy for the pseudo-rotaxane naphthalene complex (see Sect. 3.1 for details on the system). The accuracy is measured with (11) when the first walker escapes from the associated state and, for every value of N_W, is averaged over ten independent simulations. The horizontal line corresponds to the accuracy predicted by (28). After [59]

$$t_{REL} = c_d \sqrt{\frac{S^{d+2}}{\delta s^d D} \frac{\tau_G}{\beta w N_W}} \qquad (33)$$

where c_d is a constant that depends only on the dimensionality.

This relaxation time defines implicitly an upper bound for N_W. In fact, a free energy profile of depth F is filled by the N_W walkers in a time that is approximately $t_{sim} = \frac{1}{N_W} \tau_G \frac{F}{w} \left(\frac{S}{\delta s} \right)^d$. Imposing $t_{sim}/t_{corr} \gg 1$, we have

$$N_W \ll \frac{D}{S^2} \tau_G \frac{\beta F^2}{w} \left(\frac{S}{\delta s} \right)^d \frac{1}{c_d^2}. \tag{34}$$

The maximum number of walkers that can be used scales exponentially with the dimensionality of the free energy that has to be reconstructed. In practical applications, the number of walkers can be used for reconstructing a free energy in two, three and for dimensions is of the order of 10, 100 and 1000, respectively. This feature of the algorithm allows to reconstruct free energies as a function of several variables in a clock time that depends only on the maximum number of available processors.

4.2 Merging Results from Different Metadynamics Calculations

When one is interested in reconstructing a free energy surface encompassing multiple wells and barriers the overall accuracy becomes a crucial issue: for complex and slowly diffusing systems, complete convergence would be achieved only after an impractically long time.

In [44] we suggested that for the two basin case, to obtain a correct relative depth of the free energy basins metadynamics should be stopped immediately after a recrossing event through the same reactive pathway, i.e. after the system has overcome the same saddle point in the reverse direction. This, however, might provide a poor description of the saddle point region and the overall accuracy would be lower than that which can be achieved with the same set of parameters in the two basins seperately, since the diffusion time τ_S through the whole CV space will be much larger (see (30)). Moreover, if multiple walkers metadynamics is used the identification of the correct time to stop the simulations becomes even more problematic. In fact, stopping the reconstruction when the first walker recrosses the barrier becomes an arbitrary criterion when the number of walkers is very large, since recrossing can occur also due to thermal fluctuations.

In [30] and [59] we introduced an estimator of the free energy that is largely insensitive to the time at which the metadynamics run is stopped also if several local minima are present and allows reducing very significantly the error. The procedure consists in dividing the CV space into subregions Ω_k, using metadynamics to compute the free energies $F_k(s)$ in those portions. The informations are then combined by a suitable weighted histogram technique [6], to obtain the best possible estimate of the free energy on the whole region Ω. At variance with standard WHAM [8], in which the optimization procedure leads to the best possible estimator of the probability distribution, we here minimize the error on its logarithm, namely on the free energy. This can be done because in a metadynamics run it is the error on the free energy that can be directly controlled by (27). Indicating by $\epsilon_k^2(s)$ this error, the function to be minimized is:

$$L\left(F(s), c_k\right) = \int ds \sum_k \frac{(F(s) - F_k(s) - c_k)^2}{\epsilon_k^2(s)}, \tag{35}$$

where c_k are the constants that determine the alignment of the $F_k(s)$-s. The least square procedure leads to two self-consistent equations for c_k and F,

$$c_k = \int ds \frac{(F(s) - F_k(s))}{\epsilon_k^2(s)} \cdot \frac{1}{\int ds 1/\epsilon_k^2(s)}, \tag{36}$$

and

$$F(s) = \sum_k \frac{(F_k(s) + c_k)}{\epsilon_k^2(s)} \cdot \frac{1}{\sum_k 1/\epsilon_k^2(s)}, \tag{37}$$

which can be solved iteratively.

In practical applications, this technique is applied by performing a meta-dynamics reconstruction for each local minimum in the free energy surface. The walkers are forced to remain in their minimum by reflecting walls located just behind the saddle point region. In this manner the transition regions are explored in two independent metadynamics, and the overlap is sufficient to make the solution to (36) reliable.

Although this technique requires a preliminary knowledge of the topology of the free energy surface, it is extremely powerful, and should be always preferred to simple metadynamics if the free energy difference between two or more different states has to be estimated with great accuracy. For example, as we show in [59], by applying this technique it is possible to gain a factor of four in the computational time required to reproduce a two-basin free energy. The increase in performance derives simply from the fact that for the collective variables it is much more simple to diffuse in an approximately spherical basin rather than on a complex surface in which the transition between two different regions can be done only by diffusing through the bottleneck corresponding to the transition region.

4.3 Minimal Free Energy Path with Metadynamics

Another powerful strategy for extracting informations about a highly dimensional free energy surface has been introduced in [40, 41, 53]. The idea is to use metadynamics to localize the lowest free energy path (LFEP) connecting the reactant well with the product well in a high dimensional CV space. The LFEP describes the most probable reaction mechanism and can be used as a generalized reaction coordinate for describing the process. After the LFEP has been identified with a low accuracy metadynamics reconstruction, the free energy along this path is estimated using umbrella sampling [5, 7] in one dimension. This method exploits at best the capability of metadynamics to provide in a short simulation time a very reasonable estimate of the most probable reaction pathway. Still, it allows estimating *quantitatively* the free energy difference between reactants and products in a complex chemical reaction that cannot be described by any standard coordinate, since the metadynamics result is refined by one dimensional umbrella sampling.

Metadynamics is first used to reconstruct the free energy as a function of many simple coordinates at the same time. The filling rate w/τ_G and the Gaussian width δs have to be chosen in order to provide the possibility to explore completely the free energy surface in the available computational time. This requirement, in high dimensionality, necessarily implies a large error, especially if the method is used in conjunction with Car-Parrinello molecular dynamics [49].

Even if the accuracy of the reconstructed free energy is poor, it is usually possible to localize very clearly the best possible path connecting reactants and products in the CV space. The path is defined as the set of $n - 1$ points $s_1, \ldots, s_{n-1}$ that minimize

$$\sum_{\tau=1}^{n-1} F_G(s_\tau) - \sum_{\tau=1}^{n} \lambda_\tau (|s_{\tau-1} - s_\tau| - \Delta s)$$

with s_0 and s_n belonging to the reactant and product state respectively. The n Lagrange multipliers λ_i are introduced in order to enforce a constant distance $\Delta s = \frac{1}{n} \sum_{\tau'=1}^{n} |s_{\tau'-1} - s_{\tau'}|$ between two successive points. A practical method to solve this minimization problem that keeps into account also the irregularities of the reconstructed free energy is described in [53].

For example, in Fig. 9 we show a three dimensional projection of a six dimensional free energy surface for the base induced elimination (E2) and the bimolecular nucleophilic substitution (SN2) reactions between F^- and CH_3CH_2F. The collective variables used for this systems are all the six independent coordination numbers between the atoms in the systems that might change (see [41]. The red arrows depict the LFEPs connecting the reactant state ($F^- + CH_3CH_2F$) with the E2 product state and the SN2 product state. These LFEPs are then used as generalized reaction coordinates in umbrella sampling, obtaining the reaction barriers with an error below 1 kcal/mol.

The values of $F_G(s_\tau)$, $\tau = 1, \ldots, n$ provides a first estimate of the free energy along the LFEP. This potential is used as a bias to perform a one-dimensional umbrella sampling along the generalized reaction coordinate defined by the LFEP [5, 7]. The forces on the atoms of position r due to the umbrella potential are

$$f^U = -\sum_{k=1}^{d} \frac{F_G(s_{\tilde{\tau}}) - F_G(s_{\tilde{\tau}-1})}{\Delta s} \frac{(s_{\tilde{\tau}-1} - s_{\tilde{\tau}})_k}{\Delta s} \frac{\partial s_k}{\partial r} \tag{38}$$

where $s_{\tilde{\tau}}$ is the point along the path that is closest to $s(r)$.

If $F_G(s_\tau)$ is a good estimate of the one-dimensional free energy, the system, under the combined action of the microscopic Hamiltonian and of the

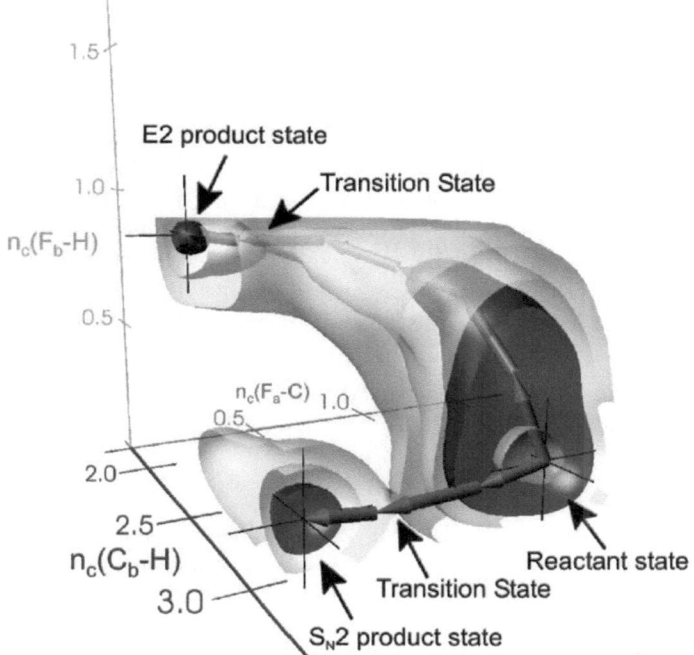

Fig. 9. Three dimensional projection of the 6D free energy surface, showing the local minima (*black crosses*) and the E2 and S_N2 reaction channels. The gray arrows depict the lowest free energy paths for the two reactions. After [41]

forces (38), will perform a diffusive motion along the generalized reaction co-ordinate defined by the path. Since the metadynamics reconstruction has been performed in several dimensions, errors of several kcal/mol can be expected, and it is usually necessary to refine the estimate of the free energy. This can be done, for example, by adding extra potentials of the form $\frac{1}{2}k\left|s - s_\tau\right|^2$ and constructing histograms for several values of τ. If the parameter k of the biasing potential is properly chosen, the histograms will overlap sufficiently, and the unbiased free energy along the path can be estimated by WHAM [7,8] with the required precision.

This approach combines many advantages of metadynamics and umbrella sampling: the often difficult problem of choosing a good one-dimensional re-action coordinate is facilitated by reconstructing the free energy as a function of several relatively simple collective variables. A good reaction coordinate, namely the LFEP, is generated *a posteriori*. The umbrella sampling corrects in a very efficient manner the errors that are unavoidable in metadynamics reconstruction performed in several dimensions.

Acknowledgments

A number of people contributed to the development of metadynamics, making the methodology useful for simulating realistic systems of increasing complexity. The results presented in this chapter are due to the combined effort of Marco Bernasconi, Mauro Boero, Davide Branduardi, Giovanni Bussi, Rosa Bulo, Matteo Ceccarelli, Davide Donadio, Bernd Ensing, Francesco Gervasio, Marcella Iannuzzi, Roman Martoňák, Cristian Micheletti, Artem Oganov, Paolo Raiteri and Andras Stirling. We would also like to thank Giovanni Ciccotti, David Chandler, Eric Vanden-Eijden and Mike Klein for several illuminating discussions and many precious suggestions. Finally, a particular acknowledgment goes to Jannis Kevrekidis, who first stimulated us to consider dimensional reduction as a tool for exploring efficiently the phase space.

References

1. W. E. and E. Vanden-Eijnden (2004) *Metastability, conformation dynamics, and transition pathways in complex systems.* Lecture notes in computational science and engineering, Ed. S. Attinger and P. Koumoutsakos, Springer Berlin Heidelberg
2. E. Carter, G. Ciccotti, J. Hynes, and R. Kapral (1989) Constrained reaction coordinate dynamics for the simulation of rare events. *Chem. Phys. Lett.* **156**, p. 472
3. M. Sprik and G. Ciccotti (1998) Free energy from constrained molecular dynamics. *J. Chem. Phys.* **109**, p. 7737
4. P. A. Bash, U. C. Singh, F. K. Brown, R. Langridge, and P. A. Kollman (1987) Free energy calculation by computer simulation. *Science* **235**, pp. 574–576
5. G. N. Patey and J. P. Valleau (1975) Monte-carlo method for obtaining interionic potential of mean force in ionic solution. *J. Chem. Phys.* **63**, pp. 2334–2339
6. A. Ferrenberg and R. Swendsen (1988) Optimized monte-carlo data-analysis. *Phys. Rev. Lett.* **61**, p. 2635
7. S. Kumar, J. M. Rosenberg, D. Bouzida, R. H. Swendsen, and P. A. Kollman (1995) Multidimensional free-energy calculations using the weighted histogram analysis method. *J. Comput. Chem.* **16**, pp. 1339–1350
8. B. Roux (1995) The calculation of the potential of mean force using computer-simulations. *Comput. Phys. Comm.* **91**, pp. 275–282
9. H. Merlitz and W. Wenzel (2002) Comparison of stochastic optimization methods for receptor-ligand docking. *Chem. Phys. Lett.* **362**, pp. 271–277
10. C. Jarzynski (1997) Nonequilibrium equality for free energy differences. *Phys. Rev. Lett.* **78**, p. 2690
11. E. Darve and A. Pohorille (2001) Calculating free energies using average force. *J. Chem. Phys.* **115**, pp. 9169–9183
12. D. Rodriguez-Gomez, E. Darve, and A. Pohorille (2004) Assessing the efficiency of free energy calculation methods *J. Chem. Phys.* **120**, p. 3563
13. L. Rosso, P. Minary, Z. Zhu, and M. Tuckerman (2002) On the use of the adiabatic molecular dynamics technique in the calculation of free energy profiles. *J. Chem. Phys.* **116**, pp. 4389–4402

14. E. Weinan, W. Ren, and E. Vanden-Eijnden (2005) Finite temperature string method for the study of rare events. *J. Phys. Chem. B* **109**, p. 6688
15. C. Dellago, P. Bolhuis, F. S. Csajka, and D. Chandler (1998) Transition path sampling and the calculation of rate constants *J. Chem. Phys.* **108**, pp. 1964–1977
16. C. Dellago, P. Bolhuis, and P. Geissler (2002) Transition path sampling. *Adv. Chem. Phys.* **123**, pp. 1–78
17. A. K. Faradjian and R. Elber (2004) Computing time scales from reaction coordinates by milestoning. *J. Chem Phys.* **120**, pp. 10880–10889
18. H. Grubmüller (1995) Predicting slow structural transitions in macromolecular systems: conformational flooding. *Phys. Rev. E* **52**, pp. 2893–2906
19. D. Cvijovic and J. Klinowski (1995) Taboo search - an approach to the multiple minima problem. *Science* **267**, pp. 664–666
20. T. Huber, A. Torda, and W. van Gunsteren (1994) Local elevation: a method for improving the searching properties of molecular dynamics simulation. *J. Comput. Aided Mol. Des.* **8**, pp. 695–708
21. A. K. N. Nakajima, J. Higo and H. Nakamura (1997) Flexible docking of a ligand peptide to a receptor protein by multicanonical molecular dynamics simulation. *Chem. Phys. Lett.* **278**, pp. 297–301
22. B. A. Heymann and H. Grubmüller (2001) Molecular dynamics force probe simulations of antibody/antigen unbinding: Entropic control and nonadditivity of unbinding forces. *Biophys. J.* **81**, pp. 1295–1313
23. H. G. and J. H. (2000) Improved tangent estimate in the nudged elastic band method for finding minimum energy paths and saddle points. *J. Chem. Phys.* **113**, pp. 9978–9985
24. R. Fletcher and M. J. D. Powell (1963) A rapidly convergent descent method for minimization. *Comput. J.* **6**, pp. 163–168
25. J. Gullingsrud, R. Braun, and K. Schulten (1999) Reconstructing potentials of mean force through time series analysis of steered molecular dynamics simulations. *J. Comp. Phys.* **151**, pp. 190–211
26. R. Elber and M. Karplus (1987) A method for determining reaction paths in large molecules – application to myoglobin. *Chem. Phys. Lett.* **139**, p. 375
27. A. Laio and M. Parrinello (2002) Escaping free energy minima. *Proc. Natl. Acad. Sci. USA* **99**, pp. 12562–12566
28. C. Theodoropoulos, Y. Qian, and I. G. Kevrekidis (2000) Coarse stability and bifurcation analysis using time-steppers: A reaction-diffusion example. *Proc. Natl. Acad. Sci. USA* **97**, pp. 9840–9843
29. I. G. Kevrekidis, C. W. Gear, and G. Hummer (2004) Equation-free: The computer-aided analysis of comptex multiscale systems. *Aiche J.* **50**(7), pp. 1346–1355
30. C. Micheletti, A. Laio, and M. Parrinello (2004) Reconstructing the density of states by history-dependent metadynamics. *Phys. Rev. Lett.* **92**, p. 170601
31. F. Wang and D. Landau (2001) Efficient, multiple-range random walk algorithm to calculate the density of states. *Phys. Rev. Lett.* **86**, p. 2050
32. G. Bussi, A. Laio, and M. Parrinello (2006) Equilibrium free energies from non-equilibrium metadynamics. *Phys. Rev. Lett.* **96**, p. 090601
33. R. Martoňák, A. Laio, and M. Parrinello (2003) Predicting crystal structures: The parrinello-rahman method revisited. *Phys. Rev. Lett.* **90**, p. 75503
34. P. Raiteri, R. Martoňák, and M. Parrinello (2005) Exploring polymorphism: the case of benzene. *Angew. Chem. Int. Ed.* **44**, pp. 3769–3773

35. R. Martoňák, A. Laio, M. Bernasconi, C. Ceriani, P. Raiteri, and M. Parrinello (2005) Simulation of structural phase transitions by metadynamics. *Zeitschrift fur Kristallographie* **220**, pp. 489–498

36. M. Iannuzzi, A. Laio, and M. Parrinello (2003) Efficient exploration of reactive potential energy surfaces using car-m. parrinelloolecular dynamics. *Phys Rev. Lett.* **90**, p. 238302

37. A. Stirling, M. Iannuzzi, A. Laio, and M. Parrinello (2004) Azulene-to-naphthalene rearrangement: The car-parrinello metadynamics method explores various reaction mechanisms. *Chem. Phys. Chem.* **5**, pp. 1558–156

38. S. Churakov, M. Iannuzzi, and M. Parrinello (2004) Ab initio study of dehydroxylation-carbonation reaction on brucite surface. *J. Phys. Chem. B* **108**, p. 11567

39. F. Gervasio, A. Laio, M. Iannuzzi, and M. Parrinello (2004) Influence of dna structure on the reactivity of the guanine radical cation. *Chem. Eur. J.* **10**, p. 4846

40. B. Ensing, A. Laio, F. Gervasio, M. Parrinello, and M. Klein (2004) A minimum free energy reaction path for the e2 reaction between fluoro ethane and a fluoride ion. *J. Am. Chem. Soc.* **126**(31), p. 9492

41. B. Ensing and M. L. Klein (2005) Perspective on the reactions between f- and ch3ch2f: The free energy landscape of the e2 and s(n)2 reaction channels. *Proc. Natl. Acad. Sci. USA* **102**, pp. 6755–6759

42. Y. Wu, J. Schmitt, and R. Car (2004) Mapping potential energy surfaces. *J. Chem. Phys.* **121**, pp. 1193–1200

43. M. Ceccarelli, C. Danelon, A. Laio, and M. Parrinello (2004) Microscopic mechanism of antibiotics translocation through a porin. *Biophys J.* **87**, pp. 58–64

44. F. L. Gervasio, A. Laio, and M. Parrinello (2005) Flexible docking in solution using metadynamics. *J. Am. Chem. Soc.* **127**, pp. 2600–2607

45. D. Branduardi, F. L. Gervasio, A. Cavalli, M. Recantini, and M. Parrinello (2005) The role of the peripheral anionic site and cation-pi interactions in the ligand penetration of the human ache gorge. *J. Am. Chem. Soc.* **127**(25), p. 9147

46. F. Zipoli, M. Bernasconi, and R. Martoňák (2004) Constant pressure reactive molecular dynamics simulations of phase transitions under pressure: The graphite to diamond conversion revisited. *Eur. Phys. J. B* **39**, p. 41

47. M. Iannuzzi and M. Parrinello (2004) Proton transfer in heterocycle crystals. *Phys. Rev. Lett.* **93**, p. 025901

48. D. Donadio, P. Raiteri, and M. Parrinello (2005) Topological defects and bulk melting of hexagonal ice. *J. Phys. Chem. B* **109**, p. 5421

49. R. Car and M. Parrinello (1985) Unified approach for molecular-dynamics and density-functional theory. *Phys. Rev. Lett.* **45**, p. 2471

50. M. Parrinello and A. Rahman (1980) Crystal structure and pair potentials: a molecular-dynamics study. *Phys. Rev. Lett.* **45**, p. 1196

51. S. Nose (1984) A unified formulation of the constant temperature molecular-dynamics methods. *J. Chem. Phys.* **81**, pp. 511–519

52. H. C. Andersen (1980) Molecular-dynamics simulations at constant pressure and-or temperature. *J. Chem. Phys.* **72**, pp. 2384–2393

53. B. Ensing, A. Laio, M. Parrinello, and M. Klein (2005) A recipe for the computation of the free energy barrier and the lowest free energy path of concerted reactions. *J. Phys. Chem. B* **109**, pp. 6676–6687

54. R. Zwanzig (1961) Memory effects in irreversible thermodynamics. *Phys. Rev.* **124**, pp. 983–992

55. H. Ottinger (1998) General projection operator formalism for the dynamics and thermodynamics of complex fluids. *Phys. Rev. E* **57**, p. 1416

56. A. Laio, A. Rodriguez-Fortea, F. L. Gervasio, M. Ceccarelli, and M. Parrinello (2005) Assessing the accuracy of metadynamics. *J. Phys. Chem. B* **109**, pp. 6714–6721

57. C. Gardiner (2004) *Handbook of Stochastic Methods.* Springer Berlin Heidelberg

58. M. Ceccarelli, F. Mercuri, D. Passerone, and M. Parrinello (2005) The microscopic switching mechanism of a [2]catenane. *J. Phys. Chem. B* **109**, pp. 17094–17099

59. P. Raiteri, A. Laio, F. L. Gervasio, C. Micheletti, and M. Parrinello (2006) Efficient reconstruction of complex free energy landscapes by multiple walkers metadynamics. *J. Phys. Chem. B* **110**, pp. 3533–3539

Transition Path Sampling Methods

C. Dellago[1], P.G. Bolhuis[2], and P.L. Geissler[3]

[1] Faculty of Physics, University of Vienna, Boltzmanngasse 5, 1090 Wien, Austria
   `Christoph.Dellago@univie.ac.at`
[2] van 't Hoff Institute for Molecular Sciences, University of Amsterdam, Nieuwe
   Achtergracht 166, 1018 WV Amsterdam, The Netherlands
   `bolhuis@science.uva.nl`
[3] Department of Chemistry, University of California at Berkeley, 94720 Berkeley,
   CA, USA
   `geissler@cchem.berkeley.edu`

Christoph Dellago

C. Dellago et al.: *Transition Path Sampling Methods*, Lect. Notes Phys. **703**, 349–391 (2006)
DOI 10.1007/3-540-35273-2_10   © Springer-Verlag Berlin Heidelberg 2006

Transition path sampling, based on a statistical mechanics in trajectory space, is a set of computational methods for the simulation of rare events in complex systems. In this chapter we give an overview of these techniques and describe their statistical mechanical basis as well as their application.

1 Introduction

This chapter will be focused on computer simulation algorithms to overcome the problem of widely disparate time scales. In doing that we concentrate on processes dominated by rare events: events that occur only rarely, but when they occur, they occur quickly. Examples of such processes include the nucleation of first order phase transitions, chemical reactions in solution and on surfaces, transport in and on solids, and protein folding to name but a few. The occurrence of rare events is related to *high potential energy barriers* or *entropic bottlenecks* (narrow but energetically flat passageways in configuration space [1]) partitioning the configuration space of the system into (meta)stable basins. In equilibrium, the system spends the bulk of its time fluctuating whithin these long-lived stable states and barriers are crossed only very rarely. In order to understand such processes in detail it is necessary to identify the relevant degrees of freedom and to separate them from orthogonal variables which may be regarded as random noise. While in principle conventional computer simulations can provide the information necessary to do that, the wide gap in time scales originating in the long times between barrier crossings is a serious computational problem.

The straightforward approach to this kind of difficulty is to follow the time evolution of the system, for instance by molecular dynamics simulation, and wait until a sufficient number of events have been observed. However, the computational requirements of such a procedure are excessive for most interesting systems. In practice, it is frequently impossible to observe a single transition of interest, let alone collect enough statistics for a microscopic resolution of the mechanism. For instance, reaction times of chemical reactions occurring in solution often exceed the second time scale. Since the simulation of molecular systems typically proceeds in steps of roughly one femtosecond, of the order of 10^{15} steps are required to observe just one transition. Such calculations are far beyond the reach of the fastest computers even in the foreseeable future.

A different strategy to approach such problems is to search for the dynamical bottlenecks through which the system passes during a transition between metastable states. If the dynamics of the system is dominated by energetic effects (as opposed to entropic effects), such bottlenecks can be identified with saddle points in the potential energy surface. In this case, saddle points are transition states, activated states from which the system can access different stable states through small fluctuations. Comparing stable states with transition states one can often infer the mechanism of the reaction. Reaction rate constants, which are very important because they are directly comparable to

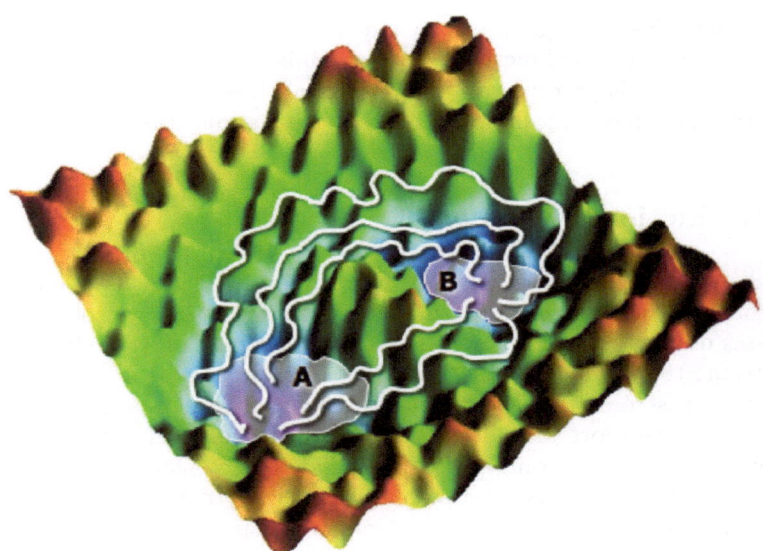

Fig. 1. Transitions pathways connecting stable states A and B on a caricature of a complex energy landscape

experimental observables, can then be determined via transition state theory (TST). If the transition state theory estimate of the reaction rate constant is not sufficiently accurate, the reactive flux formalism, in which dynamical corrections to simple transition state theory are calculated from dynamical trajectories initiated at the transition state, can provide the desired corrections [2].

While transition state theory and its modern variants can be very successful in simple (small or highly ordered) systems, complex systems with strongly non-linear potential energy landscapes require an entirely different approach. In such complex systems the saddle points in the potential energy surface cease to be characteristic points of the free energy barrier. Instead, the free energy barrier may encompass a large set of configurations some of which are stationary points but most are not (see Fig. 1). One may hope to be guided by physical intuition in the search of transition states, in effect postulating the reaction coordinate. But the relevant degrees of freedom may be highly collective and therefore difficult to anticipate. This problem can be overcome with the transition path sampling approach [3–6]. Based on a statistical mechanics of trajectories, this method focuses on those segments of the time evolution where the rare, but important events actually occur, thus avoiding long intervening waiting times between. Since in such a complex system rare transitions between stable states can occur via a multitude of different pathways, the notion of a single, well-defined reaction path, such as a zero kinetic energy path, is abandoned in favor of a large set of possibly markedly different paths: the *transition path ensemble*. Defining this ensemble does not require any a priori knowledge on how the process occurs, i.e., no definition of a reaction coordinate is necessary. Instead, it is sufficient to specify unambiguously the initial and the final state of the transition. This is a crucial feature of the method since information on the reaction coordinate is usually unavailable. Transition path sampling can therefore be used to study rare transitions between long-lived stable states in complex systems characterized by a rugged potential energy surface. Other path based methods that can be used to study rare events include the so called string method [7, 8] applicable to stochastic dynamics and methods based on the minimization of the classical action [9, 10]. Both of these approaches will be discussed in detail in other contributions to this summer school.

The transition path sampling techniques developed for the study of rare events can also be used to improve the calculation of free energies from nonequilibrium transformations on the basis of Jarzynski's theorem. In this approach equilibrium free energies are linked to the statistics of work performed during irreversible transformations. When applying these ideas in computer simulations, one has to deal with rare events (of a different kind, however) and transition path sampling can help to solve this problem. Due to space limitations we cannot comprehensively discuss all aspects of transition path sampling in this chapter. For further information the reader is referred to several recent review articles [11–15]. Of these articles, [12] is the most detailed and comprehensive.

2 The Transition Path Ensemble

Figure 1 depicts the general situation that can be addressed with the transition path sampling methodology:

- The system is (meta)stable in states A and B in the sense that if the system is put there it will remain there for a long time. We assume that states A and B can be easily characterized using an order parameter.
- The system evolves according to deterministic or stochastic equations of motion.
- The system spends most of its time either in A or B but, rarely, transition between the stable states A and B occur.
- No long-lived metastable states exist between states A and B.
- The stable states are separated by an unknown and, in general, rough energy or free energy barrier.

The goal of a transition path sampling simulation is to collect all likely transition pathways. These pathways can then be analyzed to find the transition mechanism, i.e., to identify the degrees of freedom that capture the physics of the transition and to determine how they change during the transition. Since the pathways collected with transition path sampling are fully dynamical trajectories rather than artificial pathways such as minimum energy pathways or elastic bands, it is also possible to extract kinetic information from transition path sampling simulations. The first step in the application of transition path sampling consists in the definition of an appropriate path ensemble. This is what we will do next.

2.1 Path Probability

It is convenient to discretize the continuous time evolution of the system and view a trajectory of length $\mathcal{T}$ as an ordered sequence of states:

$$x(\mathcal{T}) \equiv \{x_0, x_{\Delta t}, x_{2\Delta t}, \ldots, x_{\mathcal{T}}\} . \tag{1}$$

Consecutive states are separated by a small time increment Δt and $x_{i\Delta t}$ is a complete snapshot of the system at time $i\Delta t$. For a molecular system evolving according to Newton's equations of motion, for instance, x comprises the positions r and momenta p of all particles. Since the trajectory $x(\mathcal{T})$ results from slicing the continuous path in time, the states $x_{i\Delta t}$ are often called *time slices* (see Fig. 2).

The probability (density) to observe a given path $x(\mathcal{T})$ depends on the probability of its initial condition and the specific dynamics of the system. For Markovian processes, i.e., processes for which the probability to move from x_t to $x_{t+\Delta t}$ after one time step Δt depends only on x_t and not on the history of the system prior to t, the total path probability can be written as the product of single time step transition probabilities $p(x_t \to x_{t+\Delta t})$,

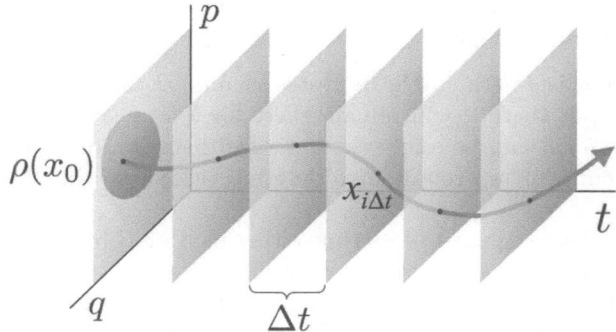

Fig. 2. A trajectory is discretized into "time slices" separated by a time increment Δt. Time slice $x_{i\Delta t}$ is a complete description of the system at time $i\Delta t$. For Newtonian and Langevin dynamics each time slice consists of the positions and momenta of all particles. Initial conditions for the trajectories are distributed according to $\rho(x_0)$

$$\mathcal{P}[x(\mathcal{T})] = \rho(x_0) \prod_{i=0}^{\mathcal{T}/\Delta t - 1} p(x_{i\Delta t} \to x_{(i+1)\Delta t}) \,. \tag{2}$$

The first factor on the right hand side of the above equation, $\rho(x_0)$, is the probability distribution of the initial conditions. The particular form of the transition probability $p(x_t \to x_{t+\Delta t})$ for different types of deterministic and stochastic dynamics will be discussed below.

2.2 Reactive Pathways

Since we want to focus on reactive pathways we now restrict the path ensemble to those pathways that start in region A, the *reactant* region, and end in region B, the *product* region. This is achieved by multiplying the path probability $\mathcal{P}[x(\mathcal{T})]$ with the appropriate characteristic functions:

$$\mathcal{P}_{AB}[x(\mathcal{T})] \equiv h_A(x_0)\mathcal{P}[x(\mathcal{T})]h_B(x_{\mathcal{T}})/Z_{AB}(\mathcal{T}) \,. \tag{3}$$

Here, $h_A(x)$ is the characteristic function of region A which is unity if the argument x is inside A and vanishes otherwise, and $h_B(x)$ is defined similarly:

$$h_{A,B}(x) = \begin{cases} 1 & \text{if} \quad x \in A, B \\ 0 & \text{if} \quad x \notin A, B \,. \end{cases} \tag{4}$$

Due to the restriction applied to the pathways in the transition path probability (3) a path not beginning in A or not ending in B (or both) has a statistical weight of zero. A path connecting A and B, on the other hand, may have a non-zero weight that depends on the unrestricted path probability $\mathcal{P}[x(\mathcal{T})]$. The transition path ensemble (3) selects only the reactive trajectories from the ensemble of all possible pathways while leaving the relative probabilities of the reactive trajectories among each other unchanged. Since by multiplication with the characteristic functions we have reduced the size of the path ensemble, the probability distribution is normalized by

$$Z_{AB}(\mathcal{T}) \equiv \int \mathcal{D}x(\mathcal{T})\, h_A(x_0)\mathcal{P}[x(\mathcal{T})]h_B(x_{\mathcal{T}}) \,. \tag{5}$$

The notation $\int \mathcal{D}x(\mathcal{T})$, familiar from path integral theory, implies a summation over all pathways:

$$\int \mathcal{D}x(\mathcal{T}) \equiv \int \cdots \int dx_0 dx_{\Delta t} dx_{2\Delta t} \cdots dx_{\mathcal{T}} \,. \tag{6}$$

The transition path ensemble (3) is a complete statistical description of all possible pathways connecting reactants with products. Pathways sampled according to this ensemble are typical trajectories which can then be analyzed to yield information about mechanisms and rates. The definition of the transition path ensemble is very general and valid for all Markovian processes. In the following we will write down the specific form of the transition path ensemble for different types of processes.

2.3 Path Probabilities for Deterministic and Stochastic Dynamics

The general transition path sampling formalism can be applied to various ensembles of pathways differing both in the distributions of initial conditions as well as in the particular transitions probabilities. The specific form of the path probability depends on the process one wants to study and is not imposed by the transition path sampling technique itself. In the following we discuss several path probabilities that frequently occur in the study of condensed matter systems.

Initial Conditions

If initial conditions are prepared by placing the system in contact with a heat bath at a specific temperature T the distribution of initial conditions is the *canonical* one,

$$\rho(x) = \exp\{-\beta\mathcal{H}(x)\}/Z \, , \tag{7}$$

where

$$Z(\beta) = \int dx \exp\{-\beta\mathcal{H}(x)\} \tag{8}$$

is the partition function of the system, $\mathcal{H}(x)$ is its Hamiltonian, and $\beta = 1/k_\mathrm{B}T$ is the inverse temperature. In other situations the energy E of the initial states is well defined requiring the use of the *microcanonical* distribution,

$$\rho(x) = \delta[E - \mathcal{H}(x)]/g(E) \, , \tag{9}$$

where

$$g(E) = \int dx\delta[E - \mathcal{H}(x)] \tag{10}$$

is the density of states. Different distributions of initial conditions can easily be incorporated into the path sampling scheme as well. Indeed, path sampling can be applied even in the case on non-equilibrium systems where the distribution of initial conditions is not known explicitly [16].

Newtonian Dynamics

Consider a classical molecular system that evolves according to Hamilton's equations of motion,

$$\dot{r} = \frac{\partial\mathcal{H}(r,p)}{\partial p}, \qquad \dot{p} = -\frac{\partial\mathcal{H}(r,p)}{\partial r} \, . \tag{11}$$

Since the time evolution of such a system is deterministic, the state of the system x_t at time t is completely determined by the state x_0 at time 0:

$$x_t = \phi_t(x_0) \, . \tag{12}$$

Here, $x = \{r, p\}$ includes the positions r and momenta p of all particles. The time dependent function $\phi_t(x_0)$ is the *propagator* of the system. For such deterministic dynamics the short time transition probability can be written in terms of a Dirac delta function:

$$p(x_t \to x_{t+\Delta t}) = \delta[x_{t+\Delta t} - \phi_{\Delta t}(x_t)] \,. \tag{13}$$

Accordingly, the path probability is given by

$$\mathcal{P}_{AB}[x(\mathcal{T})] = \rho(x_0) h_A(x_0) \prod_{i=0}^{\mathcal{T}/\Delta t - 1} \delta[x_{(i+1)\Delta t} - \phi_{\Delta t}(x_{i\Delta t})] h_B(x_{\mathcal{T}})/Z_{AB}(\mathcal{T}) \,, \tag{14}$$

where the partition function

$$Z_{AB}(\mathcal{T}) = \int dx_0 \, \rho(x_0) h_A(x_0) h_B(x_{\mathcal{T}}) \,, \tag{15}$$

normalizes the distribution. Here, all but the first state of the pathway, x_0, have been integrated out and the remaining integral is over initial conditions only. Other examples of deterministic dynamics include the extended Lagrangian dynamics of Car and Parrinello [17] and various dynamics with thermostats such as the Nose-Hoover thermostat [18] or the Gaussian isokinetic thermostat [19]. Also for these types of dynamics the above definition of the transition path ensemble applies.

Brownian Dynamics

As an example for a stochastic process consider a system evolving according to the Langevin equation in the high friction limit where inertial effects can be neglected and momenta are not required for the description of the system:

$$m\gamma\dot{r} = -\frac{\partial V(r)}{\partial r} + \mathcal{F} \,. \tag{16}$$

Here, m is the mass of the particles, $V(r)$ the potential energy of the system, γ the friction constant, and $\mathcal{F}$ is a Gaussian random force uncorrelated in time that satisfies the fluctuation dissipation theorem [20]

$$\langle \mathcal{F}(t)\mathcal{F}(0) \rangle = 2m\gamma k_B T \delta(t) \,. \tag{17}$$

Due to the random force the time evolution of the system in a short time step Δt consists of a systematic part depending on the force $-\partial V/\partial r$ and a random displacement δr [21, 22]:

$$r_{t+\Delta t} = r_t - \frac{\Delta t}{\gamma m}\frac{\partial V}{\partial r} + \delta r \,. \tag{18}$$

This random displacement is a Gaussian random variable with zero mean and a width σ that depends on the time step and the temperature [21, 22],

$$\sigma^2 = \frac{2k_B T}{m\gamma} \Delta t \ . \tag{19}$$

The short time transition probability then follows from the statistics of the random displacement:

$$p(r_t \rightarrow r_{t+\Delta t}) = \frac{1}{\sqrt{2\pi\sigma^2}} \exp \left\{ -\frac{(r_{t+\Delta t} - r_t + \frac{\Delta t}{\gamma m} \frac{\partial V}{\partial r})^2}{2\sigma^2} \right\} \ . \tag{20}$$

Appropriate expressions for the short time transition probability $p(x_t \rightarrow x_{t+\Delta t})$ can be derived also for Langevin dynamics with arbitrary friction and for Monte Carlo "dynamics" [4, 12].

2.4 Defining the Stable States A and B

While a transition path sampling simulation does not require knowledge of a *reaction coordinate* capable of characterizing the progress of a transition through the dynamical bottleneck region, a careful definition of the initial region A and the final region B is important. Regions A and B are most conveniently characterized by the value of a low dimensional *order parameter* q. Often, defining a good order parameter capable of discriminating between A and B is far from trivial. For example, to define stable states for autoionization in liquid water the connectivity of the hydrogen bond network has to be taken into account [23].

The definitions of stable states A and B, and hence of the order parameter q, have to meet several criteria. First, A and B have to be large enough to accommodate typical equilibrium fluctuations. If this is not the case important transition pathways might be missing from the transition path ensemble. Second, region A spanned by $h_A(x)$ and region B spanned by $h_B(x)$ should be located entirely within the corresponding basins of attraction. In other words, region A should not overlap with the basin of attraction of B and vice versa. The basin of attraction of a specific stable state consists of all configurations from which trajectories relax into that stable state. If this second criterion is not met, the transition path sampling algorithm most likely collects non-reactive trajectory.

Correct definition of regions A and B may require considerable trial and error experimentation. Fortunately, it is easy to detect problems associated with an inappropriate choice of order parameter. For instance, short trajectories initiated from the initial point of the unsuccessful trajectory in Fig. 3 (solid line) will most likely end with a value of q characteristic of the final state. In contrast, the probability to relax into B from the initial point of the dashed trajectory is negligible. When problems are caused by non-discriminating regions A and B, the order parameter has to be refined until correct sampling can be achieved.

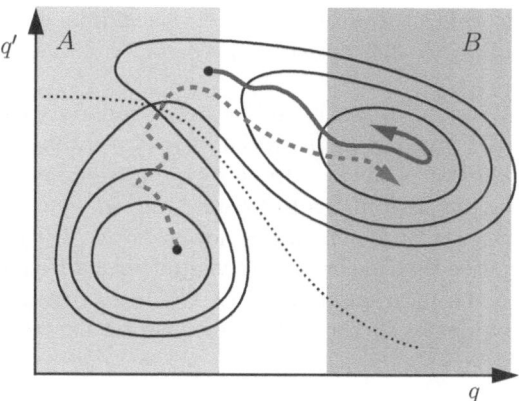

Fig. 3. In this free energy landscape the coordinate q fails to distinguish the basins of attraction of the stable states. If region A is defined by requiring that q is less than some limit (light gray region), points belonging to the basin of attraction of B are located inside A. Path starting from such points (such as the dark gray path) belong to the path ensemble but are not real reactive trajectories. For a correct definition of the stable states also the variable q' must be taken into account

3 Sampling Path Ensembles

Various deterministic and stochastic sampling techniques for path ensembles have been proposed [4–6]. Here we consider only Monte Carlo methods. It is important, however, to be aware that while the path ensemble is sampled with a Monte Carlo procedure each single pathway is a fully dynamical trajectory such as one generated by molecular dynamics.

3.1 Monte Carlo in Path Space

In a transition path sampling simulation the goal is to collect reactive trajectories according to their weight in the transition path ensemble. This can be achieved with a Monte Carlo importance sampling procedure similar to a Monte Carlo simulation of, say, an atomic fluid. In the latter case, configurations are sampled by iterating a basic step: first a new configuration is generated from an old one by, for instance, displacing a randomly selected particle by a random amount. Then, the new configuration is accepted or rejected according to how the probability of the new configuration has changed with respect to the probability of the old configuration. Usually, the Metropolis rule is used to obtain an acceptance probability that satisfies the detailed balance condition. Provided that the algorithm is ergodic (i.e. that any configuration can be reached from any other configuration in a finite number of steps), this choice guarantees that the desired distribution of configurations is sampled. Since this procedure generates a random walk that visits configurations according to their statistical weight, it is also called *importance sampling*.

Monte Carlo procedures can be applied very generally to sample probability distributions [24]. In particular, Monte Carlo techniques can also be used to sample ensembles of pathways. In this case a random walk is carried out in the space of trajectories instead of configuration space. The basic step of this procedure consists of generating a new path, $x^{(n)}(T)$, from an old one, $x^{(o)}(T)$. To guarantee that pathways are visited with a frequency proportional to their weight in the path ensemble $\mathcal{P}_{AB}[x(T)]$ we require that the forward move from the old path to the new one must be exactly balanced by the reverse move:

$$\mathcal{P}_{AB}[x^{(o)}(T)]\pi[x^{(o)}(T) \to x^{(n)}(T)]$$
$$= \mathcal{P}_{AB}[x^{(n)}(T)]\pi[x^{(n)}(T) \to x^{(o)}(T)] . \tag{21}$$

In this *detailed balance* condition $\pi[x^{(o)}(T) \to x^{(n)}(T)]$ is the probability to move from the old to the new path. According to the Monte Carlo generation and acceptance/rejection scheme, this probability is the product of a generation probability P_{gen} and an acceptance probability P_{acc}:

$$\pi[x^{(o)}(T) \to x^{(n)}(T)]$$
$$= P_{\text{gen}}[x^{(o)}(T) \to x^{(n)}(T)] \times P_{\text{acc}}[x^{(o)}(T) \to x^{(n)}(T)] . \tag{22}$$

From the detailed balance criterion (21) one obtains the following condition that must be obeyed by the acceptance probability:

$$\frac{P_{\text{acc}}[x^{(o)}(T) \to x^{(n)}(T)]}{P_{\text{acc}}[x^{(n)}(T) \to x^{(o)}(T)]} = \frac{\mathcal{P}_{AB}[x^{(n)}(T)]P_{\text{gen}}[x^{(n)}(T) \to x^{(o)}(T)]}{\mathcal{P}_{AB}[x^{(o)}(T)]P_{\text{gen}}[x^{(o)}(T) \to x^{(n)}(T)]} . \tag{23}$$

This relation can be satisfied by the Metropolis rule [25]

$$P_{\text{acc}}[x^{(o)}(T) \to x^{(n)}(T)] = \min\left\{1, \frac{\mathcal{P}_{AB}[x^{(n)}(T)]P_{\text{gen}}[x^{(n)}(T) \to x^{(o)}(T)]}{\mathcal{P}_{AB}[x^{(o)}(T)]P_{\text{gen}}[x^{(o)}(T) \to x^{(n)}(T)]}\right\} . \tag{24}$$

Since the old trajectory $x^{(o)}(T)$ is reactive, i.e. $h_A[x_0^{(o)}] = 1$ and $h_B[x_T^{(o)}] = 1$, this acceptance probability can be written as:

$$P_{\text{acc}}[x^{(o)}(T) \to x^{(n)}(T)] = h_A[x_0^{(n)}]h_B[x_T^{(n)}]$$
$$\times \min\left\{1, \frac{\mathcal{P}[x^{(n)}(T)]P_{\text{gen}}[x^{(n)}(T) \to x^{(o)}(T)]}{\mathcal{P}[x^{(o)}(T)]P_{\text{gen}}[x^{(o)}(T) \to x^{(n)}(T)]}\right\} . \tag{25}$$

In practice, the Metropolis rule is implemented in the following way. First, the new path is generated. If the new path is not reactive, it is rejected. Otherwise, the ratio $\mathcal{P}[n]P_{\text{gen}}[n \to o]/\mathcal{P}[o]P_{\text{gen}}[o \to n]$ is computed. If this ratio is larger than one, the new path is accepted. If it is smaller than one, the new path is accepted only if a random number drawn from a uniform distribution in the interval $[0, 1]$ is smaller than the ratio. Otherwise the path is rejected. In the case of a rejection the old path is retained as the current one. Repetition of this basic combination of generation and acceptance/rejection step yields a random walk through trajectory space that visits pathways according to their weight in the transition path ensemble.

3.2 Shooting and Shifting

In the preceding section we have laid out the general Monte Carlo procedure in trajectory space, but we have not yet specified how new pathways are generated from old ones. The efficiency of a transition path sampling simulation crucially depends on how this is done. A variety of schemes to do that exist [12], but so far the shooting algorithm has proven to be the most efficient one. For this reason and since the shooting algorithm is the only algorithm applicable to deterministic dynamics, we will discuss only this algorithm in detail (together with the complementary shifting procedure) and refer the reader to [12] for other path generating procedures. The rather detailed discussion of the shooting move may also serve as an illustration of how the general Monte Carlo formalism needs to be applied to derive appropriate acceptance probabilities for specific path moves.

The basic idea of the shooting algorithm is to generate pathways by using the propagation rules of the underlying dynamics. This is done by "shooting off" a new pathway from a randomly selected time slice $x_{t'}^{(o)}$ along the old path. Before shooting this time slice may be modified yielding $x_{t'}^{(n)}$ (this step is necessary for deterministic dynamics). From this perturbed time slice at time t' two new path segments are generated forward to time $\mathcal{T}$ and backward to time 0 according to the rules of the underlying dynamics. Together these two new segments form the new pathway (see Fig. 4). Since this shooting procedure is done stepwise the corresponding generation probabilities for the forward and backward path segments are:

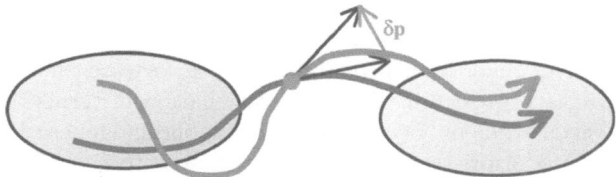

Fig. 4. In a shooting move for Newtonian dynamics a new path (dark gray) is created from an old on (light gray) by perturbing the momenta at a randomly selected time slice. From this perturbed point the equations of motion are then integrated backward and forward in time

$$P_{\text{gen}}^{\text{f}}[x^{\text{o}}(\mathcal{T}) \to x^{\text{n}}(\mathcal{T})] = \prod_{i=t'/\Delta t}^{\mathcal{T}/\Delta t - 1} p\left(x_{i\Delta t}^{(\text{n})} \to x_{(i+1)\Delta t}^{(\text{n})}\right) . \tag{26}$$

and

$$P_{\text{gen}}^{\text{b}}[x^{\text{o}}(\mathcal{T}) \to x^{\text{n}}(\mathcal{T})] = \prod_{i=1}^{t'/\Delta t} \bar{p}\left(x_{i\Delta t}^{(\text{n})} \to x_{(i-1)\Delta t}^{(\text{n})}\right) . \tag{27}$$

Since to generate the forward segment the dynamical rule of the underlying dynamics is used, (26) is just the dynamical path weight for the forward trajectory. For the backward move a time inverted transition probability $\bar{p}(x \to x')$ is used. (The choice of the backwards propagation rule is obvious for most systems, but difficulties may arise in some non-equilibrium cases [12].) Combining the forward and backward generation probabilities one obtains the generation probability for the entire shooting move:

$$P_{\text{gen}}[x^{(\text{o})}(\mathcal{T}) \to x^{(\text{n})}(\mathcal{T})] = p_{\text{gen}}[x_{t'}^{(\text{o})} \to x_{t'}^{(\text{n})}] \prod_{i=t'/\Delta t}^{\mathcal{T}/\Delta t - 1} p\left(x_{i\Delta t}^{(\text{n})} \to x_{(i+1)\Delta t}^{(\text{n})}\right)$$

$$\times \prod_{i=1}^{t'/\Delta t} \bar{p}\left(x_{i\Delta t}^{(\text{n})} \to x_{(i-1)\Delta t}^{(\text{n})}\right) , \tag{28}$$

where $p_{\text{gen}}[x_{t'}^{(\text{o})} \to x_{t'}^{(\text{n})}]$ is the probability to obtain the *shooting point* $x_{t'}^{(\text{n})}$ by modification of state $x_{t'}^{(\text{o})}$. (In the following we assume this generating probability for the shooting point to be symmetric. If it is not, this must be explicitly taken into account in the acceptance probability.) The generation probability (28) can now be used to derive an acceptance probability for pathways generated with the shooting algorithm.

$$P_{\text{acc}}[x^{(\text{o})}(\mathcal{T}) \to x^{(\text{n})}(\mathcal{T})] = h_A[x_0^{(\text{n})}]h_B[x_{\mathcal{T}}^{(\text{n})}]$$

$$\times \min\left[1, \frac{\rho\left(x_0^{(\text{n})}\right)}{\rho\left(x_0^{(\text{o})}\right)} \prod_{i=0}^{t'/\Delta t - 1} \frac{p\left(x_{i\Delta t}^{(\text{n})} \to x_{(i+1)\Delta t}^{(\text{n})}\right)}{\bar{p}\left(x_{(i+1)\Delta t}^{(\text{n})} \to x_{i\Delta t}^{(\text{n})}\right)} \times \frac{\bar{p}\left(x_{(i+1)\Delta t}^{(\text{o})} \to x_{i\Delta t}^{(\text{o})}\right)}{p\left(x_{i\Delta t}^{(\text{o})} \to x_{(i+1)\Delta t}^{(\text{o})}\right)}\right] . \tag{29}$$

In this expression, all terms stemming from the forward segment have cancelled because the generation probability of the forward shot is identical to the dynamical path probability. Such a cancellation of terms does not occur for the backwards segment, because the corresponding generation probability $\bar{p}(x \to x')$ is not a priori related in any simple way to the corresponding dynamical path probability. The acceptance probability (29), however, simplifies considerably if p and $\bar{p}$ obey a microscopic reversibility condition with respect to the distributions of initial conditions:

$$\frac{p(x \to y)}{\bar{p}(y \to x)} = \frac{\rho_0(y)}{\rho_0(x)} . \tag{30}$$

A backward generation probablity $\bar{p}$ satisfying this condition can be constructed in many cases, in particular in equilibrium situations. If condition (30) holds, the acceptance probability for the shooting move becomes:

$$P_{\text{acc}}[x^{(o)}(\mathcal{T}) \rightarrow x^{(n)}(\mathcal{T})] = h_A[x_0^{(n)}] h_B[x_{\mathcal{T}}^{(n)}] \min\left[1, \frac{\rho(x_{t'}^{(n)})}{\rho(x_{t'}^{(o)})}\right] . \qquad (31)$$

Hence, any new path connecting the stable states A and B is accepted with a probability depending only on the shooting points of the old and the new pathway. This simple acceptance rule also suggests the following algorithm. First a shooting point $x_{t'}^{(o)}$ is selected at random and modified to $x_{t'}^{(n)}$. Then, this shooting point is accepted if a random number selected from a uniform distribution in the interval $[0, 1]$ is smaller than $\rho(x_{t'}^{(n)})/\rho(x_{t'}^{(o)})$. If the shooting point is accepted, either one of the path segments is grown. If this segment reaches the appropriate stable region, the other segment is grown. The whole path is finally accepted if the boundary condition for that segment is also satisfied. If any of these steps is rejected, the procedure must be started over.

The acceptance rule (31) is very general and valid for any procedure that is microscopically reversible in the sense of (30). Cases in which its is applicable include Langevin dynamics, Monte Carlo dynamics and Newtonian dynamics [12]. Since Newtonian dynamics is the arguably most important of these (most MD simulations are based on Newton's equations of motion), we next discuss the shooting algorithm for this case with initial conditions distributed according to the equilibrium phase space density. Extensions for other kinds of deterministic dynamics such as Nose-Hoover or Gaussian isokinetic dynamics can be easily derived from the general procedure presented above.

To carry out a shooting move for Newtonian dynamics, a procedure to generate the shooting point $x_{t'}^{(n)}$ from the time slice $x_{t'}^{(o)}$ has to be specified. In general, one can do that in a symmetric way by adding a perturbation δx to $x_{t'}^{(o)}$,

$$x_{t'}^{(n)} = x_{t'}^{(o)} + \delta x . \qquad (32)$$

The perturbation δx is drawn from a symmetric distribution, such that the generation probability for the new shooting point $x_{t'}^{(n)}$ is symmetric with respect to the reverse move, $p_{\text{gen}}(x_{t'}^{(o)} \rightarrow x_{t'}^{(n)}) = p_{\text{gen}}(x_{t'}^{(n)} \rightarrow x_{t'}^{(o)})$. Symmetric generation of the shooting point can be a subtle issue, especially when there are rigid constraints. Methods for generating such displacements are discussed in [12].

In most cases, it is sufficient to change only the momentum part of the selected time slice $x_{t'}^{(o)}$ while the configurational part is left unchanged , i.e., $p_{t'}^{(n)} = p_{t'}^{(o)} + \delta p$ and $r_{t'}^{(n)} = r_{t'}^{(o)}$. However, in some cases it is advantageous to change both the configuration and momentum parts of $x_{t'}^{(o)}$ [26]. After generation of the shooting point the new trajectory is grown by applying dynamical propagation rules. For Newtonian dynamics this amounts to integrating the equations of motion. While the forward segments of the trajectory can be generated by carrying out the appropriate number of small time steps in forward direction, the backward segment has to be generated with inverted time direction, i.e., with a negative time step. Since Newton's equation of motion are time reversible, backward trajectory segments may be obtained by first inverting all momentum-like variables, and then integrating as if forward in time. Momenta in the resulting chain of states are then inverted to give the proper direction of time. To check whether this procedure satisfies the reversibility criterion (30) we consider the forward and backward transition probabilities for a short time step,

$$p(x \to y) = \delta[y - \phi_{\Delta t}(x)] \,, \tag{33}$$

and

$$\bar{p}(y \to x) = \delta[x - \phi_{-\Delta t}(y)] = \delta[x - \phi_{\Delta t}^{-1}(y)] \,. \tag{34}$$

For Dirac delta-functions the following equality is valid,

$$\delta\left[y - \phi_{\Delta t}(x)\right] = \delta\left[\phi_{\Delta t}^{-1}(y) - x\right] |\partial\phi_{\Delta t}(x)/\partial x|^{-1} \,, \tag{35}$$

where $|\partial\phi_{\Delta t}(x)/\partial x|$ is the Jacobian factor associated with the time evolution for a time step Δt. This Jacobian describes the contraction or expansion of an infinitesimal comoving volume element in phase space. Consequently, the single step forward and backward generation probabilities are related by

$$\frac{p(x \to y)}{\bar{p}(y \to x)} = \left|\frac{\partial\phi_{\Delta t}(x)}{\partial x}\right|^{-1} \,. \tag{36}$$

Since the phase space flow conserves the equilibrium distribution, we also have

$$\rho_0(\phi_{\Delta t}(x)) = \rho_0(x) \left|\frac{\partial\phi_{\Delta t}(x)}{\partial x}\right|^{-1} \,. \tag{37}$$

Hence, condition (30) holds and pathways generated with this algorithm can be accepted with the probability (31). For a microcanonical distribution of initial conditions together with Newtonian dynamics care must be taken that the perturbation of the shooting point does not change the energy. Algorithms to do that in a symmetric way are available [12]. In this case the acceptance probability further simplifies to

$$P_{\text{acc}}[x^{(o)}(\mathcal{T}) \to x^{(n)}(\mathcal{T})] = h_A[x_0^{(n)}]h_B[x_{\mathcal{T}}^{(n)}] \,. \tag{38}$$

In words, any new pathway that starts in A and ends in B is accepted. All other pathways must be rejected. Shooting moves can be complemented with shifting moves (see Fig. 5). In this computationally inexpensive move a new path is generated from old one by translating the trajectory forward or backward in time. More specifically, in a forward shifting move a trajectory

Fig. 5. In a shifting move for Newtonian dynamics a new path (*dark gray*) is created from an old on (light gray) by translating the initial point of the trajectory forward in time (such as in the picture) or backward in time

segment of a certain length is first removed from the beginning of the path. Then, a segment of the same length is grown at the end of the path by integrating the equations of motion for an appropriate number of time steps starting from the final point of the old path. As a result of this procedure the new path overlaps partially with the old one. For a backward shifting move one proceeds in an analogous way. The shifting procedure effectively translates the path in time in a way reminiscent of the "reptation" motion of a polymer in a dense melt [27].

For Hamiltonian dynamics with a canonical or microcanonical distribution of initial conditions the acceptance probability for pathways generated with the shifting algorithm is particularly simple. Provided forward and backward shifting moves are carried out with the same probability, the acceptance probability reduces to [12]

$$P_{\mathrm{acc}}[x^{(\mathrm{o})}(\mathcal{T}) \to x^{(\mathrm{n})}(\mathcal{T})] = h_A[x_0^{(\mathrm{n})}]h_B[x_t^{(\mathrm{n})}] \ . \tag{39}$$

implying that any new path that still connects regions A and B is accepted. Similarly, simple acceptance probabilities can be derived also for pathways generated with stochastic instead of deterministic dynamics [12]. Although new pathways generated by the shifting algorithm differ little from the corresponding old pathways especially in the transition region, shifting moves can improve the convergence of averages taken over the transition path ensemble.

3.3 Generating an Initial Path

To start a transition path sampling simulation with the algorithms described above an initial pathway connecting A with B must be available. Hence, generating an initial transition pathway is an important step in the application of the transition path sampling methodology. In the simplest case an initial trajectory connecting A and B can be obtained by running a long molecular dynamics (or stochastic dynamics) simulation. For most applications, however, this straightforward approach is ruled out by the rarity of the event one wants to study and an initial trajectory has to be created artificially. The specific way to generate such a trajectory is highly system dependent, but in general one produces an atypical trajectory with a low weight in the transition path ensemble $\mathcal{P}_{AB}[x(t)]$. Therefore the first part of a transition path sampling simulation starting from such a newly created trajectory consists of equilibrating the pathway towards the more important regions of trajectory space. This situation is analogous to that encountered in a conventional MC simulation of a molecular system. In that case a sufficiently long equilibration period is necessary to relax an unlikely initial configuration towards more probable regions of configuration space. Similarly, a transition path sampling simulation can start from an artificial pathway which does not even need to be a true dynamical trajectory. Then, repeated application of the Monte Carlo steps described above move the pathways towards more typical regions of pathways and the actual transition path sampling simulation can begin.

A more systematic way to create a new transition pathway is to gradually change the ensemble $\mathcal{P}_{AB}[x(t)]$ from one which includes all trajectories starting in A (without restrictions on the end point) to one which consists only of trajectories starting in A and ending in B. As will be discussed in Sect. 5, conversion of one path ensemble into another is computationally demanding and in most cases more efficient ways to generate an initial trajectory exist.

In some situations, high-temperature pathways can be used to initiate a transition path sampling simulation. Consider, for example, the folding/unfolding of a protein. At physiological temperatures, a protein in its native states unfolds only very rarely on a molecular time scale. At higher temperatures, however, unfolding occurs so quickly that it can be simulated with ordinary molecular dynamics simulations. Such a high temperature trajectory can then be used to start a transition path sampling simulation at the temperature of interest. If high temperature transition pathways are qualitatively very different from those at lower temperatures it might be necessary to carry out a systematic cooling procedure, in which the ensemble of pathways if brought to lower temperature in small steps.

In other cases, one may have some, possibly incomplete notion of a reaction coordinate. Controlling this presumed reaction coordinate one might be able to drive the system from A to B obtaining a chain of states from which shooting and shifting moves can be initiated. In our experience no general recipe exists for the generation of an initial trajectory. Rather, specific procedures have to be developed for this purpose for each application of the transition path sampling method to a new problem.

4 Reaction Mechanisms: Analyzing Transition Pathways

The path ensemble, as created by the transition path sampling methodology, is a statistically representative collection of trajectories leading from a reactant region to a product region. Further analysis of this ensemble of pathways is necessary to obtain rate constants, reaction mechanisms, reaction coordinates, transition state structures etc. In this section we will describe how to analyze the path ensemble by determining transition state ensembles, and how to test proposed reaction coordinates using *committor* distributions.

4.1 Reaction Coordinate vs. Order Parameter

The application of the transition path sampling formalism requires a proper definition of the stable states A and B. This can be achieved using so called *order parameters*, variables that discriminate configurations belonging to the reactant region from those in the product region. Such order parameters, however, are not necessarily suitable for the characterization of the reaction mechanism. To emphasize this fact we distinguish the term *order parameter* from the term *reaction coordinate*, which is a variable capable of describing the dynamical bottleneck separating products from reactants. This distinction is illustrated schematically in Fig. 6. Along a good reaction coordinate the system can be driven reversibly from one state to the other because all degrees of freedom orthogonal to the reaction coordinate equilibrate quickly. Driving the system from A to B by controlling the wrong coordinate, on the other hand, leads to large hysteresis effects because such a procedure neglects important barriers that may exist in orthogonal directions. A good reaction coordinate should capture the essence of the dynamics and allow us to predict what a trajectory starting from a given configuration will most likely do. As we will see later, this is exactly what the committor does: it tells us about the probability to relax into A or B. Thus, if a good reaction coordinate is available, the committor can be parametrized in terms of this coordinate [28].

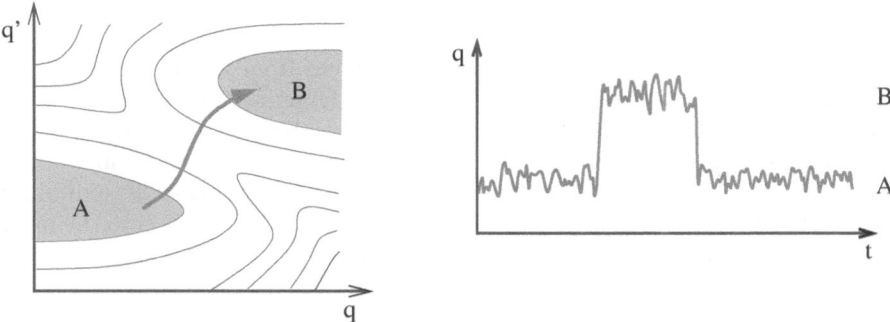

Fig. 6. Free energy landscape in which during a transition from A to B both the variable q and q' change systematically (*left hand side*). For this case, the variable q is a good order parameter and suffices to distinguish the equilibrium fluctuations in the stable states A and B. By following q as a function of time t we can detect transitions between the stables states (*right hand side*). However, this variable does not capture all essential aspects of the transition and is therefore not a good reaction coordinate. In the definition of such of a good reaction coordinate capable of describing the complete course of the reaction the variable q' can not be neglected

4.2 Committor

In the analysis of chemical dynamics the transition state plays a prominent role. Usually, the transition state is identified with saddle points of the potential energy surface where forces vanish and there is exactly one unstable mode. A system resting on a saddle point is an unstable state that can evolve either to the reactants or the products depending on which side we kick it. The concept of a transition state can be generalized and quantified for arbitrary configurations by considering the probability to proceed to reactants or products after averaging over noise and/or initial velocities. Onsager called this probability the *splitting probability* and used it to investigate ion-pair recombination [29] and in protein folding studies the term p_{fold} is used [30]. Here, we will call this relaxation probability the *committor*, because it tells us with which likelihood a certain configuration is committed to one of the two stable states (usually state B). The committor is a direct statistical indicator of the progress of the reaction. In this sense it is an ideal reaction coordinate. Parametrizing the committor in terms of small number of atomic variables to gain insight into the reaction mechanism, however, is a highly non-trivial task.

For a given configuration r the committor is defined as the fraction of trajectories started in A that reach stable state B after time t,

$$p_B(r,t) \equiv \frac{\int \mathcal{D}x(t)\, P[x(t)]\delta(r_0 - r)h_B(x_t)}{\int \mathcal{D}x(t)\, P[x(t)]\delta(r_0 - r)} \ , \tag{40}$$

where r_0 is the initial configuration of the system from which the trajectory $x(t)$ is integrated (see Fig. 7). For Newtonian dynamics and a canonical distribution of initial conditions the average in the above expression is over all momenta distributed according to a Maxwell-Boltzmann distribution. For stochastic dynamics the average extends also over noise histories. The committor is a statistical measure for how *committed* a given configuration is to the product state. A value of $p_B = 0$ indicates no commitment at all while a value of

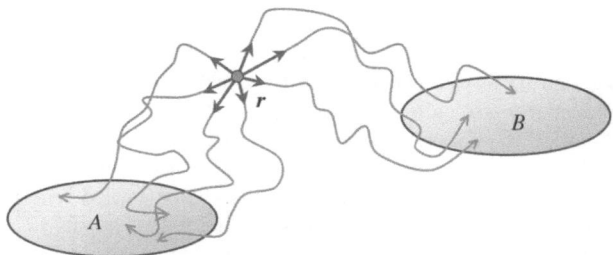

Fig. 7. For a given configuration r the committor $p_B(r)$ is calculated by initiating many trajectory with random momenta from r. Each trajectory is then propagated for a time t. The committor p_B is estimated from the fraction of trajectories that reach B in the time t

$p_B = 1$ that the configuration is fully committed to B. Configurations with $p_A \approx p_B$ are committed neither to A nor to B. Note the committor as defined above depends on time. This time has to be selected such that it is larger than the molecular time scale τ_{rxn} discussed in Sect. 5. A time-independent form of the committor can be defined by counting all trajectories originating from a given configuration r that reach state B before they reach state A [8]. In a practical calculation, only a finite sample of trajectories is available to determine the committor:

$$p_B(r,t) \approx \frac{1}{N} \sum_{i=1}^{N} h_B(x_t^{(i)}) \equiv p_B^{(N)}(r,t) . \tag{41}$$

Assuming that the N trajectories are statistically independent, the standard deviation of p_B calculated from N trajectories is:

$$\sigma = \sqrt{\langle (p_B^{(N)} - p_B)^2 \rangle} = \sqrt{\frac{p_B(1 - p_B)}{N}} . \tag{42}$$

This expression can be used to terminate a committor calculation after a certain desired accuracy has been reached. The largest number of trajectories is needed for configurations with a committor of $p_B \approx 1/2$.

4.3 The Transition State Ensemble

We define a configuration r to be a transition state (TS), if both stable states A and B are equally accessible from that configuration (see Fig. 8). In other words, r is a transition state if

$$p_A(r) = p_B(r) . \tag{43}$$

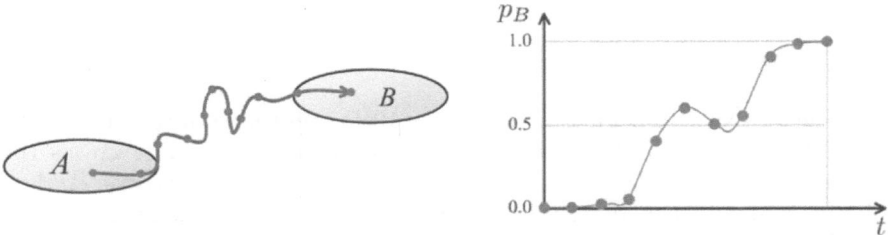

Fig. 8. The committor calculated for points along a path connecting the stable states A and B is a function which grows from $p_B = 0$ at configuration lying in and near A to $p_B = 1$ for configurations in B or nearby. In between, the committor will take the value $p_B = 0.5$ at least once but possibly several times. Configurations with $p_A = p_B = 0.5$ are equally likely to relax into either one of the stables states. We define these configurations to be the transition states

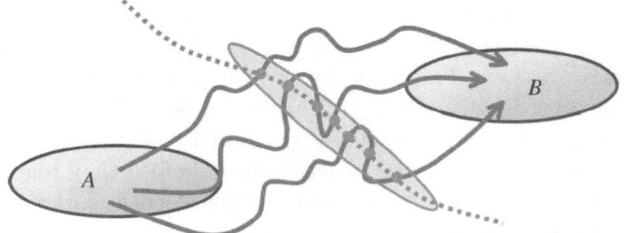

Fig. 9. All configurations on transition pathways that have a committor of $p_B = 0.5$ are members of the *transition state ensemble*. Each transition path contributes at least one configuration but possibly more to the transition state ensemble. The dashed line is the separatrix (or isocommittor 1/2 surface)

Transition states defined in this statistical way do not necessarily coincide with particular features of the potential energy surface. In certain cases, such as the diffusion of adatoms on solids at low temperature, points with $p_A(r) = p_B(r)$ may be located at or near saddle points of the potential energy surface, but in general entropic contributions play an important role in determining the location of statistically defined transition states in configuration space. For instance, the transition from a hard sphere fluid to a hard sphere solid proceeds through a purely entropic transition state [31]. Equation (43) defines a surface separating the basins of attraction of the stable states. This dividing surface is also called the *separatrix*. If all trajectories started from configuration r end either in A or in B then $p_A(r) + p_B(r) = 1$. In this case a transition state is characterized by $p_A = p_B = 1/2$. To find configurations belonging to the separatrix one can locate configurations with $p_A = p_B$ on transition pathways. The set of all such configurations is the *transition state ensemble* (see Fig. 9). This operation introduces a weight on the separatrix. Analysis of the transition state ensemble can yield important information about the reaction mechanism.

4.4 Committor Distributions

Consider the situation illustrated in Fig. 10 for proton transfer in the protonated water trimer. Suppose one has postulated the variable Δr as the reaction coordinate for the transition. Then, the question arises if Δr is a relevant and sufficient description of the reaction mechanism. In the case of the Fig. 10 it is clearly not. If one drives the transition by controlling Δr, hysteresis will occur, indicating that the variable Δr is not sufficient for a dynamical description of the reaction. Observing hysteresis is a crude way of testing reaction coordinates and a more precise procedure is called for. Calculating distributions of committors for constraint ensembles can be precisely the powerful tool we need to test the correctness of the proposed reaction coordinate. This diagnostic tool is not restricted to TPS but can be applied

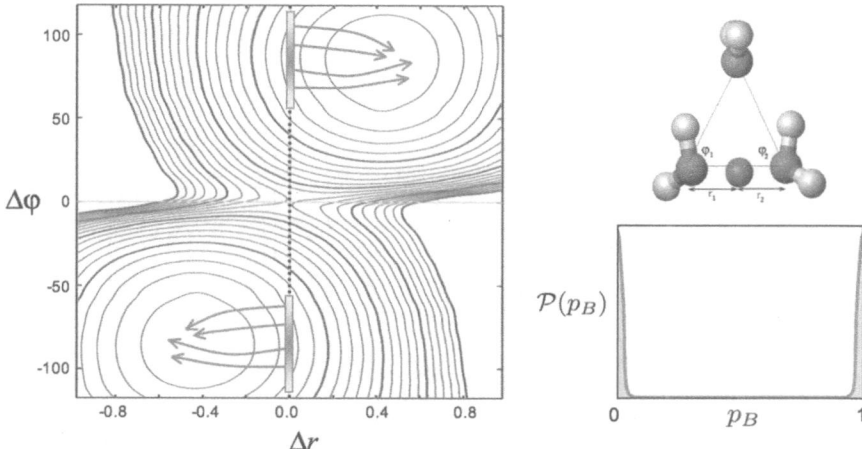

Fig. 10. If in the case of the protonated water trimer trajectories are started from configurations in which the proton coordinate $\Delta r = r_2 - r_1$ vanishes (all other variables are distributed according to the equilibrium distribution), the proton always relaxes to one of the two stable states. To which one depends on the values of the angular variable $\Delta\varphi = \varphi_2 - \varphi_1$ describing the geometry of the OOO-triangle. For initial conditions located in the lower red stripe in the free energy landscape on the left hand side, the proton will always relax into the state characterized by negative values of $\Delta\varphi$. Trajectories started in the upper red stripe will always evolve towards the other stable state with positive values of $\Delta\varphi$. The distribution $\mathcal{P}(p_B)$ obtained by repeated calculation of the committor p_B will hence show a sharp peak at $p_B = 0$ and another sharp peak at $p_B = 1$. In general such a result indicates that degrees of freedom important for the analysis have been neglected. In the case of the protonated water trimer this degree of freedom is the angular difference $\Delta\varphi$

to any case requiring a test whether or not the proposed order parameter describes the transition or is only partially correlated (if at all) with the reaction coordinate.

The first step in this procedure is the calculation of the free energy profile as a function of a variable q (in the case of the protonated trimer this is Δr). If q differs in the reactant and product region, the free energy as a function of q will most likely show a barrier with maximum at $q = q^*$. Next, configurations with $q = q^*$ are generated, for instance by imposing constraints in a molecular dynamics simulation [32]. Subsequently, commitment probabilities are computed for a representative number of configurations in this ensemble. If each of these configurations is truly in the transition state region, the committors will be around $p_B = 0.5$. If, in contrast, they are far from one half, say 0 or 1, the configurations are clearly not part of the separatrix. Hence, they cannot be transition states in the sense we discussed in the previous section.

To analyze this situation further we construct a histogram of the committors, a committor distribution $P(p_B)$,

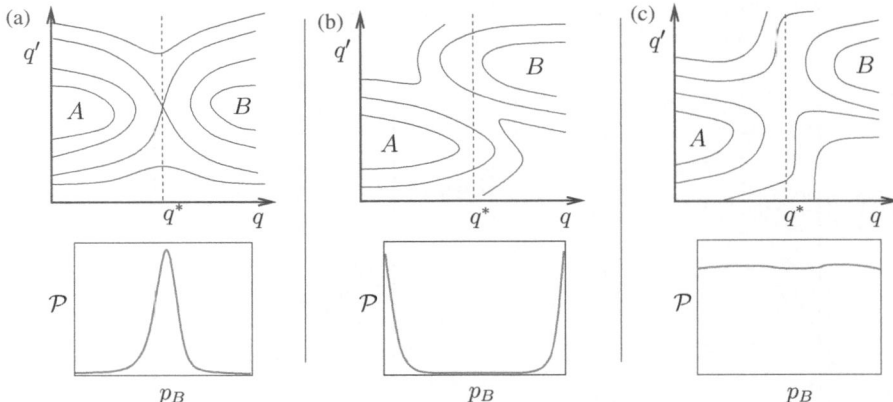

Fig. 11. Three scenarios leading to different committor distributions. (**a**) The variable q is a good reaction coordinate. Configurations constrained at $q = q^*$ produce a distribution of committors peaked at $p_B = 0.5$. (**b**) The variable q is insufficient to describe the reaction properly. As a results the committor distribution for configurations with $q = q^*$ is peaked at zero and unity. For a correct description of the transition the variable q' must be taken into account. (**c**) The transition occurs diffusively in the direction of q'. In this case the committor distribution is flat

$$P(\tilde{p}_B) = \frac{\langle \delta[\tilde{p}_B - p_B(r, t)]\delta[q^* - q(r)]\rangle}{\langle \delta[q^* - q(r)]\rangle} . \tag{44}$$

Here, $P(p_B)$ is the probability (density) for finding the committor p_B in the ensemble $q = q^*$. If this distribution is peaked around $p_B = 0.5$, the constraint ensemble $q = q^*$ is located on the separatrix and coincides with the transition state ensemble. In this case, q is a good reaction coordinate, at least in the neighborhood of the separatrix. This is illustrated in Fig. 11a. Other possible scenarios for the underlying free energy landscape result in different committor distributions, and are also illustrated in Fig. 11. The committor distribution can thus be used to estimate how far a postulated reaction coordinate is removed from the correct reaction coordinate. An application of this methodology can be found in [33], where the reaction coordinate of the crystallization of a Lennard-Jones fluid has been resolved by analysis of committor distributions.

5 Kinetics

Since pathways collected with transition path sampling are true dynamical trajectories rather than artificial sequences of states such as minimum energy pathways, they can be used to study the kinetics of the reaction. In the following we will first define reaction rate constants and explain how they are related to the microscopic dynamics of the system. At the heart of this relation lie specific time correlation functions. Then, we will discuss several methods for calculating transition rate constants within the framework of transition path sampling. These methods use slightly different expressions, but all of them rely on efficient path sampling techniques.

5.1 Reaction Rate Constants and Population Fluctuations

Consider the unimolecular reaction

$$A \rightleftharpoons B \tag{45}$$

which interconverts molecules of type A and B into each other. At a given time t there are $N_A(t)$ molecules of type A and $N_B(t)$ molecules of type B in a container of volume V. The corresponding concentrations $c_A(t) = N_A(t)/V$ and $c_B(t) = N_B(t)/V$ can change in time due to molecules undergoing the reaction. Since according to the assumption molecules can only interconvert into each other, the total number of molecules is constant, $N_A(t)+N_B(t) = N$. Provided the solution is sufficiently dilute,i.e., the molecules are independent from each other, the time evolution of the concentrations can be described on a phenomenological level by the rate equations

$$\dot{c}_A = -k_{AB}\, c_A + k_{BA}\, c_B,$$
$$\dot{c}_B = k_{AB}\, c_A - k_{BA}\, c_B\, . \tag{46}$$

Here, k_{AB} and k_{BA} are the forward and backward reaction rate constants, respectively. These are the quantities that we want to calculate with transition path sampling. To solve the kinetic equations (46) it is convenient to consider the deviation of the concentrations from their equilibrium value:

$$\Delta c_A(t) = c_A(t) - \langle c_A \rangle$$
$$\Delta c_B(t) = c_B(t) - \langle c_B \rangle\, , \tag{47}$$

where $\langle c_A \rangle$ and $\langle c_B \rangle$ are the concentrations one finds when equilibrium sets in. Since $\Delta c_A(t) = -\Delta c_B(t)$ it is sufficient to consider only $\Delta c_A(t)$. According to the analytical solution of the rate equations $\Delta c_A(t)$ decays exponentially from its initial value $\Delta c_A(0)$,

$$\Delta c_A(t) = \Delta c_A(0) \exp(-t/\tau_{\mathrm{rxn}})\, . \tag{48}$$

The *reaction time* τ_{rxn} depends on the forward and backward reaction rate constants,

$$\tau_{\mathrm{rxn}}^{-1} = k_{AB} + k_{BA}\, . \tag{49}$$

If the reaction rate constants for the forward and backward reactions differ considerably, the reaction time is dominated by the larger rate constant. The above solution of the rate equations describes how a non-equilibrium concentration decays towards its equilibrium value. The description of the reaction in terms of kinetic equations is purely phenomenological and does not make any reference to the molecular nature of the system. To calculate the kinetic coefficients k_{AB} and k_{BA} from an atomistic theory we need to compare the time evolution (48) predicted by the kinetic equations (46) with the time evolution observed in the microscopic picture. Usually, such a comparison is carried out in the framework of linear response theory, in which one consideres the response of the system to small perturbations of the concentrations [34]. Here, we connect the microscopic with the macroscopic picture following a slightly different approach. Rather than considering *small* perturbations we study the response of the system to *large* changes in the concentrations. Remarkably, both approaches yield identical expressions for the reaction rate constants. The reason for this agreement is that for the systems considered here the full response to a perturbation of arbitrary strength, just as the linear response, is fully determined by the equilibrium fluctuations of the system. This is explained in detail in the Appendix.

To derive microscopic expressions for the reaction rate constants we first use the solution of the rate equations to determine the fraction of molecules of type B at time t under the condition that initially all of the N molecules where of type A, i.e. $N_A(0) = N$ and $N_B(0) = 0$,

$$\frac{N_B(t)}{N} = \frac{\langle N_B \rangle}{N} \left(1 - e^{-t/\tau_{\mathrm{rxn}}} \right) , \tag{50}$$

where $\langle N_B \rangle$ is the number of molecules of type B in equilibrium. This fraction is nothing else than the conditional probability to find a molecule in state B at time t provided it was in state A at time 0:

$$P(B, t | A, 0) = \frac{\langle N_B \rangle}{N} \left(1 - e^{-t/\tau_{\mathrm{rxn}}} \right) . \tag{51}$$

We will now make contact with the microscopic picture by calculating this conditional probability from the microscopic dynamics of the system. Since by assumption the molecules are independent from each other, it is sufficient to follow the time evolution of a single molecule. For this molecule the characteristic functions $h_A(x)$ and $h_B(x)$ introduced earlier serve to distinguish between state A and B. The characteristic functions usually depend only on structural properties such as angles, distances or other geometric criteria, but for generality we write x, the full states of the system in phase space, as their argument. By calculating h_A and h_B for a particular state x we can tell whether the molecule is of type A or B. Here we do not assume that the molecule is either in A or in B at all times, i.e., $h_A(x) + h_B(x)$ is not necessarily unity for all configurations. But if states A and B are stable and the characteristic functions are defined properly, $h_A(x) + h_B(x) = 1$ most of the time. Only for short times, for instance during a transition from A to B, we can have $h_A(x) + h_B(x) = 0$.

To calculate the conditional probability $P(B, t|A, 0)$ from the microscopic dynamics of the system we imagine that we prepare initial conditions x_0 according to the equilibrium distribution $\rho(x_0)$ but restricted to the reactant region A,

$$\rho_A(x_0) = \rho(x_0)h_A(x_0)/Z_A , \tag{52}$$

where

$$Z_A = \int dx_0 \rho(x_0)h_A(x_0) \tag{53}$$

normalizes this restricted distribution. We now follow the time evolution of the system starting from many of such initial conditions in A and observe if the system is in B after time t by calculating $h_B(x_t)$. Averaging over all initial conditions x_0 with the appropriate weight yields the conditional probability to find the system in B at time t given that it was in A at time 0,

$$P(B, t|A, 0) = \int dx_0 \, \rho_A(x_0)h_B(x_t) = \frac{\int dx_0 \, \rho(x_0)h_A(x_0)h_B(x_t)}{\int dx \, \rho(x_0)h_A(x_0)} . \tag{54}$$

Denoting averages over the equilibrium distribution (and possibly over the noise history) by $\langle \cdots \rangle$ we can also write

$$P(B, t|A, 0) = \frac{\langle h_A(x_0)h_B(x_t) \rangle}{\langle h_A \rangle} , \tag{55}$$

where $\langle h_A(x_0)h_B(x_t) \rangle$ is a so called time correlation function and $\langle h_A \rangle$ is the average value of $h_A(x)$, i.e., the fraction of molecules of type A in equilibrium.

Equation (55), expressing the conditional probability $P(B, t|A, 0)$ in terms of a time correlation function, now permits to link the phenomenological and microscopic descriptions. By equating (51) and (55) and by noting that $\langle N_B \rangle/N = \langle h_B \rangle$ we find that

$$\frac{\langle h_A(x_0)h_B(x_t) \rangle}{\langle h_A \rangle} = \langle h_B \rangle \left(1 - e^{-t/\tau_{\mathrm{rxn}}} \right) . \tag{56}$$

While the left hand side of this equation is purely microscopic, the right hand side originates from the solution of the phenomenological rate equations. Relation (56) provides the link we were looking for.

Since the rate equations (46) treat transitions at a very coarse grained level and neglect all microscopic details of the transition, (56) cannot hold at all times. Particularly for short times deviations are expected. This does, however, not imply that the phenomenological description is incorrect but only limits its range of validity. For times that are long compared to τ_{mol}, the time scale of molecular correlations, this relation should hold. The situation is similar to that encountered in the context of diffusion. While the (phenomenological) diffusion equation predicts a linear dependence of the mean squared displacement on time, microscopic dynamics yields a quadratic dependence for short times (ballistic regime). Only after a transient time, the length of which is determined by the persistence of correlations, does linear behavior set in.

To analyze (56) further let us assume that the reaction time τ_{rxn} is much longer than the molecular time scale τ_{mol} (this is exactly the situation we are interested in). Then, for times $\tau_{\mathrm{mol}} < t \ll \tau_{\mathrm{rxn}}$ a Taylor expansion of the right hand side yields

$$\frac{\langle h_A(x_0) h_B(x_t) \rangle}{\langle h_A \rangle} = \langle h_B \rangle \frac{t}{\tau_{\mathrm{rxn}}} . \tag{57}$$

Exploiting that in equilibrium $\langle h_A \rangle k_{AB} = \langle h_B \rangle k_{BA}$ [this detailed balance condition follows from the rate (46)] and noting that $\tau_{\mathrm{rxn}}^{-1} = k_{AB} + k_{BA}$ we finally find

$$C(t) \equiv \frac{\langle h_A(x_0) h_B(x_t) \rangle}{\langle h_A \rangle} = (\langle h_A \rangle + \langle h_B \rangle) k_{AB} t \approx k_{AB} t . \tag{58}$$

Thus, the slope of the time correlation $C(t)$ in the time regime $\tau_{\mathrm{mol}} < t \ll \tau_{\mathrm{rxn}}$ is the forward reaction rate constant k_{AB}. The function $C(t)$ contains all the information needed to determine the reaction rate constant. Note that the relations obtained in this section can also be derived using linear response theory [34].

5.2 Free Energies in Path Space

To evaluate the time correlation function $C(t)$ in the transition path sampling framework we rewrite it in terms of sums over trajectories:

$$C(t) = \frac{\int \mathcal{D}x(t) \, h_A(x_0) \mathcal{P}[x(t)] h_B(x_t)}{\int \mathcal{D}x(t) \, h_A(x_0) \mathcal{P}[x(t)]} = \frac{Z_{AB}(t)}{Z_A} . \tag{59}$$

This expression can be viewed as the ratio between the partition functions for two different path ensembles: in the denominator we have the partition functions for the set of pathways that start in A and end anywhere. The partition function in the numerator is the one for the ensemble of pathways that start in A and end in B. In a sense, this ratio of partition functions measures the volume in trajectory space occupied by reactive trajectories relative to that of the trajectories starting in A but without restriction on the ending point.

This perspective suggests to determine the correlation function $C(t)$ via calculation of $\Delta F(t) = F_{AB}(t) - F_A$, the difference of the free energies related to the partition functions in the numerator and denominator. The free energy difference ΔF can also be viewed as the work $W_{AB}(t)$ necessary to convert the two ensembles *reversibly* into each other,

$$W_{AB}(t) \equiv -\ln \frac{Z_{AB}(t)}{Z_A} . \tag{60}$$

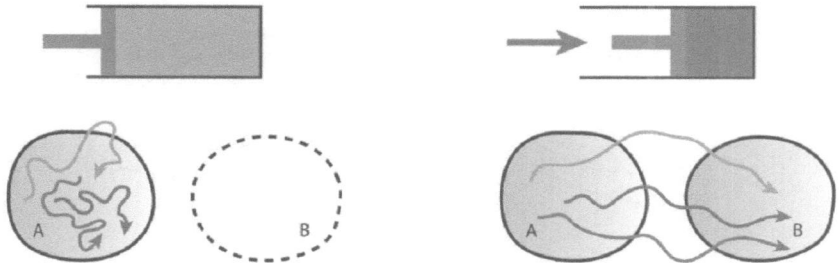

Fig. 12. Calculation of the time correlation function $C(t)$ is equivalent to determining the reversible work $W_{AB}(t)$ required to confine the endpoints of paths originating from region A into region B. This amounts to a "compression" of pathways in trajectory space

This is the work we need to "compress" the path ensemble and confine the endpoints of the paths with length t to the product region B (see Fig. 12). Note that $W_{AB}(t)$ depends on the path length t. From the free energy difference one can than immediately determine the time correlation function, $C(t) = \exp[-W_{AB}(t)]$. Due to the formal similarity of free energies of ensembles of configurations and ensembles of trajectories, standard free energies estimation methods such as thermodynamic integration or umbrella sampling can be used for the calculation of $W_{AB}(t)$ and hence of the time correlation function $C(t)$. In the following section we will discuss the application of the umbrella sampling technique [35] to this problem.

5.3 Umbrella Sampling

Umbrella sampling is a non-Boltzmann technique designed to enhance the sampling of rare but important regions of configuration space [35]. Here, we will apply it to enhance the sampling of rare but important regions in path space [36]. For this purpose we introduce an order parameter $\lambda(x)$ that distinguishes between the reactant and product region. In other words, the range of values λ takes when the system is in A, $\lambda_{\min}^A < \lambda < \lambda_{\max}^A$, does not overlap with the range of values λ takes in B, $\lambda_{\min}^B < \lambda < \lambda_{\max}^B$. Let us also assume that $\lambda_{\max}^A < \lambda_{\min}^B$. Next, we consider the probability density for finding a particular value of the parameter λ at the endpoints of pathways starting in A:

$$P_A(\tilde{\lambda}, t) = \frac{\int \mathcal{D}x(t)\, h_A(x_0) \mathcal{P}[x(t)] \delta[\tilde{\lambda} - \lambda(x_t)]}{Z_A} = \langle \delta[\tilde{\lambda} - \lambda(x_t)] \rangle_A . \quad (61)$$

By integrating this distribution over the final region we obtain the conditional probability to find the system in B at time t provided its was in A at time 0,

$$C(t) = \exp[-W_{AB}(t)] = \int_{\lambda_{\min}^B}^{\lambda_{\max}^B} d\lambda\, P_A(\lambda, t) . \quad (62)$$

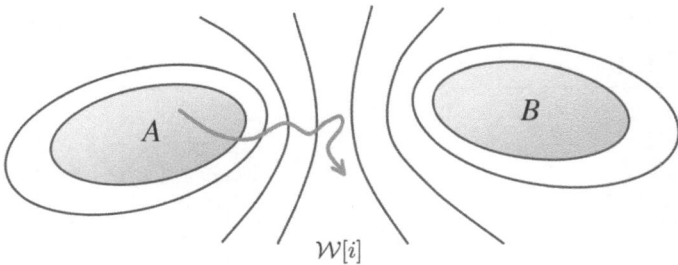

Fig. 13. In an umbrella sampling simulation for pathways configuration space is covered with a sequence of overlapping windows $\mathcal{W}[i]$. For each window a separate transition path sampling simulation for paths with starting points in A and endpoints in $\mathcal{W}[i]$ is carried out. From the order parameter distributions matched where the windows overlap the distribution $P_A(\lambda, t)$ can be calculated

In principle one could calculate the distribution (61) by generating initial conditions in A and histogram the values of λ at the endpoints of the trajectories started from these initial conditions. Since transitions from A to B are rare, however, such an approach fails to collect sufficient statistics in the λ-range corresponding to region B. Here, an umbrella sampling procedure can help. This is done by dividing the whole range of the parameter λ into overlapping windows. Each of these windows corresponds to a region $\mathcal{W}[i]$ in phase space (see Fig. 13). For each window a separate path sampling simulation is carried out for the ensemble

$$\mathcal{P}_{A\mathcal{W}[i]}[x(t)] = \rho(x_0)\mathcal{P}[x(t)]h_A(x_0)h_{\mathcal{W}[i]}(x_t) , \qquad (63)$$

in which pathways are required to start in A and end in window $\mathcal{W}[i]$. From these simulations the order parameter distribution in each window is computed:

$$P_{A\mathcal{W}[i]}(\tilde{\lambda}, t) = \frac{\int \mathcal{D}x(t)\, h_A(x_0)\mathcal{P}[x(t)]h_{\mathcal{W}[i]}(x_t)\delta[\tilde{\lambda} - \lambda(x_t)]}{\int \mathcal{D}x(t)\, h_A(x_0)h_{\mathcal{W}[i]}(x_t)}$$

$$= \langle \delta[\tilde{\lambda} - \lambda(x_t)] \rangle_{A\mathcal{W}[i]} . \qquad (64)$$

By matching these order parameter probabilities where the windows overlap one finally obtains the distribution $P_A(\lambda, t)$ over the whole order parameter range. The results of an example calculation are shown in Fig. 14. Note that through this windowing procedure it is possible to calculate the distribution $P_A(\lambda, t)$ even in the range where it is very small.

5.4 Reversible Work to Change the Path Length

With the umbrella sampling procedure described in the previous paragraph it is possible to determine the time correlation function $C(t)$ for different times

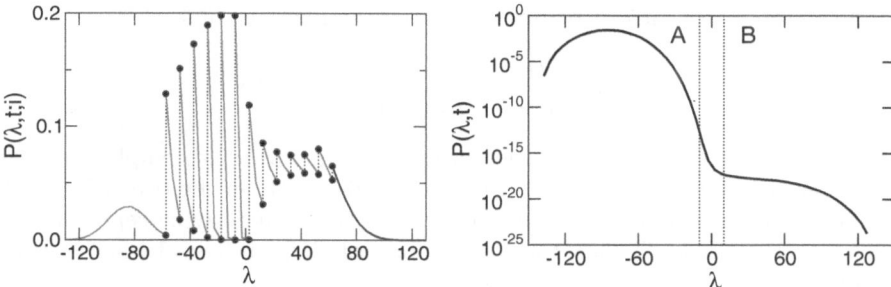

Fig. 14. Distribution of the order parameter λ in different order parameter windows for pathways of a given length t (*left hand side*). Matching the distributions in the windows where they overlap yields the distribution over the full order parameter range from the reactant region to the product region (*right hand side*). The data are from simulations of proton transfer in the protonated water trimer

t. From a time derivative determined numerically one can then calculate the transition rate constant k_{AB}. While feasible, this is an expensive procedure. It is more convenient to divide the calculation of $C(t)$, or, equivalently, of the free energy $W_{AB}(t)$, into two parts. In one part the free energy $W_{AB}(t')$ is determined for a particular time t' (the time t' can be chosen small to make the simulation more efficient). In the second part of the calculation this free energy is complemented with the free energy to change the time at which the constraint to reach B is applied:

$$W_{AB}(t) = W_{AB}(t') + \Delta W_{AB}(t; t') \,. \tag{65}$$

The free energy $\Delta W_{AB}(t, t')$ is the reversible work required to change the length of reactive trajectories and can be written in terms of the ratio of correlation functions,

$$\exp(-\Delta W_{AB}(t, t')) = \frac{C(t)}{C(t')} = \frac{\langle h_A(x_0) h_B(x_t) \rangle}{\langle h_A(x_0) h_B(x_{t'}) \rangle} \,. \tag{66}$$

This ratio can be computed efficiently for all times $t, t' < \mathcal{T}$ from a single path sampling simulation [12]:

$$\exp(-\Delta W_{AB}(t, t')) = \frac{\langle h_B(x_t) \rangle^*_{AB}}{\langle h_B(x_{t'}) \rangle^*_{AB}} \,. \tag{67}$$

where the notation $\langle h_B(x_t) \rangle^*_{AB}$ indicates an average over an ensemble of pathways starting in A but only required to visit B at some time slice along the path rather then ending in B. The correlation function $C(t)$ then is

$$C(t) = \frac{\langle h_B(x_t) \rangle^*_{AB}}{\langle h_B(x_{t'}) \rangle^*_{AB}} \times C(t') \tag{68}$$

and the rate constant can be obtained from the plateau value of

$$k(t) \equiv \frac{dC(t)}{dt} = \frac{\langle \dot{h}_B(x_t) \rangle^*_{AB}}{\langle h_B(x_{t'}) \rangle^*_{AB}} \times C(t') , \tag{69}$$

where the dot indicates a time derivative. It is important to emphasize that the time correlation function $C(t)$ is calculated exactly in the transition path sampling method and that no assumption about an underlying separation of time scales is made. Rather, the specific form of $C(t)$ can reveal whether such a separation exists.

5.5 Transition Interface Sampling and Partial Path Sampling

Transition Interface Sampling (TIS) is a path sampling technique for the calculation of reaction rate constants [37, 38]. In this method, based on the determination of effective fluxes through hypersurfaces in phase space, the use of pathways with variable length and a reduced sensitivity to recrossing events leads to an enhanced efficiency with respect to the methods described in the previous sections. For diffusive dynamics a further efficiency increase can be achieved by using the partial path transition interface sampling (PPTIS) method [38, 39]. This approach is applicable to processes such as the nucleation of a phase transition of the folding of a protein, where due to strong coupling to the environment memory of prior motion is rapidly lost [20].

Central to the derivation of the rate constant expressions used in the TIS approach is the definition of *overall* states $\mathcal{A}$ and $\mathcal{B}$ denoted by calligraphic letters. A point x in phase space is defined to be part or the overall region $\mathcal{A}$ of a trajectory started in x and followed backward in time reaches the product region A before the reactant region B (see Fig. 15). Thus the overall region $\mathcal{A}$ contains all phase space points that lie directly in A plus all phase space points that, in a sense, originate from A. Similarly x is defined to be in $\mathcal{B}$ if the respective backwards trajectory reaches B before A. Note that such overall

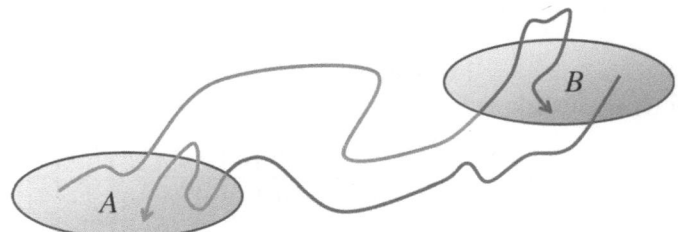

Fig. 15. Overall state $\mathcal{A}$ is defined to consist of all phase space points x whose projections into configuration space either lie in the reactant region A and all points that reach region A before region B when propagated backward in time. Overall region $\mathcal{B}$ is defined similarly. Hence, the trajectory segments colored in light gray belong to overall state $\mathcal{A}$ while those colored in dark gray belong to $\mathcal{B}$

states can be defined unambiguously only for deterministic dynamics. The overall states $\mathcal{A}$ and $\mathcal{B}$ cover the entire phase space and have a well defined, possibly highly fractal boundary between each other.

We can now use the characteristic functions $h_\mathcal{A}(x)$ and $h_\mathcal{B}(x)$ of the overall states to define the time correlation function,

$$C(t) \equiv \frac{\langle h_\mathcal{A}(x_0) h_\mathcal{B}(x_t) \rangle}{\langle h_\mathcal{A} \rangle} . \tag{70}$$

This time correlation function is the conditional probability that a phase space point in $\mathcal{A}$ at time 0 will be found in $\mathcal{B}$ at time t. This correlation function for the overall states has the advantage that for regions A and B that are sufficiently far apart recrossings of the phase space hypersurface separating the overall states are essentially eliminated and the reaction rate constant k_{AB} can be identified with the slope of $C(t)$ at time 0:

$$k_{AB} = \frac{\langle h_\mathcal{A}(x_0) \dot{h}_\mathcal{B}(x_0) \rangle}{\langle h_\mathcal{A} \rangle} = \frac{\langle \phi_{AB} \rangle}{\langle h_\mathcal{A} \rangle} . \tag{71}$$

The second part of the equations defines the *effective positive flux* $\langle \phi_{AB} \rangle$, the average flux into region B stemming from trajectories coming directly from A. (Here, "directly" means that only the first entry of the trajectory into B is counted). This expression is equivalent to the transitions state theory (TST) rate constant for transitions from the overall state $\mathcal{A}$ and $\mathcal{B}$. This approximation is valid in most cases due to the particular definition of the overall states in phase space. Since transitions from A to B (and hence also from $\mathcal{A}$ to $\mathcal{B}$) are rare, the TIS rate expression (71) cannot be evaluated by simple MD-simulation. Rather, a route similar to the umbrella sampling approach described in the previous paragraphs can be taken. For this purpose it is convenient to introduce a sequence of n non-intersecting dividing surfaces defined by $\lambda(x) = \lambda_i$ (see Fig. 16). The order parameter $\lambda(x)$ and the values $\lambda_0, \lambda_1, \cdots, \lambda_n$ are chosen such that the boundaries of regions A and B correspond to $\lambda(x) = \lambda_0$ and $\lambda(x) = \lambda_n$, respectively. For each pair of surfaces i and j one then defines the crossing probability $\mathcal{P}_A(j|i)$ as follows:

$\mathcal{P}_A(j|i)$ = Probability that a path crossing surface i for the first time after coming from region A reaches surface j before going back to A.

Using a concatenation of such transition probabilities for consecutive surfaces the TIS expression for the reaction rate constant can be rewritten as

$$k_{AB} = \frac{\langle \phi_{AB} \rangle}{\langle h_\mathcal{A} \rangle} = \frac{\langle \phi_{A1} \rangle}{\langle h_\mathcal{A} \rangle} \prod_{i=1}^{n-1} \mathcal{P}_A(i+1|i) , \tag{72}$$

where $\langle \phi_{A1} \rangle$ is the average flux through surface 1 stemming from trajectories coming directly from A. Note that the time dependence in (72) has become

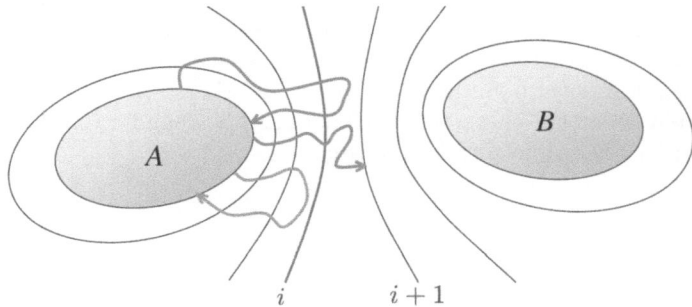

Fig. 15. In a transition interface sampling simulation to calculate the probability $\mathcal{P}_A(i+1|i)$ the shooting algorithm is used to sample an ensemble of pathways that originate in A, cross interface i and then return to A or proceed to interface $i+1$. Then, $\mathcal{P}_A(i+1|i)$ equals the fraction of the trajectories that reach interface $i+1$ rather than the surface of A. The dark gray pathways depicted are members of this ensemble but the light gray one is not because it does not cross interface i before returning to A

implicit. Instead of reaching a plateau in time as in (69), the factors appearing in (72) reach a plateau as a function of the order parameter.

In (72) the effective flux $\langle\phi_{AB}\rangle$ is expressed as a product of the effective flux through surface 1, $\langle\phi_{A1}\rangle$ and the probability that a trajectory that crosses interface 1 will reach B before A written as a product of crossing probabilities between consecutive interfaces. Each factor appearing in (72) can be determined easily with computer simulations. While the effective flux $\langle\phi_{A1}\rangle$ can be computed from a straightforward MD simulation carried out in the reactant region A, the probabilities $\mathcal{P}_A(i+1|i)$ can be calculated with path sampling simulations. More specifically, one defines an ensemble of pathways with variable length that are required to originate from A, cross surface i and then proceed to surface $i+1$ or go back to A. This ensemble can be sampled with the shooting algorithm complemented with efficiency enhancing time reversal moves [38]. For strongly diffusive dynamics the calculation of the transition probabilities $\mathcal{P}_A(i+1|i)$ can be simplified leading to a further efficiency increase. This approach, that views transitions as resulting from a sequence of uncorrelated hopping events is called partial path transition interface sampling (PPTIS) [39]. Note that the milestoning method proposed by Faradjan and Elber [40] and collaborators is very similar in spirit to PPTIS. Another method for the calculation of reaction rate constant that relies on the definition of a sequence of hypersurfaces and is based on (72) is the forward flux method of Allen, Warren, and ten Wolde [41].

5.6 Calculating Activation Energies
with Transition Path Sampling

The temperature dependence of many rate constants for chemical reactions follows the Arrhenius law [42], $k = \nu \exp(-\beta E_a)$, where ν is the so-called pre-exponential factor, E_a is the activation energy, and $\beta = 1/k_B T$ is the inverse temperature. Both the activation energy E_a and the pre-exponential factor ν can be determined from experimental data by plotting the logarithm of the reaction rate constant k as a function of β. While the negative slope of this line is the activation energy, the pre-exponential factor ν, can be determined from the intersect of the $\ln k$ vs. β curve with the y-axis. In the framework of transition state theory the activation energy E_a is the potential energy difference between reactants and the transition state, usually identified with a saddle point on the potential energy barrier separating reactants from products. The pre-exponential factor ν depends on the entropy difference between reactants and transition state and it also includes dynamical corrections.

Since the Arrhenius parameters E_a and ν are accessible experimentally, their theoretical prediction is of great interest. In principle, these parameters can be determined computationally by calculating the reaction rate constant as a function of temperature and then proceeding in a manner analogous to the analysis of the experimental data. The calculation of rate constants for reactions occurring in complex systems, however, is computationally expensive. For this reason, a direct calculation of kinetic observables, such as the activation energy, is highly desirable. This can be accomplished with a transition path sampling procedure [43]. This approach to determine activation energies does not require full rate constants calculations. Instead, the derivative of the reaction rate constant with respect to the inverse temperature β is expressed in terms of transition path averages which can be evaluated directly in a transition path sampling simulation. Since the transition path sampling method does not require any knowledge of mechanisms or transition states, this approach can be used to determine activation energies in complex systems where such knowledge is usually unavailable.

6 Rare Events in Trajectory Space

As a consequence of the second law of thermodynamics, the average work required to change an external parameter (for instance the position of a piston sliding in or out of a gas-filled cylinder) is larger than the Helmholtz free energy difference between the two equilibrium states corresponding to the initial and final value of the external parameter [34, 44]:

$$\langle W \rangle \geq \Delta F . \tag{73}$$

If the control parameter is changed reversibly, the average work $\langle W \rangle$ is equal to the free energy difference ΔF. (This is the reason that the term *reversible work* is often used for as a synonym of the equilibrium free energy difference.)

Remarkably, the so called Clausius inequality (73) can be changed into an equality by averaging the exponential $\exp(-\beta W)$ instead of the work W [45, 46]:

$$\langle \exp(-\beta W) \rangle = \exp(-\beta \Delta F) , \qquad (74)$$

Here, the angular brackets $\langle \cdots \rangle$ denote an average over an ensemble of non-equilibrium transformations initiated from states distributed according to a canonical distribution. This relation, first proven by Jarzynski in 1997, relates equilibrium free energies to the statistics of work expended along non-equilibrium trajectories.

From a computational point of view the Jarzynski equality is interesting, because it permits the calculation of free energy differences from simulations in which a control parameter is switched at arbitrary speed. To be more specific, consider a system with Hamiltonian $\mathcal{H}(x, \lambda)$ depending on the phase space point x and the control parameter λ. By changing λ continuously from its initial value λ_0 to its final value λ_1 the Hamiltonian $\mathcal{H}(x, \lambda_0)$ of the initial state is transformed into that of the final state $\mathcal{H}(x, \lambda_1)$. The free energy difference

$$\Delta F = -k_\mathrm{B} T \ln \frac{\int dx \, \exp\{-\beta \mathcal{H}(x, \lambda_1)\}}{\int dx \, \exp\{-\beta \mathcal{H}(x, \lambda_0)\}} = -k_\mathrm{B} T \ln \frac{Q_1}{Q_0} \qquad (75)$$

can be calculated by first generating initial conditions distributed according to $\exp\{-\beta \mathcal{H}(x, \lambda_0)\}/Q_0$. Then, the equations of motion of the system are integrated starting from these initial conditions. While the system evolves in time, the control parameter is switched from its initial to its final value. Along each trajectory the work necessary to switch the control parameter λ is determined and averaged according to (74).

The Jarzynski equality is exactly valid for arbitrary switching rates. This seems to suggest that free energy differences can be calculated in a computationally convenient way by averaging over short an therefore inexpensive trajectories. However, the statistical accuracy of the calculation rapidly degrades as the switching rate is increased [47–49]. The reason, familiar from applications of Zwanzig's exponential averaging approach [21,50] or Widom's particle insertion method [51], is that exponential averages are dominated by rare but important work values. As the switching rate increases and the work distribution is shifted towards larger values, fewer and fewer trajectories contribute significantly to the exponential average causing large statistical fluctuations in the estimated free energy. These statistical difficulties can easily offset the gain originating from the low computational cost of short trajectories. For straightforward fast switching simulations this statistical problem limits the switching rates to values for which the average work does not deviate from the free energy difference by more than the thermal energy $k_\mathrm{B} T$ [48]. In this regime, however, fast switching simulations are not superior to conventional methods such as umbrella sampling and thermodynamic integration [48,49].

A possible route to overcome the statistical problems occurring in a straightforward of Jarzynski's equality was recently suggested by Sun [52].

The basic idea of Sun's approach is to favor the generation of those trajectories that mostly contribute to the exponential average by using a biased sampling scheme based on transition path sampling. This work biased sampling of fast switching trajectories was implemented by Sun in a thermodynamic integration framework [52], but it can easily be adapted to an umbrella sampling simulation in path space [49,53,54]. In all these approaches the transition path sampling methods discussed in the earlier sections of this article are used to select the rare trajectories with important work values. While first results have validated this approach [49, 52–54], further research is necessary to establish whether the combination of fast switching simulations with transition path sampling algorithms yields computational methods that are competitive with conventional approaches.

Acknowledgments

This work was supported by the Austrian Science Foundation (FWF) under Grant No. P17178-N02. The authors are grateful to Jürgen Köfinger, Elisabeth Schöll-Paschinger, Harald Oberhofer, and Wolfgang Lechner for useful discussions.

Appendix: Response to Large Perturbations

In this Appendix we justify the approach followed in Sect. 5 to derive microscopic expressions for the reaction rate constants. The standard way to do this for the reaction

$$A \rightleftharpoons B \qquad (76)$$

occurring at low concentrations of species A and B is to relate the solution (48) of the phenomenological rate equations (46) with the relaxation of a nonequilibrium state prepared by applying an external perturbation. It can then be shown that in the case of weak perturbations (i.e., in the lineare response limit) the non-equilibrium relaxation has the same form as the autocorrelation of spontaneous population fluctuations:

$$\frac{\overline{N}_A(t) - \langle N_A \rangle}{\overline{N}_A(0) - \langle N_A \rangle} = \frac{\langle \delta h_A(0) \delta h_A(t) \rangle}{\langle (\delta h_A(0))^2 \rangle} , \qquad (77)$$

where $\delta h_A(x) = h_A(x) - \langle h_A \rangle$ and $h_A(t) = h_A(x_t)$. Here we show that for chemical reactions described by the rate equations (46) on a phenomenological level the assumption of a weak perturbation is unneccessary and that the same results can be obtained for perturbations of arbitrary strength. Consider a system with Hamiltonian $\mathcal{H}_0(x)$ where x represents the coordinates and momenta of all particles in the system. The system contains N molecules that can exist either in state A or in state B. The characteristic functions $h_A[q^{(i)}]$

and $h_B[q^{(i)}]$ can be used to test whether molecule i is in state A or state B. Here, $q^{(i)}$ denotes the cooordinates of all atoms in molecule i. In equilibrium, the average number of molecules of type A and B is

$$\langle N_A \rangle = N \int dx \rho_0(x) h_A[q^{(i)}] \quad \text{and} \quad \langle N_B \rangle = N \int dx \rho_0(x) h_B[q^{(i)}] \,,$$
(78)

where we have averaged over the equilibrium ensemble

$$\rho_0(x) = \frac{\exp(-\beta \mathcal{H}_0(x))}{\int dx \exp(-\beta \mathcal{H}_0(x))} \,.$$
(79)

Now imagine that a perturbation

$$\Delta \mathcal{H}(x) = -\varepsilon \sum_i^N h_A[q^{(i)}]$$
(80)

is coupled to the system and the system is allowed to equilibrate at temperature T under the action of the new Hamiltonian

$$\mathcal{H}(x) = \mathcal{H}_0(x) + \Delta \mathcal{H}(x)$$
(81)

leading to the new equilibrium distribution

$$\rho(x) = \frac{\exp(-\beta[\mathcal{H}_0(x) + \Delta \mathcal{H}(x)])}{\int dx \exp(-\beta[\mathcal{H}_0(x) + \Delta \mathcal{H}(x)])} \,.$$
(82)

For positive values of the perturbation strength ε the species A is favored with respect to species B in the new equilibrium state. If the perturbation is removed the number of molecules of type A relaxes back to its equilibrium value $\langle N_A \rangle$. The average number $\overline{N}_A(t)$ of molecules of type A present in the system at a time t after the perturbation is switched off is given by the non-equilibrium average

$$\overline{N}_A(t) = \frac{\int dx \exp(-\beta[\mathcal{H}_0(x) + \Delta \mathcal{H}(x)]) \sum_i^N h_A[q_t^{(i)}]}{\int dx \exp(-\beta[\mathcal{H}_0(x) + \Delta \mathcal{H}(x)])} \,,$$
(83)

where $q_t^{(i)}$ is the configuration of molecule i at time t and the time evolution of the system after the perturbation is removed is governed by the unperturbed Hamiltonian $\mathcal{H}_0(x)$. For the specific perturbation from (80) the nonequilibrium average is

$$\overline{N}_A(t) = \frac{\int dx \exp(-\beta \mathcal{H}_0(x)) \exp(\beta \varepsilon \sum_i^N h_A[q^{(i)}]) \sum_i^N h_A[q_t^{(i)}]}{\int dx \exp(-\beta \mathcal{H}_0(x)) \exp(\beta \varepsilon \sum_i^N h_A[q^{(i)}])} \,.$$
(84)

In the conventional treatment of unimolecular kinetics [34, 55] one evaluates this average in the limit of weak perturbation strength ε by expanding the above expression in a Taylor series. Truncating the series after the term linear in ε yields the well known linear response result:

$$\overline{N}_A(t) - \langle N_A \rangle = N\beta\varepsilon\langle\delta h_A(0)\delta h_A(t)\rangle . \tag{85}$$

Hence, the response to a weak perturbation is completely determined by the equilibrium fluctuations of the system. From the above equation an expression for the reaction rate constant can be derived [34].

The assumption of weak perturbations, however, is unecessary and the nonequilibrium average can be evaluated exactly for arbitrary perturbation strengths. To do that we first divide both numerator and denominator on the right hand side of (84) by $\int dx \exp(-\beta\mathcal{H}_0(x))$ obtaining:

$$\overline{N}_A(t) = \frac{\langle\exp(\beta\varepsilon\sum_i^N h_A[q^{(i)}])\sum_i^N h_A[q_t^{(i)}]\rangle}{\langle\exp(\beta\varepsilon\sum_i^N h_A[q^{(i)}])\rangle} . \tag{86}$$

The Boltzmann factor in the above equation originating from the perturbation can be written as a product:

$$\exp\left(\beta\varepsilon\sum_i^N h_A[q^{(i)}]\right) = \prod_i^N \exp(\beta\varepsilon h_A[q^{(i)}]) = \prod_i^N f_A[q^{(i)}, \varepsilon] , \tag{87}$$

where, to simplify the notation, we have introduced the function $f_A[q^{(i)}, \varepsilon] \equiv \exp(\beta\varepsilon h_A[q^{(i)}])$.

The nonequilibrium average thus becomes:

$$\overline{N}_A(t) = \frac{\sum_i^N \langle\prod_j^N f_A[q^{(j)}, \varepsilon]h_A[q_t^{(i)}]\rangle}{\langle\prod_j^N f_A[q^{(i)}, \varepsilon]\rangle} . \tag{88}$$

Since the solution is assumed to be sufficiently dilute such that all molecules are statistically independent from each other the numerator and the denominator in the above equation can be written as

$$\sum_i^N \left\langle\prod_j^N f_A[q^{(j)}, \varepsilon]h_A[q_t^{(i)}]\right\rangle = \sum_i^N \left(\langle f_A[q^{(i)}, \varepsilon]h_A[q_t^{(i)}]\rangle\prod_{j\neq i}\langle f_A[q^{(j)}, \varepsilon]\rangle\right)$$

$$= \left(\sum_i^N \frac{\langle f_A[q^{(i)}, \varepsilon]h_A[q_t^{(i)}]\rangle}{\langle f_A[q^{(i)}, \varepsilon]\rangle}\right)\prod_j^N\langle f_A[q^{(j)}, \varepsilon]\rangle \tag{89}$$

and

$$\left\langle\prod_j^N f_A[q^{(i)}, \varepsilon]\right\rangle = \prod_j^N\langle f_A[q^{(j)}, \varepsilon]\rangle , \tag{90}$$

respectively. Inserting these two expressions into (88) we obtain

$$\overline{N}_A(t) = \sum_i^N \frac{\langle f_A[q^{(i)}, \varepsilon] h_A[q_t^{(i)}]\rangle}{\langle f_A[q^{(i)}, \varepsilon]\rangle} = N \frac{\langle f_A[q, \varepsilon] h_A[q_t]\rangle}{\langle f_A[q, \varepsilon]\rangle} , \tag{91}$$

where we have omitted the superscripts in the last part of the equation because all of the N molecules are equivalent.

Now, the function $f_A[q, \varepsilon]$ can take only two different values depending on whether the corresponding molecule is in state A or not:

$$f_A[q, \varepsilon] = \exp(\beta \varepsilon h_A[q]) = \begin{cases} \exp(\beta \varepsilon) & \text{if } q \in A , \\ 1 & \text{if } q \notin A . \end{cases} \tag{92}$$

Therefore, we can write $f_A[q, \varepsilon]$ in terms of the characteristic function h_A:

$$f_A[q, \varepsilon] = \exp(\beta \varepsilon h_A[q]) = h_A[q](e^{\beta \varepsilon} - 1) + 1 . \tag{93}$$

Using this expression we obtain

$$\begin{aligned}
\overline{N}_A(t) &= N \frac{\langle \{(e^{\beta \varepsilon} - 1)h_A(q) + 1\} h_A(q_t)\rangle}{\langle (e^{\beta \varepsilon} - 1)h_A(q) + 1\rangle} \\
&\quad N \frac{(e^{\beta \varepsilon} - 1)\langle h_A(q)h_A(q_t)\rangle + \langle h_A\rangle}{(e^{\beta \varepsilon} - 1)\langle h_A\rangle + 1} .
\end{aligned} \tag{94}$$

Thus, an initial excess $\overline{N}_A(0)$ of species A decays to its equilibrium value $\langle N_A\rangle$ according to:

$$\overline{N}_A(t) - \langle N_A\rangle = N(e^{\beta \varepsilon} - 1)\frac{\langle \delta h_A(q)\delta h_A(q_t)\rangle}{(e^{\beta \varepsilon} - 1)\langle h_A\rangle + 1} . \tag{95}$$

where we have used that $\langle N_A\rangle = N\langle h_A\rangle$. This results implies that the response of the system to a perturbation of arbitrary strength is completely determined by its equilibrium fluctuations. Nevertheless, the response to the perturbation is a nonlinear function of the perturbation strength ε. Expansion of the above expression into a power series in $\beta \varepsilon$ and truncation after the linear term yields the familiar linear response expression of (85).

The standard linear response derivation [34] of expressions for the reaction rate constants rests on the observation that the *relative* deviation from the equilibrium average generated by a weak perturbation decays in the same way as the equilibrium fluctuations normalized by their initial value:

$$\frac{\overline{N}_A(t) - \langle N_A\rangle}{\overline{N}_A(0) - \langle N_A\rangle} = \frac{\langle \delta h_A(q)\delta h_A(q_t)\rangle}{\langle [\delta h_A(q)]^2\rangle} . \tag{96}$$

It is thus sufficient that the relaxation of the nonequilibrium population is *proportional* to the decay of equilibrium fluctuations. While the above expression (also known as Onsager's regression hypothesis) follows for weak perturbations, a linear dependence of the system's response on the perturbation strength ε it is not necessary. In fact, the exact expressions (95) derived above for perturbations of arbitrary strength also leads to the behavior described by (96). Hence, for the type of chemical dynamics considered here, Onsager's regression hypothesis remains exactly valid also for the relaxation of arbitrarily large excess populations. This result justifies the approach used in Sect. 5 (we started with all molecules of type A) to derive expressions for the reaction rate constant.

References

1. H.-X. Zhou and R. Zwanzig (1991) A rate process with an entropy barrier. *J. Chem. Phys.*, **94**, pp. 6147–6152
2. J. B. Anderson (1973) Statistical theories of chemical reactions: Distributions in the transition region. *J. Chem. Phys.*, **58**, pp. 4684–4692; C.H. Bennett (1977) Molecular Dynamics and Transition State Theory: the Simulation of Infrequent Events. In *Algorithms for Chemical Computations*, ed. R. E. Christoffersen, pp. 63–97; Washington, D.C.: Amer. Chem. Soc.; D. Chandler (1978) Statistical mechanics of isomerization dynamics in liquids and the transition state approximation. *J. Chem. Phys.*, **68**, pp. 2959–2970
3. L.R. Pratt (1986) A statistical method for identifying transition states in high dimensional problems. *J. Chem. Phys.*, **85**, pp. 5045–5048
4. C. Dellago, P.G. Bolhuis, F.S. Csajka, and D. Chandler (1998) Transition Path Sampling and the Calculation of Rate Constants. *J. Chem. Phys.*, **108**, pp. 1964–1977
5. C. Dellago, P.G. Bolhuis, and D. Chandler (1998) Efficient Transition Path Sampling: Applications to Lennard-Jones cluster rearrangements. *J. Chem. Phys.*, **108**, pp. 9236–9245
6. P.G. Bolhuis, C. Dellago, and D. Chandler (1998) Sampling ensembles of deterministic transition pathways. *Faraday Discuss.*, **110**, pp. 421–436
7. W. E, W. Ren, and E. Vanden-Eijnden (2002) String method for the study of rare events. *Phys. Rev. B*, **66**, 052301/1-4
8. W. E, W. Ren, and E. Vanden-Eijnden (2005) Finite Temperature String Method for the Study of Rare Events. *J. Phys. Chem. B*, **109**, pp. 6688-6693
9. R. Elber, A. Ghosh, A. Cardenas, and H. Stern (2004) Bridging the gap between reaction pathways, long time dynamics and calculation of rates. *Adv. Chem. Phys.*, **126**, pp. 93–129
10. D. Passerone, M. Ceccarelli, and M. Parrinello (2003) A concerted variational strategy for investigating rare events. *J. Chem. Phys.*, **118**, pp. 2025–2032
11. P. G. Bolhuis, D. Chandler, C. Dellago, and P.L. Geissler (2002) Transition Path Sampling: Throwing Ropes over Mountain Passes in the Dark. *Ann. Rev. Phys. Chem.*, **53**, pp. 291–318
12. C. Dellago, P.G. Bolhuis, and P.L. Geissler (2002) Transition Path Sampling. *Adv. Chem. Phys.*, **123**, pp. 1–78
13. C. Dellago and D. Chandler (2002) Bridging the time scale gap with transition path sampling, in *Molecular Simulation for the Next Decade*, ed. by P. Nielaba, M. Mareschal, and G. Ciccotti, Springer, Berlin. pp. 321–333
14. C. Dellago (2005) Transition Path Sampling, in *Handbook of Materials Modeling*, ed. by S. Yip, Springer, Berlin. pp. 1585–1596

15. C. Dellago (2006) Transition Path Sampling and the Calculation of Free En-
    ergies, in *Free energy calculations: Theory and applications in chemistry and
    biology*, ed. by A. Pohorille and C. Chipot, Springer, Berlin, pp. 265–294
16. G.E. Crooks and D. Chandler (2001) Efficient transition path sampling for non-
    equilibrium stochastic dynamics. *Phys. Rev. E*, **E 64**, 026109/1–4
17. R. Car and M. Parrinello (1985) Unified Approach for Molecular Dynamics and
    Density-Functional Theory. *Phys. Rev. Lett.*, **55**, pp. 2471–2474
18. S. Nose (1984) A unified formulation of the constant temperature molecular dy-
    namics methods. *J. Chem. Phys*, **81**, pp. 511-519; W.G. Hoover (1985) Canonical
    dynamics: Equilibrium phase-space distributions", *Phys. Rev. A*, **31**, pp. 1695–
    1697
19. D.J. Evans, W.G. Hoover, B.H. Failor, B. Moran, and A.J.C. Ladd (1983) Non-
    equilibrium molecular dynamics via Gauss's principle of least constraint. *Phys.
    Rev. A*, **28**, pp. 1016–2021
20. R. Zwanzig (2001) *Nonequilibrium Statistical Mechanics*, Oxford University
    Press, Oxford
21. M.P. Allen and D.J. Tildesley (1987) *Computer Simulation of Liquids*, Claren-
    don Press, Oxford
22. S. Chandrasekhar (1943) Stochastic Problems in Physics and Astronomy. *Rev.
    Mod. Phys.*, **15**, pp. 1–89
23. P.L. Geissler, C. Dellago, D. Chandler, J. Hutter, and M. Parrinello (2001)
    Autoionization in liquid water. *Science*, **291**, pp. 2121–2124
24. D.F. Landau and K. Binder (2000) *A Guide to Monte Carlo Simulations in
    Statistical Physics*, Cambridge University Press, Cambridge
25. N. Metropolis, A.W. Rosenbluth, M.N. Rosenbluth, A.H. Teller, and E. Teller
    (1953) Equation of State Calculations by Fast Computing Machines. *J. Chem.
    Phys.*, **21**, pp. 1087–1092
26. T.J.H. Vlugt, C. Dellago, and B. Smit (2000) Diffusion of Isobutane in Silicalite
    studied by Transition Path Sampling. *J. Chem. Phys.*, **113**, p. 8791
27. P.G. de Gennes (1971) Reptation of a Polymer Chain in the Presence of Fixed
    Obstacles. *J. Chem. Phys.*, **55**, pp. 572–579
28. R.B. Best and G. Hummer (2005) Reaction coordinates and rates from transition
    paths. *Proc. Nat. Acad. Sci. USA*, **102**, pp. 6732–6737
29. L. Onsager (1938) Initial Recombination of Ions. *Phys. Rev.*, **54**, pp. 554–557
30. V. Pande, A.Y. Grosberg, T. Tanaka, and E.I. Shakhnovich (1998) On the tran-
    sition coordinate for protein folding. *J. Chem. Phys.*, **108**, pp. 334–350
31. S. Auer and D. Frenkel (2001) Prediction of absolute crystal-nucleation rate in
    hard-sphere colloids. *Nature*, **409**, pp. 1020–1023
32. E.A. Carter, G. Ciccotti, J.T. Hynes, and R. Kapral (1989) Constrained reaction
    coordinate dynamics for the simulation of rare events. *Chem. Phys. Lett.*, **156**,
    pp. 472–477
33. D. Moroni, P.R. ten Wolde, and P.G. Bolhuis (2005) Interplay between Structure
    and Size in a Critical Crystal Nucleus. *Phys. Rev. Lett.*, **94**, 235703/1–4
34. D. Chandler (1987) *Introduction to Modern Statistical Mechanics*, Oxford Uni-
    versity Press, New York
35. G.M Torrie and J.P. Valleau (1974) Monte Carlo free energy estimates using
    non-Boltzmann sampling: Application to the sub-critical Lennard-Jones fluid.
    Chem. Phys. Lett., **28**, pp. 578–581
36. C. Dellago, P.G. Bolhuis and D. Chandler (1999) On the calculation of rate
    constants in the transition path ensemble. *J. Chem. Phys.*, **110**, pp. 6617–6625

37. T.S. van Erp, D. Moroni, and P.G. Bolhuis (2003) A novel path sampling method for the calculation of rate constants. *J. Chem. Phys.*, **118**, pp. 7762–7774

38. T.S. van Erp and P.G. Bolhuis (2005) Elaborating transition interface sampling methods. *J. Comp. Phys.*, **205**, pp. 157–181

39. D. Moroni, P.G. Bolhuis, and T.S. van Erp (2004) Rate constants for diffusive processes by partial path sampling. *J. Chem. Phys.*, **120**, pp. 4055–4065

40. A.K. Faradjian and R. Elber (2004) Computing time scales from reaction coordinates by milestoning. *J. Chem. Phys.*, **120**, pp. 10880–10889

41. R. J. Allen, P. B. Warren, and P. R. ten Wolde (2005) Sampling Rare Switching Events in Biochemical Networks. *Phys. Rev. Lett.*, **94**, 018104/1–4

42. P.W. Atkins (2001) *Physical Chemistry, 7th edition*, Oxford University Press, Oxford

43. C. Dellago and P.G. Bolhuis (2004) Activation energies from transition path sampling simulations. *Mol. Sim.*, **30**, pp. 795–799

44. H.B. Callen (1985) *Thermodynamics and an Introduction to Thermostatistics, 2nd edition*, John Wiley and Sons, New York

45. C. Jarzynski (1997) Nonequilibrium Equality for Free Energy Differences. *Phys. Rev. Lett.*, **78**, pp. 2690-2693

46. G.E. Crooks (1998) Nonequilibrium Measurements of Free Energy Differences for Microscopically Reversible Markovian Systems. *J. Stat. Phys.*, **90**, pp. 1481–1487

47. F. Ritort (2003) Work fluctuations and transient violations of the second law: perspectives in theory and experiments. *Sem. Poincare* **2**, pp. 193–226

48. G. Hummer (2001) Fast-growth thermodynamic integration: Error and efficiency analysis. *J. Chem. Phys.*, **114**, pp. 7330-7337

49. H. Oberhofer, C. Dellago and P.L. Geissler (2005) Biased sampling of nonequilibrium trajectories: Can fast switching simulations outperform conventional free energy calculation methods? *J. Phys. Chem. B*, **69**, pp. 6902–6915

50. R. Zwanzig (1954) High-Temperature Equation of State by a Perturbation Method. I. Nonpolar Gases. *J. Chem. Phys.*, **22**, pp. 1420–1426

51. B. Widom (1963) Some Topics in the Theory of Fluids. *J. Chem. Phys.*, **39**, pp. 2808-2812

52. S.X. Sun (2003) Equilibrium free energies from path sampling of nonequilibrium trajectories. *J. Chem. Phys.*, **118**, pp. 5769–5775

53. F.M. Ytreberg and D.M. Zuckerman (2004) Single-ensemble nonequilibrium path-sampling estimates of free energy differences. *J. Chem. Phys.*, **120**, pp. 10876–10879

54. M. Athènes (2004) A path-sampling scheme for computing thermodynamic properties of a many-body system in a generalized ensemble. *Eur. Phys. J. B*, **38**, pp. 651–663

55. D. Frenkel and B. Smit (2002) *Understanding Molecular Simulation*, Academic Press, San Diego

Sampling Kinetic Protein Folding Pathways using All-Atom Models

P.G. Bolhuis

van 't Hoff Institute for Molecular Sciences, Universiteit van Amsterdam,
Amsterdam, The Netherlands
bolhuis@science.uva.nl

Peter G. Bolhuis

P.G. Bolhuis: *Sampling Kinetic Protein Folding Pathways using All-Atom Models*, Lect. Notes
Phys. **703**, 393–433 (2006)
DOI 10.1007/3-540-35273-2_11 　　　　　　　　　© Springer-Verlag Berlin Heidelberg 2006

This chapter summarizes several computational strategies to study the kinetics of two-state protein folding using all atom models. After explaining the background of two state folding using energy landscapes I introduce common protein models and computational tools to study folding thermodynamics and kinetics. Free energy landscapes are able to capture the thermodynamics of two-state protein folding, and several methods for efficient sampling of these landscapes are presented. An accurate estimate of folding kinetics, the main topic of this chapter, is more difficult to achieve. I argue that path sampling methods are well suited to overcome the problems connected to the sampling of folding kinetics. Some of the major issues are illustrated in the case study on the folding of the GB1 hairpin.

1 Introduction

1.1 The Protein Folding Problem

Before proteins can actively function in the living cell they must fold up into a specific 3-dimensional structure, the so-called native state (see Fig. 1). Already in the 1960's it was recognized that the long linear polypeptides chains can adopt their native structure starting from the random coil state in a surprisingly short time. The famous Levinthal paradox states that if a peptide bond between amino acids can only adopt two conformations a relatively short protein of a hundred residues can have around $2^{100} \approx 10^{30}$ possible conformations. If the protein has to search conformational space randomly with approximately 1 conformer per ps, which is probably the speed limit, the time to find the native state should take more than the age of the universe. Clearly there must be some other principle at work, to make proteins fold in fractions of a second.

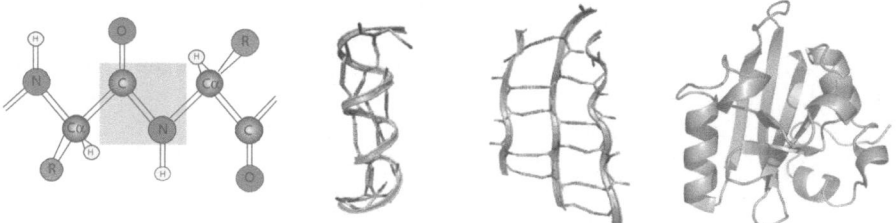

Fig. 1. Hierarchical protein structure: *Left*: Small part of the protein backbone. The peptide bond itself is marked by a shaded rectangle, R denotes one of the 20 amino acid side-chains. *Middle*: Secondary structure in the form of an alpha-helix and beta-sheet. Only the backbone is shown and highlighted by a ribbon. The hydrogen bonds are indicated by dashed lines. *Right*: Tertiary structure in the native state. The alpha-helices and beta-sheets are indicated by thick ribbons and arrows respectively

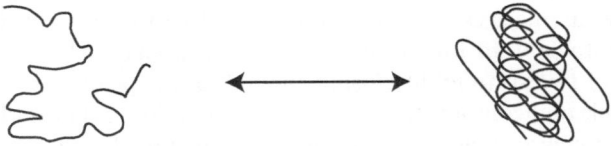

Fig. 2. Cartoon of the basic folding problem: how does the protein find its way to the folded state?

The protein folding problem can be summarized in two basic questions: 1) can we predict the structure and function of proteins from its amino acid sequence alone? 2) how does the protein find its way toward the native state (see Fig. 2)? The first question is more or less one of the holy grails of molecular biology, as it would allow the design of proteins with a specific desirable structural or enzymatic property, starting from first principles. The second question is related, because the protein's stability and folding kinetics are crucial components of the structure and function prediction. In addition, the folding properties of proteins are important for the understanding of the basis of well known diseases, such as Alzheimer's, BSE, Creuzfeld-Jacob, ALS, Huntington's, Parkinson's disease, and many cancers and cancer-related syndromes [1].

In this chapter I will discuss several molecular simulation methods that address the second question. This is not a review of the state of the entire protein folding field, because that would be a daunting task indeed, considering the $\mathcal{O}(10^4)$ papers that have been published on the subject. Instead, I refer to many excellent reviews available in the literature [1–7]. In contrast, papers on computer simulation techniques to investigate folding kinetics are less common. Nevertheless, a excellent review on simulating folding kinetics by Snow et al. recently appeared [8]. The chapter is organized as follows. The first section gives some background on protein folding, protein stability and folding kinetics. In Sect. 2 I discuss the most common protein simulation methods and models. Section 3 is devoted to free energy calculations, and Sect. 4 focuses on the kinetics of protein folding, in particular, using transition path sampling (TPS) methods. Section 5 contains an exemplary case study of beta-hairpin folding.

1.2 Stability of the Native State

The fact that proteins spontaneously fold to their native state requires a physical explanation. An unfolded or denatured protein makes many interactions with the solvent (water), and as the protein folding proceeds, these interactions are exchanged with intra-molecular non-covalent bonds, such as hydrogen bonds or salt-bridges [2]. Many hydrogen bond donors and acceptors make intra-molecular pairs to form secondary structures such as alpha helices and beta sheets (see Fig. 1). Each of these bonds adds a small energetic contribution to the stability of the native state, but added together the interaction energy of the protein is many thousands of kJ/mol in both the denatured and native state. When the entropy is also taken into account, these large values somehow almost cancel each other, and the free energy difference between unfolded and folded states is usually only of the order of a few tens of kJ/mol. This marginal stability is one of the causes for the difficulties that theoretical treatments of folding face. Both the interaction energies as well as the entropic contributions (configurational, vibrational, rotational) have to be taken into account accurately, in order to predict the folding behavior. The energy of the protein can be described by classical force fields as long as quantum mechanical processes do not play an important role, but the entropy can only be obtained from statistical mechanics, i.e. Boltzmann sampling. It is this entropic sampling that makes protein folding a hard theoretical problem.

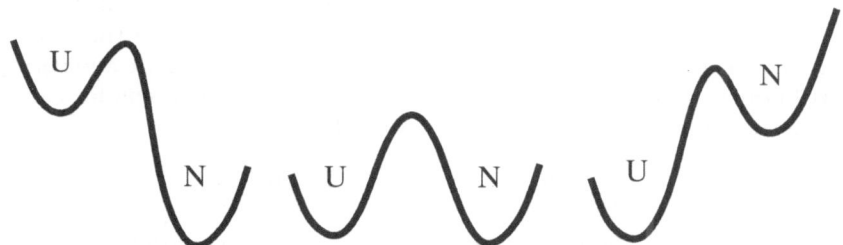

Fig. 3. Schematic free energy landscapes of two-state folders for increasing denaturant concentration (or increasing temperatures). Note that the unfolded state (U) becomes more stable with respect to the native state (N). The folding rate constant is inversely proportional to the exponent of the barrier height

1.3 Two State Folders

The marginal stability of a protein's native state means unfolding can be forced relatively easily. Usually this is done by heating, adding denaturant such as urea, changing pH or increasing pressure [2]. These conditions destabilize the native state, and/or stabilize the unfolded state. When reverting to the initial conditions the protein re-finds its native state, a property first discovered by Anfinsen [9][1]. For most single domain proteins (only tertiary structure, no quaternary structure), experimental measurements of the relaxation of the unfolded population toward equilibrium show single exponential behavior [4]: so called two state kinetics. In addition, the logarithm of the folding and unfolding rate constants are linearly dependent on the concentration of the denaturant (see Fig. 3) [2]. Such behavior indicates that only the native and denatured states are (meta)stable and all (possible) intermediate states

---

[1] Large multi-domain proteins require often molecular chaperones to fold properly. This is not considered in this review.

have a much higher free energy, and do not show significant population in the experiments. The kinetics of protein folding is hence conceptually reduced to that of a simple unimolecular chemical reaction $(A \rightarrow B)$ with a similar two state kinetic behavior [10] (except for the dependence on external parameters e.g. temperature, and pressure), as there is (supposed to be) no correlation between different protein molecules in the solution. Folding also has properties similar to the kinetics of a first order phase transition, due to the cooperative nature of the monomers [11]. For instance, the heat capacity as a function of temperature exhibits a peak around the folding temperature. Because of these similarities, it is possible to use the statistical mechanics originally developed for chemical kinetics and first order phase transitions also for the study of protein folding. These properties make two state proteins interesting systems from both experimental and theoretical points of view.

Two state kinetics does not necessarily obey the van 't Hoff-Arrhenius law, which presumes a linear relation between the logarithm of the rate constant and the inverse temperature. As for proteins both energetic and entropic contributions are important a more general applicable expression for the rate constant is given by transition state theory (TST)

$$k = \nu e^{-\Delta G^{\ddagger}/k_B T} = \nu e^{\Delta S^{\ddagger}/k_B} e^{-\Delta H^{\ddagger}/k_B T}, \tag{1}$$

where ν is a kinetic prefactor and $\Delta G^{\ddagger}, \Delta S^{\ddagger}, \Delta H^{\ddagger}$ are respectively the free energy, entropy and enthalpy difference between reactant and the transition state. As in proteins both S and H can be temperature dependent, folding can show severe non-Arrhenius behavior, even with the folding rate decreasing with temperature [2]. This non-Arrhenius behavior is beyond the scope of this chapter.

1.4 Explaining Folding with Energy Landscapes

The solution to Levinthal's paradox is to view protein folding as a one-way downhill process, guided by an energy landscape [12,13] in the form of a funnel as depicted in Fig. 4. Consider a population of unfolded denatured proteins, for which the conditions suddenly change such that the native state becomes stable again, e.g. by a temperature-jump, or by dilution of the denaturant [2]. Starting from the random coil unfolded state (schematically represented in the top of Fig. 4, with a high energy and a high entropy, and labeled with an order parameter value $Q = 0$) the hydrophobic, electrostatic, van der Waals and hydrogen bonds interactions between the residues guide the protein towards the part of phase space in which much less possible conformations are allowed. The energy of the protein decreases due to favorable interactions, but the entropy decreases as well (indicated by the horizontal space between the solid curve.). The result of this coil collapse is called the molten globule in which the protein is more or less a compact structure, but can still explore many different conformations. These molten globule states are characterized

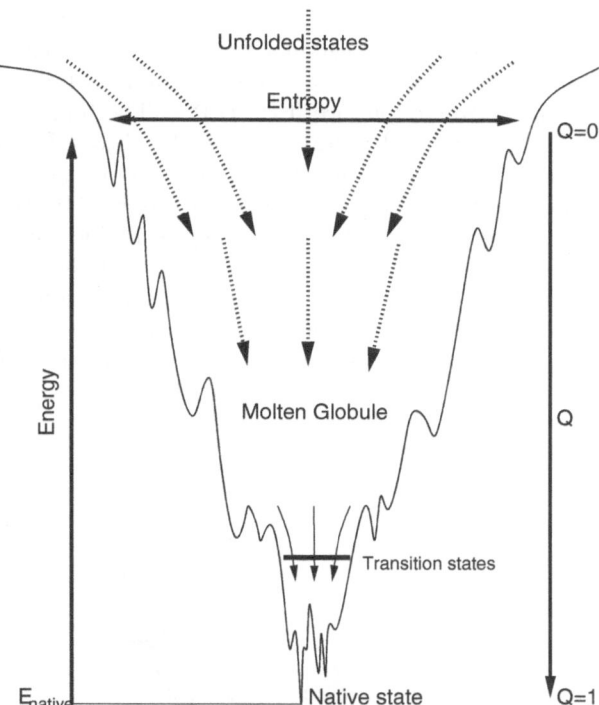

Fig. 4. The curved solid line depicts the energy landscape funnel of the folding process. The x-axis is a measure of accessible configuration space. At high energies, many configurations are possible, hence the entropy is large. The landscape guides or funnels the protein to lower energies (indicated by the arrows). The lower the energy, the more the protein is confined to fewer conformations. The global minimum corresponds to the native state. Note that the transition states have to be put in manually

with order-parameter values $0 < Q < 1$. The energy can decrease even more by the formation of secondary structure and the creation of specific backbone hydrogen bonds, salt bridges etc., thus guiding the protein into the native structure (labeled as $Q = 1$) with a very low entropy. Naturally occurring proteins have evolved such that their energy landscapes exhibits a single, stable native state, separated from the first misfolded state by a relatively large energy gap (some misfolded states are also indicated in the funnel picture, Fig. 4). In contrast, random heteropolymers will not have a single lowest energy native state but mostly exhibit a glassy energy landscape with many degenerate misfolded states [12,14].

As the balance between entropy and energy is by construction taken into account in the funnel landscape, temperature denaturation is conceptually explained. However, the down hill process does not clearly exhibit the free energy barrier required for the experimentally measured two state kinetics. Therefore, in the 1990's it was realized that the funnel landscape must have a transition state region somewhere between the molten globule and the native state (indicated in the funnel picture, Fig. 4) [12]. The funnel landscape picture still separates entropy and energy in an artificial way. It is more natural to replace the energy landscape by a free energy landscape [15], that combines the funnel concept with the experimentally observed barrier (see Fig. 5). From the unfolded state, where there are many available conformations the free energy landscape funnels the protein quickly into the compact molten globule state, where fewer conformations are allowed. The molten globule is (meta)stable and a free energy barrier has to be overcome upon folding. This barrier is caused by a reduction of entropy (fewer available conformations in

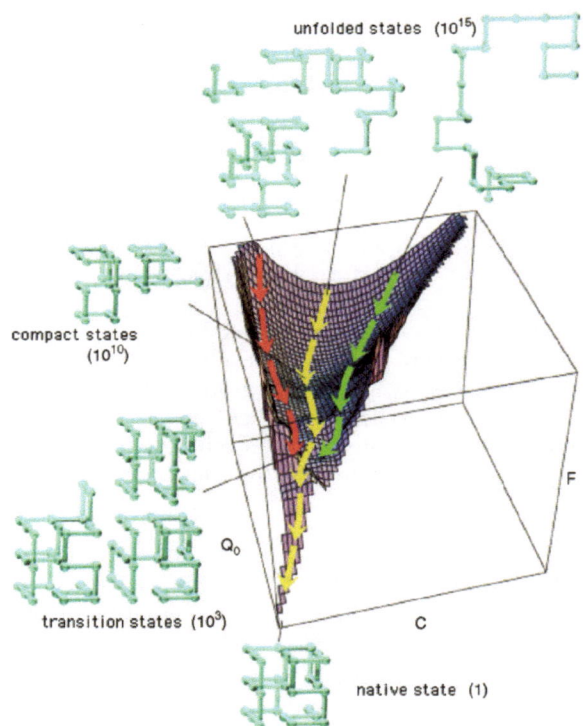

Fig. 5. Free energy landscape of a lattice model protein (see Sect. 2.2), as a function of two order parameters, the number of contacts C and the number of native contacts Q_0 (see Sect. 2.3). Unlike the energy landscape funnel picture, the free energy shows two stable states separated by a barrier (the transition state). Extended unfolded conformers quickly collapse to the molten globule, and have to overcome a barrier to folding to the native state. The funnel picture is thus reconciled with the two-state concept of a free energy barrier. Reprinted from Dinner et al., Trends Biochem. Sci. **25**, 331, (2000) with permission from Elsevier

the transition state) whereas most energetically favorable contacts have not yet been made. The nature of the barrier can be explained with two different generic models (see Fig. 6). The first is the nucleation-condensation model [5]. In the transition state, a few crucial contacts must be made to form the so-called folding nucleus. This nucleus can subsequently grow towards the stable native state, for instance, by reeling in dangling loops.

The other explanation is the diffusion-collision model in which the secondary structure forms very fast [5, 17]. These secondary structure elements then diffuse in a random walk fashion until they collide in the proper way and form the tertiary structure. Islam et al. used this model to analyze the folding of a three helix bundle proteins [17].

In the literature there is some controversy about which mechanism is prevalent. A recent paper by Gianni et al. reconciles these two views [16].

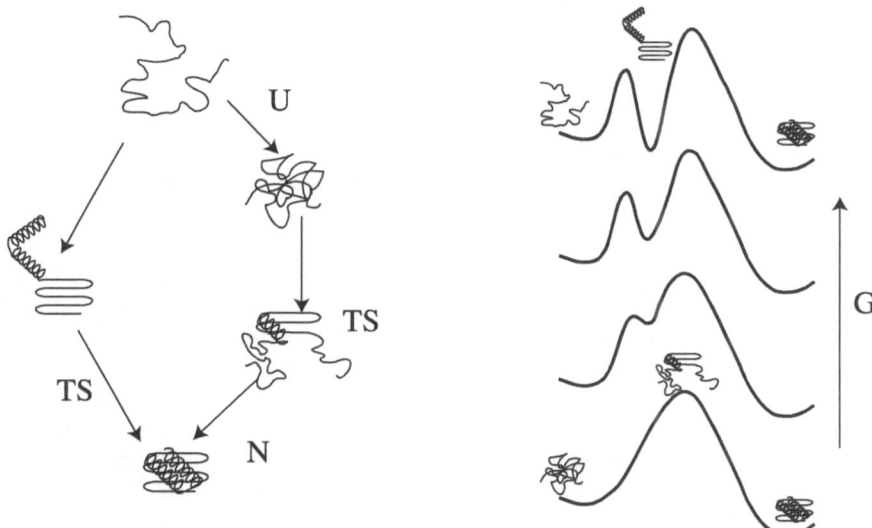

Fig. 6. (a): Schematic cartoon of the two main mechanisms of folding. The left pathway denotes the diffusion collision model, in which the secondary structure is rather stable and forms fast, after which the elements diffuse until they find the native structure. The right pathway represents the nucleation condensation mechanism, in which first a folding nucleus is formed, before the secondary and tertiary structure can fully develop into the native state. (b): Gradual change of the free energy landscape according to Gianni et al. [16]. The top landscape denotes the free energy for proteins that follow the diffusion-collision pathway, with a molten globule in which the secondary structure is stable. If the secondary structure is destabilized the free energy landscape slowly evolves into the bottom landscape, which models proteins that fold through the nucleation-condensation pathway. Note that in all cases two state behavior is observed (the left barrier is much lower than the largest barrier)

Depending on the protein, the free energy landscape differs as is illustrated in Fig. 6. For some two-state proteins, with an independently stable secondary structure, the diffusion-collision mechanism is preferred. Other proteins,for which the secondary structure is less stable on its own, fold cooperatively using the nucleation-condensation pathway. In all cases there is still two state behavior, because there is only one rate limiting barrier.

By analyzing experimental data Akmal and Munoz [18] conclude that the top of the free energy barrier is reached when the protein in its search for the native state reaches a 'critical native density', i.e. is close enough to the native state to expel the interstitial water molecules and form a folding nucleus. At that point the stabilization energy starts to overcome the decrease of conformational entropy. In addition, the expelled interstitial solvent gains translational entropy as well [19].

This schematic view is the current frame work for understanding two-state protein folding [5]. For several reviews of the extensive literature on the theory of folding, and folding statistical mechanics see e.g. [5, 12, 14].

1.5 What Determines the Folding Rate?

The values of folding rate constants for different small single domain two state proteins vary enormously, over more than 6 orders of magnitude [20]. Experiments revealed helices form in a few hundred nanoseconds, and beta-turns in a few microsecond. Apparently the folding speed limit is on the order of a microsecond [4]. On the other hand, proteins like chymotrypsin inhibitor-2 fold on the order of seconds [2]. This variation in rate begs for an explanation. Of course, the height of the free energy barrier determines the folding rate, but one would like to explain the variance based on the molecular structure. Available candidates for the rate determining factors are the energy stability gap between the ground (native) state and the first excited state, the hydrophobic collapse cooperativity, and the topology of the native state. Plaxco and coworkers argue that of these candidates the topology correlates best with the folding rate [20]. The topology is in Plaxco's view well characterized by the so-called *contact order*

$$CO = \frac{1}{NM} \sum_{i,j} \delta_{i,j} |i - j| \,, \tag{2}$$

where $\delta_{i,j} = 1$ if residues i and j are in contact and zero otherwise, M is the number of contacts N the number of residues. This contact order is larger if contacts are further apart along the sequence, and is thus supposed to capture the tertiary topology. This finding is rationalized in the so called topomer model, which states that the protein most of the time is searching for its native topology, after which the protein can fold relatively fast to the native state [20].

On the other hand, Mirny and Shakhnovich argue that contact order has not much to do with tertiary structure, and is actually dominated by local contacts. In their view, α-helices have a low contact order, while β-sheets have more distant contacts and hence a higher contact order. Indeed helical proteins fold faster than beta-proteins, providing a rather trivial correlation between rate and contact order [5]. Clearly, the role of the topology is not without controversy.

2 Simulating Proteins

2.1 Molecular Simulation

To study protein folding theoretically, simulation methods have proved indispensable. The folding transition is ultimately governed by statistical thermodynamics and hence it is paramount to use sampling methods that are able to reproduce the canonical Boltzmann distribution. Common sampling techniques are molecular dynamics (MD), Langevin or Brownian dynamics (BD) and Monte Carlo (MC).

Molecular Dynamics generates a trajectory by integration of the deterministic (Newtonian) equations of motion, usually performed with the Verlet algorithm. As the resulting trajectory contains all dynamical information, MD can be used to study kinetics. The ergodicity theorem states that in the limit of long time a straightforward MD simulation samples the micro-canonical (NVE) ensemble. Thermostats such as the Nose-Hoover thermostat, or Andersen thermostat [21] allow for canonical NVT sampling, and barostats can keep the pressure constant [22]. Detailed descriptions of MD methods can be found in [21,23].

Stochastic models, e.g. in which the solvent is replaced by an effective medium, should be described with the Langevin equation, in which friction plays the role of the degrees of freedom that were integrated out. In the high friction limit this simplifies to Brownian dynamics. The Langevin equation of motion can be integrated by a modified version of the Verlet algorithm [23], although other more accurate integrators are currently available [24].

The Monte Carlo method obtains the Boltzmann distribution through importance sampling [21]. Small trial moves are accepted or rejected according to a rule, e.g. the Metropolis criterion, that conserves detailed balance. Basic Metropolis Monte Carlo samples the canonical distribution, but implementation of the constant pressure ensemble is straightforward by allowing for volume changes with the proper acceptance rule [21] The Monte Carlo technique is discussed in detail in several other chapters in this edition. More excellent accounts of the MC methodology can be found in [21,25]. In principle, Monte Carlo can only access thermodynamics and does not provide kinetic information. In some cases, with several additional assumptions certain Monte Carlo schemes can be viewed as solutions to the Fokker-Planck equation, and thus can be used to study dynamics [26].

The relatively small system size feasible in molecular simulation can have a large effect on the results [21, 23]. To minimize such finite size effects, an infinite system is mimicked by applying periodic boundary conditions. These conditions effectively create an infinite number of copies of the system in each direction. The cubic box is simplest to implement, but more complex tilings of space such as the truncated octahedron, or the dodecahedron results in smaller system sizes because they require fewer solvent molecules, and hence are more efficient [21, 22].

2.2 Models for Proteins

Atomistic Models

The success of any molecular simulation method relies on the potential energy function for the system of interest, also known as force fields [27]. In case of proteins, several (semi)empirical atomistic force fields have been developed over the years, of which ENCAD [28,29], AMBER [30], CHARMM [31], GROMOS [32], and OPLSAA [33] are the most well known. In principle, the force field should include the electronic structure, but for most except the smallest systems the calculation of the electronic structure is prohibitively expensive, even when using approximations such as density functional theory.[2] Instead, most potential energy functions are (semi)empirical classical approximations of the Born-Oppenheimer energy surface.

Because the Verlet algorithm in MD requires the analytical derivative of the energy, the force, at every integration time step, the potential functional form is often taken as simple as possible. Most force fields express the energy as a sum over bond, angle, torsional angle and non-bonded interaction terms. The bond between two neighboring atoms is approximated by a harmonic spring, as is the bending potential. The torsion potential, (or dihedral for four consecutive atoms), is approximated by a periodic function based on the cosine of the torsion angle. The non-bonded van der Waals attraction is modeled by the 6-12 Lennard-Jones (LJ) potential and the electrostatic interaction by a Coulomb potential. The force fields differ in the specific values of the parameters and how they are derived. In all cases the van der Waals parameters are derived using empirical data. In this fitting procedure the water model (e.g. TIP3P [34] or SPC/E [35]) is included. For the electrostatic term an important ingredient is the value of the (partial) charges on the atoms, which has to be fitted to empirical or ab-initio data as well. Polarization is not included in most classical forcefields, although this is certainly an active direction of research in the force field community [36].

Whereas most potential terms decay to zero within the box-size, the long-ranged nature of the Coulomb force necessitates the inclusion of the interaction of a charged atom not only with all other atoms in the box, but with

---

[2] We do not consider the QM/MM methods where classical forcefields are combined with electronic structure calculations.

all atoms in all periodic images as well. This makes the explicit calculation of the electrostatic interaction the most expensive part of the energy evaluation. The Ewald summation method has been developed to approximate this infinite sum efficiently using Fourier transformation. Special techniques such as Particle Mesh Ewald (PME) [22], particle-particle particle-mesh (P^3M) [22] improve the efficiency further by applying fast Fourier algorithms, and clever separation of the real space and reciprocal space parts.

Existing MD packages employ several additional tricks to speed up the simulations. The time-step is governed by the fastest vibration in the system: the bond vibrations. The time-step and hence the efficiency can be increased by constraining these fast degrees of freedom using algorithms such as SHAKE [23], RATTLE [37], or LINCS [38]. Another strategy is to use multiple time-step integration schemes, in which the long ranged interactions are not updated each time-step [39]. Implementation of Verlet neighbor list and cell lists dramatically speed up the evaluation of non-bonded van der Waals interactions [23]. The LJ potential is usually truncated at a cutoff-radius and a switching function is applied to keep the derivatives continuous.

The development of force fields has made enormous progress over the past decades, and (some of) the current force fields can make reasonably accurate predictions [8]. Nevertheless, further improvement of force fields as well as careful comparisons to experiment are required.

Implicit Solvent

The above mentioned strategy for molecular simulation of complex systems using an atomistic forcefield is popular, partly because it is considered a reasonable approach to reality and partly because it is fairly straightforward to implement. In particular, many software packages are available that can perform molecular dynamics and/or Monte Carlo [30, 31, 40]. One might rightly ask the question why folding poses such a computational problem, if these accurate potentials and software packages are available?. The main reason is that the system size (>10000 atoms)and the time scale (>1 μs) needed to deal with folding of a small protein in aqueous solution, is still beyond current reach for current computer power. To give an idea how expensive it is to go beyond the microsecond level for atomistic simulation; the first microsecond folding simulation was performed in 1998 by the Kollman group, who needed about 4 months on a 256 node supercomputer to create a single trajectory of a 36-residue Villin headpiece [41]. At present, the systems sizes and timescales that molecular dynamics can handle are still roughly limited to 10^5 atoms and 1 μs.

Most protein folding simulations using explicit solvent consist of 80 or more percent water, and it turns out that the calculation of the water interaction indeed also takes more than 80% of the CPU time. Some MD packages improve on this by using a special routine for the water interaction [42]. Nevertheless, it seems a waste that most of the computer time is spent on solvent molecules that apparently have hardly anything to do with the process of interest. Therefore, in the so-called implicit solvent models all water molecules have been replaced by an effective potential. Simply scaling the electrostatic interaction alone by e.g. an effective distance dependent dielectric constant cannot faithfully reproduce the role of water as it neglects polarization, hydrogen bonding and the hydrophobic interaction. Therefore more sophisticated implicit solvent models such as the effective energy function (EEF1) [43] and the Generalized Born/surface area (GB/SA)model [44] were developed. EEF1 is based on the solvation free energies of the amino acid side-chains, and treats electrostatic interaction via a distance-dependent dielectric constant. The GB/SA model takes into account both the solute-solvent electrostatic polarization (GB) as well as the solvent accessible surface of the solute (SA), approximating the cost of the creation of a solvent interface and the van der Waals interactions.

Even sophisticated implicit solvent models cannot completely reproduce the effects of the solvent and there is much debate whether or not it is allowed to replace the solvent by a effective forcefield without losing valuable information [8]. Comparison between implicit and explicit solvent shows significant differences [45–47]. Nevertheless, the GB/SA implicit solvent is very popular, because it is computationally much cheaper than explicit solvent force fields and can reach much longer timescales (up to microseconds) [8].

Coarse-Grained Models

For most but the smallest fastest folding proteins even the implicit solvent atomistic models are too expensive. While implicit solvent models can reach the microsecond timescale, most proteins take much longer to fold. To reach these time scales, one can coarse-grain the atomistic models. The strategy behind coarse-graining is to squint your eyes and view the protein as a string of blobs, instead of individual atoms. These blobs can be of the size of a few atoms, or of the size of the residue (amino acid) or even larger. The blobs are concatenated in order to maintain the topology of the protein. The trick is to find the proper effective interaction between these blobs, such that most or all of the interesting protein properties are reproduced. There are many flavors of coarse-grained potentials. The Go-model, originally introduced in the 70's [48] is among the most popular. In this model the protein is represented by a chain of beads, interacting via effective potentials, designed such that the model protein folds in the a priori chosen native state. These potentials have been used for off-lattice as well as lattice models, and can be sampled by MC, Langevin dynamics and MD [49]. Go-models have been crucial in the debate on the "topology as rate determinant" [5], and the "diffusion-collision versus nucleation-condensation" discussion [5, 17].

A more ab initio coarse-graining approach is that by Scheraga and co-workers [50], who model the protein as a chain of beads representing the alpha carbon atoms, with an additional bead for the side chain. By integrating out the degrees of freedom in an all-atom model of the protein, these authors obtain a effective potential able to reproduce the properties of the polypeptide. Application of Langevin dynamics allows for prediction of protein structure without having to rely on knowledge of the native state [50, 51].

Although each of these models have their own merits, it is currently not clear how much predictive power these coarse-grained potentials have. Nevertheless, coarse-graining is a valid way of studying protein folding, and in the near future might have the much sought after power to predict novel folds.

Lattice Models

In some theoretical studies, requiring many folding/unfolding events for different sequences, even these highly coarse-grained models would be computationally too expensive. In order to simplify the protein models further, in the last two decades minimal models emerged that can reproduce folding behavior [15, 52, 53]. Lattice models such as the self avoiding walk (SAW) were already employed in the polymer field to investigate scaling behavior, swelling and collapse [54]. The SAW model assumes that the protein is a linear heteropolymer that lives on a cubic lattice, where one residue can only occupy one lattice site at a time. If only excluded volume would be taken into account, the model would resemble a polymer in a good solvent [54]. Including attraction and repulsion between residues at neighboring sites induces protein folding behavior. These attractions and repulsions are given by the energy function

$$E = \sum_{i,j} C_{ij} S_{ij} , \tag{3}$$

where i, j are the residue indexes, S is the interaction energy matrix and C is the contact map, with $C_{i,j} = 1$, if i and j are neighbors and 0 otherwise. Early models only made a difference between hydrophilic and hydrophobic interactions, where hydrophobic residues attract each other and hydrophilic are repulsive. The simple HP model just assigns a negative energy to hydrophobic interactions and positive to hydrophilic interactions [52,53]. A slightly more realistic and often used interaction matrix is that of Miyazawa and Jernigan [55] who extracted from many PDB database protein structures the frequency of contacts for each amino acid pair. Based on this frequency and using a quasi chemical approximation, they constructed the effective free energy of each amino acid pair in a 20×20 interaction matrix.

Unfortunately, a given existing (natural) protein sequence will not fold into its native state using this matrix. Instead, these highly simplified models are mostly used to address theoretical issues, such as the stability of the native state in general, the temperature dependence of the folding rate, and protein evolution. Sampling the Boltzmann distribution with Monte Carlo techniques leads to insight in the relative thermodynamic stability of the collapsed and extended states. In principle, however, one could use the entire machinery of advanced Monte Carlo techniques developed to sample polymers, e.g. Configurational Biased Monte Carlo and recoil growth [21], but one would lose all kinetic information. In contrast, by only allowing for local moves, such as the crankshaft and corner moves, it is possible to extract some kinetic information [15].

A random sequence (or even a naturally occurring protein) lattice polymer would probably collapse or become trapped in a glassy state, but would almost certainly not behave like a two state folder. Instead one must design the lattice protein sequence to fold into the desired native topology, by minimizing the total energy E of this fixed structure through mutation of the sequence, while maintaining the sequence heterogeneity (to prevent ending up with a homopolymer) [5, 56]. A minimal energy sequence does not yet guarantee two state folding behavior. An additional criterion is the previously mentioned 'energy gap': all other configurations than the native state must have a much higher energy to avoid getting trapped into a misfolded state.

The random energy model greatly simplifies the analysis of general folding behavior of lattice heteropolymers by assuming that the interaction energies S_{ij} are Gaussian distributed. The details of this analysis goes beyond the scope of this chapter, but can be found in [12, 14, 57].

In summary, lattice models of proteins can be employed to test theories, and find generic folding behavior, including substrate induced folding, protein binding, disorder-order transitions associated with signaling proteins and even translocation [5, 58, 59]. However, although they give global insight into the statistical mechanics of folding they are not probable candidates for giving molecular insight into the folding problem, simply because there is no molecular information in these models, other than the linearity of the polymer and the interaction matrix.

2.3 Order Parameters

Molecular simulations result in sequences of configurations of the protein. These configurations consists of many atoms, beads, particles or lattice sites. One might inspect these trajectories visually, but in general, it is desirable to analyze the data more quantitatively. To do so, one has to reduce a 3N-dimensional configuration to a number of low dimensional order parameters that contain physical information on the state of the system. For proteins many such order parameters exist. In this section I will discuss a few.

The number of native contacts n_c is an important and often used parameter in protein folding descriptions. A contact is made when the α-carbons of non-adjacent residues are within a 6 Å distance. A native contact is a contact that also occurs in a reference configuration representing the native state. This reference configuration can be taken from, for instance, the Protein Database (PDB) or from simulations (e.g. the most likely structure, minimum free energy, etc).[3] The number of native contacts can be evaluated for arbitrary configurations and measures the similarity between the configuration of interest

---

[3] Note that in general a PBD structure of a certain protein obtained by crystallography will not correspond that protein's solution structure. One should therefore

and the native state. Often n_c is presented as the fraction ρ of the maximum possible native contacts. It is also common to construct a $N \times N$ matrix of all residue-residue contacts, the so-called contact map (as in Sect. 2.2).

Another measure of the native state similarity is the root mean square deviation (RMSD) with respect to a reference (native) state given by

$$\sigma_{RMSD}^2 = \frac{\sum_i^N m_i(\mathbf{r}_i - \mathbf{r}_i^{\text{ref}})^2}{\sum_i^N m_i} \tag{4}$$

where m_i denotes the mass of the ith atom, $\mathbf{r}_i$ and $\mathbf{r}_i^{\text{ref}}$ are the position of atom i in the structure of interest and the reference structure, respectively. This function should equal zero for two identical protein configurations, but in general it will not, because of the offset by translation and rotation. Therefore, it is necessary to minimize this function with respect to translation and rotation. The RMSD can accurately characterize the native reference state. For instance, a configuration with an RMSD value below 2.5 Å is considered a close approximation of the native state. However, RMSD quickly deteriorates as a measure for configurations away from the native state. An intermediate state, a transition state and an denatured state, can easily have similar native RMSD values.

The native state of a protein often contains intra-protein hydrogen bonds. A common definition of a hydrogen bond is when the distance between the acceptor (A) and donor (D) is smaller than 3.5 Å, and the D-H-A angle is larger than 150°. Of special interest are the backbone hydrogen bonds, as they stabilize the secondary and tertiary structure. These bonds are formed between a carbonyl oxygen and an amide hydrogen in the protein back bone (see Fig. 1). If a backbone hydrogen bond also occurs in the native reference state it is considered a native hydrogen bond. Again, the number of native backbone hydrogen bonds is an order parameter that reveals the similarity between a configuration and the native state.

The above described order parameters are effective for the native state. How about the unfolded or denatured state? It is much harder to characterize this state, as there is not a single reference configuration to compare to. One would like to use a nonspecific order parameter, such as the radius of gyration R_g. This parameter, often used for polymers, is defined as [54]

$$R_g^2 = \frac{1}{6N} \frac{\sum_i^N m_i(\mathbf{r}_i - \mathbf{r}_{CM})^2}{\sum_i^N m_i} \tag{5}$$

where $\mathbf{r}_{CM}$ denotes the center of mass of the polymer, and m_i the mass of the atom. R_g is naturally expected to be smaller in the native state than in the unfolded. R_g is often calculated using the alpha carbon atoms alone, although it is of course also possible to use all atoms in the protein. Alternatively, one

---

be careful with using the x-ray structures as a reference for simulation in solvent. A better option might be to use NMR-structures.

might be specifically interested in the hydrophobic residues. R_g^{hc} gives an indication of the size of the hydrophobic core, and hence of the progress of the hydrophobic collapse.

Besides R_g one can, for instance, look at the degree of solvation, e.g. through the solvent accessible surface (SAS), the surface of the protein that is accessible to a solvent molecule. This surface is estimated by rolling a hard sphere of the size of a solvent molecule over the protein surface. The SAS is generally larger in the denatured states (molten globule and unfolded states). It also has the advantage it can be calculated both for implicit and explicit solvent. In case of an explicit solvent, a more physical intuitive solvation parameter might be the number of first shell solvent molecules around the protein, or even the number of hydrogen bonds between the protein and the solvent.

Naturally, it is possible to construct a multitude of other parameters for specific systems, based on e.g. dihedral angles or atom distances in salt bridges.

3 Sampling the Thermodynamics of Folding

3.1 Free Energy

The free energy of protein folding is a key quantity in computational studies of the thermodynamic behavior of the folding process. In particular, one is interested in the free energy difference ΔG between the (meta) stable states that the protein can adopt, because it is related to the experimental equilibrium constant K. In two state folders, the most important states are the native and the denatured state. All other metastable states are assumed not well populated, i.e. have a much higher free energy, compared to the thermal energy. Of course, these additional metastable states can be interesting from a mechanistic viewpoint, but they do not show up in thermodynamic and kinetic measurements.

A second interesting quantity is the height of the folding free energy barrier that the protein has to overcome upon folding or unfolding. The barrier height gives insight in the stability of the native state, and yields an estimate for the rate constant. This height can be related to experimental phi analysis [2]. Phi analysis uses sequence mutations to influence the folding rate constants and thus the barrier height, and can give insight in the structure of the transition states.

The most straightforward way to obtain free energy differences by computer simulation is to express the free energy as a function of the set of order parameters λ_i. Up to a constant this Landau free energy [21] is given by

$$\beta F(\{\lambda_i\}) = -\ln P(\{\lambda_i\}) + const \tag{6}$$

where $F(\{\lambda_i\})$ is the probability to find the system at certain value of the set of order parameters λ_i. This probability is obtained by integrating out all degrees of freedom in the configurational partition function, except the set of order parameters $\{\lambda_i\}$ that are of interest:

$$P(\{\lambda_i\}) = \frac{\int d\mathbf{r}^N \exp(-\beta U(\mathbf{r}^N)) \prod_i(\delta(\lambda_i - \lambda_i(\mathbf{r}^N)))}{\int d\mathbf{r}^N \exp(-\beta U(\mathbf{r}^N))} \tag{7}$$

Here $\beta = 1/k_B T$ is the inverse temperature, $\mathbf{r}^N$ denotes the coordinate vector for all N particles, U is the potential energy, and $\lambda_i(\mathbf{r}^N)$ the instantaneous value of the order parameter λ_i, and $\delta(x)$ is the Dirac delta function. Note that this expression can be used irrespective of the simulation method as long as one samples the canonical distribution. Both straightforward MC and MD can thus serve to find the free energy by simply histogramming the values of the order parameters [21]. Of course, sampling problems arise when the free energy barriers between the stable states are high enough to make the transition region unlikely. In that case one has to rely on additional algorithms to enhance statistics. Moreover, as I will discuss in more detail the next section, the calculated height of the barrier depends very much on the chosen set of order parameters $\{\lambda_i\}$ to project the free energy on.

3.2 Biased Sampling

One of the most often used methods in simulation of free energy barriers is the Umbrella Sampling (US) technique where one includes a biasing "umbrella" potential W in the partition function [21]. Introducing this bias in the partition function leads to a different sampling distribution:

$$P_{bs}(\{\lambda_i\}) = \frac{\int d\mathbf{r}^N \exp(-\beta U(\mathbf{r}^N) + W(\{\lambda_i(\mathbf{r}^N)\}) \prod_i(\delta(\lambda_i - \lambda_i(\mathbf{r}^N)))}{\int d\mathbf{r}^N \exp(-\beta U(\mathbf{r}^N) + W(\{\lambda(\mathbf{r}^N)\})} \tag{8}$$

The true free energy has to be "unbiased" by simply subtracting the biasing potential

$$\beta F(\{\lambda_i\}) = -\ln P_{bs}(\{\lambda_i\}) - W(\{\lambda_i\}) + const \tag{9}$$

A biasing function that is exactly the negative of the free energy would yield optimal statistics as it would lead to a uniform $P_{bs}(\{\lambda_i\})$. However, to obtain this function requires either an iterative method, or a priori knowledge of the barrier [21]. Therefore, a more practical method is to choose the function W such that one biases the distribution towards the unlikely region. For instance, one could use a simple harmonic potential $W(\lambda) = c(\lambda - \lambda^*)^2$, where c is the force constant, and λ^* denotes the fixed value of an order parameter that characterizes the transition region. Usually, this λ^* is chosen at several values, so that many US simulations can be run in parallel. All histograms can then be "glued" together, e.g. using the weighted histogram analysis method [6, 60]. Multiple dimensions might be used in this approach, although histograms in more than two dimension are usually unpractical. An alternative to US is the metadynamics method which allows for sampling up

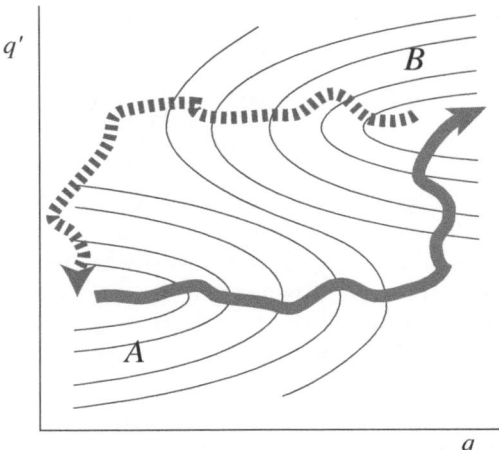

Fig. 7. Generic free energy landscape illustrating the hysteresis problem. If q is the only order parameter used to describe the reaction, but other parameters such as q' are also important slow variables, hysteresis might occur. *Solid curve*: starting from the initial state A, the ensemble is slowly biased along q, until the transition to B suddenly occurs. *Dashed curve*: When the simulation starts in B and q is slowly reduced a different part of phase space is sampled, leading to hysteresis

to 6 dimensions [61]. Other recently developed methods focusing on multi-dimensional biasing functions include flooding [62], hyperdynamics [63], and multicanonical sampling [64].

The Umbrella Sampling technique assumes there is a fast sampling of all degrees of freedom perpendicular to the biasing variables. If this is not case, and if one of these variables exhibit slow sampling, hysteresis might occur (see Fig. 7). Such hysteresis becomes apparent when one samples the unfolding landscape first by moving λ from the folded to the unfolded state, and subsequently back to the folded state. If other slow variables are important, the free energy landscape is poorly sampled, and the results for the unfolding and the folding sequence do not agree. Shea et al. [6] circumvented this problem partly by taking high temperature unfolding trajectories as an initial ensemble, around which they probed the free energy landscape at 298 K. Apparently, the high temperature unfolding trajectories were close enough to the room temperature trajectories that this approach led to meaningful results. It is however not a priori given that such umbrella sampling gives the correct Boltzmann distribution. For instance, if the high temperature trajectories visit parts of phase space that are not representative for the folding, such an approach fails, because US only locally samples the free energy, and will not explore far away parts of phase space. Therefore, in the last decades, alternative methods such as parallel tempering methods have gained popularity, which should, in principle, suffer less from this problem. Notwithstanding the limitations, umbrella sampling is extremely versatile and has been used in many MD and MC studies [6, 21].

3.3 Replica Exchange/Parallel Tempering

Parallel Tempering (PT), a.k.a. the Replica Exchange Method (REM), is a Monte Carlo scheme invented to sample free energy landscapes with many local minima [21, 65]. The basic idea of PT/REM is to heat up the system to high temperature periodically in a simulation and then lower the temperature again to the temperature of interest. In this way, a glassy system trapped in a local minimum at low temperature can escape this minimum at a higher temperature. At first sight one might think that simply alternating periods of low temperature (e.g. around room temperature) with high temperature might solve the problem of getting stuck in local minima, but such a scheme does not obey detailed balance, and hence does not conserve the canonical distribution. Restoration of detailed balance is possible by running two simulation (MD or MC) in parallel, one at high and one at low temperature, and occasionally trying to exchange configurations. Considering both systems simultaneously, the Boltzmann distribution is proportional to $\exp(-\beta_{low}U_{low} - \beta_{high}U_{high})$. The Metropolis acceptance rule for exchanging the two systems is then [21]

$$P_{acc} = \min[1, \exp((\beta_{high} - \beta_{low})(U_{high} - U_{low}))] . \tag{10}$$

This acceptance probability is in general exponentially small for large temperature gaps and large system sizes, which both lead to huge energy differences. To improve the acceptance ratio, in replica exchange/parallel tempering many copies at temperatures in between the low and high temperature are inserted, such that the exchange probability is reasonable. The temperature distribution can be chosen such that the swapping probability is optimal. The Metropolis algorithm for many replicas becomes

$$P_{acc}(i \leftrightarrow j) = \min[1, \exp((\beta_i - \beta_j)(U_i - U_j))] . \tag{11}$$

The parallel tempering method is very useful for rugged energy landscapes, and if one does not know the order parameters to bias in. However, although the PT/REM in principle conserves the Boltzmann distribution, its sampling efficiency relies on the fact that at high temperature the transition of interest happens spontaneously within the simulation time per replica, so that the phase space is adequately sampled. During its random walk through temperature space every replica should visit the highest and the lowest temperature many times. In addition, the exchange between the temperatures should be not too fast, as the systems have to be able to adapt to the new temperature. The convergence of PT/REM in case of protein folding in explicit solvent is still a controversial issue [66]. One reason for slow convergence is the large energy of the explicit solvent. Berne and coworkers recently introduced an adapted PT/REM scheme that only takes into account the potential energy of the solute, which allows a reduction of the number of required replicas and hence leads to enhancement of the sampling [67].

It is also possible to combine the umbrella sampling and PT/REM by again adding a biasing function that is now a function of both the order parameter set $\{\lambda_i\}$ and the temperature T. The advantage of such a combination might be that even though the dynamics at high temperature might be faster, it is not a priori certain that all relevant states are well sampled at high temperature. This is in particular a problem for folding processes in which one state is stabilized by entropy, and the other by energetic interaction. Including a biasing function might be able to improve that, but, of course, assumes one already has a proper order parameter to bias in [68].

The PT methods can also be combined with Frenkel's recently proposed scheme of waste recycling [69]. In this scheme Monte Carlo trial moves are not just discarded as useless but actually contribute to the average. It is straightforward to extend this to PT [68], improving the statistics on the free energy dramatically. Note however, that the waste recycling scheme does not improve the sampling itself, but only the accuracy of the histograms.

As a last note, the PT/REM method is not limited to temperature exchange other parameters can be used for the replica exchange [21].

4 Sampling the Kinetics of Protein Folding

4.1 Direct Molecular Dynamics Methods

In addition to thermodynamics, computational methods can in principle provide insight in the kinetics of folding. Although lattice models have been used to address generic questions about folding kinetics [52], all-atom MD with explicit water is the natural choice for the study of the folding kinetics of small proteins in aqueous solution, as it contains the true dynamics (at the accuracy level of the forcefield of course). However, because current computational power does not allow for exploration of time scales on the order of microseconds, it seems not possible to investigate the kinetics of even the fastest folding proteins by direct MD.

If one is interested in how often a protein passes through the transition region, one could start a MD simulation in a stable state, and just wait for it to make a transition over the barrier. The average of the time it takes to cross the barrier is called the mean first passage time (MFPT) τ_{MFPT}. The MFPT is simply related to the rate constant by $k = 1/\tau_{\mathrm{MFPT}}$. In general, it is extremely costly to calculate MFPT for atomistic models with explicit solvent by brute force. Moreover, even if a folding event takes place within a microsecond of MD simulation time, it is only one possible pathway out of the many available to the system, while many folding events are needed for an accurate estimate of the rate. Therefore, reproduction of many folding and unfolding events using direct MD, Langevin/Brownian dynamics or MC is currently only possible for coarse-grained models or atomistic models with implicit solvent.

Parallel Replica

One possible solution to the problem of long timescales is to run many trajectories (replicas) in parallel with different initial conditions [70]. When one replica makes a transition to another basin of attraction, all other replicas are also moved to the other stable basin [71], after which the procedure restarts. This approach is extremely effective in the distributed computing scheme pioneered by Pande et al. [14]. This scheme initiates trajectories from a stable state at a large number of independent processors and uses the fact that for high free energy barriers the distribution of passage times follows a Poisson distribution. Spawning 10000 trajectories from the reactant state with different initial conditions is likely to yield successful passages after only a few ns, even when the MFPT is on the order of microseconds. Of course, the other 9990+ trajectories will have yielded no information on the barrier crossing, making this an extremely costly scheme. Also, it is not a priori obvious whether these fast successful events are a representative ensemble of the transition trajectories.

High Temperature Trajectories

The difficulties involving long folding trajectories simplify if one considers the unfolding instead of the folding process. Assuming reversibility, the unfolding pathways of a stable protein should be similar to the folding pathways. At room temperature (298 K) the simulation of the unfolding is almost as hard as the folding process. However, by increasing the temperature the unfolding happens spontaneously in an short (ns) time regime, because the thermal energy can overcome the barrier and the unfolded state is entropically favorable. Much about the mechanics can already be deduced from these high temperature unfolding trajectories [72–74].

4.2 What is the Reaction Coordinate in Folding?

Rare Events

The long mean first passage times that make straightforward MD very costly, indicate the existence of high free energy barriers between folded and unfolded states (and possible intermediates). The crossing of such high free energy barriers is a rare but important event. Several methods enable the computation of rare event kinetics, such as the transition state theory (TST) based Bennett-Chandler approach [75, 76], which expresses the rate constant as a product of two factors: the equilibrium probability to be on the barrier, and a kinetic prefactor (cf. eqn. 1). The first factor is given by the FE difference between the transition state region and the stable state, and can be calculated by, for instance, US. The second factor, the transmission coefficient, is obtained by firing off many trajectories from the top of the barrier [75, 76]. The TST-based methods are based on the FE as a function of reaction coordinates. Unfortunately, they can therefore suffer from the previously mentioned hysterisis problems for high dimensional complex processes (e.g. folding) when it is difficult to find the correct reaction coordinates for such a process.

The Reaction Coordinate Problem

The reaction coordinate problem that theorists face when studying folding kinetics is far from trivial. Because the barrier is high it is not well populated, both in experiments as well as in simulations. However, to obtain insight in the process it is important to know how far the transition has progressed: e.g. when the transition state has been reached. The problem is to find a description of such a reaction coordinate in terms of a low dimensional order parameter. In a case of simple the reaction (e.g. between two atoms) the distance between atoms might play the role of a good reaction coordinate. For large distances the system is in its reactant state. When the distance shortens the free energy increases until the top of the barrier has been reached. Subsequently, the free energy decreases again and the system is in the product state. In this example, the reaction coordinate is an extremely simple order parameter: the distance. In protein folding the situation is more complex. As many atoms play a role in the folding process, no a priori obvious reaction coordinate can be found. Although some of the native state order parameters, such as the fraction of native contacts ρ or RMSD (see previous sections), might seem good candidates for the reaction coordinate, this is not necessarily the case.

Suppose as is depicted in Fig. 8 and discussed in more detail in the TPS chapter, that the two true ingredients of the reaction coordinate are given by two order parameters, q and q'. All other degrees of freedom are fast variables just supplying thermal noise, or friction. The free energy landscape might then look like the top panel of Fig. 8. This free energy landscape has a single transition region, or dynamical bottleneck through which almost all trajectories between the initial and final stable state must pass. On inspection, it is clear that these trajectories first make a change in q' before changing q. This clearly indicates that q' is an important ingredient for describing the transition state. But q' does not distinguish the stable states as both states have the same average value of q'. Obviously, the reaction coordinate must be a combination of both q and q'. However, in general, the situation is less straightforward. Suppose that one has only access to q, as it can distinguish the stable states, whereas q' is an yet unknown quantity. Then, as illustrated in the bottom panel of Fig. 8 the free energy as a function of q shows a single barrier, indicating the expected two state behavior. The top of this barrier does not correspond to the transition state region, but is already in the basin of attraction of the final state. On the other hand, the true TS region is located

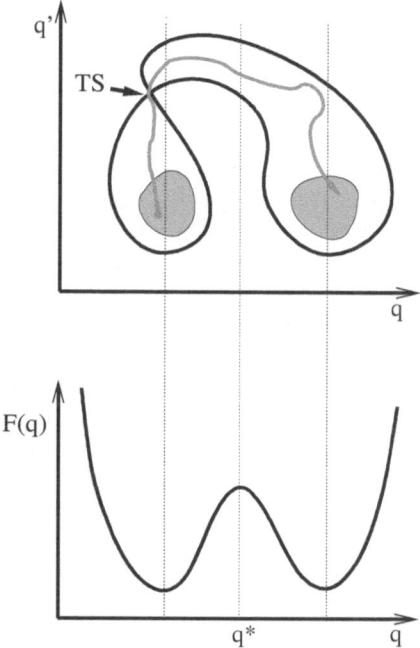

Fig. 8. Generic free energy landscapes can illustrate the reaction coordinate problem in complex systems. If both q and q' are important ingredients of the reaction coordinate (*top left panel*), leaving out one of them, e.g. q', will lead to a wrong prediction of the transition state region (*bottom panel*), and hence a wrong mechanism and a statistically inaccurate rate constant. See text for more details

within the q boundaries of the initial state. Only taking into account the q order parameter leads to hysteresis (similar to the illustration in Fig. 7). In protein folding similar hysteresis problems occur. The native and unfolded state parameters such as R_g, n_c etc, can most likely not serve as reaction coordinates, since there will be other unknown parameters that are important to describe the folding transition.

Many Folding Pathways: the Search for the Transition State Ensemble

As mentioned in the introduction, there is not a single folding pathway. The flexible protein might choose from many different pathways via many different low populated intermediates or transition states, all leading eventually to the native state. Hence, there is not only one transition state, but instead a whole ensemble of states: the transition state ensemble (TSE). To explore the

TSE computationally requires a definition or criterion of a transition state[4]. Usually a TS is defined as a saddle point on the potential energy landscape. However, this makes only sense at very low temperature, when entropy is not important. Besides, in complex systems, the number of saddle points in the energy landscape grows exponentially with the number of degrees of freedom [77]. Clearly, it makes no sense to count all of these saddle points. What about saddle points in the free energy landscape? While these might seem good candidates for the TSE at first sight, the discussion in the previous section clarifies that such an approach is very dangerous, if one misses out an important ingredient of the reaction coordinate. Therefore, Du et al. proposed a more operational definition of a transition state [78]. Consider a very long trajectory traveling many times between two stable states. By construction, this trajectory must pass the TS region many times. Next, define transition states as those configurations that lie precisely on the border of the basin of attraction of the initial (unfolded) and final (native) state. To identify a transition state, one can take a configuration from the long trajectory, and start many new short trajectories from it (with randomized Maxwell-Boltzmann velocities at the temperature of interest). When the folding probability p_{fold}, i.e. the fraction of trajectories that make it to the native state, equals 0.5 this configuration is labeled a TS. The probability p_{fold} is also called the commitment probability or *committor*, as in general it gives the probability for commitment to the final product (native state).

Mapping the Kinetics on Reaction Coordinates

It is tempting to use a free energy surface as a function of a reaction coordinate, in combination with a Langevin model, in order to extract all kinetic information. However, a simple diffusion model on a free energy surface can only describe the kinetics of complex systems as long as all the proper reaction coordinates (slow variables) have been identified. The above discussion in the previous section clearly shows, that it is not easy to establish if this goal has been reached. In addition, a diffusion model requires the correct effective diffusion constant. When these requirements are met, it is possible to apply the idea of mapping the kinetics on a 1 dimensional effective reaction coordinate (see e.g. [79]) In Sect. 5 I discuss the case of the folding of a simple secondary structure element, in which these requirements are not so easily fulfilled.

4.3 Path Sampling

Path sampling tries to circumvent the problems related to defining a reaction coordinate by constructing dynamical pathways from the initial and final

---

[4] Experimentally the TSE is investigated by phi-analysis, which is beyond the scope of this paper [2].

states, of which much more is known. Already in the 1980's Pratt [80] introduced the notion of sampling Markovian chains of states. Such chains lead from an initial state via a number of intermediate configurations to a known final state. These configurations are connected with each other through transition probabilities. When the transition probabilities correspond to the underlying dynamics, an ensemble of these chains of state is equivalent to the path ensemble: the collection of true dynamical unbiased trajectories connecting the initial and final stable states. By changing configurations on these pathways one can obtain new paths and thus sample the path ensemble.

Several different methods have been developed to sample the path ensemble. For instance, Elber et al. have devised a number of methods to study protein folding based on the stochastic Onsager-Machlup action formalism [81–83]. In this approach, every path's action corresponds to the weight of the path in the ensemble, and thus can be sampled through a Monte Carlo scheme. Although such approach is elegant and efficient, the large time-steps involved make a proper correspondence between the transition probabilities and the underlying dynamics problematic. Hence, a quantitative treatment of kinetics is beyond reach for these methods.

Transition Path Sampling

Similar in spirit is the transition path sampling (TPS) method, which is explained thoroughly in another chapter in this book. TPS has proved to be a viable method for studying processes separated by high free energy barriers in complex environments [77, 84]. Starting from an existing initial transition pathway, TPS gathers a collection of paths connecting a reactant with a product stable region by employing the Monte Carlo shooting algorithm. While one is free to choose any MC algorithm to sample the path ensemble, the shooting move turns out to be extremely efficient [77, 84]. The shooting algorithm changes the momenta of a randomly chosen time slice on an existing a path, and determines a new trajectory from this time slice by integrating the equations of motion backward and forward in time using a standard MD integrator. When the new trajectory connects the initial with the final region, the new path is accepted, otherwise the old path is retained. Repeating this procedure with different time slices results in a random walk through path space and a collection of transition paths: the *path ensemble*. Because detailed balance is obeyed, the paths are properly sampled according to their statistical weight. Subsequent analysis of the path ensemble gives an unbiased insight in the mechanism of the reaction. The major advantage of TPS is that one does not have to impose reaction coordinates on the system, but rather extracts these from the simulation results.

The TPS shooting algorithm might run into problems for long deterministic trajectories over rough energy landscapes so that the old and the new trajectories will have diverged completely before the other basin of attraction is reached. In that case, a small change in momenta will cause the forward as well as the backward trajectory to return to the same stable region. These non-reactive trajectories will make up the majority of the trial moves. We can improve the efficiency dramatically by making use of a stochastic TPS scheme [85]. In the stochastic TPS algorithm one can accept a forward or backward shot independent of each other [77]. Stochastic trajectories are generated by coupling the system to a heat bath using a thermostat similar to the Andersen thermostat [86]. The frequency of the coupling is chosen low enough to obtain realistic mechanisms and rate constants, while still having a quick but controlled divergence from the old path.

Further efficiency is obtained by making the path length variable [87]. As only shots from the barrier itself are useful in stochastic path sampling, it is natural to stop with the integration of the equation of motion once one reaches a stable state, provided that the trajectory is then really committed to a stable state [87, 88].

During the sampling the path length fluctuates, but remains bounded, because of the Monte Carlo acceptance rule [87]. In previous TPS studies, shifting moves were used to enhance statistics in the correlation functions for the calculation of rate constants. We do not have to perform such shifting moves as these have no effect on the path ensemble. The enhancement of statistics the shifting moves provide, is taken care of in the TIS rate expression described below [87].

Transition Interface Sampling

Conventional TST methods such as the Bennett-Chandler technique [75, 76] are not efficient to calculate the rate constant for complex transitions with unknown reaction coordinate. If the order parameters used to estimate the barrier do not correspond to the reaction coordinates, the constraining techniques to calculate the free energy barrier (such as umbrella sampling) can lead to unrealistic free energy profiles and extreme hysteresis. Even if the free energy barrier is properly sampled, the barrier will likely be lower than the real barrier, and the calculation of a statistically accurate transmission coefficient will be exceedingly difficult [77, 87]. Within the TPS framework, however, it is possible to calculate the rate constant accurately and efficiently using the transition interface sampling (TIS) method [87]. In short, the TIS method calculates the flux to leave a stable initial state and, subsequently, the conditional probability for trajectories leaving the initial state to reach the final state. The basis of the method is to map the entire phase space onto a foliation of interfaces characterized by a one-dimensional order parameter $\lambda(x)$, which is a function of the system configuration x. The forward rate constant k_{AB} then follows from the following equation

$$k_{AB} = f_A P(\lambda_B | \lambda_A) , \qquad (12)$$

where the first factor is the flux f_A to leave region A (defined by $\lambda(x) < \lambda_A$) and the second factor is the conditional probability $P(\lambda_B | \lambda_A)$ to reach region

B (defined by order parameter $\lambda(x) > \lambda_B$) once the surface λ_A defining region A is passed. The flux factor can be measured by starting an MD simulation in stable state A. The interface λ_A will be crossed often, resulting in statistically accurate value for f_A. In contrast, the value of $P(\lambda_B|\lambda_A)$ is very low for a high barrier, and cannot be measured directly. The statistics can be considerably improved by employing a biased path sampling scheme. By sampling paths with the constraint that it comes from region A, crosses an interface λ_i and then either goes to B or returns to A, we can measure the probability $P(\lambda|\lambda_A)$ to reach values of λ larger than λ_i. These probabilities are binned in a histogram, and after having performed several simulations for different interfaces λ_i, these histograms are joined into a master curve, from which the value of $P(\lambda_B|\lambda_A)$ can be extracted. A more detailed description of TIS can be found in [87].

Partial Paths and Milestoning

Path sampling seems to be a promising way to study folding mechanics and kinetics [88]. However, for larger proteins the folding pathways might become prohibitively long and diffusive. Some recently developed methods make use of the diffusivity of paths and the corresponding loss of correlation. For instance, the partial path TIS technique (PPTIS) takes advantage of this loss of correlation by describing the rare event as a Markovian hopping process between the interfaces, with corresponding hopping probabilities. If the loss of correlation between three consecutive interfaces is justified, this method exactly gives the kinetics [89].

Similar in spirit is the Milestoning [90] method by Elber and coworkers, who assume that the diffusion of interest occurs through a "tube" in configuration space, and translate the rare process into a non-Markovian hopping between configuration space hyperplanes, the so-called "milestones" (which are in fact rather similar to the TIS interfaces, except that they do not form a foliation). The kinetics is obtained from starting an equilibrium ensemble on a milestone, and measuring the time distribution needed to reach the next milestone. The distribution can subsequently be used to construct the kinetics. The assumption is that there is an equilibrium (Boltzmann) distribution on each milestone.

Markovian State Models and Stochastic Road Maps

The PPTIS method coarse-grains the transition pathways to a one-dimensional Markovian state model[5]. In principle, there is no need to limit oneself to only one dimension. It is possible to coarse-grain the folding pathways using a Markovian state network model, also called stochastic road map [91–93] or

---

[5] The Milestoning method also coarse-grains pathways in a one-dimensional model, but is non-Markovian.

equilibrium kinetic network [94]. Here, one assumes that the pathways can be described by a Markov chain of hopping events on a network of (metastable) states. The only ingredients going in to the model are the configurations of such metastable states and the transition probabilities between them. In this sense, the model can be viewed as coarse-graining using Kinetic Monte Carlo [95]. The advantage is that for such a model it is straightforward and computationally efficient to calculate rate constants and committor distributions. Of course, the difficult part is to find the description of metastable states and the transition probabilities between these state. Recently, Pande and coworkers have shown how to use transition path ensembles to obtain such a Markovian state model, by clustering the conformations and measuring the transition probabilities [91]. Krivov and Karplus used a similar network approach to study the β-hairpin of the next section in an implicit solvent [94].

5 Case Study: the GB1 β-Hairpin

In this section we apply some of the above techniques on the 16 residue C-terminal fragment (41–56) of protein G-B1 (sequence GEWTYDDATK-TFTVTE, structure given in Fig. 9) which in the last decade has become a model system to investigate β sheet formation. Seminal fluorescence experiments by Eaton and coworkers [96, 97] revealed a two state kinetics between the folded and unfolded hairpin, with a relaxation time of 6 microseconds. This work inspired many simulation studies on the β-hairpin using either simplified models [98–100], full atom models in implicit solvent [94, 101, 102] or in an explicit solvent [74, 83, 103–107]. The first investigations, high temperature molecular dynamics (MD) in explicit solvent by Pande and Rokhsar [74] and multicanonical Monte Carlo sampling in implicit solvent by Dinner et al. [101], demonstrated that the folding takes place via a number of discrete steps (see Fig. 9). After the initial β turn formation the hydrophobic residues in the peptide collapse into a hydrophobic core (H-state), followed by formation of the backbone hydrogen bonds (F and N states). Other simulations, notably by Garcia et al. [105] and Zhou et al. [106], determined the β hairpin free energy landscape in explicit solvent.

Straightforward MD shows that at 300 K there is a large barrier to unfolding, preventing the system to unfold or fold spontaneously in an accessible simulation time scale [74, 103, 105, 106]. The rate determining step in the folding process seems to be the H-F transition: the formation of the backbone hydrogen bonds. I have applied TPS techniques to the GB1-β hairpin's H-F transition in explicit solvent at room temperature in order to study the kinetic pathways and obtain the folding rates [85, 88].

The TSE was determined by computing the commitment probabilities, p_A (folding) and p_B (unfolding) [88]. One typical transition state configuration is depicted in Fig. 10 Interestingly, the backbone is already in the hairpin shape. This agrees with the observation that the rate determining step in the

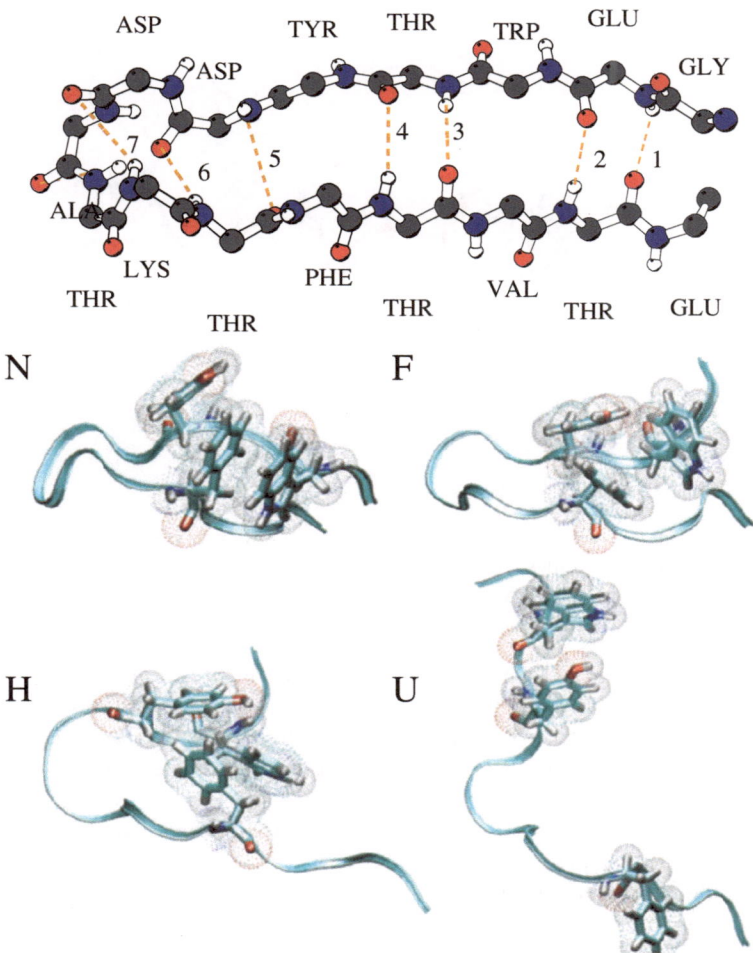

Fig. 9. *Top*: Native PDB structure of the β hairpin. The side chains are left out. The 7 backbone hydrogen bonds are indicated by dotted lines, and counted from the tail. *Bottom*: Spatial 3D structures of the several (meta) stable states. The backbone is represented by a ribbon, the hydrophobic core in stick model with dots to indicate the size of the atoms. All other residues and solvent molecules are left out

folding process is the final stabilizing transition to the F state. Configurations in the TSE are characterized by the fact that the native backbone hydrogen bonds are not yet formed and a strip of water molecules is in between the two strands, forming hydrogen bonds to the backbone donors and acceptors. This seems to be the overall generic picture for all configurations in the transition state ensemble. It supports the hypothesis brought forward by several authors

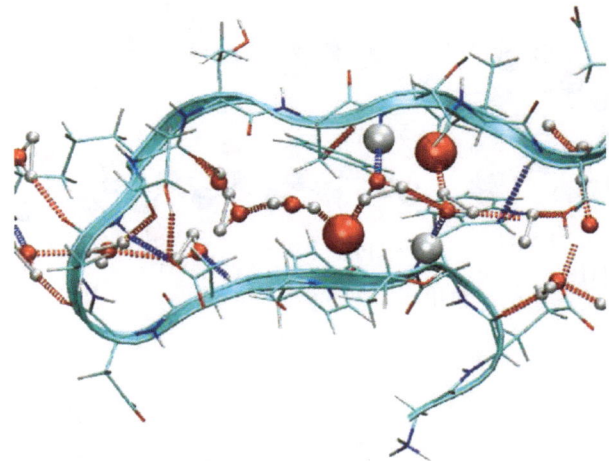

Fig. 10. Typical F-H transition state configuration. The two strands are separated by a strip of water molecules. In particular, the important backbone hydrogen bonds 3-4 (indicated by large spheres) are bridged by water molecules

that the solvent plays the role of lubricant in the final stages of folding [6, 108–110], bringing together the native backbone hydrogen bonds. The TSE results also indicates that a major cause for the folding barrier might be the loss of conformational entropy of the protein and translational entropy of the bound solvent molecules in the transition state [18, 19].

Figure 11a shows the free energy from the replica exchange method simulations (see Sect. 3.3) for $T = 295$ K in the $N_{hb} - R_g$ plane. The number of backbone hydrogen bonds N_{hb} and the hydrophobic core radius of gyration R_g, were previously thought to be the most important order parameters for describing the folding transition [106]. The transition state saddle point region seems to be located at $R_g \approx 4.7$ and a $N_{hb} \approx 0.5$, where the free energy difference between the saddle point and the F state minimum is $\Delta F = F_{TS} - F_F \approx 1.5 k_B T$, where k_B is Boltzmann's constant.[6] However, if we plot the TSE in the same plane, it is located around $R_g \approx 4.7$ and $N_{hb} = 0$. This suggests that the TSE is not at the saddle point of the free energy landscape, and that the order parameters N_{hb} and R_g do not entirely describe the kinetic pathways of folding. In retrospect, this is not extremely surprising, considering the nature of the transition states in Fig. 10. I also plotted in the same plane part of the $p_B < 0.2$ ensemble, and a part of the $p_B > 0.9$ ensemble in the picture to give an indication of the basin of attraction and commitment to the stable states.

---

[6] As we will see later, this very low value is caused by the choice of order parameters for the FE projection.

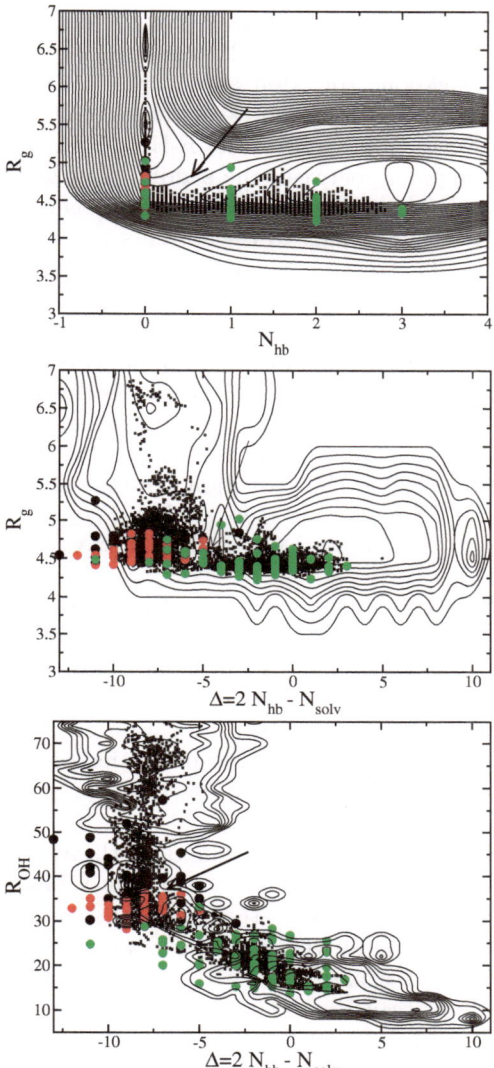

Fig. 11. Representations of the folding event in 3 different order parameter planes. The free energy landscape from replica exchange is given by thin solid contour lines separated by $0.2\ k_BT$. A few smoothed paths in the F-H ensemble are denoted by a scatter plot (*small dots*). Each dot represents a time slice along a path. Also given are the different committor ensembles: $p_B < 0.2$ light gray, $0.4 < p_B < 0.6$ in dark gray and $p_B > 0.9$ in black. The apparent transition state saddle points in the FE landscape are indicated by *arrows*

The TSE structure in Fig. 10 suggests that the degree of backbone solvation plays an important role in the F-H transition. To introduce a measure for the degree of backbone solvation as an order parameter we can define $\Delta = 2N_{hb} - N_{solv}$ where N_{solv} denotes the number of backbone-solvent hydrogen bonds. However, representing the F-H transition in the $\Delta - R_g$ FE landscape, the discrepancy between the saddle point and the TSE becomes even more clear, indicating that not only all hydrogen bonds must be broken but also the hydrogen bonds to the solvents must be formed before the transition can occur. This discrepancy reveals the R_g of the hydrophobic core is also not a very good descriptor of the F-H transition, as in both states the R_g value is similar. Figure 10 suggests the inter-strand proximity might be a better indicator. The sum of distances between oxygen and hydrogen of the native backbone hydrogen-bonds R_{OH} captures this inter-strand distance. Figure 11c shows the TSE in the $R_{OH} - \Delta$ plane, and here it is clear that besides the hydrogen bond parameter Δ also R_{OH} first has to reach a certain threshold ($R_{OH} \approx 30$) before the transition can take place. Arguably, the TS saddle region in the FE landscape is located at $R_{OH} \approx 30$ and $\Delta \approx -9$. In that case the R_{OH} appears to be a reasonable reaction coordinate for the F-H transition.

Clearly the TSE does not correspond with the saddle points in all the FE landscapes. One might think that this discrepancy could be caused by insufficient sampling of the free energy or transition paths. However, the free energy minima are well characterized (e.g. in the $(N_{hb}-R_g)$-plane), and better sampling will not change the qualitative location of the saddle points. The discrepancy is due to fact that the FE is a projection of all degrees of freedom on the chosen order parameters. This projection can obscure the location of true TS [77].

The general picture is then that the kinetic pathways to unfolding go from a state in which hydrogen bonds 3 and 4 are tightly bound, via a transition state in which a strip of water bridges the backbone strands and form hydrogen bonds to the backbone. Only then the paths are really committed to the H-state. The folding takes place in the reverse fashion. In the H-state, the hairpin has to find the transition state conformation in which the backbone hydrogen bonds 3–4 are bridged by water, before they are actually formed. Hydrogen bonds 3–4 are formed first as they are driven together by the nearby hydrophobic core. The expulsion of bridging waters might also be caused by the vicinity of the core.

Using $\lambda = R_{OH}$ I measured the crossing probability $P(\lambda_B|\lambda_A)$ by conducting transition interface sampling (TIS) simulations for both the folding (H-F) and unfolding (F-H) rate limiting step. The rate constants were, respectively, $k_{H-F} = 0.20\,\mu s^{-1}$, and $k_{F-H} = 0.4\,\mu s^{-1}$, both in reasonable agreement with experiment. The values of the rate constants also agree with previous theoretical estimates in implicit solvent by Zagrovic et al. [102]. One might conclude that the explicit solvent apparently is not required to determine the folding rate, but it clearly is important for the mechanism and transition state ensemble.

The TIS results indicate that there is a free energy barrier of more than $10\,k_BT$ between the F and H states. However, the REM free energies only show a barrier of approximately 3–4 k_BT. This value is similar to the barrier found in the free energy landscapes of [106]. The large discrepancy between the rate constant and FE results is caused by an overlap between the F and H states, thus apparently lowering the barriers in the projection on not only the R_{OH}, but also the other used representations. It follows that free energy landscapes cannot always be trusted to give correct barrier heights. The low barrier also shows that TST based methods [75,76] cannot be used to calculate rates, as they will result in extremely low transmission coefficients. Using a one dimensional high friction Langevin description on the fitted one-dimensional free energy landscape the rate constant was about a factor 500 too high with respect to the experimental and TIS rates. This also suggests that the true free energy barrier should be about $k_BT\ln 500 \approx 6k_BT$ higher [88].

6 Outlook

In these notes I hope to have given an overview of the computational challenges posed by protein folding. While much progress has been made in the past years, especially the development of increasingly accurate force fields, in conjunction with novel methodology that allows for more efficient sampling of phase space, much is still unknown. Path sampling methods allow the prediction of mechanistic kinetic details, that cannot be obtained otherwise. However, for large proteins the computational effort due to both system size and long time-scales becomes prohibitive. Thus, there is still much room for improvement, in particular by using coarse-grained models based on atomistic detailed force-fields, and by applying rare event methods. Transition path sampling related methods provide one way, but coarse-graining in time, e.g. by Markovian state models will also enable sampling the folding pathways of larger proteins. In this way, computer simulation might lead to a better understanding of how proteins fold to their native state and why they sometimes misfold [1].

References

1. C. M. Dobson (2004) Principles of protein folding, misfolding and aggregation.*Semin. Cell. Dev. Biol.* **15**, pp. 3–16
2. A. Fersht 1999 Structure and Mechanism in Protein Science, Freeman, New York
3. M. S. Cheung, L. L. Chavez, J. N. Onuchic (2004) The energy langscape for protein folding and possible connections to function. *Polymer* **45**, pp. 547–55

4. J. Kubelka, J. Hofrichter, W. A. Eaton (2004) The protein folding 'speed limit'. *Curr. Opin. Struc. Biol.* **14**, pp. 76–88

5. L. Mirny, E. Shakhnovich (2001) Protein folding theory: From lattice to all-atom models. *Annu. Rev. Biophys. Biom.* **30**, pp. 361–396

6. J. E. Shea, C. L. Brooks (2001) From folding theories to folding proteins: A review and assessment of simulation studies of protein folding and unfolding. *Annu. Rev. Phys. Chem.* **52** pp. 499–535

7. S. Gnanakaran, H. Nymeyer, J. Portman, K. Y. Sanbonmatsu, A. E. Garcia (2003) Peptide folding simulations. *Curr. Opin. Struc. Biol.* **13**, pp. 168–174

8. C. D. Snow, E. J. Sorin, Y. M. Rhee, V. Pande (2005) How well can simulation predict protein folding kinetics and thermodynamics? *Annu. Rev. Biophys. Biomol. Struct.* **34**, pp. 43–69

9. C. B. Anfinsen (1973) Principles that govern the folding of protein chains., Science **181**, pp. 223–230

10. D. Chandler, (1987) Introduction to Modern Statistical Mechanics, Oxford University Press, New York

11. A. Grosberg (2004) Statistical mechanics of protein folding: some outstanding problems, in: N. Attig, K.Binder, H. Grubmüller, K. Kremer (Eds.), Computational Soft Matter: from Synthetic Polymers to Proteins, Vol. 23 of NIC Series, Graphische Betriebe, Jülich, pp. 375–399

12. J. N. Onuchic, Z. Luthey-Schulten, P. G. Wolynes (1997) Theory of protein folding: The energy landscape perspective. *Annu. Rev. Phys. Chem.* **48**, pp. 545–600

13. D. Wales (2003) Energy Landscapes, Cambridge University Press, Cambridge

14. V. S. Pande, A. Y. Grosberg, T. Tanaka (2000) Heteropolymer freezing and design: Towards physical models of protein folding. *Rev. Mod. Phys.* **72**, pp. 259–314

15. A. R. Dinner, A. Sali, L. J. Smith, C. M. Dobson, M. Karplus (2000) Understanding protein folding via free-energy surfaces from theory and experiment. *Trends Biochem. Sci.* **25**, pp. 331–339

16. S. Gianni, N. R. Guydosh, F. Khan, T. D. Caldas, U. Mayor, G. W. N. White, M. L. DeMarco, V. Daggett, A. R. Fersht (2003) Unifying features in protein-folding mechanisms. *P. Natl. Acad. Sci. USA* **100**, pp. 13286–13291

17. S. Islam, M. Karplus, D. Weaver (2002) Application of the diffusion-collision model to the folding of three-helix bundle proteins. *J. Mol. Biol.* **318**, pp. 199–215

18. A. Akmal, V. Munoz (2004) The nature of the free energy barriers to two-state folding. *Proteins: Struc. Funct. Bio.* **47**, pp. 142–152

19. Y. Harano, M. Kinoshita (2005) Translational-entropy gain of solvent upon protein folding. *Biophys. J.* **89**, pp. 2701–2710

20. B. Gillespie, K. W. Plaxco (2004) Using protein folding rates to test protein folding theories. *Annu. Rev. Biochem.* **73**. pp. 837–859

21. D. Frenkel, B. Smit (2002) Understanding molecular simulation. 2nd ed., Academic Press, San Diego, CA

22. J. Norberg, L. Nilsson (2003) Advances in biomolecular simulations: methodology and recent applications. *Q. Rev. Biophys.* **36**, pp. 257–306

23. M. P. Allen, D. J. Tildesley (1987) Computer Simulation of Liquids. Oxford University Press, Oxford

24. A. Ricci, G. Ciccotti. (2003) Algorithms for brownian dynamics. *Mol. Phys.* **101**, pp. 1927–1931

25. K. Binder, D. Heermann (2002) Monte Carlo simulation in statistical physics, Springer, Berlin
26. K. Kikuchi, M. Yoshida, T. Maekawa, H. Watanabe (1991) Metropolis Monte-Carlo method as a numerical technique to solve the Fokker-Planck equation. *Chem. Phys. Lett.* **185**, pp. 335–338
27. W. Wang, O. Donini, C. M. Reyes, P. A. Kollman (2001) Biomolecular simulations: Recent developments in force fields, simulations of enzyme catalysis, protein-ligand, protein-protein, and protein-nucleic acid noncovalent interactions. *Annu. Rev. Biophys. Biom.* **30**, pp. 211–243
28. M. Levitt, (1983) Molecular dynamics of native protein .1. Computer simulation of trajectories. *J. Mol. Biol.* **168**, pp. 595–620
29. M. Levitt, M. Hirschberg, R. Sharon, V. Daggett (1995) Potential-energy function and parameters for simulations of the molecular-dynamics of proteins and nucleic-acids in solution. *Comput. Phys. Commun.* **91**, pp. 215–231
30. W. Cornell, P. Cieplak, C. I. Bayly, I. R. Gould, M. K. M. (1995) A 2nd generation force-field for the simulation of proteins, nucleic-acids, and organic-molecules. *J. Am. Chem. Soc.* **117**, pp. 5179–5197
31. A. D. MacKerell Jr., D. Bashford, M. Bellott, R. Dunbrack Jr., J. Evanseck, M. Field, S. Fischer, J. Gao, H. Guo, S Ha et al. (1998) All-atom empirical potential for molecular modeling and dynamics studies of proteins. *J. Phys. Chem. B* **102**, pp. 3586–3616
32. W. van Gunsteren (1987) H. Berendsen, Gromos-87 manual, *Biomos BV*, Groningen, The Netherlands
33. W. L. Jorgensen, D. S. Maxwell, J. Tirado-Rives (1996) Development and testing of the opls all-atom force field on conformational energetics and properties of organic liquids. *J. Am. Chem. Soc.* **118**, pp. 11225–11236
34. J. Jorgensen, W.L.and Chandrasekhar, J. Madura, R. Impey, M. Klein (1983) Comparison of simple potential functions for simulating liquid water. *J. Chem. Phys.* **79**, pp. 926–935
35. H. J. C. Berendsen, J. P. M. Postma, W. F. van Gunsteren, J. Hermans (1981) Intermolecular Forces, D. Reidel Publishing Company, Dordrecht, Ch. Interaction models for water in relation to protein hydration, pp. 331–342
36. J. Banks, G. Kaminski, R. Zhou, D. Mainz, B. Berne, R. Friesner (1999) Parametrizing a polarizable force field from ab initio data. i. the fluctuating point charge model. *J. Chem. Phys.* **110**, pp. 741–754
37. H. Andersen (1983) Rattle: a "velocity" version of the shake algorithm for molecular dynamics. *J. Comput. Phys.* **52**, pp. 24–34
38. B. Hess, B. Bekker (1997) H. J. C. Berendsen, J. G. E. M. Fraaije, LINCS: a linear constraints solver for molecular simulations. *J. Comp. Chem.* **18**, pp. 1463–1472
39. M. Tuckerman, G. Martyna, B. Berne (1992) Reversible multiple time scale molecular dynamics. *J. Chem. Phys.* **97**, pp. 1990–2001
40. B. Lindahl, E. Hess, D. van der Spoel (2001) Gromacs 3.0: a package for molecular simulation and trajectory analysis. *J. Mol. Mod.* **7**, pp. 306–317
41. Y. Duan, P. A. Kollman (1998) Pathways to a protein folding intermediate observed in a 1-microsecond simulation in aqueous solution. *Science* **282**, pp. 740–744
42. S. Miyamoto, P. A. Kollman (1997) SETTLE: an analytical version of the SHAKE and the RATTLE algorithms for rigid water molecules. *J. Comp. Chem.* **13**, pp. 952–962

43. T. Lazaridis, M. Karplus (1999) Effective energy function for proteins in solution. *Prot. Struct. Func. Gen.* **35** , pp. 133–152

44. D. Qiu, P. S. Shenkin, F. P. Hollinger, W. C. Still (1997) The GB/SA continuum model for solvation. a fast analytical method for the calculation of approximate born radii. *J. Phys. Chem.* A **101**, pp. 3005–3014

45. M. Y. Shen, K. F. Freed (2002) Long time dynamics of met-enkephalin: Comparison of explicit and implicit solvent models. *Biophys. J.* **82**, pp. 1791–1808

46. H. Nymeyer, A. E. Garcia (2003) Simulation of the folding equilibrium of alpha-helical peptides: A comparison of the generalized born approximation with explicit solvent. *P. Natl. Acad. Sci.* USA **100**, pp. 13934–13939

47. R. H. Zhou, B. J. Berne (2002) Can a continuum solvent model reproduce the free energy landscape of a beta-hairpin folding in water? *P. Natl. Acad. Sci.* USA **99**, pp. 12777–12782

48. H. Taketomi, Y. Ueda, N. Go (1975) Studies on protein folding, unfolding and fluctuations by computer-simulation. 1. effect of specific amino-acid sequence represented by specific inter-unit interactions. *Int. J. Pept. Protein Res.* **7**, p. 445

49. C. Clementi, A. E. Garcia, J. N. Onuchic (2003) Interplay among tertiary contacts, secondary structure formation and side-chain packing in the protein folding mechanism: All-atom representation study of protein L. *J. Mol. Biol.* **326**, pp. 933–954

50. A. Liwo, M. Khalili, H. A. Scheraga (2005) Ab initio simulations of protein-folding pathways by molecular dynamics with the united-residue model of polypeptide chains. *P. Natl. Acad. Sci.* USA **102**, pp. 2362–2367

51. S. Oldziej, C. Czaplewski, A. Liwo, M. Chinchio, M. Nanias, J. A. Vila, M. Khalili, Y. A. Arnautova, A. Jagielska, M. Makowski, H. D. Schafroth, R. Kazmierkiewicz, D. R. Ripoll, J. Pillardy, J. A. Saunders, Y. K. Kang, K. D. Gibson, H. A. Scheraga, (2005) Physics-based protein-structure prediction using a hierarchical protocol based on the unres force field: Assessment in two blind tests. *P. Natl. Acad. Sci.* USA **102**, pp. 7547–7552

52. K. A. Dill, S. Bromberg, K.Yue, K. M. Fiebig, D. P. Yee, P. Thomas, H. S. Chan (1995) Principles of protein-folding - a perspective from simple exact models. *Prot. Science* **4**, pp. 561–602

53. H. S. Chan, K. A. Dill (1998) Protein folding in the landscape perspective: Chevron plots and non-arrhenius kinetics. *Proteins* **30**, pp. 2–33

54. P. de Gennes (1979) Scaling Concepts in Polymer Physics, Cornell University Press, Ithaca NY

55. S. Miyazawa, R. Jernigan (1985) Estimation of effective interresidue contact energies from protein crystal-structures - quasi-chemical approximation. *Macromolecules* **18**, p. 534

56. I. Coluzza, H. G. Muller, D. Frenkel (2003) Designing refoldable model molecules. *Phys. Rev.* E **68**, p. 046703

57. S. S. Plotkin, J. N. Onuchic (2002) Understanding protein folding with energy landscape theory - part ii: Quantitative aspects. *Q. Rev. Biophys.* **35**, pp. 205–286

58. I. Coluzza, D. Frenkel (2005) Designing specificity of protein-substrate interactions. *Phys. Rev.* E **70**, p. 051917

59. I. Coluzza, S. van der Vies, D. Frenkel (2006) Translocation boost protein-folding efficiency of double-barreled chaperonins. *Biophys. J.* **90**, pp. 3375–3381

60. A. Ferrenberg, R. Swendsen (1989) Optimized monte-carlo data-analysis. *Phys. Rev. Lett.* **63**, pp. 1195–1198

61. A. Laio, M. Parrinello (2002) Escaping free-energy minima. *P. Natl. Acad. Sci. USA* **99**, pp. 12562-12567

62. H. Grubmüller (1995) Predicting slow structural transitions in macromolecular systems - conformational flooding. *Phys. Rev. E* **52**, pp. 2893–2906

63. A. Voter (1997) Hyperdynamics: Accelerated molecular dynamics of infrequent events. *Phys. Rev. Lett.* **78**, pp. 3908–3911

64. B. Berg, T. Neuhaus (1992) Multicanonical ensemble - a new approach to simulate 1st-order phase-transitions. *Phys. Rev. Lett.* **68**, pp. 9–12

65. A. Mitsutake, Y. Sugita, Y. Okamoto (2001) Generalized-ensemble algorithms for molecular simulations of biopolymers. *Biopolymers* **60**, pp. 96–123

66. E. J. Sorin, V. S. Pande (2005) Exploring the helix-coil transition via all-atom equilibrium ensemble simulations. *Biophys. J.* **88**, pp. 2472–2493

67. P. Liu, B. Kim, R. Friensner, B. Berne (2005) Replica exchange with solute tempering: A method for sampling biological systems in explicit water. *P. Natl. Acad. Sci. USA* **102**, pp. 13749–13754

68. I. Coluzza, D. Frenkel (2005) Virtual-move parallel tempering. *Phys. Chem. Phys* **6**, pp. 1779–1783

69. D. Frenkel (2004) Speed-up of monte carlo simulations by sampling of rejected states. *P. Natl. Acad. Sci. USA* **101**, pp. 17571–17575

70. P. Ferrara, J. Apostolakis, A. Caflisch (2000) Thermodynamics and kinetics of folding of two model peptides investigated by molecular dynamics simulations. *J.Phys. Chem. B.* **104**, pp. 5000–5010

71. V. S. Pande, I. Baker, J. Chapman, S. P. Elmer, S. Khaliq, S. M. Larson, Y. M. Rhee, M. R. Shirts, C. D. Snow, E. J. Sorin, B. Zagrovic (2003) Atomistic protein folding simulations on the submillisecond time scale using worldwide distributed computing. *Biopolymers* **68**, pp. 91–109

72. D. A. C. Beck, V. Dagget (2004) Methods for molecular dynamics simulation of protein folding/unfolding in solution. *Methods* **34**, pp. 112–120

73. N. Ferguson, R. Day, C. M. Johnson, M. D. Allen, V. Dagget, A. Fersht (2005) Simulation and experiment at high temperatures: Ultrafast folding of a thermophilic protein by nucleation-condensation. *J. Mol. Biol.* **347**, pp. 855–870

74. V. S. Pande, D. S. Rokhsar (1999) Molecular dynamics simulations of unfolding and refolding of a beta-hairpin fragment of protein G. *P. Natl. Acad. Sci. USA* **96**, pp. 9062–9067

75. D. Chandler (1978) Statistical mechanics of isomerization dynamics in liquids and the transition state. *J. Chem. Phys.* **68**, pp. 2959–2970

76. C. H. Bennett (1977) Molecular dynamics and transition state theory: the simulation of infrequent events, in: R. Christofferson (Ed.), Algorithms for Chemical Computations, ACS Symposium Series No. 46, American Chemical Society, Washington, D.C., pp. 63–97

77. C. Dellago, P. G. Bolhuis, P. L. Geissler (2002) Transition path sampling. *Adv. Chem. Phys.* **123**, pp. 1–78

78. R. Du, V. S. Pande, A. Y. Grosberg, T. Tanaka, E. S. Shakhnovich (1998) On the transition coordinate for protein folding. *J. Chem. Phys.* **108**, pp. 334–350

79. Y. Rhee, V. Pande (2005) One-dimensional reaction coordinate and the corresponding potential of mean force from commitment probability distribution. *J. Phys. Chem. B* **109** , pp. 6780–6786

80. L. R. Pratt (1986) A statistical-method for identifying transition-states in high dimensional problems. *J. Chem. Phys.* **85**, p. 5045

81. R. Olender, R. Elber (1996) Calculation of classical trajectories with a very large time step: Formalism and numerical examples. *J. Chem. Phys.* **105**, pp. 9299–9315

82. R. Elber, A. Ghosh, A. Cardenas, H. Stern (2003) Bridging the gap between long time trajectories and reaction pathways. *Adv. Chem. Phys.* **126**, pp. 93–129

83. P. Eastman, N. Gronbech-Jensen, S. Doniach (2001) Simulation of protein folding by reaction path annealing. *J. Chem. Phys.* **114**, pp. 3823–3841

84. P. G. Bolhuis, D. Chandler, C. Dellago, P. L. Geissler (2002) Transition path sampling: Throwing ropes over rough mountain passes, in the dark. *Annu. Rev. Phys. Chem.* **53**, pp. 291–318

85. P. G. Bolhuis (2003) Transition path sampling on diffusive barriers. *J. Phys.-Condens. Mat.* **15**, pp. S113–S120

86. H. C. Andersen (1980) Molecular dynamics simulations at constant pressure and/or temperature. *J. Chem. Phys.* **72**, pp. 2384–2389

87. T. S. van Erp, D. Moroni, P. G. Bolhuis (2003) A novel path sampling method for the calculation of rate constants. *J. Chem. Phys.* **118**, pp. 7762–7774

88. P. G. Bolhuis (2005) Kinetic pathways of beta-hairpin (un)folding in explicit solvent. *Biophys. J.* **88**, pp. 50–61

89. D. Moroni, P. G. Bolhuis, T. S. van Erp (2004) Rate constants for diffusive processes by partial path sampling. *J. Chem. Phys.* **120**, pp. 4055–4065

90. A. Faradjian, R. Elber (2004) Computing time scales from reaction coordinates by milestoning. *J. Chem. Phys.* **120**, pp. 10880–10889

91. N. Singhal, C. D. Snow, V. S. Pande (2004) Using path sampling to build better markovian state models: Predicting the folding rate and mechanism of a tryptophan zipper beta hairpin. *J. Chem. Phys.* **121**, pp. 415–425

92. N. M. Amato, G. Song (2002) Using motion planning to study protein folding pathways. *J. Comput. Biol.* **9**, pp. 149–168

93. N. M. Amato, K. A. Dill, G. Song (2003) Using motion planning to map protein folding landscapes and analyze folding kinetics of known native structures. *J. Comput. Biol.* **10**, pp. 239–255

94. S. V. Krivov, M. Karplus (2004) Hidden complexity of free energy surfaces for peptide (protein) folding. *P. Natl. Acad. Sci. USA* **101**, pp. 14766–14770

95. A. B. Bortz, M. H. Kalos, J. L. Lebowitz (1975) New algorithm for monte-carlo simulation of ising spin systems. *J. Comput. Phys.* **17**, p. 10

96. V. Munoz, P. A. Thompson, J. Hofrichter, W. A. Eaton (1997) Folding dynamics and mechanism of beta-hairpin formation. *Nature* **390**, pp. 196–199

97. V. Munoz, E. R. Henry, J. Hofrichter, W. A. Eaton (1998) A statistical mechanical model for beta-hairpin kinetics. *P. Natl. Acad. Sci. USA* **95**, pp. 5872–5879

98. A. Kolinski, B. Ilkowski, J. Skolnick (1999) Dynamics and thermodynamics of beta-hairpin assembly: Insights from various simulation techniques. *Biophys. J.* **77**, pp. 2942–2952

99. D. K. Klimov, D. Thirumalai (2000) Mechanisms and kinetics of beta-hairpin formation. *P. Natl. Acad. Sci. USA* **97**, pp. 2544–2549

100. G. H. Wei, P. Derreumaux, N. Mousseau (2004) Complex folding pathways in a simple beta-hairpin. *Prot. Struct. Func. Bio.* **56**, pp. 464–474

101. A. R. Dinner, T. Lazaridis, M. Karplus (1999) Understanding beta-hairpin formation. *P. Natl. Acad. Sci. USA* **96**, pp. 9068–9073

102. B. Zagrovic, E. Sorin, V. S. Pande (2001) Beta-hairpin folding simulations in atomistic detail using an implicit solvent model. *J. Mol. Biol.* **313**, pp. 151–169

103. D. Roccatano, A. Amadei, A. Di Nola, H. J. C. Berendsen (1999) A molecular dynamics study of the 41-56 beta-hairpin from B1 domain of protein G. *Protein Sci.* **8**, pp. 2130–2143

104. B. Y. Ma, R. Nussinov (2000) Molecular dynamics simulations of a beta-hairpin fragment of protein G: Balance between side-chain and backbone forces. *J. Mol. Biol.* **296**, pp. 1091–1104

105. A. E. Garcia, K. Y. Sanbonmatsu (2001) Exploring the energy landscape of a beta hairpin in explicit solvent. *Proteins* **42**, pp. 345–354

106. R. H. Zhou, B. J. Berne, R. Germain (2001) The free energy landscape for beta hairpin folding in explicit water. *P. Natl. Acad. Sci.* USA **98**, pp. 14931–14936

107. J. Tsai, M. Levitt (2002) Evidence of turn and salt bridge contributions to beta-hairpin stability: MD simulations of C-terminal fragment from the B1 domain of protein G. *Biophys. Chem.* **101**, pp. 187–201

108. F. B. Sheinerman, C. L. Brooks (1998) Calculations on folding of segment B1 of streptococcal protein G. *J. Mol. Biol.* **278**, pp. 439–456

109. F. B. Sheinerman, C. L. Brooks (1998) Molecular picture of folding of a small alpha/beta protein. *P. Natl. Acad. Sci.* USA **95**, pp. 1562–1567

110. M. S. Cheung, A. E. Garcia, J. N. Onuchic (2002) Protein folding mediated by solvation: Water expulsion and formation of the hydrophobic core occur after the structural collapse. *P. Natl. Acad. Sci.* USA **99**, pp. 685–690

Calculation of Classical Trajectories with Boundary Value Formulation

R. Elber

Department of Computer Science, Cornell University, 4130 Upson Hall, Ithaca NY 14853
ron@cs.cornell.edu

Ron Elber

Ron Elber: *Calculation of Classical Trajectories with Boundary Value Formulation*, Lect. Notes Phys. **703**, 435–451 (2006)
DOI 10.1007/3-540-35273-2_12   © Springer-Verlag Berlin Heidelberg 2006

An algorithm to compute classical trajectories using boundary value formulation is presented and discussed. It is based on an optimization of a functional of the complete trajectory. This functional can be the usual classical action, and is approximated by discrete and sequential sets of coordinates. In contrast to initial value formulation, the pre-specified end points of the trajectories are useful for computing rare trajectories. Each of the boundary-value trajectories ends at desired products. A difficulty in applying boundary value formulation is the high computational cost of optimizing the whole trajectory in contrast to the calculation of one temporal frame at a time in initial value formulation.

1 Introduction

Calculation of classical trajectories of molecular systems is a powerful tool for studying the thermodynamics and the kinetics of biophysical problems. Indeed the availability of numerous computer packages and journals devoted to such simulations suggests that these studies are extremely popular, and are influencing many fields. It is therefore no surprise that a number of chapters in this book are devoted to the use of classical dynamics in condensed matter physics.

The emphasis in other chapters of this book is on initial value solution of classical equations of motion (e.g. the Newton's equations). The Newton's equations are second order differential equations – $M\ddot{X} = -dU/dX$, where X $(X \in R^{3N})$ is the coordinate vector. Throughout this chapter X is assumed to be a Cartesian vector, M is a $3N \times 3N$ (diagonal) mass matrix, and U is the potential energy. A widely used algorithm that employs the coordinates and the velocities (V) at a specific time and integrates the equations of motion in small time steps is the Verlet algorithm [1]:

$$
\begin{aligned}
X_{i+1} &= X_i + V_i \cdot \Delta t - \left(\Delta t^2/2M\right) \cdot (dU/dX_i) \\
V_{i+1} &= V_i - \left(\Delta t^2/2M\right) \cdot (dU/dX_i + dU/dX_{i+1})
\end{aligned}
\tag{1}
$$

where i is the index of the integration step, Δt is the step size, and the division by the matrix, means the inverse. There are two limitations of this popular initial value formulation that are worth mentioning: (i) The time step, Δt, must be small to maintain numerical stability, and (ii) It is difficult to end a trajectory solved with an initial value approach in a pre-determined and desired final state.

The small size of the time step restricts the calculation to rapid processes. Even for fast molecular events (nanoseconds) millions of femtosecond steps are required to reach the nanosecond time scale. Longer time scales are important for simulating many biophysical events and are not reachable with routine calculations (e.g. rapid protein folding occurs at the microseconds and milliseconds time scales).

The second limitation is more subtle. In principle, if the final state is strongly attractive with a large radius of convergence, we anticipate that the trajectory will quickly stop at the end state, making the sampling of reactive trajectories relatively easy. This is a key idea behind the calculation of the transmission coefficient [2] and the transition path sampling [3]. These techniques initiate trajectories at improbable configurations that are intermediates between the reactants and products. By integrating the trajectory in the forward and backward directions to the two attractive end states it is possible to generate a complete reactive trajectory. Nevertheless, if the desired final states are not strong attractors it becomes difficult to sample reactive trajectories.

A complementary approach to the initial value formulation is the calculation of trajectories as a solution of a boundary value problem. In this approach the two end points are given as input to the calculations. It is therefore obvious that the trajectories will end at the pre-set interesting configurations. This simple construction solves the second limitation on initial value calculations that we mentioned above. Of course if the end points are not known and only the beginning configuration is available (e.g. the protein folding problem), then the initial value approach is the only viable option.

The second limitation is harder to tackle rigorously, however as we discuss in Sects. 3.3 and 3.4, the boundary value approach can be used to suggest approximate solutions with much larger time steps. Increasing the time step is much harder to achieve with initial value formulation.

2 Basic Theory

We consider the classical action S [4] for a collection of N particles, defined by

$$S = \int_0^t \left(\frac{1}{2} \dot{X}^t M \dot{X} - U(X) \right) d\tau \equiv \int_0^t (T - U) d\tau \equiv \int_0^t L d\tau \qquad (2)$$

The coordinates are stored in the vector X (X^t denotes a transposed vector) and throughout this manuscript we use Cartesian coordinates only. A dot denotes a time derivative. The mass matrix M is diagonal, T is the kinetic energy, U is the potential energy, and L is the Lagrangian. We seek trajectories such that the total time, t, and the end points of the trajectories, $X(0)$ and $X(t)$, are fixed, and the action is stationary with respect to path variations. With the above conditions the Newton's equations of motion are obtained by a standard variation of the classical path [4]. Let $\eta(\tau)$ be an arbitrary displacement vector from a path, $X(\tau)$. The stationary condition of the action is obtained from the expression below

$$S\left[X\left(\tau\right)+\eta\left(\tau\right)\right] - S\left[X\left(\tau\right)\right] = \int_{0}^{t} \left(\frac{1}{2}\left(\dot{X}^{t}+\dot{\eta}^{t}\right)M\left(\dot{X}+\dot{\eta}\right) - U\left(X+\eta\right)\right)d\tau$$

$$- \int_{0}^{t}\left(\frac{1}{2}\left(\dot{X}^{t}\right)M\left(\dot{X}\right) - U\left(X\right)\right)d\tau \qquad (3)$$

To the first order in the displacement η we have

$$S\left[X\left(\tau\right)+\eta\left(\tau\right)\right] - S\left[X\left(\tau\right)\right] \cong \int_{0}^{t}\left(\dot{X}^{t}M\dot{\eta} - [dU\left(X\right)/dX]^{t}\cdot\eta\right)d\tau = 0 \quad (4)$$

Integrating by parts $\int_{0}^{t}\dot{X}^{t}M\dot{\eta}\cdot d\tau = [\dot{X}^{t}M\eta]_{0}^{t} - \int_{0}^{t}\ddot{X}^{t}M\eta\cdot d\tau$, and setting the variation on the boundaries to zero ($\eta\left(0\right)=\eta\left(t\right)=0$) we have for an arbitrary variation η

$$S\left[X\left(\tau\right)+\eta\left(\tau\right)\right] - S\left[X\left(\tau\right)\right] \cong - \int_{0}^{t}\left(\ddot{X}^{t}\cdot M + [dU\left(X\right)/dX]^{t}\right)\cdot\eta d\tau = 0$$

$$\Rightarrow M\ddot{X} + \frac{dU}{dX} = 0 \qquad (5)$$

This is (of course) the expression for the Newton's equations of motion. The above derivation is well known and can be found in any textbook on mechanics (e.g. Landau and Lifshitz [4]). It shows the equivalence of the variation principle and the usual differential form of the equations of motion. It is also the basis for introducing initial value numerical solvers, using a finite difference to represent the second derivatives with respect to time, for example

$$M\frac{X_{i+1} + X_{i-1} - 2X_{i}}{\Delta t^{2}} + \frac{dU}{dX_{i}} = 0 \qquad (6)$$

We obtain an initial value solver by exchanging variables on both sides of the equation to give $X_{i+1} = 2X_{i} - X_{i-1} - M^{-1}\Delta t^{2}\frac{dU}{dX}$. This equation can be solved in steps. Given X_{i-1} and X_{i} we can produce X_{i+1}, and so on.

It is useful to introduce another analytical variant of the classical action that suggests a different perspective on the boundary value formulation of classical mechanics and a corresponding initial value solver. We consider a system of a constant total energy $E = (T+U)$. We write $U = E - T$ and substitute in (2) $S = \int_{0}^{t}\left(2T - E\right)d\tau = \int_{0}^{t}2T\cdot d\tau - Et$. Define $Q = M^{1/2}X$, mass weighted coordinates, which we now use in the expression for the kinetic energy $\int_{0}^{t}\dot{X}^{t}M\dot{X}\cdot d\tau = \int_{0}^{t}\dot{Q}^{2}\cdot d\tau$. The last expression can also be written as $\int_{Q(0)}^{Q(t)}\dot{Q}dQ$ ($dQ \equiv \dot{Q}d\tau$). Collecting the pieces, we now have a new expression for the action

$$S = \int\limits_{Q(0)}^{Q(t)} \dot{Q} dQ - Et \tag{7}$$

Note that $\dot{Q}$ and dQ are parallel vectors. Therefore their inner product $\dot{Q} \cdot dQ$ can be written as $|\dot{Q}| \cdot |dQ|$ which allows for the final twist of (7). We return to the expression of the energy and write

$$E = \frac{1}{2}\dot{Q}^t \dot{Q} + U(Q) \;\Rightarrow\; \left|\dot{Q}\right| = \sqrt{2(E - U(Q))} \tag{8}$$

Our new expression for the action therefore becomes

$$S = \int\limits_{Q(0)}^{Q(t)} \sqrt{2(E - U(Q))} dQ - Et \tag{9}$$

The boundary conditions of (9) are the same as of the usual classical action in (2) (fixed end points and total time). However, separation of variables is now self-evident. The second term on the right hand side of the equation (Et) is independent of the coordinates (the energy is a constant). Similarly the first term is independent of time. In fact, it is not necessary to write $dQ = \dot{Q} d\tau$. Any parameterization of the trajectory with respect to a scalar variable s which is monotonic between the points $Q(0)$ and $Q(t)$ will do ($dQ = (dQ/ds)\,ds$). It means that variation of the integral part of the right hand side of (9) with respect to the coordinates can be done with no reference to time. We therefore consider a variation with the abbreviated action [5]

$$S_{abbrev} = \int\limits_{Q_{init}}^{Q_{final}} \sqrt{2(E - U(Q))} dQ \tag{10}$$

One should keep in mind that the new abbreviated action poses no constraint on time. Another way of thinking about the elimination of time is to shift the total energy by a constant, C, such that $E' = E - C = 0$. In this case (9) becomes identical to (10). Note that the potential energy U is shifted in a similar way to the total energy $U' = U - C$. This keeps the value of $E - U$ unchanged. The time is arbitrary and is no longer fixed as in the usual definition of the action. The time can be computed in retrospect once the stationary path is obtained using the standard classical mechanics formula $t = \int_{Q_{init}}^{Q_{final}} \frac{d|Q|}{\sqrt{2(E-U(Q))}}$ (the integration is over the previously determined classical path in $3N$ space).

The variation of S_{abbrev} is done in a similar way to the variation of the usual classical action. The parameterization of Q ($Q(s)$) we have in mind is in terms of the arc length $ds = \sqrt{dQ^t dQ}$

$$S_{abbrev}\left[Q + \eta\right] - S_{abbrev}\left[Q\right] = \int\limits_{Q_{init}}^{Q_{final}} \sqrt{2\left(E - U\left(Q + \eta\right)\right)}d\sqrt{\left(Q + \eta\right)^t\left(Q + \eta\right)}$$

$$- \int\limits_{Q_{init}}^{Q_{final}} \sqrt{2\left(E - U\left(Q\right)\right)}\sqrt{dQ^t \cdot dQ} \qquad (11)$$

The expansion of the length element to the first order in η is as follows

$$\sqrt{d\left(Q + \eta\right)^t \cdot d\left(Q + \eta\right)} \cong \sqrt{dQ^t \cdot dQ + 2d\eta^t \cdot dQ} = ds\left(1 + \frac{d\eta^t}{ds} \cdot \frac{dQ}{ds}\right)$$

where $\frac{dQ}{ds} \equiv e_Q$ is a unit vector in the direction of the path Q.

We now have

$$S_{abbrev}\left[Q + \eta\right] - S_{abbrev}\left[Q\right] \cong \int\limits_0^{s_f} \frac{-\eta^t \cdot dU/dQ}{\sqrt{2\left(E - U\left(Q\right)\right)}}ds + \int\limits_0^{s_f} \sqrt{2\left(E - U\left(Q\right)\right)}\frac{d\eta}{ds}\frac{dQ}{ds}ds$$

$$= \int\limits_0^{s_f}\left(\frac{-\eta^t \cdot dU/dQ}{\sqrt{2\left(E - U\left(Q + \eta\right)\right)}} + \frac{\left(e_Q^t \cdot dU/dQ\right)\left(\eta^t \cdot e_Q\right)}{\sqrt{2\left(E - U\left(Q + \eta\right)\right)}} + \sqrt{2\left(E - U\left(Q\right)\right)}\frac{de_Q^t}{ds} \cdot \eta\right)ds$$

and we finally write an equation of motion in s which is the arc-length of the classical path

$$\frac{d^2Q}{ds^2} = -\frac{1}{2\left(E - U\left(Q\right)\right)}\left(\frac{dU}{dQ} - \left(e_Q^t \cdot \frac{dU}{dQ}\right)\left(e_Q\right)\right) \qquad (12)$$

This equation is similar to the Newton's equation (remember that Q is weighted by the mass) except that the component of the force in the direction perpendicular to the path is projected out and the mass is replaced by twice the kinetic energy. From the differential equation a finite difference formula for Q as a function of s can be obtained similarly to the initial difference formula we have for X as a function of the time t. This equation is not so popular with initial value solvers since the term $E - U$ can go to zero, or becomes (numerically) even negative causing significant implementation problems.

If we multiply (10) by e_Q^t from the left we obtain

$$e_Q^t \cdot \frac{d^2Q}{ds^2} = \frac{dQ}{ds}\frac{d^2Q}{ds^2} = \frac{1}{2}\frac{d}{ds}\left(\frac{dQ}{ds}\right)^2 = 0 \rightarrow |e_Q| = \text{constant}$$

The last result is however not new since (by definition!) $\left|\frac{dQ}{ds}\right| \equiv |e_Q| = 1$. Hence, the equations of motion (12) are only for the component of the coordinate vector that is perpendicular to the path. The motion along the path is determined by the definition of s.

3 Numerical Algorithm

The two equations of motion we discussed so far can be found in classical mechanics text books (the differential equations of motion as a function of the arch-length can be found in [5]). Amusingly, the usual derivation of the initial value equations starts from boundary value formulation while a numerical solution by the initial value approach is much more common. Shouldn't we try to solve the boundary value formulation first? As discussed below the numerical solution for the boundary value representation is significantly more expensive, which explains the general preference to initial value solvers. Nevertheless, there is a subset of problems for which the boundary value formulation is more appropriate. For example boundary value formulation is likely to be efficient when we probe paths connecting two known end points.

An algorithm to solve the boundary value formulation uses the local stationary conditions on the action and seeks a simultaneous solution for the coordinates at all times. Consider a set of discrete coordinates that we use to approximate the trajectory $\{X_1, ..., X_N\}$ where the initial and final coordinate sets of the trajectory, X_1 and X_N, are fixed. We have $N - 2$ (possibly non-linear) coupled equations for the coordinates of the remaining time slices based on the local stationary conditions

$$M\frac{2X_i - X_{i+1} - X_{i-1}}{\Delta t^2} + \frac{dU}{dX_i} \equiv r_i = 0 , \quad i = 2, \ldots, N - 1 \qquad (13)$$

The set of unknown are the X_i-s, and the goal is to set all the r_i-s (the residuals) to zero. The same condition is obtained if we first discretize the classical action and consider a stationary condition of the discrete classical action as a function of the intermediate coordinates

$$S = \sum_{i=1}^{N-1} \left[M\frac{(X_{i+1} - X_i)^2}{2\Delta t^2} - U(X_i) \right]^2 \Delta t \qquad (14)$$

The difference (13) is recovered from (14) by the requirement that the action will be stationary with respect to all intermediate coordinates X_i, $i = 2, \ldots, N - 1$, i.e. $\partial S / \partial X_i = 0$

3.1 The Action is not a Minimum

It is important to emphasize that the stationary solution of the action functional (the classical trajectory) is not necessarily a minimum. This means that a straightforward minimization of S as a function of all the intermediate coordinates is not possible in the general case. The action can be optimized if the matrix of its second derivatives with respect to the intermediate coordinates is definite (positive or negative) to support direct minimization of S or $-S$. As a simple example that S need not be definite, we consider a one-dimensional case and compute the second derivatives of the discrete action in (14) with respect to the intermediate coordinates

$$\frac{\Delta t}{M}\frac{\partial^2 S}{\partial X_j \partial X_k} = 2\delta_{j,k} - \delta_{j,k+1} - \delta_{j,k-1} - \frac{\Delta t^2}{M}\frac{\partial^2 U}{\partial X_j \partial X_k}\delta_{j,k} \qquad (15)$$

The last term on the right hand side of the equation is diagonal in the time slices. For stable systems (convex potentials) such as the harmonic oscillator, $(U(X) = \frac{1}{2}k(X - X_{eq})^2)$, it is negative. The first three terms form a tri-diagonal matrix proportional to a discrete representation of the operator $\frac{d^2}{dt^2}$. The eigenvectors of this operator are of the form: $A\exp\left(-i\lambda j\right)$ where A and λ are constants and $i \equiv \sqrt{-1}$. The eigenvalues are $2\left(1 - \cos\left(\lambda\right)\right)$. The eigenvalues of the tri-diagonal matrix vary from zero to four (the precise spread depends on the boundary conditions and the discretization of time). On the other hand the last diagonal term (the potential derivatives) tends to be negative in stable systems. Therefore the sum of the time derivative operator and the diagonal term has the potential of creating a non-definite matrix. This is indeed the case for numerous applications. We cannot be more specific since the relative size of the two terms is related to concrete choice of potential and to the total time under consideration.

A transition from a definite to a non-definite matrix occurs even for the simple case of the harmonic oscillator as a function of the total time. In general, the shorter the time, the smaller the term with the potential derivatives and the matrix as a whole is more positive. At sufficiently long time we expect some of the eigenvalues to reverse their sign and become negative, making the matrix indefinite.

3.2 Minimization of a Target Function

The above discussion suggests that we cannot minimize the action directly to compute classical trajectories. We focus instead on the residuals (13). One boundary value formulation that we frequently use minimizes the squares of the residual vectors, i.e., we define a target function T that we wish to minimize as a function of all the intermediate coordinates $X_j, j = 2, ..., N-1$.

$$T = \sum_{j=2,...,N-1} r_j^t \cdot r_j \qquad (16)$$

The global minimum of T $(T = 0)$ is our approximation to a classical trajectory for a finite step size Δt. A typical calculation of a trajectory will start with an initial guess, say X_j^0 $j = 2, \ldots, N-1$ which in many applications we set to a straight line in Cartesian space, i.e. $X_j = X_1 + \frac{j-1}{N-1} \cdot (X_N - X_1)$. The optimization proceeds by optimization of the target function. For example using over-damped dynamics, we solve the following coupled stochastic equations

$$\gamma\frac{dX_j}{d\tau} = -\nabla T_j + R_j , \quad j = 2, ...N - 1$$
$$\langle R_i \rangle = 0 \quad \langle R_i(0) R_i(\tau) \rangle = 4\gamma k_B \theta \delta(\tau) \tag{17}$$

with the initial condition $\{X_j(\tau = 0) = X_j^0\}$. In the above equation τ is a fictitious time (with units of $1/\gamma$ - fictitious friction) that indexes the relaxation of the whole trajectory until the residuals are fully optimized (set to zero). R_i is a white-noise random force chosen to satisfy the dissipation-fluctuation theorem, the relation in the second line of (17). In the simulation we gradually decrease the temperature, $k_B\theta$, from a high initial value to zero. Note that this annealing is for the whole classical trajectories and we simulated simultaneously changes in $N - 2$ structures. This is more expensive that initial value formulation in which one structure is changing at a time. However, the algorithm of (17) brings us to a desired product and is considerably more stable as a function of integration step size.

There are important differences between the initial value formulation and the boundary value approach. Initial value solutions are based on interpolation forward in time one coordinate set after another. The boundary value approach is based on minimization of a target function of the whole trajectory. Minimization (and the study of a larger system) is more expensive in the boundary value formulation compared to initial value solver. However, the calculations of state to state trajectories and the abilities to use approximations (next section), make it a useful alternative for a large number of problems.

3.3 Large Step in Time

It is instructive to consider the solution of the coupled (13) for the harmonic oscillator demonstrating an important difference between the initial value solver and boundary value formulation. For simplicity we consider a harmonic oscillator with two degrees of freedom $Y^{(1)} \equiv X^{(1)} - X^{(1eq)}$ and $Y^{(2)} \equiv X^{(2)} - X^{(2eq)}$ where $X^{(1eq)}$ and $X^{(2eq)}$ are the equilibrium positions. The corresponding force constants are $c^{(1)}$ and $c^{(2)}$ respectively. We are particularly interested in the case of $c^{(1)} \gg c^{(2)}$.

$$M\frac{2Y_j^{(k)} - Y_{j+1}^{(k)} - Y_{j-1}^{(k)}}{\Delta t^2} - c^{(k)}Y_j^{(k)} = r_j^{(k)} = 0 , \tag{18}$$
$$j = 2, ..., N - 1 , \quad k = 1, 2$$

The index k refers to different degrees of freedom and the index j to time slices.

It is possible to solve the coordinates $Y_j^{(k)}$ at different times as independent variables (like we do in practical calculation with boundary value formulation). However, for the stability analysis below it is useful to consider a solution of the type $Y_j^{(k)} = A_k \exp(i\lambda^{(k)}j)$, we have

$$A^{(k)} \left[2 - 2\cos\left(\lambda^{(k)}\right) - \frac{c^{(k)} \cdot \Delta t^2}{M} \right] = 0 \text{ or } \cos\left(\lambda^{(k)}\right) = \left(1 - \frac{\Delta t^2 c^{(k)}}{2M}\right).$$

For a small Δt we can find an appropriate solution for $\lambda^{(k)}$. Substituting it in the expression for $Y_j^{(k)}$ provides a sound approximation to the time evolution of the coordinates. The amplitudes $A^{(k)}$ and its conjugate solution are determined from the boundary conditions.

If the time step is not small (such that $\Delta t^2 > 2M/c^{(1)}$) an initial value solver becomes unstable. A solution to the finite difference equation can be found in which $\lambda^{(1)}$ is imaginary, and the cosine function is replaced by a hyperbolic cosine. This solution is not only a poor approximation but it is also numerically unstable. Using initial value solvers the above solution leads to exponential (unbound) growth of the coordinate as a function of time $(Y_j^{(1)} = A\exp(\lambda^{(1)}j))$.

Since we assume different force constants, we can still have an accurate and stable solution for the second mode if $\Delta t^2 \ll 2M/c^{(2)}$. However, even a small coupling between the degrees of freedom (as expected in practice) will make the solution for the two coordinates unstable. In typical simulations instability developed in one or a few coordinates rapidly spreads to the rest of the degrees of freedom and cannot be ignored.

Interestingly the behavior of the fast mode is different in boundary value formulation, resulting in a stable solution. In boundary value formulation we have two end coordinates with sensible values that are hard to fit to an exponentially growing function. A solution of the form $Y(t) = A\exp(\lambda t)$ is not likely to satisfy boundary value conditions with $Y(0)$ and $Y(t)$ bound by $a \cdot A$ where a is of order of 1. The minimization of the target function T finds a likely minimum for the residuals (and the trajectory) in which the amplitude is zero ($A = 0$) for all intermediate points. The solution is (of course) not exact since rapid oscillations have been removed. However, it is numerically stable which is surprising for a step size comparable or even larger than the (fast) oscillation period. It is also expected to be quite accurate if the coupling between the fast and the slow modes is rather weak [6] as is the case (for example) for fast bond vibrations in comparison to the slow overall diffusion of biological molecules.

Setting to zero rapid oscillations that are difficult to integrate accurately is one model of the dynamics. Another model is to use mathematical white noise, attempting to capture the influence of the rapid oscillations on the slow modes. The finite difference equation is replaced by

$$M\frac{2Y_j^{(k)} - Y_{j+1}^{(k)} - Y_{j-1}^{(k)}}{\Delta t^2} - c^{(k)}Y_j^{(k)} = r_j^{(k)} = R_j^k, \quad j = 2, \ldots, N-1. \quad (19)$$

The deviations from zero are modeled as Gaussian white noise such that $\langle R_j^k \rangle = 0$ and $\langle R_j^k R_m^k \rangle = b\delta_{jm}$ where b is a constant. This procedure was used in the past to sample and optimize trajectories [7,8].

We noted above that if the coupling between the fast and slow modes is significant then the filtering may impact the slow modes as well. Consider an activated process, a transition from a deep minimum A to another deep minimum B passing over a significant energy barrier. This is a classical problem which is deceptively simple. The transition is slow on the average since the process is rare. However, both the vibrations within the well, and the time of transition over the barrier are fast. The application of the "filter" mentioned above will remove all the rapid oscillations. The filtered discrete trajectory will have some points "sitting" at the minimum of A and some points "sitting" at the minimum of B. Hence, all the intrinsic dynamics are washed away if the integration time step, Δt, is greater than both the local vibration inverse frequency and the transition time scale. This is regardless of the additional (hidden) time scale in the system (the fact that the process is rare). It is therefore clear that the filtering of high frequencies in time is a useful approach to describe dynamics of slow modes, as long as slow modes with interesting motions and relatively weak coupling to the fast modes can be found in the system.

The length formulation of the classical action is better suited to describe transitions between the wells with large integration steps. This is since the overall distance between the states A and B is large in the length representation (even if it is short in time). The large overall distance is captured by the arc length formulation even if the typical integration step in length is not small. This is to be contrasted with a large time step that misses in this case the rapid transition time.

3.4 Large Step in Length

An interesting large step limit can be found in the length formulation. Consider the finite-difference residual-equation as a function of the arc-length.

$$\frac{Q_{s+1} + Q_{s-1} - 2Q_s}{\Delta s^2} =$$
$$-\frac{1}{2\left(E - U\left(Q_s\right)\right)}\left(\frac{dU}{dQ_s} - \left(\left(\frac{Q_{s+1} - Q_s}{\Delta s}\right)^t \cdot \frac{dU}{dQ}\right) \cdot \frac{Q_{s+1} - Q_s}{\Delta s}\right) \quad (20)$$

with the additional requirement $|e_Q| \equiv \left|\frac{dQ}{ds}\right| \simeq \left|\frac{Q_{s+1} - Q_s}{\Delta s}\right| = 1$. Consider next the limit of a large step Δs or a small curvature such that (note that the forces are independent of the step size Δs)

$$2\left(E - U\left(Q_s\right)\right)\frac{Q_{s+1} + Q_{s-1} - 2Q_s}{\Delta s^2} \ll \frac{dU}{dQ_s} \quad (21)$$

We also require that the finite difference approximation to e_Q is a sound approximation to the path slope. Under these circumstances the residual equations are dominated by the potential derivatives and can be approximately solved as

$$\left(\frac{dU}{dQ_s} - \left(\left(\frac{Q_{s+1} - Q_s}{\Delta s}\right)^t \cdot \frac{dU}{dQ}\right) \cdot \frac{Q_{s+1} - Q_s}{\Delta s}\right) = 0 \; s = 2, \ldots, N-1 \quad (22)$$

and $\left|\dfrac{Q_{s+1} - Q_s}{\Delta s}\right| = 1.$

This is an identical equation for the minimum energy path [9, 10] or the so-called steepest descent path. The first implementation of an algorithm to compute minimum energy paths based on the above formula was the LUP (Locally Updated Planes) method [9] that did not include the constraint on the displacement size. This is formally correct since different parameterizations of the path are possible, but may lead to numerical problems in which the distances between the intermediates grow without control. This was adjusted to produce more stable algorithms by the Nudge Elastic Band approach [11] and later by the String method [10].

In summary the length formulation provides physically meaningful results at the small (exact result) and the large step (the minimum energy path) limits.

3.5 The Gauss and Onsager-Machlup Actions

We have started from expressions for the classical action, replaced them by a local differential equation, and by a local difference equation which we solve either "as is" using relaxation methods (a topic that was not discussed in this manuscript) or by minimizing a target function T ($T = \sum_{i=2,\ldots,N-1} r_i^2$). Can we "close the cycle" and find the continuous limit of the target function? The target function G can be easily written down in the limit of a small time step providing a functional of the path

$$G\left[X\left(\tau\right)\right] = \int_0^t \left(M\frac{d^2 X}{d\tau^2} + \frac{dU\left(X\right)}{dX}\right)^2 d\tau \quad (23)$$

The functional above was used already by Gauss [12] to study classical trajectories (which explains our choice of the action symbol). Onsager and Machlup used path integral formulation to study stochastic trajectories [13]. The origin of their trajectories is different from what we discussed so far, which are mechanical trajectories. However, the functional they derive for the most probable trajectories, $O\left[X\left(\tau\right)\right]$ is similar to the equation above:

$$O\left[X\left(\tau\right)\right] = \int_0^t \left(M\frac{d^2 X}{dt^2} - \gamma\frac{dX}{dt} + \frac{dU}{dX}\right)^2 d\tau \quad (24)$$

A dissipative (friction) term is the difference between the Onsager and the Gauss actions. In principle the friction can be set to zero which makes the optimal paths identical in the two actions. The optimal paths are obtained when the actions are zeroes.

These functionals are different from the usual classical action (see (2) and (10)). It is of interest to examine the variation of the Gauss action and its stationary solutions. It is clear that the global minimum of all paths of the Gauss action is when the differential equations of motion (Newton's law) are satisfied. Nevertheless, the possibility of alternative stationary solutions cannot be dismissed. This has practical ramifications since it is the Gauss action that we approximate when we minimize the sum of the residuals in (18). To the first order we have

$$
\begin{aligned}
& G\left[X\left(\tau\right)+\eta\right]-G\left[X\left(\tau\right)\right] \\
& \cong \int_0^t 2\left(M\frac{d^2X}{dt^2}+\frac{dU}{dX}\right)\left(M\frac{d^2}{dt^2}\eta+\frac{d^2U}{dX^2}\eta\right)dt
\end{aligned}
\tag{25}
$$

Integrating by part and setting η and $d\eta/dt$ to zero at the boundaries ($\tau=0$ and $\tau=t$) we obtain the following Euler Lagrange equations

$$
M\frac{d^2}{dt^2}\left(M\frac{d^2X}{dt^2}+\frac{dU}{dX}\right)+\frac{d^2U}{dX^2}\cdot\left(M\frac{d^2X}{dt^2}+\frac{dU}{dX}\right)=0 .
\tag{26}
$$

We define $C\equiv\left(M\frac{d^2X}{dt^2}+\frac{dU}{dX}\right)$, which allows us to write (26) more compactly $M\frac{d^2C}{dt^2}+\frac{d^2U}{dX^2}C=0$. The Newtonian trajectories are the special case in which C is equal to zero on the boundaries and everywhere else. There are however other solutions that may satisfy the variational principle and the boundary conditions. For illustration purposes we consider a one-dimensional harmonic oscillator for which $\frac{d^2U}{dX^2}=K$, the spring force constant. In this particular example C itself becomes a harmonic oscillator $C=A\cos(\sqrt{\frac{K}{M}}t+\varphi)\equiv A\cos(\omega t+\phi)$ where A,ω and φ are constants.

Substituting back the solution of C in the Euler Lagrange equations for X we obtain the equation for a forced harmonic oscillator.

$$
M\frac{d^2X}{dt^2}+KX=A\cos\left(\omega t+\varphi\right)
$$

Since the frequency of the external force is exactly on resonance with the free oscillator, the amplitude of the forced solution is unbound. Hence, the second stationary trajectory of the Gauss action (in addition to the Newtonian trajectory) is a solution with a growing amplitude (and energy) as a function of time. This solution is unlikely to be present in a boundary value formulation in which both end points have physically sound values. Note that the energy growth here has a different origin when compared to the numerical instability induced by a large step in the difference equation (Sect. 12.3.3).

The existence of multiple solutions (stationary paths of the action) suggests why in the discrete representation we need to use a global minimizer like simulated annealing. On the other hand the last illustration for the harmonic oscillator suggests that the behavior of the alternative trajectories is markedly different from the Newtonian's trajectory and therefore relatively easy to detect.

4 Examples

In the last ten years or so, a number of studies addressing numerous molecular processes of physical and biophysical origins were investigated using the boundary value techniques that were presented here. With the time formulation we have studied a conformational transition in glycosyltransferase [14]. This study was (and still is) the largest system for which such a reactive trajectory was computed. More recent time formulation studies include ion permeation through the gramicidin channel [15], and an isomerization of a Lennard-Jones cluster [16]. Other studies of biophysical systems in the length formulation include folding of protein A [17], cytochrome C [18] and a helix formation of a short peptide [19]. Some of the calculations were performed for problems accessible to straightforward molecular dynamics which makes it possible to compare mechanisms. Other calculations were extended to domains of time inaccessible to the initial value approach making it possible to explore and examined in atomic details problems we were not able to study in the past.

5 Summary

We have discussed numerous boundary value formulations of classical mechanics and the corresponding discrete versions that led to convenient numerical procedures. The boundary value approach has the following advantages compared to initial value formulation: (i) the initial and the final points are fixed, enabling the study of transition mechanisms between pre-specified states, (ii) A large time step is possible while maintaining the stability of the solution, and (iii) alternative representation (time, arc-length) are possible, adding to the flexibility of the calculation and interpretation. The disadvantage compared to initial value formulation is the much larger problem we have at our hand (a complete trajectory is optimized starting from an initial guess as compared to computing one structure at the time with initial value formulation). Another disadvantage is the requirement for two end points. When two end points are not available, the initial value formulation is preferred. Comparing the time and the length formulation with a large integration step, the length formulation is more appropriate for activated processes with a large integration step.

Acknowledgements

This research is supported by NIH grant GM059796. I am grateful to Stacey Shirk for editorial assistance.

References

1. L. Verlet (1967) Computer experiments on classical fluids I. Thermodynamics properties of Lennard Jones molecules. *Phys. Rev.* **98**, p. 159
2. D. Chandler (1978) Statistical mechanics of isomerization dynamics in liquids and transition-state approximation. *J. Chem. Phys.* **68**, pp. 2959–2970
3. C. Dellago, P. G. Bolhuis, and D. Chandler (1999) On the calculation of reaction rates in the transition path ensemble. *J. Chem. Phys.* **110**, pp. 6617–6625
4. L. D. Landau and E. M. Lifshitz (2000) *Mechanics*, third edition, Butterworth-Heinenann, Oxford, Chap. 1
5. L. D. Landau and E. M. Lifshitz (2000) *Mechanics*, third edition, Butterworth-Heinenann, Oxford, pp. 140–142
6. R. Olender and R. Elber (1996) Calculation of classical trajectories with a very large time step: Formalism and numerical examples. *J. Chem. Phys.* **105**, pp. 9299–9315
7. R. Elber, J. Meller and R. Olender (1999) A stochastic path approach to compute atomically detailed trajectories: Application to the folding of C peptide. *J. Phys. Chem. B* **103**, pp. 899–911
8. K. Siva and R. Elber (2003) Ion permeation through the gramicidin channel: Atomically detailed modeling by the stochastic difference equation. *Proteins, Structure, Function and Genetics* **50**, pp. 63–80
9. A. Ulitksy and R. Elber (1990) A new technique to calculate the steepest descent paths in flexible polyatomic systems. *J. Chem. Phys.* **96**, p. 1510
10. E. Weinan, R. Weiqing, and E. Vanden-Eijnden (2002) String method for the study of rare events. *Physical Review B* **66**, p. 52301
11. H. Jonsson, G. Mills, and K. W. Jacobsen (1998) Nudge Elastic Band Method for Finding Minimum Energy Paths of Transitions, in Classical and Quantum Dynamics in Condensed Phase Simulations. Edited by B.J. Berne, G. Ciccotti, D.F. Coker, World Scientific, p. 385.
12. C. Lanczos (1970) The variational principles of mechanics. University of Toronto Press
13. L. Onsager and S. Machlup (1953) *Phys. Rev.* **91**, p. 1505; *ibid.* (1953); **91**, p. 1512
14. J. C. M. Uitdehaag, B. A. van der Veen, L. Dijkhuizen, R. Elber, and B. W. Dijkstra (2001) Enzymatic circularization of a malto-octaose linear chain studied by stochastic reaction path calculations on cyclodextrin glycosyltransferase. *Proteins Structure Function and Genetics* **43**, pp. 327–335
15. K. Siva and R. Elber (2003) Ion permeation through the gramicidin channel: Atomically detailed modeling by the Stochastic Difference Equation. *Proteins Structure Function and Genetics* **50**, pp. 63–80
16. D. Bai and R. Elber, Calculation of point-to-point short time and rare trajectories with boundary value formulation. *J. Chemical Theory and Computation,* 2, 484–494(2006)

17. A. Ghosh, R. Elber, and H. Scheraga (2002) An atomically detailed study of the folding pathways of Protein A with the Stochastic Difference Equation. *Proc. Natl. Acad. Sci.* **99**, pp. 10394–10398
18. A. Cárdenas and R. Elber (2003) Kinetics of Cytochrome C Folding: Atomically Detailed Simulations. *Proteins, Structure Function and Genetics* **51**, pp. 245–257
19. A. Cárdenas and R. Elber (2003) Atomically detailed simulations of helix formation with the stochastic difference equation. *Biophysical Journal,* **85**, pp. 2919–2939

Transition Path Theory

E. Vanden-Eijnden

Courant Institute of Mathematical Sciences, New York University New York, NY 10012
eve2cims.nyu.edu

Eric Vanden-Eijnden

E. Vanden-Eijnden: *Transition Path Theory*, Lect. Notes Phys. **703**, 453–493 (2006)
DOI 10.1007/3-540-35273-2_13

1 Introduction

The dynamical behavior of many systems arising in physics, chemistry, biology, etc. is dominated by rare but important transition events between long lived states. For over 70 years, transition state theory (TST) has provided the main theoretical framework for the description of these events [17,33,34]. Yet, while TST and evolutions thereof based on the reactive flux formalism [1,5] (see also [30,31]) give an accurate estimate of the transition rate of a reaction, at least in principle, the theory tells very little in terms of the *mechanism* of this reaction. Recent advances, such as transition path sampling (TPS) of Bolhuis, Chandler, Dellago, and Geissler [3,7] or the action method of Elber [15,16], may seem to go beyond TST in that respect: these techniques allow indeed to sample the ensemble of reactive trajectories, i.e. the trajectories by which the reaction occurs. And yet, the reactive trajectories may again be rather uninformative about the mechanism of the reaction. This may sound paradoxical at first: what more than actual reactive trajectories could one need to understand a reaction? The problem, however, is that the reactive trajectories by themselves give only a very indirect information about the *statistical properties* of these trajectories. This is similar to why statistical mechanics is not simply a footnote in books about classical mechanics. What is the probability density that a trajectory be at a given location in state-space conditional on it being reactive? What is the probability current of these reactive trajectories? What is their rate of appearance? These are the questions of interest and they are not easy to answer directly from the ensemble of reactive trajectories. The right framework to tackle these questions also goes beyond standard equilibrium statistical mechanics because of the nontrivial bias that the very definition of the reactive trajectories imply – they must be involved in a reaction. The aim of this chapter is to introduce the reader to the probabilistic framework one can use to characterize the mechanism of a reaction and obtain the probability density, current, rate, etc. of the reactive trajectories.

Since our results are rather general, it is useful to set the stage somewhat abstractly. We shall consider a system whose state-space is $\Omega \subseteq \mathbb{R}^n$ and denote by $x(t)$ the current state of the system in Ω at time t. For instance, $x(t)$ may be the set of instantaneous positions and momenta of the atoms of a molecular system. We assume that the system is ergodic (in the sense of (3) below) and that we have observed an infinitely long equilibrium trajectory, $\{x(t) : -\infty < t < \infty\}$. Out of this observation, we wish to understand the mechanism of reaction between a *reactant state* A and a *product state* B, here thought of as two disjoint regions in Ω, $A \subset \Omega$ and $B \subset \Omega$ with $A \cap B = \emptyset$. (Of course our final aim is to bypass the need of actually observing $\{x(t) : -\infty < t < \infty\}$ and find more efficient ways to understand the reaction, but it is useful to formulate the questions this way.) Since the system is ergodic, the trajectory $\{x(t) : -\infty < t < \infty\}$ will go infinitely many times from A to B, and each time the reaction happens. This reaction involves *reactive trajectories* defined as follows: given the trajectory $\{x(t) : -\infty < t < \infty\}$, we say that the reactive pieces of it are those pieces during which $x(t)$ is out of both A and B and such that it came out of A last and will go to B next. To formalize things, given a trajectory $\{x(t) : -\infty < t < \infty\}$, let

$$
\begin{aligned}
t_{AB}^+(t) &= \text{smallest } t' \geq t \text{ such that } x(t') \in A \cup B \\
t_{AB}^-(t) &= \text{largest } t' \leq t \text{ such that } x(t') \in A \cup B
\end{aligned}
\tag{1}
$$

Then

$$\text{ensemble of reactive trajectories} = \{x(t) : t \in R\}$$
where $t \in R$ if and only if $x(t) \notin A \cup B$, $x(t_{AB}^+(t)) \in B$ and $x(t_{AB}^-(t)) \in A$
$$\tag{2}$$

and each continuous piece of trajectory going from A to B in the ensemble (2) is a specific reactive trajectory (see Fig. 1).

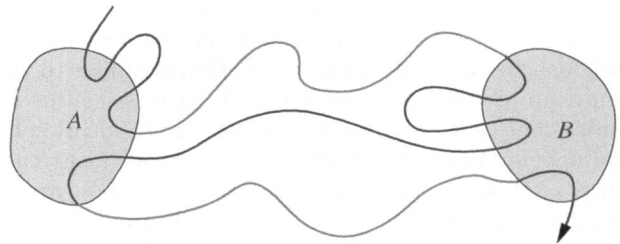

Fig. 1. Schematic representation of the reactant state A, the product state B, a piece of an equilibrium trajectory (shown in black) and the two reactive trajectories along it (shown in light gray)

Our objective is to understand the statistical properties of the reactive trajectories in the ensemble (2). We will try to do so under minimum assumptions about the dynamics of $x(t)$, but we have to require the following from the start. First we require that the dynamics be *Markov*, i.e. given $x(t)$, its future $\{x(t') : t' > t\}$ and its past $\{x(t') : t' < t\}$ are statistically determined. Second, that it be *ergodic* with respect to some equilibrium probability density $m(x)$, i.e. given a suitable observable $F(x)$ and a generic trajectory $\{x(t) : -\infty < t < \infty\}$, we have

$$
\lim_{T \to \infty} \frac{1}{2T} \int_{-T}^{T} F(x(t))dt = \int_{\Omega} F(x)m(x)dx
\tag{3}
$$

Ergodicity guarantees that the trajectory oscillates infinitely often between A and B, so that the number of reactive trajectories in the ensemble (2) is infinite and this ensemble is statistically independent of which particular trajectory $\{x(t) : -\infty < t < \infty\}$ one chooses to generate it. In other words, we try to understand the generic way the reaction happens at equilibrium, by opposition to the way it may happen once if one prepares the system in some specific initial state.

We then ask the following:

1. What is the *probability density function* of reactive trajectories, i.e. the probability density $m_{AB}(x)$ that the trajectory be at point $x \in \Omega \setminus (A \cup B)$ at time t conditional on this trajectory being reactive at time t? We stress that $m_{AB}(x)$ is not simply $m(x)$ properly renormalized on $\Omega \setminus (A \cup B)$ since a trajectory may be in $\Omega \setminus (A \cup B)$ by leaving B, or by leaving A and re-entering this set afterwards without visiting B in between (in both cases it is not considered as reactive). From (2) $x(t)$ belongs to the set of reactive trajectories if and only if $\chi_{\Omega \setminus (A \cup B)}(x(t))\chi_B(x(t^+_{AB}(t)))\chi_A(x(t^-_{AB}(t))) = 1$ (here and below, given any set C, $\chi_C(x) = 1$ if $x \in C$ and $\chi_C(x) = 0$ otherwise). Therefore, by ergodicity, $m_{AB}(x)$ can be defined as the probability density such that for any suitable function $F(x)$, we have (compare (3))

$$\lim_{T \to \infty} \frac{\int_{-T}^T F(x(t))\chi_{\Omega \setminus (A \cup B)}(x(t))\chi_B(x(t^+_{AB}(t)))\chi_A(x(t^-_{AB}(t)))dt}{\int_{-T}^T \chi_{\Omega \setminus (A \cup B)}(x(t))\chi_B(x(t^+_{AB}(t)))\chi_A(x(t^-_{AB}(t)))dt}$$
$$= \int_{\Omega \setminus (A \cup B)} F(x)m_{AB}(x)dx \tag{4}$$

2. What is the *probability current* of reactive trajectories? This probability current is the vector field $J_{AB}(x)$ defined in $\Omega \setminus (A \cup B)$ which is such that, given any surface $S \subset \Omega \setminus (A \cup B)$ which is the boundary of a region Ω_S, the surface integral of $J_{AB}(x)$ over S gives the probability flux of reactive trajectories across S. More precisely,

$$\lim_{s \to 0^+} \frac{1}{s} \lim_{T \to \infty} \frac{1}{2T} \int_{-T}^T \left(\chi_{\Omega_S}(x(t))\chi_{\Omega \setminus \Omega_S}(x(t+s)) \right.$$
$$\left. - \chi_{\Omega \setminus \Omega_S}(x(t))\chi_{\Omega_S}(x(t+s)) \right)$$
$$\times \chi_A(x(t^-_{AB}(t)))\chi_B(x(t^+_{AB}(t+s)))dt \tag{5}$$
$$= \int_S \hat{n}_S(x) \cdot J_{AB}(x)d\sigma_S(x)$$

where $\hat{n}_S(x)$ is the unit normal on S pointing outward Ω_S and $d\sigma_S(x)$ is the surface element on S. The integral at the left hand-side of (5) counts the balance of how many reactive trajectories go in and out of Ω_S in the infinitesimal interval $[t, t+s]$ and averages this count over time. Therefore the current $J_{AB}(x)$ is the true indicator of the mechanism of the reaction since, roughly, it measures the average flow of the reactive trajectories at a given point $x \in \Omega \setminus (A \cup B)$: in particular, the streamlines of $J_{AB}(x)$ indicate the average pathway of the reaction as they allow to define regions (or tubes) joining A to B which contribute to a specific percentage of the reaction.

3. What is the *rate* of the reaction, i.e. what is the mean frequency k_{AB} of transitions from A to B? If N_T^R is the number of reactive trajectories observed during the time interval $[-T, T]$ in the ensemble (2), the rate is the limit

$$k_{AB} = \lim_{T \to \infty} \frac{N_T^R}{2T}. \tag{6}$$

The answers to these questions and some extra ones are given below. This chapter is a summary of the material presented originally in [9–14, 26–28].

2 Probability Density of Reactive Trajectories

What is the probability density $m_{AB}(x)$ that a trajectory be at point $x \in \Omega \setminus (A \cup B)$ at time t conditional on it being reactive at time t? To answer this question, let us first derive the probability density $m_R(x)$ that there be a reactive trajectory at point x at time t; in terms of the trajectory $\{x(t) : -\infty < t < \infty\}$, $m_R(x)$ can be defined as the density such that for any suitable function $F(x)$ we have

$$\lim_{T \to \infty} \frac{1}{2T} \int_{-T}^{T} F(x(t)) \chi_{\Omega \setminus (A \cup B)}(x(t)) \chi_B(x(t_{AB}^+(t))) \chi_A(x(t_{AB}^-(t))) dt$$

$$= \int_{\Omega \setminus (A \cup B)} F(x) m_R(x) dx \tag{7}$$

Observe that this expression involves the integral at numerator in (4) divided by $2T$ in the limit as $T \to \infty$. Using Markovianity, $m_R(x)$ is the probability density that any trajectory (reactive or not) be at x at time t, times the probability $P_R(x)$ that the trajectory be reactive (i.e. that it came from A last and will go to B next) conditional on it being at x at time t. The probability $P_R(x)$ can be expressed in terms of the forward and backward *committor functions*, defined as follows. Given $x \in \Omega \setminus (A \cup B)$, the forward committor function is

$$q_+(x) = \text{probability}$$
$$\text{that } x(t) \text{ goes next to } B \text{ rather than } A \text{ given that } x(0) = x \quad (8)$$

whereas the backward commitor function is

$$q_-(x) = \text{probability}$$
$$\text{that } x(t) \text{ came last from } A \text{ rather than } B \text{ given that } x(0) = x \quad (9)$$

Note that since the dynamics need not be time-reversible, $q_-(x) \neq 1 - q_+(x)$ in general.[1] In terms of $q_+(x)$ and $q_-(x)$, we simply have $P_R(x) = q_+(x)q_-(x)$

---

[1] $q_+(x) = 1 - q_-(x)$ if and only if the dynamics is time-reversible, which is here understood in the following sense. By definition, time-reversible Markov processes

and therefore we deduce that the probability density that there be a reactive trajectory at point x at time t is

$$m_R(x) = q_+(x)q_-(x)m(x) \tag{10}$$

Note that the integral of $m_R(x)$ over $\Omega \setminus (A \cup B)$ is in general less than one since it gives the probability the trajectory be reactive at time t:

$$
\begin{aligned}
Z_{AB} &= \int_{\Omega \setminus (A \cup B)} q_+(x)q_-(x)m(x)dx \\
&= \lim_{T \to \infty} \frac{1}{2T} \int_{-T}^{T} \chi_{\Omega \setminus (A \cup B)}(x(t))\chi_B(x(t_{AB}^+(t)))\chi_A(x(t_{AB}^-(t)))dt
\end{aligned}
\tag{11}
$$

which is also the limit as $T \to \infty$ of the integral at denominator in (4) divided by $2T$. Therefore Z_{AB} is the factor by which $m_R(x)$ needs to be divided to account for the extra condition that the trajectory be reactive at time t and answers our original question about $m_{AB}(x)$:

$$m_{AB}(x) = Z_{AB}^{-1}m_R(x) = Z_{AB}^{-1}q_+(x)q_-(x)m(x) , \tag{12}$$

This relation between $m_{AB}(x)$, $Z_{AB}(x)$ and $m_R(x)$ as well as the expression (10) for $m_R(x)$ in a simpler setting were first derived in [20].

Since the dynamics is Markov by assumption, the committor functions $q_+(x)$, $q_-(x)$ and hence $m_R(x)$ and $m_{AB}(x)$ from (10) and (12) are well-defined. However, these functions could be quite nasty. For instance, if the dynamics is deterministic (say, Hamiltonian), $q_+(x)$ and $q_-(x)$ are either 1 or 0, depending on whether the trajectory passing through x goes next to B or not, and came last from A or not. So, to avoid pathologies, we will assume that the dynamics is not deterministic and $q_+(x)$ and $q_-(x)$ are smooth functions taking value in $[0, 1]$. Later, we will give conditions under which this assumption holds true and give closed equations for $q_+(x)$ and $q_-(x)$ (see (23) and (24) below).

The density $m_{AB}(x)$ gives a first interesting indicator of the mechanism of the reaction since it allows e.g. to determine the proportion of time that the trajectory spends in any region $C \subseteq \Omega \setminus (A \cup B)$ while it is reactive. This proportion of time is simply given by (4) with $F(x) = \chi_C(x)$:

$$\int_C m_{AB}(x)dx \tag{13}$$

---

are such that, at equilibrium, the joint probability density that the system be at x at time t and at y at time $t + s$ with $s > 0$ is the same as the joint probability density that the system be at y at time t and at x at time $t + s$. Note that this definition is the standard one in stochastic processes theory, but it is different from what time-reversibility means in the context of Hamiltonian systems. In particular, an Hamiltonian system is *not* time-reversible with our definition, since one is not allowed to revert the directions of the momenta.

Since the reactive trajectories will in general slow down near the dynamical bottlenecks of the reaction, this allows one to identify the *transition state regions* roughly as the regions where $m_{AB}(x)$ is peaked. Observe however that these regions can be multiple (i.e. there may be more than one dynamical bottleneck for a reaction) and quite wide (i.e. the dynamical bottleneck may be a rather extended region in state-space).

Quite interestingly, the above argument gives also the rate of the reaction k_{AB} in terms of Z_{AB} and the *mean reaction time* t_{AB} defined as

$$t_{AB} = \lim_{T \to \infty} \frac{1}{N_T^R} \int_{-T}^{T} \chi_{\Omega \setminus (A \cup B)}(x(t)) \chi_B(x(t_{AB}^+(t))) \chi_A(x(t_{AB}^-(t))) dt \quad (14)$$

Indeed, letting

$$T_T^R = \int_{-T}^{T} \chi_{\Omega \setminus (A \cup B)}(x(t)) \chi_B(x(t_{AB}^+(t))) \chi_A(x(t_{AB}^-(t))) dt \quad (15)$$

be the total time the trajectory is reactive in $[-T, T]$ and combining (6), (11) and (14), one sees that

$$\begin{aligned} k_{AB} &= \lim_{T \to \infty} \frac{N_T^R}{2T} = \lim_{T \to \infty} \frac{T_T^R}{2T} \frac{N_T^R}{T_T^R} \\ &= \frac{\lim_{T \to \infty} T_T^R / 2T}{\lim_{T \to \infty} T_T^R / N_T^R} = Z_{AB}/t_{AB} \end{aligned} \quad (16)$$

An alternative to that expression will be given below (see (43)) in terms of $m(x)$, $q_+(x)$ and $q_-(x)$. Note that (16) is not fully explicit since we have not expressed t_{AB} in terms of $m(x)$, $q_+(x)$ and $q_-(x)$ (this will also be done below, see (44)). A formula equivalent to (16) was first derived in [20].

Finally, notice that $m_{AB}(x)$ restricted to any dividing surface S and properly normalized to this surface gives the probability density that a reactive trajectory crosses S at point $x \in S$. In other words, if t_j^S with $j \in \mathbb{Z}$ is the set of all times at which the reactive trajectories in the ensemble (2) cross S, i.e.

$$\{t_j^S : j \in \mathbb{Z}\} \text{ is the set of all times such that } x(t_j^S) \in S \text{ and } t_j^S \in R \quad (17)$$

then for any suitable function $F(x)$

$$\lim_{N \to \infty} \frac{1}{2N} \sum_{j=-N}^{N} F(x(t_j^S)) = Z_S^{-1} \int_S F(x) q_+(x) q_-(x) m(x) d\sigma_S(x) \quad (18)$$

where $Z_S = \int_S q_+(x) q_-(x) m(x) d\sigma_S(x)$ and $d\sigma_S(x)$ is the surface element on S.[2]

---

[2] Formula (17) and (18) assume that we can count the crossing times t_j^S which is not possible in general if the dynamics of $x(t)$ is governed by a stochastic differential equation such as (19). However, (17) and (18) can be generalized to this case, and the claim remains valid: $m_{AB}(x)$ restricted to any dividing surface S and properly normalized to this surface gives the probability density that the reactive trajectories cross S at point $x \in S$.

3 Probability Current of Reactive Trajectories

To proceed further, we must make some additional assumptions about the equation of motion governing the evolution of $x(t)$. We shall suppose that $x(t)$ satisfies the following *stochastic differential equation*

$$\dot{x}(t) = b(x(t)) + \sqrt{2}\,\sigma(x(t))\,\eta(t) \tag{19}$$

where $b(x) = (b_1(x), \dots, b_n(x))^T \in \mathbb{R}^n$ is the drift vector,

$$\sigma(x) = \begin{pmatrix} \sigma_{11}(x) & \cdots & \sigma_{1n}(x) \\ \vdots & \ddots & \vdots \\ \sigma_{n1}(x) & \cdots & \sigma_{nn}(x) \end{pmatrix} \in \mathbb{R}^n \times \mathbb{R}^n \tag{20}$$

is the square root of the diffusion tensor, and $\eta(t) = (\eta_1(t), \dots, \eta_n(t))^T \in \mathbb{R}^n$ is an n dimensional white-noise, i.e. a Gaussian process with mean 0 and covariance $\langle \eta_i(t)\eta_j(s) \rangle = \delta_{ij}\delta(t - s)$. An important example of (19) which arises as a special case of this equation is the *Langevin equation*

$$\begin{cases} \dot{r}_i = v_i \\ \mu_i \dot{v}_i = -\dfrac{\partial V(r)}{\partial r_i} - \gamma v_i + \sqrt{2k_B T \gamma}\,\eta_i(t) \end{cases} \tag{21}$$

where $(r, v) = (r_1, \dots, r_m, v_1, \dots, v_m)^T \in \mathbb{R}^m \times \mathbb{R}^m$ are the position and velocities of the particles, $V(r)$ is the potential, $-\nabla V = -(\partial V/\partial r_1, \dots, \partial V/\partial r_m)^T$ is the force, γ is the friction coefficient and μ_i is the mass of the particle. (21) can be put in the form of (19) by writing $x = (r, v)^T$, $b(x) = (v, -\mu^{-1}\nabla V(r) - \gamma\mu^{-1}v)^T$, etc.

We also need some background material about (19). If $m(x)$ denotes the *equilibrium probability density function* of $x(t)$, i.e. the probability density to find a trajectory (reactive or not) at position x at time t, $m(x)$ satisfies the (steady) *forward Kolmogorov equation* (also known as Fokker-Planck equation)

$$0 = -\sum_{i=1}^{n} \frac{\partial}{\partial x_i}(b_i(x)m(x)) + \sum_{i,j=1}^{n} \frac{\partial^2}{\partial x_i \partial x_j}(a_{ij}(x)m(x)) \tag{22}$$

where $a(x) = a^T(x) = \sigma(x)\sigma^T(x) \in \mathbb{R}^n \times \mathbb{R}^n$ is the nonnegative-definite diffusion tensor (to avoid the problems with the deterministic dynamics discussed in Sect. 2, we assume that at least some of the entries $a_{ij}(x)$ are nonzero). If (22) has a unique solution such that $\int_\Omega m(x)dx = 1$, then the process defined by (19) is ergodic, i.e. it satisfies (3). In addition, it can be shown that the forward committor function $q_+(x)$ satisfies the *backward Kolmogorov equation* associated with (19):

$$\begin{cases} 0 = \displaystyle\sum_{i=1}^{n} b_i(x)\frac{\partial q_+(x)}{\partial x_i} + \sum_{i,j=1}^{n} a_{ij}(x)\frac{\partial^2 q_+(x)}{\partial x_i \partial x_j} \\ q_+(x)|_{x \in \partial A} = 0, \qquad q_+(x)|_{x \in \partial B} = 1, \end{cases} \tag{23}$$

whereas the backward committor function $q_-(x)$ satisfies the backward Kolmogorov equation associated with the time-reversed process:

$$\begin{cases} 0 = \sum_{i=1}^{n} b_i^R(x) \dfrac{\partial q_-(x)}{\partial x_i} + \sum_{i,j=1}^{n} a_{ij}(x) \dfrac{\partial^2 q_-(x)}{\partial x_i \partial x_j} \\ q_-(x)|_{x \in \partial A} = 1, \qquad q_-(x)|_{x \in \partial B} = 0 \end{cases} \tag{24}$$

where

$$b_i^R(x) = -b_i(x) + \frac{2}{m(x)} \sum_{j=1}^{n} \frac{\partial}{\partial x_i}(a_{ij}(x)m(x)) \tag{25}$$

The operator acting on $q_+(x)$ at the right hand-side of (23) is called the *generator* of the process, and it plays a very important role. In particular, for any suitable $F(x)$ we have

$$\lim_{t \to 0^+} \frac{1}{t}\left(\mathbf{E}_x F(x(t)) - F(x)\right) = \sum_{i=1}^{n} b_i(x) \frac{\partial F(x)}{\partial x_i} + \sum_{i,j=1}^{n} a_{ij}(x) \frac{\partial^2 F(x)}{\partial x_i \partial x_j} \tag{26}$$

where $x(t)$ denotes the solution of (19) in a given realization of the white-noise $\eta(t)$ and $\mathbf{E}_x$ denotes the expectation conditional on $x(0) = x$ over the ensemble of solutions generated with different realizations of the white-noise The first term at the right hand side of (26) is similar to the one one would get for the ordinary differential equation $\dot{x}(t) = b(x(t))$ by chain rule: $\dot{F}(x(t)) = \sum_{i=1}^{n} \dot{x}_i(t)\partial F(x(t))/\partial x_i = \sum_{i=1}^{n} b_i(x(t))\partial F(x(t))/\partial x_i$. The second term at the right hand-side of (26) arises because of the presence of the white-noise in (19). The derivation of (23), (24) and (26) from the definition of $q_+(x)$, $q_-(x)$ and $x(t)$ is beyond the scope of the present chapter but it can be found in any elementary textbook on probability and stochastic processes theory like e.g. chapters in 4 and 5 in [8] (another accessible reference is [18]).

Going back to the problem of the probability current of reactive trajectories, notice first that (22) can be written in the form of a continuity equation,

$$0 = -\sum_{i=1}^{n} \frac{\partial J_i(x)}{\partial x_i} \tag{27}$$

where $J(x) = (J_1(x), \dots, J_n(x))^T \in \mathbb{R}^n$ is the equilibrium *probability current*

$$J_i(x) = b_i(x)m(x) - \sum_{j=1}^{n} \frac{\partial}{\partial x_j}(a_{ij}(x)m(x)) \tag{28}$$

Of course, this current accounts for what the trajectory does, irrespective of whether it is reactive or not. In particular, if Ω_S is the region enclosed by the surface S, then the equilibrium current $J(x)$ is such that (compare (5))

$$\lim_{s \to 0^+} \frac{1}{s} \lim_{T \to \infty} \frac{1}{2T} \int_{-T}^{T} \left(\chi_{\Omega_S}(x(t))\chi_{\Omega \setminus \Omega_S}(x(t+s)) \right.$$
$$\left. - \chi_{\Omega \setminus \Omega_S}(x(t))\chi_{\Omega_S}(x(t+s)) \right) dt \qquad (29)$$
$$= \int_S \hat{n}_S(x) \cdot J(x) d\sigma_S(x) = 0$$

where $\hat{n}_S(x)$ is the unit normal on S pointing outward Ω_S and $d\sigma_S(x)$ is the surface element on S (below it is actually shown how to derive (28) from (29)). This is not what we want. For instance, the equilibrium current (28) is identically zero for processes that are time-reversible in the sense of the footnote on page 458.[3] But even for these processes, there must be a probability current of reactive trajectories since, by construction, these trajectories flow from A to B.

To determine what is the probability current of reactive trajectories let us go back to the definition (5). Taking the limit as $T \to \infty$ in this expression using ergodicity gives

$$\lim_{s \to 0^+} \frac{1}{s} \left(\int_{\Omega_S} m(x)q_-(x)\mathbf{E}_x\left(q_+(x(s))\chi_{\Omega \setminus \Omega_S}(x(s))\right) dx \right.$$
$$\left. - \int_{\Omega \setminus \Omega_S} m(x)q_-(x)\mathbf{E}_x\left(q_+(x(s))\chi_{\Omega_S}(x(s))\right) dx \right) \qquad (30)$$
$$= \int_S \hat{n}_S(x) \cdot J_{AB}(x) d\sigma_S(x)$$

where $\mathbf{E}_x$ denotes expectation conditional on $x(0) = x$ as in (26). Taking the limit as $s \to 0^+$ can now be done using (26) but this operation is somewhat tricky because of the presence of the discontinuous functions $\chi_{\Omega_S}(x)$ and $\chi_{\Omega \setminus \Omega_S}(x)$ over which the generator in (26) arising in the limit must act. The spatial derivatives on $\chi_{\Omega_S}(x)$ and $\chi_{\Omega \setminus \Omega_S}(x)$ bring delta distributions concentrated on the boundary S of Ω_S, and so intuitively in the limit as $s \to 0^+$, the left hand-side of (30) should reduce to a surface integral over S. This is indeed the case, but since the argument leading to this conclusion is somewhat technical we defer it till the end of this section. The result is that (30) gives

$$\int_S \sum_{i=1}^{n} \hat{n}_{S,i}(x) \left(q_-(x)q_+(x)J_i(x) + q_-(x)m(x) \sum_{j=1}^{n} a_{ij}(x)\frac{\partial q_+(x)}{\partial x_j} \right.$$
$$\left. -q_+(x)m(x) \sum_{j=1}^{n} a_{ij}(x)\frac{\partial q_-(x)}{\partial x_j} \right) d\sigma_S(x)$$
$$= \int_S \hat{n}_S(x) \cdot J_{AB}(x) d\sigma_S(x) \qquad (31)$$

---

[3] For processes satisfying a stochastic differential equation such as (19), it can be shown that the time-reversibility is precisely equivalent to the *detailed balance condition* $J(x) = 0$ which is, in fact, a constraint on $b(x)$ and $a(x)$ since $m(x)$ is determined by these coefficients through (22).

where $J(x)$ is the probability current (28). Since this relation must hold for any surface S, we deduce that the *probability current of reactive trajectories* is

$$J_{AB,i}(x) = q_-(x)q_+(x)J_i(x)$$
$$+ q_-(x)m(x)\sum_{j=1}^n a_{ij}(x)\frac{\partial q_+(x)}{\partial x_j} - q_+(x)m(x)\sum_{j=1}^n a_{ij}(x)\frac{\partial q_-(x)}{\partial x_j} \tag{32}$$

Derivation of (31).

The proper way to take the limit as $s \to 0^+$ and avoid ambiguities on how to interpret the derivatives of $\chi_{\Omega_S}(x)$ and $\chi_{\Omega \backslash \Omega_S}(x)$ is to *mollify* these functions, that is, replace them by functions varying rapidly on S but smooth, then let $s \to 0^+$ and finally remove the mollification. Let then $f_\delta(x)$ be a smooth function which is 1 in Ω_S at a distance δ from S, 0 out of Ω_S at a distance δ from S and varies rapidly but smoothly from 0 to 1 in the strip of size 2δ around S. Thus (30) is the limit as $\delta \to 0$ of

$$I_\delta = \lim_{s\to 0^+} \frac{1}{s}\int_\Omega \Big(m(x)q_-(x)f_\delta(x)\mathbf{E}_x\big(q_+(x(s))(1-f_\delta(x(s)))\big)$$
$$- m(x)q_-(x)(1-f_\delta(x))\mathbf{E}_x\big(q_+(x(s))f_\delta(x(s)))\big)\Big)dx \tag{33}$$

Inserting

$$0 = -m(x)q_-(x)f_\delta(x)\big(q_+(x)(1-f_\delta(x))\big) + m(x)q_-(x)(1-f_\delta(x))\big(q_+(x)f_\delta(x)\big)$$

under the integral then letting $s \to 0^+$ using (26), (33) gives

$$I_\delta = \int_\Omega \Big(m(x)q_-(x)f_\delta(x)L\big(q_+(x)(1-f_\delta(x))\big)$$
$$- m(x)q_-(x)(1-f_\delta(x))L\big(q_+(x)f_\delta(x)\big)\Big)dx \tag{34}$$

where L is a short hand notation for the generator defined in (26): for any suitable $F(x)$

$$LF(x) = \sum_{i=1}^n b_i(x)\frac{\partial F(x)}{\partial x_i} + \sum_{i,j=1}^n a_{ij}(x)\frac{\partial^2 F(x)}{\partial x_i \partial x_j} \tag{35}$$

Expanding the integrand in (34), several terms cancel and we are simply left with

$$I_\delta = -\int_\Omega m(x)q_-(x)L\big(q_+(x)f_\delta(x)\big)dx \tag{36}$$

Using the explicit form (35) for L and expanding, this is

$$I_\delta = - \int_\Omega m(x) q_-(x) \left(f_\delta(x) L q_+(x) + \sum_{i,j=1}^n a_{ij}(x) \frac{\partial}{\partial x_i} \left(q_+(x) \frac{\partial f_\delta(x)}{\partial x_j} \right) \right.$$
$$\left. + \sum_{i=1}^n \frac{\partial f_\delta(x)}{\partial x_i} \left(b_i(x) q_+(x) + \sum_{j=1}^n a_{ij}(x) \frac{\partial q_+(x)}{\partial x_j} \right) \right) dx \tag{37}$$

By (23), $L q_+(x) = 0$ and integrating by parts the second term in the parenthesis under the integral, we arrive at

$$I_\delta = - \int_\Omega \sum_{i=1}^n \frac{\partial f_\delta(x)}{\partial x_i} \left(q_+(x) q_-(x) J_i(x) + q_-(x) m(x) \sum_{j=1}^n a_{ij}(x) \frac{\partial q_+(x)}{\partial x_j} \right.$$
$$\left. - q_+(x) m(x) \sum_{j=1}^n a_{ij}(x) \frac{\partial q_-(x)}{\partial x_j} \right) dx \tag{38}$$

where $J(x)$ is the probability current (28). Now let $\delta \to 0$ and recall that for any suitable $F(x) = (F_i(x), \ldots, F_n(x))^T$

$$\lim_{\delta \to 0} \int_\Omega \sum_{i=1}^n \frac{\partial f_\delta(x)}{\partial x_i} F_i(x) dx = - \lim_{\delta \to 0} \int_\Omega f_\delta(x) \sum_{i=1}^n \frac{\partial F_i(x)}{\partial x_i} dx$$
$$= - \int_{\Omega_S} \sum_{i=1}^n \frac{\partial F_i(x)}{\partial x_i} dx \tag{39}$$
$$= - \int_S \sum_{i=1}^n \hat{n}_{S,i}(x) F_i(x) d\sigma_S(x)$$

where the first equality follows by integration by parts, the second by definition of $f_\delta(x)$, and the third by the divergence theorem. Using (39), we conclude that the limit of (38) as $\delta \to 0$ is the surface integral at the left hand-side of (31), and we are done.

4 Reaction Rate, Streamlines of the Current and Transition Tubes

Using (22), (23) and (24), it is easy to see that $J_{AB}(x)$ is a divergence free vector field in $\Omega \setminus (A \cup B)$. Indeed, after some straightforward algebraic manipulations, one arrives at

$$\sum_{i=1}^n \frac{\partial J_{AB,i}(x)}{\partial x_i} = -q_+(x) q_-(x) \times \text{RHS(22)} - m(x) q_-(x) \times \text{RHS(23)}$$
$$- m(x) q_+(x) \times \text{RHS(24)} = 0 \tag{40}$$

where RHS stands for right hand-side. As a result, by the divergence theorem, the probability flux across any closed surface $S \subset \Omega \setminus (A \cup B)$ is zero,

$$\int_{S} \hat{n}_S(x) \cdot J_{AB}(x) d\sigma_S(x) = \int_{\Omega_S} \mathrm{div} J_{AB}(x) dx = 0 \quad \text{(closed } S) \tag{41}$$

where Ω_S is the volume whose boundary is S, $\hat{n}_S(x)$ is the unit normal to S pointing outward Ω_S, $d\sigma_S(x)$ is the surface element on S and $\mathrm{div} J_{AB}(x) = 0$ denotes the divergence of the current that we calculated in (40). For the same reason, the probability flux across any dividing surface S, i.e. any surface which separates Ω into two pieces, one which contains A and the other B, is constant. This is consistent with the fact that there are no source nor sink of reactive trajectories in $\Omega \setminus (A \cup B)$: every reactive trajectory that leaves A must reach B. By construction, the constant probability flux across any dividing surface is simply the *reaction rate*,

$$k_{AB} = \int_{S} \hat{n}_S(x) \cdot J_{AB}(x) d\sigma_S(x) \quad \text{(dividing } S), \tag{42}$$

where $\hat{n}_S(x)$ is the unit normal to S pointing toward B. This quantity is the limit defined in (6), i.e. it gives the *exact* mean frequency at which the reactive trajectories are observed within a given trajectory.

The expression (42) for the rate can be simplified and transformed into a volume integral over $\Omega \setminus (A \cup B)$:

$$k_{AB} = \int_{\Omega \setminus (A \cup B)} m(x) \sum_{i,j=1}^{n} a_{ij}(x) \frac{\partial q_+(x)}{\partial x_i} \frac{\partial q_+(x)}{\partial x_j} dx \tag{43}$$

(Equivalently, we could use $q_-(x)$ instead of $q_+(x)$ in this expression.) Since the derivation of (43) from (42) involves a few technical steps we postpone it till the end of this section. To the best of our knowledge, this expression was first given in [14]. It provides an exact alternative to the expression for the reaction rate given by TST and the reactive flux formalism (see Sect. 5). Finally, notice that (11) and (43) can be combined with (16) to give the following expression for the *mean reaction time* t_{AB} defined in (14):

$$t_{AB} = \frac{\int_{\Omega \setminus (A \cup B)} q_+(x) q_-(x) m(x) dx}{\int_{\Omega \setminus (A \cup B)} m(x) \sum_{i,j=1}^{n} a_{ij}(x) (\partial q_+/\partial x_i)(\partial q_+(x)/\partial x_j) dx} \tag{44}$$

There is more than the reaction rate that can be extracted from the probability current of reactive trajectories. First, notice that on ∂A where $q_+(x) = 0$ and $q_-(x) = 1$, we have

$$\hat{n}_{\partial A}(x) \cdot J_{AB}(x) = m(x) \sum_{i,j=1}^{n} \hat{n}_{\partial A,i}(x) a_{ij}(x) \frac{\partial q_+(x)}{\partial x_j}$$

$$= m(x) |\nabla q_+(x)|^{-1} \sum_{i,j=1}^{n} a_{ij}(x) \frac{\partial q_+(x)}{\partial x_i} \frac{\partial q_+(x)}{\partial x_j} \geq 0. \tag{45}$$

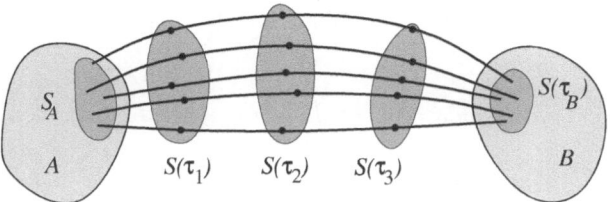

Fig. 2. Schematic of the streamlines of the probability current of reactive trajectories out of $S_A \subset \partial A$, and the pushed forward region $S(s)$ at "times" $0 < \tau_1 < \tau_2 < \tau_3 < s_A$. The collection of these regions, $\{S(\tau) : \tau \geq 0\} = \cup_{\tau \geq 0} S(\tau)$ forms a tube carrying a given percentage of the probability flux of reactive trajectories

where $\hat{n}_{\partial A}(x)$ is the unit normal on ∂A pointing outward A which can be expressed as $\hat{n}_{\partial A}(x) = |\nabla q_+(x)|^{-1} \nabla q_+(x)$. The last inequality in (45) follows from the fact that $a(x)$ is a nonnegative definite tensor. Similarly on ∂B where $q_+(x) = 1$ and $q_-(x) = 0$, we have

$$\hat{n}_{\partial B}(x) \cdot J_{AB}(x) = m(x) \sum_{i,j=1}^{n} \hat{n}_{\partial B,i}(x) a_{ij}(x) \frac{\partial q_-(x)}{\partial x_j}$$

$$= m(x) |\nabla q_-(x)|^{-1} \sum_{i,j=1}^{n} a_{ij}(x) \frac{\partial q_-(x)}{\partial x_i} \frac{\partial q_-(x)}{\partial x_j} \geq 0. \tag{46}$$

where $\hat{n}_{\partial B}(x)$ is the unit normal on ∂B pointing inward B which can be expressed as $\hat{n}_{\partial B}(x) = -|\nabla q_-(x)|^{-1} \nabla q_-(x)$. Equations (45) and (46) shows that the probability current of reactive trajectories point outward A everywhere on ∂A and inward B everywhere on ∂B, as it should since every reactive trajectory connects A to B by construction. Now suppose that we identify a region on ∂A, say $S_A \subset \partial A$, which is such that $X\%$ of the probability current goes out of A through S_A, i.e.

$$\int_{S_A} \hat{n}_{\partial A}(x) \cdot J_{AB}(x) d\sigma_{\partial A}(x) = \frac{X}{100} \int_{\partial A} \hat{n}_{\partial A}(x) \cdot J_{AB}(x) d\sigma_{\partial A}(x) \equiv \frac{X}{100} k_{AB} \tag{47}$$

We can then push forward the surface S_A, using the artificial dynamics

$$\frac{dx(\tau)}{d\tau} = J_{AB}(x(\tau)), \qquad x(0) \in S_A \subset \partial A. \tag{48}$$

This equation defines the *streamlines* of $J_{AB}(x)$ (see Fig. 2). Eventually, every streamline $x(\tau)$ must reach B for some $\tau_B > 0$ and we will terminate the streamlines on ∂B by assuming that $x(\tau) = x(\tau_B) \in \partial B$ for $\tau \geq \tau_B$ (notice that τ_B depends on $x(0) \in \partial A$ and may be different for every streamline). Using the divergence theorem again, the probability flux through the surface $S(\tau) = \cup_{x(0) \in S_A} x(\tau)$ (that is, the push-forward of the surface S_A along the streamlines), is equal to the probability flux through the surface $S_A \equiv S(0)$, i.e.

$$\int_{S_A} \hat{n}_{\partial A}(x) \cdot J_{AB}(x) d\sigma_{\partial A}(x) = \int_{S(\tau)} \hat{n}_{S(\tau)}(x) \cdot J_{AB}(x) d\sigma_{S(\tau)}(x). \quad (49)$$

As a result, the region

$$\{S(\tau) : \tau \geq 0\} = \bigcup_{\tau \geq 0} S(\tau) \quad (50)$$

defines one or more *transition tubes* in $\Omega \backslash (A \cup B)$ which connect A and B and which carries $X\%$ of the probability flux of reactive trajectories (see Fig. 2).

Derivation of (43).

To check that (43) gives the rate, let $S_+(z) = \{q_+(x) = z\}$ be the (forward) isocommittor surface with committor value $z \in [0, 1]$, and consider the integral

$$A(z) = \int_{S_+(z)} m(x) \sum_{i,j=1}^{n} \hat{n}_{S_+(z),i}(x) a_{ij}(x) \frac{\partial q_+(x)}{\partial x_j} d\sigma_{S_+(z)}(x) \quad (51)$$

where $\hat{n}_{S_+(z)}(x)$ is the unit normal to $S_+(x)$ pointing toward B and $d\sigma_{S_+(z)}(x)$ is the surface element on $S_+(z)$. Since $S_+(0) \equiv \partial A$, is easy to see from (42) and (45) that:

$$A(0) = \int_{\partial A} m(x) \sum_{i,j=1}^{n} \hat{n}_{\partial A,i}(x) a_{ij}(x) \frac{\partial q_+(x)}{\partial x_j} d\sigma_{\partial A}(x) \equiv k_{AB} \quad (52)$$

Next, we show that $A(z) = A(0) = k_{AB}$ for all $z \in [0, 1]$. Using the Dirac delta function we can express $A(z)$ as

$$A(z) = \int_{\Omega} m(x) \sum_{i,j=1}^{n} \frac{\partial q_+(x)}{\partial x_i} a_{ij}(x) \frac{\partial q_+(x)}{\partial x_j} \delta(q_+(x) - z) dx \quad (53)$$

and hence

$$\begin{aligned}
\frac{dA(z)}{dz} &= -\int_{\Omega} m(x) \sum_{i,j=1}^{n} \frac{\partial q_+(x)}{\partial x_i} a_{ij}(x) \frac{\partial q_+(x)}{\partial x_j} \delta'(q_+(x) - z) dx \\
&= -\int_{\Omega} m(x) \sum_{i,j=1}^{n} \frac{\partial q_+(x)}{\partial x_i} a_{ij}(x) \frac{\partial}{\partial x_j} \delta(q_+(x) - z) dx
\end{aligned} \quad (54)$$

Integrating by parts, this gives

$$\frac{dA(z)}{dz} = \int_\Omega m(x) \sum_{i,j=1}^n a_{ij}(x) \frac{\partial^2 q_+(x)}{\partial x_i \partial x_j} \delta(q_+(x) - z) dx$$

$$+ \int_\Omega \sum_{i,j=1}^n \frac{\partial q_+(x)}{\partial x_i} \frac{\partial}{\partial x_j} (a_{ij}(x) m(x)) \delta(q_+(x) - z) dx$$

$$= - \int_\Omega m(x) \sum_{i=1}^n b_i(x) \frac{\partial q_+(x)}{\partial x_i} \delta(q_+(x) - z) dx$$

$$+ \int_\Omega \sum_{i,j=1}^n \frac{\partial q_+(x)}{\partial x_i} \frac{\partial}{\partial x_j} (a_{ij}(x) m(x)) \delta(q_+(x) - z) dx$$

(55)

where in the second step we used (23). Using the definition (28) for the equilibrium current $J(x)$, the two integrals in the last equality can be recombined into

$$\frac{dA(z)}{dz} = - \int_\Omega \sum_{i=1}^n \frac{\partial q_+(x)}{\partial x_i} J_i(x) \delta(q_+(x) - z) dx$$

$$= - \int_{S_+(z)} \sum_{i=1}^n n_{S_+(z),i}(x) J_i(x) d\sigma_{S_+(z)}(x) = 0$$

(56)

where in the last equality we use the fact that the probability flux of the regular (by opposition to reactive) trajectories through any surface is zero at equilibrium. Equation (56) implies that $A(z) = A(0) = k_{AB}$ for all $z \in [0, 1]$ as claimed. Hence, $\int_0^1 A(z) dz = k_{AB}$ which from (53) gives

$$k_{AB} = \int_0^1 \int_\Omega m(x) \sum_{i,j=1}^n \frac{\partial q_+(x)}{\partial x_j} a_{ij}(x) \frac{\partial q_+(x)}{\partial x_j} \delta(q_+(x) - z) dx dz$$

$$= \int_{\Omega \setminus (A \cup B)} m(x) \sum_{i,j=1}^n \frac{\partial q_+(x)}{\partial x_j} a_{ij}(x) \frac{\partial q_+(x)}{\partial x_j} dx$$

(57)

This is (43).

5 Comparison with Transition State Theory (TST) and Transition Path Sampling (TPS)

A recent account of transition state theory (TST) has been given in [31] (see also [30]). Here we shall content ourselves with briefly summarizing TST (by which we mean the modern version of the theory which accounts for dynamical corrections) and contrasting it with the present approach. The main result of TST is an expression for the reaction rate; this expression is equivalent to (42) but different from it and less general (it only applies to specific cases of (19), such as the Langevin equation (21), and only with specific dividing surfaces).

These restrictions arise because TST is based on a two-step procedure [1, 5]: first one measures the frequency of crossing of a dividing surface S by the trajectory (this gives the bare TST rate of S), then one corrects this frequency to account only for the last crossing by each reactive trajectories (this gives the transmission coefficient of the surface S, which is the coefficient by which the bare TST rate of S must be multiplied to get the actual reaction rate).[4] This two-step procedure is in general not possible with (19) because every reactive trajectory cross any dividing surface infinitely many times, hence the bare TST rate is infinite, the transmission coefficient 0, and their product undetermined. In addition, while the transmission coefficient can be expressed in terms of the functions $q_+(x)$ and $q_-(x)$ (see [31]), within TST these functions are not defined through (23) and (24) (though, of course, they could) but rather estimated via running trajectories from the dividing surface S itself and following them till they reach A or B (which is a step which may prove rather inefficient numerically). Finally, and most importantly, since TST is based on a single dividing surface, this precludes TST from giving any information about the probability density and current of the reactive trajectories away from this surface. Summarizing: TST is oblivious to the probabilistic framework discussed here and, as a result, it gives no information about the mechanism of the reaction.

Transition path sampling (TPS) is discussed in detail in another chapter of this book, and here too we will content ourselves with a few remarks. TPS is based on the observation that the probability density to observe a piece of trajectory $\{x(t) : 0 \leq t \leq T\}$ satisfying (19) is proportional to

$$\exp\Big(-\frac{1}{4}\int_0^T \sum_{i=1}^n \Big(\sum_{j=1}^n \sigma_{ij}^{-1}(x(t))(\dot{x}_j(t) - b_j(x(t)))\Big)^2 dt\Big). \qquad (58)$$

where $\sigma^{-1}(x)$ is the inverse of the tensor $\sigma(x)$ defined in (20) (assuming it to be invertible). Therefore (58) allows one to design Metropolis Monte-Carlo schemes in trajectory space, as was suggested earlier by Pratt [25]. If the constraints that $x(0) = x_0 \in A$ with x_0 distributed according to the equilibrium density $m(x)$ restricted and normalized to A and $x(T) \in B$ are added, then these Monte-Carlo schemes permit to sample trajectories which, by construction, start in A and end in B without making any a priori assumption about their behavior in between. The big advantage of TPS is that, if A and B are long-lived (as is the case in situations of interest), by definition these trajectories would be much more difficult to observe by direct simulation of (19). In addition, these trajectories belong to an ensemble (the TPS ensemble) which is close to the ensemble of reactive trajectories defined in (2), except for the additional constraint that only the reactive trajectories which make the transition from A to B in less than T are accounted for. In principle, this bias can

---

[4] The fact that TST can be viewed this way may not be readily apparent from the traditional exposition of TST, but it is clear from the viewpoint on TST taken in [30, 31] (see also [5, 32]).

be minimized by taking T large enough, though this may leads to additional difficulties both conceptual (when T is large, a piece of trajectory may go back and forth between A and B more than once which requires to re-weight each of these carefully in the ensemble) and practical (the longer the pieces of trajectories, the costlier they are to handle). On the other hand, TPS does not use nor rely on the probabilistic framework discussed in this chapter and as a result it offers no direct way to estimate the probability density of the reactive trajectories, their probability current or their rate. As we know, these quantities depends on the committor functions $q_-(x)$ and $q_+(x)$. In TPS, these functions must be estimated *a posteriori*, via the processing of the trajectories in the TPS ensemble, which is a nontrivial operation (see e.g. [22] where this was attempted on the example of alanine dipeptide). Nevertheless, using the probabilistic framework discussed in this chapter in conjunction with TPS may prove useful and is certainly worth further considerations (for some results in this direction see [2, 20]).

6 The Situations with Localized Transition Tubes

The results presented so far indicates that the isocommittor functions $q_-(x)$ and $q_+(x)$ are essential to understand the mechanism of a reaction. However these results do not say how to compute $q_-(x)$ and $q_+(x)$, except via the solution of (24) and (23) which is a formidable task, even numerically, when the dimensionality of the system is large (that is, in any situation of interest). In this section, we show that the transition tube carrying most of the probability flux of reactive trajectories can be identified under the assumption that this tube is localized, in a sense made precise below. As show in Sect. 7, this is a way to make practical the probabilistic framework presented so far: while standard numerical methods based on finite difference or finite element are inappropriate to determine $q_-(x)$ and $q_+(x)$, under the assumptions of this section one can develop algorithms to estimate these functions locally inside the tubes carrying most of the probability flux of reactive trajectories, i.e. where they matter most.

For simplicity, we will consider first the time-reversible situation, when

$$0 = J_i(x) = b_i(x)m(x) - \sum_{j=1}^{n} \frac{\partial}{\partial x_j}(a_{ij}(x)m(x)) \qquad (59)$$

holds (see the footnote on p. 463). An important example which is *not* time-reversible is that of the Langevin equation (21) (recall that reverting the momenta is not allowed in our definition of time-reversibility, see the footnote on p. 458): how to deal with this example will be briefly explained in Sect. 6.5. (59) used in (25) implies that $b^R(x) = b(x)$ and as a result $q_-(x) = 1 - q_+(x)$ since (24) and (23) are identical when $b^R(x) = b(x)$ except for the boundary conditions. Denoting $q_+(x)$ by $q(x)$ to stress that (59) holds, it is then easy to see that (23) can be written as

$$\begin{cases} 0 = \displaystyle\sum_{i,j=1}^{n} \frac{\partial}{\partial x_i}\left(a_{ij}(x)m(x)\frac{\partial q(x)}{\partial x_j}\right) \\ q(x)|_{x\in\partial A} = 0, \qquad q(x)|_{x\in\partial B} = 1. \end{cases} \qquad (60)$$

In addition the probability density of reactive trajectories (12) reduces to

$$m_{AB}(x) = Z_{AB}^{-1}q(x)(1 - q(x))m(x) \qquad (61)$$

where $Z_{AB} = \int_{\Omega\setminus(A\cup B)} q(x)(1 - q(x))m(x)dx$, and the probability current of reactive trajectories (32) to

$$J_{AB,i}(x) = \sum_{j=1}^{n} a_{ij}(x)m(x)\frac{\partial q(x)}{\partial x_j} \qquad (62)$$

6.1 The Localized Tube Assumption

Let $S(z) = \{x : q(x) = z\}$, $z \in [0,1]$, be the family of *isocommittor surfaces* which foliate $\Omega\setminus(A\cup B)$. The results in this section are based on the following *local transition tube assumption* (see Fig. 3):[5]

There exist a family of regions $C(z) \subset S(z)$, $z \in [0,1]$, such that $\cup_{z\in[0,1]}C(z) \equiv \{C(z) : z \in [0,1]\}$ forms a tube in $\Omega \setminus (A \cup B)$ joining A and B with the following properties: $C(z)$ contributes most to the integral of $m(x)$ on $S(z)$, i.e.

$$\int_{S(z)} m(x)d\sigma_{S(z)}(x) \approx \int_{C(z)} m(x)d\sigma_{S(z)}(x), \qquad (63)$$

and inside each $C(z)$ we have

$$a_{ij}(x) \approx cst_{ij}, \qquad \frac{\partial q(x)}{\partial x_i} \approx cst_i \qquad x \in C(z), \quad i,j = 1,\ldots,n \quad (64)$$

(63) says that the equilibrium probability density restricted to $S(z)$ is actually concentrated on $C(z)$; (64) says that the diffusion tensor $a_{ij}(x)$ is approximately constant in $C(z)$ (the reason for this assumption will become clear shortly) and that the isocommittor surfaces $S(z)$ are locally planar inside the tube. We will denotes these planes by $P(z)$, $z \in [0,1]$, so that $S(z) \approx P(z)$

---

[5] Observe that it may be necessary to introduce more than one tube $\cup_{z\in[0,1]}C(z)$ if the reaction can proceed by several localized channels. The arguments below can be generalized to these situations, but we will not consider them here for simplicity.

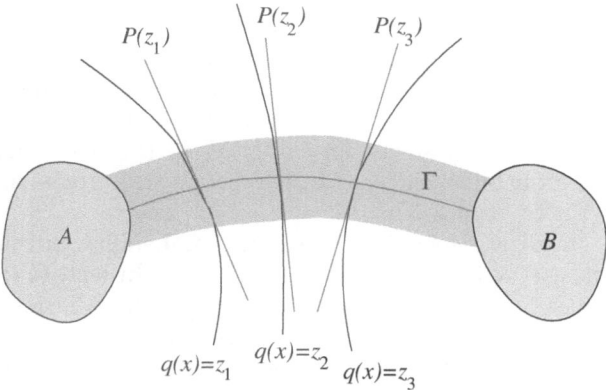

Fig. 3. Schematic representation of tube $\cup_{z\in[0,1]}C(z)$ connecting A and B (shown in *dark grey*), the isocommittor surfaces $S(z)$ where $q(x) = z$, the planes $P(z)$ which approximate $S(z)$ inside the tube, and the curve $\Gamma = \{\varphi(z) : z \in [0,1]\}$ whose geometric location is that of a streamline of the probability current of reactive trajectories inside the tube

inside $\cup_{z\in[0,1]}C(z)$. Equation (63) implies that inside each $P(z)$ there must be a region $D(z) \subset P(z)$ which is such that $D(z) \approx C(z)$ and contributes most to the integral of $m(x)$ on $P(z)$, i.e.

$$\int_{D(z)} m(x)d\sigma_{P(z)}(x) \approx \int_{P(z)} m(x)d\sigma_{P(z)}(x) . \tag{65}$$

Observe that the localized assumption does not necessarily mean that the region $C(z)$ in $S(z)$ (or equivalently $D(z)$ in $P(z)$) is small: beside depending on how $a_{ij}(x)$ varies, its extension in various directions actually depends on how much the isocommittor surface $q(x) = z$ is curved in these directions (the more planar it is in one direction, the wider the region can be in this direction).

The main claim of this section is that, if the localized tube assumption holds, then the tube $\cup_{z\in[0,1]}C(z) \approx \cup_{z\in[0,1]}D(z)$ is the tube carrying most of the probability flux of reactive trajectories and it can be determined by an algorithmic procedure which is much simpler than solving (23) and (24) for $q_+(x)$ and $q_-(x)$.

To see this, recall that $m_{AB}(x)$ restricted and properly re-normalized to any dividing surface S is the probability density that the reactive trajectories cross S at $x \in S$ (see (18) in Sect. 2). By (61) $m_{AB}(x)$ restricted and re-normalized to the isocommittor surface $S(z) = \{x : q(x) = z\}$ is the same as $m(x)$ restricted and re-normalized to $S(z)$ since $m_{AB}(x) = Z_{AB}^{-1}z(1-z)m(x)$ in $S(z)$ by construction and the constant factor $Z_{AB}^{-1}z(1-z)$ is absorbed by the re-normalization. Therefore, $m(x)$ restricted and normalized to the isocommittor surface $S(z) = \{x : q(x) = z\}$ is also the probability density that the reactive trajectories cross $S(z)$ at $x \in S(z)$. As a result (63) implies that the reactive trajectories cross $S(z)$ mostly in the region $C(z)$. Since this is true for all $z \in [0,1]$, this implies that the reactive trajectories remain preferably in the tube $\cup_{z \in [0,1]} C(z)$. As a result, most of the probability flux of reactive trajectories must go through $\cup_{z \in [0,1]} C(z)$ since there cannot be any significant probability current of the reactive trajectories in regions that they do not visit to begin with.

It follows that the streamlines of the the probability current of reactive trajectories must stay inside the tube $\cup_{z \in [0,1]} C(z)$. In fact, (64) implies that these streamlines form a bundle of quasi-parallel curves inside $\cup_{z \in [0,1]} C(z)$, the collection of which actually represent the tube $\cup_{z \in [0,1]} C(z)$. This is a consequence of the equation (48) for the streamlines, which in the present time-reversible context is explicitly

$$\frac{dx_i(\tau)}{d\tau} = m(x(\tau)) \sum_{j=1}^{n} a_{ij}(x(\tau)) \frac{\partial q(x(\tau))}{\partial x_j}. \tag{66}$$

Together with assumption (64) which says that the vector $(\partial q/\partial x_1, \ldots, \partial q/\partial x_{r_r})^T$ evaluated at any $x \in C(z) \approx D(z) \subset P(z)$ is approximately parallel to the unit normal $\hat{n}(z)$ to the plane $P(z)$, (66) implies that the streamlines solution of (66) cross every $C(z) \approx D(z)$ at approximately the same angle (for instance, if $a(x)$ was the identity matrix, (66) would imply that each streamline is perpendicular to each plane).

Remarkably, the constraint that the streamlines must form a bundle of quasi-parallel curves inside the tube gives us an extra constraint on this tube which permits to determine its geometric location. To see this, it is convenient to re-express (66) as a geometric constraint on the streamlines inside the tube by viewing these streamlines as a collection of curves $\Gamma_x = \{\varphi^x(z) : z \in [0,1]\}$, where $x \in C(0) \subset \partial A$ indexes the curves and

$$\varphi^x(z) = \{x(\tau) : x(0) = x \in C(0) \subset \partial A, \tau \in [0, \tau_B]\} \cap C(z) \tag{67}$$

is the location in $C(z)$ where the streamline starting from $x \in C(0) \subset \partial A$ intersects $C(z)$ (recall that $\tau_B = \tau_B(x)$ is the time at which the streamline starting from $x(0) = x \in \partial A$ reaches ∂B, i.e. $x(\tau_B) \in \partial B$). By construction of Γ_x, we can now represent the tube as

$$\bigcup_{z \in [0,1]} C(z) \approx \bigcup_{z \in [0,1]} D(z) \approx \bigcup_{x \in C(0) \subset \partial A} \Gamma_x \tag{68}$$

and from (66) it follows that each $\varphi^x(z)$ must satisfy

$$\frac{d\varphi_i^x(z)}{dz} \text{ parallel to } \sum_{j=1}^{n} a_{ij}(\varphi^x(z)) \frac{\partial q(\varphi^x(z))}{\partial x_j} \tag{69}$$

Since the unit normal $\hat{n}(z)$ to $P(z)$ is approximately parallel to $(\partial q/\partial x_1, \ldots, \partial q/\partial x_n)^T$ evaluated at $\varphi^x(z)$, (69) can be written as

$$\frac{d\varphi_i^x(z)}{dz} \text{ parallel to } \sum_{j=1}^{n} a_{ij}(\varphi^x(z))\hat{n}_j(z) \tag{70}$$

Similar to (66), this equation is a constraint on the angle at which each of the curves $\Gamma_x = \{\varphi^x(z) : z \in [0,1]\}$ intersects each of the planes $P(z)$. Taken together with (63) and (65), (70) is the additional constraint which determines the geometric location tube $\cup_{z \in [0,1]} C(z) \approx \cup_{z \in [0,1]} D(z)$. Let us see why.

Identification of the Tube

Assume that the localized assumption holds and suppose that we give ourself one specific curve $\Gamma = \{\varphi(z) : z \in [0,1]\}$ joining some point in ∂A to some point on ∂B; view Γ as the tentative location of a streamline. Knowing this curve, we can then construct the collection of planes $P(z)$, $z \in [0,1]$, along the curve which are such that $\varphi(z) \in P(z)$ and the unit normal to $P(z)$ satisfies (this is a re-writing of (70) with $\varphi^x(z) \equiv \varphi(z)$):

$$\hat{n}_i(z) \text{ parallel to } \sum_{j=1}^{n} a_{ij}^{-1}(\varphi(z))\frac{d\varphi_j(z)}{dz} \tag{71}$$

Once we have the planes $P(z)$, $z \in [0,1]$, we can also construct the bundle of all the curves Γ_x satisfying (70) (one of the curves Γ_x is the initial Γ.) Observe that this bundle may be bigger than the one in (67) and (68) since we have not specified the regions $D(z)$ yet. In fact, the only thing that limits the extension of this bundle at this point is that it must be defined in the region where the planes $P(z)$ do not intersect, that is in the union of the regions $Q(z) \subset P(z)$, $z \in [0,1]$, around Γ which are such that any point x in $Q(z)$ is closest to $\varphi(z)$ than to any other point along Γ:

$$Q(z) = \{x : x \in P(z) \text{ and } |x - \varphi(z)| = \min_{z' \in [0,1]} |x - \varphi(z')|\} \tag{72}$$

Now look at the regions $D(z)$ in the planes $P(z)$ which concentrate the probability, i.e. the smallest regions in $P(z)$ where (65) holds to some prescribed accuracy. If the initial curve Γ that we choose is a streamline, then we will have

$$\bigcup_{z \in [0,1]} D(z) \subseteq \bigcup_{z \in [0,1]} Q(z), \quad \text{i.e. } D(z) \subseteq Q(z) \text{ for all } z \in [0,1] \tag{73}$$

In addition, consistent with (68), any curve in the bundle of curves in $\cup_{z \in [0,1]} Q(z)$ will either be in the tube $\cup_{z \in [0,1]} D(z)$ or out of it:

either: $\Gamma_x = \{\varphi^x(z) : z \in [0,1]\} \in \cup_{z \in [0,1]} D(z)$,

   i.e. if $\varphi^x(0) \in D(0)$ then $\varphi^x(z) \in D(z)$ for all $z \in [0,1]$,

or: $\Gamma_x = \{\varphi^x(z) : z \in [0,1]\} \notin \cup_{z \in [0,1]} D(z)$,

   i.e. if $\varphi^x(0) \notin D(0)$ then $\varphi^x(z) \notin D(z)$ for all $z \in [0,1]$

$$\tag{74}$$

(Since the second equality in (68) is only supposed to be satisfied approximately to some prescribed accuracy, it is allowed to fiddle things a little bit at the frontier of $\cup_{z \in [0,1]} D(z)$ in function of what the curves do at this frontier.) If (73) and (74) hold, then $\cup_{z \in [0,1]} D(z) \approx \cup_{x \in D(0)} \Gamma_x \approx \cup_{z \in [0,1]} C(z)$ is the transition tube.

In contrast, if the initial curve Γ that we choose is not an actual streamline, then the regions $D(z)$, $z \in [0,1]$ obtained via (65) will not be consistent with the bundle of curves Γ_x obtained via (70): either some of the regions $D(z)$ will be at least partially out of the possible regions $Q(z)$, i.e. we will have $D(z) \not\subseteq Q(z)$ for some $z \in [0,1]$ and (73) will fail, or some of the curves in the bundle will enter or leave the tube $\cup_{z \in [0,1]} D(z)$ and (74) will fail, or both.

Summarizing: we need to adjust the initial curve Γ in order to get collection of planes $P(z)$ $z \in [0,1]$ via (71), a collection of regions $D(z) \subset P(z)$, $z \in [0,1]$ via (65) and $Q(z) \subset P(z)$, $z \in [0,1]$ via (72), and an associated bundle of curves via (70) which are such that both (73) and (74) hold. On the other hand, once this collection of $D(z) \subset P(z)$, $z \in [0,1]$ and the associated bundle have been identified, then the tube $\cup_{z \in [0,1]} D(z) \approx \cup_{x \in D(0)} \Gamma_x \approx \cup_{z \in [0,1]} C(z)$ is the transition tube. In fact, such an adjustment procedure is algorithmic in nature, i.e. it can be turned into an numerical scheme to determine the tube, as explained in Sect. 7.

Finally, observe that the adjustment procedure described above determines the geometric location of the tube and the bundle of curves Γ_x, but not their parametrization, i.e. even if we have localized the tube and the bundle of curves Γ_x, we still do not know where are $D(z)$ and $\varphi^x(z)$ along Γ_x: this is important since $D(z)$ is the local approximation of the isocommittor surface $S(z) = \{q(x) = z\}$. We handle this question in the next section.

6.2 Tube Parametrization and Committor Function

In this section, we assume that the geometric location of the tube $\cup_{z \in [0,1]} D(z) \approx \cup_{z \in [0,1]} C(z)$ has been determined by the adjustment procedure described at the end of Sect. 6.1 and that so has the bundle of curves Γ_x which are the geometric location of the streamlines inside the tube. For simplicity we will denote by Γ a specific curve within the bundle (it is not important which one in particular), and we ask how to parametrize it as $\Gamma = \{\varphi(z), z \in [0,1]\}$ where, as in (67), $\varphi(z)$ is the location along Γ where Γ intersects $D(z)$. Recall that this is important, because $D(z) \approx C(z) \subset S(z)$, i.e. it is a local approximation of the isocommittor surface $S(z) = \{q(x) = z\}$: in other words, $q(x) \approx q(\varphi(z)) \equiv z$ if $x \in D(z)$.

We will proceed as follows: let $\phi(s)$ be the parametrization of Γ by arc-length, i.e. $\Gamma = \{\phi(s) : s \in [0, L_\Gamma]\}$ where L_Γ is the length of Γ, and let $\bar{P}(s)$ be the plane which intersects Γ at point $\phi(s)$ in such a way that the unit normal to $\bar{P}(s)$, which we shall denote by $\hat{\nu}(s)$, satisfies the equivalent of (70): $d\phi_i/ds$ is parallel to $\sum_{j=1}^n a_{ij}(\phi(s))\hat{\nu}_j(s)$. Since by construction $|d\phi/ds| = 1$, where $|d\phi/ds|$ is the Euclidean norm of the vector $(d\phi_1/ds, \ldots, d\phi_n/ds)^T$, we have

$$\frac{d\phi_i(s)}{ds} = \frac{\sum_{j=1}^n a_{ij}(\phi(s))\hat{\nu}_j(s)}{|a(\phi(s))\hat{\nu}(s)|} \tag{75}$$

where $|a(\phi(s))\hat{\nu}(s)|$ is the Euclidean norm of

$$(\sum_{j=1}^n a_{1,j}(\phi(s))\hat{\nu}_j(s), \ldots, \sum_{j=1}^n a_{n,j}(\phi(s))\hat{\nu}_j(s))^T.$$

Since Γ is assumed to be known, all these quantities are known and what remains to be done is relate $\varphi(z)$ to $\phi(s)$, $\hat{n}(z)$ to $\hat{\nu}(s)$ and $P(z)$ to $\bar{P}(s)$, i.e. find the function $f(z)$ such that $\varphi(z) = \phi(f(z))$, $\hat{n}(z) = \hat{\nu}(f(z))$ and $P(z) = \bar{P}(f(z))$.

To do so, recall that $z = q(\varphi(z)) = q(\phi(f(z)))$. Differentiating with respect to z, we deduce that

$$\begin{aligned}
1 &= \frac{df(z)}{dz} \sum_{i=1}^n \frac{d\phi_i(f(z))}{ds} \frac{\partial q(\phi(f(z)))}{\partial x_i} \\
&= \frac{df(z)}{dz} \frac{\sum_{i,j=1}^n a_{ij}(\phi(f(z)))\hat{\nu}_j(f(z))\partial q(\phi(f(z)))/\partial x_i}{|a(\phi(f(z)))\hat{\nu}(f(z))|}
\end{aligned} \tag{76}$$

where we used (75) in the second step. This equation is not closed because we do not know $\partial q/\partial x_i$. To get the latter, let $\Omega(z)$ be the region enclosed between ∂A and $S(z)$ and integrate (60) over $\Omega(z)$. Using the divergence theorem, this gives

$$\begin{aligned}
0 &= \int_{\Omega(z)} \sum_{i,j=1}^n \frac{\partial}{\partial x_i} \left(a_{ij}(x)m(x)\frac{\partial q(x)}{\partial x_j} \right) \\
&= -\int_{\partial A} \sum_{i,j=1}^n \hat{n}_{\partial A,i}(x)a_{ij}(x)\frac{\partial q(x)}{\partial x_j}m(x)d\sigma_{\partial A}(x) \\
&\quad + \int_{S(z)} \sum_{i,j=1}^n \hat{n}_{S(z),i}(x)a_{ij}(x)\frac{\partial q(x)}{\partial x_j}m(x)d\sigma_{S(z)}(x)
\end{aligned} \tag{77}$$

where $\hat{n}_{\partial A}(x)$ is the unit normal to ∂A pointing outward A and $d\sigma_{\partial A}(x)$ is the surface element on ∂A. From (42) and (62), both integrals at the right-hand side of (77) are the reaction rate k_{AB}, so this equation can be written as

$$k_{AB} = \int_{S(z)} \sum_{i,j=1}^n \hat{n}_{S(z),i}(x)a_{ij}(x)\frac{\partial q(x)}{\partial x_j}m(x)d\sigma_{S(z)}(x) \tag{78}$$

So far, we have not made any approximation, i.e. (78) is exact. Let us now use the localized tube assumption, i.e. (63), (64) and (65): it implies that

$$k_{AB} \approx \sum_{i,j=1}^{n} \hat{n}_i(z) a_{ij}(\varphi(z)) \frac{\partial q(\varphi(z))}{\partial x_j} Z_{P(z)}$$

$$= \sum_{i,j=1}^{n} \hat{\nu}_i(f(z)) a_{ij}(\phi(f(z))) \frac{\partial q(\phi(f(z)))}{\partial x_j} Z_{\bar{P}(f(z))} \tag{79}$$

where given any plane P, Z_P is defined as

$$Z_P = \int_P m(x) d\sigma_P(x) \tag{80}$$

(79) gives the information on $\partial q/\partial x_i$ that we were missing in (76): combining these two equations we arrive at the following equation for $f(z)$:

$$\frac{df(z)}{dz} = \frac{1}{k_{AB}} Z_{\bar{P}(f(z))} |a(\phi(f(z)))\hat{\nu}(f(z))| \tag{81}$$

This differential equation actually determines both $f(z)$ and k_{AB}. Indeed it is a first order differential equation, but with two boundary conditions for $f(z)$: $f(0) = 0$ (since $\phi(0) = \varphi(0)$) and $f(1) = L_\Gamma$ (since $\phi(L_\Gamma) = \varphi(1)$). Due to the special structure of (81) (namely because k_{AB} enters as an overall multiplicative factor at the right hand-side), we can actually imposes these two boundary conditions, which gives $f(z)$ and fixes k_{AB} as well.

Once $f(z)$ is known, using $\varphi(z) = \phi(f(z))$ and (75), we have

$$\frac{d\varphi_i(z)}{dz} = \frac{df(z)}{dz} \frac{d\phi_i(f(z))}{ds} = \frac{df(z)}{dz} \frac{\sum_{j=1}^{n} a_{ij}(\varphi(z))\hat{n}_j(z)}{|a(\varphi(z))\hat{n}(z)|} \tag{82}$$

This equation gives the proportionality factor that was missing in (70).

6.3 Reaction Rate and Probability to be Reactive

Once the locations of $\varphi(z)$ and the associated planes $P(z)$ along Γ are known, we can give simple expressions for the various quantities of interest in terms of these objects.

First, we can re-express the rate k_{AB} in terms of $\varphi(z)$ and $\hat{n}(z)$ alone as follows. Using $\phi(f(z)) = \varphi(z)$ and $\hat{\nu}(f(z)) = \hat{n}(z)$ in (76), this equation can be written as

$$1 = \frac{df(z)}{dz} \frac{\sum_{i,j=1}^{n} a_{ij}(\varphi(z))\hat{n}_j(z)\partial q(\varphi(z))/\partial x_i}{|a(\varphi(z))\hat{n}(z)|} \tag{83}$$

where $|a(\varphi(z))\hat{n}(z)|$ denotes the Euclidean norm of

$$(\textstyle\sum_{j=1}^{n} a_{1,j}(\varphi(z))\hat{n}_j(z), \ldots, \sum_{j=1}^{n} a_{n,j}(\varphi(z))\hat{n}_j(z))^T.$$

Solving (83) in $\sum_{i,j=1}^{n} a_{ij}(\varphi(z))\hat{n}_j(z)\partial q(\varphi(z))/\partial x_i$ and using the result in the first equality in (79) gives

$$k_{AB} \approx \frac{|a(\varphi(z))\hat{n}(z)|Z_{P(z)}}{df(z)/dz} \tag{84}$$

Multiplying both members by $df(z)/dz$ and integrating on $z \in [0,1]$ using $f(0) = 0$ and $f(1) = L_\Gamma$, leads to

$$L_\Gamma k_{AB} \approx \int_0^1 |a(\varphi(z))\hat{n}(z)|Z_{P(z)}dz \tag{85}$$

Since $L_\Gamma = \int_0^1 |d\varphi/dz|dz$, where $|d\varphi/dz|$ denotes the Euclidean norm of the vector $(d\varphi_1/dz, \ldots, d\varphi_n/dz)^T$, we arrive at the following expression for k_{AB}:

$$k_{AB} \approx \frac{\int_0^1 |a(\varphi(z))\hat{n}(z)|Z_{P(z)}dz}{\int_0^1 |d\varphi/dz|dz} \tag{86}$$

Proceeding similarly, we can express Z_{AB} in terms of $\varphi(z)$ and $\hat{n}(z)$. Observe that

$$\begin{aligned}
Z_{AB} &= \int_{\Omega \setminus (A \cup B)} q(x)(1 - q(x))m(x)dx \\
&= \int_0^1 z(1 - z)\left(\int_{S(z)} \frac{m(x)}{|\nabla q(x)|} d\sigma_{S(z)}(x) \right) dz \\
&\approx \int_0^1 z(1 - z)\frac{Z_{P(z)}}{|\nabla q(\varphi(z))|} dz
\end{aligned} \tag{87}$$

where we used (63) and (64) in the last step. Since $z = q(\varphi(z))$, we have

$$1 = \sum_{i=1}^n \frac{d\varphi_i(z)}{dz}\frac{\partial q(\varphi(z))}{\partial x_i} = |\nabla q(\varphi(z))| \sum_{i=1}^n \frac{d\varphi_i(z)}{dz}\hat{n}_i(z) \tag{88}$$

Solving this equation for $|\nabla q(\varphi(z))|$ and using the result in (87) gives

$$Z_{AB} \approx \int_0^1 z(1 - z) \sum_{i=1}^n \frac{d\varphi_i(z)}{dz}\hat{n}_i(z)Z_{P(z)}dz \tag{89}$$

6.4 Working in Collective Variables

In Sect. 6.1 we assumed that the local transition tube assumption was valid in the original variables x. This is a strong assumption which may fail in many situations of interest. On the other hand, it is reasonable to assume that the range of validity of the local transition tube assumption may be extended in some suitable set of collective variables which are sufficient to describe the reaction. This is the situation that we describe in this section.

Suppose that $q(x)$ depends of x only through a set of *collective variables*, $(\theta_1(x), \ldots, \theta_N(x))$, i.e.

$$q(x) \approx Q(\theta_1(x), \ldots, \theta_N(x)), \tag{90}$$

for some unknown function $Q(\theta_1, \ldots, \theta_N)$. The collective variables can go all the way from being a set of variables introduced to account for some symmetries (like e.g. translation or rotation), in which case $N < n$ but $N \approx n$ and there is no approximation in (90), to being a much smaller set of variables (like e.g torsion angles, bond distances, etc.), chosen because one has reason to believe a priori that this set of variables is large enough to describe the reaction. In this second case, $N \ll n$, but the there is no reason to assume that N is small, i.e. there could still be many collective variables $(\theta_1(x), \ldots, \theta_N(x))$, so the approximation made in (90) is not as restrictive as it may look at first sight.

What is the equation for $Q(\theta_1, \ldots, \theta_n)$ if one assumes that (90) holds? Let us go back to (60): to incorporate (90) in (60), observe that (60) is the Euler-Lagrange equation associated with the minimization of [6]

$$I(q) = \int_\Omega \sum_{i,j=1}^n a_{ij}(x) \frac{\partial q(x)}{\partial x_i} \frac{\partial q(x)}{\partial x_j} m(x) dx \tag{91}$$

over all $q(x)$ satisfying $q(x)|_{x \in A} = 0$ and $q(x)|_{x \in B} = 0$. Consistent with (90), we can minimize (91) over all functions $q(x)$ in the form of (90) and get a good approximation of the committor function. Using (90) in (91), this object function can be reduced to the following object function for Q:

$$\bar{I}(Q) = \int_\Omega \sum_{k,l=1}^N \sum_{i,j=1}^n a_{ij}(x) \frac{\partial \theta_k(x)}{\partial x_i} \frac{\partial \theta_l(x)}{\partial x_j} \left(\frac{\partial Q(\theta^*)}{\partial \theta_k} \frac{\partial Q(\theta^*)}{\partial \theta_l} \right)_{\theta(x)=\theta^*} m(x) dx$$

$$= \int_{\Omega_\theta} \sum_{k,l=1}^N A_{kl}(\theta^*) \frac{\partial Q(\theta^*)}{\partial \theta_k} \frac{\partial Q(\theta^*)}{\partial \theta_l} M(\theta^*) d\theta^* \tag{92}$$

where Ω_θ, is the image of Ω by the map $\theta : \mathbb{R}^n \mapsto \mathbb{R}^N$,

$$M(\theta^*) = \int_\Omega m(x) \delta(\theta_1(x) - \theta_1^*) \cdots \delta(\theta_N(x) - \theta_N^*) dx \tag{93}$$

is the equilibrium probability density of the collective variable $\theta(x)$ and

$$A_{kl}(\theta^*) = M^{-1}(\theta^*) \int_\Omega \sum_{i,j=1}^n a_{ij}(x) \frac{\partial \theta_k(x)}{\partial x_i} \frac{\partial \theta_l(x)}{\partial x_j}$$

$$\times m(x) \delta(\theta_1(x) - \theta_1^*) \cdots \delta(\theta_N(x) - \theta_N^*) dx \tag{94}$$

$$\equiv \left\langle \sum_{i,j=1}^n a_{ij}(x) \frac{\partial \theta_k(x)}{\partial x_i} \frac{\partial \theta_l(x)}{\partial x_j} \right\rangle_{\theta(x)=\theta^*}.$$

---

[6] Observe that the object function (91) may be used as a starting point to implement other approximations than (90). Observe also that as a direct consequence of (43)), $k_{AB} = \min I(q)$ over all $q(x)$ satisfying $q(x)|_{x \in A} = 0$ and $q(x)|_{x \in B} = 1$, i.e. the actual $q(x)$ solution of (60) is the one that minimizes the reaction rate.

is the conditional average of $\sum_{i,j=1}^n a_{ij}(x)(\partial\theta_k/\partial x_i)(\partial\theta_l/\partial x_j)$ in $\theta(x) = \theta^*$. (92) must be minimized over all functions $Q(\theta)$ satisfying $Q(\theta)|_{\theta \in A_\theta} = 0$ and $Q(\theta)|_{\theta \in B_\theta} = 1$ where A_θ and B_θ are the images of A and B, respectively, by the map $\theta : \mathbb{R}^n \mapsto \mathbb{R}^N$

Equation (92) is simpler than (60) but otherwise structurally identical to it. In particular, the Euler-Lagrange equation associated with minimizing (92) has the form of a backward Kolmogorov equation:

$$\begin{cases} 0 = \sum_{k,l=1}^N \frac{\partial}{\partial\theta_k}\left(A_{kl}(\theta)M(\theta)\frac{\partial Q(\theta)}{\partial\theta_l}\right) \\ Q(\theta)|_{\theta \in \partial A_\theta} = 0, \qquad Q(\theta)|_{\theta \in \partial B_\theta} = 1. \end{cases} \tag{95}$$

Therefore (95) instead of (60) can be taken as a starting point for what we did in Sect. 6.1. Observe that (95) is the backward Kolmogorov equation associated with the stochastic differential equation (compare (19))

$$\dot{\theta}(\tau) = B(\theta(\tau)) + \sqrt{2}\bar{\sigma}(\theta(\tau))\eta(\tau) \tag{96}$$

where $\eta_k(\tau)$ is a white-noise satisfying $\langle\eta_k(\tau)\eta_l(\tau')\rangle = \delta_{kl}\delta(\tau - \tau')$, $\bar{\sigma}(\theta)$ is such that $\sum_{l=1}^N \bar{\sigma}_{kl}(\theta)\bar{\sigma}_{k'l}(\theta) = A_{kk'}(\theta)$ and

$$B_k(\theta) = \frac{1}{M(\theta)}\sum_{l=1}^N \frac{\partial}{\partial\theta_l}(M(\theta)A_{kl}(\theta)) \tag{97}$$

The time τ in (96) is an artificial time whose relation with the physical time t in (19) is unspecified by our argument. In other words, (96) allows one to understand the mechanism of the reaction but not its timing or rate. To obtain the latter, we must go back to the object function in (92) and observe that $k_{AB} = \min_{q(x)} I(q) \approx \min_{Q(\theta)} \bar{I}(Q)$, i.e. within approximation (90) we have

$$k_{AB} = \int_{\Omega_\theta} \sum_{k,l=1}^N A_{kl}(\theta)\frac{\partial Q(\theta)}{\partial\theta_k}\frac{\partial Q(\theta)}{\partial\theta_l}M(\theta)d\theta \tag{98}$$

where $Q(\theta)$ is the solution of (95).

6.5 The Case of the Langevin Dynamics

Consider now the Langevin equation in (21). This is an important example which is not time-reversible, i.e. $q_+(r, v) \neq 1 - q_-(r, v)$ (in fact it is easy to see that $q_+(r, v) = 1 - q_-(r, -v)$ since, in order to know what happens when we revert time, we need to revert the momenta). The backward Kolmogorov equation (23) in the present case is

$$0 = \sum_{i=1}^{m} \left(v_i \frac{\partial q_+(r,v)}{\partial r_i} - \mu_i^{-1} \frac{\partial V(r)}{\partial r_i} \frac{\partial q_+(r,v)}{\partial v_i} \right)$$
$$+ \gamma \sum_{i=1}^{m} \left(-\mu_i^{-1} v_i \frac{\partial q_+(r,v)}{\partial v_i} + \mu_i^{-2} k_B T \frac{\partial^2 q_+(r,v)}{\partial v_i^2} \right) \tag{99}$$

plus boundary conditions. To deal with this case, let us assume that $q_+(x,v)$ can be approximated by a function independent of the momenta v and depending on the positions only through the collective variables $(\theta_1(r), \ldots, \theta_N(r))$, i.e.

$$q_+(r,v) \approx Q(\theta_1(r), \ldots, \theta_N(r)) \tag{100}$$

for some unknown function $Q(\theta_1, \ldots, \theta_N)$. (Observe that since $q_-(r,v) = 1 - q_+(r,-v)$, we would then also have $q_-(r,v) \approx 1 - Q(\theta_1(r), \ldots, \theta_N(r))$ for the same function Q as in the time-reversible case.) To incorporate the approximation (100) into (99), observe that the solution of this equation minimizes the object function

$$I(q_+) = \int_{\Omega \setminus (A \cup B)} \left| \sum_{i=1}^{m} \left(v_i \frac{\partial q_+(r,v)}{\partial r_i} - \mu_i^{-1} \frac{\partial V(r)}{\partial r_i} \frac{\partial q_+(r,v)}{\partial v_i} \right) \right.$$
$$\left. + \gamma \sum_{i=1}^{m} \left(-\mu_i^{-1} v_i \frac{\partial q_+(r,v)}{\partial v_i} + \mu_i^{-2} k_B T \frac{\partial^2 q_+(r,v)}{\partial v_i^2} \right) \right|^2 e^{-\beta H(r,v)} dr dv \tag{101}$$

where $H(r,v) = \frac{1}{2} \sum_{i=1}^{m} \mu_i v_i^2 + V(r)$ is the Hamiltonian. Now insert the ansatz (99) into (101). It can be checked that (101) then reduces precisely to (92) with $m(x) = Z^{-1} e^{-\beta V(r)}$ and $a_{ij}(x) = \mu_i^{-1} \delta_{ij}$. In other words, we can again apply all what we did before in Sects. 6.1–6.4 to the present case.

It should be stressed, however, that the approximation in (100) may be adequate to describe the mechanism of the transition but it does not allow us to compute the rate of the reaction. Indeed, from (43), the reaction rate in the present case is

$$k_{AB} = Z_H^{-1} k_B T \gamma \int_{\Omega \setminus (A \cup B)} \sum_{i=1}^{m} \mu_i^{-2} \left(\frac{\partial q_+(r,v)}{\partial v_i} \right)^2 e^{-\beta H(r,v)} dr dv \tag{102}$$

where $Z_H = \int_{\mathbb{R}^m \times \mathbb{R}^m} e^{-\beta H(r,v)} dr dv$ ($Z_H^{-1} e^{-\beta H(r,v)}$ is the equilibrium probability density function of the Langevin equation (21)). Thus (102) depends on the corrections to (100) which are not accounted for by this approximation. Observe that this does not make (100) inconsistent: the rate depends on the derivatives of $q(r,v)$, which may be more difficult to approximate that $q(r,v)$ itself. How is this possible is illustrated by the following example.

Remark: the high friction limit of (99).

Suppose that one is interested in a situation where the friction coefficient in (99) is big, $\gamma \gg 1$. In this case (99) can be solved by singular perturbation techniques: here we used the techniques developed in [24] where the proofs of the claims that we make below can be found. Look for a solution of (99) in the form of

$$q_+(r, v) = q_0(r, v) + \gamma^{-1} q_1(r, v) + \gamma^{-2} q_2(r, v) + \cdots \tag{103}$$

Inserting this ansatz into (99) and equating equal powers of γ leads to the hierarchy:

$$
\begin{cases}
\displaystyle\sum_{i=1}^{m} \left(-\mu_i^{-1} v_i \frac{\partial q_0(r, v)}{\partial v_i} + \mu_i^{-2} k_B T \sum_{i=1}^{n} \frac{\partial^2 q_0(r, v)}{\partial v_i^2} \right) = 0, \\[2.5ex]
\displaystyle\sum_{i=1}^{m} \left(-\mu_i^{-1} v_i \frac{\partial q_1(r, v)}{\partial v_i} + \mu_i^{-2} k_B T \sum_{i=1}^{n} \frac{\partial^2 q_1(r, v)}{\partial v_i^2} \right) \\[2.5ex]
\qquad = -\displaystyle\sum_{i=1}^{m} \left(v_i \frac{\partial q_0(r, v)}{\partial r_i} - \mu_i^{-1} \frac{\partial V(r)}{\partial r_i} \frac{\partial q_0(r, v)}{\partial v_i} \right), \\[2.5ex]
\displaystyle\sum_{i=1}^{m} \left(-\mu_i^{-1} v_i \frac{\partial q_2(r, v)}{\partial v_i} + \mu_i^{-2} k_B T \sum_{i=1}^{m} \frac{\partial^2 q_2(r, v)}{\partial v_i^2} \right) \\[2.5ex]
\qquad = -\displaystyle\sum_{i=1}^{m} \left(v_i \frac{\partial q_1(r, v)}{\partial r_i} - \mu_i^{-1} \frac{\partial V(r)}{\partial r_i} \frac{\partial q_1(r, v)}{\partial v_i} \right), \\[2ex]
\cdots
\end{cases}
\tag{104}
$$

The first equation implies that $q_0(r, v)$ belongs to the null-space of the operator at the left hand-side, $L_v \equiv \sum_{i=1}^{m} (-\mu_i^{-1} v_i \partial/\partial v_i + \mu_i^{-2} k_B T \partial^2/\partial v_i^2)$. It can be shown [24] that this requires that $q_0(r, v)$ be a function of r only, i.e. $q_0(r, v) = Q(r)$. Since the null-space of the operator L_v is non-trivial, this operator is not invertible and the second and the third equations in (104) require solvability conditions. To understand how these solvability conditions come about, multiply both sides of the second equation in (104) by $(2\pi k_B T)^{-m/2} e^{-\frac{1}{2}\beta|v|_\mu^2}$ where $|v|_\mu^2 = \sum_{i=1}^{m} \mu_i v_i^2$, and integrate on v. Integrating by parts at the left hand-side using

$$0 = \sum_{i=1}^{m} \left(\mu_i^{-1} \frac{\partial}{\partial v_i} \left(v_i e^{-\frac{1}{2}\beta|v|_\mu^2} \right) + \mu_i^{-2} k_B T \frac{\partial^2}{\partial v_i^2} e^{-\frac{1}{2}\beta|v|_\mu^2} \right) \tag{105}$$

we arrive at the condition (using $\partial q_0/\partial v_i = \partial Q/\partial v_i = 0$)

$$0 = -\int_{\mathbb{R}^n} (2\pi k_B T)^{-m/2} e^{-\frac{1}{2}\beta|v|_\mu^2} \sum_{i=1}^{m} v_i \frac{\partial Q(r)}{\partial r_i} dv \tag{106}$$

This condition is the solvability condition for the second equation in (104); our argument above shows that it is a necessary condition in order that this equation be solvable and it can be proven [24] that this condition is also sufficient to that end. Equation (106) is automatically satisfied since

$\partial Q/\partial r_i$ can be pulled out of the integral (it is independent of v) and $\int_{\mathbb{R}^m} (2\pi k_B T)^{-m/2} e^{-\frac{1}{2}\beta|v|^2_\mu} v_i dv = 0$. This means that one can solve the equation for q_1: it is easy to see that

$$q_1(r, v) = \sum_{i=1}^{n} \mu_i v_i \frac{\partial Q(r)}{\partial r_i}. \tag{107}$$

is a solution to this equation and it can be shown [24] that this solution is in fact unique. Inserting this expression into the equation for q_2 in (104) gives:

$$\sum_{i=1}^{m} \left(-\mu_i^{-1} v_i \frac{\partial q_2(r, v)}{\partial v_i} + \mu_i^{-2} k_B T \frac{\partial^2 q_2(r, v)}{\partial v_i^2} \right)$$
$$= -\sum_{i,j=1}^{n} \mu_i v_i v_j \frac{\partial^2 Q(r)}{\partial r_i \partial r_j} + \sum_{i=1}^{n} \frac{\partial V(r)}{\partial r_i} \frac{\partial Q(r)}{\partial r_i}. \tag{108}$$

The solvability condition for this equation is also obtained by multiplying both sides by $(2\pi k_B T)^{-m/2} e^{-\frac{1}{2}\beta|v|^2_\mu}$ and integrating over v. Integrating by parts at the left hand-side and using (105), we arrive at

$$0 = \int_{\mathbb{R}^n} (2\pi k_B T)^{-n/2} e^{-\frac{1}{2}\beta|v|^2_\mu} \left(-\sum_{i,j=1}^{n} \mu_i v_i v_j \frac{\partial^2 Q(r)}{\partial r_i \partial r_j} + \sum_{i=1}^{n} \frac{\partial V(r)}{\partial r_i} \frac{\partial Q(r)}{\partial r_i} \right) dv$$
$$= \sum_{i=1}^{n} \left(-k_B T \frac{\partial^2 Q(r)}{\partial r_i^2} + \frac{\partial V(r)}{\partial r_i} \frac{\partial Q(r)}{\partial r_i} \right), \tag{109}$$

This equation is the limiting equation for $Q(r)$: it is the backward Kolmogorov equation associated with the overdamped equation obtained from (21) in the limit as $\gamma \to \infty$ (the overdamped dynamics is considered in Sect. 6.6). On the other hand, inserting (107) into (103) gives

$$q_+(r, v) = Q(r) + \gamma^{-1} \sum_{i=1}^{n} \mu_i v_i \frac{\partial Q(r)}{\partial r_i} + O(\gamma^{-2}) \tag{110}$$

This expression is consistent with (100) to leading order in γ, but it is the next order correction which gives the rate since $Q(r)$ is independent of v and (102) depends on the gradient of $q_+(r, v)$ with respect to v. In fact, using (110) in (102) and integrating out the momenta, we see that

$$k_{AB} = Z^{-1} k_B T \gamma^{-1} \int_{\mathbb{R}^m \setminus (A_r \cup B_r)} \left(\sum_{i=1}^{n} \frac{\partial Q(r)}{\partial r_i} \right)^2 e^{-\beta V(r)} dr \tag{111}$$

where A_r and B_r are the projections of A and B in configuration space.

Of course, in general one is not interested in the high friction limit in the original Cartesian space. However, assuming that $q(r, v) \approx Q(\theta_1(r), \dots, \theta_N(r))$, it may be reasonable to suppose that the dynamics is overdamped at the level of the θ_j's. In this case a generalization of (111) where one would take $Q(r) \approx Q(\theta_1(r), \ldots, \theta_N(r))$ may be appropriate provided that one can decide what γ is. On the other hand, it may be safer to estimate the rate via other techniques, for instance by estimating t_{AB} by some other means (e.g. using reactive trajectories generated by TPS or obtained by initiating trajectories from the isocommittor $\frac{1}{2}$ surface) and using (16) with Z_{AB} given by (89) (the latter expression being dominated by the leading order term $q(r, v) \approx Q(\theta_1(r), \ldots, \theta_N(r))$ since (11) depends on q itself and not on its gradient).

6.6 Remark: the Role of the Minimum Energy Path

Consider the *overdamped dynamics* when (19) takes the form

$$\gamma \dot{r}_i(t) = -\frac{\partial V(r(t))}{\partial r_i} + \sqrt{2k_B T \gamma}\, \eta_i(t) \tag{112}$$

As explained in Sect. 6.5, this equation arises from (21) in the high friction limit as $\gamma \gg 1$. The equilibrium probability density associated with (112) is

$$m(r) = Z^{-1} e^{-\beta V(r)} \quad \text{where} \quad Z = \int_{\mathbb{R}^m} e^{-\beta V(r)} dr \tag{113}$$

In addition, $q(r)$ satisfies (60), which in the present context is (this is nothing but (109))

$$\begin{cases} 0 = -\nabla V(r) \cdot \nabla q(r) + k_B T \, \Delta q(r) \\ q(r)|_{r \in \partial A} = 0, \qquad q(r)|_{r \in \partial B} = 1 \,, \end{cases} \tag{114}$$

where $\nabla q = (\partial q/\partial r_1, \ldots, \partial q/\partial r_m)^T$, the dot denotes the standard inner product and $\Delta q = \sum_{i=1}^m \partial^2 q/\partial r_i^2$. In the present case the calculations in Sect. 6.1 indicates that the curves $\Gamma_x = \{\varphi^x(z) : z \in [0,1]\}$ must be perpendicular to each plane $P(z)$ (this is (70)). Quite interestingly, if one makes the additional assumption that the temperature is so small that the density $e^{-\beta V(r)}$ in (113) is strongly peaked at the minimum of $V(r)$ in $P(z)$, then each $D(z) \approx C(z)$ shrinks to a single point located at the minimum of $V(r)$ in $P(z)$; hence the tube $\cup_{z \in [0,1]} C(z) \approx \cup_{z \in [0,1]} D(z)$ shrinks to a single curve, $\Gamma = \{\varphi(z) : z \in [0,1]\}$, which is such that

$$\varphi(z) = \text{minimum of } V(r) \text{ in } P(z) \,. \tag{115}$$

This relation requires that

$$\hat{n}_i(z) \quad \text{parallel to} \quad \frac{\partial V(\varphi(z))}{\partial r_i} \tag{116}$$

Combining this equation with (70), which in the present context reads

$$\frac{d\varphi(z)}{dz} \quad \text{parallel to} \quad \hat{n}_i(z) \tag{117}$$

we therefore deduce that

$$\{\varphi(z) : z \in [0,1]\} = \text{curve } \Gamma \text{ such that } \nabla V(r) \text{ along } \Gamma \text{ is parallel to } \Gamma \tag{118}$$

This equation is sometimes written as $0 = [\nabla V(r)]^{\perp}$, which is supposed to hold everywhere along Γ and where $[\cdot]^{\perp}$ denotes the projection perpendicular to Γ.

The curve defined by (118) is a very well-known object: it is the *minimum energy path* (MEP) which connects two minima of $V(r)$ via a saddle point. How, why and by whom the MEP was first introduced in the context of molecular dynamics is not clear (to the author at least). The argument above indicates that it is the relevant object that concentrates most of the probability current of the reactive trajectories in its vicinity in the case of the overdamped dynamics when the temperature is small and the potential is sufficiently smooth (otherwise, if $V(r)$ has many critical points, (118) has many solutions, none of which taken alone is relevant). It is however not clear when such situations arise, and the MEP may often prove irrelevant.

7 Some Computational Aspects

In this section, we briefly discuss the computational aspects of the theory discussed so far. As explained in Sect. 6, the committors functions $q_+(x)$ and $q_-(x)$ are the key objects to determine. Whereas this cannot be done by solving (23) and (24) directly (in most applications of practical interest these equations are simply too large for standard numerical techniques such as finite differences or finite elements), $q_+(x)$ and $q_-(x)$ can be determined locally in the situations where the localized tube assumption discussed in Sect. 6.1 (or its equivalent in collective variables, see Sects. 6.4 and 6.5) is valid. This is the essence of the *string method* which we briefly discuss now. Since the focus of the present chapter is mainly the theory, we will be rather sketchy here and content ourselves with indicating how the ideas presented in Sect. 6 naturally lead to the string method: for the details on how to implement this algorithm, we refer the reader to the original references [9–12, 23, 26–28].

As explained in Sect. 6.1, if the localized tube assumption is valid, it is possible to identify this tube by moving a single curve, Γ, viewed as a candidate for the location of a streamline in the tube, to get a collection of planes $P(z)$ $z \in [0,1]$ via (71), a collection of regions $D(z) \subset P(z)$, $z \in [0,1]$ via (65) and $Q(z) \subset P(z)$, $z \in [0,1]$ via (72), and an associated bundle of curves via (70) which are such that both (73) and (74) hold. Now, since the location of the tube depends at least in part on the equilibrium probability density $m(x)$, it is natural to simplify the procedure described at the end of Sect. 6.1 and try to estimate a priori (that is, guess) the location of a specific streamline inside the tube based on the value of $m(x)$ inside the tube. There are two natural a priori estimates for the location of the streamline: the first leads to the zero-temperature string method and is explained in Sect. 7.1; the second leads to the finite temperature string method and is explained in Sect. 7.2. Both methods can be generalized to work in collective variables, as explained in Sect. 7.3. As shown below, making an a priori estimate for the location of the streamline simplifies the actual identification of the tube, though it introduces an additional assumption beyond the localized tube assumption of Sect. 6.1 and hence is somewhat more restrictive.

7.1 The Zero-Temperature String Method

The first natural a priori estimate for the location of a specific streamline $\Gamma = \{\varphi(z) : z \in [0, 1]\}$ inside the tube is to take for $\varphi(z)$ the location in $C(z)$ where $m(x)$ is maximum, i.e. take

$$\varphi(z) = \text{location of the maximum of } m(x) \text{ in } C(z). \tag{119}$$

Since $C(z) \approx D(z) \subset P(z)$, this implies that $\varphi(z)$ satisfies

$$\frac{\partial m(\varphi(z))}{\partial x_i} \text{ parallel to } \hat{n}_i(z) \tag{120}$$

This equation and (70) can then be combined into the following single equation for the curve:

$$\{\varphi(z) : z \in [0, 1]\} = \text{curve } \Gamma \text{ such that } a(x)\nabla m(x) \text{ along } \Gamma \text{ is parallel to } \Gamma \tag{121}$$

where $a(x)\nabla m(x)$ is the vector

$$(\textstyle\sum_{j=1}^{n} a_{1,j}(x)\partial m/\partial x_j, \ldots, \sum_{j=1}^{n} a_{n,j}(x)\partial m/\partial x_j)^T$$

Using (59), (121) can also be written as

$$\{\varphi(z) : z \in [0, 1]\} = \text{curve } \Gamma \text{ such that } b(x)$$
$$- \text{div } (a(x)) \text{ along } \Gamma \text{ is parallel to } \Gamma \tag{122}$$

where $\text{div}(a(x))$ is the vector with components

$$(\text{div}(a(x)))_i = \sum_{j=1}^{n} \frac{\partial a_{ij}(x)}{\partial x_j} \tag{123}$$

Of course, even if the localized tube assumption is valid (and hence the tube can be identified by the adjustment procedure described at the end of Sect. 6.1), since (119) may not be consistent with being the location of a streamline in the tube, there is no guarantee that this curve will lead to regions $D(z)$ defined in the planes $P(z)$ via (65) and an associated bundle of curves Γ_x via (70) such that both (73) and (74) are valid. In particular, if $m(x)$ has many critical points (local minima, maxima, etc.) (119) may lead to a curve which is too wiggly, in which case the potential regions $Q(z) \subset P(z)$ where the bundle lives (see (72)) may be too small to contain the regions $D(z) \subset P(z)$. But again, this does not mean that the localized tube assumption is invalid, simply that the a priori estimate (119) is to restrictive. How to improve upon (119) is explained in Sect. 7.2.

For now, however, let us consider a situation where (119) is valid and the problem reduces to determining the curve Γ satisfying (122). To do so, let Γ_0 be an initial guess for Γ. We wish to find a sequence of curves, Γ_0, Γ_1, etc. such that Γ_n converges towards the solution of (122) as $n \to \infty$. To write the iterative procedure which gives Γ_{n+1} from Γ_n, it is convenient to represent this curve parametrically as $\Gamma_n = \{\phi_n(\alpha) : \alpha \in [0,1]\}$ where α is some parameter not necessarily related to the parameter z used in $\varphi(z)$ in (122). How to choose and fix the parametrization of Γ_n will be explained in a moment, but for time being it suffices to recall that, once $\Gamma = \lim_{n\to\infty} \Gamma_n$ is found, no matter how it is parametrized, the actual $\varphi(z)$ entering (122) can be obtained from $\phi(\alpha) = \lim_{n\to\infty} \phi_n(\alpha)$ by the procedure explained in Sect. 6.2 and k_{AB} and Z_{AB} by the procedure explained in Sect. 6.3.

Now recall that (122) is the combination of (70) and (120). This suggests to use the following iterative procedure. Given $\Gamma_n = \{\phi_n(\alpha) : \alpha \in [0,1]\}$, construct the planes $P_n(\alpha)$ with unit normal $\hat{n}_n(\alpha)$ satisfying (this is the equivalent of (70))

$$\hat{n}_{n,i}(\alpha) \text{ parallel to } \sum_{j=1}^{n} a_{ij}^{-1}(\phi_n(\alpha)) \frac{\partial \phi_{n,j}(\alpha)}{\partial \alpha} \tag{124}$$

where $a_{ij}^{-1}(x)$ are the entries of the inverse of the diffusion matrix $a(x)$. Next find the maximum of $m(x)$ in each $P_n(\alpha)$, which can be done e.g. by running the dynamical equation

$$\dot{x}_i^{\alpha}(t) = \frac{\partial m(x^{\alpha}(t))}{\partial x_i} - \hat{n}_{n,i}(\alpha) \sum_{j=1}^{n} \hat{n}_{n,j}(\alpha) \frac{\partial m(x^{\alpha}(t))}{\partial x_j}, \qquad x^{\alpha}(0) = \varphi_n(\alpha) \tag{125}$$

The right hand side of this equation is the component in $P_n(\alpha)$ of the vector $\nabla m(x) = (\partial m/\partial x_1, \ldots, \partial m/\partial x_n)^T$, and the steady state solution of (125) is where this component is zero, i.e. $\nabla m(x)$ is parallel to $\hat{n}_n(\alpha)$, consistent with (120). Now construct the curve Γ_{n+1} out of the solutions of (125), i.e. take $\phi_{n+1}(\alpha) = x^{\alpha}(\Delta t)$ for some $\Delta t > 0$. If $\Gamma_{n+1} = \Gamma_n$ within required accuracy stop; if $\Gamma_{n+1} \neq \Gamma_n$ repeat. This iteration can be shown to converge, at least if the initial curve Γ_0 is close enough to the actual one satisfying (122) and Δt is small enough. Finally, once this curve has been identified, construct the associated regions $D(z)$ via (65), $Q(z)$ via (72) and the associated bundle of curves via (70). Then check whether (73) and (74) are satisfied. In practice, the regions $D(z)$ and $Q(z)$ (or part of the latter at least) can be identified by running some constrained version of (19) such that $x(t)$ stays in the plane and samples the density $m(x)$ restricted to this plane (which can be done by generalizing the idea of the blue moon sampling [6, 29] to (19), see [4]): $D(z)$ is then the region in which the constrained trajectory is most likely to stay, and each region $D(z)$ is contained in the region $Q(z)$ if the region $D(z)$ do not intersect with any other region $D(z')$, $z' \neq z$. If (73) and (74) are not satisfied, then either the localized tube assumption is invalid or the curve defined by (122) is not a streamline in the tube (the latter reason of failure can be excluded or eliminated by using (70) and (127) instead of (122), see below).

The above iterative procedure is straightforward to implement except maybe for the following catch. In practice the curves Γ_n must be discretized, i.e. one does not work with $\{\phi_n(\alpha) : \alpha \in [0,1]\}$ but rather with a collection of representative points or *images* along this curve, $\{\phi_n(\alpha_j) : j = 1, \ldots, J\}$ for some $J \in \mathbb{N}$. If we naively apply the iterative procedure outlined above to these discretized curves, then the representative points $\{\phi_n(\alpha_j) : j = 1, \ldots, J\}$ may end up clustering in certain regions along the continuous curve while leaving other regions along this curve under-resolved. This may lead to numerical instabilities. This difficulty is easy to solve: at each iteration, we can interpolate a continuous curve along the images $\{\phi_n(\alpha_j) : j = 1, \ldots, J\}$, then redistribute new images evenly along this interpolated curve and restart with the planes associated with the redistributed images rather than the original ones. This way, it is guaranteed that the images always remain evenly spaced and no numerical instability arises. The iterative procedure just described, including the reparametrization step is, in essence, the *zero-temperature string method*, introduced in [9, 26] and further developed and used in [10, 27]. [7] The zero-temperature string method is a generalization of the nudged elastic band method [19, 21].

7.2 The Finite Temperature String Method

Suppose now that (122) is a curve leading to regions $D(z)$ defined in the planes $P(z)$ via (65) and an associated bundle of curves Γ_x via (70) such that (73) and (74) are not satisfied. This means that the a priori estimate (119) is not consistent with being the location of a streamline and must be improved upon. A second natural a priori estimate for a streamline is to take the mean of $m(z)$ restricted to $C(z)$, i.e

$$\varphi(z) = Z_{C(z)}^{-1} \int_{C(z)} x m(x) d\sigma_{S(z)}(x) \tag{126}$$

---

[7] The zero-temperature string method was primarily formulated as a time-continuous evolution of a curve $\Gamma(t)$ rather than the time-discrete evolution as explained here, but this difference is not essential.

where $Z_{C(z)} = \int_{C(z)} m(x)d\sigma_{C(z)}(x)$. Since $P(z)$ approximates $S(z)$ inside the tube $\cup_{z\in[0,1]} C(z)$, from (63) it follows that the equation above can also be written as

$$\varphi(z) \approx Z_{P(z)}^{-1} \int_{P(z)} x m(x)d\sigma_{P(z)}(x) \tag{127}$$

Again, there is no guarantee that (127) will lead to regions $D(z)$ defined in the planes $P(z)$ via (65) and an associated bundle of curves Γ_x via (70) such that (73) and (74) are satisfied. If (127) fails to do the job because this curve is too wiggly again, then other guesses should be used that lead to smoother curves, or one should go back to the full procedure based on moving a whole bundle of curves as outlined in Sect. 6.1 (which, by the way, has not been done yet).

Assuming that (126) is valid, it is possible to determine the curve Γ satisfying (70) and (127) instead of (122) via a straightforward generalization of the iterative procedure described in Sect. 7.1. Instead of using (125) to find the maximum of $m(x)$ in each $P_n(\alpha)$, one uses a constrained version of (19) to sample $m(x)$ and get the mean of this density in each plane. The new curve $\Gamma_{n+1} = \{\phi_{n+1}(\alpha) : \alpha \in [0,1]\}$ is then constructed by taking for $\phi_{n+1}(\alpha)$ a weighted average between $\phi_n(\alpha)$ and the mean position in $P_n(\alpha)$. Except for this modification, the iterative procedure is conducted as above, including the reparametrization step at each iteration, and in essence it is the *finite-temperature string method* introduced in [11] and further developed in [12,28]. At the end of the iteration, when the curve satisfying both (70) and (127) has been identified, one has again to check whether it leads to regions $D(z)$ defined in the planes $P(z)$ via (65) and an associated bundle of curves Γ_x via (70) such that (73) and (74) are satisfied. If not, one has to go into other procedures to estimate a priori the location of Γ inside the tube, e.g. by smoothing the curve defined by (127). Eventually, if none of these work, we may have to conclude that the localized tube assumption is invalid, at least in the original variables, and look for collective variables in which it may be satisfied.

7.3 The String Method in Collective Variables

To generalize the zero- or the finite-temperature string methods to work in collective variables (Sects. 6.4 and 6.5), it suffices to do the following. At each iteration, it is now required to estimate the tensor $A_{kl}(\theta)$ defined in (94) and determine the gradient of the density $M(\theta)$ defined in (93) to be used in the equivalent of (125) (in the zero-temperature version) or sample with respect to this density to identify the equivalent of (127) (in the finite-temperature version). This requires to introduce an additional loop of sampling of blue moon type, now in the original ensemble with density $m(x)$ under the additional constraint that $\theta(x) = \theta^*$. For details, see [23].

8 Outlook

In this chapter, we have shown why the recent transition path theory (TPT) offers the correct probabilistic framework to understand the mechanism by which rare events occur by analyzing the statistical properties of the reactive trajectories involved in these events. The main results of TPT are the probability density of reactive trajectories and the probability current (and associated streamlines) of reactive trajectories, which also allows one to compute the probability flux of these trajectories and the rate of the reaction. It was also shown that TPT is a constructive theory: under the assumption that the reaction channels are local, TPT naturally leads to algorithms that allow to identify these channels in practice and compute the various quantities that TPT offers.

Acknowledgments

I am grateful to Weinan E and Weiqing Ren: the results reported here are part of a joint research project with them. I also thank very warmly Giovanni Ciccotti who helped me to clarify things when I was confused (and often unaware of it) and to whom the presentation owns a lot (the good parts anyway; I am sole responsible for the rest). I also thank David Chandler, Ron Elber, Christof Schuette, Alexander Fischer, Luca Maragliano, Phillip Metzner and Paul Maragakis for many interesting discussions. This work was partially supported by NSF grants DMS02-09959 and DMS02-39625, and by ONR grant N00014-04-1-0565.

References

1. C. H. Bennett (1977) In   *Algorithms for Chemical Computation*, eds. A. S. Nowick and J. J. Burton ACS Symposium Series No. 46, **63**
2. R. B. Best, and G. Hummer (2005) Reaction coordinates and rates from transition paths. *Proc. Natl. Acad. Sci. USA* **102**, p. 6732
3. P. G. Bolhuis, D. Chandler, C. Dellago, and P. Geissler (2002) Transition path sampling: Throwing ropes over rough mountain passes, in the dark. *Ann. Rev. Phys. Chem.* **59**, p. 291
4. G. Ciccotti, R. Kapral, and E. Vanden-Eijnden (2005) Blue moon sampling, vectorial reaction coordinates, and unbiased constrained dynamics. *Chem. Phys. Chem.* **6**, p. 1809
5. D. Chandler (1978) Statistical-Mechanics of isomerization dynamics in liquids and transition-state approximation. *J. Chem. Phys.* **68**, p. 2959
6. E. A. Carter, G. Ciccotti, J. T. Hynes, and R. Kapral (1989) Constrained reaction coordinate dynamics for the simulation of rare event. *Chem. Phys. Lett.* **156**, p. 472
7. C. Dellago, P. G. Bolhuis, and P. L. Geissler (2002) Transition Path Sampling. *Advances in Chemical Physics* **123**, p. 1

8. R. Durrett (1996) *Stochastic Calculus.* CRC Press
9. W. E, W. Ren and E. Vanden-Eijnden (2002) String method for the study of rare event. *Phys. Rev. B* **66**, 052301
10. W. E, W. Ren and E. Vanden-Eijnden (2003) Energy landscape and thermally activated switching of submicron-sized ferromagnetic element. *J. App. Phys.* **93**, p. 2275
11. W. E, W. Ren and E. Vanden-Eijnden (2005) Finite temperature string method for the study of rare events. *J. Phys. Chem. B* **109**, p. 6688
12. W. E, W. Ren and E. Vanden-Eijnden (2005) Transition pathways in complex systems: Reaction coordinates, isocommittor surfaces, and transition tubes. *Chem. Phys. Lett.* **413**, p. 242
13. W. E and E. Vanden-Eijnden (2004) Metastability, conformation dynamics, and transition pathways in complex systems. In: *Multiscale Modelling and Simulation*, eds. S. Attinger and P. Koumoutsakos (LNCSE **39**, Springer Berlin Heidelberg
14. W. E and E. Vanden-Eijnden (2006) Towards a Theory of Transition Paths. *J. Stat. Phys.* **123**, p. 503
15. R. Elber, A. Ghosh, and A. Cárdenas (2002) Long time dynamics of complex systems. *Account of Chemical Research* **35**, p. 396
16. R. Elber, A. Ghosh, A. Cárdenas, and H. Stern (2003) Bridging the gap between reaction pathways, long time dynamics and calculation of rates. *Advances in Chemical Physics* **126**, p. 93
17. H. Eyring (1935) The activated complex in chemical reactions. *J. Chem. Phys.* **3**, p. 107
18. C. W. Gardiner (1997) *Handbook of Stochastic Methods for Physics, Chemistry, and the Natural Sciences*, Springer Berlin Heidelberg
19. G. Henkelman and H. Jónsson (2000) Improved tangent estimate in the nudged elastic band method for finding minimum energy paths and saddle points. *J. Chem. Phys.* **113**, p. 9978
20. G. Hummer (2004) From transition paths to transition states and rate coefficients. *J. Chem. Phys.* **120**, p. 516
21. H. Jónsson, G. Mills, and K. W. Jacobsen (1998) Nudged Elastic Band Method for Finding Minimum Energy Paths of Transitions. In: *Classical and Quantum Dynamics in Condensed Phase Simulations*, ed. by: B. J. Berne, G. Ciccoti, and D. F., Coker, World Scientific
22. A. Ma, A. R. Dinner (2005) Automatic Method for Identifying Reaction Coordinates in Complex Systems. *J. Phys. Chem. B* **109**, p. 6769
23. L. Maragliano, A. Fischer, E. Vanden-Eijnden and G. Ciccotti (2006) String method in collective variables: Minimum free energy paths and isocommittor surfaces. *J. Chem. Phys.* **125**, 024106
24. G. Papanicolaou (1976) Probabilistic problems and methods in singular perturbation. *Rocky Mountain Math. J* **6**, p. 653
25. L. R. Pratt (1986) A statistical-method for identifying transition-states in high dimensional problems. *J. Chem. Phys.* **9**, p. 5045
26. W. Ren (2002) *Numerical Methods for the Study of Energy Landscapes and Rare Events.* PhD thesis, New York University
27. W. Ren (2003) Higher order string method for finding minimum energy paths. *Comm. Math. Sci.* **1**, p. 377

28. W. Ren, E. Vanden-Eijnden, P. Maragakis, and W. E. (2005) Transition pathways in complex systems: Application of the finite-temperature string method to the alanine dipeptide. *J. Chem. Phys.* **123**, 134109

29. M. Sprik and G. Ciccotti (1998) Free energy from constrained molecular dynamics. *J. Chem Phys.* **109**, p. 7737

30. F. Tal and E. Vanden-Eijnden (2006) Transition state theory and dynamical corrections in ergodic systems. *Nonlinearity* **19**, p. 501

31. E. Vanden-Eijnden and F. Tal (2005) Transition state theory: Variational formulation, dynamical corrections, and error estimates. *J. Chem. Phys.* **123**, 184103

32. T. S. van Erp and P. G. Bolhuis (2005) Elaborating transition interface sampling method. *J. Comp. Phys.* **205**, p. 157

33. E. Wigner (1938) The transition state method. *Trans. Faraday Soc.* **34**, p. 29

34. T. Yamamoto (1960) Quantum statistical mechanical theory of the rate of exchange chemical reactions in the gas phase. *J. Chem Phys.* **33**, p. 281

Multiscale Modelling in Molecular Dynamics: Biomolecular Conformations as Metastable States*

E. Meerbach, E. Dittmer, I. Horenko, and C. Schütte

Institut für Mathematik II, Freie Universität Berlin, Arnimallee 26, 14195 Berlin, Germany
schuette@math.fu-berlin.de

Illia Horenko

* Supported by the DFG research center MATHEON "Mathematics for key technologies" (FZT 86) in Berlin

E. Meerbach et al.: *Multiscale Modelling in Molecular Dynamics: Biomolecular Conformations as Metastable States*, Lect. Notes Phys. **703**, 495–517 (2006)
DOI 10.1007/3-540-35273-2_14

We report on a novel approach to the automatic identification of metastable states from long term simulation of complex molecular systems. The new approach is based on a hierarchical concept of metastability: metastable states are understood as subsets of state or configuration space from which the dynamics exits only very rarely; subsets with the smallest exit probabilities are of most interest, their further decomposition then may reveal subsets from which exiting is less but comparably difficult for the system under investigation. The article gives a survey of the theoretical foundation of the approach and its algorithmic realization that generalizes the well-known concept of Hidden Markov Models. The performance of the resulting algorithm is illustrated by an application to a 100 ns simulation of penta-alanine with explicit water. We demonstrate that the resulting metastable states allow to reveal the conformation dynamics of the molecule.

1 Introduction

The macroscopic dynamics of typical biomolecular systems is mainly characterized by the existence of biomolecular conformations which can be understood as metastable geometrical large scale structures, i.e., geometries which are persistent for long periods of time. On the longest time scales biomolecular dynamics is a kind of flipping process between these conformations, while on closer inspection it exhibits a rich temporal multiscale structure. Recent research seems to indicate that the conformations with the most pronounced persistency can be understood as metastable or "almost invariant" sets in state or configuration space [1, 2]. In many applications a Markovian picture is an appropriate description of the dynamics since typical correlation times in the system are sufficiently smaller than the waiting times between conformation changes (and thus much smaller than the timescale the effective description is intended to cover). In other words, the effective or macroscopic dynamics is given by a Markov jump process that hops between the metastable sets while the dynamics within these sets might be mixing on time scales that are smaller than the typical waiting time between the hops.

While the problem of computing the transition rates or the transition pathways between two *given* conformations attracted a lot of attention recently [3–7], the problem of efficient algorithmic identification of the most persistent conformations of a given system still is a challenging open problem. Recently there have been several *set-oriented* approaches to this problem [1, 8, 9]. These approaches are based on the construction of a transition matrix that describes transition probabilities between sets in the state space of the system. Via analysis of this transition matrix metastable sets are obtained. [2, 10]. For higher dimensional systems this always requires coarse graining of the state space, that is a partition of the state space in disjoint sets to avoid the curse of dimensionality, which that has to be designed carefully since the resulting metastable sets are unions of the sets from the partition.

Even more recently, alternative approaches have been introduced that apply appropriate Hidden Markov models (HMMs) to the identification problem [11–13].

We will first explain the background of these two approaches and comment on their relation. In the second part of this contribution we will discuss the identification of the most persistent conformations of penta-alanine as a numerical example. Other examples can be found in [13, 14].

2 Identification of Metastable States

2.1 Dynamics and Statistics

In classical molecular dynamics, a molecular system with a fixed number of N atoms is given by a state vector $(q, p) \in \mathbf{X} = \mathbb{R}^{3N} \times \mathbb{R}^{3N}$, where $q \in \mathbb{R}^{3N}$ denotes the position vector and $p \in \mathbb{R}^{3N}$ the momentum vector. The dynamical behavior, given a specified potential energy function V, a mass matrix M and initial conditions (q_0, p_0), is described by the Hamilton's equations

$$\dot{q} = M^{-1}p \,, \qquad\qquad q(0) = q_0 \,, \qquad\qquad (1)$$
$$\dot{p} = -\nabla_q V(q) \,, \qquad\qquad p(0) = p_0 \,. \qquad\qquad (2)$$

Equation (1) models an energetically closed system, whose total energy, given by the Hamiltonian

$$H(q, p) \;=\; \frac{1}{2}\, p^T M^{-1} p + V(q), \qquad\qquad (3)$$

is preserved under the dynamics.

It is well known that for every smooth function $\mathcal{F} : \mathbb{R} \to \mathbb{R}$ the probability measure $\mu(\mathrm{d}x) \propto \mathcal{F}(H)(x)\mathrm{d}x$ is invariant wrt. the Markov process X_t given by the solution of the Hamiltonian system (1). The most frequent choice is the canonical density or *canonical ensemble*

$$f(x) \;\propto\; \exp(-\beta H(x)) \qquad\qquad (4)$$

for some constant $\beta > 0$ that can be interpreted as the inverse temperature. The associated measure $\mu(\mathrm{d}x) \propto f(x)\mathrm{d}x$ is called the *canonical measure*. The canonical ensemble is often used in modeling experiments on molecular systems that are performed under the conditions of constant volume and temperature $\mathcal{T} = \frac{1}{k_B \beta}$, where k_B is Boltzmann's constant. Obviously, a single solution of the Hamiltonian system (1) can never be ergodic wrt. the canonical measure, since it conserves the internal energy H, as defined in (3). There are several approaches in the construction of (stochastic) dynamical systems that allow sampling of the canonical ensemble by means of long-term simulation. Most deterministic methods reduce to the construction of a Hamiltonian system in some slightly extended state space $\hat{\mathbf{X}}$, whose projection onto the lower dimensional state space $\mathbf{X}$ of positions and momenta generates a sampling according to (4). One of the most prominent examples is the Nosé-Hoover thermostat [15]. There are also non-deterministic methods. Amongst them, for example, are the well-known Langevin dynamics models, as well as Hybrid Monte Carlo approaches, cf. [2].

2.2 Metastability and the Transfer Operator Approach

Each of the optional dynamical models mentioned above involves a *homogeneous Markov process* $X_t = \{X_t\}_{t \in \mathcal{T}}$ in either continuous or discrete time on some state space $\mathbf{X}$. The motion of X_t is given in terms of the *stochastic transition function*

$$p(t, x, A) = \mathbb{P}[X_{t+s} \in A \mid X_s = x], \tag{5}$$

for every $t, s \in \mathcal{T}$, $x \in \mathbf{X}$ and $A \subset \mathbf{X}$. We write $X_0 \sim \mu$, if the Markov process X_t is initially distributed according to the probability measure μ and denote the corresponding probability function of the process by $\mathbb{P}_\mu$. A Markov process X_t admits an *invariant probability measure* μ, or μ is invariant wrt. X_t, if $\int_{\mathbf{X}} p(t, x, A)\mu(\mathrm{d}x) = \mu(A)$. In the following we always assume that the invariant measure of the process under investigation exists and is unique. A Markov process is called *reversible* wrt. an invariant probability measure μ if $\int_A p(t, x, B)\mu(\mathrm{d}x) = \int_B p(t, x, A)\mu(\mathrm{d}x)$ for every $t \in \mathcal{T}$ and $A, B \subset \mathbf{X}$.

Metastability of some subset of the state space is characterized by the property that the Markov process is likely to remain within the subset for a long period of time, until it exits and a transition to some other region of the state space occurs. There are in fact several related but different definitions of metastability in the literature (see, e.g., [16–20]); we will focus on the so-called ensemble concept introduced in (6), for a comparison with, e.g., the exit time concept, see [2].

The *transition probability* $p(t, B, C)$ from a subset $B \subset \mathbf{X}$ to another subset $C \subset \mathbf{X}$ within the time span t is defined as the conditional probability

$$p(t, B, C) = \mathbb{P}_\mu[X_t \in C \mid X_0 \in B] = \frac{\mathbb{P}_\mu[X_t \in C \text{ and } X_0 \in B]}{\mathbb{P}_\mu[X_0 \in B]}. \tag{6}$$

This may be rewritten as

$$p(t, B, C) = \frac{1}{\mu(B)} \int_B p(t, x, C)\, \mu(\mathrm{d}x). \tag{7}$$

In other words, the transition probability quantifies the dynamical fluctuations within the stationary ensemble μ. A subset $B \subset \mathbf{X}$ is called *metastable* on the time scale $\tau > 0$ if

$$p(\tau, B, B^c) \approx 0, \quad \text{or equivalently,} \quad p(\tau, B, B) \approx 1,$$

where $B^c = \mathbf{X} \setminus B$ denotes the complement of B.

The objective of the transfer operator approach is an identification of a *decomposition of the state space into metastable subsets* and the corresponding "flipping dynamics" between these sub-states. By a decomposition d = $\{D_1, \ldots, D_m\}$ of the state space **X** we mean a collection of subsets $D_k \subset \mathbf{X}$ with the following properties: (1) positivity $\mu(D_k) > 0$ for every k, (2) disjointness up to null sets, and (3) the covering property $\cup_{k=1}^{m} \overline{D_k} = \mathbf{X}$.

The *metastability of a decomposition* d is defined as the sum of the metastabilities of its subsets, supposed that the time scale τ of interest is fixed. Then, for each arbitrary decomposition $\mathrm{d}_m = \{D_1, \ldots, D_m\}$ of the state space **X** into m sets we define its metastability measure by

$$\mathrm{meta}(\mathrm{d}_m) = \sum_{j=1}^{m} p(\tau, D_j, D_j).$$

For given m the *optimal metastable decomposition* into m sets can then be defined as that decomposition into m sets which maximizes the functional meta. This means in particular that the appropriate number m of metastable subsets must be identified. Both the determination of m and the identification of the metastable subsets can be achieved via spectral analysis of the so-called transfer operator.

Transfer Operator

The *semigroup of propagators* or forward transfer operators $P^\tau : L^r(\mu) \to L^r(\mu)$ with t ersetzen durch $\tau \in \mathcal{T}$ and $1 \le r < \infty$ is characterized as follows:

$$\int_A P^\tau v(y) \, \mu(\mathrm{d}y) = \int_{\mathbf{X}} v(x)p(\tau, x, A)\mu(\mathrm{d}x) \qquad (8)$$

for $A \subset \mathbf{X}$. As μ is the invariant probability measure of the considered Markov process, the characteristic function $\mathbf{1_X}$ of the entire state space is an invariant density of P^τ, i.e., $P^\tau \mathbf{1_X} = \mathbf{1_X}$. Furthermore, P^τ is a Markov operator, i.e., P^τ conserves both norm $\|P^\tau v\|_1 = \|v\|_1$ and positivity $P^\tau v \ge 0$ if $v \ge 0$, which is a simple consequence of the definition. Due to (8), the semigroup of propagators mathematically models the evolution of sub-ensembles in time. In our algorithmic strategy we furthermore exploit self-adjointness of the propagator which is inherited from reversiblity of the underlying dynamic and results in a real-valued spectrum (cp. Theorem 1).

The *key idea of the transfer operator approach* wrt. the identification of metastable decompositions can be described as follows:

Metastable subsets can be detected via eigenvalues of the propagator P^τ close to its maximal eigenvalue $\lambda = 1$; moreover they can be identified by exploiting the corresponding eigenfunctions. In doing so, the number of metastable subsets in the metastable decomposition is equal to the number of eigenvalues close to 1, including $\lambda = 1$ and counting multiplicity.

This strategy was first proposed by Dellnitz and Junge [9] for discrete dynamical systems with weak random perturbations, and has been successfully applied to molecular dynamics in different contexts [1,2,21,22]. The key idea requires the following two additional *conditions on the propagator* P^τ:
(C1) The essential spectral radius of P^τ is less than one, i.e., $r_{ess}(P^\tau) < 1$.
(C2) The eigenvalue $\lambda = 1$ of P is simple and dominant, i.e., $\eta \in \sigma(P^\tau)$ with $|\eta| = 1$ implies $\eta = 1$.

Consider for example: (1) high-friction Langevin processes, and (2) (Nose-Hoover) constant temperature molecular dynamics. For both cases the dynamics is reversible and the transfer operator is self-adjoint. For type (1) examples, conditions (C1) and (C2) are known to be satisfied under rather weak condition on the potential [2]. For type (2) examples, it is unknown whether or not the conditions are satisfied; however, it is normally assumed in molecular dynamics that they are valid for realistically complex systems in solution.

The next result [2, 23] justifies the above key idea:

Theorem 1. *Let $P^\tau : L^2(\mu) \to L^2(\mu)$ denote a reversible propagator satisfying (C1) and (C2). Then P^τ is self–adjoint with spectrum of the form*

$$\sigma(P^\tau) \subset [a,b] \cup \{\lambda_m\} \cup \ldots \cup \{\lambda_2\} \cup \{1\}$$

with $-1 < a \le b < \lambda_m \le \ldots \le \lambda_1 = 1$ and λ_i isolated, eigenvalues that are counted according to their finite multiplicities. Denote by $v_m, \ldots, v_1$ the corresponding eigenfunctions, normalized to $\|v_k\|_2 = 1$. Let Q be the orthogonal projection of $L^2(\mu)$ onto $\mathrm{span}\{1_{D_1}, \ldots, 1_{D_m}\}$, where $\mathrm{d}_m = \{D_1, \ldots, D_m\}$ is an arbitrary partition of the state space. Then the metastability of d_m can be bounded from above by

$$p(\tau, D_1, D_1) + \ldots + p(\tau, D_m, D_m) \le 1 + \lambda_2 + \ldots + \lambda_m ,$$

while it is bounded from below according to

$$1 + \kappa_2\lambda_2 + \ldots + \kappa_m\lambda_m + c \le p(\tau, D_1, D_1) + \ldots + p(\tau, D_m, D_m) ,$$

where $\kappa_j = \|Qv_j\|^2_{L^2(\mu)}$ and $c = a\left((1 - \kappa_2) + \ldots + (1 - \kappa_n)\right)$.

Theorem 1 highlights the strong relation between a decomposition of the state space into metastable subsets and a *Perron cluster* of dominant eigenvalues close to 1. It states that the metastability of an arbitrary decomposition d_m cannot be larger than $1+\lambda_2+\ldots+\lambda_m$, while it is at least $1+\kappa_2\lambda_2+\ldots+\kappa_m\lambda_m+c$, which is close to the upper bound whenever the dominant eigenfunctions $v_2, \ldots, v_m$ are almost constant on the metastable subsets $D_1, \ldots, D_m$ implying $\kappa_j \approx 1$ and $c \approx 0$. The term c can be interpreted as a correction that is small whenever $a \approx 0$ or $\kappa_j \approx 1$. It is demonstrated in [23] that the lower and upper bounds are sharp and asymptotically exact.

There is an important message contained in the last theorem: *metasta-bility analysis has to be hierarchical*. Whenever we approximate the optimal metastable decomposition d_2 of the state space into, say, two sets, we should always be aware that there could be a decomposition d_3 into three sets for which meta(d_3) is almost as large as meta(d_2). For example, one or both of the two subsets in d_2 could decompose into two or several metastable subsets from which it is comparably difficult to exit for the system under investigation.

However, whenever there is a gap in the spectrum of the transfer operator after m dominant eigenvalues, then the results of, e.g., [16,21] tell us that any decomposition into more than m sets will be associated with a significantly larger drop in metastability as measured by the function meta.

Example

The easiest nontrivial example is a time-discrete Markov chain on a discrete state space. For example, take the chain with state space $S = \{1, 2, 3, 4\}$ and one-step transition probabilities as illustrated in Fig. 1.

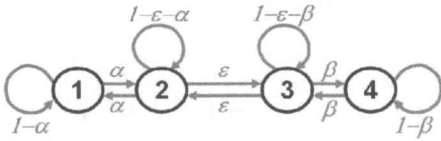

Fig. 1. Markov chain with states 1,2,3,4. The numbers on the arrows linking the states are the one-step transition probabilities. Any transition that is not represented by an arrow is not allowed

This chain has a unique invariant measure, $\mu = (0.25, 0.25, 0.25, 0.25)$, and is reversible. Its transition matrix is given by the following stochastic matrix:

$$\mathcal{T} = \begin{pmatrix} 1 - \alpha & \alpha & 0 & 0 \\ \alpha & 1 - \alpha - \epsilon & \epsilon & 0 \\ 0 & \epsilon & 1 - \beta - \epsilon & \beta \\ 0 & 0 & \beta & 1 - \beta \end{pmatrix}.$$

In this specific case $\mathcal{T}$ is the transfer operator of the process.

Let us consider two cases defined in Table 1 below: In case 1, we obviously have a metastable decomposition into the subsets $\{1, 2\}$ and $\{3, 4\}$ with metastability measure 0.995 while any further decomposition significantly lowers the metastability measure. Table 2 shows that the spectrum of the transfer operator exhibits a corresponding gap after the first two eigenvalues, and that the upper bound from our theorem, $(\lambda_1 + \lambda_2)/2 = 0.995$, is a very good approximation of meta($\{1, 2\}, \{3, 4\}$). The second eigenvector $v_2 = (0.51, 0.49, -0.49, -0.50)$ clearly exhibits almost constant levels on the two sets $\{1, 2\}$ and $\{3, 4\}$ of the metastable decomposition. In case 2,

Table 1. Different parameter sets for the Markov chain considered herein, and metastability measures of the two different decompositions discussed in the text

case	α	β	ϵ	meta($\{1,2\},\{3,4\}$)	meta($\{1\},\{2\},\{3,4\}$)
1	0.25	0.4	0.01	0.995	0.828
2	0.05	0.4	0.01	0.995	0.962

Table 2. Different parameter sets for the Markov chain considered herein, and spectrum of the associated transfer operators

case	α	β	ϵ	λ_1	λ_2	λ_3	λ_4
1	0.25	0.4	0.01	1.000	0.990	0.495	0.195
2	0.05	0.4	0.01	1.000	0.991	0.895	0.195

the decreased value of α introduces an additional, milder metastability that separates state 1 from state 2. We can also see this in the spectrum, see Table 2. Still, $\{1,2\}$ and $\{3,4\}$ is the metastable decomposition into two sets with metastability measure 0.995, and associated second eigenvector $v_2 = (0.55, 0.45, -0.49, -0.50)$. But this time, this is the top level of the hierarchy of metastable decompositions only: we can further decompose the set $\{1,2\}$ and the resulting decomposition $\{1\}$, $\{2\}$ and $\{3,4\}$ into three sets also has a significantly high metastability measure of 0.962. For both decompositions, the upper bounds on the metastability measure computed from the eigenvalues are very close to the true values.

2.3 Discretization and PCCA

In the typical case the dynamical process under investigation lives on a continuous state space such that the transfer operator does not have the form of a nice stochastic matrix. Therefore, discretization of the transfer operator is needed to yield a stochastic matrix with which one can proceed as in the example above.

Let $\chi = \{\chi_1, \ldots, \chi_n\} \subset L^2(\mu)$ denote a set of *non–negative* functions that are a partition of unity, i.e., $\sum_{k=1}^{n} \chi_k = \mathbf{1_X}$. The *Galerkin projection* $\Pi_n : L^2(\mu) \to S_n$ onto the associated finite dimensional ansatz space $S_n = \text{span}\{\chi_1, \ldots, \chi_n\}$ is defined by

$$\Pi_n v = \sum_{k=1}^{n} \frac{\langle v, \chi_k \rangle_\mu}{\langle \chi_k, \chi_k \rangle_\mu} \, \chi_k \, .$$

Application of the Galerkin projection to $P^\tau v = \lambda v$ yields an eigenvalue problem for the discretized propagator $\Pi_n P^\tau \Pi_n$ acting on the finite-dimensional space S_n. The matrix representation of this finite dimensional operator is given by the $n \times n$ *stochastic transition matrix* $\mathcal{T} = (\mathcal{T}_{kl})$, whose entries are given by

$$\mathcal{T}_{kl} = \frac{\langle P^\tau \chi_k, \chi_l \rangle_\mu}{\langle \chi_k, \chi_k \rangle_\mu} . \tag{9}$$

The transition matrix inherits the main properties of the transfer operator: it is a stochastic matrix with invariant measure given by the invariant measure μ of P^τ, it is reversible if P^τ is self-adjoint, and (if the discretization is fine enough) it also exhibits a Perron cluster of eigenvalues that approximates the corresponding Perron cluster of P^τ, and with eigenvectors that approximate the dominant eigenvectors of P^τ [2]. It thus allows to compute the metastable sets of interest by computation of the dominant eigenvectors of $\mathcal{T}$ and by realization of the identification strategy proposed on page 500 based on these (discrete) eigenvectors. This has led to the construction of an aggregation technique called "Perron Cluster Cluster Analysis" (PCCA) [8, 10, 24].

If $x_0, \ldots, x_N$ denote a time series obtained from a realization of the Markov process with time stepping τ, then the entries of $\mathcal{T}$ can be approximated from the relative transition rates computed by means of this time series:

$$\mathcal{T}_{kl} \approx \mathcal{T}_{kl}^{(N)} = \frac{\sum_{j=1}^N \chi_k(x_j) \cdot \chi_l(x_{j+1})}{\sum_{j=1}^N \chi_k(x_j)} . \tag{10}$$

Although it looks extremely simple, using equation (10) algorithmically may become problematic. There are two main reasons for potential difficulties.

Trapping Problem

The *rate of convergence* of $\mathcal{T}_{kl}^{(N)} \to \mathcal{T}_{kl}$ depends on the smoothness of the partition functions χ_k as well as on the mixing properties of the Markov process [25]. The latter property is crucial here: The convergence is geometric with a rate constant $\lambda_1 - \lambda_2 = 1 - \lambda_2$ where λ_2 denotes the second largest eigenvalue (in modulus). In the case of metastability with λ_2 being very close to $\lambda_1 = 1$, we will have dramatically slow convergence. However we will *not* go into the depth of the discussion on overcoming the trapping problem, but instead assume in all of the following that we have already generated or can directly generate a time series that is "long enough" in the sense that it contains statistically significant information about more than one –if not all– interesting metastable states of the system under consideration. The interested reader may refer to the vast literature [26–28].

Curse of Dimension

Any discretization of the transfer operator will suffer from the curse of dimension whenever it is based on a uniform partition of all of the hundreds or thousands of degrees of freedom in a typical biomolecular system. Fortunately, chemical observations reveal that –even for larger biomolecules– the curse of dimensionality can be circumvented by exploiting the hierarchical structure of the dynamical and statistical properties of biomolecular systems:

only relatively few *essential degrees of freedom* may be needed to describe the conformational transitions; furthermore, the canonical density has a rich spatial multiscale structure induced by the rich structure of the potential energy landscape, which again underlines the necessity of a hierarchical approach.

2.4 Relaxations, Transitions, and Effective Dynamics

Assume that we successfully identified a metastable decomposition into the sets $D_1, \ldots, D_m$ for a given lag time τ. Due to our above results the dynamics is jumping from sets D_k to set D_j with probability $p(\tau, D_k, D_j)$ during time τ. Then, it is an intriguing idea to describe the effective dynamics of the system by means of the Markov chain with discrete states $D_1, \ldots, D_m$ and transition matrix $P = (P_{kj})$ with $P_{kj} = p(\tau, D_k, D_j)$. This "effective dynamics" is Markovian and thus cannot take into account that there may be memory in the system that is much longer than the time span τ used to compute the metastable decomposition.

In order to categorize the obtained Markovian model for the effective dynamics more precisely, let us denote the typical (mean) exit time from D_j to D_k by T_{jk}, and the typical relaxation timescale within D_j by τ_j (that is, when the system enters D_j at $t = 0$ it has lost almost all of its memory at $t = \tau_j$).

The simplest case is that we have τ being comparable to the largest τ_j and $\tau_j \ll \min_k T_{jk}$ for all $j = 1, \ldots, m$. Then the above construction is a good model of the effective dynamics, and the system on average samples its restricted invariant density $\mu|_{D_j}$ in D_j before exiting from D_j.

The more complicated case is that still $\tau \ll \min_k T_{jk}$ but no longer $\tau_j \ll \min_k T_{jk}$ for all j. Then the above construction constitutes a misleading model for the effective dynamics.

Comparable problems appear for other types of coarse graining. In the above, the coarse graining is given by the metastable decomposition. But we could also try to realize the transfer operator approach, by means of discretizing only a subspace instead of the full state space. For example the subspace spanned by the essential degrees of freedom, or the torsion angles space. As the Markov property does not hold for projections of Markov processes in general, we have to be aware that the process on the (torsion angles) subspace might not longer be Markovian.

2.5 The Hidden Markov Model Approach

We are now going to consider the case of a given timeseries $(O_t)_{t=t_1, \ldots, t_N}$, with constant sampling time $\tau = t_{j+1} - t_j$. Here, the O_t do not necessarily denote the state of the molecule at time t but rather some low-dimensional observable, for example, some or all torsion angles or the set of essential degrees of freedom (if this should be available). We assume that there is an unknown metastable decomposition, say, into m sets $D_1, \ldots, D_m$. Using the notation introduced above, we furthermore assume $\tau \ll \min_k T_{jk}$ but do not specify the relation between τ and the τ_j. We then can premise that, at any time t, the system is in one of the metastable states D_{j_t} to which we simply refer by j_t in the following. However, the time series (j_t) is *hidden*, i.e., not known in advance, while the series (O_t) is called the output series or the *observed* sequence.

This design can be represented by a Hidden Markov Model (HMM). A HMM abstractly consists of two related stochastic processes: a hidden process j, that fulfills the Markov property and an observed process O_t that depends on the state of the hidden process j_t at time t. A HMM is fully specified by the initial distribution π, the rate matrix R of the hidden Markov process j, as well as by the law that governs the observable O_t depending on the respective hidden state j_t.

In the standard versions of HMMs the observables are i.i.d. random variables with stationary distributions that depend on the respective hidden states [13]. Within the scope of molecular dynamics this means, that one considers the simple case where τ is comparable to the τ_j and $\tau_j \ll \min_k T_{jk}$, i.e., the process samples the restricted invariant density before exiting from a metastable state, and the sampling time of the time series is long enough to assume statistical independence between steps. Nevertheless, if this is not the case, only a slight modification of the model structure is required to represent the relaxation behavior: Instead of i.i.d. random variables one can use an Ornstein-Uhlenbeck (OU) process as a model for the output behavior in each hidden state. The HMM then gets the form [11]:

$$\dot{Y}_t = -DV^{(q)}(Y_t) + \sigma^{(j_t)}\dot{W}_t, \tag{11}$$

$$j_t \ : \ \mathbf{R}^1 \to \{1, 2, \ldots, m\} \,, \tag{12}$$

where j_t are the realizations of the hidden Markov process with discrete state space, W_t is standard "white noise", and $\{V^{(j)}, \sigma^{(j)}\}$ is a set of the state-specific model parameters with harmonic potentials $V^{(j)}$ of the form

$$V^{(j)}(Y) = \frac{1}{2}(Y - \mu^{(j)})^T D^{(j)}(Y - \mu^{(j)}) + V_0^{(j)} \,, \tag{13}$$

where $\mu^{(j)}$ and $D^{(j)}$ are *equilibrium position* and Hesse-matrix of the OU process within conformation j. This process is therefore given by the parameters $\Theta^{(j)} = (\mu^{(j)}, D^{(j)}, \sigma^{(j)})$. Since the output process is specified by a stochastic differential equation we will refer to this model modification as HMMSDE . Its entire parameter set will be denoted $\Theta = (\Theta^{(1)}, \ldots, \Theta^{(m)}, R)$ in the following, where R denotes the rate matrix of the Markov chain in (12).

Assume for a moment that the hidden state is fixed, i.e. $j_t = j$. Then, the evolution of a probability density $\rho(t, Y|j)$ under the dynamics given by (11) can be obtained as the solution of the corresponding Fokker Planck equation:

$$\partial_t \rho = \triangle_Y V^{(j)}(Y)\rho + \nabla_Y V^{(j)}(x) \cdot \nabla_Y \rho + \nabla_Y \cdot B^{(j)}\nabla_Y \rho, \tag{14}$$

$$\rho(t = 0, Y|j) = \rho_0(Y|j) \,, \tag{15}$$

where $B^{(j)} = (\sigma^{(j)})^2 \in \mathbf{R}^1$ denotes the variance of the white noise (for $\mathbf{R}^d$ it is a positive definite selfadjoint matrix). In the subsequent we will denote the partial differential operator on the RHS of (14) by $L_{\Theta^{(j)}}$. Then, the solution of (14) can be written as $\rho(t, Y|j) = (\exp(tL_{\Theta^{(j)}})\rho_0)(Y)$.

We have the following aim: For a given observed sequence $O = (O_t)$ determine those parameters in the dynamical model (11 & 12) for which the *probability that O is an output of these parameters is maximal*. To this end, one has to know the probability $p(O, j|\Theta)$ of the observed sequence O and a specific hidden sequence j for given parameters Θ. First assume that the dynamics (11) yields output $Y_1 = O_{t_{k+1}}$ at time $t = t_k + \tau = t_{k+1}$ after starting at $Y_0 = O_{t_k}$ while remaining in $j = j_{t_k}$ during the evolution form t_k to t_{k+1}. Due to the above this probability obviously is

$$\rho(O_{t_{k+1}}|j, O_{t_k}) = \Big(\exp(\tau L_{\Theta^{(j)}})\delta_{O_{t_k}} \Big)(O_{t_{k+1}}) , \qquad (16)$$

where δ_O denotes the Dirac measure supported at state O. With this the total probability that the observed sequence O and a given hidden sequence j is an output of (11), parametrized by Θ, is

$$p(O, j|\Theta) = \pi(j_{t_0})\nu(O_{t_0}|j_{t_0}) \prod_{k=1}^{N-1} \mathcal{T}(j_{t_k}, j_{t_{k+1}})\rho(O_{t_{k+1}}|j_{t_k}, O_{t_k}) , \quad (17)$$

where $\mathcal{T} = \exp(\tau R)$ denotes the transition matrix of the Markov jump process, specified in (12), in time τ and π and ν are initial distributions that have to chosen in addition.

In putting everything together, we have to face these algorithmic problems: (1) determine the optimal parameters Θ by maximizing the probability $p(O, j|\Theta)$ –this is a nonlinear global optimization problem–, (2) determine the optimal sequence of hidden metastable states $j = (j_t)$ for given optimal parameters, and (3) determine the number of important metastable states; which we, up to now, simply assumed to be identical with the number of hidden states.

The above formulation of the two first problems seems to contain a considerable contradiction: How can we determine optimal parameters without knowing the optimal hidden sequence? Fortunately the solution is already available from the standard HMM framework: The parameter optimization is carried out by the Expectation Maximization (EM) algorithm that *iteratively* determines the optimal parameters Θ_* via maximizing the expectation

$$Q(\Theta; \Theta_k) = \mathbb{E}\Big(\log p(O, j|\Theta) \,|\, O, \Theta_k \Big) \qquad (18)$$

of the complete probability $p(O, q|\lambda)$ wrt. the hidden sequence j given the observation sequence and the current parameter estimate Θ_k. It is a classical result [29] (Chap. 4.2) that (18) can be rewritten as a sum over all hidden sequences:

$$Q(\Theta; \Theta_k) = \sum_{j=(j_t)} p(O, j|\Theta_k) \log\left(p(O, j|\Theta)\right) . \tag{19}$$

The expectation-step of the EM algorithm evaluates the expectation value Q based on the given parameter estimate Θ_k, while the maximization-step determines the refined parameter set $\Theta_{k+1} = \text{argmax}_\Theta Q(\Theta; \Theta_k)$. The expectation step is standard but the maximization step can also be realized algorithmically, see [11] or [12] for different realizations.

For the identification of the optimal sequence of hidden metastable states we can use the well-known Viterbi algorithm [30], which exploits dynamic programming techniques to resolve in a recursive manner the optimization problem

$$j_* = \text{argmax } p(O, j|\Theta_*) .$$

The obtained optimal sequence j_* is called "Viterbi path". For technical details see [11].

The parameter fitting step requires the specification of the number of hidden states, which, whenever the hidden states should be metastable states, is in general not apriori known. One policy to overcome this problem is to assume a sufficient large number of hidden states, perform the parameter fitting and conduct a further aggregation of the resulting transition matrix. This can be done by Perron cluster cluster analysis (PCCA), e.g., by the spectral properties of the resulting transition matrix $\mathcal{T}$ as proposed in the transfer operator approach (we will illustrate this procedure on an example in the next section), see [11] for details.

As the numerical effort of the used algorithmic process scales wrt. to the dimension of the observable d as $\mathcal{O}(d^3)$ the three steps above provide a tool for metastability analysis only on low-dimensional observables. But this obstacle can be circumvented by first applying the algorithm *separately* to several low-dimensional projections. We end up with an aggregated Viterbi path j_* for each projection (for example, when the observed sequence contains all peptide angles along a polypeptide chain, we could apply HMMSDE to the peptide angles of all single amino acid units of the sequence first). Then, simple combinatorics allows to combine these Viterbi paths into a "combined" Viterbi path, that contains every occurred state combination. A transition matrix for the full-dimensional system can be obtained by counting relative frequencies in the combined Viterbi path and another aggregation via PCCA finally allows for the identification of metastable sets.

We want to underline two important points on this algorithmic scheme: (1) The philosophy of HMM models gives a justification to work on low-dimensional projections of an observable, because the observable is not meant to specify the occupied metastable state at a certain time, but to reflect different states by different dynamical behavior which seems quite reasonable. (2) We not only obtain an optimal sequence of hidden metastable states, but also optimal parameters for a simple, but physically motivated, reduced model (11 & 12) of the dynamics. We think that the extraction of such models is quite important to gain more insight in the mechanisms behind metastable behavior.

3 Numerical Example: Penta-Alanine

As illustration we demonstrate the performance of the proposed algorithmic procedure in application to the analysis of a peptide molecule. The global (secondary) structure of a peptide is determined by the so-called peptide angles. For each alanine amino acid we have to consider two of these backbone torsion angles. These peptide angles pairs can not take arbitrary values due to steric interaction, but will adopt values in definite regions, belonging to various secondary structures. In Fig. 2 the backbone torsion angles of penta-alanine are shown. As usual the pair of angles belonging to the same amino acid residue is labelled by Φ and Ψ. Illustration of quantities as functions of Φ and Ψ are called Ramachandran plots. The Ramachandran plot in Fig. 2 exhibits the values that a pair of peptide angles typically takes in certain secondary structures.

Our analysis is based upon a time series of the 10 backbone torsion angles of penta-alanine, extracted from the long time simulation that has already been discussed in [31] (courtesy of G. Stock, Frankfurt). The simulation was done in explicit water using a thermostat of $300\,\mathrm{K}$ over an interval of $100\,\mathrm{ns}$, while the coordinates were written out every 0.1ps, resulting in 1000000 data points. Figure 10 below shows a histogram Ramachandran plot of the entire time series for each $\Phi\backslash\Psi$ pair. It reveals that the fifth angle pair has a substantial different behavior than the other pairs. This will not be our concern here since we just take the timeseries for demonstration of the algorithmic procedure of extracting information about metastable states from a given time series.

As penta-alanine is a short peptide it will not have a stable β-sheet conformation, but as we will see in our analysis it exhibits a stable α-helix conformation and several other conformations which can be characterized by certain flexibility patterns.

The first step of our analysis consists in analyzing each of the peptide angles *separately* by using the HMMSDE techniques to determine the Viterbi path for each of the angles. Beforehand an initial guess must specify the number of hidden states, which we set to 4. As result each peptide angle is represented by a discrete time series with 4 states, the Viterbi path of this angle, assigning each instance in the time series to a state of a hidden Markov process. Fig. 3 provides an example. Note the fundamental difference to a direct transfer operator approach, where we have to specify a box discretization of the ten dimensional state space.

This way we obtain 10 Viterbi paths. In the next step these are now combined to pairwise Viterbi paths by simple superposition of the Viterbi paths which belong to a $\Phi\backslash\Psi$ combination. This produces five *pairwise* Viterbi

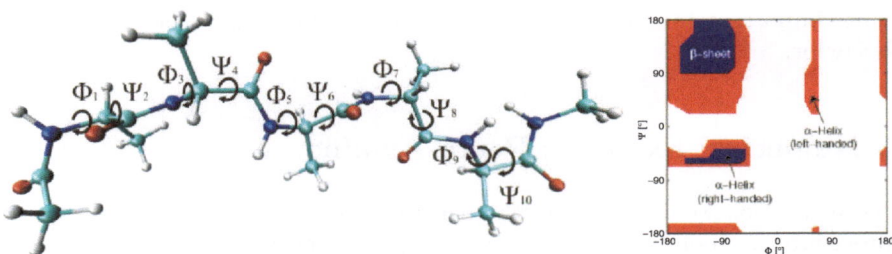

Fig. 2. *Left:* The Penta-alanine peptide in ball-and-stick representation. The ten peptide angles determining the secondary structure are marked by $\Phi_1, \Psi_2, \ldots, \Phi_9, \Psi_{10}$. *Right:* Ramachandran plot, showing the energetically preferred regions of a $\Phi\backslash\Psi$ pair with the associated secondary structures (simplefied plot due to [32])

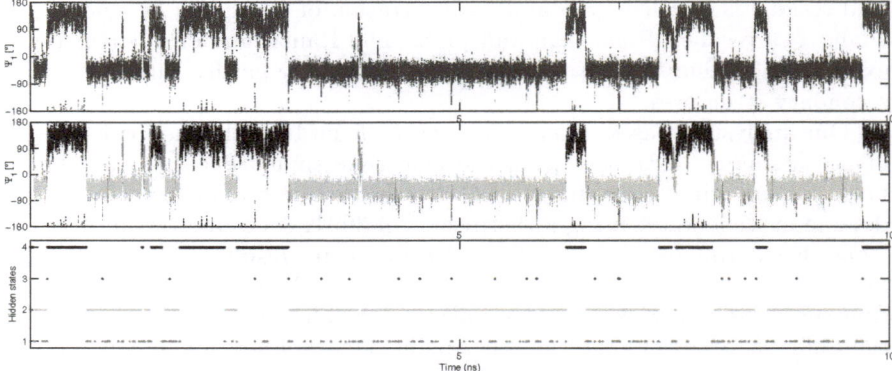

Fig. 3. Results of HMMSDE for the Ψ_2 time series. *Top:* Illustration of the first 10 ns of the Ψ_2 time series. *Middle:* The same picture colored according to the association with the assumed 4 hidden states. *Bottom:* The Viterbi path displayed as a discrete time series which specifies only the hidden states

paths, each with 16 states. Each of these Viterbi paths now is understood as the output time series of a Markov process with discrete state space. For each pair the corresponding stochastic transition matrix can be computed by counting the transitions between different states. Following the transfer operator approach the spectra of these transition matrices contain information about metastability in the dynamics of each peptide angle pair. For example the eigenvalues of the transition matrix T_4, extracted from the fourth pairwise Viterbi path (time lag $\tau = 0.1$ ps), are

$\lambda_1(T_4)$	$\lambda_2(T_4)$	$\lambda_3(T_4)$	$\lambda_4(T_4)$	$\lambda_5(T_4)$	$\lambda_6(T_4)$	$\lambda_7(T_4)$	$\lambda_8(T_4)$
1	0.998	0.994	0.991	0.979	0.975	0.959	0.947

$\lambda_9(T_4)$	$\lambda_{10}(T_4)$	$\lambda_{11}(T_4)$	$\lambda_{12}(T_4)$	$\lambda_{13}(T_4)$	$\lambda_{14}(T_4)$	$\lambda_{15}(T_4)$	$\lambda_{16}(T_4)$
0.9337	0.898	0.8798	0.839	0.176	0.154	0.139	-0.001

which suggests four metastable subsets for this pairwise Viterbi path. Obviously other interpretations of the spectrum are also reasonable, as one could argue that the clearest gap occurs after the twelfth eigenvalue. This is more or less a decision of how much detail one wants to or can afford to preserve at this stage of analysis, but it turns out that if we turn to the global analysis these details are filtered out anyway.

The next step is to cluster all pairwise Viterbi paths according to the structure of the eigenvectors belonging to the dominant eigenvalues of the associated transition matrix. This yields five *clustered* pairwise Viterbi paths with 4 or 5 states each, allocating each instance in the corresponding $\Phi\backslash\Psi$ pair time series to a *metastable* hidden set. Plotting this information in the form of a Ramachandran plot reveals a similar, although not equal, structure for each $\Phi\backslash\Psi$ pair, see Fig. 4. Note again that the different metastable sets are not disjoint sets in the $\Phi\backslash\Psi$ plane, as the HMMSDE analysis assumes the given data to be a projection of some hidden full process.

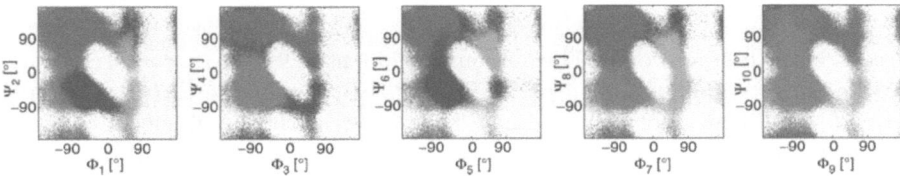

Fig. 4. Ramachandran plots of the five $\Phi\backslash\Psi$ pairs colored according to the associated metastable sets. The numbers of metastable sets differ for different $\Phi\backslash\Psi$ pairs (it is from left to right 5,4,5,4,4)

By another superposition of the five clustered pairwise Viterbi paths a *global* Viterbi path is obtained. This path contains 1114 different states, due to the fact that the states of the clustered pairwise Viterbi paths can be combined in any way giving a theoretical maximum of $5 \cdot 4 \cdot 5 \cdot 4 \cdot 4 = 1600$ possible global states (of which only 1114 actually occur). Setting up the transition matrix again yields a sparse stochastic matrix in which more than 99% of the entries are equal to zero.

It is instructive to compare the eigenvalues of transition matrices obtained for different lag times τ. That is, we do not count transitions on a timescale of $0.1ps$ which means to observe transitions from one instance of the time series to the next, but count transitions on a timescale of, say, $1ps$ which is between every tenth step in the global Viterbi path. For all time lags, Fig. 5 clearly indicates two dominant eigenvalues after which we find a gap, followed by other gaps after 4, 9, or 16 eigenvalues. This yields 2, respectively 4, 9, or 16 metastable sets. To avoid confusion we call these metastable sets (molecular) conformations.

To gain more insight into the metastability analysis we will now compare the results of the procedure based on the first two eigenvectors with the results

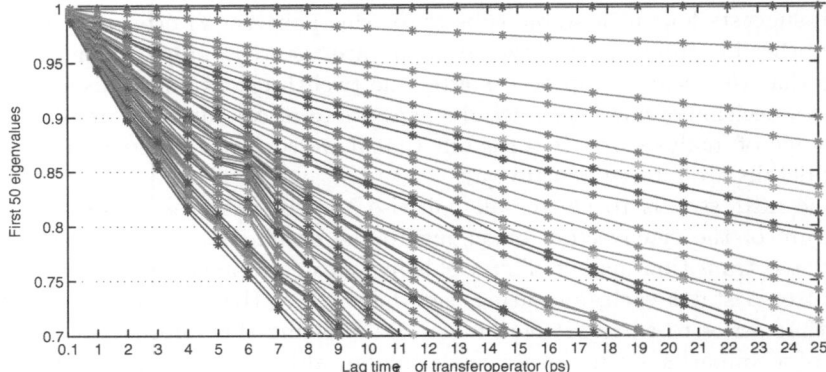

Fig. 5. Illustration of the 50 largest eigenvalues of the transition matrix obtained from the global Viterbi path vs. lag time τ, e.g. a lag time of 10 ps means that transition are considered that occur from time x to time $x+10$ ps. On clearly observes that the structure of the spectrum does not depend on τ

based on the first four ones. The outcome will identify 2, resp. 4, conformations in the discrete global Viterbi path. This allows to associate each data point of the ten dimensional peptide angle time series with one of the 2, resp. 4, conformations.

In Fig. 6 the results based on the first two eigenvectors are displayed ($M = 2$). Each of the two resulting conformations is represented by five histogram plots that belong to the five $\Phi\backslash\Psi$ pairs. These histograms are based on the assignment of data points to the conformation. Comparing the positions of the histogram peaks in the first conformation with the classifications given in Fig. 2 shows that this conformation corresponds to an α-helix structure. In contrast, the other conformation allows no assignment to a specific secondary structure, as every angle pair is very flexible and adopts regions of the α-helix structure and the β-strand structure.

Redoing the analysis with the $M = 4$ leading eigenvectors yields that the α-helix structure is still identified as a conformation, while the other conformation of the previous analysis splits-up into three conformations. Each of these three conformations can be uniquely described by the dynamical behavior of the peptide angle pairs, some of them are fixed to α-helix, resp. β-strand regions, while others remaining flexible in the sense that they alternate between these regions, see Fig. 7.

Taking more leading eigenvectors into account would resolve more flexible angle pairs by separating α-helix and β-strand parts, but it is important to note that this means resolving metastability on a faster timescale, cf. Fig. 5. As illustration we show representatives of each conformation $C_1, \ldots, C_4$ resulting from the $M = 4$ analysis in Fig. 8, This figure also includes the conditional transition probabilities $p(\tau, C_j, C_k)$ of the Markov switching process for lag time $\tau = 0.1$ps. Note that in accordance with our definition of metastability

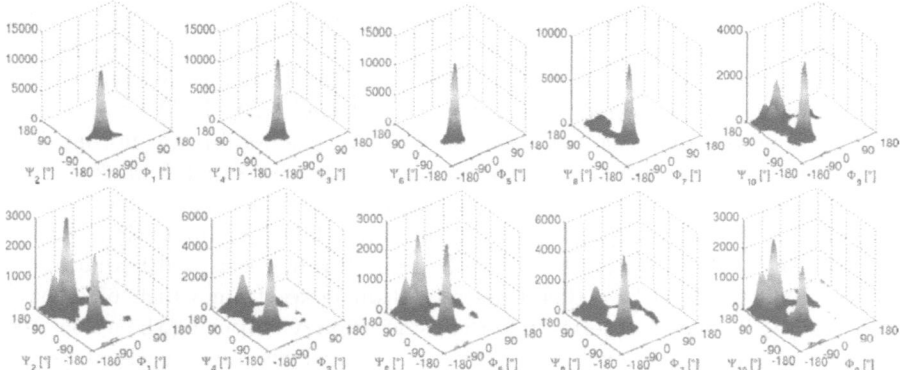

Fig. 6. Histogram plots for the two conformations (as resulting from the $M = 2$ analysis) displayed in the Ramachandran plane of each of the five $\Phi\backslash\Psi$ pairs. *Top row:* This conformation constitutes a clear helix structure. *Bottom row:* This conformation corresponds to no clear secondary structure

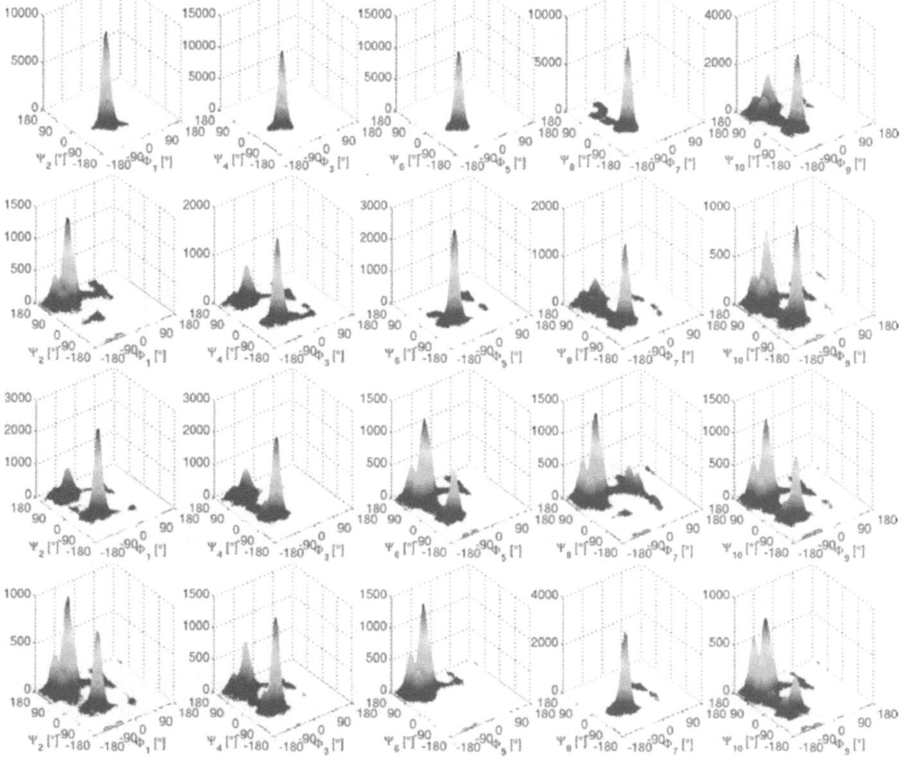

Fig. 7. Conformations as resulting from $M = 4$ analysis represented as histogram plots displayed in the Ramachandran plane of each of the five $\Phi\backslash\Psi$ pairs. *Top:* The helix structure of the first conformation is unchanged compared to the $M = 2$ analysis. *Below:* The other conformations exhibit mixed structures, with some angle pairs fixed to α-helix or β-strand regions while others are flexible

the conditional probability to stay within one conformation is nearly 1 for each conformation.

Finally we want to try to verify the results of the HMMSDE based procedure by comparison with a direct transfer operator based analysis. We will do this by reducing the dimensionality of the system considerable by noting that the relevant dynamical information is contained in the first four Ψ angles. Therefore we can reduce the ten-dimensional to a four-dimensional peptide angle space by skipping the other dimensions. The four-dimensional space can be partitioned directly by discretizing each dimension in 10 equidistant boxes. This yields $10^4 = 10000$ discretization boxes from which 6551 have been visited by the time series under consideration. Computing the associated transition matrix and evaluating the dominant spectrum is easily feasible, particular as it is a sparse matrix. The results are obtained by analyzing this transition matrix, based upon a direct partition of the state space, are similar to the results we obtained with HMMSDE . Without giving details, we indicate

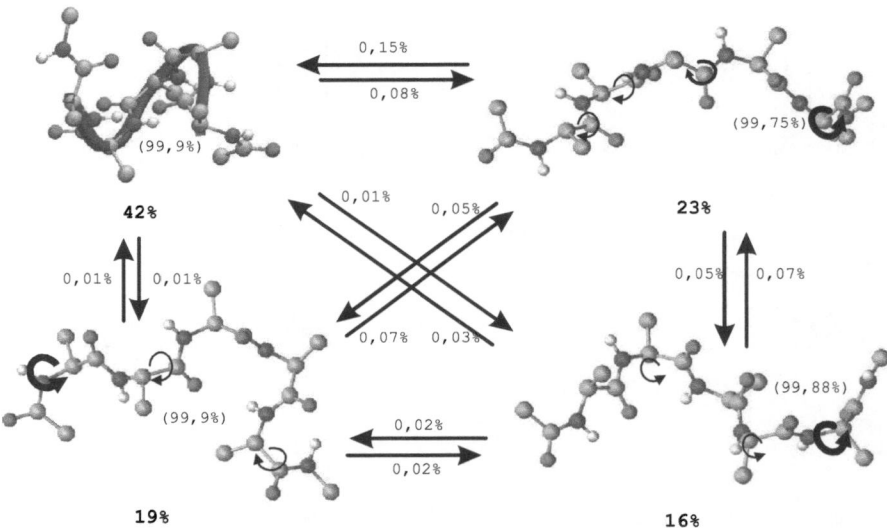

Fig. 8. Representatives of the four conformations obtained in the $M = 4$ analysis and the conditional transition probabilities between them (lag time $\tau = 0.1$ps). Fat numbers indicating the statistical weight of each conformation, numbers in brackets the conditional probability to stay within a conformation. Flexibility in peptide angles is marked with arrows, cf. Fig. 7. Note that the transition matrix relating to this picture is not symmetric but reversible. *Top left:* For the helix conformation the backbone is colored blue for illustrative purpose. It should be obvious from Fig. 5 that for significantly larger lag time τ only two eigenvalues will correspond to metastability such that only the helical conformation and a mixed flexible and partially unfolded one remain with significantly high conditional probability to stay within

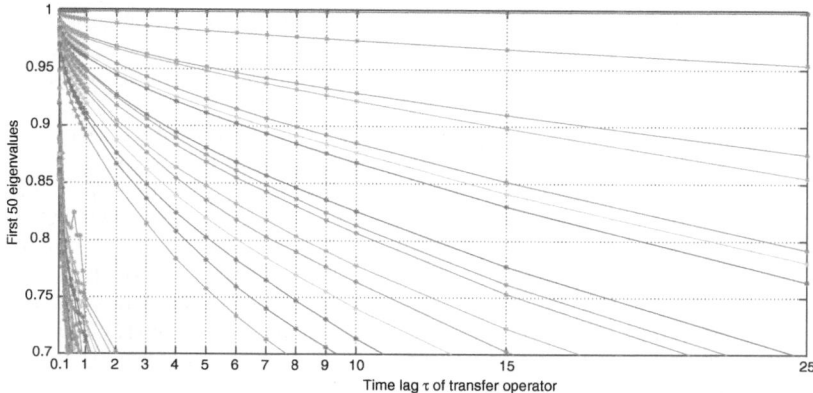

Fig. 9. The 50 largest eigenvalues of the transition matrix obtained from a direct discretization of the $\Psi_2, \Psi_4, \Psi_6, \Psi_8$-subspace versus the lag time τ

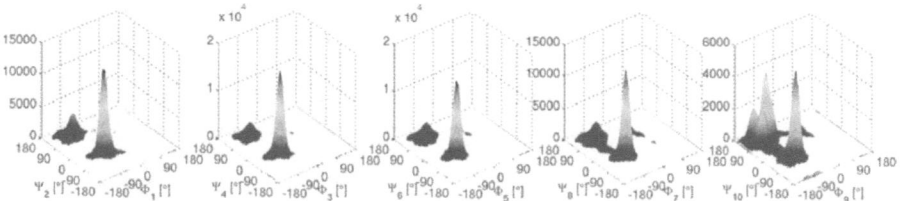

Fig. 10. Pairwise Ramachandran plots of the histogram of the entire time series

this by showing the eigenvalues plotted against different lag times τ in Fig. 9, which reveals a very similar spectral structure as Fig. 5.

Acknowledgements

We are indebted to G. Stock for the courtesy of making the simulation data of penta-alanine available to us.

References

1. C. Schütte, A. Fischer, W. Huisinga, and P. Deuflhard (1999) A direct approach to conformational dynamics based on hybrid Monte Carlo. *J. Comput. Phys. Special Issue on Computational Biophysics* **151**, pp. 146–168
2. C. Schütte and W. Huisinga (2003) Biomolecular conformations can be identified as metastable sets of molecular dynamics. In *Handbook of Numerical Analysis* (P. G. Ciaret and J.-L. Lions, eds.), Computational Chemistry, North-Holland
3. D. Chandler (1998) Finding transition pathways: Throwing ropes over rough montain passes, in the dark, in *Classical and Quantum Dynamics in Condensed*

Phase Simulations (B. Berne, G. Ciccotti, and D. Coker, eds.), Singapure: World Scientific, pp. 51–66

4. W. E, W. Ren, and E. Vanden-Eijnden (2002) String method for the study of rare events. *Phys. Rev. B* **66**, p. 052301

5. W. E, W. Ren, and E. Vanden-Eijnden (2005) Finite temperature string method for the study of rare events. *J. Phys. Chem. B* **109**, pp. 6688–6693

6. A. Laio and M. Parrinello (2002) Escaping free-energy minima. *Proceedings of the National Academy of the United States of America* **99**, pp. 12562–23566

7. R. Elber and M. Karplus (1987) Multiple conformational states of proteins: A molecular dynamics analysis of Myoglobin. *Science* **235**, pp. 318–321

8. P. Deuflhard, W. Huisinga, A. Fischer, and C. Schütte (2000) Identification of almost invariant aggregates in reversible nearly uncoupled Markov chains. *Lin. Alg. Appl.* **315**, pp. 39–59

9. M. Dellnitz and O. Junge (1999) On the approximation of complicated dynamical behavior. *SIAM J. Num. Anal.* **36**(2), pp. 491–515

10. P. Deuflhard and M. Weber (2005) Robust Perron cluster analysis in conformation dynamics. *Lin. Alg. Appl.* **398**, pp. 161–184

11. I. Horenko, E. Dittmer, A. Fischer, and C. Schütte, Automated model reduction for complex systems exhibiting metastability. Submitted to Multiscale Modeling and Simulation

12. I. Horenko, E. Dittmer, and C. Schütte (2005) Reduced stochastic models for complex molecular systems. *Computing and Visualization in Science* **9**, pp. 89–102

13. A. Fischer, S. Waldhausen, I. Horenko, E. Meerbach, and C. Schütte (2004) Identification of biomolecular conformations from incomplete torsion angle observations by hidden Markov models. *Journal of computational Physics* (submitted)

14. I. Horenko, E. Dittmer, F. Lankas, J. Maddocks, P. Metzner, and C. Schütte (2005) Macroscopic dynamics of complex metastable systems: Theory, algorithms, and application to b-dna. *J. Appl. Dyn. Syst., submitted*

15. S. D. Bond and B. B. L. Benedict J. Leimkuhler (1999) The Nosé–Poincaré method for constant temperature molecular dynamics. *JCP* **151**(1), pp. 114–134

16. A. Bovier, M. Eckhoff, V. Gayrard, and M. Klein (2001) Metastability in stochastic dynamics of disordered mean–field models. *Probab. Theor. Rel. Fields* **119**, pp. 99–161

17. E. B. Davies (1982) Metastable states of symmetric Markov semigroups I. *Proc. London Math. Soc.* **45**(3), pp. 133–150

18. C. Schütte, W. Huisinga, and P. Deuflhard (2001) Transfer operator approach to conformational dynamics in biomolecular systems. In *Ergodic Theory, Analysis, and Efficient Simulation of Dynamical Systems* (B. Fiedler, ed.), Springer Berlin Heidelberg, pp. 191–223

19. G. Singleton (1984) Asymptotically exact estimates for metatstable Markov semigroups. *Quart. J. Math. Oxford* **35**(2), pp. 321–329

20. W. E and E. Vanden-Eijnden (2005) Metastability, conformation dynamics, and transition pathways in complex systems preprint

21. W. Huisinga, S. Meyn, and C. Schütte (2004) Phase transitions and metastability in Markovian and molecular systems. *Ann. Appl. Probab.* **14**(1), pp. 419–458

22. C. Schütte and W. Huisinga (2000) On conformational dynamics induced by Langevin processes. In *EQUADIFF 99 – International Conference on Differential Equations* (B. Fiedler, K. Gröger, and J. Sprekels, eds.), vol. 2, (Singapore), pp. 1247–1262, World Scientific

23. W. Huisinga and B. Schmidt (2005) Metastability and dominant eigenvalues of transfer operators. In *New Algorithms for Macromolecular Simulation* (C. Chipot, R. Elber, A. Laaksonen, B. Leimkuhler, A. Mark, T. Schlick, C. Schütte, and R. Skeel, eds.), vol. **49** of *Lecture Notes in Computational Science and Engineering*, Springer, to appear

24. M. Weber (2004) Improved Perron cluster analysis. ZIB-Report, (Zuse Institute, Berlin, pp. 03–04

25. P. Lezaud (2001) Chernoff and Berry–Esséen inequalities for Markov processes. *ESIAM: P & S* **5**, pp. 183–201

26. B. J. Berne and J. E. Straub (1997) Novel methods of sampling phase space in the simulation of biological systems. *Curr. Opinion in Struct. Biol.* **7**, pp. 181–189

27. D. M. Ferguson, J. I. Siepmann, and D. G. Truhlar, eds. (1999) *Monte Carlo Methods in Chemical Physics*, vol. **105** of *Advances in Chemical Physics*. New York: Wiley

28. A. Fischer, C. Schütte, P. Deuflhard, and F. Cordes (2002) Hierarchical uncoupling-coupling of metastable conformations. In *Computational Methods for Macromolecules: Challenges and Applications* (T. Schlick and H. H. Gan, eds.), vol. **24** of *Lecture Notes in Computational Science and Engineering*, Springer Berlin Heidelberg, pp. 235–259

29. J. A. Bilmes (1998) A gentle tutorial of the EM algorithm and its application to parameter estimation for Gaussian mixture and Hidden Markov Models. tech. rep., International Computer Science Institute, Berkeley

30. A. J. Viterbi (1967) Error bounds for convolutional codes and an asymptotically optimum decoding algorithm. *IEEE Trans. Informat. Theory* **IT-13**, pp. 260–269

31. Y. Mu, P. H. Nguyen, and G. Stock (2004) Energy landscape of a small peptide revealed by dihedral angle principal component analysis. *Proteins: Structure, Function, and Bioinformatics* **58**(1), pp. 45–52

32. G. N. Ramachandran and V. Sasiskharan (1968) Conformations of polypeptides and proteins. *Advan. Prot. Chem.* **23**, pp. 283–427

Transport Coefficients
of Quantum-Classical Systems

R. Kapral[1] and G. Ciccotti[2]

[1] Chemical Physics Theory Group, Department of Chemistry, University of
Toronto, Toronto, ON M5S 3H6, Canada
rkapral@chem.utoronto.ca
[2] Dipartimento di Fisica, Università "La Sapienza", Piazzale Aldo Moro, 2, 00185
Roma, Italy
giovanni.ciccotti@roma1.infn.it

Raymond Kapral

R. Kapral and G. Ciccotti: *Transport Coefficients of Quantum-Classical Systems*, Lect. Notes
Phys. **703**, 519–551 (2006)
DOI 10.1007/3-540-35273-2_15

1 Introduction

Quantum mechanics provides us with the most fundamental description of natural phenomena. In many instances classical mechanics constitutes an adequate approximation and it is widely used in simulations of both static and dynamic properties of many-body systems. Often, however, quantum effects cannot be neglected and one is faced with the task of devising methods to simulate the behavior of the quantum system.

The computation of the equilibrium properties of quantum systems is a challenging problem. The simulation of dynamical properties, such as transport coefficients, presents additional problems since the solution of the quantum equations of motion for many-body systems is even more difficult. This fact has prompted the development of approximate methods for dealing with such problems.

The topic of this chapter is the description of a quantum-classical approach to compute transport coefficients. Transport coefficients are most often expressed in terms of time correlation functions whose evaluation involves two aspects: sampling initial conditions from suitable equilibrium distributions and evolution of dynamical variables or operators representing observables of the system. The schemes we describe for the computation of transport properties pertain to quantum many-body systems that can usefully be partitioned into two subsystems, a quantum subsystem S and its environment $\mathcal{E}$. We shall be interested in the limiting situation where the dynamics of the environmental degrees of freedom, in isolation from the quantum subsystem S, obey classical mechanics.

We show how the quantum-classical evolution equations of motion can be obtained as an approximation to the full quantum evolution and point out some of the difficulties that arise because of the lack of a Lie algebraic structure. The computation of transport properties is discussed from two different perspectives. Transport coefficient formulas may be derived by starting from an approximate quantum-classical description of the system. Alternatively, the exact quantum transport coefficients may be taken as the starting point of the computation with quantum-classical approximations made only to the dynamics while retaining the full quantum equilibrium structure. The utility of quantum-classical Liouville methods is illustrated by considering the computation of the rate constants of quantum chemical reactions in the condensed phase.

2 Wigner Formulation of Quantum Statistical Mechanics

We begin our discussion with a survey of the quantum dynamics and linear response theory expressions for quantum transport coefficients. Since we wish to make a link to a partial classical description, the use of Wigner transforms provides a means to introduce a phase space description of the quantum system. Consequently, our formulation of quantum mechanics will be carried out in this Wigner transform framework.

2.1 Dynamics

The von Neumann equation, or quantum mechanical Liouville equation,

$$\frac{\partial \hat{\rho}(t)}{\partial t} = -\frac{i}{\hbar}[\hat{H}, \hat{\rho}(t)] \equiv -i\hat{L}\hat{\rho}(t) \;, \tag{1}$$

where $\hat{H}$ is the Hamiltonian of the system and $\hat{L}$ is the quantum Liouville operator, specifies the time evolution of the density matrix $\hat{\rho}$. The formal solution of (1) is

$$\hat{\rho}(t) = e^{-i\hat{L}t}\hat{\rho}(0) = e^{-i\hat{H}t/\hbar}\hat{\rho}(0)e^{i\hat{H}t/\hbar} \;. \tag{2}$$

We may take the Wigner transform of the quantum Liouville equation to obtain an alternate formulation of the equation of motion. The Wigner transforms of the density matrix and an operator $\hat{A}$ are defined, respectively, by [1]

$$\rho_W(\mathcal{X}) = (2\pi\hbar)^{-\mathcal{N}} \int d\mathcal{Z}\, e^{i\mathcal{P}\cdot\mathcal{Z}/\hbar} \left\langle \mathcal{R} - \frac{\mathcal{Z}}{2} |\hat{\rho}| \mathcal{R} + \frac{\mathcal{Z}}{2} \right\rangle \;, \tag{3}$$

and

$$\begin{aligned}
A_W(\mathcal{X}) &= \int d\mathcal{Z}\, e^{i\mathcal{P}\cdot\mathcal{Z}/\hbar} \left\langle \mathcal{R} - \frac{\mathcal{Z}}{2} |\hat{A}| \mathcal{R} + \frac{\mathcal{Z}}{2} \right\rangle \\
&= \int d\mathcal{Z}\, e^{-i\mathcal{P}\cdot\mathcal{Z}/\hbar} \left\langle \mathcal{R} + \frac{\mathcal{Z}}{2} |\hat{A}| \mathcal{R} - \frac{\mathcal{Z}}{2} \right\rangle \;,
\end{aligned} \tag{4}$$

where $\mathcal{N}$ is the coordinate space dimension of the system and we use calligraphic symbols to denote phase space variables for the entire system, $\mathcal{X} = (\mathcal{R}, \mathcal{P})$. (Later in this chapter we shall make a distinction between phase space variables for the entire system and those for the $\mathcal{S}$ and $\mathcal{E}$ subsystems. We let $\mathcal{X} = (x, X)$ where $x = (r, p)$ and $X = (R, P)$ for the $\mathcal{S}$ and $\mathcal{E}$ subsystems, respectively.) Taking the Wigner transform of (1) we find

$$\frac{\partial}{\partial t}\rho_W(\mathcal{X}, t) = -\frac{i}{\hbar}\left[(\hat{H}\hat{\rho}(t))_W(\mathcal{X}) - (\hat{\rho}(t)\hat{H})_W(\mathcal{X})\right] \;. \tag{5}$$

To proceed, we must evaluate the Wigner transform of a product of operators. This calculation is given in several reviews [2] but we sketch it here for completeness. Letting $\mathcal{Q} = \mathcal{R} + \mathcal{Z}/2$ and $\mathcal{Q}' = \mathcal{R} - \mathcal{Z}/2$, we may invert the relation (4) to obtain

$$\langle \mathcal{Q}|\hat{A}|\mathcal{Q}'\rangle = (2\pi\hbar)^{-\mathcal{N}} \int d\mathcal{P}\, e^{i\mathcal{P}\cdot(\mathcal{Q}-\mathcal{Q}')/\hbar} A_W((\mathcal{Q}+\mathcal{Q}')/2, \mathcal{P}) \;. \tag{6}$$

If we introduce the Fourier transform of $A_W(\mathcal{Q}, \mathcal{P})$,

$$A_W(\mathcal{Q}, \mathcal{P}) = \int d\sigma \, d\tau \, e^{i(\sigma \mathcal{Q} + \tau \mathcal{P})/\hbar} \alpha(\sigma, \tau) \,, \tag{7}$$

into (6) we find,

$$\langle \mathcal{Q}|\hat{A}|\mathcal{Q}'\rangle = \int d\sigma \, e^{i\sigma(\mathcal{Q}+\mathcal{Q}')/(2\hbar)} \alpha(\sigma, \mathcal{Q}' - \mathcal{Q}) \,. \tag{8}$$

Writing the Wigner transform of a product of operators as

$$(\hat{A}\hat{B})_W(\mathcal{R}, \mathcal{P}) = \int d\mathcal{Z} d\mathcal{R}' \, e^{i\mathcal{P}\mathcal{Z}/\hbar} \left\langle \mathcal{R} - \frac{\mathcal{Z}}{2} \Big| \hat{A} \Big| \mathcal{R}' \right\rangle \left\langle \mathcal{R}' \Big| \hat{B} \Big| \mathcal{R} + \frac{\mathcal{Z}}{2} \right\rangle \,, \tag{9}$$

and inserting (8) and its analog for the the operator $\hat{B}$ with Fourier coefficients β we find

$$\begin{aligned}
(\hat{A}\hat{B})_W(\mathcal{R}, \mathcal{P}) &= \int d\mathcal{Z} d\mathcal{R}' d\sigma d\sigma' \, e^{i\mathcal{P}\mathcal{Z}/\hbar} e^{i\sigma(\mathcal{R}+\mathcal{R}'-\mathcal{Z}/2)/(2\hbar)} e^{i\sigma'(\mathcal{R}'+\mathcal{R}+\mathcal{Z}/2)/(2\hbar)} \\
&\quad \times \alpha(\sigma, \mathcal{R}' - \mathcal{R} + \mathcal{Z}/2)\beta(\sigma', \mathcal{R} - \mathcal{R}' + \mathcal{Z}/2) \\
&= \int d\tau d\tau' d\sigma d\sigma' \, \alpha(\sigma, \tau) \\
&\quad \times e^{i(\sigma\mathcal{R}+\tau\mathcal{P})/\hbar} e^{i(\sigma'\tau-\sigma\tau')/(2\hbar)} e^{i(\sigma'\mathcal{R}+\tau'\mathcal{P})/\hbar} \beta(\sigma', \tau') \,.
\end{aligned} \tag{10}$$

In the second line we made the change of variables $\tau = \mathcal{R} - \mathcal{R}' + \mathcal{Z}/2$ and $\tau' = \mathcal{R}' - \mathcal{R} + \mathcal{Z}/2$. Finally, we note that

$$\begin{aligned}
e^{i(\sigma\mathcal{R}+\tau\mathcal{P})/\hbar} &e^{i(\sigma'\tau-\sigma\tau')/(2\hbar)} e^{i(\sigma'\mathcal{R}+\tau'\mathcal{P})/\hbar} \\
&= e^{i(\sigma\mathcal{R}+\tau\mathcal{P})/\hbar}(1 + i(\sigma'\tau - \sigma\tau')/2\hbar + \dots)e^{i(\sigma'\mathcal{R}+\tau'\mathcal{P})/\hbar} \\
&= e^{i(\sigma\mathcal{R}+\tau\mathcal{P})/\hbar}(1 + \hbar\Lambda/2i + \dots)e^{i(\sigma'\mathcal{R}+\tau'\mathcal{P})/\hbar} \\
&= e^{i(\sigma\mathcal{R}+\tau\mathcal{P})/\hbar} e^{\hbar\Lambda/2i} e^{i(\sigma'\mathcal{R}+\tau'\mathcal{P})/\hbar} \\
&= e^{i(\sigma'\mathcal{R}+\tau'\mathcal{P})/\hbar} e^{-\hbar\Lambda/2i} e^{i(\sigma\mathcal{R}+\tau\mathcal{P})/\hbar} \,,
\end{aligned} \tag{11}$$

where Λ is the negative of the Poisson bracket operator, $\Lambda = \overleftarrow{\nabla}_{\mathcal{P}} \cdot \overrightarrow{\nabla}_{\mathcal{R}} - \overleftarrow{\nabla}_{\mathcal{R}} \cdot \overrightarrow{\nabla}_{\mathcal{P}}$, where the direction of an arrow indicates the direction in which the operator acts. Inserting this result into (10) and using the definition in (7) we find the result $(\hat{A}\hat{B})_W = A_W(\mathcal{X})e^{\frac{\hbar\Lambda}{2i}} B_W(\mathcal{X}) = B_W(\mathcal{X})e^{\frac{-\hbar\Lambda}{2i}} A_W(\mathcal{X})$ for the Wigner transform of a product of operators. The second equality follows from the equality in the last two lines of (11). Using these results, the Wigner transform of the quantum Liouville equation can be written as,

$$\begin{aligned}
\frac{\partial}{\partial t} \rho_W(\mathcal{X}, t) &= -\frac{i}{\hbar}\left(H_W(\mathcal{X})e^{\frac{\hbar\Lambda}{2i}} \rho_W(\mathcal{X}, t) - \rho_W(\mathcal{X}, t)e^{\frac{\hbar\Lambda}{2i}} H_W(\mathcal{X}) \right) \\
&= -\frac{i}{\hbar}\left(H_W(\mathcal{X})\left(e^{\frac{\hbar\Lambda}{2i}} - e^{\frac{-\hbar\Lambda}{2i}}\right) \rho_W(\mathcal{X}, t) \right) \\
&= -\frac{2}{\hbar} H_W(\mathcal{X}) \sin\left(\frac{\hbar\Lambda}{2}\right) \rho_W(\mathcal{X}, t) \equiv -iL_W(\mathcal{X})\rho_W(\mathcal{X}, t).
\end{aligned} \tag{12}$$

Here, the Wigner transformed Hamiltonian is $H_W(\mathcal{X}) = \mathcal{P}^2/2M + V_W(\mathcal{R})$. The last line of this equation defines iL_W, the Wigner form of quantum Liouville operator, $iL_W = \frac{2}{\hbar} H_W(\mathcal{X}) \sin\left(\frac{\hbar\Lambda}{2}\right)$.

A similar calculation may be carried out for the time evolution of an observable. Starting with the Heisenberg equation of motion for a dynamical variable $\hat{B}$,

$$\frac{d\hat{B}(t)}{dt} = \frac{i}{\hbar}[\hat{H}, \hat{B}(t)] , \tag{13}$$

and taking the Wigner transform, we obtain,

$$\frac{d}{dt} B_W(\mathcal{X}, t) = iL_W(\mathcal{X})B_W(\mathcal{X}, t) . \tag{14}$$

The classical limit of these equations of motion is easily taken by retaining only the term independent of $\hbar$ in the Liouville operator $iL_W = \frac{2}{\hbar} H_W(\mathcal{X}) \sin\left(\frac{\hbar\Lambda}{2}\right) = H_W(\mathcal{X})\Lambda + O(\hbar)$. Using this result we find the classical Liouville equation for the density matrix,

$$\frac{\partial}{\partial t}\rho(\mathcal{X}, t) = -H(\mathcal{X})\Lambda\rho(\mathcal{X}, t) = \{H(\mathcal{X}), \rho(\mathcal{X}, t)\}$$
$$\equiv -iL_{cl}(\mathcal{X})\rho(\mathcal{X}, t) , \tag{15}$$

with a similar evolution equation for a dynamical variable,

$$\frac{d}{dt}B(\mathcal{X}, t) = iL_{cl}(\mathcal{X})B(\mathcal{X}, t) , \tag{16}$$

where we have dropped the subscript W on the density matrix and Wigner transformed operators in this classical limit.

2.2 Response Theory and Time Correlation Functions

Equilibrium time correlation function expressions for transport properties can be derived using linear response theory [3]. Linear response theory can be carried out directly on the Wigner transformed equations of motion to obtain the transport properties as correlation functions involving Wigner transformed quantities. Alternatively, we may carry out the linear response analysis in terms of abstract operators and insert the Wigner representation of operators in the final form for the correlation function. We use the latter route here.

In linear response theory, it is assumed that a time dependent external force $F(t)$ couples to an observable $\hat{A}$ (self-adjoint operator) and the response of the system to linear order in the external force is computed. More specifically, the Hamiltonian in the presence of the external force is $\hat{H}(t) = \hat{H} - \hat{A}F(t)$, and the evolution equation for the density matrix is

$$\frac{\partial\hat{\rho}(t)}{\partial t} = (i\hbar)^{-1}[\hat{H}(t), \hat{\rho}(t)] = -(i\hat{L} - i\hat{L}_A F(t))\hat{\rho}(t) , \tag{17}$$

where $i\hat{L}_A \equiv (i/\hbar)[\hat{A}, \ ]$.

Given that the system was in thermal equilibrium in the distant past, the solution of this equation to linear order in the external force is [3]

$$\hat{\rho}(t) = \hat{\rho}_e^Q + \int_{-\infty}^t dt'\, e^{-i\hat{L}(t-t')} i\hat{L}_A \hat{\rho}_e^Q F(t') \, . \tag{18}$$

The canonical equilibrium density matrix is $\hat{\rho}_e^Q = Z_Q^{-1} \exp(-\beta\hat{H})$ and $Z_Q = \mathrm{Tr}\exp(-\beta\hat{H})$ is the partition function. The response of the system to the external force in a property B is given by the average value of the operator $\hat{B}$ using the density matrix at time t (assuming, without loss of generality, that $\hat{B}$ has an average value of zero in equilibrium),

$$\overline{B(t)} = \mathrm{Tr}\hat{B}\hat{\rho}(t) = \frac{i}{\hbar}\int_{-\infty}^t dt'\, \mathrm{Tr}\hat{B}(t-t')[\hat{A},\hat{\rho}_e^Q]F(t')$$

$$= \frac{i}{\hbar}\int_{-\infty}^t dt'\, \mathrm{Tr}[\hat{B}(t-t'),\hat{A}]\hat{\rho}_e^Q F(t') \equiv \int_{-\infty}^t dt'\, \phi_{BA}(t-t')F(t') \tag{19}$$

with the response function defined by

$$\phi_{BA}(t) = \left\langle \frac{i}{\hbar}[\hat{B}(t),\hat{A}] \right\rangle_Q \, , \tag{20}$$

where the angle brackets denote a quantum canonical equilibrium average, $\langle\cdots\rangle_Q = \mathrm{Tr}\cdots\hat{\rho}_e^Q$.

The response function may be written in an equivalent form by using the quantum mechanical operator identity, [3]

$$\frac{i}{\hbar}[\hat{A},\hat{\rho}_e^Q] = \int_0^\beta d\lambda\, \hat{\rho}_e^Q \dot{\hat{A}}(-i\hbar\lambda) \, , \tag{21}$$

in the second line of (19) to obtain

$$\phi_{BA}(t) = \int_0^\beta d\lambda\, \mathrm{Tr}\dot{\hat{A}}(-i\hbar\lambda)\hat{B}(t)\hat{\rho}_e^Q \, . \tag{22}$$

Transport properties are typically expressed as time integrals of flux-flux correlation functions. Letting $\hat{B} = \dot{\hat{A}} \equiv \hat{j}^A$ be the flux corresponding to the operator $\hat{A}$, the quantum expression for a transport coefficient takes the general form,

$$\lambda_A \propto \int_0^\infty dt\, \langle \hat{j}^A; \hat{j}^A(t) \rangle_Q \, , \tag{23}$$

where

$$\beta\langle \hat{j}^A; \hat{j}^A(t) \rangle_Q \equiv \left\langle \frac{i}{\hbar}[\hat{j}^A(t),\hat{A}] \right\rangle_Q = \int_0^\beta d\lambda\, \mathrm{Tr}\hat{j}^A(-i\hbar\lambda)\hat{j}^A(t)\hat{\rho}_e^Q \, , \tag{24}$$

defines the Kubo transformed correlation function.

We may now insert the definitions of the operators in terms of their Wigner transforms, as discussed in the previous subsection, to obtain equivalent representations of the transport coefficient expressions. Noting that $\text{Tr}\hat{A}\hat{B} = \int d\mathcal{X}\, A_W(\mathcal{X})B_W(\mathcal{X})$ and the rule for the Wigner transform of a product of operators, we can write

$$\left\langle \frac{i}{\hbar}[\hat{j}^A(t), \hat{A}]\right\rangle_Q = \frac{i}{\hbar}\int d\mathcal{X}\left(j_W^A(t)e^{\frac{\hbar A}{2i}}A_W - A_W e^{\frac{\hbar A}{2i}}j_W^A(t)\right)\rho_{We}^Q$$

$$= \frac{2}{\hbar}\int d\mathcal{X}\left(j_W^A(t)\sin(\hbar A/2)A_W\right)\rho_{We}^Q . \tag{25}$$

Using these results and carrying out this program, we find that a transport coefficient can be written in the following equivalent forms,

$$\lambda_A \propto \frac{1}{\beta}\int_0^\infty dt \int d\mathcal{X}\,\frac{2}{\hbar}\left(j_W^A(t)\sin(\hbar A/2)A_W\right)\rho_{We}^Q ,$$

$$= \frac{1}{\beta}\int_0^\infty dt \int_0^\beta d\lambda \int d\mathcal{X}\left(j_W^A(-i\hbar\lambda)e^{\hbar A/2i}j_W^A(t)\right)\rho_{We}^Q . \tag{26}$$

We see that the correlation functions have a rather complex form when expressed in terms of Wigner transformed variables, involving exponential operators of the Poisson bracket operator.

The classical limits of the correlation function expressions are easily obtained by taking the $\hbar \to 0$ limit of these equations. We find,

$$\lambda_A \propto \frac{1}{\beta}\int_0^\infty dt \int d\mathcal{X}\left(j^A(t)Aj^A\right)\rho_e = \int_0^\infty dt \int d\mathcal{X}\, j^A j^A(t)\rho_e , \tag{27}$$

where ρ_e is the classical canonical equilibrium density matrix.

3 Partial Wigner Representation of Quantum Mechanics

As discussed in the Introduction, our interest is in quantum-classical systems where the environmental degrees of freedom can be treated classically. The above formulation of quantum dynamics and quantum statistical mechanics in the Wigner representation suggests that we consider another formulation of quantum mechanics based on a *partial* Wigner representation where only the degrees of freedom in the $\mathcal{E}$ subsystem are Wigner transformed [4]. We now sketch how this program can be carried out.

In order to distinguish between the subsystem $\mathcal{S}$ and environment $\mathcal{E}$ variables we use the notation $\mathcal{R} = (r, R)$, $\mathcal{P} = (p, P)$ and $\mathcal{X} = (r, R, p, P)$ where the lower case symbols refer to the subsystem and the upper case symbols refer to the bath. Again, calligraphic symbols are used to denote variables for the entire system, $\mathcal{S} \bigoplus \mathcal{E}$. The Hamiltonian is the sum of the kinetic energy operators of the subsystem and bath and the potential energy of the entire system, $\hat{H} = \hat{P}^2/2M + \hat{p}^2/2m + \hat{V}(\hat{q}, \hat{Q})$. We suppose that the coordinate space dimension of $\mathcal{S}$ is n and that of $\mathcal{E}$ is N with $\mathcal{N} = n + N$.

The partial Wigner transformation [4] of the density matrix with respect to the subset of Q coordinates is

$$\hat{\rho}_W(R, P) = (2\pi\hbar)^{-N} \int dz e^{iP \cdot z/\hbar} \left\langle R - \frac{z}{2} |\hat{\rho}| R + \frac{z}{2} \right\rangle . \tag{28}$$

In this representation the quantum Liouville equation is,

$$\begin{aligned}
\frac{\partial \hat{\rho}_W(R, P, t)}{\partial t} &= -\frac{i}{\hbar} \left((\hat{H}\hat{\rho})_W - (\hat{\rho}\hat{H})_W \right) \\
&= -\frac{i}{\hbar} \left(\hat{H}_W e^{\hbar\Lambda/2i} \hat{\rho}_W(t) - \hat{\rho}_W(t) e^{\hbar\Lambda/2i} \hat{H}_W \right) ,
\end{aligned} \tag{29}$$

where the partially Wigner transformed Hamiltonian is

$$\hat{H}_W(R, P) = \frac{P^2}{2M} + \frac{\hat{p}^2}{2m} + \hat{V}_W(\hat{q}, R) , \tag{30}$$

and Λ is again the negative of the Poisson bracket operator, but now operating only on the Wigner variables of $\mathcal{E}$. Note also that the partially Wigner transformed density matrix and operators are still operators in the Hilbert space of $\mathcal{S}$ and not just phase space functions when full Wigner transforms are taken. We may rewrite the quantum Liouville equation in a more compact form [5, 6]

$$\frac{\partial \hat{\rho}_W(R, P, t)}{\partial t} = -i\hat{L}_W \hat{\rho}_W(t) \equiv -(\hat{H}_W, \hat{\rho}_W(t))_Q . \tag{31}$$

by defining the quantum Liouville operator and quantum bracket in this partial Wigner representation. As might be expected, the Heisenberg equation of motion for a partially transformed operator takes the form,

$$\frac{d\hat{A}_W(R, P, t)}{dt} = i\hat{L}_W \hat{A}_W(t) \equiv (\hat{H}_W, \hat{A}_W(t))_Q . \tag{32}$$

The partial Wigner transform of a product of operators satisfies the associative product rule,

$$\begin{aligned}
(\hat{A}\hat{B}\hat{C})_W &= \left(\left(\hat{A}_W e^{\hbar\Lambda/2i} \hat{B}_W \right) e^{\hbar\Lambda/2i} \hat{C}_W \right) \\
&= \left(\hat{A}_W e^{\hbar\Lambda/2i} \left(\hat{B}_W e^{\hbar\Lambda/2i} \hat{C}_W \right) \right) .
\end{aligned} \tag{33}$$

The time evolution of a quantum operator $\hat{C} = \hat{A}\hat{B}$, which is the product of two operators, can be written in the partial Wigner representation as

$$\hat{C}_W(t) = \hat{A}_W(t) e^{\hbar\Lambda/2i} \hat{B}_W(t) . \tag{34}$$

Quantum mechanics in the partial Wigner representation is exact and the partially Wigner transformed quantum bracket satisfies the Jacobi identity,

$$(\hat{A}_W, (\hat{B}_W, \hat{C}_W)_Q)_Q + (\hat{C}_W, (\hat{A}_W, \hat{B}_W)_Q)_Q + (\hat{B}_W, (\hat{C}_W, \hat{A}_W)_Q)_Q = 0 \, , \tag{35}$$

consistent with the Lie algebraic structure of quantum dynamics.

4 Quantum-Classical Dynamics

We saw that it was a simple matter to take the classical limit of the quantum Liouville equation in Wigner transformed form by retaining only those terms in the Liouville operator that were independent of $\hbar$; i.e., taking the $\hbar \to 0$ limit of the Liouville operator. We cannot follow so simple a procedure to construct a quantum-classical equation of motion since $\hbar$ appears in the evolution operator arising from the $\mathcal{S}$ degrees of freedom and in the exponential operator involving the Poisson bracket operator coming from the partial Wigner transform over the $\mathcal{E}$ subsystem degrees of freedom. We now describe how to construct approximate evolution equations.

4.1 Dynamics

While one can imagine various routes to obtain the quantum-classical equations of motion, one way to disentangle these different contributions is to suppose that the particles comprising $\mathcal{E}$ have mass M while those of $\mathcal{S}$ have mass m, with $m \ll M$. In this circumstance we can scale variables so that distances are measured on the scale characteristic of the light particles, $\lambda_m = (\hbar^2/m\epsilon_0)^{1/2}$, with ϵ_0 some characteristic energy of the system, and velocities of both light and heavy particles are scaled to be comparable using momentum units $p_m = (m\lambda_m/t_0) = (m\epsilon_0)^{1/2}$ and $P_M = (M\epsilon_0)^{1/2}$, respectively. Here $t_0 = \hbar/\epsilon_0$ is the chosen time unit. This scaling is reminiscent of that used to derive Langevin equations from the Liouville equation in the theory of Brownian motion [7].

Let energy be measured in units of ϵ_0 and time in units of $t_0 = \hbar/\epsilon_0$. In scaled units $\hat{q}' = \hat{q}/\lambda_m$, $R' = R/\lambda_m$, $\hat{p}' = \hat{p}/p_m$, $P' = P/P_M$ and $t' = t/t_0$, we have,

$$\frac{\partial \hat{\rho}'_W(R', P', t)}{\partial t'} = -i\Big(\hat{H}'_W e^{\mu \Lambda'/2i} \hat{\rho}'_W(t') - \hat{\rho}'_W(t') e^{\mu \Lambda'/2i} \hat{H}'_W\Big)$$

$$\approx -i\Big(\hat{H}'_W(1 + \frac{\mu\Lambda'}{2i})\hat{\rho}'_W(t') - \hat{\rho}'_W(t')(1 + \frac{\mu\Lambda'}{2i})\hat{H}'_W\Big) \, . \tag{36}$$

To obtain the second approximate equality we expanded the right hand side to first order in the small parameter $\mu = (m/M)^{1/2}$. Returning to unscaled units we have the quantum-classical Liouville equation,

$$\frac{\partial \hat{\rho}_W(R,P,t)}{\partial t} = -\frac{i}{\hbar}[\hat{H}_W, \hat{\rho}_W(t)] + \frac{1}{2}\left(\{\hat{H}_W, \hat{\rho}_W(t)\} - \{\hat{\rho}_W(t), \hat{H}_W\}\right)$$

$$= -(\hat{H}_W, \hat{\rho}_W(t)) = -i\hat{\mathcal{L}}\hat{\rho}_W(t) , \qquad (37)$$

that gives the time evolution of the quantum-classical density matrix $\hat{\rho}_W(R,P,t)$ [4, 8–15]. The last two equalities in (37) define the quantum-classical bracket and quantum-classical Liouville operator [4, 5]. In (37), the coupling between the quantum subsystem and bath appears in both terms in the quantum-classical Liouville operator. The quantum character manifests itself in the Poisson bracket terms since the quantum operators do not commute and their order must be respected.

Using a similar procedure we may derive the equation of motion for an observable, $\hat{A}_W(R,P,t)$ [4]

$$\frac{d\hat{A}_W(R,P,t)}{dt} = (\hat{H}_W, \hat{A}_W(t)) = i\hat{\mathcal{L}}\hat{A}_W(t) , \qquad (38)$$

which is the quantum-classical analog of the Heisenberg equation of motion. This equation manifestly conserves energy.

In contrast to quantum mechanical and classical brackets, the quantum-classical bracket does not satisfy the Jacobi identity since [5]

$$(\hat{A}_W, (\hat{B}_W, \hat{C}_W)) + (\hat{C}_W, (\hat{A}_W, \hat{B}_W)) + (\hat{B}_W, (\hat{C}_W, \hat{A}_W)) \neq 0 . \qquad (39)$$

Consequently, quantum-classical dynamics does not possess a Lie algebraic structure and this leads to pathologies in the general formulation of quantum-classical dynamics and statistical mechanics as we shall see below [5, 6].

Furthermore, the evolution of a composite operator in quantum-classical dynamics cannot be written exactly in terms of the quantum-classical evolution of its constituent operators, but only to terms $\mathcal{O}(\hbar)$. To see this consider the action of the quantum-classical Liouville operator on the composite operator $\hat{C}_W = \hat{B}_W(1 + \hbar\Lambda/2i)\hat{A}_W$. A straightforward calculation shows that

$$i\hat{\mathcal{L}}\hat{C}_W = (i\hat{\mathcal{L}}\hat{B}_W)\left(1 + \frac{\hbar\Lambda}{2i}\right)\hat{A}_W + \hat{B}_W\left(1 + \frac{\hbar\Lambda}{2i}\right)(i\hat{\mathcal{L}}\hat{A}_W) + \mathcal{O}(\hbar) . \qquad (40)$$

It follows that

$$\hat{C}_W(t) = e^{i\hat{\mathcal{L}}t}\hat{C}_W = (1 + i\hat{\mathcal{L}}t + (i\hat{\mathcal{L}})^2\frac{t}{2!} + \ldots)\left(\hat{B}_W\left(1 + \frac{\hbar\Lambda}{2i}\right)\hat{A}_W + \mathcal{O}(\hbar^2)\right)$$

$$= \hat{B}_W\left(1 + \frac{\hbar\Lambda}{2i}\right)\hat{A}_W + \mathcal{O}(\hbar^2)$$

$$+ t\left((i\hat{\mathcal{L}}\hat{B}_W)\left(1 + \frac{\hbar\Lambda}{2i}\right)\hat{A}_W + \hat{B}_W\left(1 + \frac{\hbar\Lambda}{2i}\right)(i\hat{\mathcal{L}}\hat{A}_W) + \mathcal{O}(\hbar)\right)$$

$$+ \frac{t^2}{2!}\left(((i\hat{\mathcal{L}})^2\hat{B}_W)\left(1 + \frac{\hbar\Lambda}{2i}\right)\hat{A}_W + 2(i\hat{\mathcal{L}}\hat{B}_W)\left(1 + \frac{\hbar\Lambda}{2i}\right)(i\hat{\mathcal{L}}\hat{A}_W)\right.$$

$$\left. + \hat{B}_W\left(1 + \frac{\hbar\Lambda}{2i}\right)((i\hat{\mathcal{L}})^2\hat{A}_W) + \mathcal{O}(\hbar)\right) + \ldots . \qquad (41)$$

Resumming the series of operators we find,

$$\hat{C}_W(t) = \left(e^{i\hat{\mathcal{L}}t}\hat{B}_W\right)\left(1 + \frac{\hbar\Lambda}{2i}\right)\left(e^{i\hat{\mathcal{L}}t}\hat{A}_W\right) + \mathcal{O}(\hbar)$$

$$= \hat{B}_W(t)\left(1 + \frac{\hbar\Lambda}{2i}\right)\hat{A}_W(t) + \mathcal{O}(\hbar) . \tag{42}$$

While these features lead to some pathologies in the formulation of quantum-classical dynamics and statistical mechanics, the violations are in terms of higher order in $\hbar$ for the bath (or, better, the mass ratio μ), so that for systems where quantum-classical dynamics is likely to be applicable the numerical consequences are often small. We remark that almost all quantum-classical schemes suffer from these problems, although these deficiencies are often not highlighted.

The quantum-classical Liouville equation may be expressed in any convenient basis. In particular, the adiabatic basis vectors, $|\alpha; R\rangle$, are given by the solutions of $\hat{h}_W|\alpha; R\rangle = E_\alpha(R)|\alpha; R\rangle$, where $\hat{h}_W = \frac{\hat{p}^2}{2m} + \hat{V}_W(\hat{q}, R)$. We take an Eulerian view of the dynamics so that the adiabatic basis vectors are parameterized by the time-independent values of the bath coordinates R. In this basis, the Liouville operator has matrix elements [4],

$$i\mathcal{L}_{\alpha\alpha',\beta\beta'} = (i\omega_{\alpha\alpha'} + iL_{\alpha\alpha'})\delta_{\alpha\beta}\delta_{\alpha'\beta'} - J_{\alpha\alpha',\beta\beta'}$$

$$\equiv i\mathcal{L}^0_{\alpha\alpha'}\delta_{\alpha\beta}\delta_{\alpha'\beta'} - J_{\alpha\alpha',\beta\beta'} , \tag{43}$$

where $\omega_{\alpha\alpha'}(R) = (E_\alpha(R) - E_{\alpha'}(R))/\hbar$ is a frequency determined by the difference in energies of adiabatic states and $iL_{\alpha\alpha'}$ is the Liouville operator for classical evolution under the mean of the Hellmann-Feynman forces for adiabatic states α and α',

$$iL_{\alpha\alpha'} = \frac{P}{M} \cdot \frac{\partial}{\partial R} + \frac{1}{2}\left(F_W^\alpha + F_W^{\alpha'}\right) \cdot \frac{\partial}{\partial P} , \tag{44}$$

where $F_W^\alpha = -\langle\alpha; R|\frac{\partial\hat{V}_W(\hat{q}, R)}{\partial R}|\alpha; R\rangle$ is the Hellmann-Feynman force for state α. The operator $J_{\alpha\alpha',\beta\beta'}$ accounts for non-adiabatic transitions and corresponding changes of the bath momentum. It is given by

$$J_{\alpha\alpha',\beta\beta'} = -\frac{P}{M} \cdot d_{\alpha\beta}\left(1 + \frac{1}{2}S_{\alpha\beta} \cdot \frac{\partial}{\partial P}\right)\delta_{\alpha'\beta'}$$

$$-\frac{P}{M} \cdot d^*_{\alpha'\beta'}\left(1 + \frac{1}{2}S^*_{\alpha'\beta'} \cdot \frac{\partial}{\partial P}\right)\delta_{\alpha\beta} , \tag{45}$$

where $d_{\alpha\beta} = \langle\alpha; R|\nabla_R|\beta; R\rangle$ is the nonadiabatic coupling matrix element and $S_{\alpha\beta} = \Delta E_{\alpha\beta}\hat{d}_{\alpha\beta}(\frac{P}{M} \cdot \hat{d}_{\alpha\beta})^{-1}$ with $\Delta E_{\alpha\beta}(R) = E_\alpha(R) - E_\beta(R)$.

4.2 Transport Properties

We may easily carry out a linear response theory derivation of transport properties based on the quantum-classical Liouville equation that parallels the derivation for quantum systems outlined above. We suppose the quantum-classical system with Hamiltonian $\hat{H}_W$ is subjected to a time dependent external force that couples to the observable $\hat{A}_W$, so that the total Hamiltonian is

$$\hat{\mathcal{H}}_W(t) = \hat{H}_W - \hat{A}_W F(t) \,. \tag{46}$$

The evolution equation for the density matrix takes the form

$$\frac{\partial \hat{\rho}_W(t)}{\partial t} = -(i\hat{\mathcal{L}} - i\hat{\mathcal{L}}_A F(t))\hat{\rho}_W(t) \,, \tag{47}$$

where $i\hat{\mathcal{L}}_A$ has a form analogous to $i\hat{\mathcal{L}}$ with $\hat{A}_W$ replacing $\hat{H}_W$, $i\hat{\mathcal{L}}_A = (\hat{A}_W,\ )$. The formal solution of this equation is found by integrating from t_0 to t,

$$\hat{\rho}_W(t) = e^{-i\hat{\mathcal{L}}(t-t_0)}\hat{\rho}_W(t_0) + \int_{t_0}^t dt'\, e^{-i\hat{\mathcal{L}}(t-t')} i\hat{\mathcal{L}}_A \hat{\rho}_W(t') F(t') \,. \tag{48}$$

The next step in the calculation is to choose $\hat{\rho}_W(t_0)$ to be the equilibrium density matrix, $\hat{\rho}_{We}$. One of the differences between quantum and quantum-classical response theories appears at this stage. In quantum mechanics, the quantum canonical equilibrium density is $\hat{\rho}_e^Q = Z_Q^{-1} \exp(-\beta\hat{H})$ which, when expressed in terms of the partial Wigner transform, can be written as

$$\hat{\rho}_{We}^Q(R, P) = (2\pi\hbar)^{-N} \int dZ e^{iP \cdot Z/\hbar} \left\langle R - \frac{Z}{2} | \hat{\rho}_e^Q | R + \frac{Z}{2} \right\rangle \,. \tag{49}$$

The density matrix $\hat{\rho}_{We}^Q(R, P)$ is not stationary under quantum-classical dynamics. Instead, the equilibrium density of a quantum-classical system has to be determined by solving the equation $i\hat{\mathcal{L}}\hat{\rho}_{We} = 0$. An explicit solution of this equation has not been found although a recursive solution, obtained by expressing the density matrix $\hat{\rho}_{We}$ in a power series in $\hbar$ or the mass ratio μ, can be determined. While it is difficult to find the full solution to all orders in $\hbar$, the solution is known analytically to $\mathcal{O}(\hbar)$. When expressed in the adiabatic basis, the result is [5]·

$$\rho_{We}^{\alpha\alpha'} = \rho_{We}^{(0)\alpha} \Big(\delta_{\alpha\alpha'} - i\hbar \frac{P}{M} \cdot d_{\alpha\alpha'} \Big(\frac{\beta}{2}(1 + e^{-\beta E_{\alpha'\alpha}}) $$
$$+ \frac{1}{E_{\alpha\alpha'}}(1 - e^{-\beta E_{\alpha'\alpha}}) \Big)(1 - \delta_{\alpha\alpha'}) \Big) + \mathcal{O}(\hbar^2) \,. \tag{50}$$

If the partial Wigner transform of the exact canonical quantum equilibrium density in (49) is expressed in the adiabatic basis and is expanded to linear order in $\hbar$, we obtain the same result as in (50), indicating that the quantum-classical expression is exact to this order.

Using the quantum-classical form for $\hat{\rho}_{We}$, which, by construction, is invariant under quantum-classical dynamics, the first term on the right hand side of (48) reduces to $\hat{\rho}_{We}$ and is independent of t_0. We may assume that the system with Hamiltonian $\hat{H}_W$ is in thermal equilibrium at $t_0 = -\infty$, and with this boundary condition, to first order in the external force, (48) is

$$\hat{\rho}_W(t) = \hat{\rho}_{We} + \int_{-\infty}^{t} dt' \, e^{-i\hat{\mathcal{L}}(t-t')} i\hat{\mathcal{L}}_A \hat{\rho}_{We} F(t') \, . \tag{51}$$

Then, computing $\overline{B_W(t)} = \mathrm{Tr}' \int dR dP \, \hat{B}_W \hat{\rho}_W(t)$ to obtain the response function, we find

$$\begin{aligned}
\overline{B_W(t)} &= \int_{-\infty}^{t} dt' \, \mathrm{Tr}' \int dR dP \, \hat{B}_W e^{-i\hat{\mathcal{L}}(t-t')} i\hat{\mathcal{L}}_A \hat{\rho}_{We} F(t') \\
&= \int_{-\infty}^{t} dt' \, \langle (\hat{B}_W(t-t'), \hat{A}_W) \rangle F(t') \equiv \int_{-\infty}^{t} dt' \, \phi_{BA}^{QC}(t-t') F(t') \, .
\end{aligned} \tag{52}$$

Thus, the quantum-classical form of the response function is

$$\phi_{BA}^{QC}(t) = \langle (\hat{B}_W(t), \hat{A}_W) \rangle = \mathrm{Tr}' \int dR dP \, \hat{B}_W(t) (\hat{A}_W, \hat{\rho}_{We}) \, . \tag{53}$$

where, in writing the second equality in (53), we have used cyclic permutations under the trace and integrations by parts.

Given the response function, an expression for a transport coefficient can be obtained by taking $\hat{B}_W = \dot{\hat{A}}_W = i\mathcal{L}\hat{A}_W \equiv \hat{j}_W^A$. The quantum-classical analog of the expression for a quantum mechanical transport coefficient in (23) is given by

$$\lambda_A \propto \int_0^{\infty} dt \, \langle (\hat{j}_W^A(t), \hat{A}_W) \rangle = \int_0^{\infty} dt \, \mathrm{Tr}' \int dR dP \, \hat{j}_W^A(t) (\hat{A}_W, \hat{\rho}_{We}) \, . \tag{54}$$

This correlation function expression involves both quantum-classical dynamical evolution of observables and quantum-classical expressions for the equilibrium density.

5 Quantum-Classical Approximations for Quantum Correlation Functions

Rather than carrying out a linear response derivation to obtain correlation function expressions for transport coefficients based on the quantum-classical equations of motion, in this section we show how transport coefficients can be obtained by a different route. We take as a starting point the quantum mechanical expression for a transport coefficient and consider a limit where the *dynamics* is approximated by quantum-classical dynamics [17, 18]. The advantage of this approach is that the full quantum equilibrium structure can be retained while still performing the evolution of observables using quantum-classical dynamics, which is computationally more tractable than full quantum evolution.

The quantum mechanical expression for a transport property was given in (23) and its generalization to a time-dependent transport coefficient, defined as the finite time integral of a general flux-flux correlation function involving the fluxes of operators $\hat{A}$ and $\hat{B}$, is

$$\lambda_{AB}(t) = \int_0^t dt' \langle \hat{j}_A ; \hat{j}_B(t') \rangle_Q = \langle \dot{\hat{A}} ; \hat{B}(t) \rangle = \frac{1}{\beta} \left\langle \frac{i}{\hbar} [\hat{B}(t), \hat{A}] \right\rangle_Q . \quad (55)$$

It is convenient in simulations to determine the transport coefficient from the plateau value of $\lambda_{AB}(t)$ [3]. Considering the second equality in (55) in detail, we can write the transport coefficient $\lambda_{AB}(t)$ as,

$$
\begin{aligned}
\lambda_{AB}(t) &= \frac{1}{\beta Z_Q} \int_0^\beta d\lambda \, \mathrm{Tr} \left(\dot{\hat{A}} \, (-i\hbar\lambda) \hat{B}(t) e^{-\beta \hat{H}} \right) \\
&= \frac{1}{\beta Z_Q} \int_0^\beta d\lambda \, \mathrm{Tr} \left(\dot{\hat{A}} \, e^{\frac{i}{\hbar} \hat{H}(t + i\hbar\lambda)} \hat{B} e^{-\frac{i}{\hbar} \hat{H}(t + i\hbar\lambda)} e^{-\beta \hat{H}} \right) \\
&= \frac{1}{\beta Z_Q} \int_0^\beta d\lambda \int \prod_{i=1}^4 d\mathcal{Q}_i \langle \mathcal{Q}_1 | \dot{\hat{A}} | \mathcal{Q}_2 \rangle \langle \mathcal{Q}_2 | e^{\frac{i}{\hbar} \hat{H}(t + i\hbar\lambda)} | \mathcal{Q}_3 \rangle \langle \mathcal{Q}_3 | \hat{B} | \mathcal{Q}_4 \rangle \\
&\quad \times \langle \mathcal{Q}_4 | e^{-\frac{i}{\hbar} \hat{H}(t + i\hbar\lambda)} e^{-\beta \hat{H}} | \mathcal{Q}_1 \rangle , \quad (56)
\end{aligned}
$$

where the last line follows from introducing a coordinate representation $\{\mathcal{Q}\} = \{q\}\{Q\}$ of the operators (recall that calligraphic symbols are used to denote variables for the entire system, subsystem plus bath). Making a change of variables, $\mathcal{Q}_1 = \mathcal{R}_1 - \mathcal{Z}_1/2$, $\mathcal{Q}_2 = \mathcal{R}_1 + \mathcal{Z}_1/2$, etc., and then expressing the matrix elements in terms of the Wigner transforms of the operators using (6), we have [18]

$$
\begin{aligned}
\lambda_{AB}(t) &= \frac{1}{\beta} \int_0^\beta d\lambda \int d\mathcal{X}_1 d\mathcal{X}_2 (\dot{A})_W(\mathcal{X}_1) B_W(\mathcal{X}_2) \frac{1}{(2\pi\hbar)^{2\mathcal{N}} Z_Q} \\
&\quad \times \int d\mathcal{Z}_1 d\mathcal{Z}_2 e^{-\frac{i}{\hbar}(\mathcal{P}_1 \cdot \mathcal{Z}_1 + \mathcal{P}_2 \cdot \mathcal{Z}_2)} \left\langle \mathcal{R}_1 + \frac{\mathcal{Z}_1}{2} \left| e^{\frac{i}{\hbar} \hat{H}(t + i\hbar\lambda)} \right| \mathcal{R}_2 - \frac{\mathcal{Z}_2}{2} \right\rangle \\
&\quad \times \left\langle \mathcal{R}_2 + \frac{\mathcal{Z}_2}{2} \left| e^{-\beta \hat{H} - \frac{i}{\hbar} \hat{H}(t + i\hbar\lambda)} \right| \mathcal{R}_1 - \frac{\mathcal{Z}_1}{2} \right\rangle . \quad (57)
\end{aligned}
$$

We define the spectral density by [19, 20],

$$
\begin{aligned}
W(\mathcal{X}_1, \mathcal{X}_2, t) &= \frac{1}{(2\pi\hbar)^{2\mathcal{N}} Z_Q} \int d\mathcal{Z}_1 d\mathcal{Z}_2 e^{-\frac{i}{\hbar}(\mathcal{P}_1 \cdot \mathcal{Z}_1 + \mathcal{P}_2 \cdot \mathcal{Z}_2)} \\
&\quad \times \left\langle \mathcal{R}_1 + \frac{\mathcal{Z}_1}{2} \left| e^{\frac{i}{\hbar} \hat{H}t} \right| \mathcal{R}_2 - \frac{\mathcal{Z}_2}{2} \right\rangle \left\langle \mathcal{R}_2 + \frac{\mathcal{Z}_2}{2} \left| e^{-\beta \hat{H} - \frac{i}{\hbar} \hat{H}t} \right| \mathcal{R}_1 - \frac{\mathcal{Z}_1}{2} \right\rangle , (58)
\end{aligned}
$$

which, for real t, satisfies the property

$$W(\mathcal{X}_1, \mathcal{X}_2, t)^* = W(\mathcal{X}_2, \mathcal{X}_1, -t) . \tag{59}$$

Letting

$$\overline{W}(\mathcal{X}_1, \mathcal{X}_2, t) = \frac{1}{\beta} \int_0^\beta d\lambda W(\mathcal{X}_1, \mathcal{X}_2, t + i\hbar\lambda) , \tag{60}$$

which satisfies the same symmetry property as (59), we can write the transport coefficient as

$$
\begin{aligned}
\lambda_{AB}(t) &= \int d\mathcal{X}_1 d\mathcal{X}_2 (\dot{A})_W(\mathcal{X}_1) B_W(\mathcal{X}_2) \overline{W}(\mathcal{X}_1, \mathcal{X}_2, t) \\
&= \int d\mathcal{X}_1 d\mathcal{X}_2 (iL_W(\mathcal{X}_1) A_W(\mathcal{X}_1)) B_W(\mathcal{X}_2) \overline{W}(\mathcal{X}_1, \mathcal{X}_2, t) \\
&= -\int d\mathcal{X}_1 d\mathcal{X}_2 A_W(\mathcal{X}_1) B_W(\mathcal{X}_2) (iL_W(\mathcal{X}_1) \overline{W}(\mathcal{X}_1, \mathcal{X}_2, t)).
\end{aligned} \tag{61}
$$

From (61), the time evolution of $\overline{W}(\mathcal{X}_1, \mathcal{X}_2, t)$ may be defined by[3]

$$\frac{\partial}{\partial t} \overline{W}(\mathcal{X}_1, \mathcal{X}_2, t) = iL_W(\mathcal{X}_1) \overline{W}(\mathcal{X}_1, \mathcal{X}_2, t) . \tag{62}$$

Using these results, the transport coefficient expression can be written as

$$\lambda_{AB}(t) = -\int d\mathcal{X}_1 d\mathcal{X}_2 A_W(\mathcal{X}_1) B_W(\mathcal{X}_2) \left(\frac{\partial}{\partial t} \overline{W}(\mathcal{X}_1, \mathcal{X}_2, t) \right). \tag{63}$$

This transport coefficient expression is exact. The full quantum equilibrium structure is contained in the initial value $\overline{W}(\mathcal{X}_1, \mathcal{X}_2, 0)$, which is proportional to the integral over λ of

$$W(\mathcal{X}_1, \mathcal{X}_2, i\hbar\lambda) = \frac{1}{(2\pi\hbar)^{2N} Z_Q} \int d\mathcal{Z}_1 d\mathcal{Z}_2 e^{-\frac{i}{\hbar}(\mathcal{P}_1 \cdot \mathcal{Z}_1 + \mathcal{P}_2 \cdot \mathcal{Z}_2)}$$
$$\times \left\langle \mathcal{R}_1 + \frac{\mathcal{Z}_1}{2} \left| e^{-\hat{H}\lambda} \right| \mathcal{R}_2 - \frac{\mathcal{Z}_2}{2} \right\rangle \left\langle \mathcal{R}_2 + \frac{\mathcal{Z}_2}{2} \left| e^{-(\beta-\lambda)\hat{H}} \right| \mathcal{R}_1 - \frac{\mathcal{Z}_1}{2} \right\rangle . \tag{64}$$

Its time evolution is given by full quantum mechanics in the Wigner representation. In order to obtain a computationally tractable form, we consider a limit where the time evolution of $\overline{W}(\mathcal{X}_1, \mathcal{X}_2, t)$ is approximated by quantum-classical dynamics.

For future reference, we remark that the evolution equation can be written in other forms using the symmetry relation (59). Taking complex conjugates of both sides of (62) and using the fact that $(iL_W)^* = iL_W$ gives

---

[3] We may also obtain this equation of motion directly by differentiating the definition of W in (58)

$$\frac{\partial}{\partial t}\overline{W}(\mathcal{X}_2, \mathcal{X}_1, -t) = iL_W(\mathcal{X}_1)\overline{W}(\mathcal{X}_2, \mathcal{X}_1, -t) \ . \tag{65}$$

If we then exchange variables $\mathcal{X}_1 \leftrightarrow \mathcal{X}_2$ and $t \leftrightarrow -t$ we get [18]

$$\frac{\partial}{\partial t}\overline{W}(\mathcal{X}_1, \mathcal{X}_2, t) = -iL_W(\mathcal{X}_2)\overline{W}(\mathcal{X}_1, \mathcal{X}_2, t) \ . \tag{66}$$

We shall make use of these various equivalent forms of the evolution to write the reaction rate coefficient expression in a form that is most convenient for simulation.

5.1 Quantum-Classical Evolution Equation for W

The Wigner form of the quantum evolution operator $iL_W(\mathcal{X}_1)$ in (62) for the equation of motion for $\overline{W}(\mathcal{X}_1, \mathcal{X}_2, t)$ can be rewritten in a form that is convenient for the passage to the quantum-classical limit. Recalling that the system may be partitioned into $\mathcal{S}$ and $\mathcal{E}$ subspaces, the Poisson bracket operator Λ can be written as the sum of Poisson bracket operators acting in each of these subspaces as $\Lambda(\mathcal{X}_1) = \Lambda(x_1) + \Lambda(X_1)$. Thus, we may write

$$\begin{aligned}
iL_W(\mathcal{X}_1) &= \frac{2}{\hbar}H_W(\mathcal{X}_1)\sin(\hbar\Lambda(\mathcal{X}_1)/2) \\
&= \frac{2}{\hbar}H_W(\mathcal{X}_1)\sin(\hbar\Lambda(x_1)/2 + \hbar\Lambda(X_1)/2) \\
&= \frac{2}{\hbar}H_W(\mathcal{X}_1)\Big(\sin(\hbar\Lambda(x_1)/2)\cos(\hbar\Lambda(X_1)/2) \\
&\quad + \cos(\hbar\Lambda(x_1)/2)\sin(\hbar\Lambda(X_1)/2)\Big) \ . \tag{67}
\end{aligned}$$

If we introduce scaled coordinates as defined in Sect. 4.1 and expand the evolution operator to first order in the mass ratio μ, the resulting expression for the quantum-classical evolution operator in unscaled coordinates is[4]

$$i\mathcal{L}(\mathcal{X}_1) = \frac{2}{\hbar}H_W(\mathcal{X}_1)\sin(\hbar\Lambda(x_1)/2) + H_W(\mathcal{X}_1)\cos(\hbar\Lambda(x_1)/2)\Lambda(X_1) \ . \tag{68}$$

With this approximate expression for iL_W, (62) becomes the quantum-classical evolution equation for the spectral density function,

$$\frac{\partial}{\partial t}\overline{W}(\mathcal{X}_1, \mathcal{X}_2, t) = i\mathcal{L}(\mathcal{X}_1)\overline{W}(\mathcal{X}_1, \mathcal{X}_2, t) \ . \tag{69}$$

A similar analysis can be carried out on (66) to obtain a quantum-classical evolution equation involving the operator in $\mathcal{X}_2$ phase space coordinates.

---

[4] $i\mathcal{L}(\mathcal{X})$ is the Wigner transform of $i\hat{\mathcal{L}}(X)$ defined earlier in (37) over the $\mathcal{S}$ subsystem degrees of freedom.

5.2 Adiabatic Basis

Since Wigner transformed quantum mechanics is difficult to compute, we express the $\mathcal{S}$ subsystem degrees of freedom in an adiabatic basis rather than in a Wigner representation. To this end, we first observe that $A_W(\mathcal{X}_1)$ can be written as

$$A_W(\mathcal{X}_1) = \int dz_1 \, e^{\frac{i}{\hbar} p_1 \cdot z_1} \left\langle r_1 - \frac{z_1}{2} | \hat{A}_W(X_1) | r_1 + \frac{z_1}{2} \right\rangle, \tag{70}$$

where $\hat{A}_W(X_1)$ is the *partial* Wigner transform of $\hat{A}$. We may now express the subsystem operators in the adiabatic basis to obtain,

$$A_W(\mathcal{X}_1) = \sum_{\alpha_1 \alpha_1'} \int dz_1 \, e^{\frac{i}{\hbar} p_1 \cdot z_1} \left\langle r_1 - \frac{z_1}{2} | \alpha_1; R_1 \right\rangle A_W^{\alpha_1 \alpha_1'}(X_1) \left\langle \alpha_1'; R_1 | r_1 + \frac{z_1}{2} \right\rangle, \tag{71}$$

where $A_W^{\alpha_1 \alpha_1'}(X_1) = \langle \alpha_1; R_1 | \hat{A}_W(X_1) | \alpha_1'; R_1 \rangle$.

Starting with (63) and inserting the expression (71) for $A_W(\mathcal{X}_1)$ and its analog for $B_W(\mathcal{X}_2)$, the time-dependent transport coefficient expression becomes

$$\lambda_{AB}(t) = - \sum_{\alpha_1, \alpha_1', \alpha_2, \alpha_2'} \int \prod_{i=1}^{2} dX_i \, A_W^{\alpha_1 \alpha_1'}(X_1) B_W^{\alpha_2 \alpha_2'}(X_2)$$

$$\times \frac{\partial}{\partial t} \overline{W}^{\alpha_1' \alpha_1 \alpha_2' \alpha_2}(X_1, X_2, t), \tag{72}$$

where

$$W^{\alpha_1' \alpha_1 \alpha_2' \alpha_2}(X_1, X_2, t) = \int \prod_{i=1}^{2} dx_i dz_i \, e^{\frac{i}{\hbar}(p_1 \cdot z_1 + p_2 \cdot z_2)} \left\langle r_1 - \frac{z_1}{2} | \alpha_1; R_1 \right\rangle$$

$$\times \left\langle \alpha_1'; R_1 | r_1 + \frac{z_1}{2} \right\rangle \left\langle r_2 - \frac{z_2}{2} | \alpha_2; R_2 \right\rangle \left\langle \alpha_2'; R_2 | r_2 + \frac{z_2}{2} \right\rangle W(\mathcal{X}_1, \mathcal{X}_2, t), \tag{73}$$

Performing integrals over subsystem $\mathcal{S}$ coordinates, this expression may also be written in the more explicit form,

$$W^{\alpha_1' \alpha_1 \alpha_2' \alpha_2}(X_1, X_2, t) = \int \prod_{i=1}^{2} dZ_i e^{-\frac{i}{\hbar}(P_1 \cdot Z_1 + P_2 \cdot Z_2)} \frac{1}{Z_Q} \frac{1}{(2\pi\hbar)^{2N}}$$

$$\times \left\langle \alpha_1'; R_1 \left| \left\langle R_1 + \frac{Z_1}{2} | e^{\frac{i}{\hbar} \hat{H} t} | R_2 - \frac{Z_2}{2} \right\rangle | \alpha_2; R_2 \right\rangle \right.$$

$$\times \left\langle \alpha_2'; R_2 \left| \left\langle R_2 + \frac{Z_2}{2} \left| e^{-\frac{i}{\hbar} \hat{H} t - \beta \hat{H})} \right| R_1 - \frac{Z_1}{2} \right\rangle | \alpha_1; R_1 \right\rangle \right.. \tag{74}$$

Equation (73) shows how $W(\mathcal{X}_1, \mathcal{X}_2, t)$ may be related to its matrix elements $W^{\alpha_1' \alpha_1 \alpha_2' \alpha_2}(X_1, X_2, t)$ in the adiabatic basis. Similarly,

$$W(\mathcal{X}_1, \mathcal{X}_2, t) = \sum_{\alpha_1'\alpha_1\alpha_2'\alpha_2} \int \prod_{i=1}^{2} dz_i \; e^{-\frac{i}{\hbar}(p_1 \cdot z_1 + p_2 \cdot z_2)} \frac{1}{(2\pi\hbar)^{2n}} \left\langle r_1 + \frac{z_1}{2} |\alpha_1'; R_1 \right\rangle$$

$$\times \left\langle \alpha_1; R_1 | r_1 - \frac{z_1}{2} \right\rangle \left\langle r_2 + \frac{z_2}{2} |\alpha_2'; R_2 \right\rangle \left\langle \alpha_2; R_2 | r_2 - \frac{z_2}{2} \right\rangle W^{\alpha_1'\alpha_1\alpha_2'\alpha_2}(X_1, X_2, t) \,.$$

$$(75)$$

relates W to its matrix elements $W^{\alpha_1'\alpha_1\alpha_2'\alpha_2}$.

The quantum-classical evolution equation for $W^{\alpha_1'\alpha_1\alpha_2'\alpha_2}(X_1, X_2, t)$ may now be obtained from (69) by taking matrix elements of this equation using the definitions given above. We obtain [17,18],

$$\frac{\partial}{\partial t} \overline{W}^{\alpha_1'\alpha_1\alpha_2'\alpha_2}(X_1, X_2, t) = \sum_{\beta_1'\beta_1} i\mathcal{L}_{\alpha_1'\alpha_1,\beta_1'\beta_1}(X_1) \overline{W}^{\beta_1'\beta_1\alpha_2'\alpha_2}(X_1, X_2, t) \,. (76)$$

The quantum-classical evolution operator $i\mathcal{L}_{\alpha'\alpha,\beta'\beta}$ appearing in this equation is the same as that already defined in (43). We shall also have occasion to use (66) expressed in terms of matrix elements in the analysis presented below. Using (66), an equivalent form of the evolution equation is,

$$\frac{\partial}{\partial t} \overline{W}^{\alpha_1'\alpha_1\alpha_2'\alpha_2}(X_1, X_2, t) = -\sum_{\beta_2'\beta_2} i\mathcal{L}_{\alpha_2'\alpha_2,\beta_2'\beta_2}(X_2) \overline{W}^{\alpha_1'\alpha_1\beta_2'\beta_2}(X_1, X_2, t) \,. (77)$$

5.3 Transport Coefficient

Using these results, we may now obtain a form for transport coefficients which is convenient for simulation. We use the equality in (76), insert this into (72), and move the evolution operator $i\mathcal{L}(X_1)$ onto the $A_W(X_1)$ dynamical variable making use of integration by parts and cyclic permutations under the trace. We find

$$\lambda_{AB}(t) = \sum_{\alpha_1,\alpha_1',\alpha_2,\alpha_2'} \int \prod_{i=1}^{2} dX_i \; (i\mathcal{L}(X_1)A_W(X_1))^{\alpha_1\alpha_1'}$$

$$\times B_W^{\alpha_2\alpha_2'}(X_2) \overline{W}^{\alpha_1'\alpha_1\alpha_2'\alpha_2}(X_1, X_2, t). \quad (78)$$

Next, we use the equality in (77) and formally solve the equation to obtain $\overline{W}(X_1, X_2, t) = e^{-i\mathcal{L}(X_2)t} \overline{W}(X_1, X_2, 0)$. Finally we substitute this form for $\overline{W}(X_1, X_2, t)$ into (78) and move the evolution operator to the dynamical variable $B_W(X_2)$. In the adiabatic basis, the action of the propagator $e^{-i\mathcal{L}(X_2)t}$ on $\hat{B}_W(X_2)$ is

$$B_W^{\alpha_2\alpha_2'}(X_2, t) = \sum_{\beta_2\beta_2'} \left(e^{-i\mathcal{L}(X_2)t} \right)_{\alpha_2\alpha_2',\beta_2\beta_2'} B_W^{\beta_2\beta_2'}(X_2) \,. \quad (79)$$

The final expression for a transport coefficient is

$$\lambda_{AB}(t) = \sum_{\alpha_1, \alpha_1', \alpha_2, \alpha_2'} \int \prod_{i=1}^{2} dX_i \, (i\mathcal{L}(X_1) A_W(X_1))^{\alpha_1 \alpha_1'}$$

$$\times B_W^{\alpha_2 \alpha_2'}(X_2, t) \overline{W}^{\alpha_1' \alpha_1 \alpha_2' \alpha_2}(X_1, X_2, 0). \tag{80}$$

This equation can serve as the basis for the computation of transport properties for quantum-classical systems. The evolution of dynamical variables is carried out using quantum-classical Liouville dynamics as in the quantum-classical linear response expression (54). However, in contrast to (54), full quantum equilibrium effects are incorporated in the initial value of $\overline{W}$, which, from its definition in (60), depends on the quantity,

$$W^{\alpha_1' \alpha_1 \alpha_2' \alpha_2}(X_1, X_2, i\hbar\lambda) = \int \prod_{i=1}^{2} dZ_i e^{-\frac{i}{\hbar}(P_1 \cdot Z_1 + P_2 \cdot Z_2)} \frac{1}{Z_Q} \frac{1}{(2\pi\hbar)^{2N}}$$

$$\times \left\langle \alpha_1'; R_1 \middle| \left\langle R_1 + \frac{Z_1}{2} \middle| e^{-\hat{H}\lambda} \middle| R_2 - \frac{Z_2}{2} \right\rangle \middle| \alpha_2; R_2 \right\rangle$$

$$\times \left\langle \alpha_2'; R_2 \middle| \left\langle R_2 + \frac{Z_2}{2} \middle| e^{\hat{H}(\lambda - \beta)} \middle| R_1 - \frac{Z_1}{2} \right\rangle \middle| \alpha_1; R_1 \right\rangle. \tag{81}$$

6 Simulation Algorithms

Thus far we have focused on the formal development of quantum-classical dynamics and the derivation of expressions for transport coefficients which utilize this dynamics. We now turn to a discussion of how quantum-classical Liouville dynamics can be simulated for arbitrary many-body systems.

Various schemes have been proposed for the solution of the quantum-classical Liouville equation [13, 21–24]. Here we describe the sequential short-time algorithm that represents the solution in an ensemble of surface-hopping trajectories [25, 26].

6.1 Momentum-Jump Approximation

Before describing the simulation algorithm, it is useful to discuss an approximation to the operator J, defined in (45), which is responsible for both quantum transitions and associated momentum changes in the bath. This operator is difficult to evaluate because it involves derivatives with respect to bath particle momenta. We make an approximation to the $(1 + (S_{\alpha\beta}/2) \cdot (\partial/\partial P))$ operator in J so that its action on any function of the momenta yields the function evaluated at a shifted momentum value [4, 6, 26, 27].

Here we show the steps leading to this momentum-jump approximation[5]. Since $S_{\alpha\beta} = \Delta E_{\alpha\beta} \hat{d}_{\alpha\beta} (\frac{P}{M} \cdot \hat{d}_{\alpha\beta})^{-1}$, with $\Delta E_{\alpha\beta} = E_\alpha - E_\beta$, we may write

---

[5] We present the derivation in a simple form where the masses of all bath particles are the same. The general case for different bath masses has also been derived [27].

$$\left(1 + \frac{1}{2} S_{\alpha\beta} \cdot \frac{\partial}{\partial P}\right) = 1 + \frac{1}{2} \Delta E_{\alpha\beta} M \frac{1}{(P \cdot \hat{d}_{\alpha\beta})} \frac{\partial}{\partial (P \cdot \hat{d}_{\alpha\beta})}$$

$$= 1 + \Delta E_{\alpha\beta} M \frac{\partial}{\partial [(P \cdot \hat{d}_{\alpha\beta})^2]} \tag{82}$$

Consider the action of the operator on any function $f(P)$ of the momentum. We obtain,

$$\left(1 + \Delta E_{\alpha\beta} M \frac{\partial}{\partial [(P \cdot \hat{d}_{\alpha\beta})^2]}\right) f(P) \approx e^{\Delta E_{\alpha\beta} M \partial / \partial (P \cdot \hat{d}_{\alpha\beta})^2} f(P)$$

$$= e^{\Delta E_{\alpha\beta} M \partial / \partial [(P \cdot \hat{d}_{\alpha\beta})^2]} f\left(\hat{d}_{\alpha\beta}^\perp (P \cdot \hat{d}_{\alpha\beta}^\perp) + \hat{d}_{\alpha\beta} \mathrm{sgn}(P \cdot \hat{d}_{\alpha\beta}) \sqrt{(P \cdot \hat{d}_{\alpha\beta})^2}\right)$$

$$= f\left(\hat{d}_{\alpha\beta}^\perp (P \cdot \hat{d}_{\alpha\beta}^\perp) + \hat{d}_{\alpha\beta} \mathrm{sgn}(P \cdot \hat{d}_{\alpha\beta}) \sqrt{(P \cdot \hat{d}_{\alpha\beta})^2 + \Delta E_{\alpha\beta} M}\right)$$

$$= f(P + \Delta P) . \tag{83}$$

In the first line of this equation we made the main assumption that the first two terms on the left hand side could be approximated by the exponential of the operator. In the second line we wrote the momentum vector as a sum of its components along $\hat{d}_{\alpha\beta}$ and perpendicular to $\hat{d}_{\alpha\beta}^\perp$, and in the penultimate line we used the fact that the exponential operator is a translation operator in the variable $(P \cdot \hat{d}_{\alpha\beta})^2$. In the last line the momentum jump ΔP is given by

$$\Delta P = \hat{d}_{\alpha\beta} \left(\mathrm{sgn}(P \cdot \hat{d}_{\alpha\beta}) \sqrt{(P \cdot \hat{d}_{\alpha\beta})^2 + \Delta E_{\alpha\beta} M} - (P \cdot \hat{d}_{\alpha\beta})\right) . \tag{84}$$

The use of this approximation leads to a representation of the dynamics in terms of energy-conserving trajectories.

This approximation may be useful even beyond its strict domain of validity. Non-adiabatic transitions are likely to occur when adiabatic potential energy surfaces lie close in energy so that $\Delta E_{\alpha\beta}$ is small and the non-adiabatic coupling matrix element $d_{\alpha\beta}$ is large. The momentum jump approximation will be valid if $P \cdot d_{\alpha\beta}$ is not too small. If $\Delta E_{\alpha\beta}$ is large, the prefactor of $(1 + \frac{1}{2} S_{\alpha\beta} \cdot \frac{\partial}{\partial P})$, $P \cdot d_{\alpha\beta}/M$, is usually small and the contributions to the evolution coming from the J factors carry a small weight.

6.2 Short-Time Sequential Propagation Algorithm

An operator $\hat{A}_W(R, P, t)$ in quantum-classical dynamics has the formal solution $\hat{A}_W(t) = \exp(i\hat{\mathcal{L}}t) \hat{A}_W(0)$. The propagator, $\exp(i\hat{\mathcal{L}}t)$, can be decomposed into a composition of propagators in time segments of arbitrary length. The evolution of a dynamical variable over any time interval can then be obtained by the successive application of evolution operators in the small time segments [25].

Suppose we are interested in determining the time evolution over a time interval $(0, t)$. We first divide this interval into N segments such that the jth segment has length $\Delta t_j = t_j - t_{j-1} = \Delta t$. We may then write the propagator in the adiabatic basis as

$$(e^{i\hat{\mathcal{L}}t})_{\alpha_0\alpha_0',\alpha_N\alpha_N'} = \sum_{(\alpha_1\alpha_1')\dots(\alpha_{N-1}\alpha_{N-1}')} \prod_{j=1}^{N} (e^{i\hat{\mathcal{L}}(t_j-t_{j-1})})_{\alpha_{j-1}\alpha_{j-1}',\alpha_j\alpha_j'} , \quad (85)$$

where $(\alpha_0\alpha_0') \equiv (\alpha\alpha')$. Using this expression for the propagator we have

$$A_W^{\alpha\alpha'}(R, P, t) = \sum_{(\alpha_1\alpha_1')\dots(\alpha_N\alpha_N')} \left[\prod_{j=1}^{N} (e^{i\hat{\mathcal{L}}(t_j-t_{j-1})})_{\alpha_{j-1}\alpha_{j-1}',\alpha_j\alpha_j'}\right] A_W^{\alpha_N\alpha_N'}(R, P) .$$

$$(86)$$

Given this form of a time-dependent observable, the simulation algorithm exploits the structure of the propagator in the short-time segments.

In any time segment $(t_j - t_{j-1})$, using the decomposition of the quantum-classical Liouville operator into diagonal $i\mathcal{L}^0$ and off-diagonal J parts in (43), the quantum-classical propagator may be written in Dyson form as

$$\left(e^{i\hat{\mathcal{L}}(t_j-t_{j-1})}\right)_{\alpha_{j-1}\alpha_{j-1}',\alpha_j\alpha_j'} = e^{i\mathcal{L}^0_{\alpha_{j-1}\alpha_{j-1}'}(t_j-t_{j-1})}\delta_{\alpha_{j-1}\alpha_j}\delta_{\alpha_{j-1}'\alpha_j'}$$

$$- \sum_{\alpha_l\alpha_l'}\int_{t_{j-1}}^{t_j} d\tau_1 e^{i\mathcal{L}^0_{\alpha_j\alpha_j'}(\tau_1-t_{j-1})} J_{\alpha_{j-1}\alpha_{j-1}',\alpha_l\alpha_l'} \left(e^{i\hat{\mathcal{L}}(t_j-\tau_1)}\right)_{\alpha_l\alpha_l',\alpha_j\alpha_j'} .$$

$$(87)$$

Taking the time interval Δt to be small enough, we can approximate the propagator in an interval by the first order term in the Dyson expression as,

$$(e^{i\hat{\mathcal{L}}(t_j-t_{j-1})})_{\alpha_{j-1}\alpha_{j-1}',\alpha_j\alpha_j'} \approx e^{i\mathcal{L}^0_{\alpha_{j-1}\alpha_{j-1}'}\Delta t}\left(\delta_{\alpha_j\alpha_j',\alpha_{j-1}\alpha_{j-1}'} - \Delta t J_{\alpha_{j-1}\alpha_{j-1}',\alpha_j\alpha_j'}\right)$$

$$= \mathcal{W}_{\alpha_{j-1}\alpha_{j-1}'}(t_{j-1}, t_j)e^{iL_{\alpha_{j-1}\alpha_{j-1}'}\Delta t}$$

$$\times \left(\delta_{\alpha_j\alpha_j',\alpha_{j-1}\alpha_{j-1}'} - \Delta t J_{\alpha_{j-1}\alpha_{j-1}',\alpha_j\alpha_j'}\right) , \quad (88)$$

where the phase factor $\mathcal{W}_{\alpha_{j-1}\alpha_{j-1}'}(t_{j-1}, t_j) = e^{i\omega_{\alpha_{j-1}\alpha_{j-1}'}(t_j-t_{j-1})}$ associated with time segment (t_j, t_{j-1}) arises from the action of $\exp(i\mathcal{L}^0_{\alpha_{j-1}\alpha_{j-1}'}(t_j - t_{j-1}))$ on any phase space function. More specifically, the action of $\exp(i\mathcal{L}^0_{\alpha\alpha'} (t_2 - t_1)) = \exp(-(i\omega_{\alpha\alpha'} + iL_{\alpha\alpha'})(t_2 - t_1))$ on any function $f_{\alpha\alpha'}(R, P)$ may be computed explicitly in terms of time-reversed trajectories starting at the phase point (R, P) at time t_2 and terminating at another phase point at time t_1, $(t_1 < t_2)$. In particular we let

$$\tilde{R}_{t_1,\alpha\alpha'}^{t_2} = e^{-iL_{\alpha\alpha'}(t_2-t_1)}R ,$$

$$\tilde{P}_{t_1,\alpha\alpha'}^{t_2} = e^{-iL_{\alpha\alpha'}(t_2-t_1)}P , \quad (89)$$

be the time-reversed trajectory that starts at (R, P) at time t_2 and ends at $(\tilde{R}^{t_2}_{t_1,\alpha\alpha'}, \tilde{P}^{t_2}_{t_1,\alpha\alpha'})$ at time t_1. Using the analog of the Dyson identity in (87) for the propagator

$$e^{-i\mathcal{L}^0_{\alpha\alpha'}t} = e^{-(i\omega_{\alpha\alpha'}+iL_{\alpha\alpha'})} = e^{-iL_{\alpha\alpha'}t} - \int_0^t dt'\, e^{-iL_{\alpha\alpha'}t'} i\omega_{\alpha\alpha'}(R) e^{-i\mathcal{L}^0_{\alpha\alpha'}(t-t')} ,$$

$$(90)$$

solving the equation by iteration and resumming, we may show that

$$e^{-(i\omega_{\alpha\alpha'}+iL_{\alpha\alpha'})(t_2-t_1)} f_{\alpha\alpha'}(R, P) = e^{-i\int_{t_1}^{t_2} d\tau \omega_{\alpha\alpha'}(\tilde{R}_{\tau,\alpha\alpha'})} f_{\alpha\alpha'}(\tilde{R}^{t_2}_{t_1,\alpha\alpha'}, \tilde{P}^{t_2}_{t_1,\alpha\alpha'})$$

$$\equiv \mathcal{W}_{\alpha\alpha'}(t_1, t_2) f_{\alpha\alpha'}(\tilde{R}^{t_2}_{t_1,\alpha\alpha'}, \tilde{P}^{t_2}_{t_1,\alpha\alpha'}) . \quad (91)$$

Now that we know how to evolve the dynamics within a small time segment, we can decide to construct a Monte-Carlo-style stochastic algorithm to account for the quantum transitions that arise from the action of J. At the end of each time segment, the system either may remain in the same pair of adiabatic states or make a transition to a new pair of states. More specifically, for an initial pair of quantum states, $(\alpha_0 \alpha'_0)$, the phase point (R, P) is evolved for a time Δt to a new value $(R_{\Delta t}, P_{\Delta t})$ (here we use a simplified notation for the time-evolved phase points in the interval Δt) using the classical propagator $e^{iL_{\alpha_0 \alpha'_0} \Delta t}$ and the phase factor $\mathcal{W}_{\alpha_0 \alpha'_0}$ is computed. With probability $1/2$, one chooses whether the transition $\alpha_0 \to \alpha_1$ or $\alpha'_0 \to \alpha'_1$ occurs. The states α_1 and α'_1 are chosen uniformly from the set of allowed final states; the weight $w_{\alpha_0 \alpha'_0, \alpha_1 \alpha'_1}$ associated with the final state is the number of allowed final states. Once $(\alpha_1 \alpha'_0)$ or $(\alpha_0 \alpha'_1)$ is chosen, the nonadiabatic coupling matrix element $d_{\alpha_0 \alpha_1}$ (or $d_{\alpha'_0 \alpha'_1}$) is computed at $R_{\Delta t}$ and the probability, π, of a nonadiabatic transition is given by

$$\pi = \left| \frac{P_{\Delta t}}{M} \cdot d_{\alpha_0 \alpha_1}(R_{\Delta t}) \right| \Delta t \left(1 + \left| \frac{P_{\Delta t}}{M} \cdot d_{\alpha_0 \alpha_1}(R_{\Delta t}) \right| \Delta t \right)^{-1} . \quad (92)$$

If the transition is rejected, then

$$A_W^{\alpha_0 \alpha'_0}(R, P, \Delta t) = \mathcal{W}_{\alpha_0 \alpha'_0}(\Delta t) A_W^{\alpha_0 \alpha'_0}(R_{\Delta t}, P_{\Delta t}) \frac{1}{1 - \pi} . \quad (93)$$

If the transition is accepted, then, using the momentum jump approximation, we translate the momentum $P_{\Delta t}$ to $\tilde{P}_{\Delta t} = P_{\Delta t} + \Delta P$ where ΔP is defined in (84). We then write

$$A_W^{\alpha_0 \alpha'_0}(R, P, \Delta t) = \mathcal{W}_{\alpha_0 \alpha'_0}(\Delta t) A_W^{\alpha_1 \alpha'_0}(R_{\Delta t}, \tilde{P}_{\Delta t}) \Delta t$$

$$\times \frac{P_{\Delta t}}{M} \cdot d_{\alpha_0 \alpha_1}(R_{\Delta t}) \frac{1}{\pi} w_{\alpha_0 \alpha'_0, \alpha_1 \alpha'_1} . \quad (94)$$

From (84) we see that if $\Delta E_{\alpha\beta} < 0$ (an upward transition from $\alpha \to \beta$) and $(\bar{P} \cdot \tilde{d}_{\alpha\beta})^2 < |\Delta E_{\alpha\beta}|$ so that there is insufficient kinetic energy from bath momenta along $\bar{\bar{d}}_{\alpha\beta}$ for the quantum transition to occur, the argument of the square root is negative leading to imaginary momentum changes. In this case, the quantum transition does not occur and the trajectory is continued adiabatically. The total energy of the system is conserved along a quantum-classical surface-hopping trajectory when the momentum-jump approximation is used, even if the transition is to a pair of coherently coupled surfaces.

7 Chemical Reaction Rates

In this section we illustrate applications of the formalism and simulation method by calculating the rate constants of activated chemical reactions. The calculations are carried out by evaluating time-correlation reactive-flux expressions for the rate. The general transport coefficients we derived earlier can easily be specialized to this case. Quantum reaction rates, such as proton and electron transport processes, very appropriately fall into the category of systems that can be studied using quantum-classical dynamics since the light particle (proton or electron) being transferred must be treated quantum mechanically but dynamics of the environment in which the transfer takes place (condensed phase polar solvent or large biomolecule) can often be treated classically to a good approximation.

7.1 Reactive-Flux Correlation Functions

For a quantum mechanical system in thermal equilibrium undergoing a transformation $A \rightleftharpoons B$, a rate constant k_{AB} may be calculated from the time-dependent reactive-flux correlation function [28],

$$k_{AB}(t) = \frac{1}{n_A^{eq}} \int_0^t dt' \langle \dot{\hat{N}}_A; \dot{\hat{N}}_B \rangle = \frac{1}{\beta n_A^{eq}} \left\langle \frac{i}{\hbar}[\hat{N}_B(t), \hat{N}_A] \right\rangle, \qquad (95)$$

where $\hat{N}_A$ is the A species operator, n_A^{eq} is the equilibrium density of species A, and $\dot{\hat{N}}_A = (i/\hbar)[\hat{H}, \hat{N}_A]$ is the flux of $\hat{N}_A$ with Hamiltonian $\hat{H}$, with an analogous expression for $\dot{\hat{N}}_B$.[6] We may now directly apply the general formula, (80), derived earlier for the quantum-classical limit of quantum time correlation function, to obtain an expression for the reaction rate in the form,

$$k_{AB}(t) = \frac{1}{n_A^{eq}} \sum_{\alpha\alpha'} \int dX N_{BW}^{\alpha\alpha'}(X,t) W_{A'}^{\alpha'\alpha}\left(X, \frac{i\hbar\beta}{2}\right), \qquad (96)$$

---

[6] The time evolution of the reactive flux is given by projected dynamics [28] but in simulations we may replace projected dynamics by ordinary dynamics and insert absorbing states in the reactant and product regions to yield well defined plateau values. Such a procedure will be accurate provided there is a sufficient time scale separation between the relaxation time for reactive events and other microscopic relaxation times in the system.

where we used the high temperature approximation

$$\overline{W}\left(\mathcal{X}_1, \mathcal{X}_2, 0\right) = W\left(\mathcal{X}_1, \mathcal{X}_2, \frac{i\hbar\beta}{2}\right) + \mathcal{O}(\beta^2) . \tag{97}$$

The matrix elements of $W_{A'}$ in the adiabatic basis are given by

$$W_{A'}^{\alpha'\alpha}\left(X, \frac{i\hbar\beta}{2}\right) = \sum_{\alpha_1\alpha_1'} \int dX' \left(i\mathcal{L}(X')N_{AW}(X')\right)^{\alpha_1\alpha_1'} W^{\alpha_1'\alpha_1\alpha'\alpha}\left(X', X, \frac{i\hbar\beta}{2}\right),$$
$$\tag{98}$$

where

$$
\begin{aligned}
W^{\alpha_1'\alpha_1\alpha'\alpha}\left(X', X, \frac{i\hbar\beta}{2}\right) &= \frac{1}{(2\pi\hbar)^{2N} Z_Q} \int dZ dZ' e^{-\frac{i}{\hbar}(P\cdot Z + P'\cdot Z')} \\
&\times \left\langle \alpha'; R \left| \left\langle R + \frac{Z}{2} \left| e^{-\frac{\beta}{2}\hat{H}} \right| R' - \frac{Z'}{2} \right\rangle \right| \alpha_1; R' \right\rangle \\
&\times \left\langle \alpha_1'; R' \left| \left\langle R' + \frac{Z'}{2} \left| e^{-\frac{\beta}{2}\hat{H}} \right| R - \frac{Z}{2} \right\rangle \right| \alpha; R \right\rangle .
\end{aligned}
\tag{99}
$$

This rate coefficient expression involves quantum-classical evolution of the matrix element $N_{BW}^{\alpha\alpha'}(X, t)$ but retains the full quantum equilibrium structure of the system.

7.2 Proton Transfer

As an example, we consider the calculation of the rate constant for a proton transfer reaction ($AH\text{-}B \rightleftharpoons A^- \text{-} H^+ B$) in a hydrogen-bonded complex (AHB) dissolved in a polar solvent. The model we consider [29] has potential parameters chosen to describe proton transfer in a slightly strongly hydrogen-bonded phenol (A) trimethylamine (B) complex in a methyl chloride liquid-state solvent. At the equilibrium $A - B$ separation, $R_{AB} = 2.7$ Å, the proton potential energy function in the AHB complex has two minima, the deeper minimum corresponding to the stable covalent state and the shallower minimum corresponding to the metastable ionic state. Additional details of the calculations can be found in [27]. This model has been studied often using different methods [30–35].

Proton transfer dynamics in polar liquids can be monitored by the solvent polarization, $\Delta E(R)$, [36, 37]

$$\xi(R) = \Delta E(R) = \sum_{i,a} z_a e \left(\frac{1}{|\mathbf{R}_i^a - s|} - \frac{1}{|\mathbf{R}_i^a - s'|} \right), \tag{100}$$

where $z_a e$ is the charge on solvent atom a, and s and s' are two points within the complex, one at the center of mass and the other displaced from the center of mass, which correspond to the minima of the bare hydrogen bonding

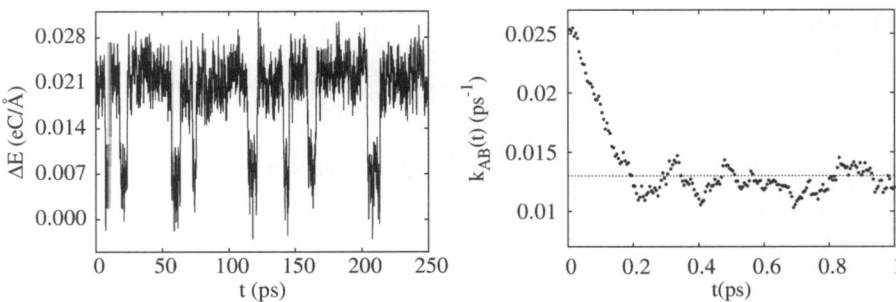

Fig. 1. (*Left*) Time series of the solvent polarization (ΔE) for a ground state adiabatic trajectory. (*Right*) The rate coefficient, $k_{AB}(t)$, as a function of time. The dotted line indicates the plateau value k_{AB}

potential. The sums run over all solvent molecules i and atoms a. The solvent polarization is an example of a reaction coordinate for a quantum rate process that depends solely on the environmental coordinates. In Fig. 1 (left panel) we see that ΔE tracks the hops of the proton between the reactant/covalent state ($\Delta E \approx 0.005\ eC/\text{Å}$) and the product/ionic state ($\Delta E \approx 0.0225\ eC/\text{Å}$). The complex spends more time in the ionic configuration than in the covalent configuration since electrostatic interactions with the polar solvent preferentially stabilize the ionic configuration of the complex. In the absence of the polar solvent, the complex is primarily found in the covalent configuration.

Since we consider systems at approximately room temperature and the dynamics of the solvent and complex atoms can be accurately captured using classical mechanics, a high temperature/classical approximation may be made to $\overline{W}^{\alpha'_1 \alpha\ \alpha'_2 \alpha_2}(X, X')$ to obtain,

$$\overline{W}^{\alpha'_1 \alpha_1 \alpha'_2 \alpha_2}(X, X') = \frac{e^{-\beta\left(\frac{P^2}{2M} + E_{\alpha'_1}(R)\right)}}{(2\pi\hbar)^{\mathcal{N}} Z_Q} \delta_{\alpha'_1 \alpha_2} \delta_{\alpha'_2 \alpha_1} \delta(X - X')\,, \quad (101)$$

where $Z_Q = (2\pi\hbar)^{-\mathcal{N}} \sum_\alpha \int dRdP e^{-\beta\left(\frac{P^2}{2M} + E_\alpha(R)\right)}$. Using this approximation for the spectral density function, the rate constant for this proton transfer reaction can easily be written. Taking $\Delta E(R)$ as the reaction coordinate, the A and B species variables can be defined as $\hat{N}_A = \theta(\Delta E(R) - \Delta E^\ddagger)$ and $\hat{N}_B = \theta(\Delta E^\ddagger - \Delta E(R))$, respectively. The time-dependent rate constant is given by

$$k_{AB}(t) = \frac{-1}{n_A^{eq}} \sum_\alpha \int dRdP\ \Delta\dot{E}(R) N_B^{\alpha\alpha}(R, P, t)\delta(\Delta E(R) - \Delta E^\ddagger)\rho_{W_e}^{\alpha\alpha}\,. (102)$$

The time derivative of the solvent polarization can be written as $\Delta\dot{E}(R) = \frac{P}{M} \cdot \nabla_R \Delta E(R)$. The canonical equilibrium distribution is given by $\rho_{W_e}^{\alpha\alpha} = Z_0^{-1} e^{-\beta H_W^\alpha}$, with $Z_0 = \sum_\alpha \int dRdP\ e^{-\beta H_W^\alpha}$. Equation (102) provides a well-defined formula involving initial sampling from the barrier top $\Delta E = \Delta E^{\ddagger}$ and quantum-classical time evolution of $N_B^{\alpha\alpha}(R, P, t)$.

In Fig. 1 (right panel), we plot the time-dependent rate coefficient obtained from an average over 16000 trajectories. We see that $k_{AB}(t)$ falls quickly from its initial transition state theory value in a few tenths of a picosecond to a plateau from which the rate constant can be extracted. The decrease in the rate coefficient from its transition state theory value is due to recrossing by the trajectory of the barrier top before the system reaches a metastable state. The value of k_{AB} obtained from the plateau is $k_{AB} = 0.013$ ps^{-1}. The adiabatic rate constant is $k_{AB}^{ad} = 0.019$ ps^{-1}, indicating that nonadiabatic effects influence the proton transfer rate.

7.3 Quantum Sampling of the Reaction Coordinate

In the previous proton transport example the equilibrium structure of the polarization reaction coordinate, which depends on the positions of all particles in the environment, was treated classically. We now consider a simpler reaction model to examine the effect on the reaction rate of treating equilibrium sampling of the reaction coordinate quantum mechanically [38]. This example further illustrates formalism developed in Sect. 5 where the quantum equilibrium structure embodied in the spectral density function W is combined with quantum-classical dynamics to compute the transport property. The imaginary time propagators in W can, in principle, be computed using quantum path integral methods [39] or other approximations such as linearization methods [40–42]. By treating the initial sampling of the reaction coordinate quantum mechanically, we show that the time-dependent rate coefficient has an initial value of zero in agreement with the full quantum rate expression [28], and its asymptotic value, which gives the rate constant, differs from that obtained using classical initial sampling.

We consider a system in which only one coordinate, R_0, is directly coupled to the quantum subsystem and this coordinate serves as the reaction coordinate, $\xi(R) = R_0$. The coordinate R_0 is, in turn, coupled to a bath. The A and B species operators may be defined as $\hat{N}_{AW} = \theta(-R_0)$ and $\hat{N}_{BW} = \theta(R_0)$, where θ is the Heaviside function and the dividing surface is located at $\xi^{\ddagger} = R_0^{\ddagger} = 0$. For this choice of species variable, $W_{A'}^{\alpha'\alpha}(X, \frac{i\hbar\beta}{2})$ defined in (98), can be simplified by performing the integrations over the X' coordinates to yield,

$$W_{A'}^{\alpha'\alpha}\left(X, \frac{i\hbar\beta}{2}\right) = \frac{1}{(2\pi\hbar)^N Z_Q} \frac{i\hbar}{M_0} \int dZ dZ_0' (d\delta(Z_0')/dZ_0') e^{-\frac{i}{\hbar}P\cdot Z}$$

$$\times \left\langle \alpha'; R_0 \left| \left\langle R + \frac{Z}{2} \left| e^{-\frac{\beta}{2}\hat{H}} \right| -\frac{Z_0'}{2} \right\rangle \left\langle \frac{Z_0'}{2} \left| e^{-\frac{\beta}{2}\hat{H}} \right| R - \frac{Z}{2} \right\rangle \right| \alpha; R_0 \right\rangle .$$

$$(103)$$

The adiabatic eigenstates depend only on R_0 since the bath is coupled directly only to this coordinate. In the results sketched below we use an approximate analytical expression for this quantity. The details of this calculation and the approximations employed to obtain the result can be found in [38]. As a result of this analysis we find

$$
W_{A'}^{\alpha'\alpha}\left(X, \frac{i\hbar\beta}{2}\right) = \frac{1}{2\pi\hbar Z_Q}\frac{1}{\cos^2 u}\sqrt{\frac{2M_0 u'}{\beta\hbar^2\pi}}e^{-\frac{2M_0 u'}{\beta\hbar^2}R_0^2}
$$
$$
\times \frac{P_0}{M_0}e^{-\frac{\beta P_0^2}{2M_0 u'}}F_{\alpha'\alpha}(R_0)\rho_b(P_b, R_b; R_0), \qquad (104)
$$

where $u' \equiv u\cot u$ and $\rho_b(P_b, R_b; R_0)$ is proportional to the Wigner transform of the canonical equilibrium density matrix for the bath in the field of the R_0 coordinates,

$$
\rho_b(P_b, R_b; R_0) = \frac{1}{(2\pi\hbar)^{N-1}}\int dZ_b e^{-\frac{i}{\hbar}P_b\cdot Z_b}\left\langle R_b + \frac{Z_b}{2}\left|e^{-\beta\hat{H}_{b(n)}}\right|R_b - \frac{Z_b}{2}\right\rangle.
$$
$$
(105)
$$

The function $F_{\alpha'\alpha}(R_0)$ is defined by

$$
F_{\alpha'\alpha}(R_0) = e^{-\beta\varepsilon_\alpha(R_0)}\left(\delta_{\alpha'\alpha} + \frac{1}{2}\left(1 - \frac{\beta P_0^2}{M_0 u'}\right)\frac{i\hbar}{P_0}d_{\alpha'\alpha}O_{\alpha'\alpha}\right), \qquad (106)
$$

with $O_{\alpha'\alpha}(R_0) = (1 - e^{-\frac{\beta}{2}\varepsilon_{\alpha'\alpha}(R_0)})^2$ and $\varepsilon_{\alpha'\alpha} = \varepsilon_{\alpha'} - \varepsilon_\alpha$, where the $\varepsilon_\alpha(R_0)$ are related to the $E_\alpha(R_0)$ introduced earlier by $\varepsilon_\alpha(R_0) = E_\alpha(R_0) + \frac{1}{2}M_0\omega^{\ddagger 2}R_0^2$ with $\omega^\ddagger$ is the frequency at the barrier top. Note that in contrast to a classical treatment of the reaction coordinate, initial sampling of R_0 is no longer restricted to the barrier top.

As an application of this formalism, we consider a two-level quantum system coupled to a classical bath as a simple model for a transfer reaction in a condensed phase environment. The Hamiltonian operator of this system, expressed in the diabatic basis $\{|L\rangle, |R\rangle\}$, has the matrix form [43]

$$
\mathbf{H} = \begin{pmatrix} V_n(R_0) + \hbar\gamma_0 R_0 & -\hbar\Omega \\ -\hbar\Omega & V_n(R_0) - \hbar\gamma_0 R_0 \end{pmatrix}
$$
$$
+ \left(\frac{P_0^2}{2M_0} + \sum_{j=1}^{N}\frac{P_j^2}{2M_j} + \sum_{j=1}^{N}\frac{M_j}{2}\omega_j^2\left(R_j - \frac{c_j}{M_j\omega_j^2}R_0\right)^2\right)\mathbf{I}.
$$
$$
(107)
$$

In this model, a two-level system is coupled to a classical nonlinear oscillator with a mass M_0 and phase space coordinates (R_0, P_0). This coupling is given by $\hbar\gamma_0 R_0$. The nonlinear oscillator, which has a quartic potential energy function $V_n(R_0) = aR_0^4/4 - M_0\omega^{\ddagger 2}R_0^2/2$, is then bilinearly coupled

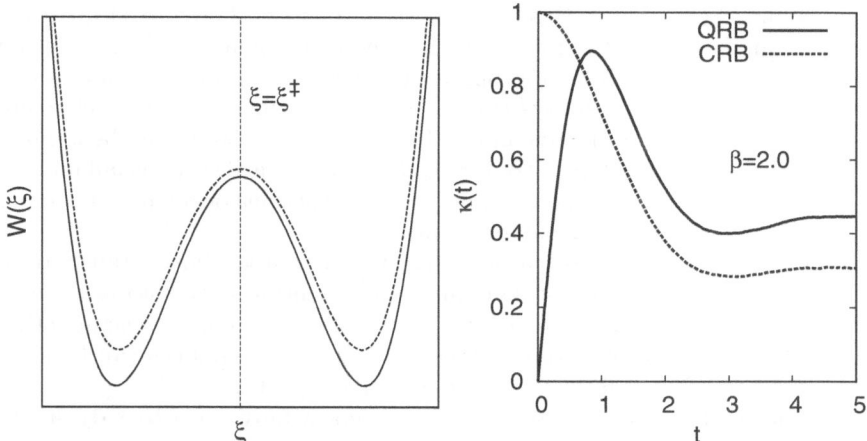

Fig. 2. (*Left*) A schematic illustration of the free energy W along a reaction coordinate ξ for weak coupling to the reaction coordinate. The dotted line at $\xi = \xi^{\ddagger}$ indicate the position of the barrier top. (*Right*) Comparison of the time-dependent transmission coefficients k_{AB}/k_{AB}^{TST} for quantum (QRB) and classical (CRB) sampling of the reaction coordinate. Parameters values: $\beta = 2$, $\gamma_0 = 0.1$, $\Omega = 0.1$, and $\xi = 3$

to a bath of N independent harmonic oscillators. The bath harmonic oscillators labelled $j = 1, \ldots, N$ have masses M_j and frequencies ω_j. The bilinear coupling is characterized by an Ohmic spectral density [44, 45], $J(\omega) = \pi \sum_{j=1}^{N} (c_j^2/(2M_j\omega_j^2))\delta(\omega - \omega_j)$, where $c_j = (\xi\hbar\omega_0 M_j)^{1/2}\omega_j$, $\omega_j = -\omega_c \ln(1 - j\omega_0/\omega_c)$ and $\omega_0 = \frac{\omega_c}{N}(1 - e^{-\omega_{max}/\omega_c})$, with ω_c a cut-off frequency.

Figure 2 (left panel) shows the energy profile for a two-level system weakly coupled to the reaction coordinate. Both the ground and excited state surfaces have two minima separated by a high barrier at $\xi(R_0) = \xi^{\ddagger}$. The right panel of this figure compares the time dependent rate coefficients for quantum (QRB) and classical (CRB) treatments of the reaction coordinate for a moderately low temperature ($\beta = 2$). At $t = 0$, the CRB result for the time-dependent transmission coefficient, $\kappa(t) = k_{AB}/k_{AB}^{TST}$, where k_{AB}^{TST} is determined from a classical treatment of the reaction coordinate [38], is non-zero and equal to unity. The QRB results for the time-dependent transmission coefficient are zero at $t = 0$, which is expected for quantum rate processes [28]. Quantum effects are pronounced and we see that the QRB formulation yields a larger rate constant than the CRB treatment. This enhancement of the quantum rate has also been observed in other studies [39, 46, 47].

8 Conclusion

The computation of quantum mechanical transport properties for many-body systems remains one of the most challenging problems in condensed matter physics and chemistry. In this chapter we have shown how mixed quantum-classical methods can be used to tackle these problems. Such a quantum-classical description will be appropriate if the system of interest can usefully be partitioned into two subsystems, one of which whose quantum character must be retained while the dynamics of the other subsystem may be approximated by classical mechanics. We saw that many physical systems fall into this category, notably proton and electron transfer reactions occurring in complex condensed phase and other environments.

Using the quantum-classical formulations presented above, transport coefficients can be calculated by sampling from quantum initial states or, more approximately, from quantum-classical initial states combined with quantum-classical dynamics of observables. Quantum-classical dynamics can be simulated in terms of ensembles of surface-hopping trajectories.

Quantum-classical Liouville dynamics is not a fully consistent dynamical theory and there is scope for further development [48]. For systems where a quantum-classical decomposition of the system is appropriate, the numerical consequences of such inconsistencies are likely to be small, as confirmed by calculations on model systems. For quantum systems bilinearly coupled to harmonic baths the computation of transport properties is exact if quantum-classical Liouville dynamics is used in conjunction with sampling from quantum equilibrium distributions. Thus, although not completely free from difficulties, quantum-classical Liouville dynamics has provided a more systematic and complete underpinning to quantum-classical surface-hopping schemes and is a formalism that can be used to investigate the dynamical properties of quantum systems coupled to large and complex many-body environments. Future developments in this area are likely to yield other algorithms and insight into the quantum dynamics of open systems.

Acknowledgments

This work was supported in part by a grant from the Natural Sciences and Engineering Research Council of Canada. We would also like to thank Steve Nielsen, Donal MacKernan, Alessandro Sergi, Gabriel Hanna and Hyojoon Kim whose work was described in this chapter.

References

1. E. Wigner (1932) On the Quantum Correction for Thermodynamic Equilibrium. *Phys. Rev.* **40**, pp. 749–759
2. K. Imre, E. Ozizmir, M. Rosenbau, and P. F. Zweifel (1967) Wigner method in quantum statistical mechanics. *J. Math. Phys.* **8**, pp. 1097–1108; M. Hillery, R. F. O'Connell, M. O. Scully, and E. P. Wigner (1984) Distribution functions in physics: Fundamentals. *Phys. Repts.* **106**, pp. 121–167

3. R. Kubo (1957) Statistical-mechanical theory of irreversible processes. 1. General theory and simple applications to magnetic and conduction problems. *J. Phys. Soc.* (Japan) **12**, 570–586; R. Kubo (1966) The fluctuation-dissipation theorem. *Repts. Prog. Phys.* **29**, pp. 255–284
4. R. Kapral and G. Ciccotti (1999) Mixed Quantum-Classical Dynamics. *J. Chem. Phys.* **110**, pp. 8919–8929
5. S. Nielsen, R. Kapral, and G. Ciccotti (2001) Statistical Mechanics of Quantum-Classical Systems. *J. Chem. Phys.* **115**, pp. 5805–5815
6. R. Kapral and G. Ciccotti (2002) A statistical mechanical theory of quantum dynamics in classical environments. In Bridging Time Scales: Molecular Simulations for the Next Decade, ed by P. Nielaba, M. Mareschal, and G. Ciccotti, Springer Berlin Heidelberg, pp 445–472
7. P. Mazur and I. Oppenheim (1970) Molecular theory of Brownian motion. *Physica* **50**, pp. 241–258
8. V. I. Gerasimenko (1982) Correlation-less equations of motion of quantum-classical systems. *Repts. Acad. Sci. Ukr. SSR* **10**, pp. 64–67; V. I. Gerasimenko (1982) Dynamical equation of quantum-classical systems. *Theor. Math. Phys.* **50**, pp. 49–55
9. I. V. Aleksandrov (1981) The statistical dynamics of a system consisting of a classical and a quantum subsystem. *Z. Naturforsch.* **36a**, pp. 902–908
10. W. Boucher and J. Traschen (1988) Semiclassical physics and quantum fluctuations. *Phys. Rev. D* **37**, pp. 3522–3532
11. W. Y. Zhang and R. Balescu (1988) Statistical-mechanics of a spin-polarized plasma. *J. Plasma Phys.* **40**, pp. 199–213; *ibid.* (1988) Kinetic-equation, spin hydrodynamics and collisional depolarization rate in a spin-polarized plasma. *J. Plasma Phys.* **40**, pp. 215–234
12. O. V. Prezhdo and V. V. Kisil (1997) Mixing quantum and classical mechanics. *Phys. Rev. A* **56**, pp. 162–175
13. C. C. Martens and J.-Y. Fang (1996) Semiclassical-limit molecular dynamics on multiple electronic surfaces. *J. Chem. Phys.* **106**, pp. 4918–4930; A. Donoso, and C. C. Martens (1998) Simulation of Coherent Nonadiabatic Dynamics Using Classical Trajectories. *J. Phys. Chem. A* **102**, pp. 4291–4300
14. I. Horenko, C. Salzmann, B. Schmidt, and C. Schütte (2002) Quantum-classical Liouville approach to molecular dynamics: Surface hopping Gaussian phase-space packets. *J. Chem. Phys.* **117**, pp. 11075–11088
15. Q. Shi and E. Geva (2004) A derivation of the mixed quantum-classical Liouville equation from the influence functional formalism. *J. Chem. Phys.* **121**, pp. 3393–3404
16. S. Nielsen, R. Kapral, and G. Ciccotti (2000) Statistical Mechanics of Quantum-Classical Systems. *J. Chem. Phys.* **112**, pp. 6543–6553
17. A. Sergi and R. Kapral (2004) Quantum-Classical Limit of Quantum Correlation Functions. *J. Chem. Phys.* **121**, pp. 7565–7576
18. H. Kim and R. Kapral (2005) Transport Properties of Quantum-Classical Systems. *J. Chem. Phys.* **122**, 214105
19. V. S. Filinov, Y. V. Medvedev, and V. L. Kamskyi (1995) Quantum dynam-ics and wigner representation of quantum-mechanics. *Mol. Phys.* **85**, pp. 711–726; V. S. Filinov (1996) Wigner approach to quantum statistical mechanics and quantum generalization molecular dynamics method. Part 1. *Mol. Phys.* **88**, pp. 1517–1528

20. V. S. Filinov (1996) Wigner approach to quantum statistical mechanics and quantum generalization molecular dynamics method. Part 2. *Mol. Phys.* **88**, pp. 1529–1540
21. C. C. Wan and J. Schofield (2000) Mixed quantum-classical molecular dynamics: Aspects of the multithreads algorithm. *J. Chem. Phys.* **113**, pp. 7047–7054
22. C. C. Wan and J. Schofield (2002) Solutions of mixed quantum-classical dynamics in multiple dimensions using classical trajectories. *J. Chem. Phys.* **116**, pp. 494–506
23. M. Santer, U. Manthe, and G. Stock (2001) Quantum-classical Liouville description of multidimensional nonadiabatic molecular dynamics. *J. Chem. Phys.* **114**, pp. 2001–2012
24. I. Horenko, M. Weiser, B. Schmidt, and C. Schütte (2004) Fully adaptive propagation of the quantum-classical Liouville equation. *J. Chem. Phys.* **120**, pp. 8913–8923
25. D. MacKernan, G. Ciccotti, and R. Kapral (2002) Sequential Short-Time Propagation of Quantum-Classical Dynamics. *J. Phys. Condens. Matt.* **14**, pp. 9069–9076
26. A. Sergi, D. Mac Kernan, G. Ciccotti, and R. Kapral (2003) Simulating quantum dynamics in classical environments. *Theor. Chem. Acc.* **110**, pp. 49–58
27. G. Hanna and R. Kapral (2005) Quantum-Classical Liouville Dynamics of Nonadiabatic Proton Transfer. *J. Chem. Phys.* **122**, 244505
28. R. Kapral, S. Consta, L. McWhirter (1998) Chemical rate laws and rate constants. In *Classical and Quantum Dynamics in Condensed Phase Simulations*, ed by B. J. Berne, G. Ciccotti, D. F. Coker World Scientific, Singapore pp. 583–616
29. H. Azzouz and D. Borgis (1993) A quantum molecular-dynamics study of proton-transfer reactions along asymmetrical H bonds in solution. *J. Chem. Phys.* **98**, pp. 7361–7374
30. S. Hammes-Schiffer and J. C. Tully (1994) Proton transfer in solution: Molecular dynamics with quantum transitions. *J. Chem. Phys.* **101**, pp. 4657–4667
31. R. P. McRae, G. K. Schenter, B. C. Garrett, Z. Svetlicic, and D. G. Truhlar (2001) Variational transition state theory evaluation of the rate constant for proton transfer in a polar solvent. *J. Chem. Phys.* **115**, pp. 8460–8480
32. D. Antoniou and S. D. Schwartz (1999) A molecular dynamics quantum Kramers study of proton transfer in solution. *J. Chem. Phys.* **110**, pp. 465–472
33. D. Antoniou and S. D. Schwartz (1999) Quantum proton transfer with spatially dependent friction: Phenol-amine in methyl chloride. *J. Chem. Phys.* **110**, pp. 7359–7364
34. S. Y. Kim and S. Hammes-Schiffer (2003) Molecular dynamics with quantum transitions for proton transfer: Quantum treatment of hydrogen and donoracceptor motions. *J. Chem. Phys.* **119**, pp. 4389–4398
35. T. Yamamoto and W. H. Miller (2005) Path integral evaluation of the quantum instanton rate constant for proton transfer in a polar solvent. *J. Chem. Phys.* **122**, 044106
36. P. M. Kiefer and J. T. Hynes (2004) Adiabatic and nonadiabatic proton transfer rate constants in solution. *Solid State Ionics*, **168**, pp. 219–224
37. D. Laria, G. Ciccotti, and M. Ferrario et al. (1992) Molecular-Dynamics Study of Adiabatic Proton-Transfer Reactions in Solution. *J. Chem. Phys.* **97**, pp. 378–388

38. H. Kim and R. Kapral (2005) Nonadiabatic quantum-classical reaction rates with quantum equilibrium structure. *J. Chem. Phys.* **123**, 194108

39. M. Topaler and N. Makri (1994) Quantum rates for a double well coupled to a dissipative bath: Accurate path integral results and comparison with approximate theories. *J. Chem. Phys.* **101**, pp. 7500–7519

40. S. Bonella and D. F. Coker (2005) LAND-map, a linearized approach to nonadiabatic dynamics using the mapping formalism. *J. Chem. Phys.* **122**, 194102

41. H. Kim and P. J. Rossky (2002) Evaluation of Quantum Correlation Functions from Classical Data. *J. Phys. Chem. B* **106**, pp. 8240–8247

42. J. A. Poulsen, G. Nyman, and P. J. Rossky (2003) Practical evaluation of condensed phase quantum correlation functions: A FeynmanKleinert variational linearized path integral method. *J. Chem. Phys.* **119**, pp. 12179–12193

43. A. Sergi and R. Kapral (2003) Quantum-Classical Dynamics of Nonadiabatic Chemical Reactions. *J. Chem. Phys.* **118**, pp. 8566–8575

44. N. Makri and K. Thompson (1998) Semiclassical influence functionals for quantum systems in anharmonic environments. *Chem. Phys. Lett.* **291**, pp. 101–109; *ibid.* (1999) Influence functionals with semiclassical propagators in combined forwardbackward time. *J. Chem. Phys.* **110**, pp. 1343–1353; N. Makri (1999) The Linear Response Approximation and Its Lowest Order Corrections: An Influence Functional Approach. *J. Phys. Chem. B* **103**, pp. 2823–2829

45. D. McKernan, R. Kapral, and G. Ciccotti (2002) Surface-Hopping Dynamics of a Spin-Boson System. *J. Chem. Phys.* **116**, pp. 2346–2353

46. H. B. Wang, X. Sun, and W. H. Miller (1998) Semiclassical approximations for the calculation of thermal rate constants for chemical reactions in complex molecular systems. *J. Chem. Phys.* **108**, pp. 9726–9736

47. E. Rabani, G. Krilov, and B. J. Berne (2000) Quantum mechanical canonical rate theory: A new approach based on the reactive flux and numerical analytic continuation methods. *J. Chem. Phys.* **112**, pp. 2605–2614

48. V. V. Kisil (2005) A quantum-classical bracket from p-mechanics. *Europhys. Lett.* **72**, pp. 873–879

Linearized Path Integral Methods
for Quantum Time Correlation Functions

D.F. Coker[1] and S. Bonella[2]

[1] Department of Chemistry, Boston University, 590 Commonwealth Avenue,
Boston, MA 02215, USA
coker@bu.edu
[2] NEST Scuola Normale Superiore, Piazza dei Cavalieri 7, It-56126 Pisa
s.bonella@sns.it

David Coker

D.F. Coker and S. Bonella: *Linearized Path Integral Methods for Quantum Time Correlation Functions*, Lect. Notes Phys. **703**, 553–590 (2006)
DOI 10.1007/3-540-35273-2_16

We review recently developed approximate methods for computing quantum time correlation functions based on linearizing the phase of their path integral expressions in the difference between paths representing the forward and backward propagators. Our focus here will be on problems that can be partitioned into two subsystems: One that is best described by a few discrete quantum states such as the high frequency vibrations or electronic states of molecules, and the other subsystem, "the bath", composed of the remaining degrees of freedom that will be described by a continuous representation. The general theory will first be developed and applied to model condensed phase problems. Approximations to the theory will be then made enabling applications to large scale realistic systems.

1 Introduction

According to the principles of statistical mechanics, the prediction and analysis of experimental responses of physical systems to general external perturbations requires the evaluation of appropriate time dependent average values or correlation functions. The calculation of these quantities for systems in which quantum effects are important demands the solution of the time dependent Schrödinger equation for the appropriate Hamiltonian. For a generic Hamiltonian, such a task can not be accomplished exactly except for the very simplest of systems with relatively few degrees of freedom. Consequently, the development of approximate numerical methods for calculating time-dependent quantum properties is an active, and growing field at the forefront of chemistry and physics.

In this chapter we review a new class of approximate methods for calculating quantum time correlation functions for systems that show non-adiabatic effects. In non-adiabatic phenomena, the coupling, for example, between nuclear and electronic motions in a molecule, or interactions with the environment, can induce transitions among the different states of the electronic subsystem. These transitions can change the products of a chemical reaction or of a scattering process by opening different reaction channels, they can modify the relaxation path and the final state of a photo-excited molecule, or influence the time scale for dissociation or recombination of molecules in the presence of solvent. In fact, the ability to analyze and predict the rich variety of behaviors produced by non-adiabatic phenomena is crucial for understanding a wide range of interesting physical and chemical processes. Proton and electron transfer in solution, electron-hole recombination dynamics, reactive scattering on surfaces, environmentally controlled chemical reactions, are just a few examples.

Here we discuss the theory with reference to "electronic" and "nuclear" degrees of freedom with applications such as those mentioned above in mind. However, the conceptual scheme applies whenever the degrees of freedom can be partitioned into two coupled groups: one best described by a discrete set of (perhaps many body) states, the other, a "bath" of continuous degrees of freedom. For example, quantum effects for high frequency vibrations can be viewed in this non-adiabatic framework. The transitions now occur between discrete nuclear vibrational states on a given electronic potential surface driven by the coupled dynamics of the continuous solvent variables. Vibrational relaxation and dephasing of molecules in solutions can be described in this way.

When dealing with non-adiabatic dynamics, a fruitful strategy to approximate the fully quantum propagation involves taking a classical limit for the evolution of the nuclear degrees of freedom in the presence of a set of quantum states. This kind of approach profits from the powerful simulation techniques available for evolving classical degrees of freedom, with the added benefit that the dimensionality of the quantum part of the problem is reduced to that of the electronic subset.

A delicate point in such mixed quantum-classical schemes is the self-consistency between the classical and quantum motions. The time-dependent variation of the electronic Hamiltonian arising from the nuclear dynamics results in electronic transitions. These quantum transitions, on the other hand, influence the forces on the nuclei, often with dramatic dynamical consequences.

Several quantum-classical approaches have been developed to include an appropriate coupling mechanism between the classical and quantum variables. For example, in the Ehrenfest method (see [1] and references therein), effective Schrödinger equations are obtained for the slow (nuclear) and fast (electronic) variables. According to these equations, each set of variables moves in a mean field potential determined by averaging over the complementary set. The classical limit for the motion of the nuclei is invoked to describe the evolution of the full system in terms of classical nuclear trajectories evolving on a single time dependent potential energy surface given by the expectation value of the electronic Hamiltonian.

Though it has been successfully applied, for example to describe energy transfer processes at metal surfaces, the Ehrenfest method fails when it becomes important to monitor different paths for different electronic states rather than a trajectory determined by an average over the different surfaces. This problem is particularly serious if one is interested in studying state specific nuclear pathways, such as those present in scattering events, or those determining low probability products in a chemical reaction.

An alternative mixed quantum-classical evolution scheme which does not suffer from this limitation is Surface Hopping [1]. This method, although based on heuristic arguments applied to the coupled channel equation more than on a rigorous classical limit, maintains a multiconfiguration picture of the system, and is able to describe complex non-adiabatic phenomena such as proton transfer in solution, or the branching into different product channels of photo-excited chemical reactions in clusters and condensed phase environments [1–7].

In Surface Hopping, an ensemble of classical nuclear trajectories is propagated. Each moves under the Hellmann-Feynman force of a particular electronic quantum state except for occasional switches from one state to another. The occurrence of such "hops" is determined by a stochastic mechanism triggered by the characteristics of the electronic states. For each specification of the initial conditions, the algorithm produces an ensemble of initially identical trajectories which will branch into many paths during the dynamics, with each path evolving on different potential energy surfaces for different periods of time as determined by the coupling with the electronic quantum evolution. In order to ensure the correct statistical weight of any average, the switching algorithm is constructed so that, at any instant of time and for a sufficiently large ensemble of nuclear paths, the fraction of trajectories assigned to each state is approximately equal to the population of that state. Though based on convincing physical arguments, the most popular hopping algorithms still suffer from a certain degree of arbitrariness. This manifests itself in various well known ambiguities of Surface Hopping methods, such as the handling of the so called forbidden hops (necessary for these techniques to maintain microscopic reversibility [8]), which bring into question their general reliability.

A different mixed quantum-classical procedure can be derived which, though still based on a set of classical surface switching nuclear trajectories, derives the hopping mechanism from a rigorous classical limit of the quantum evolution of the nuclear degrees of freedom. In this case, the attention is focused on the properties of the density matrix of the mixed system, and the language used is that of the Wigner formulation of quantum mechanics [9–15]. The Wigner-Lioville equation for a mixed quantum-classical system is derived and its solution obtained in the form of an iterative series that can, in principle, be integrated numerically. Though theoretically more sound, this method is less efficient than Surface Hopping from a numerical point of view as it requires summing a potentially slow converging integral series to determine the time evolution of the density matrix. Further, the probability to hop among the different electronic states associates weights to the different classical trajectories. These weights are not always well behaved and numerical methods to accumulate their effects efficiently are still at the early stages of development.

In recent work [16, 17] we presented a new mixed quantum-classical method, which we call LAND-Map (Linearized approach to non-adiabatic dynamics in the mapping formulation), for calculating correlation functions. The method couples the linearization ideas put forth by various workers [18–26] with the mapping description of non-adiabatic transitions [27–31].

Linearization methods start from a path integral representation of the forward and backward propagators in expressions for time correlation function, and combine them to describe the overall time evolution of the system in terms of a set of classical trajectories whose initial conditions are sampled from a quantity related to the Wigner transform of the quantum density operator. The linearized expression for a correlation function provides a powerful tool for describing systems in the condensed phase. The rapid decay of correlation functions for such systems enables reliable results to be obtained using a representation of the dynamics strictly valid only for relatively short times. It is interesting to note that the formal result used in these calculations has also emerged in the literature as an approximation to the time evolution of the Wigner equivalent of a correlation function [32], and is known in the semiclassical context as the linearized semiclassical initial value (SC-IVR) representation [26, 33–39].

In order to extend the linearization scheme to non-adiabatic dynamics it is convenient to represent the role of the discrete electronic states in terms of operators that simplify the evolution of the quantum subsystem with out changing its effect on the classical bath. A way to do this was first suggested by Miller, McCurdy and Meyer [28, 29] and has more recently been revisited by Thoss and Stock [30, 31]. Their method, known as the mapping formalism, represents the electronic degrees of freedom and the transitions between different states in terms of positions and momenta of a set of fictitious harmonic oscillators. Formally the approach is exact, but approximations (e.g. semi-classical, linearized SC-IVR, etc.) must be made for its numerical implementation.

LAND-Map takes advantage of the general quadratic nature of the degrees of freedom that describe electronic transitions in the quantum mapping propagator to evolve them exactly and to obtain a new mixed quantum-classical scheme by applying the linearization procedure only to the nuclear variables. The resulting coupled nuclear and electronic evolution is quite different from the propagation schemes used in related methods, but still requires only the integration of purely classical equations of motion with initial conditions sampled from a Wigner density. Sampling initial conditions for the nuclear variables from the quantum Wigner density for general many-body systems is as yet an unsolved problem. However, several promising approximate methods for incorporating nuclear quantum effects in the initial equilibrium density in this way have been presented [18–24].

The chapter is organized as follows: A detailed derivation of the linearized non-adiabatic algorithm for calculating time correlation functions is first presented. The method is then compared to other non-adiabatic techniques by examining the differences and similarities in the calculation of the properties of a simple, but instructive, model system. As an example of the implementation of LAND-map in more challenging cases, we describe calculations on the spin-boson model (a two level quantum system coupled to a bath of harmonic oscillators). We also explore the adiabatic limit of the theory (i.e. no electronic transitions resulting from nuclear dynamics) and apply it to compute the velocity and position autocorrelation functions for electronic diffusion in a realistic model of a metal-molten salt solution. The applications presented in this chapter do not require a general approach for sampling the quantum Wigner distribution and rather test the accuracy of the linearized approximate dynamics. The chapter is concluded by mentioning various questions of general implementation including initial condition sampling and efficient averaging procedures.

2 Theory

2.1 System-Bath Representation of Quantum Time Correlation Functions

Let us consider a system of nuclei and electrons described by the Hamiltonian

$$\hat{H} = \frac{\hat{P}^2}{2M} + \hat{h}_{el}(\hat{R}, \hat{p}, \hat{r}) \tag{1}$$

The quantum time correlation function of two operators of the system is defined as

$$\langle \hat{A}\hat{B}(t) \rangle = Tr\{\hat{\rho}\hat{A}e^{\frac{i}{\hbar}\hat{H}t}\hat{B}e^{-\frac{i}{\hbar}\hat{H}t}\} \tag{2}$$

For a system in thermal equilibrium the density operator is

$$\hat{\rho} = \frac{e^{-\beta\hat{H}}}{Z} \tag{3}$$

In (1) $\hat{R}$ and $\hat{P}$ refer to nuclear variables, while $\hat{r}$ and $\hat{p}$ describe the electronic degrees of freedom.

The correlation function can be written in a basis set $|R\alpha\rangle = |R\rangle|\alpha\rangle$ chosen as the tensor product of the coordinate representation for the nuclear degrees of freedom, and a nuclear coordinate independent electronic basis. In this chapter we shall refer to this electronic representation as the diabatic basis. (We refer the interested reader to reference [40] for the development of this approach with a more general electronic representation). By inserting resolutions of the identity in this basis, one obtains

$$\langle \hat{A}\hat{B}(t) \rangle = \sum_{\alpha\beta,\alpha'\beta'} \int dR_0 dR_N d\tilde{R}_0 d\tilde{R}_N \langle R_0\alpha|\hat{\rho}\hat{A}|\tilde{R}_0\alpha'\rangle \tag{4}$$

$$\langle \tilde{R}_0\alpha'|e^{\frac{i}{\hbar}\hat{H}t}|\tilde{R}_N\beta'\rangle\langle\tilde{R}_N\beta'|\hat{B}|R_N\beta\rangle\langle R_N\beta|e^{-\frac{i}{\hbar}\hat{H}t}|R_0\alpha\rangle$$

Notice that the electronic Hamiltonian is in general non-diagonal in the coordinate-diabatic state representation, and its matrix elements are given by

$$h_{\alpha\beta}(\hat{R}) = \langle\alpha|\hat{h}_{el}(\hat{R}, \hat{p}, \hat{r})|\beta\rangle \tag{5}$$

The expression above for the time correlation function can be read as a sequence of propagations and measurements. Reading from the right, the first propagator evolves the system from the nuclear configuration R_0 and the electronic state α to configuration R_N and state β in a time t. At this time a measurement, i.e. the evaluation of a given matrix element of operator $\hat{B}$,

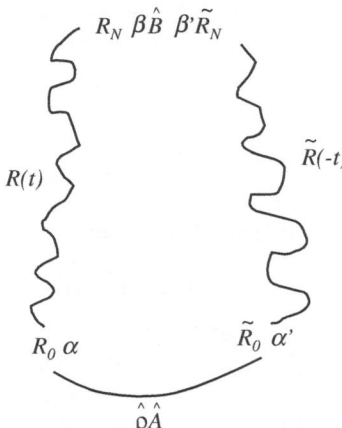

Fig. 1. Diagramatic representation of the paths involved in this general correlation function expression

is performed and then the time evolution is resumed. As a result of this measurement, the second segment of the propagation may start from a different nuclear configuration $\tilde{R}_N$, and a different electronic state β'. This part of the propagation takes us to a new configuration $\tilde{R}_0$ and a new electronic state α' in a time $-t$ and is sometimes referred to as the backward evolution leg. The trace operation in the definition of the correlation function can then be completed by evaluating the matrix element of the product of the operator $\hat{A}$ with the density operator $\hat{\rho}$ between the final state $|\alpha'\tilde{R}_0\rangle$, and the initial state $|\alpha R_0\rangle$. A diagramatic representation of the paths involved in this general correlation function expression is displayed in Fig. 1.

Because of the off-diagonal nature of the electronic Hamiltonian in the diabatic basis, transitions among different electronic states can, and in general will, occur during the forward and backward time evolutions. As a first step towards a computationally convenient expression for the non-adiabatic correlation function, we shall make use of the mapping Hamiltonian formalism to account for these transitions [28–31].

2.2 Quantum Subsystem Mapping Hamiltonian Representation

The core of the idea is to replace the evolution of the electronic subsystem with the evolution of a system of fictitious harmonic oscillators by means of two mapping relations. The first involves the states and is defined by

$$|\alpha\rangle \rightarrow |m_\alpha\rangle = |0_1, \ldots, 1_\alpha, ..0_n\rangle \tag{6}$$

This prescription maps the Hilbert space spanned by the original n diabatic states into one coinciding with a subspace of n-oscillators of unit mass and at most one quantum of excitation in a single specific oscillator.

The second mapping relation acts on the electronic Hamiltonian operator. This quantity can be rewritten in the diabatic basis as

$$\hat{h}_{el} = \sum_{\lambda,\lambda'} h_{\lambda,\lambda'}(\hat{R})|\lambda\rangle\langle\lambda'| \tag{7}$$

The mapping substitutes the diadic operators as follows

$$|\lambda\rangle\langle\lambda'| \rightarrow \hat{a}_\lambda^\dagger \hat{a}_{\lambda'} \tag{8}$$

where $\hat{a}$ and $\hat{a}^\dagger$ are creation and annihilation operators of mapping oscillator excitation. These operators can be expressed in terms of the positions and momenta of the n oscillators, for example

$$\hat{a}_{\lambda'} = \frac{1}{\sqrt{2\hbar}}(\hat{q}_{\lambda'} + i\hat{p}_{\lambda'}) \tag{9}$$

With these prescriptions the electronic Hamiltonian becomes

$$\hat{h}_{el} \rightarrow \hat{h}_m(\hat{R}) = \frac{1}{2}\sum_\lambda h_{\lambda,\lambda}(\hat{R})(\hat{q}_\lambda^2 + \hat{p}_\lambda^2 - \hbar) + \frac{1}{2}\sum_{\lambda,\lambda'} h_{\lambda,\lambda'}(\hat{R})(\hat{q}_{\lambda'}\hat{q}_\lambda + \hat{p}_{\lambda'}\hat{p}_\lambda) \tag{10}$$

The mapping provides the following equality

$$\langle R_N\beta|e^{-\frac{i}{\hbar}\hat{H}t}|R_0\alpha\rangle = \langle R_N m_\beta|e^{-\frac{i}{\hbar}\hat{H}_m t}|R_0 m_\alpha\rangle \tag{11}$$

where $\hat{H}_m = \hat{P}^2/2M + \hat{h}_m(\hat{R})$.

Using the above we may substitute the forward and backward propagators in the correlation function expression and obtain

$$\langle \hat{A}\hat{B}(t)\rangle = \sum_{\alpha\beta,\alpha'\beta'} \int dR_0 dR_N d\tilde{R}_0 d\tilde{R}_N \langle R_0\alpha|\hat{\rho}\hat{A}|\tilde{R}_0\alpha'\rangle \tag{12}$$

$$\times \langle \tilde{R}_0 m_{\alpha'}|e^{\frac{i}{\hbar}\hat{H}_m t}|R_N m_{\beta'}\rangle\langle \tilde{R}_N\beta'|\hat{B}|R_N\beta\rangle\langle R_N m_\beta|e^{-\frac{i}{\hbar}\hat{H}_m t}|R_0 m_\alpha\rangle$$

2.3 Bath Subsystem Phase Space Path Integral Representation

We next introduce a discretized path integral representation for the nuclear part of the propagators, and choose to do so in a hybrid momentum-coordinate representation [18,41]. This can be accomplished, for example for the forward propagator, by first using the identity

$$\langle R_N m_\beta|e^{-\frac{i}{\hbar}\hat{H}_m t}|R_0 m_\alpha\rangle = \langle R_N m_\beta|e^{-\frac{i}{\hbar}\hat{H}_m \epsilon}\ldots e^{-\frac{i}{\hbar}\hat{H}_m \epsilon}|R_0 m_\alpha\rangle \tag{13}$$

with $\epsilon = t/N$, (N integer), to write the total time evolution of the system as a sequence of short time propagations. We then introduce a set of resolutions of the identity in the nuclear coordinate representation and write the above expression as

$$\langle R_N m_\beta | e^{-\frac{i}{\hbar}\hat{H}_m \epsilon} \dots e^{-\frac{i}{\hbar}\hat{H}_m \epsilon} | R_0 m_\alpha \rangle = \int \prod_{k=1}^{N-1} dR_k \langle R_N m_\beta | e^{-\frac{i}{\hbar}\hat{H}_m \epsilon} | R_{N-1} \rangle$$
$$\times \langle R_{N-1} | e^{-\frac{i}{\hbar}\hat{H}_m \epsilon} | R_{N-2} \rangle \dots \langle R_1 | e^{-\frac{i}{\hbar}\hat{H}_m \epsilon} | R_0 m_\alpha \rangle \quad (14)$$

Using the Trotter approximation for the exponential of a sum of non-commuting operators allows separation of the kinetic and potential contributions to the propagation in each matrix element in the above expression, so that

$$\langle R_k | e^{-\frac{i}{\hbar}\hat{H}_m \epsilon} | R_{k-1} \rangle \approx \langle R_k | e^{-\epsilon \frac{i}{\hbar}\hat{P}^2/2M} e^{-\frac{i}{\hbar}\hat{h}_m(\hat{R})\epsilon} | R_{k-1} \rangle \quad (15)$$

Equation (15) becomes exact in the $\epsilon \to 0$ limit, and can be made more explicit by evaluating the kinetic term by inserting a complete set of states in the nuclear momentum representation to obtain, for example,

$$\langle R_k | e^{-\epsilon \frac{i}{\hbar}\hat{P}^2/2M} e^{-\frac{i}{\hbar}\hat{h}_m(\hat{R})\epsilon} | R_{k-1} \rangle = \int dP_k \langle R_k | e^{-\epsilon \frac{i}{\hbar}\hat{P}^2/2M} | P_k \rangle$$
$$\times \langle P_k | e^{-\frac{i}{\hbar}\hat{h}_m(\hat{R})\epsilon} | R_{k-1} \rangle$$
$$= \int \frac{dP_k}{2\pi\hbar} e^{-\epsilon \frac{i}{\hbar}P_k^2/2M} e^{-\frac{i}{\hbar}\epsilon \hat{h}_m(R_{k-1})} e^{\frac{i}{\hbar}P_k(R_k - R_{k-1})} \quad (16)$$

where we have used $\langle R_k | P_k \rangle = e^{\frac{i}{\hbar}P_k R_k}/\sqrt{2\pi\hbar}$, and we took advantage of the diagonal nature of $\hat{h}_m$ in the nuclear coordinate representation. Notice that the mapping Hamiltonian is still an operator in the mapping variables subspace. Combining the analogous representation of the other short time propagators, we finally obtain that the forward propagator has the following path integral form

$$\langle R_N m_\beta | e^{-\frac{i}{\hbar}\hat{H}_m t} | R_0 m_\alpha \rangle = \int \prod_{k=1}^{N-1} dR_k \frac{dP_k}{2\pi\hbar} \frac{dP_N}{2\pi\hbar} e^{\frac{i}{\hbar}S} \quad (17)$$
$$\times \langle m_\beta | e^{-\frac{i}{\hbar}\epsilon \hat{h}_m(R_N)} \dots e^{-\frac{i}{\hbar}\epsilon \hat{h}_m(R_0)} | m_\alpha \rangle$$

where

$$S = \epsilon \sum_{k=1}^{N} \left[P_k \frac{(R_k - R_{k-1})}{\epsilon} - \frac{P_k^2}{2M} \right] \quad (18)$$

A similar expression can be obtained for the backward propagator.

2.4 Quantum Subsystem Mapping Transition Amplitude

We can now simplify the above expression by examining the transition amplitude $\langle m_\beta | e^{-\frac{i}{\hbar}\epsilon \hat{h}_m(R_N)} \dots e^{-\frac{i}{\hbar}\epsilon \hat{h}_m(R_0)} | m_\alpha \rangle$ between the mapping states m_α and

m_β for a given nuclear path $(R_0, R_1, \ldots, R_N)$. As a consequence of the mapping, the electronic Hamiltonian in (10) is a quadratic function of the position and momentum operators of the oscillators. Such a Hamiltonian still contains the nuclear coordinates which appear as a set of (time dependent) external parameters influencing the evolution of the electronic degrees of freedom. For any given specification of the nuclear path, the quadratic nature of the electronic Hamiltonian allows us to obtain an exact expression for the mapping transition amplitude. One way to achieve this goal is to use a semi-classical form for the mapping propagator. A particularly convenient choice is provided by the Herman-Kluk or coherent state representation [42, 43] of the propagator. With this approach the function being propagated is represented in terms on an over complete set of fixed width complex Gaussian basis functions whose center position, phase, and weight are treated as dynamical parameters. A semi-classical analysis gives that the position and phase of these basis functions follow classical trajectories and the weight of the contribution of each basis function, in this limit, is related to the stability of the corresponding classical trajectory with respect to variations in initial Gaussian position and momentum parameters. Semi-classical theory is exact for quadratic Hamiltonians, so after some manipulations [16] we find the transition amplitude can be written as

$$\langle m_\beta | e^{-\frac{i}{\hbar}\epsilon\hat{h}_m(R_N)} \ldots e^{-\frac{i}{\hbar}\epsilon\hat{h}_m(R_0)} | m_\alpha \rangle = \qquad (19)$$

$$\int dq_0 dp_0 \frac{(2\gamma q_{\beta t} + \frac{i}{\hbar}p_{\beta t})}{2\sigma_{pq}} e^{-\frac{i}{2\hbar\sigma_{pq}}\sum_\lambda q_{\lambda t}p_{\lambda t}} e^{-\frac{1}{2}\sum_\lambda(\frac{q_{\lambda t}^2}{\sigma_q^2}+\frac{p_{\lambda t}^2}{\sigma_p^2})} c_t e^{\frac{i}{\hbar}s_t}$$

$$\times \frac{(2\gamma q_{\alpha 0} - \frac{i}{\hbar}p_{\alpha 0})}{2\sigma_{pq}} e^{-\frac{1}{2}\sum_\lambda(\frac{q_{\lambda 0}^2}{\sigma_q^2}+\frac{p_{\lambda 0}^2}{\sigma_p^2})} e^{\frac{i}{2\hbar\sigma_{pq}}\sum_\lambda q_{\lambda 0}p_{\lambda 0}}$$

Here γ is the width of the coherent states [42] used in the representation of the semi-classical propagator, $\sigma_q^2 = (\gamma + 1/2)/\gamma$, $\sigma_p^2 = 2\hbar^2(\gamma + 1/2)$, and $\sigma_{pq} = \gamma + 1/2$. Further in this semi-classical approach, as outlined above, (q_t, p_t) (where, for example, $q_t = (q_{1t}, \ldots, q_{nt})$) is the end point in phase space of a classical trajectory which starts at (q_0, p_0) and evolves according to the Hamiltonian

$$h(R) = \frac{1}{2}\sum_\lambda h_{\lambda,\lambda}(R)(q_\lambda^2 + p_\lambda^2 - \hbar) + \frac{1}{2}\sum_{\lambda\neq\lambda'} h_{\lambda,\lambda'}(R)(q_{\lambda'}q_\lambda + p_{\lambda'}p_\lambda)$$

s_t is the action (see (23)), and

$$c_t = \left[\det\frac{1}{2}\left(\frac{\partial \mathbf{q}_t}{\partial \mathbf{q}_0} + \frac{\partial \mathbf{p}_t}{\partial \mathbf{p}_0} - 2i\hbar\gamma\frac{\partial \mathbf{q}_t}{\partial \mathbf{p}_0} + \frac{i}{2\hbar\gamma}\frac{\partial \mathbf{p}_t}{\partial \mathbf{q}_0}\right)\right]^{\frac{1}{2}} \qquad (20)$$

is the square root of the determinant of the complex matrix which measures the stability of the mapping variable trajectories with respect to variations in the initial conditions.

Mapping Variable Evolution

At this stage, the mapping amplitude is expressed entirely in terms of quantities that can be calculated by propagating an ensemble of classical trajectories for the mapping variables and accumulating appropriate complex weights. Because of the parametric dependence of the Hamiltonian on the set of nuclear positions $\{R\}$, the explicit form of the evolution of the mapping variables cannot be determined in general. However, a number of useful properties of this dynamics can be obtained directly from the corresponding Hamilton's equations

$$\frac{dq_\beta}{dt} = h_{\beta,\beta}(R)p_\beta + \sum_{\lambda \neq \beta} h_{\beta,\lambda}(R)p_\lambda \tag{21}$$

$$\frac{dp_\beta}{dt} = -h_{\beta,\beta}(R)q_\beta - \sum_{\lambda \neq \beta} h_{\beta,\lambda}(R)q_\lambda$$

These properties allow us to further simplify the expression of the mapping amplitude and we shall now describe them in some detail to prepare for the crucial step in our linearization approach. First of all, it can immediately be shown that the quantity

$$\sum_\lambda (q_\lambda^2 + p_\lambda^2) \tag{22}$$

is a constant of the motion. This result will be used below, together with appropriate choice of the arbitrary width parameter, and units ($\hbar = 1$), to combine the different time Gaussian terms appearing in (19). Next, notice that the action s_t is given by

$$s_t = \int_0^t d\tau \left\{ \sum_\lambda p_{\lambda\tau} \frac{dq_{\lambda\tau}}{d\tau} - h(R_\tau) \right\} \tag{23}$$

$$= \int_0^t d\tau \left\{ \frac{1}{2} \sum_\lambda h_{\lambda,\lambda}(R_\tau)(p_{\lambda\tau}^2 - q_{\lambda\tau}^2 + \hbar) \right.$$

$$\left. + \frac{1}{2} \sum_{\lambda \neq \lambda'} h_{\lambda,\lambda'}(R_\tau)(p_{\lambda'\tau}p_{\lambda\tau} - q_{\lambda'\tau}q_{\lambda\tau}) \right\}$$

On the other hand, from the evolution equations, we find

$$\frac{d}{d\tau} \sum_\lambda p_{\lambda\tau}q_{\lambda\tau} = \left\{ \sum_\lambda h_{\lambda,\lambda}(R_\tau)(p_{\lambda\tau}^2 - q_{\lambda\tau}^2) \right.$$

$$\left. + \sum_{\lambda \neq \lambda'} h_{\lambda,\lambda'}(R_\tau)(p_{\lambda'\tau}p_{\lambda\tau} - q_{\lambda'\tau}q_{\lambda\tau}) \right\} \tag{24}$$

The exponents coming from the action and the phases of the initial and final mapping states in (19) can then be combined to obtain

$$s_t - \int_0^t d\tau \frac{d}{d\tau} \sum_\lambda p_{\lambda\tau} q_{\lambda\tau} = \frac{1}{2} \int_0^t d\tau \sum_\lambda h_{\lambda\lambda}(R_\tau) \tag{25}$$

$$+ \frac{1}{2\hbar} \left(1 - \frac{1}{\sigma_{pq}} \right) \int_0^t d\tau \left\{ \sum_\lambda h_{\lambda,\lambda}(R_\tau)(p_{\lambda\tau}^2 - q_{\lambda\tau}^2) \right.$$

$$\left. + \sum_{\lambda \neq \lambda'} h_{\lambda,\lambda'}(R_\tau)(p_{\lambda'\tau} p_{\lambda\tau} - q_{\lambda'\tau} q_{\lambda\tau}) \right\}$$

The choice of the value of the coherent state width γ is arbitrary, since this parameter does not affect the mathematical properties of the Gaussian basis set which determine the form of the semi-classical propagator. In fact, the value of this quantity is usually chosen in practical implementations so as to facilitate the numerical convergence. In the following, we shall set $\gamma = 1/2$ since with this choice (25) simplifies considerably and becomes

$$s_t - \int_0^t d\tau \frac{d}{d\tau} \sum_\lambda p_{\lambda\tau} q_{\lambda\tau} = \frac{1}{2} \int_0^t d\tau \sum_\lambda h_{\lambda\lambda}(R_\tau) \tag{26}$$

Next, let us consider the following matrix

$$\left(\frac{\partial \mathbf{q}_t}{\partial \mathbf{q}_0} + \frac{\partial \mathbf{p}_t}{\partial \mathbf{p}_0} - 2i\hbar\gamma \frac{\partial \mathbf{q}_t}{\partial \mathbf{p}_0} + \frac{i}{2\hbar\gamma} \frac{\partial \mathbf{p}_t}{\partial \mathbf{q}_0} \right) = \frac{\partial \eta_t}{\partial \eta_0} \tag{27}$$

where we define

$$\eta = \mathbf{q} + i\mathbf{p} , \tag{28}$$

and use the above choice for γ, and select a system of units so that $\hbar = 1$. The time evolution of the complex vector η can be readily obtained from Hamilton's equations for the mapping momenta and coordinates, and is given by

$$\frac{d\eta_t}{dt} = -ih_{el}(R_t)\eta_t \tag{29}$$

where h_{el} is the electronic Hamiltonian matrix. As long as care is taken in preserving the time-ordering of the operators, this equation can be integrated in a time-stepping fashion as

$$\eta_t = e^{-i\epsilon h_{el}(R_N)} \dots e^{-i\epsilon h_{el}(R_0)} \eta_0 \tag{30}$$

so that

$$\frac{\partial \eta_t}{\partial \eta_0} = e^{-i\epsilon h_{el}(R_N)} \dots e^{-i\epsilon h_{el}(R_0)} \tag{31}$$

Substituting this expression in (20), and using

$$\det[e^A] = e^{\mathrm{Tr}A} \tag{32}$$

where A is a general matrix, we obtain

$$c_t \sim e^{-\frac{i}{2}\int_0^t d\tau \sum_\lambda h_{\lambda\lambda}(R_\tau)} \tag{33}$$

This result can be combined with the action's surviving contribution to the phase in (19) to rewrite the semi-classical amplitude as

$$\langle m_\beta| e^{-\frac{i}{\hbar}\epsilon\hat{h}_m(R_N)} \dots e^{-\frac{i}{\hbar}\epsilon\hat{h}_m(R_1)} |m_\alpha\rangle \sim \int dq_0 dp_0 (q_{\beta t} + ip_{\beta t})(q_{\alpha 0} - ip_{\alpha 0})$$
$$\times e^{-\frac{1}{2}\sum_\lambda (q_{\lambda 0}^2 + p_{\lambda 0}^2)} \tag{34}$$

Here, as mentioned above, we have used (22) to combine the two Gaussians in (19) in a single Gaussian function evaluated at zero time. The symbol $\sim$ in this equation indicates that some irrelevant constants have been dropped, not that there are approximations in this result. Below we replace it with an equality as the constants which normalize the propagators divide out when the appropriately normalized correlation function is computed. The semi-classical amplitude can be conveniently re-expressed by introducing a polar representation of the complex polynomials identifying the initial and final occupied mapping states, thus

$$\langle m_\beta| e^{-\frac{i}{\hbar}\epsilon\hat{h}_m(R_N)} \dots e^{-\frac{i}{\hbar}\epsilon\hat{h}_m(R_0)} |m_\alpha\rangle = \int dq_0 dp_0 r_{t,\beta}(\{R_k\}) e^{-i\Theta_{t\beta}(\{R_k\})}$$
$$\times r_{0\alpha} e^{i\Theta_{0,\alpha}} G_0 \tag{35}$$

where $G_0 = e^{-\frac{1}{2}\sum_\lambda (q_{0,\lambda}^2 + p_{0,\lambda}^2)}$,

$$r_{t,\beta}(\{R_k\}) = \sqrt{q_{t,\beta}^2(\{R_k\}) + p_{t,\beta}^2(\{R_k\})} \tag{36}$$

and

$$\Theta_{t,\beta}(\{R_k\}) = \tan^{-1}\left(\frac{p_{t,\beta}(\{R_k\})}{q_{t,\beta}(\{R_k\})}\right) \tag{37}$$

A more explicit form for the phase in (35) can be obtained if one observes that

$$\Theta_{t,\beta}(\{R_k\}) = \tan^{-1}\left(\frac{p_{0,\beta}}{q_{0,\beta}}\right) + \int_0^t d\tau \left[\frac{d}{d\tau}\tan^{-1}\left(\frac{p_{\tau,\beta}(R_\tau)}{q_{\tau,\beta}(R_\tau)}\right)\right] \tag{38}$$

$$= \tan^{-1}\left(\frac{p_{0,\beta}}{q_{0,\beta}}\right) + \int_0^t d\tau h_{\beta,\beta}(R_\tau)$$

$$+ \int_0^t d\tau \sum_{\lambda \neq \beta} \left[h_{\beta,\lambda}(R_\tau)\frac{(p_{\tau\beta}p_{\tau\lambda} + q_{\tau\beta}q_{\tau\lambda})}{(p_{\tau\beta}^2 + q_{\tau\beta}^2)}\right]$$

$$= \tan^{-1}\left(\frac{p_{0,\beta}}{q_{0,\beta}}\right) + \int_0^t \theta_\beta(R_\tau)d\tau$$

Equation (38) defines the function $\theta_\beta(R)$, and in going from the first to the second line of this equation, the evolution equations of the mapping variables have been used.

2.5 Linearization of Bath Subsystem Path Integrals: A Trajectory Based Expression for the Correlation Function

Having derived a suitable expression for the transition amplitude, we now return to the time correlation function. Substituting (35) in the path integral expression for the propagator, (17), and using the latter and its analog for the backward evolution in (12) we obtain

$$\langle \hat{A}\hat{B}(t)\rangle = \sum_{\alpha\beta,\alpha'\beta'} \int dR_0 d\tilde{R}_0 \int \prod_{k=1}^{N} dR_k \frac{dP_k}{2\pi\hbar} \int \prod_{k=1}^{N} d\tilde{R}_k \frac{d\tilde{P}_k}{2\pi\hbar} e^{\frac{i}{\hbar}(S-\tilde{S})} \quad (39)$$

$$\times \int dq_0 dp_0 d\tilde{q}_0 d\tilde{p}_0 \langle R_0\alpha|\hat{\rho}\hat{A}|\tilde{R}_0\alpha'\rangle \langle \tilde{R}_N\beta'|\hat{B}|R_N\beta\rangle$$

$$\times r_{t,\beta}(\{R_k\})e^{-i\Theta_{t\beta}(\{R_k\})}r_{0\alpha}e^{i\Theta_{0,\alpha}}G_0$$

$$\times \tilde{r}_{t,\beta'}(\{\tilde{R}_k\})e^{i\tilde{\Theta}_{t\beta'}(\{\tilde{R}_k\})}\tilde{r}_{0\alpha'}e^{-i\tilde{\Theta}_{0,\alpha'}}\tilde{G}_0$$

Here, and in the following we use a shorthand notation for the mapping state indices such that α stands for m_α, for example.

All the formal manipulations performed so far are exact, and the nuclear evolution is still described at the full quantum level. To proceed to a computable expression [16–21], we now change bath subsystem variables to mean, $\bar{R}_k = (R_k + \tilde{R}_k)/2$, and difference, $Z_k = R_k - \tilde{R}_k$, coordinates (with similar transformation for the bath momenta, $\bar{P}_k = (P_k + \tilde{P}_k)/2$ and $Y_k = P_k - \tilde{P}_k$) and Taylor series expand the phase in (39) in the difference variables. Truncating this expansion to linear order we obtain the following approximate expression for the correlation function

$$\langle \hat{A}\hat{B}(t)\rangle = \sum_{\alpha\beta,\alpha'\beta'} \int dq_0 dp_0 d\tilde{q}_0 d\tilde{p}_0 \tilde{r}_{0\alpha'} e^{-i\tilde{\Theta}_{0,\alpha'}} \tilde{G}_0 r_{0\alpha} e^{i\Theta_{0,\alpha}} G_0 \quad (40)$$

$$\times \int d\bar{R}_0 dZ_0 \int \prod_{k=1}^{N} d\bar{R}_k \frac{d\bar{P}_k}{2\pi\hbar} \int \prod_{k=1}^{N} dZ_k \frac{dY_k}{2\pi\hbar}$$

$$\times \left\langle \bar{R}_0 + \frac{Z_0}{2}\alpha|\hat{\rho}\hat{A}|\bar{R}_0 - \frac{Z_0}{2}\alpha'\right\rangle e^{-i\bar{P}_1 Z_0}$$

$$\times \left\langle \bar{R}_N - \frac{Z_N}{2}\beta'|B|\bar{R}_N + \frac{Z_N}{2}\beta\right\rangle e^{i\bar{P}_N Z_N}$$

$$\times e^{-i\epsilon[\nabla\theta_\beta(\bar{R}_N)+\nabla\tilde{\theta}_{\beta'}(\bar{R}_N)]/2\}Z_N}$$

$$\times r_{t,\beta}(\{\bar{R}_k\})\tilde{r}_{t,\beta'}(\{\bar{R}_k\})e^{-i\epsilon\sum_{k=1}^{N}[\theta_\beta(\bar{R}_k)-\tilde{\theta}_{\beta'}(\bar{R}_k)]}$$

$$\times e^{-i\epsilon\sum_{k=1}^{N-1}\{(\bar{P}_{k+1}-\bar{P}_k)/\epsilon+[\nabla\theta_\beta(\bar{R}_k)+\nabla\tilde{\theta}_{\beta'}(\bar{R}_k)]/2\}Z_k}$$

$$\times e^{-i\epsilon\sum_{k=1}^{N}\{\bar{P}_k/M-(\bar{R}_k-\bar{R}_{k-1})/\epsilon\}Y_k}$$

The integrals over the end-point difference coordinates Z_0 and Z_N can be performed defining the Wigner transformed operators

$$(\hat{\rho}\hat{A})_W^{\alpha,\alpha'}(\bar{R}_0, \bar{P}_1) = \int dZ_0 \left\langle \bar{R}_0 + \frac{Z_0}{2}\alpha \left| \hat{\rho}\hat{A} \right| \bar{R}_0 - \frac{Z_0}{2}\alpha' \right\rangle e^{-i\bar{P}_1 Z_0} \quad (41)$$

and in the limit of $\epsilon \to 0$

$$(\hat{B})_W^{\beta',\beta}(\bar{R}_N, \bar{P}_N) = \int dZ_N \left\langle \bar{R}_N + \frac{Z_N}{2}\beta' \left| \hat{B} \right| \bar{R}_N - \frac{Z_N}{2}\beta \right\rangle e^{-i\bar{P}_N Z_N} \quad (42)$$

All integrals over the difference coordinates, Z_k, and difference momenta, Y_k, for $0 < k < N$ can also be performed as they are integral representations of δ-functions. Thus the linearized approximation for the time correlation function can finally be expressed as

$$\begin{aligned}
\langle \hat{A}\hat{B}(t)\rangle = \sum_{\alpha\beta,\alpha'\beta'} \int d\bar{R}_0 dq_0 dp_0 d\tilde{q}_0 d\tilde{p}_0 \int \prod_{k=1}^{N} d\bar{R}_k \frac{d\bar{P}_k}{2\pi\hbar} (\hat{\rho}\hat{A})_W^{\alpha,\alpha'}(\bar{R}_0, \bar{P}_1) \\
\times (\hat{B})_W^{\beta',\beta}(\bar{R}_N, \bar{P}_N) G_0 \tilde{G}_0 r_{0\alpha} e^{i\theta_0,\alpha} \tilde{r}_{0\alpha'} e^{-i\tilde{\theta}_0,\alpha'} \\
\times r_{t,\beta}(\{\bar{R}_k\}) \tilde{r}_{t,\beta'}(\{\bar{R}_k\}) e^{-i\epsilon \sum_{k=1}^{N}(\theta_\beta(\bar{R}_k) - \tilde{\theta}_{\beta'}(\bar{R}_k))} \\
\times \prod_{k=1}^{N-1} \delta\left(\frac{\bar{P}_{k+1} - \bar{P}_k}{\epsilon} - F_k^{\beta,\beta'}\right) \prod_{k=1}^{N} \delta\left(\frac{\bar{P}_k}{M} - \frac{\bar{R}_k - \bar{R}_{k-1}}{\epsilon}\right)
\end{aligned} \quad (43)$$

where

$$\begin{aligned}
F_k^{\beta,\beta'} = -\frac{1}{2}\left\{\nabla_{\bar{R}_k} h_{\beta,\beta}(\bar{R}_k) + \nabla_{\bar{R}_k} h_{\beta',\beta'}(\bar{R}_k)\right\} \\
-\frac{1}{2}\sum_{\lambda\neq\beta} \nabla_{\bar{R}_k} h_{\beta,\lambda}(\bar{R}_k) \left\{\frac{(p_{\beta k}p_{\lambda k} + q_{\beta k}q_{\lambda k})}{(p_{\beta k}^2 + q_{\beta k}^2)}\right\} \\
-\frac{1}{2}\sum_{\lambda\neq\beta'} \nabla_{\bar{R}_k} h_{\beta',\lambda}(\bar{R}_k) \left\{\frac{(\tilde{p}_{\beta' k}\tilde{p}_{\lambda k} + \tilde{q}_{\beta' k}\tilde{q}_{\lambda k})}{(\tilde{p}_{\beta' k}^2 + \tilde{q}_{\beta' k}^2)}\right\}
\end{aligned} \quad (44)$$

The product of δ-functions in (43) amounts to a time-stepping prescription in which the mean path evolves classically. As the motion of the mapping variables is already classical, the calculation of the time correlation function has been reduced to a two step procedure: (1) Sampling a set of initial conditions for the bath variables from a probability distribution related to the partial Wigner transform of the density times the operator $\hat{A}$, i.e. the factor $(\hat{\rho}\hat{A})_W^{\alpha,\alpha'}(\bar{R}_0, \bar{P}_1)$ in (43), and a Gaussian distribution, $\tilde{G}_0 G_0$, for the mapping subsystem variables; (2) Integration of a set of coupled classical equations of motion for the mapping and bath variables. The first of these tasks can be accomplished for complex systems only approximately using, for example, recently developed local harmonic methods [18–21]. The second task is straightforward. However, depending on the specific term of the correlation function which is being evaluated, the forces in (44) are determined by different time dependent linear combinations of pairs of diagonal, and off-diagonal elements of the quantum subsystem Hamiltonian. The diagonal terms are identified by the *final* states in the propagators appearing in the original expression for the correlation function, while the off-diagonal terms are responsible for the feedback between bath motion and changes in the quantum subsystem state amplitudes. The latter are affected by the bath propagation through the parametric dependence of the classical counterpart of (10).

The key simplifying approximation that transforms the exact expression for the time correlation in (39) to the computationally tractable approximate expression in (43) is the truncation of the Taylor series expansion of the integrand phase to linear order in the difference phase space path. This corresponds to the assumption that the most important contributions to the correlation function result from forward and backward bath subsystem paths that remain close to one another throughout the evolution period. This assumption is justified at $t = 0$ if the product of the initial density and initial observable is localized in bath subsystem phase space. For example, this will be the case for a system in thermal equilibrium at sufficiently high temperature. If the matrix elements of the operator measured at the final time are localized in the phase space representation, as will be the case for many experimentally relevant observables, the assumption will similarly be valid at the terminal point. Under these circumstances, a way to ensure end point proximity is to have the forward and backward paths stay close to one another at all times. If the total propagation time is short the above prescription includes all possible pairs of paths connecting the phase space regions isolated by the operators. At longer times and for general potentials different pairs of paths may become stationary points for the action phase difference in (39) with the specified boundary conditions. The linearization approximation neglects contributions from such pairs of different paths and is therefore expected to provide a less reliable description of the dynamics at these longer times. The hope is that interesting correlation functions for condensed phase systems decay sufficiently rapidly that dynamics on a time scale reliably represented by the linearization approximation is sufficient to yield useful results.

2.6 Comments on Implementation and Comparison with Other Methods for Non-Adiabatic Dynamics

The implementation of the algorithm outlined above is somewhat delicate due to our use of the polar representation of the complex Hermite polynomials that project onto the final states β or β'. When the complex polynomial is zero, the phase is ill-defined. This is reflected in the expression of the force in (44) by the apparent singularity in the off-diagonal terms. The existence of a divergence in the force, however, depends on the behavior of the gradients of the off-diagonal terms of the electronic Hamiltonian. As they usually are, or go to, zero very rapidly in regions of zero population of the final state, it is possible to remedy the problem by a careful implementation of the algorithm. Furthermore, since the weight of such trajectories is rigorously zero thanks to the amplitudes r_β and $\tilde{r}_{\beta'}$ in (43), the effect of this apparent pathology in the phase is limited to the integration of the evolution equations.

A few further comments are in order: All the propagations are classical and local in time, two features which simplify the numerical task considerably. The mapping variables evolve according to (21), while the classical trajectory for the nuclear variables is determined by the forces in (44). As such the overall evolution of the two coupled dynamical subsystems is not governed by a single Hamiltonian. This is in marked contrast to the usual semi-classical approach where the mapping Hamiltonian is differentiated to obtain a classically consistent system of equations of motion for all degrees of freedom [36, 44, 45]. The method we present here is also significantly different from the Pechukas semi-classical formulation of non-adiabatic dynamics [46]. The Pechukas transition amplitude and trajectory must be determined by self-consistent iteration. With the mapping formulation, however, we have an explicit form for the transition amplitude, (35), which can be integrated as the nuclear trajectory is advanced making the approach local in time and straightforward to implement.

The two distinct evolutions in the LAND-map approach are reminiscent of the situation in Surface Hopping or mean field non-adiabatic methods where different dynamical prescriptions apply to the quantum and classical subsystems (see for example [1]). To highlight the differences between the linearized path integral, mean field, and Surface Hopping approaches let us consider a 1D bath coupled to a simple two level quantum subsystem. We will specialize our example to a popular model, the simple avoided crossing, 1D scattering problem of the so-called Tully canon [47], the potentials for which are displayed in Fig. 2. A wave packet starting in the left asymptotic region in diabatic electronic state 1 and traveling towards the right with sufficient momentum will pass through the avoided crossing region, splitting into two packets, one on each surfaces emerging to the right of the crossing with different packet momenta depending on the asymptotic energy gap between the surfaces. The amount of splitting depends on the coupling between the states and the incoming packet momentum or incident scattering energy. If the scattering energy is too low the packet will simply reflect off the barrier in the ground state surface. Suppose, for example, we measure the time dependent expectation value of the bath coordinate.

With the LAND-map approach, due to our focus in this example on computing bath subsystem operators which are diagonal in the quantum subsystem states, and because we chose the initial density to be localized in state 1, the correlation function in (43) contains only two terms: ($\alpha = \alpha' = 1$ and $\beta = \beta' = 1$), and ($\alpha = \alpha' = 1$ and $\beta = \beta' = 2$). For this situation the system starts out with a phase space distribution determined by the value $\alpha = \alpha' = 1$ and evolves along trajectories whose asymptotic properties are governed by forces obtained from (44) that depend on the value of $\beta = \beta'$.

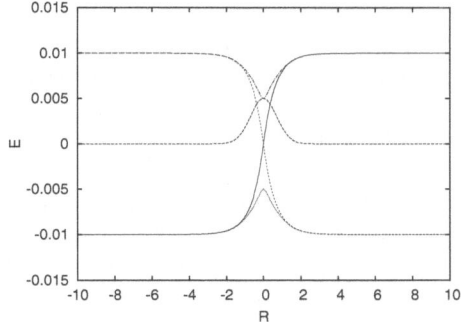

Fig. 2. Simple 1D scattering avoided crossing model. Energies and distances are in atomic units. Solid and short dashed are the two diagonal diabats, long dashed curve is the off-diagonal diabat, dotted, and dash dot are the adiabatic energy surfaces

In the first of the above terms, before and after the trajectories cross the coupling region the force is $-\nabla_R h_{11}$. This force will be modified by the off-diagonal element, $-\nabla_R h_{12}$, which, in our model, is only active in the coupling region, and, as given in (44), this additional off-diagonal force contribution is modulated by the time dependence of the mapping variables obtained by solving the auxiliary equations in (21). For the second term in the correlation function, however, the asymptotic force on the trajectory is $-\nabla_R h_{22}$ and the off-diagonal modifying force, active only in the coupling region, is $-\nabla_R h_{21}$. Thus, a LAND-map calculation of the bath coordinate correlation function of this two state problem involves propagating two sets of trajectories. Each set starts from the same initial distribution of positions and momenta of the bath and mapping oscillators but one set evolves with forces F_t^{11}, and the other with F_t^{22} as given by (44).

The two sets of trajectories carry time dependent weights which, from (43), have the form

$$W_{\alpha\alpha'\beta\beta'}(t) = r_{0\alpha}\tilde{r}_{0\alpha'}e^{i(\theta_{0,\alpha}-\tilde{\theta}_{0,\alpha'})}$$
$$\times r_{t,\beta}\tilde{r}_{t\beta'}e^{-i\epsilon\int_0^t d\tau(\theta_{\tau,\beta}-\tilde{\theta}_{\tau,\beta'})} \tag{45}$$

In general, the initial mapping variables must be sampled from Gaussian distributions and the complex polynomials in these variables which appear in (34) are written in polar form to give the weights in (45). The integrals over the initial mapping variables appearing in the correlation function expression can be reliably approximated by steepest descent [48] to give so-called "focused initial conditions" in which the initial values for the "occupied state" mapping variables are set to $(q_{occ}, p_{occ}) = (1, 1)$ while the unoccupied state mapping variables are set to zero $(q_{un}, p_{un}) = (0, 0)$. In our example this means that the weights of the ($\alpha = \alpha' = 1$ and $\beta = \beta' = 1$) trajectories start out at $W_{1111}(0) = 1$ and relax as the coupled system-bath dynamics is evolved and the states mix. The weights of the ($\alpha = \alpha' = 1$ and $\beta = \beta' = 2$) class of trajectories, on the other hand, are initially $W_{1122}(0) = 0$ and the contributions from these trajectories grow as these trajectories pass through the coupling region.

The characteristics of the two sets of trajectories and their weights differ considerably. Trajectories riding initially on the lower left diabat, the $(\alpha\alpha'\beta\beta') = (1111)$ class, will climb the wall in this surface and, as they enter the coupling region, they will be additionally accelerated and decelerated by the off-diagonal forces whose effects are modulated by the time dependent amplitude term $(p_{\beta t}p_{\alpha t} + q_{\beta t}q_{\alpha t})/(p_{\beta t}^2 + q_{\beta t}^2)$. For this class of trajectories, however, the modulation term is unity throughout so the dynamics is completely conservative. Their contribution to the correlation function will be accumulated with weights determined by (45) which depend parametrically on initial bath subsystem phase space point $(\bar{R}_0, \bar{P}_1)$ from which the trajectory was launched. The combination of weights and location for this class of trajectories gives a packet that emerges in the higher energy state on the right hand side of the coupling region, having slowed appropriately after climbing the barrier. The dynamics of the (1122) class of trajectories, on the other hand, is completely different. As mentioned above these trajectories are initiated with the same phase space distribution but they start their evolution with zero weight and riding on the upper left hand potential surface. As they enter the coupling region their weights grow and the off-diagonal force matrix elements begin to make time dependent contributions modulated by the amplitude term $(p_{2t}p_{1t} + q_{2t}q_{1t})/(p_{2t}^2 + q_{2t}^2)$. These trajectories *do not* conserve energy. To reproduce the wave packet moving with the correct momentum on the lower right surface, the off-diagonal forces need to dissipate the energy of these "ghost" trajectories. The implicitly coupled bath and mapping oscillator dynamics is responsible for both this dissipation and growth of trajectory weight.

Let us now compare the salient features of the other non-adiabatic approaches mentioned in the introduction to LAND-map in the context of the example problem outlined above. We first consider the Surface Hopping techniques that have been developed by Tully and others [1]. These methods are generally formulated in the instantaneous adiabatic eigenstate representation of the quantum subsystem. They differ from the LAND-map approach in that trajectories are only evolved with forces derived from a single adiabatic surface at any instant. This is true except in regions where the "hops" between adiabatic surfaces take place to represent the quantum transitions. For each surface hopping trajectory this occurs at points, and impulsive off-diagonal forces act to conserve surface hopping trajectory energy. An auxiliary time dependent quantum subsystem Schrödinger equation for the adiabatic basis set expansion coefficients is solved along each bath subsystem trajectory and the solutions are used to hopping probabilities. A stochastic model that branches trajectories from one surface to another is employed in an attempt to capture the effect of quantum state mixing on the bath subsystem dynamics. The classical trajectories and the quantum evolution are not influenced, as in the LAND-map approach, by the measurement one is performing. By their very nature Surface Hopping methods generate approximate densities of trajectories in different quantum subsystem states that can only be used to compute time dependent expectation values of operators that are diagonal in these states.

In general the ingredients underlying Surface Hopping methods are not derivable from a formal starting point but rather are motivated on strong physical grounds. The key advantage of these methods is that the trajectories carry unit weight rather than the phase factors associated with trajectories in the LAND-map approach. The implementation of Surface Hopping methods is thus numerically very efficient. A single ensemble sampling the initial distribution generates populations of trajectories in all possible final states by stochastic branching into representatives of the most important states as time advances. Various heuristic models have also been developed to describe the dephasing or decoherence of the off-diagonal density matrix elements due to bath subsystem packets following different paths as they move away from transition points on the coupled surfaces [49,50]. This "importance sampling" philosophy is fundamentally different from that underlying the LAND-map approach in which we propagate trajectories even if they have vanishing weights in some regions and allow all phase weighted contributions to interfere.

Let us now illustrate the nature of the Surface Hopping trajectories in the calculation of bath position expectation for the model in Fig. 2. In the limit of strongly localized coupling regions (so the adiabatic surfaces and any choice of coupled crossing diabatic potential curves look similar, except for state label switches at a small number of points), surface hopping methods can be viewed as essentially "stitching" together pieces of LAND-map trajectories from the different classes outlined in the above discussion. The initially occupied state amplitude diminishes as a (1111) LAND-map trajectory approaches the localized crossing region. As a (1122) trajectory approaches the crossing, however, its amplitude increases and, due to the non-conservative nature of the dynamics, its excess energy starts to fall. In the localized crossing region there will be overlap between these two LAND-map trajectory ensemble populations. Surface Hopping can be viewed as providing a mechanism whereby (1111) trajectories whose amplitudes are falling fast are replaced in the Surface Hopping trajectory ensemble by "rising star" (1122) trajectories. This replacement is based on a completely heuristic algorithm which terminates the (1111) trajectory and assumes that there must be a (1122) trajectory in this emerging ensemble that can be used as a replacement. A physical model of the population transfer is introduced to infer the properties of the replacement (1122) trajectory without ever having evolved any such trajectories explicitly.

In this light surface hopping methods are akin to the LAND-map approach except they throw away a lot of information in the interest of developing an efficient method to model the approximate dynamics of the diagonal densities. The multiple spawning methods developed by Martinez and coworkers [51–55] work in a similar fashion but the replacement trajectories that are spawned in new emerging regions of importance carry more quantum information, since in this approach they are frozen Gaussian wave packets.

The methods outlined above can be contrasted with the Ehrenfest approach in which the bath follows a single mean field trajectory satisfying classical-like equations of motion $\dot{R}(t) = P(t)/M$ and $\dot{P}(t) = F_{MF} = -\langle\psi(t)|\partial\hat{h}_{el}/\partial R|\psi(t)\rangle$. The mean field force on the trajectory is determined by the time dependent quantum subsystem state $|\psi(t)\rangle = \sum_\alpha c_\alpha(t)|\alpha\rangle$ which satisfies the auxiliary Schrödinger equation $i\hbar\frac{\partial}{\partial t}|\psi(t)\rangle = \hat{h}_{el}(R(t))|\psi(t)\rangle$. The composite system energy, obtained as $E(t) = P(t)^2/2M + \langle\psi(t)|\hat{h}_{el}(R(t))|\psi(t)\rangle$, is conserved along a trajectory. For our two state model problem the force is a time dependent linear combination of the gradients of the diabatic potentials and coupling matrix elements of the form $F_{MF} = -(c_1^2(t)\nabla_R h_{11} + c_2^2(t)\nabla_R h_{22} + 2\mathrm{Re}\{c_1^*(t)c_2(t)\nabla_R h_{12}\})$.

Unlike the LAND-map or Surface Hopping approaches outlined above, where expectation values are built of contributions from multiple terms, each associated with a different class of trajectories, with the mean field approach the time dependent expectation is obtained from the properties of a single trajectory. Before the coupling region in our example $c_1 = 1$ and $c_2 = 0$ and the mean field trajectory moves over the lower left hand surface in much the same way as the important LAND-map or Surface Hopping trajectories in this region. However, after the coupling region the mean field force is some time independent linear combination of diagonal force elements determined by asymptotic values of the evolved coefficients. With the LAND-map or Surface Hopping methods, on the other hand, when the trajectories emerge from the coupling region they experience forces purely from a single surface. With the mean field approach the the two packets that scatter on the different surfaces in the real, full quantum treatment of the problem, whose identity is maintained with both the LAND-map and Surface Hopping methods, are replaced by an unphysical single trajectory moving on the average surface.

Finally, we note that the linearization procedure can be applied by treating both the bath and mapping variables on the same dynamical footing. In this approach the propagators are linearized in the difference between forward and backward paths in ALL degrees of freedom [22, 36]. Initial phase space conditions for all variables are sampled from the Wigner transform of the full system-bath density. The evolution of the bath and mapping variables is derived from a single system-bath Hamiltonian, $H(P, R, \{p_\lambda, q_\lambda\}) = P^2/2M + h(R, \{p_\lambda, q_\lambda\})$ where h has the same form as in (20). The force on the bath variables is thus

$$F_L = -\frac{1}{2}\sum_\lambda \nabla_R h_{\lambda,\lambda}(R)(q_\lambda^2 + p_\lambda^2 - \hbar) - \frac{1}{2}\sum_{\lambda\neq\lambda'} \nabla_R h_{\lambda,\lambda'}(R)(q_{\lambda'}q_\lambda + p_{\lambda'}p_\lambda) \quad (46)$$

With this approach $(q_\lambda^2 + p_\lambda^2 - \hbar)$ plays the role of the state occupation so asymptotically in our example the force on the bath is some constant linear combination of the forces from the different surfaces determined by the full system dynamics evolved through the coupling region. These populations may in general differ from the mean field amplitudes, c_i, but the asymptotic linear combination form of F_L poses the same problems for this approach as outlined above for the Ehrenfest force, F_{MF}.

Note also that the populations multiplying the gradients of the diagonal matrix elements of the electronic Hamiltonian in (46) can become negative during the propagation. These terms can then result in unstable trajectories due to inversion of the diabatic surfaces which compromise the convergence of the fully linearized algorithm for long times [56, 57]. Given the form of the forces in the LAND-map approach (see (44)), this method does not suffer from these problems.

2.7 The Adiabatic Limit

In many applications it is often reasonable to suppose that the bath subsystem dynamics causes slow mixing of the quantum subsystem states. If the relevant experimental measurements involve time scales shorter than the quantum subsystem mixing time, one can proceed as if the bath dynamics occurs in a single quantum subsystem state. This is the adiabatic approximation and in this limit (43) can be simplified by making the following substitutions: $\alpha = \beta$ (the forward path begins and ends in the same quantum subsystem state), and similarly for the backward path we have $\alpha' = \beta'$. Thus the adiabatic approximation to the correlation function is obtained as

$$\langle \hat{A}\hat{B}(t) \rangle = \sum_{\alpha,\alpha'} \int d\bar{R}_0 dq_0 dp_0 d\tilde{q}_0 d\tilde{p}_0 \int \prod_{k=1}^{N} d\bar{R}_k \frac{d\bar{P}_k}{2\pi\hbar} (\hat{\rho}\hat{A})_W^{\alpha,\alpha'}(\bar{R}_0, \bar{P}_1) \quad (47)$$

$$\times (\hat{B})_W^{\alpha',\alpha}(\bar{R}_N, \bar{P}_N) G_0 \tilde{G}_0 r_{0\alpha} e^{i\theta_{0,\alpha}} \tilde{r}_{0\alpha'} e^{-i\bar{\theta}_{0,\alpha'}}$$

$$\times r_{t,\alpha}(\{\bar{R}_k\}) \tilde{r}_{t,\alpha'}(\{\bar{R}_k\}) e^{-i\epsilon \sum_{k=1}^{N}(\theta_\alpha(\bar{R}_k) - \bar{\theta}_{\alpha'}(\bar{R}_k))}$$

$$\times \prod_{k=1}^{N-1} \delta\left(\frac{\bar{P}_{k+1} - \bar{P}_k}{\epsilon} - F_k^{\alpha,\alpha'}\right) \prod_{k=1}^{N} \delta\left(\frac{\bar{P}_k}{M} - \frac{\bar{R}_k - \bar{R}_{k-1}}{\epsilon}\right)$$

From the mapping variable equations of motion in (21) we see that the evolution of these variables becomes uncoupled when the off-diagonal Hamiltonian matrix elements $h_{\beta,\lambda}(R)$ for $\beta \neq \lambda$ are small compared to the diagonal elements $h_{\beta,\beta}(R)$. This constitutes the criterion for adiabatic dynamics. These equations then give simple, constant amplitude harmonic oscillator dynamics, independent of bath subsystem path, for the mapping variables. In this case, the expression for the mapping transition amplitude in (35) simplifies as follows

$$\langle m_\alpha | e^{-\frac{i}{\hbar}\epsilon \hat{h}_m(R_N)} \dots e^{-\frac{i}{\hbar}\epsilon \hat{h}_m(R_1)} | m_\alpha \rangle = e^{-i\int_0^t \theta_\alpha(R_\tau)d\tau}$$

$$\times \int dq_0 dp_0 |r_{0,\alpha}|^2 G_0 \quad (48)$$

The Gaussian integrals can then be performed to give an irrelevant multiplicative constant. Further, from (44) it follows that, in the adiabatic limit, the forces that determine the classical-like bath subsystem trajectories according to the δ-functions in (47) become

$$F_k^{\alpha,\alpha'} = -\frac{1}{2}\left\{\nabla_{\bar{R}_k}h_{\alpha,\alpha}(\bar{R}_k) + \nabla_{\bar{R}_k}h_{\alpha',\alpha'}(\bar{R}_k)\right\} \qquad (49)$$

and the final adiabatic correlation function expression is

$$\langle\hat{A}\hat{B}(t)\rangle = \sum_{\alpha,\alpha'}\int d\bar{R}_0\int\prod_{k=1}^{N}d\bar{R}_k\frac{d\bar{P}_k}{2\pi\hbar}(\hat{\rho}\hat{A})_W^{\alpha,\alpha'}(\bar{R}_0,\bar{P}_1) \qquad (50)$$

$$\times(\hat{B})_W^{\alpha',\alpha}(\bar{R}_N,\bar{P}_N)e^{-i\epsilon\sum_{k=1}^{N}(\theta_\alpha(\bar{R}_k)-\bar{\theta}_{\alpha'}(\bar{R}_k))}$$

$$\times\prod_{k=1}^{N-1}\delta\left(\frac{\bar{P}_{k+1}-\bar{P}_k}{\epsilon}-F_k^{\alpha,\alpha'}\right)\prod_{k=1}^{N}\delta\left(\frac{\bar{P}_k}{M}-\frac{\bar{R}_k-\bar{R}_{k-1}}{\epsilon}\right)$$

The above result has been obtained using a representation of the quantum subsystem which is independent of bath subsystem coordinates (a diabatic representation). Although, in the presence of state mixing, the detailed structure of the coupling mechanism between the two subsystems in the evolution equation does depend on the choice of the basis set for the quantum subsystem [40], in the adiabatic limit the form of the time correlation function derived above remains valid. All that is required to adapt (50) to a given specification of the quantum subsytem basis set is to use the diagonal elements of the electronic Hamiltonian in the chosen representation when evaluating the phase and in the expression for the force. So, for example, if an adiabatic basis set provides a better representation of the quantum subsystem states for the problem under investigation, the adiabatic time correlation function is given by

$$\langle\hat{A}\hat{B}(t)\rangle = \sum_{\alpha,\alpha'}\int d\bar{R}_0\int\prod_{k=1}^{N}d\bar{R}_k\frac{d\bar{P}_k}{2\pi\hbar}(\hat{\rho}\hat{A})_W^{\alpha,\alpha'}(\bar{R}_0,\bar{P}_1) \qquad (51)$$

$$\times(\hat{B})_W^{\alpha',\alpha}(\bar{R}_N,\bar{P}_N)e^{-i\epsilon\sum_{k=1}^{N}(E_\alpha(\bar{R}_k)-E_{\alpha'}(\bar{R}_k))}$$

$$\times\prod_{k=1}^{N-1}\delta\left(\frac{\bar{P}_{k+1}-\bar{P}_k}{\epsilon}-F_{ak}^{\alpha,\alpha'}\right)\prod_{k=1}^{N}\delta\left(\frac{\bar{P}_k}{M}-\frac{\bar{R}_k-\bar{R}_{k-1}}{\epsilon}\right)$$

with

$$F_{ak}^{\alpha,\alpha'} = -\frac{1}{2}\left\{\nabla_{\bar{R}_k}E_\alpha(\bar{R}_k) + \nabla_{\bar{R}_k}E_{\alpha'}(\bar{R}_k)\right\} \qquad (52)$$

and $E_\alpha(R)$ is an eigenvalue of the electronic Hamiltonian matrix at nuclear configuration R. An alternative derivation of this result for the adiabatic basis set is given in reference [58].

3 Applications

In this section we present some applications of the LAND-map approach for computing time correlation functions and time dependent quantum expectation values for realistic model condensed phase systems. These representative applications demonstrate how the methodology can be implemented in general and provide challenging tests of the approach. The first test application is the spin-boson model where exact results are known from numerical path integral calculations [59–62]. The second system we study is a fully atomistic model for excess electronic transport in metal – molten salt solutions. Here the potentials are sufficiently reliable that findings from our calculations can be compared with experimental results.

3.1 The Spin-Boson Model: A Non-Adiabatic Problem

Here we apply the LAND-map approach to compute of the time dependent average population difference, $\Delta(t) = \langle \hat{\sigma}_z(t) \rangle$, between the spin states of a spin-boson model. Here $\hat{\sigma}_z = [|1\rangle\langle 1| - |2\rangle\langle 2|]$. Within the limits of linear response theory, this model describes the dissipative dynamics of a two level system coupled to an environment [59,63–65]. The environment is represented by an infinite set of harmonic oscillators, linearly coupled to the quantum subsystem. The characteristics of the system-bath coupling are completely described by the spectral density $J(\omega)$. In the following, we shall restrict ourselves to the case of an Ohmic spectral density

$$J(\omega) = \xi \omega e^{-\omega/\omega_c} \tag{53}$$

with an exponential cutoff. In this expression, ω_c is the frequency at which the spectral density has a maximum. This model has been extensively used both in exact [59–62] and approximate [36,66–68] calculations. Depending on the temperature and on the value of the Kondo parameter ξ, (also known as the friction) which determines the strength of the coupling between the system and the bath, the Ohmic case exhibits a variety of behaviors ranging from non-equilibrium coherent oscillation of the population difference to overdamped relaxation.

In actual numerical implementations of the method, the infinite harmonic bath is replaced by a finite set of N oscillators. Makri and coworkers showed that as few as ten bath modes are sufficient to represent the harmonic bath reliably [60]. With an appropriate choice of units [68], the spin-boson Hamiltonian can then be written as

$$\hat{H} = -\Omega \hat{\sigma}_x + \sum_{j=1}^{N} \left[\frac{P_j^2}{2} + \frac{1}{2} \omega_j'^2 R_j^2 - c_j R_j \hat{\sigma}_z \right] \tag{54}$$

where $\hat{\sigma}_z$ and $\hat{\sigma}_x$ are the Pauli matrices with $\pm\hbar$ on the diagonal, and $\hbar$ in the off diagonal entries respectively, and $2\hbar\Omega$ is the gap between the two levels in the bare system. As usual we work in units where $\hbar = 1$, and all frequencies have been scaled by the cutoff frequency, so for example $\omega'_j = \omega_j / \omega_c$. Below we shall remove the primes from the notation, but scaled variables are used throughout. Thus, in reduced units, the strength of the system-bath coupling is

$$c_j = \omega_j \sqrt{\xi \omega_0} \tag{55}$$

To implement the linearized path integral formulation for time correlation functions the initial density operator must be Wigner transformed in the bath variables while it remains an operator in the quantum subsystem space. In the calculations presented below we assume that the system and bath do not interact initially. Consequently total probability density at $t = 0$ is of the form

$$\hat{\rho}_{sb} = \rho(\mathbf{R}_0, \mathbf{P}_1)|1\rangle\langle 1| \tag{56}$$

In these calculations we prepare the bath in thermal equilibrium. The Wigner transform of the bath probability density can then be computed analytically and is given by

$$\rho(\mathbf{R}_0, \mathbf{P}_1) = \prod_{j=1}^{N} \frac{\tanh\left(\beta\omega_j/2\right)}{\pi} e^{-\frac{\tanh(\beta\omega_j/2)}{\omega_j}\left(\frac{P_{j0}^2}{2} + \frac{\omega_j^2 R_{j0}^2}{2}\right)} \tag{57}$$

In all the calculations described here we have used "focused" initial conditions [48] rather that sampling the full mapping distribution see Sect. 2.6.

For the spin-boson problem the off-diagonal elements of the Hamiltonian defined in (54) are independent of bath configuration so the forces on the bath variables defined in (44) contain contributions only from the diagonal diabatic surfaces. Further, since the observations we make in the computation of $\Delta(t)$ involve only electronic state populations, the operator $B_{\beta'\beta}$ is diagonal. Thus our bath trajectories exhibit classical motion over the individual bare diabatic surfaces in this case.

In the first set of calculations we present, see Fig. 3, the parameters in the spin-boson Hamiltonian were chosen so as to investigate the performance of the linearized non-adiabatic correlation function approach in the weak coupling regime, over a broad range of temperatures. We fix the adiabatic gap, 2Ω, to be close to the peak of the spectral density so there are plenty of modes to couple to, albeit relatively weakly. The specific values of the parameters were taken from the literature [36, 64, 67] and are detailed in the figure caption. It is well known that, for fixed friction and adiabatic gap, varying the temperature in the spin-boson model results in transition from coherent population transfer at a frequency determined by the gap, at low temperature, to overdamped, incoherent evolution at high temperature. Such behavior is indeed observed in the figure, which shows both exact results (symbols) and the results of our LAND-map calculations (solid lines).

The comparison suggests that the linearized technique presented in this chapter works over a wide range of temperatures in the weak coupling regime.

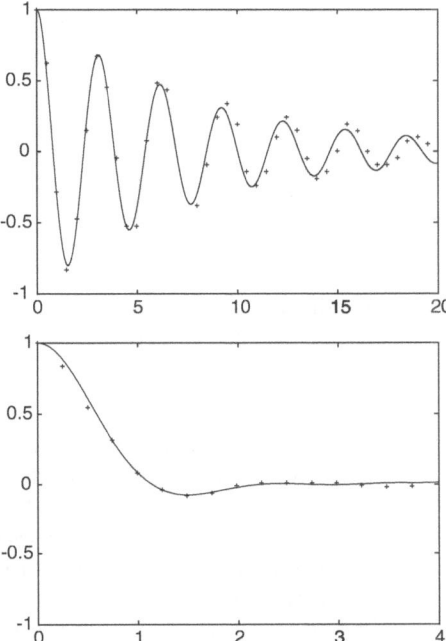

Fig. 3. Electronic population difference, $\Delta(t)$ as a function of Ωt, under various system-bath conditions. In all panels $\Omega/\omega_c = 0.4$ and $\xi = 0.09$. Upper panel has $\beta\hbar\omega_c = 12.5$. Lower panel has $\beta\hbar\omega_c = 0.25$. The solid curves are obtained with our LAND-map method and the points are exact results [36, 64, 67]

From the discussion in the theory section we might have expected that the linearization procedure should provide more reliable results at higher temperature where the initial density is dominated by diagonal terms in the bath variables. By this argument, at low temperatures where the off-diagonal terms in the density become appreciable, contributions from differing forward and backward paths should be more important and linearization should break down. Indeed, the lowest temperature results show the largest deviations and, as expected, these are only apparent at the longest times since the differences between forward and backward paths grow as the paths become longer. Nevertheless the worst discrepancy amounts to little more than a small phase shift in the coherent population transfer. Our approach actually becomes exact for this problem in the limit of vanishing friction so the fact that the Kondo parameter is small, but non-negligible, probably contributes to our linearized approximation working so well at low temperatures. The fact that the surfaces are largely harmonic in this problem also helps us. At higher temperatures the coherent transfer is rapidly damped and our linearized approximation recovers this decay remarkably well.

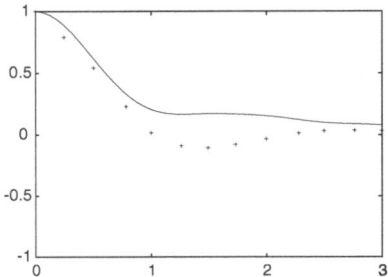

Fig. 4. Electronic population difference, $\Delta(t)$ with $\Omega/\omega_c = 1/3$, $\beta\hbar\omega_c = 3$, and $\xi = 0.5$. Solid curve is obtained with our LAND-map method and the points are exact results [61]

In Fig. 4 we show results for a spin-boson system at low temperature and large Kondo parameter where the linearization approximation is expected to do most poorly. Indeed, in this situation we see a significant discrepancy between the results of our linearized calculations and exact values. The linearized path integral approximation overemphasizes the effect of the friction, and underestimates the importance of the coherent dynamics. Thus the exact result oscillates around zero while the linearized approximate result is overdamped and shows slow incoherent decay.

The trends described above are in agreement with previous studies employing alternative forms of linearization. These approaches [22, 36, 67] all involve using the mapping Hamiltonian in some way and require surprisingly large ensembles of trajectories to converge (typically tens to hundreds of thousands). Furthermore all these calculations suffer from some pathologies (unstable trajectories resulting from motion on inverted potential energy surfaces, for example [22, 57]) which require *ad hoc* remedies and can affect the reliability of the results at long times. Calculations on the spin-boson problem have also been performed using the mixed quantum-classical Wigner-Liouville equation [68]. This method has the advantage of being exact for the Hamiltonian in (54) so that it is in principle possible to use it to study the system under any physical conditions. However, the algorithms developed so far for implementing this mixed quantum-classical propagation are plagued with numerical instabilities that make convergence problematic at long times and high friction.

LAND-map, in contrast both to the Wigner-Liouville methods and other linearization approaches, converges extremely rapidly with number of trajectories. The results in all these figures were obtained with as few as 500 trajectories, a situation which is quite encouraging in view of potential application to condensed phase problems.

3.2 Electronic Transport in a Metal-Molten Salt Solution:
An Adiabatic Calculation

The behavior of an excess electron in dilute metal-molten salt solutions has been the subject of many experimental and theoretical studies [69–72]. The details of the model we employ are exactly the same as the early calculations of Selloni and coworkers [71, 72]. Specifically, our simulations have been performed on a periodically repeated system of 32 K^+ cations, 31 Cl^- anions, and 1 electron. The mass density was set to $\rho = 1.52 \times 10^3$ kg/m^3. The temperature we use here is $T = 1800$ K.

The adiabatic approximate correlation function result of (51) has been used in calculations of the excess electron diffusion constant employing a basis of instantaneous adiabatic states. To verify the reliability of the hypothesis of slow mixing of electronic states on the time scale of the electronic transport, we monitor the time evolution of the expansion coefficients of the electronic wave function in the chosen basis along a typical nuclear trajectory generated with forces from the ground electronic state. Comparing the results in the top two panels of Fig. 5 reveals that the ground state population relaxes very slowly while significant electron displacement occurs. It is important to observe that the reliability of the adiabatic correlation function expression, for a given problem, depends on the choice of basis. In the bottom panel of Fig. 5 the lowest nine adiabatic state energies are shown as functions of time along the trajectory. The time evolution of the ground state energy is characterized by a sequence of wells separated by barriers (see, for example, the "avoided crossing" feature between 0.6 and 0.7 ps). As we see from the large amplitude electronic displacement (middle panel) that occurs when the dynamics takes the system over a ground state barrier, successive wells may be associated with the electronic distribution being localized in different regions of the fluid. Had we chosen to represent the quantum subsystem in terms of a basis of functions localized in these different regions, the behavior of the populations of these states would have been dramatically different from that of the adiabatic basis set. In fact, it would have involved rapid changes at rates comparable to the rate of electron displacement since, with this representation, different states would have been required to represent the quantum subsystem localization in different regions.

To obtain the diffusion constant, D, we consider two alternative equilibrium time correlation function approaches. First, D can be obtained from the long time limit of the slope of the time-dependent mean square displacement of the electron from its starting position. The quantum expression for this estimator is

$$D = \frac{1}{6} \lim_{t \to \infty} \frac{d}{dt} \langle (\hat{\mathbf{r}}(0) - \hat{\mathbf{r}}(t))^2 \rangle \tag{58}$$

This average is not in the form of a time dependent correlation function, but it can be cast as such when a periodic version of the position operator is introduced due to the periodic boundary conditions employed in the

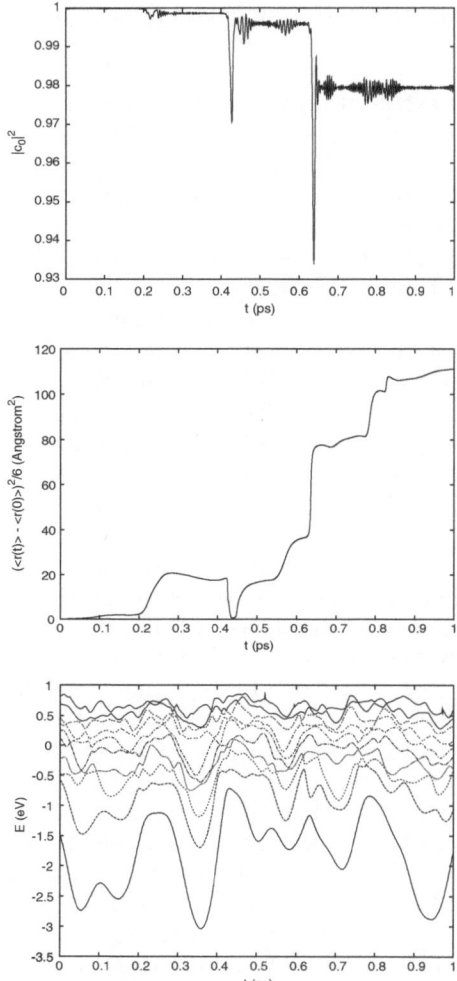

Fig. 5. Evolution of $|c_0|^2$ (*upper panel*) and the electron's mean square displacement (*middle panel*) along a 1ps trajectory at $T = 1800\,\mathrm{K}$. The lower panel shows the lowest 10 eigenvalues along the same trajectory

calculation [73]. The necessary steps are detailed in [58]. In the following we shall use the long-time limit expression for the periodic estimator introduced in that paper. This expression is based on a stationary phase analysis of the time correlation function in (51) which is valid if the measurement involves the asymptotic behavior at long times of operators with non-zero diagonal elements in the chosen basis set. Under these circumstances, the phase factor in the energy gap restricts the double sum over basis states to only diagonal contributions. Further we approximate the Wigner transform of the product of the initial bath subsystem density times the operator at the initial time which appears in (51) with its high temperature limit. Finally we assume that the temperature is sufficiently low on the scale of electronic excited state energies that we need only include contributions from the ground state. With these restrictions the sum over diagonal terms is truncated and the Wigner transform of the canonical density matrix for the bath subsystem is replaced by the classical Boltzmann probability distribution with partition function given by

$$Z_0 = \int dR(0) \int dP(0) e^{-\beta\{P^2(0)/2M + E_0(R(0))\}} \tag{59}$$

These manipulations yield the following approximation for the diffusion constant

$$D_{ad} = \frac{1}{6} \lim_{t \to \infty} \frac{d}{dt} \frac{1}{Z_0} \int dR(0) \int dP(0) e^{-\beta\{P^2(0)/2M + E_0(R(0))\}}$$
$$\times \left(\langle r \rangle_{\Phi_0(R(0))} - \langle r \rangle_{\Phi_0(R(t))} \right)^2 \tag{60}$$

where $\langle r \rangle_{\Phi_0(R(t))} = \langle \Phi_0(R(t)) | \hat{r} | \Phi_0(R(t)) \rangle$ is the average value of the periodic version of the position operator introduced by Resta [73]

$$\langle \hat{r} \rangle = \frac{L}{2\pi} \mathrm{Im}[\ln \langle \Phi_0 | e^{i\frac{2\pi}{L}\hat{r}} | \Phi_0 \rangle] \tag{61}$$

Here L is the unit cell size of the periodic system.

The average of the mean square displacement over ≈ 300 nuclear trajectories is shown in Fig. 6. The long time limiting value of the slope of the curve gives $D_{ad} = 1.9 \pm 0.5 \times 10^{-3}$ cm^2 s^{-1}. For comparison the experimental diffusion constant is $D_{expt} = 3 \times 10^{-3}$ cm^2 s^{-1} at $T = 1300$ K.

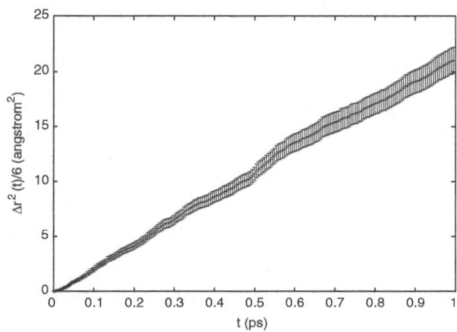

Fig. 6. Ground state mean square displacement estimator for excess electron diffusion at $T = 1800$ K

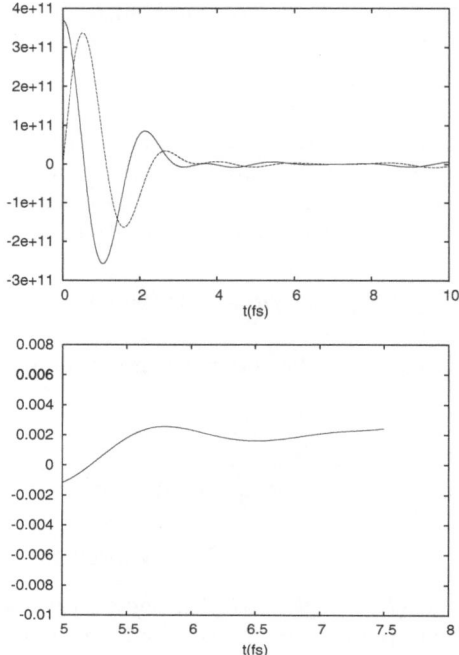

Fig. 7. Real (*solid*) and imaginary (*dashed*) parts of the velocity autocorrelation at $T = 1800\,\mathrm{K}$ (*upper panel*). Here the y-axis units are $\mathrm{m^2 s^{-2}}$. Cumulative time integral of velocity autocorrelation function (*lower panel*). Here the y-axis units are $\mathrm{cm^2 s^{-1}}$

The second approach we use to estimate D employs the following expression

$$D_v = \frac{1}{3} \int_0^\infty \mathrm{Re}[C_{vv}(t)]dt \tag{62}$$

where

$$C_{vv}(t) = \langle \hat{v} \cdot \hat{v}(t) \rangle = \mathrm{Tr}\{\hat{\rho}\hat{v}\hat{v}(t)\} \tag{63}$$

is the velocity autocorrelation function.

In Fig. 7 we present the real and imaginary parts of the velocity autocorrelation function calculated using (51). The diagonal velocity matrix elements in the instantaneous adiabatic electronic basis set are rigorously zero so the only finite contributions to our estimator come from off-diagonal terms. From (51) these contributions are computed using trajectories moving over the mean of the adiabatic potential surfaces associated with the two different states involved in the non-zero off-diagonal velocity matrix elements. In addition, (51) indicates that each of these off-diagonal contributions must be weighted by a phase factor which oscillates at a frequency determined by the time dependent energy gap between this pair of states. These phase weighted contributions must be added for each pair of different adiabatic states included in our representation since precise details of both the short and long time history of the velocity autocorrelation function are necessary if its integral is to be computed.

Similar to the situation for the position based estimator, we include only contributions from the ground state in the thermal average. Convergence is achieved by including off-diagonal elements associated with as few as 9 excited states.

In Fig. 7 we also display the integral of the velocity autocorrelation function versus the upper limit of the time integration in (62). The figure shows oscillations around a plateau value indicating that the decay of the velocity autocorrelation function is sufficiently rapid and that the interference effects among the different terms in the sum are captured reliably. Our estimate for the diffusion coefficient with this approach is $2 \pm 0.5 \times 10^{-3}$ cm^2 s^{-1}, in good agreement with the position based estimator discussed above. Despite the sizable error bars this agreement confirms that our approximation for the correlation function is accurate enough to preserve the Green-Kubo relation between the mean square displacement and velocity autocorrelation function expressions for D.

4 Conclusion

In this chapter we have presented a linearized path integral approach to non-adiabatic dynamics and demonstrated it usefulness in applications to realistic model condensed phase systems. There are two remaining aspects to the general implementation of the methods outlined here that need to be addressed: First, a means of sampling initial conditions from the Wigner transform of the initial density times an operator needs to be developed. In the example applications presented here we either use a high temperature approximation for the initial distribution (this is reliable for our electron transport in metal-molten salt solution calculations), or the problem is simple enough that the initial distribution can be obtained analytically (this is the case in the spin-boson application). At this point there are no methods for sampling general Wigner transformed distributions however some approximate techniques have been presented that are based on local Gaussian approximations to the distribution [18,19,21–24] and these can often be reliable. The second issue that needs to be addressed for the general implementation of LAND-map is the efficient accumulation of averages over the complex phase factor weighted distribution of trajectories generated by these methods. A possible strategy rewrites the slowly convergent average in terms of cumulant expansions. These methods have been applied in the context of the metal-molten salt electron transport calculation outlined in Sect. 3 with promising results [74].

In spite of these remaining numerical difficulties, the method has proved more efficient than most of its competitors. It also provides an original theoretical framework to analyze the roles of trajectories and weights associated with mixed quantum-classical calculations. Since the formal steps in the derivation do not introduce *ad hoc* ingredients in the algorithm the linearization approximation has some advantages over less rigorous methods such as the Surface Hopping or mean field approaches. Comparing these, and other methods may prove interesting, and assessing their relative merits may offer insight into new approaches for solving lingering algorithmic concerns.

Acknowledgements

We acknowledge support for this research from the National Science Foundation under grant number CHE-0316856 as well a the Petroleum Research Fund administered by the American Chemical Society grant number 39180-AC6.

References

1. J. C. Tully (1998) Mixed quantum classical dynamics: Mean field and surface hopping. In G. Ciccotti B. Berne, and D. Coker, editors, *Classical and quantum dynamics in condensed phase simulations*. World Scientific, Dordrecht, p. 489
2. N. Yu, C. J. Margulis, and D. F. Coker (2001) Influence of solvation environment on excited state avoided crossings and photo-dissociation dynamics. *J. Phys. Chem. B* **105**, p. 6728
3. C. J. Margulis and D. F. Coker (1999) Nonadiabatic molecular dynamics simulations of the photofragmentation and geminate recombination dynamics in size-selected $I_2^- \cdot (CO_2)_n$ cluster ions. *J. Chem. Phys.* **110**, p. 5677
4. V. S. Batista and D. F. Coker (1997) Nonadiabatic molecular dynamics simulations of the photofragmentation and geminate recombination dynamics of I_2^- in size selected Ar_n clusters. *J. Chem. Phys.* **106**, p. 7102
5. V. S. Batista and D. F. Coker (1997) Nonadiabatic molecular dynamics simulations of ultrafast pump-probe experiments on I_2 in solid rare gases. *J. Chem. Phys.* **106**, p. 6923
6. V. S. Batista and D. F. Coker (1996) Nonadiabatic molecular dynamics simulation of photodissociation and geminate recombination of I_2 in liquid xenon. *J. Chem. Phys.* **105**, p. 4033
7. P. J. Rossky (1998) Nonadiabatic quantum dynamics simulation using classical baths. In G. Ciccotti B. Berne, and D. Coker, editors, *Classical and quantum dynamics in condensed phase simulations*. World Scientific, Dordrecht, p. 515
8. P. V. Parandeker and J. C. Tully (2005) Mixed quantum-classical equilibrium. *J. Chem. Phys.* **122**, p. 094102
9. R. Kapral and G. Ciccotti (1999) Mixed quantum-classical dynamics. *J. Chem. Phys.* **110**, p. 8919
10. S. Nielsen, R. Kapral, and G. Ciccotti (2000) Mixed quantum-classical surface hopping dynamics. *J. Chem. Phys.* **112**, p. 6543
11. S. Nielsen, R. Kapral, and G. Ciccotti (2001) Statistical mechanics of quantum-classical systems. *J. Chem. Phys.* **115**, p. 5805

12. A. Sergi and R. Kapral (2004) Quantum-classical limit of quantum correlation functions. *J. Chem. Phys.* **121**, p. 7565

13. C. C. Martens and J.-Y. Fang (1997) Semiclassical-limit molecular dynamics on multiple electronic surfaces. *J. Chem. Phys.* **106**, p. 4918

14. A. Donoso and C. C. Martens (2000) Semiclassical multistate Liouville dynamics in the adiabatic representation. *J. Chem. Phys.* **112**, p. 3980

15. C. C. Martens and A. Donoso (1997) Simulation of coherent nonadiabatic dynamics using classical trajectories. *J. Phys. Chem. A* **102**, p. 4291

16. S. Bonella and D. F. Coker (2005) Land-map, a linearized approach to nonadiabatic dynamics using the mapping formalism. *J. Chem. Phys.* **122**, p. 194102

17. S. Bonella, D. Montemayor, and D. F. Coker (2005) Linearized path integral approach for calculating nonadiabatic time correlation functions. *Proc. Natl. Acad. Sci.* **102**, pp. 6715–6719

18. J. A. Poulsen, G. Nyman, and P. J. Rossky (2003) Practical evaluation of condensed phase quantum correlation functions: A Feynman-Kleinert variational linearized path integral method. *J. Chem. Phys.* **119**, p. 12179

19. J. A. Poulsen, G. Nyman, and P. J. Rossky (2004) Determination of the Van Hove spectrum of liquid He(4): An application of Feynman-Kleinert linearized path integral methodology. *J. Phys. Chem. A* **108**, p. 8743

20. J. A. Poulsen, G. Nyman, and P. J. Rossky (2005) Static and dynamic quantum effects in molecular liquids: A linearized path integral description of water. *Proc. Natl. Acad. Sci.* **102**, p. 6709

21. Q. Shi and E. Geva (2003) A relationship between semiclassical and centroid correlation functions. *J. Chem. Phys.* **118**, p. 8173

22. Q. Shi and E. Geva (2004) A semiclassical generalized quantum master equation for an arbitrary system-bath coupling. *J. Chem. Phys.* **120**, p. 10647

23. Q. Shi and E. Geva (2003) Semiclassical theory of vibrational energy relaxation in the condensed phase. *J. Phys. Chem. A* **107**, p. 9059

24. Q. Shi and E. Geva (2003) Vibrational energy relaxation in liquid oxygen from semiclassical molecular dynamics. *J. Phys. Chem. A* **107**, p. 9070

25. M. F. Herman and D. F. Coker (1999) Classical mechanics and the spreading of localized wavepackets in condensed phase molecular systems. *J. Chem. Phys.* **111**, p. 1801

26. R. Hernandez and G. Voth (1998) Quantum time correlation functions and classical coherence. *Chem. Phys. Lett.* **223**, p. 243

27. H. D. Meyer and W. H. Miller (1979) A Classical analog for electronic degrees of freedom in nonadiabatic collision processes. *J. Chem. Phys.* **70**, p. 3214

28. W. H. Miller and C. W. McCurdy (1978) Classical trajectory model for electronically nonadiabatic collision phenomena. A classical analog for electronic degrees of freedom. *J. Chem. Phys.* **69**, p. 5163

29. C. W. McCurdy, H. D. Meyer, and W. H. Miller (1979) Classical model for electronic degrees of freedom in nonadiabatic collision processes: Pseudopotential analysis and calculations for $F(^2P_{1/2})+H^+,Xe \rightarrow F(^2P_{3/2})+H^+,Xe$. *J. Chem. Phys.* **70**, p. 3177

30. G. Stock and M. Thoss (1997) Semiclassical description of nonadiabatic quantum dynamics. *Phys. Rev. Lett.* **78**, p. 578

31. G. Stock and M. Thoss (1999) Mapping approach to the semiclassical description of nonadiabatic quantum dynamics. *Phys. Rev. A* **59**, p. 64

32. G. Ciccotti, C. Pierleoni, F. Capuani, and V. S. Filinov (1999) Wigner approach to the semiclassical dynamics of a quantum many-body system: the dynamic scattering function of ^{4}He. *Comp. Phys. Commun.* **121**, p. 452

33. X. Sun and W. H. Miller (1997) Mixed semiclassical-classical approaches to the dynamics of complex molecular systems. *J. Chem. Phys.* **106**, p. 916

34. H. Wang, X. Sun, and W. H. Miller (1998) Semicalssical approximations for the calculation of thermal rate constants for chemical reactions in complex molecular systems. *J. Chem. Phys.* **108**, p. 9726

35. X. Sun, H. B. Wang, and W. H. Miller (1998) On the semiclassical description of quantum coherence in thermal rate constants. *J. Chem. Phys.* **109**, p. 4190

36. X. Sun, H. B. Wang, and W. H. Miller (1998) Semiclassical theory of electronically nonadaibatic molecular dynamics: Results of a linearized approximation to the initial value representation. *J. Chem. Phys.* **109**, p. 7064

37. W. H. Miller (2001) The semiclassical initial value representation: A potentially practical way for adding quantum effects to classical molecular dynamics simulations. *J. Phys. Chem. A* **105**, p. 2942

38. H. Wang, M. Thoss, and W. H. Miller (2001) Generalized forward-backward initial value representation for the calculation of correlation functions in complex systems. *J. Chem. Phys.* **114**, p. 9220

39. S. Zhang and E. Pollak (2003) Quantum dynamics for dissipative systems : A numerical study of the Wigner-Fokker-Planck equation. *J. Chem. Phys.* **118**, p. 4357

40. D. F. Coker and S. Bonella (2006) Linearized non-adiabatic dynamics in the adiabatic representation. In David Micha and Irene Burghardt, editors, *Quantum dynamics of complex molecular systems*. Springer-Verlag, Berlin, p. 307

41. H. Kleinert (2004) *Path Integrals in Quantum Mechanics, Statics, Polymer Physics and Financial Markets*. World Scientific, Singapore

42. M. F. Herman and E. Kluk (1984) A semiclassical justification for the use of non-spreading wavepackets in dynamics calculations. *Chem. Phys.* **91**, p. 27

43. W. H. Miller (2002) An alternate derivation of the herman-kluk (coherent state) semiclassical initial value representation of the time evolution operator. *Mol. Phys.* **100**, p. 397

44. X. Sun and W. H. Miller (1997) Semiclassical initial value representation for electronically nonadaibatic molecular dynamics. *J. Chem. Phys.* **106**, p. 6346

45. G. Stock and M. Thoss (2005) Classical description of nonadiabatic quantum dynamics. *Adv. Chem. Phys.* **131**, p. 243

46. P. Pechukas (1969) Time-dependent semiclassical scattering theory. II. Atomic collisions. *Phys. Rev.* **181**, p. 174

47. J. C. Tully (1990) Molecular dynamics with electronic transitions. *J. Chem. Phys.* **93**, p. 1061

48. S. Bonella and D. F. Coker (2003) Semi-classical implementation of the mapping Hamiltonian approach for non-adiabatic dynamics: Focused initial distribution sampling. *J. Chem. Phys.* **118**, p. 4370

49. O. Prezhdo B. Schwartz, E. Bittner, and P. Rossky (1996) Quantum decoherence and the isotope effect in condensed phase nonadiabatic molecular dynamics simulations. *J. Chem. Phys.* **104**, p. 5942

50. E. Bittner and P. Rossky (1997) Decoherent histories and non adiabatic quantum molecular dinamics simulations. *J. Chem. Phys.* **107**, p. 8611

51. M. Ben-Nun and T. J. Martinez (1998) Nonadiabatic molecular dynamics: Validation of the multiple spawning method for a multidimensional problem. *J. Chem. Phys.* **108**, p. 7244

52. M. Ben-Nun and T. J. Martinez (2000) A multiple spawning approach to tunneling dynamics. *J. Chem. Phys.* **112**, p. 6113

53. M. Ben-Nun, J. Quenneville, and T. J. Martinez (2000) Ab initio multiple spawning: Photochemistry from first principles quantum molecular dynamics. *J. Phys. Chem. A* **104**, p. 5161

54. M. D. Hack, A. M. Wensmann, D. G. Truhlar, M. Ben-Nun, and T. J. Martinez (2001) Comparison of full multiple spawning, trajectory surface hopping and converged quantum mechanics for electronically nonadiabatic dynamics. *J. Chem. Phys.* **115**, p. 1172

55. M. Ben-Nun and T. J. Martinez (2002) Ab initio quantum molecular dynamics. *Adv. Chem. Phys.* **121**, p. 439

56. S. Bonella and D. F. Coker (2001) A semi-classical limit for the mapping Hamiltonian approach to electronically non-adiabatic dynamics. *J. Chem. Phys.* **114**, p. 7778

57. S. Bonella and D. F. Coker (2001) Semi-classical implementation of mapping Hamiltonian methods for general non-adiabatic problems. *Chem. Phys.* **268**, p. 323

58. S. Causo, G. Ciccotti, D. Montemayor, S. Bonella, and D.F. Coker (2005) An adiabatic linearized path integral approach for quantum time correlation functions: Electronic transport in metal-molten salt solutions. *J. Phys. Chem. B* **109**, p. 6855

59. C. H. Mak and D. Chandler (1991) Coherent-incoherent transition and relaxation in condensed phase tunneling systems. *Phys. Rev. A* **44**, p. 2352

60. M. Topaler and N. Makri (1994) Quantum rates for a double well coupled to a dissipative bath: Accurate path integral results and comparison with approximate theories. *J. Chem. Phys.* **101**, p. 7500

61. K. Thompson and N. Makri (1999) Influence functionals with semiclassical propagators in combined forward-backward time. *J. Chem. Phys.* **110**, p. 1343

62. K. Thompson and N. Makri (1998) Semiclassical influence functionals for quantum systems in anharmonic environments. *Chem. Phys. Lett.* **291**, p. 101

63. A. J. Leggett, S. Chakravarty, A. T. Dorsey, M. P. A. Fisher, A. Garg, and W. Zwerger (1987) Dynamics of the dissipative two state system. *Rev. Mod. Phys.* **59**, p. 1

64. R. Egger and C. H. Mak (1994) Low temperature dynamical simulation of spin-boson systems. *Phys. Rev. B* **50**, p. 15210

65. D. G. Evans, A. Nitzan, and M. A. Ratner (1998) Photoinduced electron transfer in mixed valence compounds: Beyond the golden rule regime. *J. Chem. Phys.* **108**, p. 6387

66. G. Stock (1995) A semiclassical self-consistent-field approach to dissipative dynamics: The spin-boson problem. *J. Chem. Phys.* **103**, p. 1561

67. A. Golosov and D. R. Reichman (2001) Classical mapping approaches for nonadiabatic dynamics: Short time analysis. *J. Chem. Phys.* **114**, p. 1065

68. D. Mac Kernan, G. Ciccotti, and R. Kapral (2002) Surface hopping dynamics of a spin boson system. *J. Chem. Phys.* **116**, p. 2346

69. M. A. Bredig (1964) *Molten Salt Chemistry.* Interscience, New York, NY

70. W. Freyland, K. Garbade, and E. Pfeiffer (1983) Optical study of electron localization approaching a polarization catastrophe in liquid K_x-KCl_{1-x}. *Phys. Rev. Lett.* **51**, p. 1304
71. E. S. Fois, A. Selloni, and M. Parrinello (1989) Approach to metallic behavior in metal-molten-salt solutions. *Phys. Rev.* **39**, p. 4812
72. A. Selloni, P. Carnevali, R. Car, and M. Parrinello (1987) Localization, hopping, and diffusion of electrons in molten salts. *Phys. Rev. Lett.* **59**, p. 823
73. R. Resta (1988) Quantum-mechanical position operator in extended systems. *Phys. Rev. Lett.* **80**, p. 1800
74. S. Causo, G. Ciccotti, S. Bonella, and R. Vuillemier. (2006) An adiabatic linearized path integral approach for quantum time correlation functions II: A cumulant expansion method for improving convergence. To appear in *J. Phys. Chem. B*.

Ensemble Optimization Techniques
for Classical and Quantum Systems

S. Trebst[1] and M. Troyer[2]

[1] Microsoft Research and Kavli Institute for Theoretical Physics, University of
California, Santa Barbara, CA 93106, USA
trebst@comp-phys.org
[2] Theoretische Physik, ETH Zürich, 8093 Zürich, Switzerland
troyer@comp-phys.org

Matthias Troyer

S. Trebst and M. Troyer: *Ensemble Optimization Techniques for Classical and Quantum Systems*, Lect. Notes Phys. **703**, 591–640 (2006)
DOI 10.1007/3-540-35273-2_17   © Springer-Verlag Berlin Heidelberg 2006

We present a review of extended ensemble methods and ensemble optimization techniques. Extended ensemble methods, such as multicanonical sampling, broad histograms, or parallel tempering aim to accelerate the simulation of systems with large energy barriers, as they occur in the vicinity of first order phase transitions or in complex systems with rough energy landscapes, such as spin glasses or proteins. We present a recently developed feedback algorithm to iteratively achieve an *optimal* ensemble, with the fastest equilibration and shortest autocorrelation times. In the second part we review time-discretization free world line representations for quantum systems, and show how any algorithm developed for classical systems, such as local updates, cluster updates or the extended and optimized ensemble methods can also be applied to quantum systems. An overview over the methods is followed by a selection of typical applications.

1 Introduction

In this chapter we will review recent developments in the simulation of lattice (and continuum) models by classical and quantum Monte Carlo simulations. Unbiased numerical methods are required to obtain reliable results for classical and quantum lattice model when interactions or fluctuations are strong, especially in the vicinity of phase transitions, in frustrated models and in systems where quantum effects are important. For classical systems, molecular dynamics or the Monte Carlo method are the methods of choice since they can treat large systems.

Both Monte Carlo and molecular dynamics simulations slow down in the vicinity of phase transitions or in disordered systems with rough energy landscapes, since the time scales to tunnel through energy barriers can become prohibitively long. Here Monte Carlo simulations have an advantage over molecular dynamics, since in Monte Carlo simulations both the dynamics and the ensemble can be changed to achieve faster tunneling through the energy barriers. Using modern sampling algorithms, such as cluster updates, extended ensemble methods or parallel tempering strategies most classical magnets can be efficiently simulated, with the computational effort scaling with a low power of the system size, and usually linear in system size. The notable exceptions are spin glasses, known to be nondeterministic-polynomially (NP) hard in more than two space dimensions [1] and where most likely no polynomial-time algorithm can exist [2].

In the first part of this chapter we will give a short overview of Monte Carlo simulations for classical lattice models in Sect. 2.1 and will then review the extended and optimized ensemble methods in Sect. 3. We will focus the discussion on a recently developed algorithm to iteratively achieve an *optimal* ensemble, with the fastest equilibration and shortest autocorrelation times.

For quantum magnets, quantum Monte Carlo (QMC) methods are also the method of choice whenever they are applicable. Over the last decade

efficient algorithms for classical Monte Carlo simulations have been generalized to quantum systems and systems with millions of quantum spins have
been simulated [3]. In Sect. 4 we will present modern time-discretization free
world line representations for quantum lattice models. They faithfully map
the quantum system to an equivalent classical system with one more dimension. Efficient Monte Carlo algorithms developed for classical systems can also
be applied to quantum systems, using these world line representations, as we
will show in Sect. 5.

Unfortunately, in contrast to classical magnets, QMC methods are efficient
only for non-frustrated magnets and for bosonic systems. Fermionic degrees
of freedom or frustration in quantum systems usually lead to the "negative
sign problem", when the weights of some configurations become negative [4].
These negative weights cannot be directly interpreted as probabilities in the
Monte Carlo process and lead to cancellation effects in the sampling. As a
consequence the statistical errors grow exponentially with inverse temperature
and system size and the QMC methods are restricted to small systems and
not too low temperatures.

2 The Monte Carlo Method for Classical Lattice Models

2.1 The Metropolis Algorithm

We start with a short review of the Monte Carlo method for calculating integrals of the form

$$\langle O \rangle = \frac{\int_\Omega \mathrm{d}x W(x) O(x)}{\int_\Omega \mathrm{d}x W(x)} \ , \tag{1}$$

where Ω is a discrete or continuous configuration space and $W(x)$ a not necessarily normalized weight function. We want to sample this integral in a Monte
Carlo process by creating a sequence $\{x_i\}$ of N configurations, where each
configuration is drawn according to the normalized probability distribution
function

$$P(x) = \frac{W(x)}{\int_\Omega \mathrm{d}x W(x)} \ . \tag{2}$$

Under the assumption of uncorrelated samples x_i we can then estimate the
expectation value (1) by the sample mean

$$\langle O \rangle \approx \frac{1}{N} \sum_{i=1}^{N} O(x_i) \ , \tag{3}$$

within a statistical error

$$\Delta = \sqrt{\frac{\mathrm{Var}O}{N}} = \sqrt{\frac{\langle O^2 \rangle - \langle O \rangle^2}{N}} \ . \tag{4}$$

Since we will, in general, not have a direct algorithm to create samples x_i according to the distribution $P(x_i)$ we will use a Markov process in which starting from an initial configuration x_0 a Markov chain of configuration is generated:

$$x_0 \to x_1 \to x_2 \to \ldots \to x_n \to x_{n+1} \to \ldots . \tag{5}$$

A transition matrix T_{xy} gives the transition probabilities of going from configuration x to configuration y in one step of the Markov process. As the sum of probabilities of going from configuration x to any other configuration is one, the columns of the matrix T are normalized:

$$\sum_y T_{xy} = 1 . \tag{6}$$

A consequence is that the Markov process conserves the total probability. Another consequence is that the largest eigenvalue of the transition matrix T is 1 and the corresponding eigenvector with only positive entries is the equilibrium distribution which is reached after a large number of Markov steps.

We want to determine the transition matrix T so that we asymptotically reach the desired probability $P(x)$ for a configuration i. A set of sufficient conditions is:

1. **Ergodicity:** It has to be possible to reach any configuration $\mathbf{x}$ from any other configuration $\mathbf{y}$ in a finite number of Markov steps. This means that for all $\mathbf{x}$ and $\mathbf{y}$ there exists a positive integer $n < \infty$ such that $(T^n)_{\mathbf{xy}} \neq 0$.

2. **Detailed balance:** The probability distribution $p_x^{(n)}$ changes at each step of the Markov process:

$$\sum_x p_x^{(n)} T_{xy} = p_{\mathbf{y}}^{(n+1)} , \tag{7}$$

but converges to the equilibrium distribution p_x. This equilibrium distribution p_x is an eigenvector with left eigenvalue 1 and the equilibrium condition

$$\sum_x p_x T_{xy} = p_y \tag{8}$$

must be fulfilled. It is easy to see that the detailed balance condition

$$\frac{W_{\mathbf{xy}}}{W_{\mathbf{yx}}} = \frac{p_{\mathbf{y}}}{p_x} \tag{9}$$

is sufficient.

The simplest Monte Carlo algorithm is the Metropolis algorithm [5] which can be outlined as follows:

- Starting with a configuration $x = x_i$ propose a new configuration y with an a-priori probability A_{xy}.

- Calculate the acceptance ratio

$$P_{xy} = \min\left(1, \frac{A_{yx}W(y)}{A_{xy}W(x)}\right) \tag{10}$$

and accept the proposed configuration with probability P_{xy}. To do so we draw a uniform random number u in the interval $[0, 1[$ and choose $x_{i+1} = y$ if $u < P_{xy}$ and $x_{i+1} = x$ otherwise.
- Measure the quantity O for the new configuration x_{i+1} no matter whether the proposed configuration was accepted or not.

Since the samples created in this Markov chain are correlated (we only do small changes at each step), equation (4) for the statistical error needs to be modified to

$$\Delta = \sqrt{\frac{\mathrm{Var}\, O}{N}(1 + 2\tau_O)} \tag{11}$$

where τ_O is the integrated autocorrelation time of $O(x_i)$ in the Markov chain.

2.2 The Local Update Metropolis Algorithm for the Ising Model

We will next apply this Metropolis algorithm to simulations of the Ising ferromagnet with Hamilton function

$$H = -J \sum_{\langle i,j \rangle} \sigma_i \sigma_j - g\mu_B h \sum_{i=1}^{N} \sigma_i \, , \tag{12}$$

where J is the exchange constant, h the magnetic field, g the Landé g-factor, μ_B the Bohr magneton, and N the total number of spins. The sum runs over all pairs of nearest neighbors i and j and $\sigma_i = \pm 1$ is the value of the Ising spin at site i.

To calculate the value of an observable, such as the mean magnetization at an inverse temperature $\beta = 1/k_B T$ with T being the temperature and k_B the Boltzmann constant, we need to evaluate

$$\langle m \rangle = \sum_c m(c) \exp(-\beta E(c))/Z, \tag{13}$$

where

$$m(c) = \frac{1}{N} \sum_{i=1}^{N} \sigma_i \tag{14}$$

is the magnetization of the configuration c, $E(c)$ the energy of the configuration,

$$P(c) = \exp(-\beta E(c)) \tag{15}$$

the Boltzmann weight and

$$Z = \sum_c P(c) \tag{16}$$

the partition function, normalizing the weights.

As discussed above, Monte Carlo sampling can be performed on this sum using the Metropolis method. The simplest types of updates are local spin flips:

1. Pick a random site i. The a-priori probabilites A_{xy} are all just $1/N_{sites}$ for a system with N_{sites} spins.
2. Calculate the energy cost ΔE for flipping the spin at site i: $\sigma_i \to -\sigma_i$
3. Flip the spin with the Metropolis probability $\min[1, \exp(-\beta \Delta E)]$. If rejected, keep the original spin value.
4. Perform a measurement independent of whether the spin flip was accepted or rejected.

The same local update algorithm can be applied to systems with longer-range interactions and with coupling constants that vary from bond to bond. For more complex classical models, such as Heisenberg models, local updates will no longer consist of simple spin flips, but of arbitrary rotations of the local spin vectors.

2.3 Critical Slowing Down and Cluster Update Algorithms

Local update algorithms are easy to implement and work well away from phase transitions. Problems arise in the vicinity of continuous (second order) phase transitions, where these algorithms suffer from "critical slowing down" [6] and at first order phase transitions where there is a tunneling problem through free energy barriers.

At second order phase transitions the correlation length ξ diverges upon approaching the phase transition, and this causes the autocorrelation times τ_O to also diverge as

$$\tau_O \propto \min(L, \xi)^z \tag{17}$$

with a dynamical critical exponent of $z \approx 2$. L is the linear extent of the system. The origin of critical slowing down is the fact that close to the critical temperature large ordered domains of linear extent ξ are formed and the single spin updates are not effective in changing these large domains. The value $z \approx 2$ can be understood considering that the time for a domain wall to move a distance ξ by a random walk scales as ξ^2. The solution to critical slowing down are cluster updates, flipping carefully selected clusters of spins instead of single spins. Cluster update algorithms were originally invented by Swendsen and Wang for the Ising model [6] and soon generalized to $O(N)$ models, such as the Heisenberg model [7]. These cluster update algorithms are discussed in text books on classical Monte Carlo simulations and in computational physics text books. While most cluster algorithms require spin-inversion invariance and thus do not allow for external magnetic fields, extensions to spin models in magnetic fields have been proposed [8,9]. An open source implementation of local and cluster updates for Ising, Potts, XY and Heisenberg models is available through the ALPS (Applications and Libraries for Physics Simulations) project [10] at the web page `http://alps.comp-phys.org/`.

The tunneling problem at first order phase transitions and for disordered systems, where tunneling times often diverge exponentially can be overcome using extended ensemble methods, which are the main topic of the next chapter.

3 Extended Ensemble Methods

3.1 First Order Phase Transitions and the Multicanonical Ensemble

While cluster updates can solve critical slowing down at second order phase transitions they are usually inefficient at first order phase transitions and in frustrated systems. Let us consider a first order phase transition, such as in a two-dimensional q-state Potts model with Hamilton function

$$H = -J \sum_{\langle i,j \rangle} \delta_{\sigma_i \sigma_j} \, , \tag{18}$$

where the spins σ_i can now take the integer values $1, \ldots, q$. For $q > 4$ this model exhibits a first order phase transition, accompanied by exponential slowing down of conventional local update algorithms. The exponential slowdown is caused by the free energy barrier between the two coexisting metastable states at the first order phase transition.

This barrier can be quantified by considering the energy histogram

$$H_{\text{canonical}}(E) \propto g(E) P_{\text{Boltzmann}}(E) = g(E) \exp(-\beta E) \, , \tag{19}$$

which is the probability of encountering a configuration with energy E during the Monte Carlo simulation. Here

$$g(E) = \sum_{c} \delta_{E, E(c)} \tag{20}$$

is the density of states. Away from first order phase transitions, $H_{\text{canonical}}(E)$ has approximately Gaussian shape, centered around the mean energy. At first order phase transitions, where the energy jumps discontinuously the histogram $H_{\text{canonical}}(E)$ develops a double-peak structure. The minimum of $H_{\text{canonical}}(E)$ between these two peaks, which the simulation has to cross in order to go from one phase to the other, becomes exponentially small upon increasing the system size. This leads to exponentially large autocorrelation times.

This tunneling problem at first-order phase transitions can be relieved by extended ensemble techniques which aim at broadening the sampled energy space. Instead of weighting a configuration c with energy $E = E(c)$ using the Boltzmann weight $P_{\text{Boltzmann}}(E) = \exp(-\beta E)$ more general weights $P_{\text{extended}}(E)$ are introduced which define the extended ensemble. The configuration space is explored by generating a Markov chain of configurations

$$c_1 \rightarrow c_2 \rightarrow \ldots \rightarrow c_i \rightarrow c_{i+1} \rightarrow \ldots \, , \tag{21}$$

where a move from configuration c_1 to c_2 is accepted with probability

$$P_{\text{acc}}(c_1 \rightarrow c_2) = \min\left(1, \frac{P(c_2)}{P(c_1)}\right) = \min\left(1, \frac{W_{\text{extended}}(E_2)}{W_{\text{extended}}(E_1)}\right) . \tag{22}$$

In general, the extended weights are defined in a single coordinate, such as the energy, thereby projecting the random walk in configuration space to a random walk in energy space

$$E_1 = E(c_1) \rightarrow E_2 \rightarrow \ldots \rightarrow E_i \rightarrow E_{i+1} \rightarrow \ldots \, . \tag{23}$$

For this random walk in energy space a histogram can be recorded which has the characteristic form

$$H_{\text{extended}}(E) \propto g(E) W_{\text{extended}}(E) \, , \tag{24}$$

where the density of states $g(E)$ is fixed for the simulated system.

One choice of generalized weights is the multicanonical ensemble [11, 12] where the weight of a configuration c is defined as $W_{\text{multicanonical}}(c) \propto 1/g(E(c))$. The multicanonical ensemble then leads to a flat histogram in energy space

$$H_{\text{multicanonical}}(E) \propto g(E) W_{\text{multicanonical}}(E) = g(E)\frac{1}{g(E)} = \text{const.}, \tag{25}$$

removing the exponentially small minimum in the canonical distribution. After performing a simulation, measurements in the multicanonical ensemble are reweighted by a factor $W_{\text{Boltzmann}}(E)/W_{\text{multicanonical}}(E)$ to obtain averages in the canonical ensemble.

3.2 The Wang-Landau Algorithm

Since the density of states and thus the multicanonical weights are not known initially, a scalable algorithm to estimate these quantities is needed. The Wang-Landau algorithm [13, 14] is a simple but efficient iterative method to obtain good approximations of the density of states $g(E)$ and the multicanonical weights $W_{\text{multicanonical}}(E) \propto 1/g(E)$.

The algorithm starts with a (very bad) estimate of the density of states $g(E) = 1$ for all energies which is iteratively improved by a modification factor f in the following loop:

- Start with $g(E) = 1$ and a modification factor $f \approx \exp(1)$.
- Repeat
  - Reset a histogram of energies $H(E) = 0$.
  - Perform simulations until the histogram of energies $H(E)$ is "flat":
    - Pick a random site and propose a local update, e.g. by flipping the spin at the site, which changes the current configuration c to a new configuration c', and the energy from $E = E(c)$ to $E' = E(c')$.
    - Approximating multicanoncal weights with the current estimate of the density of states the update is accepted with probability $\min[1, g(E)/g(E')]$.
    - Increase the histogram at the current value of E: $H(E) \leftarrow H(E)+1$
    - Increase the estimate $g(E)$ at the current value of E: $g(E) \leftarrow fg(E)$.
  - Once $H(E)$ is "flat" and has "sufficient statistics", reduce $f \leftarrow \sqrt{f}$.
- Stop once f is sufficiently small, e.g. $f \approx \exp(10^{-6})$.

Only a few lines of code need to be changed in the local update algorithm for the Ising model, but a few remarks are helpful:

1. The initial value for f needs to be carefully chosen, $f = \exp(1)$ is only a rough guide. A good choice is picking the initial f such that $f^{N_{\text{sweeps}}}$ is approximately the total number of states $\sum_E g(E)$ (e.g. 2^N for an Ising model with N sites).
2. Checking for flatness of the histogram (e.g. the minimum is at least 80% of the mean) should be done only after a reasonable number of sweeps N_{sweeps}. One sweep is defined as one attempted update per site.
3. The flatness criterion is quite arbitrary. In order to ensure convergence of the estimated $g(E)$ it should be extended to enforce sufficient statistics, e.g. by requiring that each histogram entry is at least of the order of $1/\sqrt{\ln f}$ as pointed out in Refs. [15, 16].
4. The density of states $g(E)$ can become very large and easily exceed 10^{10000}. In order to obtain such large numbers the *multiplicative increase* $g(E) \leftarrow fg(E)$ is essential. A naive additive guess $g(E) \leftarrow g(E) + f$ would never be able to reach the large numbers needed.
5. Since $g(E)$ is so large, we only store its *logarithm*. The update step is thus $\ln g(E) \leftarrow \ln g(E) + \ln f$.

At the end, the density of states $g(E)$ needs to be normalized. Either a known ground state degeneracy (e.g. $g(E_{\text{GS}}) = 2$ in the Ising ferromagnet) or a known total number of states (e.g. $\sum_E g(E) = 2^N$ in the Ising model with N spins) or a combination of the two (e.g. $g(E_{\text{GS}}) \cdot \sum_E g(E) = 2^{N+1}$ for the Ising ferromagnet) can be used to normalize $g(E)$.

Besides overcoming the exponentially suppressed tunneling problem at first order phase transitions, the Wang-Landau algorithm calculates the generalized density of states $g(E)$ in an iterative procedure. The knowledge of the density of states $g(E)$ then allows the direct calculation of the free energy from the partition function (16). The internal energy, entropy, specific heat and other thermal properties are easily obtained as well, by differentiating the free energy. By additionally measuring the averages $A(E)$ of other observables A as a function of the energy E, thermal expectation values can be obtained at arbitrary inverse temperatures β by performing just a single simulation:

$$\langle A(\beta)\rangle = \frac{\sum_E A(E)g(E)e^{-\beta E}}{\sum_E g(E)e^{-\beta E}} \; . \tag{26}$$

3.3 Markov Chains and Random Walks in Energy Space

The multicanonical ensemble and Wang-Landau algorithm both project a random walk in high-dimensional configuration space onto a one-dimensional random walk in energy space where all energy levels are sampled equally often. It is important to note that the random walk in configuration space, (21), is a biased Markovian random walk, while the projected random walk in energy space, (23), is non-Markovian, as memory is stored in the configuration. This becomes evident as the system approaches a phase transition in the random walk: While the energy no longer reflects from which side the phase transition is approached, the current configuration may still reflect the actual phase the system has visited most recently. In the case of the two-dimensional ferromagnetic Ising model, the order parameter for a given configuration at the critical energy $E_c \sim -1.41N$ will reveal whether the system is approaching the transition from the magnetically ordered (lower energies) or disordered side (higher energies).

This loss of information in the projection of the random walk in configuration space has important consequences for the random walk in energy space. Most strikingly, the local diffusivity of a random walker in energy space, which for a diffusion time t_D can be defined as

$$D(E, t_D) = \langle (E(t) - E(t + t_D))^2\rangle / t_D \tag{27}$$

is *not* independent of the location in energy space. This is illustrated in Fig. 1 for the Ising ferromagnet. Below the phase transition around $E \sim -1.41\,N$ a clear minimum evolves in the local diffusivity. In this region large ordered domains are formed and by moving the domain boundaries through local spin flips only small energy changes are induced resulting in a suppressed local diffusivity in energy space.

Because of the strong energy dependence of the local diffusivity the simulation of a multicanonical ensemble sampling all energy levels equally often turns out to be suboptimal [17]. The performance of flat-histogram algorithms can be quantified for classical spin models such as the ferromagnet where the number of energy levels is given by $[-2N, +2N]$ and thereby scales with the number of spins N in the system. When measuring the typical round-trip time between the two extremal energies for multicanonical simulations, these round-trip times τ are found to scale like

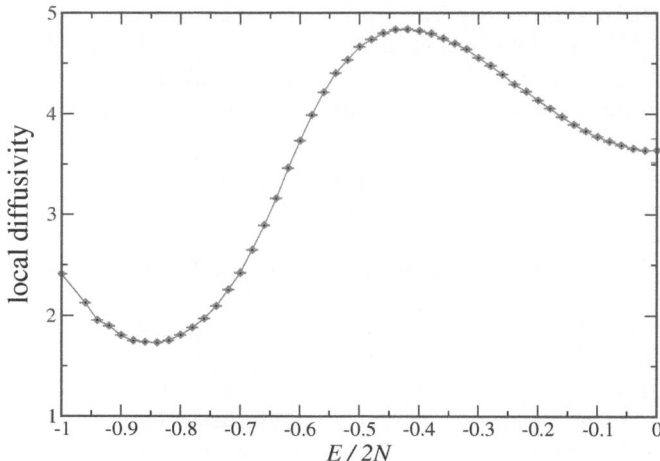

Fig. 1. Local diffusivity $D(E, t_D) = \langle (E(t) - E(t + t_D))^2 \rangle / t_D$ of a random walk sampling a flat histogram in energy space for the two-dimensional ferromagnetic Ising model. The local diffusivity strongly depends on the energy with a strong suppression below the critical energy $E_c \approx -1.41\,N$

$$\tau \sim N^2 L^z \,, \tag{28}$$

showing a power-law deviation from the N^2-scaling behavior of a completely unbiased random walk. Here z is a critical exponent describing the slowdown of a multicanonical simulation in the proximity of a phase transition [17,18]. The value of z strongly depends on the simulated model and the dimensionality of the problem. In two dimensions the exponent increases from $z = 0.74$ for the ferromagnet as one introduces competing interactions leading to frustration and disorder. The exponent becomes $z = 1.73$ for the fully frustrated Ising model which is defined by a Hamilton

$$H = \sum_{\langle i,j \rangle} J_{ij} \sigma_i \sigma_j \,, \tag{29}$$

where the spins around any given plaquette of four spins are frustrated, e.g. by choosing the couplings along three bonds to be $J_{ij} = -1$ (ferromagnetic) and $J_{ij} = +1$ (antiferromagnetic) for the remaining bond. For the spin glass where the couplings J_{ij} are randomly chosen to be $+1$ or -1 exponential scaling ($z = \infty$) is found [17, 19]. Increasing the spatial dimension for the ferromagnet the exponent is found to decrease as $z \approx 1.81, 0.74$ and 0.44 for dimension $d = 1, 2$ and 3 and z vanishes for the mean-field model in the limit of infinite dimensions [18].

3.4 Optimized Ensembles

The observed polynomial slowdown for the multicanonical ensemble poses the question whether for a given model there is an optimal choice of sampling energies, $H_{\text{optimal}}(E)$ and corresponding weights $W_{\text{optimal}}(E)$, which eliminates the slowdown. To address this question an adaptive feedback algorithm has recently been introduced that iteratively improves the weights in an extended ensemble simulationt leading to further improvements in the efficiency of the algorithm by several orders of magnitude [20]. The scaling for the optimized ensemble is found to scale like $O([N \ln N]^2)$ thereby reproducing the behavior of an unbiased Markovian random walk up to a logarithmic correction.

At the heart of the algorithm is the idea to maximize a current j of walkers that move from the lowest energy level, E_-, to the highest energy level, E_+, or vice versa, in an extended ensemble simulation by varying the weights $W_{\text{extended}}(E)$. To measure the current a label is added to the walker that indicates which of the two extremal energies the walker has visited most recently. The two extrema act as "reflecting" and "absorbing" boundaries for the labeled walker: e.g., if the label is plus, a visit to E_+ does not change the label, so this is a "reflecting" boundary. However, a visit to E_- does change the label, so the plus walker is absorbed at that boundary. The behavior of the labeled walker is *not* affected by its label except when it visits one of the extrema and the label changes.

For the random walk in energy space, two histograms are recorded, $H_+(E)$ and $H_-(E)$, which for sufficiently long simulations converge to steady-state distributions which satisfy $H_+(E) + H_-(E) = H(E) = W(E)g(E)$. For each energy level the fraction of random walkers which have label "plus" is then given by $f(E) = H_+(E)/H(E)$. The above-discussed boundary conditions dictate $f(E_-) = 0$ and $f(E_+) = 1$.

The steady-state current to first order in the derivative is

$$j = D(E)H(E)\frac{df}{dE} , \qquad (30)$$

where $D(E)$ is the walker's diffusivity at energy E. There is no current if $f(E)$ is constant, since this is equilibrium. Therefore the current is to leading order proportional to df/dE. Rearranging the above equation and integrating on both sides, noting that j is a constant and f runs from 0 to 1, one obtains

$$\frac{1}{j} = \int_{E_-}^{E_+} \frac{dE}{D(E)H(E)} . \qquad (31)$$

To maximize the current and thus the round-trip rate, this integral must be minimized. However, there is a constraint: $H(E)$ is a probability distribution and must remain normalized which can be enforced with a Lagrange multiplier:

$$\int_{E_-}^{E_+} dE \left(\frac{1}{D(E)H(E)} + \lambda H(E) \right) . \qquad (32)$$

To minimize this integrand, the ensemble, that is the weights $W(E)$ and thus the histogram $H(E)$ are varied. At this point it is assumed that the dependence of $D(E)$ on the weights can be neglected.

The optimal histogram, $H_{\text{optimal}}(E)$, which minimizes the above integrand and thereby maximizes the current j is then found to be

$$H_{\text{optimal}}(E) \propto \frac{1}{\sqrt{D(E)}} \, . \tag{33}$$

Thus for the optimal ensemble, the probability distribution of sampled energy levels is simply inversely proportional to the square root of the local diffusivity.

The optimal histogram can be approximated in a feedback loop of the form

- Start with some trial weights $W(E)$, e.g. $W(E) = 1/g(E)$.
- Repeat
    - Reset the histograms $H(E) = H_+(E) = H_-(E) = 0$.
    - Simulate the system with the current weights for N sweeps:
        - Updates are accepted with probablity $\min[1, W(E')/W(E)]$.
        - Record the histograms $H_+(E)$ and $H_-(E)$.
    - From the recorded histogram an estimate of the local diffusivity is obtained as

$$D(E) \propto \frac{1}{H(E)\frac{df}{dE}} \, , \quad f(E) = \frac{H_+(E)}{H(E)} \, , \qquad H(E) = H_+(E) + H_-(E) \, .$$

    - Define new weights as

$$W_{\text{optimized}}(E) = W(E) \sqrt{\frac{1}{H(E)} \cdot \frac{df}{dE}} \, .$$

    - Increase the number of sweeps for the next iteration $N \leftarrow 2N$.
- Stop once the histogram $H(E)$ has converged.

Again the implementation of this feedback algorithm requires to change only a few lines of code in the original local update algorithm for the Ising model. Some additional remarks are useful:

1. In contrast to the Wang-Landau algorithm the weights $W(E)$ are modified only after a batch of N sweeps, thereby ensuring detailed balance between successive moves at all times.
2. The initial value of sweeps N should be chosen large enough that a couple of round trips are recorded, thereby ensuring that steady state data for $H_+(E)$ and $H_-(E)$ are measured.
3. The derivative df/dE can be determined by a linear regression, where the number of regression points is flexible. Initial batches with the limited statistics of only a few round trips may require a larger number of regression points than subsequent batches with smaller round-trip times and better statistics.

4. Similar to the multicanonical ensemble the weights $W(E)$ can become very large, and storing the logarithms may be advantageous. The reweighting then becomes $\ln W_{\text{optimized}}(E) = \ln W(E) + (\ln \frac{df}{dE} - \ln H(E))/2$.

At the end of the simulation, the density of states can be estimated from the recorded histogram as $g(E) = H_{\text{optimized}}(E)/W_{\text{optimized}}(E)$ and normalized as described above.

Figure 2 shows the optimized histogram for the two-dimensional ferromagnetic Ising model. The optimized histogram is no longer flat, but a peak evolves at the critical region around $E_c \approx -1.41\ N$ of the transition. The feedback of the local diffusivity reallocates resources towards the bottlenecks of the simulation which have been identified by a suppressed local diffusivity.

The scaling of round-trip times is shown in Fig. 3 for the two-dimensional fully frustrated Ising model. The power-law slowdown of round-trip times for the flat-histogram ensemble $O(N^2 L^{1.73})$ is reduced to a logarithmic correction $O([N \ln N]^2)$ for the optimized ensemble in comparison to a completely unbiased random walk with $O(N^2)$-scaling. This scaling improvement results in a speedup by a nearly two orders of magnitude already for a system with some 128×128 spins.

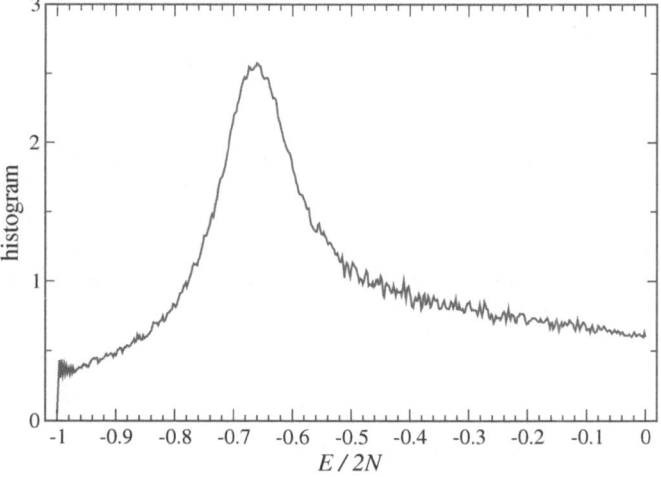

Fig. 2. Optimized histograms for the two-dimensional ferromagnetic Ising model. After the feedback of the local diffusivity a peak evolves near the critical energy of the transition $E_c \approx -1.41\ N$. The feedback thereby shifts additional resources towards the bottleneck of the simulation which were identified by a suppressed local diffusivity

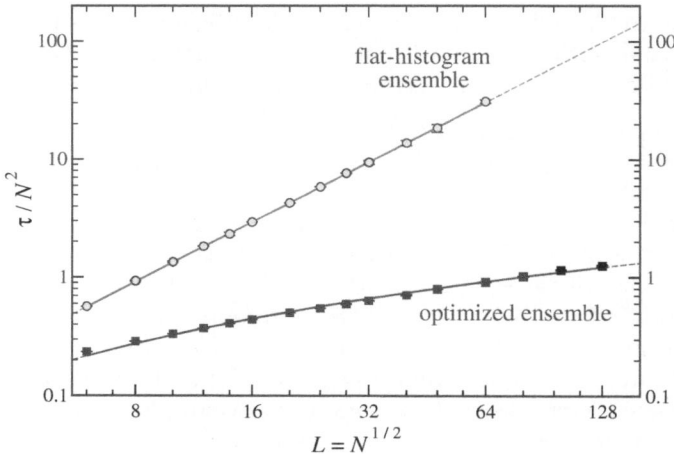

Fig. 3. Scaling of round-trip times for a random walk in energy space sampling a flat histogram (*open squares*) and the optimized histogram (*solid circles*) for the two-dimensional fully frustrated Ising model. While for the multicanonical simulation a power-law slowdown of the round-trip times $O(N^2 L^z)$ is observed, the round-trip times for the optimized ensemble scale like $O([N \ln N]^2)$ thereby approaching the ideal $O(N^2)$-scaling of an unbiased Markovian random walk up to a logarithmic correction

3.5 Simulation of Dense Fluids

Extended ensembles cannot only be defined as a function of energy, but in arbitrary reaction coordinates $\boldsymbol{R}$ onto which a random walk in configuration space can be projected. The generalized weights in these reaction coordinates $W_{\text{extended}}(\boldsymbol{R})$ are then used to bias the random walk along the reaction coordinate by accepting moves from a configuration c_1 with reaction coordinate $\boldsymbol{R}_1$ to a configuration c_2 with reaction coordinate $\boldsymbol{R}_2$ with probability

$$p_{\text{acc}}(c_1 \to c_2) = p_{\text{acc}}(\boldsymbol{R}_1 \to \boldsymbol{R}_2) = \min\left(1, \frac{W_{\text{extended}}(\boldsymbol{R}_2)}{W_{\text{extended}}(\boldsymbol{R}_1)}\right) . \tag{34}$$

The generalized weights $W_{\text{extended}}(\boldsymbol{R})$ can again be chosen in such a way that similar to a multicanonical simulation a flat histogram is sampled along the reaction coordinate by setting the weights to be inversely proportional to the density of states defined in the reaction coordinates, that is $W_{\text{extended}}(\boldsymbol{R}) \propto 1/g(\boldsymbol{R})$.

An optimal choice of weights can be found by measuring the local diffusivity of a random walk along the reaction coordinates and applying the feedback method to shift weight towards the bottlenecks in the simulation. This generalized ensemble optimization approach has recently been illustrated for the simulation of dense Lennard-Jones fluids close to the vapor-liquid equilibrium [21]. The interaction between particles in the fluid is described by a pairwise Lennard-Jones potential of the form

$$\Phi_{\mathrm{LJ}}(R) = 4\epsilon \left[\left(\frac{\sigma}{R}\right)^{12} - \left(\frac{\sigma}{R}\right)^{6} \right] , \tag{35}$$

where ϵ is the interaction strength, σ a length parameter, and R the distance between two particles. It is this distance R between two arbitrarily chosen particles in the fluid that one can use as a new reaction coordinate for a projected random walk. Defining an extended ensemble with weights $W_{\mathrm{extended}}(R)$ and recording a histogram $H(R)$ during a simulation will then allow to calculate the pair distribution function $g(R) = H(R)/W_{\mathrm{extended}}(R)$. The pair distribution function $g(R)$ is closely related to the potential of mean force (PMF)

$$\Phi_{\mathrm{PMF}}(R) = -\frac{1}{\beta} \ln g(R) , \tag{36}$$

which describes the average interaction between two particles in the fluid in the presence of many surrounding particles.

For high particle densities and low enough temperatures shell structures will form in the fluid which are reminiscent of the hexagonal lattice of the solid structure at very low temperatures. These shell structures are revealed by a sinusoidal modulation in the PMF as illustrated in the lower panel of Fig. 4 for the case of a two-dimensional fluid. Thermal equilibration between the shells is suppressed by entropic barriers which form between the shells. Again, one can ask what probability distribution, or histogram, should be sampled along the reaction coordinate, in this case the radial distance R, so that equilibration between the shells is improved. Measuring the local diffusivity for a random walk along the radial distance R in an interval $[R_{\mathrm{min}}, R_{\mathrm{max}}]$ and subsequently applying the feedback algorithm described above optimized histograms $H(R)$ are found which are plotted in Fig. 4 for varying temperatures [21]. The feedback algorithm again shifts additional weight in the histogram towards the bottleneck of the simulation, in this case towards the barriers between the shells. Interestingly, additional peaks emerge in the optimized histogram as the temperature is lowered towards the vapor-liquid equilibrium. The minima between these peaks points to additional meta-stable configurations which occur at these low temperatures, namely interstitial states which occur as the shells around two particles merge as detailed in [21].

This example illustrates that for some simulations the local diffusivity and optimized histogram *itself* are very sensitive measures that can reveal interesting phenomena which are otherwise hard to detect in a numerical simulation. In general, a strong modulation of the local diffusivity for the random walk along a given reaction coordinate is a good indicator that the reaction coordinate itself is a good choice that captures some interesting physics of the problem.

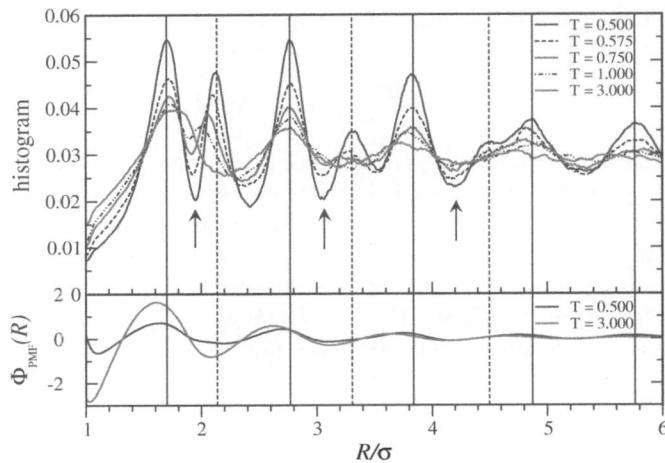

Fig. 4. Optimized histograms for the simulation of dense Lennard-Jones fluids

3.6 Parallel Tempering

The simulation of frustrated systems suffers from a similar tunneling problem as the simulation of first order phase transitions: local minima in energy space are separated by barriers that grow with system size. While the multicanonical or optimized ensembles do not help with the NP-hard problems faced by spin glasses, they are efficient in speeding up simulations of frustrated magnets without disorder.

An alternative to these extended ensembles for the simulation of frustrated magnets is the "parallel tempering" or "replica exchange" Monte Carlo method [22–25]. Instead of performing a single simulation at a fixed temperature, simulations are performed for M replicas at a set of temperatures $T_1, T_2, \ldots, T_M$. In addition to standard Monte Carlo updates at a fixed temperature, exchange moves are proposed to swap two replicas between adjacent temperatures. These swaps are accepted with a probability

$$\min[1, \exp(\Delta\beta\Delta E)], \tag{37}$$

where $\Delta\beta$ is the difference in inverse temperatures and ΔE the difference in energy between the two replicas.

The effect of these exchange moves is that a replica can drift from a local free energy minimum at low temperatures to higher temperatures, where it is easier to cross energy barriers and equilibration is fast. Upon cooling (by another sequence of exchanges) it can end up in a different local minimum on time scales that are much shorter compared to a single simulation at a fixed low temperature. This random walk of a single replica in temperature space is the analog of the random walk in energy space discussed for the extended ensemble techniques. The complement of the statistical ensemble, defined by the weights $W_{\text{extended}}(E)$, is the particular choice of temperature points in the temperature set $\{T_1, T_2, \ldots, T_M\}$ for the parallel tempering simulation. The probability of sampling any given temperature T in an interval $T_i < T < T_{i+1}$ can then be approximated by $H(T) \propto 1/\Delta T$, where $\Delta T = T_{i+1} - T_i$ is the length of the temperature interval around the temperature T. This probability distribution $H(T)$ is the equivalent to the histogram $H(E)$ in the extended ensemble simulations. The ensemble optimization technique introduced above can thus be reformulated to optimize the temperature set in a parallel tempering simulation in such a way that the rate of round-trips between the two extremal temperatures, T_1 and T_M respectively, is maximized [26, 27].

Starting with an initial temperature set $\{T_1, T_2, \ldots, T_M\}$ a parallel tempering simulation is performed where each replica is labeled either "plus" or "minus" indicating which of the two extremal temperatures the respective replica has visited most recently. This allows to measure a current of replicas diffusing from the highest to the lowest temperature by recording two histograms, $h_+(T)$ and $h_-(T)$ for each temperature point. The current j is then given by

$$j = D(T) H(T) \frac{df}{dT} \, , \tag{38}$$

where $D(T)$ is the local diffusivity for the random walk in temperature space, and $f(T) = h_+(T)/(h_+(T) + h_-(T))$ is the fraction of random walkers which have visited the highest temperature T_M most recently. The probability distribution $H(T)$ is normalized, that is

$$\int_{T_1}^{T_M} H(T) \, dT = C \int_{T_1}^{T_M} \frac{dT}{\Delta T} = 1 \, , \tag{39}$$

where C is a normalization constant. Rearranging (38) the local diffusivity $D(T)$ of the random walk in temperature space can be estimated as

$$D(T) \propto \frac{\Delta T}{df/dT} \, . \tag{40}$$

Analog to the argument for the extended ensemble in energy space the current j is maximized by choosing a probability distribution

$$H_{\text{optimal}}(T) \propto \frac{1}{\sqrt{D(T)}} \propto \sqrt{\frac{1}{\Delta T} \frac{df}{dT}} \tag{41}$$

which is inversely proportional to the square root of the local diffusivity. The optimized temperature set $\{T_1', T_2', \ldots, T_M'\}$ is then found by choosing the n-th temperature point T_n' such that

$$\int_{T_1'}^{T_n'} H_{\text{optimal}}(T) \, dT = \frac{n}{M} \, , \tag{42}$$

where M is the number of temperature points in the original temperature set, and the two extremal temperatures $T_1' = T_1$ and $T_M' = T_M$ remain unchanged. Similarly to the algorithm for the ensemble optimization this feedback of the local diffusivity should be iterated until the temperature set is converged.

Figure 5 illustrates the optimized temperature sets for the Ising ferromagnet obtained by several iterations of the above feedback loop. After feedback of the local diffusivity temperature points accumulate near the critical temperature $T_c = 2.269$ of the transition. This is in analogy to the optimized histograms for the extended ensemble simulations where resources where shifted towards the critical energy of the transition, see Fig. 2.

It is interesting to note that for the optimized temperature set the acceptance rates for swap moves are not independent of the temperature. Around the critical temperature, where temperature points are accumulated by the feedback algorithm, the acceptance rates are higher as at higher/lower temperatures, where the density of temperature points becomes considerably smaller after feedback. The almost Markovian scaling behavior for the optimized random walks in either energy or temperature space is thus generated by a problem-specific statistical ensemble which is characterized neither by a flat histogram nor flat acceptance rates for exchange moves, but by a characteristic

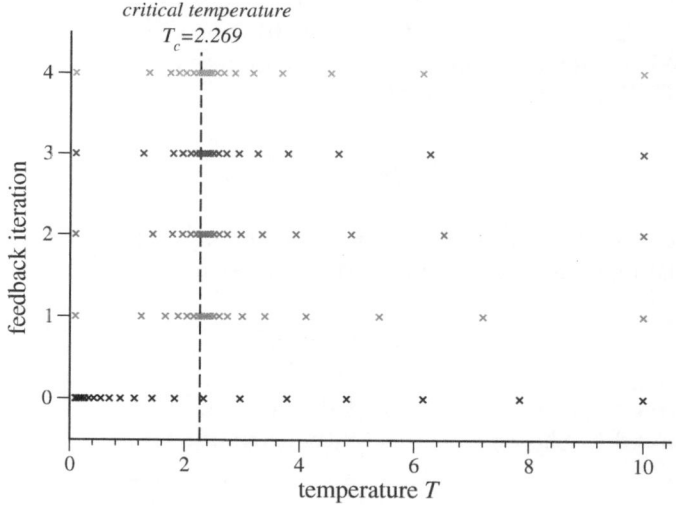

Fig. 5. Optimized temperature sets for the two-dimensional Ising ferromagnet. The initial temperature set with 20 temperature points is determined by a geometric progression for the temperature interval $[0.1, 10]$. After feedback of the local diffusivity the temperature points accumulate near the critical temperature $T_c = 2.269$ of the phase transition (*dashed line*). Similar to the ensemble optimization in energy space the feedback of the local diffusivity relocates resources towards the bottleneck of the simulation

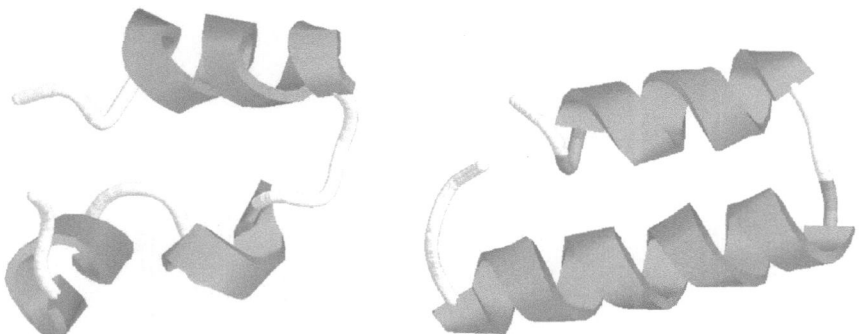

Fig. 6. Low-energy structure of the 36-residue chicken villin headpiece sub-domain HP-36. On the left the structure determined in NMR experiments is shown. The right panel shows the lowest-energy configuration found in a feedback-optimized all-atom parallel tempering simulation using the ECEPP/2 force field and an implicit solvent model. The root-mean square deviation of this structure to the structure on the left is $R_{RMSD} = 3.8$ Å

probability distribution which concentrates resources at the minima of the measured local diffusivity.

3.7 Optimized Parallel Tempering Simulations of Proteins

The feedback-optimized parallel tempering technique [26] outlined in the previous section has recently been applied to study the folding of the 36-residue chicken villin headpiece sub-domain HP-36 [27]. Since HP-36 is one of the smallest proteins with well-defined secondary and tertiary structure [28] and at the same time with 596 atoms still accessible to numerical simulations, it has recently attracted considerable interest as an example to test novel numerical techniques, including molecular dynamics [29,30] and Monte Carlo [31,32] methods. The experimentally determined structure [28] which is deposited in the Protein Data Bank (PDB code 1vii) is illustrated in the left panel of Fig. 6.

Applying an all-atom parallel tempering simulation of the protein HP-36 in the ECEPP/2 force field [33] using an implicit solvent model [34] the authors of [27] have measured the diffusion of labeled replicas in temperature space. The simulated temperature interval is chosen such that at the lowest temperature $T_{min} = 250$ K the protein is in a folded state and the highest temperature $T_{max} = 1000$ K ensures that the protein can fully unfold for the simulated force field. The measured local diffusivity for the random walk between these two extremal temperatures is shown in Fig. 7. A very strong modulation of the local diffusivity is found along the temperature. Note the logarithmic scale of the ordinate. The pronounced minimum of the local diffusivity around $T \approx 500$ K points to a severe bottleneck in the simulation which by measurements of the specific heat has been identified as the helix-coil

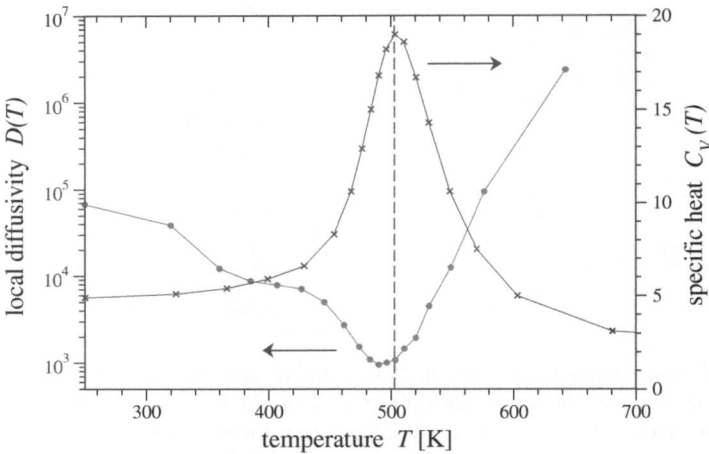

Fig. 7. Local diffusivity (*solid circles*) of the random walk in temperature space for a parallel tempering simulation of the 36-residue villin headpiece sub-domain HP-36. The diffusivity shows a strong modulation along the temperature, note the logarithmic scale of the ordinate. Slightly below the helix-coil transition around $T \approx 500$ K which is identified by a maximum in the specific heat (crosses, right ordinate) there is a strong suppression of the diffusivity

transition [31]. Above this transition the protein is in an extended unordered configuration, while below the helix-coil transition the protein is characterized by high helical content [31]. The shoulder in the local diffusivity in the temperature region 350 K $\leq T \leq 490$ K points to a second bottleneck in the simulation, possibly caused by competing low-energy configurations with high helical content.

An optimized temperature set for the parallel tempering simulation of HP-36 in the ECEPP/2 force field can then be found by feeding back the local diffusivity applying the algorithm outlined above. Results for a temperature set with 20 temperature points are illustrated in Fig. 8 for an initial temperature set which similar to a geometric progression concentrates temperature points at low temperatures [27]. After the feedback temperature points concentrate around the bottleneck of the simulation, primarily around the helix-coil transition at $T \approx 500$ K and in the temperature regime 350 K $\leq T \leq 490$ K below the transition where a shoulder in the local diffusivity was found.

In [27] it was demonstrated that by using the optimized temperature set in the simulations the low-energy configurations equilibrated considerably faster than in previous parallel tempering simulations [31]. As a consequence, the low energy structures are more compact and the configuration with lowest energy illustrated in Fig. 6 shows a root-mean square deviation to the experimentally determined structure of $R_{\mathrm{RMSD}} = 3.7$ Å. This deviation from the native structure is similar to results found by large-scale molecular dynamics

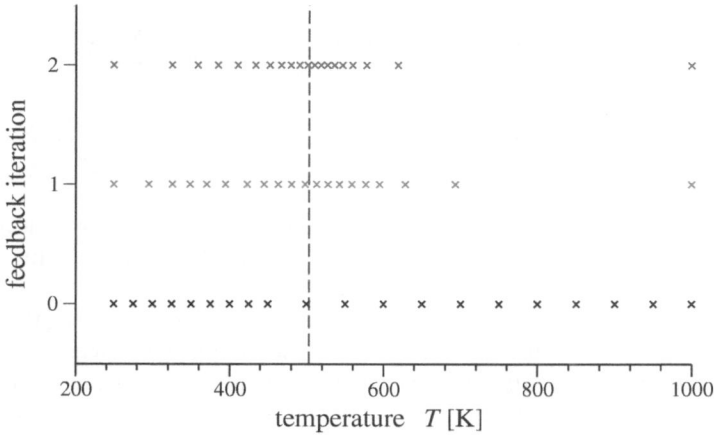

Fig. 8. Optimized temperature sets with 20 temperature points for the parallel tempering simulation of the 36-residue protein HP-36. The initial temperature set covers a temperature range 250 K $\leq T \leq$ 1000 K and concentrates temperature points at low temperatures similar to a geometric progression. After the feedback of the local diffusivity temperature points accumulate around the helix-coil transition at $T \approx 500$ K where the strong suppression of the local diffusivity points to a severe bottleneck

simulations [30] with a different force field. However, employing the optimized temperature set the Monte Carlo simulations consumed only a fraction of one percent of the computing time used for the molecular dynamics simulations. A detailed discussion of the application of the feedback method in the study of proteins is given in [27].

3.8 Simulation of Quantum Systems

Extended ensemble methods, such as the multicanonical ensemble, Wang-Landau sampling or parallel tempering can also be generalized to quantum systems [35, 36], as we will show in the next two sections.

4 Quantum Monte Carlo World Line Algorithms

4.1 The $S = 1/2$ Quantum XXZ Model

In this section we will generalize the Monte Carlo methods described in section 2 for classical spin systems to quantum spin systems. As an example we will use the spin-1/2 quantum Heisenberg or XXZ models with Hamiltonian

$$H = \sum_{\langle i,j \rangle} \left[J_z S_i^z S_j^z + J_{xy} \left(S_i^x S_j^x + S_i^y S_j^y \right) \right] - h \sum_{i=1}^{N} S_i^z \qquad (43)$$

$$= \sum_{\langle i,j \rangle} \left[J_z S_i^z S_j^z + \frac{J_{xy}}{2} \left(S_i^+ S_j^- + S_i^- S_j^+ \right) \right] - h \sum_{i=1}^{N} S_i^z$$

where S_i^α are spin $S = 1/2$ operators fulfilling the standard commutation relations and in the second line we have replaced S_i^x and S_i^y by the spin raising and lowering operators S_i^+ and S_i^-.

The case $J_{xy} = 0$ corresponds to the classical Ising model (12) up to a change in sign: while in classical Monte Carlo simulations (where there is no difference in the thermodynamics of the ferromagnet and the antiferromagnet) a positive exchange constant J denotes the ferromagnet, the convention for quantum systems is usually opposite with a positive exchange constant denoting the antiferromagnet. The other limit $J_z = 0$ corresponds to the quantum XY-model, while $J_z = J_{xy}$ is the Heisenberg model.

4.2 Representations

The basic problem for Monte Carlo simulations of quantum systems is that the partition function is no longer a simple sum over classical configurations as in (16) but an operator expression

$$Z = \mathrm{Tr} \exp(-\beta H) , \qquad (44)$$

where H is the Hamilton operator and the trace Tr goes over all states in the Hilbert space. Similarly the expression for an observable like the magnetization is an operator expression:

$$\langle m \rangle = \frac{1}{Z} \mathrm{Tr} \left[m \exp(-\beta H) \right] , \qquad (45)$$

and the Monte Carlo method cannot directly be applied except in the classical case where the Hamilton operator H is diagonal and the trace reduces to a sum over all basis states. The first step of any QMC algorithm is thus the mapping of the quantum system to an equivalent classical system

$$\langle m \rangle = \frac{1}{Z} \mathrm{Tr} \left[m \exp(-\beta H) \right] = \sum_c m(c) W(C) , \qquad (46)$$

where the sum goes over configurations c in an artificial classical system (e.g. a system of world lines), $m(c)$ will be the value of the magnetization or another observable as measured in this classical system and $W(C)$ the weight of the classical configuration. We will now present two different but related methods for this mapping, namely continuous time path integrals and the stochastic series expansion.

The Path-Integral Representation

The path-integral formulation of a quantum systems goes back to [37], and forms the basis of most QMC algorithms. Instead of following the historical route and discussing the Trotter-Suzuki (checkerboard) decomposition [38,39] for path integrals with discrete time steps $\Delta\tau$ we will directly describe the continuous-time formulation used in modern codes.

The starting point is a time-dependent perturbation expansion in imaginary time to evaluate the density matrix operator $\exp(-\beta H)$. Using a basis in which the S^z operators are diagonal we follow [40] and split the Hamiltonian $H = H_0 + V$ into a diagonal term H_0, containing the S^z term and an offdiagonal perturbation V, containing the exchange terms $(J_{xy}/2)(S_i^+ S_j^- + S_i^- S_j^+)$. In the interaction representation the time-dependent perturbation is $V(\tau) = \exp(\tau H_0)V \exp(-\tau H_0)$ and the partition function can be represented as:

$$
\begin{aligned}
Z = \mathrm{Tr}\exp(-\beta H) &= \mathrm{Tr}\left[\exp(-\beta H_0)T\exp\int_0^\beta d\tau V(\tau)\right], \\
&= \mathrm{Tr}\left[\exp(-\beta H_0)\left(1 - \int_0^\beta d\tau_1 V(\tau_1) + \frac{1}{2}\int_0^\beta d\tau_1 \int_{\tau_1}^\beta d\tau_2 V(\tau_1)V(\tau_2) + \ldots\right)\right] \\
&= \sum_i \langle i|\left[\exp(-\beta H_0)\left(1 - \int_0^\beta d\tau_1 V(\tau_1) + \frac{1}{2}\int_0^\beta d\tau_1 \int_{\tau_1}^\beta d\tau_2 V(\tau_1)V(\tau_2) + \ldots\right)\right]|i\rangle,
\end{aligned}
\tag{47}
$$

where the symbol T denotes time-ordering of the exponential and in the last line we have replaced the trace by a sum over a complete set of basis states $|i\rangle$, that are eigenstates of the local S^z operators. Note that, in contrast to a real time path integral, the imaginary time path integral always converges on finite systems of N spins at finite temperatures β, and the expansion can be truncated at orders $n \gg \beta J_{xy} N$.

Equation (47) is now just a classical sum of integrals and can be evaluated by Monte Carlo sampling in the intitial states $|i\rangle$, the order of the perturbation n and the times τ_i ($i = 1, \ldots, n$). This is best done by considering a graphical world line representation of the partition function (47) shown in Fig. 9. The zero-th order terms in the sum $\sum_i \langle i| \exp(-\beta H_0)|i\rangle$ are given by straight world lines shown in Fig. 9a. First order terms do not appear since the matrix elements $\langle i|V|i\rangle$ are zero for the XXZ model. The first non-trivial terms appear in second order with two exchanges, as shown in Fig. 9b. A general configuration of higher order is depicted in Fig. 10a.

Since the XXZ Hamiltonian commutes with the z-component of total spin $\sum_i S_i^z$, the total magnetization is conserved and all valid configurations are represented by closed world lines as shown in Figs. 9 and 10. Models that break this conservation of magnetization, such as general XYZ models with different couplings in all directions, models with transverse fields coupling to S_i^x or higher spin models with single ion anisotropies $(S_i^x)^2$ or $(S_i^y)^2$ will in addition contain configurations with broken world line segments.

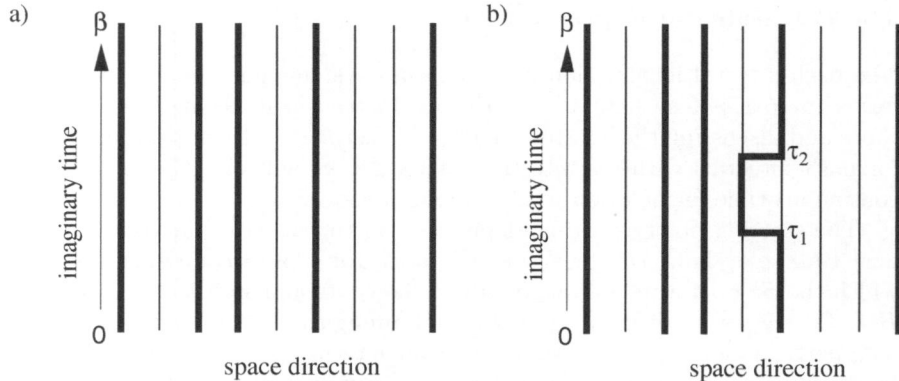

Fig. 9. Examples of simple world line configurations in imaginary time for a quantum spin model. Up-spins are shown by bold lines and down spins by thin lines. (**a**) a configuration in 0-th order perturbation theory where the spins evolve according to the diagonal term $\exp(-\beta H_0)$ and the weight is given by the classical Boltzmann weight of H_0. (**b**) a configuration in second order perturbation theory with two exchanges at times τ_1 and τ_2. Its weight is given by the matrix elements of the exchange processes and the classical Boltzmann weight of H_0 of the spins

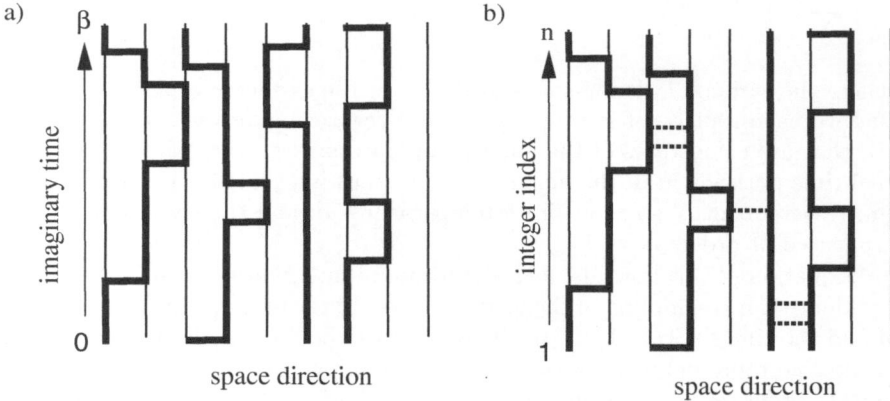

Fig. 10. Examples of world line configurations in (**a**) a path-integral representation where the time direction is continuous and (**b**) the stochastic series expansion (SSE) representation where the "time" direction is discrete. Since the SSE representation perturbs not only in offdiagonal terms but also in diagonal terms, additional diagonal terms are present in the representation, indicated by *dashed lines*

The Stochastic Series Expansion Representation

An alternative representation is the stochastic series expansion (SSE) [41], a generalization of Handscomb's algorithm [42] for the Heisenberg model. It starts from a Taylor expansion of the partition function in orders of β:

$$Z = \mathrm{Tr}\exp(-\beta H) = \sum_{n=0}^{\infty} \frac{\beta^n}{n!} \mathrm{Tr}(-H)^n \tag{48}$$

$$= \sum_{n=0}^{\infty} \frac{\beta^n}{n!} \sum_{\{i_1,\ldots i_n\}} \sum_{\{b_1,\ldots b_n\}} \langle i_1| - H_{b_1}|i_2\rangle\langle i_2| - H_{b_2}|i_3\rangle \cdots \langle i_n| - H_{b_n}|i_1\rangle \, ,$$

where in the second line we decomposed the Hamiltonian H into a sum of single-bond terms $H = \sum_b H_b$, and again inserted complete sets of basis states. We end up with a similar representation as (47) and a related world-line picture shown in Fig. 10b.

The key difference is that the SSE representation is a perturbation expansion in *all* terms of the Hamiltonian, while the path-integral representation perturbs only in the off-diagonal terms. Although the SSE method thus needs higher expansion orders for a given system, this disadvantage is compensated by a simplification in the algorithms: only integer indices of the operators need to be stored instead of continuous time variables τ_i. Except in strong magnetic fields or for dissipative quantum spin systems [43, 44] the SSE representation is thus the preferred representation for the simulation of quantum magnets.

The Negative Sign Problem

While the mapping from the quantum average to a classical average in (46) can be performed for any quantum system, it can happen in frustrated quantum magnets, that some of the weights $W(C)$ in the quantum system are negative, as is shown in Fig. 11.

Since Monte Carlo sampling requires positive weights $W(C) > 0$ the standard way of dealing with the negative weights of the frustrated quantum magnets is to sample with respect to the unfrustrated system by using the absolute values of the weights $|W(C)|$ and to assign the sign, $s(c) \equiv \mathrm{sign}\, W(C)$ to the quantity being sampled:

$$\langle m \rangle = \frac{\sum_c m(c)W(C)}{\sum_c W(C)} = \frac{\sum_c m(c)s(c)|W(C)| / \sum_c |W(C)|}{\sum_c s(c)|W(C)| / \sum_c |W(C)|} \equiv \frac{\langle ms \rangle'}{\langle s \rangle'}. \tag{49}$$

While this allows Monte Carlo simulations to be performed, the errors increase exponentially with the particle number N and the inverse temperature β. To see this, consider the mean value of the sign $\langle s \rangle = Z/Z'$, which is just the ratio of the partition functions of the frustrated system $Z = \sum_c W(C)$ with weights $W(C)$ and the unfrustrated system used for sampling with

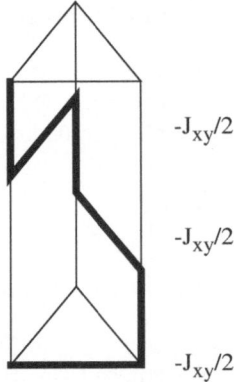

$-J_{xy}/2$

$-J_{xy}/2$

$-J_{xy}/2$

Fig. 11. Example of a frustrated world line configuration in a Heisenberg quantum antiferromagnet on a triangle. The closed world line configuration contains three exchange processes, each contributing a weight proportional to $-J_{xy}/2$. The overall is proportional to $(-J_{xy}/2)^3$ and is negative, causing a negative sign problem for the antiferromagnet with $J_{xy} > 0$

$Z' = \sum_c |W(C)|$. As the partition functions are exponentials of the corresponding free energies, this ratio is an exponential of the differences Δf in the free energy densities: $\langle s \rangle = Z/Z' = \exp(-\beta N \Delta f)$. As a consequence, the relative error $\Delta s / \langle s \rangle$ increases exponentially with increasing particle number and inverse temperature:

$$\frac{\Delta s}{\langle s \rangle} = \frac{\sqrt{(\langle s^2 \rangle - \langle s \rangle^2)/M}}{\langle s \rangle} = \frac{\sqrt{1 - \langle s \rangle^2}}{\sqrt{M} \langle s \rangle} \sim \frac{\exp(\beta N \Delta f)}{\sqrt{M}}. \tag{50}$$

Here M is the number of uncorrelated Monte Carlo samples. Similarly the error for the numerator increases exponentially and the time needed to achieve a given relative error scales exponentially in N and β.

It was recently shown that the negative sign problem is NP-hard, implying that almost certainly no solution for this exponential scaling problem exists [4]. Given this exponential scaling of quantum Monte Carlo simulations for frustrated quantum magnets, the QMC method is best suited for nonfrustrated magnets and we will restrict ourselves to these sign problem free cases in the following.

Measurements

Physical observables that can be measured in both the path-integral representation and the SSE representation include, next to the energy and the specific heat, any expectation value or correlation function that is diagonal in the basis set $\{|i\rangle\}$. This includes the uniform or staggered magnetization in the z direction, the equal time correlation functions and structure factor of the

z-spin components and the z-component uniform and momentum-dependent susceptibilites.

Offdiagonal operators, such as the magnetization in the x- or y-direction, or the corresponding correlation functions, structure factors and susceptibilities require an extension of the sampling to include configurations with broken world line segments. These are hard to measure in local update schemes (described in Sect. 4.3) unless open world line segments are already present when the Hamiltonian does not conserve magnetization, but are easily measured when non-local updates are used (see Sects. 4.4 and 4.5).

The spin stiffness ρ_s can be obtained from fluctuations of the winding numbers of the world lines [45], a measurement which obviously requires non-local moves that can change these winding numbers.

Dynamical quantities are harder to obtain, since the QMC representations only give access to imaginary-time correlation function. With the exception of measurements of spin gaps, which can be obtained from an exponential decay of the spin-spin correlation function in imaginary time, the measurement of real-time or real-frequency correlation functions requires an ill-posed analytical continuation of noisy Monte Carlo data, for example using the Maximum Entropy Method [46–48].

Thermodynamic quantities that cannot be expressed as the expectation value of an operator, such as the free energy or entropy cannot be directly measured but require an extended ensemble simulation, discussed in Sect. 5.

4.3 Local Updates

To perform a quantum Monte Carlo simulation on the world line representation, update moves that are ergodic and fulfill detailed balance are required. The simplest types of moves are again local updates. Since magnetization conservation prohibits the breaking of world lines, the local updates need to move world lines instead of just changing local states as in a classical model.

A set of local moves for a one-dimensional spin-1/2 model is shown in Fig. 12 [49,50]. The two required moves are the insertion and removal of a pair of exchange processes (Fig. 12(a)) and the shift in time of an exchange process (Fig. 12(b)). Slightly more complicated local moves are needed for higher-dimensional models, for example to allow world lines to wind around elementary squares in a square lattice [51]. Since these local updates cannot

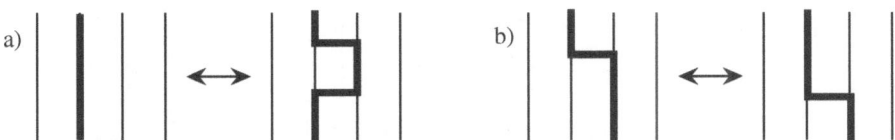

Fig. 12. Examples of local updates of world lines: (**a**) a pair of exchange processes can be inserted or removed; (**b**) an exchange process is moved in imaginary time

change global properties, such as the number of world lines (the magnetization) or their spatial winding, they need to be complemented with global updates [51].

While the local update world line and SSE algorithms enable the simulation of quantum systems they suffer from critical slowing down at second order phase transitions. Even worse, changing the spatial and temporal winding numbers usually has an exponentially small acceptance rate. While the restriction to zero spatial winding can be viewed as a boundary effect, changing the temporal winding number and thus the magnetization is essential for simulations at fixed magnetic fields.

4.4 Cluster Updates and the Loop Algorithm

The ergodicity problems of purely local updates and the critical slowing down observed also in quantum systems require the use of cluster updates. The loop algorithm [52] and its continuous time version [53], are generalizations of the classical cluster algorithms [6, 7] to quantum systems. They not only solve the problem of critical slowing down, but can also change the magnetization and winding numbers efficiently, avoiding the ergodicity problem of local updates. While the loop algorithm was initially developed for the path-integral representation it can also be applied to simulations in the SSE representation.

Since there exist extensive recent reviews of the loop algorithm [54,55], we will only outline the loop algorithm here. It constructs clusters of spins, similar to the Swendsen-Wang [6] clusters of the classical Ising model (Sect. 2.3). Upon applying the cluster algorithms to world lines in QMC we have to take into account that – in systems with conserved magnetization – the world lines may not be broken. This implies that a single spin cannot be flipped by itself, but, as shown in Fig. 13, connected world line segments of spins must be flipped together. These world line segments form a closed loop, hence the name "loop algorithm".

While the loop algorithm was originally developed only for spin-1/2 models it has been generalized to higher spin models [56–59] and anisotropic spin

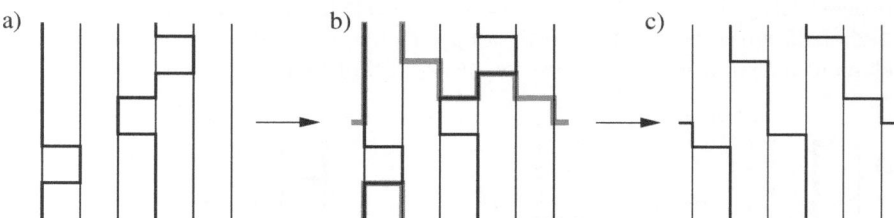

Fig. 13. A loop cluster update: (**a**) world line configuration before the update, where the world line of an up-spin is drawn as a *thick line* and that of a down-spin as a *thin line*; (**b**) world line configuration and a loop cluster (*grey line*); (**c**) the world line configurations after all spins along the loop have been flipped

models [60]. Since an efficient open-source implementation of the loop algorithm is available (see Sect. 4.6) we will not discuss further algorithmic details but refer interested readers to the reviews [54, 55].

4.5 Worm and Directed Loop Updates

The Loop Algorithm in a Magnetic Field

As successful as the loop algorithm is, it is restricted – as most classical cluster algorithms – to models with spin inversion symmetry. Terms in the Hamiltonian which break this spin-inversion symmetry, such as a magnetic field, are not taken into account during loop construction. Instead they enter through the acceptance rate of the loop flip, which can be exponentially small at low temperatures.

As an example consider two $S = 1/2$ quantum spins in a magnetic field:

$$H = J\mathbf{S}_1\mathbf{S}_2 - g\mu_B h(S_1^z + S_2^z) \tag{51}$$

In a field $g\mu_B h = J$ the singlet state $1/\sqrt{2}(|\uparrow\downarrow\rangle - |\downarrow\uparrow\rangle)$ with energy $-3/4J$ is degenerate with the triplet state $|\uparrow\uparrow\rangle$ with energy $1/4J - h = -3/4J$. As illustrated in Fig. 14a), we start from the triplet state $|\uparrow\uparrow\rangle$ and propose a loop shown in Fig. 14b). The loop construction rules, which ignore the magnetic field, propose to flip one of the spins and go to the intermediate configuration $|\uparrow\downarrow\rangle$ with energy $-1/4J$ shown in Fig. 14c). This move costs potential energy $J/2$ and thus has an *exponentially small acceptance rate* $\exp(-\beta J/2)$. Once we accept this move, immediately many small loops are built, exchanging the spins on the two sites, and gaining exchange energy $J/2$ by going to the spin singlet state. A typical world line configuration for the singlet is shown in Fig. 14d). The reverse move has the same exponentially small probability, since the probability to reach a world line configuration without any exchange term [Fig. 14c)] from a spin singlet configuration [Fig. 14d)] is exponentially small.

This example clearly illustrates the reason for the exponential slowdown: in a first step we *lose all potential energy*, before *gaining it back in exchange energy*. A faster algorithm could thus be built if, instead of doing the trade in one big step, we could trade potential with exchange energy in small pieces, which is exactly what the worm algorithm does.

The Worm Algorithm

The worm algorithm [40] works in an extended configuration space, where in addition to closed world line configurations one open world line fragment (the "worm") is allowed. Formally this is done by adding a source term to the Hamiltonian which for a spin model is

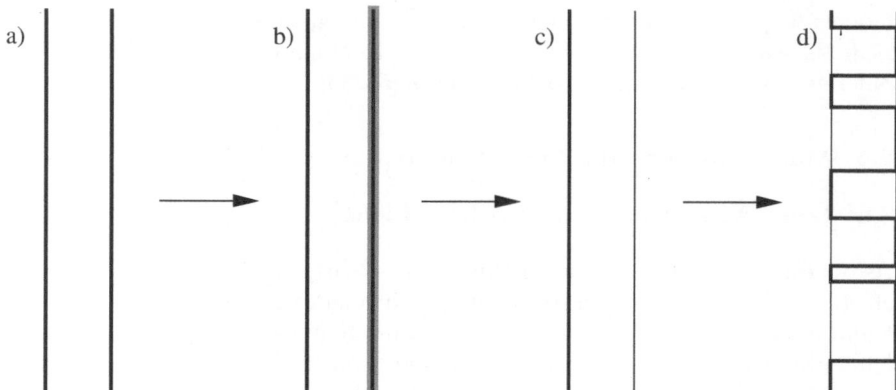

Fig. 14. A loop update for two antiferromagnetically coupled spins in a magnetic field with $J = g\mu_B h$. (**a**) Starting from the triplet configuration $|\uparrow\uparrow\rangle$, (**b**) a loop is constructed, proposing to go to (**c**), the intermediate configuration $|\uparrow\downarrow\rangle$, which has an exponentially small acceptance rate, and finally into configurations like (**d**) which represent the singlet state $1/\sqrt{2}(|\uparrow\downarrow\rangle - |\downarrow\uparrow\rangle)$. As in the previous figure a *thick line* denotes an up-spin and a *thin line* a down-spin

$$H_{\text{worm}} = H - \eta \sum_i (S_i^+ + S_i^-) \, . \tag{52}$$

This source term allows world lines to be broken with a matrix element proportional to η. The worm algorithm now proceeds as follows: a worm (i.e. a world line fragment) is created by inserting a pair (S_i^+, S_i^-) of operators at nearby times, as shown in Fig. 15a,b). The ends of this worm are then moved randomly in space and time [Fig. 15c)], using local Metropolis or heat bath updates until the two ends of the worm meet again as in Fig. 15d). Then an update which removes the worm is proposed, and if accepted we are back in a configuration with closed world lines only, as shown in Fig. 15e). This algorithm is straightforward, consisting just of local updates of the worm ends in the extended configuration space but it can perform nonlocal changes. A worm end can wind around the lattice in the temporal or spatial direction and that way change the magnetization and winding number.

In contrast to the loop algorithm in a magnetic field, where the trade between potential and exchange energy is done by first losing all of the potential energy, before gaining back the exchange energy, the worm algorithm performs this trade in small pieces, never suffering from an exponentially small acceptance probability. While it is not as efficient as the loop algorithm in zero magnetic field (the worm movement follows a random walk while the loop algorithm can be interpreted as a self-avoiding random walk), the big advantage of the worm algorithm is that it remains efficient in the presence of a magnetic field.

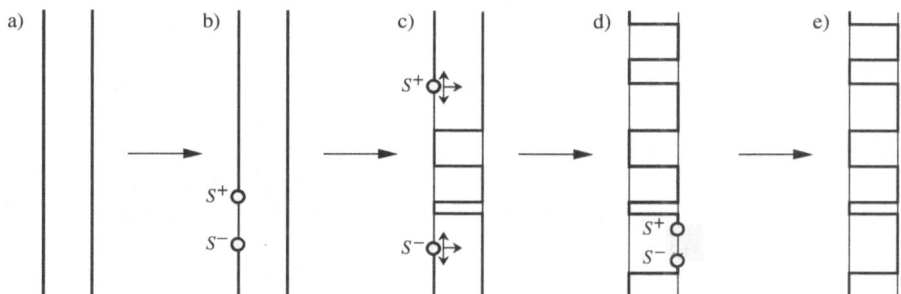

Fig. 15. A worm update for two antiferromagnetically coupled spins in a magnetic field with $J = g\mu_B h$. (**a**) starting from the triplet configuration $|\uparrow\uparrow\rangle$ a worm is constructed in (**b**) by inserting a pair of S^+ and S^- operators. (**c**) these "worm end" operators are then moved by local updates until (**d**) they meet again, when a move to remove them is proposed, which leads to the closed world line configuration (**e**). As in the two previous figures a *thick line* denotes an up-spin and a *thin line* a down-spin

The Directed Loop Algorithm

Algorithms with a similar basic idea as the worm algorithm in the path-integral representations are the operator-loop update [61,62] and the directed-loop algorithms [63] which can be formulated in both an SSE and a world-line representation. Like the worm algorithm, these algorithms create two world line discontinuities, and move them around by local updates. The main difference to the worm algorithm is that here these movements do not follow an unbiased random walk but have a preferred direction, always trying to move away from the last change, which further speeds up the simulations.

4.6 Open Source Implementations: the ALPS Project

The loop, worm and directed loop algorithms can be used for the simulation of a wide class of quantum magnets. They are of interest not only to theoretical physicists, but also to experimentalists who want to fit experimental measurements to theoretical models. The wide applicability of these methods has led to the publication of open-source versions of these algorithms as part of the ALPS project (Algorithms and Libraries for Physics Simulations) [10] on the web page `http://alps.comp-phys.org/`.

4.7 Applications

We will finally present typical applications of the above algorithms by reviewing a small and necessarily biased selection.

The loop algorithm has been applied to a wide range of problems, ranging from purely theoretical questions to experimental data fitting. Below we list a selection of applications that provide an overview over the possibilities of the loop algorithm. The first simulation using the loop algorithm was an accurate determination of the ground state properties (staggered magnetization, spin stiffness and spin wave velocity) of the square-lattice spin-1/2 quantum Heisenberg antiferromagnet [64]. In a similar spirit the uniform susceptibility, correlation length and spin gap of spin ladder models [65, 66] and integer spin chains [59] was calculated, confirming the presence of a spin gapped ground state in even-leg spin ladders and integer spin chains.

As the loop algorithm is efficient also at critical points, it has been used in the first high accuracy simulations of the critical properties of quantum phase transitions by studying the Néel to quantum paramagnet transition in two-dimensional quantum spin systems [67], for a determination of the low-temperature asymptotic scaling of two-dimensional quantum Heisenberg antiferromagnets [3, 58, 68], and for accurate calculations of the Néel temperature of anisotropic quasi-one and quasi-two dimensional antiferromagnets [69].

The loop algorithm is not only restricted to toy models, but can be applied to realistic models of quantum magnets. Comparisons to experimental measurements are done by fitting simulation data to experimental measurements, as for alternating chain compounds [70], spin ladder materials [71] or frustrated square lattice antiferromagnets [72]. In the latter material the sign problem due to frustration limits the accuracy. As an example we show in Fig. 16 the good quality of a fit of QMC data to experimental measurements on the spin ladder compound $SrCu_2O_3$.

Another interesting application is to simulate realistic models for quantum magnets, using exchange constants calculated by ab-initio methods. Comparing these ab-initio QMC data to experimental measurements, as done for a series of vanadates [73] and for ladder compounds [71] allows to quantitatively check the ab-initio calculations.

The worm and directed loop algorithms are applied when magnetic fields are present. Typical examples include the calculation of magnetization curves of quantum magnets [74], the determination of the first order nature of the spin flop transition in two dimensions [75] and the calculation of phase diagrams of dimerized quantum magnets in a magnetic field [76].

5 Extended Ensemble Methods for Quantum Systems

In this section we will present generalizations of extended ensemble simulations to world line quantum Monte Carlo simulations, in particular:

- histogram reweighting,
- parallel tempering,
- extended ensemble methods.

Histogram reweighting allows to extract information at a temperature different than (but close to) the temperature at which the simulation is performed. This is especially useful when studying critical phenomena, where a

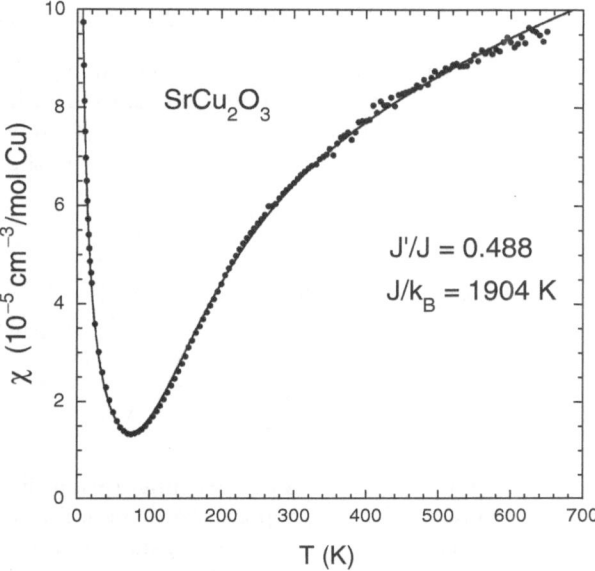

Fig. 16. Fits of experimental measurements of the uniform susceptibility of $SrCu_2O_3$ to the results of QMC simulations, determining a coupling $J \approx 1904\,K$ along the chains of the ladder and a ratio $J'/J \approx 0.488$ for the inter-chain to intra-chain coupling

single simulation can provide information for the whole critical region around the phase transition.

Parallel tempering and the extended ensemble methods (such as multi-canonical simulations and Wang-Landau sampling) speed up simulations at and below phase transitions and are especially useful at first order phase transitions or for frustrated systems. Note, however, that frustrated quantum spin systems generally suffer from the negative sign problem. Since the negative sign problem arises as a property of the representation and does not depend on the ensemble or the updates, the scaling will remain exponential even when using improved sampling algorithms, in contrast to classical simulations where extended ensemble algorithms and parallel tempering can dramatically speed up the simulations.

Another advantage of extended ensemble simulations is the ability to directly calculate the density of states and from it thermodynamic properties such as the entropy or the free energy that are not directly accessible in canonical simulations. In the following we will again use quantum magnets as concrete examples. A generalization to bosonic and fermionic models will always be straightforward.

5.1 Generalizing Extended Ensembles to Quantum Systems

Since simulations of quantum systems suffer from the same problems as classical simulations, the extension of these generalized sampling schemes to quantum systems is highly desired. The extension is not immediately obvious since the partition function of a quantum system cannot be cast in the classical form

$$Z = \sum_E g(E)e^{-\beta E} \tag{53}$$

unless the complete spectrum of the Hamilton operator H is known.

Instead of a representation like (53) we will aim for a generalized representation of the form

$$Z = \sum_c W(c) = \sum_\lambda g(\lambda)p(\mu, \lambda) , \tag{54}$$

where λ describes values of properties Λ of the configuration that are sampled, μ are external parameters such as temperature or coupling constants, and $p(\mu, \lambda)$ is the weight of that configuration. The generalized density of states $g(\lambda)$ is a sum over all configurations c with the property $\Lambda(c) = \lambda$

$$g(\lambda) = \sum_c \delta_{\Lambda(c),\lambda}\tilde{W}(c) , \tag{55}$$

where the reduced weight $\tilde{W}(c) = W(c)/p(\mu, \lambda)$ of a configuration c shall not depend on the parameters μ. By defining as

$$A(\lambda) = \frac{1}{g(\lambda)} \sum_c \delta_{\Lambda(c),\lambda}A(c)\tilde{W}(c) \tag{56}$$

the "microcanonical" average of A for configurations with $\Lambda(c) = \lambda$ we can obtain thermal averages $\langle A(\mu)\rangle$ at arbitrary parameters μ:

$$\langle A(\mu)\rangle = \frac{\sum_\lambda A(\lambda)g(\lambda)p(\mu, \lambda)}{\sum_\lambda g(\lambda)p(\mu, \lambda)} . \tag{57}$$

In a classical simulation we might choose the desired property as the energy: $\lambda = E$, the external parameter the inverse temperature: $\mu = \beta$, the weight the Boltzmann weight $p(\beta, E) = \exp(-\beta E)$. The reduced weight $\tilde{W}(c) = 1$ and, hence, $g(E)$ the standard density of states.

The generalized notation makes sense even for a classical system. Consider, for example, an Ising antiferromagnet in a magnetic field:

$$H = J \sum_{\langle i,j \rangle} \sigma_i \sigma_j - h \sum_i \sigma_i . \tag{58}$$

If we are interested in properties at a fixed inverse temperature β as a function of the magnetization we can choose as external parameter the magnetic field

$\mu = h$ and as property of the system the magnetization $\lambda = M = \sum_i \sigma_i$, giving a representation:

$$Z = \sum_M g(M)e^{\beta h M}. \tag{59}$$

If we are interested in properties as a function of both temperature and magnetization we might pick a two-dimensional representation. As external parameters we choose the inverse temperature and magnetic field $\mu = (\beta, h)$. The corresponding properties of the system are $\lambda = (E_J, M)$, where the magnetic energy E_J is defined as

$$E_J = J \sum_{\langle i,j \rangle} \sigma_i \sigma_j . \tag{60}$$

This gives a representation

$$Z = \sum_{E_J, M} g(E_J, M)e^{-\beta E_J + \beta h M} . \tag{61}$$

Continuous Time Path Integrals

To apply generalized sampling schemes to quantum systems in the path integral representation we cast (47) into the form

$$Z = \int dE_0 \sum_{n=0}^{\infty} g(E_0, n)\beta^n e^{-\beta E_0} , \tag{62}$$

where the diagonal energy contribution E_0 is the value of the diagonal part of the Hamiltonian H_0 in each configuration.

Comparing to (54) we have as control parameter the inverse temperature $\mu = \beta$ and need two properties of the configuration $\lambda = (E_0, n)$.

As in classical systems we might be interested in the dependence on a magnetic field h instead of the temperature, and rewrite (47) in a form very similar to the classical one as

$$Z = \int dM g(M)e^{\beta h M} , \tag{63}$$

where the magnetization M of a configuration is defined as

$$M = \frac{1}{\beta} \left[(\tau_1 + \beta - \tau_n) \left\langle i_1 | \sum_r S_r^z | i_1 \right\rangle + \sum_{i=2}^{n} (\tau_i - \tau_{i-1}) \left\langle i_i | \sum_r S_r^z | i_i \right\rangle \right]. \tag{64}$$

Like in the classical systems, similar expression, can be derived for the dependency on any parameters of interest.

Stochastic Series Expansion

In the stochastic series expansion a one-dimensional representation is sufficient to calculate properties as a function of the temperature:

$$Z = \sum_{n=0}^{\infty} g(n)\beta^n, \tag{65}$$

which is just the high temperature expansion of the partition function.

5.2 Histogram Reweighting

In a classsical system the thermal average of a quantity A at an inverse termperature β

$$\langle A(\beta) \rangle = \frac{1}{Z} \sum_c A_c e^{-\beta E_c}, \tag{66}$$

where A_c is the measurement of the observable A in the configuration c and E_c the energy of that configuration is usually estimated by the sample mean in a Monte Carlo simulation

$$\langle A(\beta) \rangle \approx \overline{A} = \frac{1}{M} \sum_i A_{c_i}. \tag{67}$$

This sampling scheme gives results only for the inverse temperature β, but actually there is much more information available than just the simple average (67). For example, in the search for a phase transition a range of temperatures needs to be explored and information at a nearby inverse temperature $\beta' \approx \beta$ can be obtained from a simulation performed at β. This is done by reweighting the configurations sampled with the Boltzmann weight $p_c = \exp(-\beta E_c)$ to obtain averages for the Boltzmann weight $p'_c = \exp(-\beta' E_c)$:

$$\langle A(\beta') \rangle \approx \frac{\sum_i A_{c_i} p'_{c_i}/p_{c_i}}{\sum_i p'_{c_i}/p_{c_i}} = \frac{\sum_i A_{c_i} e^{-\Delta\beta E_{c_i}}}{\sum_i e^{-\Delta\beta E_{c_i}}}, \tag{68}$$

where $\Delta\beta = \beta' - \beta$.

Instead of storing the full time series of measurements $\{A_{c_i}\}$ and energies $\{E_{c_i}\}$, it is sufficient to store a histogram $H(E)$, counting how often the energy level E occurs in the time series $\{E_{c_i}\}$, and the average $A(E)$ of all the measurements performed on configurations with energy E [77, 78]. Since the histogram $H(E)$ is a statistical estimator for the product $g(E)p(E)$, the average $\langle A(\beta') \rangle$ can be calculated from (57) as a sum over all energies

$$\langle A(\beta') \rangle \approx \frac{\sum_E H(E)A(E)e^{-\Delta\beta E}}{\sum_E H(E)e^{-\Delta\beta E}}. \tag{69}$$

In a model with continuous energy spectrum, such as the Heisenberg model, the energy range is divided into discrete bins of width ΔE and the histograms are constructed for these bins.

Histogram reweighting works well only if the configurations sampled at the inverse temperature β are also relevant at β', requiring that $\Delta\beta$ is small. Otherwise the errors become too large since there will not be sufficient entries in $H(E)$ for the energies E important at β'.

Multiple histograms obtained at different temperatures can be used to broaden the accessible temperature range [77, 78].

Generalized Histogram Reweighting

Histogram reweighting can not only be performed in the temperature, but also in any of the coupling constants. Using the generalized representation (54), we can calculate expectation values at coupling constant $\mu' \approx \mu$ from simulation performed for coupling constants μ by reweighting as

$$\langle A(\mu') \rangle \approx \frac{\sum_\lambda H(\lambda) A(\lambda) p(\mu', \lambda)/p(\mu, \lambda)}{\sum_\lambda H(\lambda) p(\mu', \lambda/p(\mu, \lambda)}, \tag{70}$$

since the recorded histogram $H(\lambda)$ is an estimator for $g(\lambda) p(\mu, \lambda)$.

For example, to investigate a phase transition as a function of the magnetic field h in an Ising antiferromagnet with Hamilton function (58) one would construct a histogram $H(M)$ of the magnetization $M = \sum_i \sigma_i$, and store the averages of the energy $E(M)$ and any observable $A(M)$ as a function of magnetization. Reweighting to a new field strength $h' = h + \Delta h$ is then easily done:

$$\langle A(h') \rangle \approx \frac{\sum_M H(M) A(M) e^{\beta \Delta h M}}{\sum_M H(M) e^{\beta \Delta h M}}. \tag{71}$$

Similar expressions are readily derived for quantum systems. Here we only give the expressions for temperature reweighting in the path integral representation:

$$\langle A(\beta') \rangle \approx \frac{\int dE_0 \sum_{n=0}^\infty H(n, E_0) A(n, E_0) e^{-\Delta\beta E_0} (\beta'/\beta)^n}{\int dE_0 \sum_{n=0}^\infty H(n, E_0) e^{-\Delta\beta E_0} (\beta'/\beta)^n}, \tag{72}$$

and the SSE representation:

$$\langle A(\beta') \rangle \approx \frac{\sum_{n=0}^\Lambda H(n) A(n) (\beta'/\beta)^n}{\sum_{n=0}^\Lambda H(n) (\beta'/\beta)^n}. \tag{73}$$

The integral in the path integral equation is again replaced by a sum over entries in a binned energy histogram.

Since the histogram $H(\lambda)$ is strongly peaked around the thermal expectation values for the observable Λ at a given set of parameters μ, the density of states $g(\lambda) = H(\lambda)/p(\mu, \lambda)$ can be accurately estimated only in a small region of phase space. Consequently, histogram reweighting can only be used to estimate averages at nearby parameters $\mu' \approx \mu$, where the same states are relevant. In order to explore larger parameter regions, parallel tempering or generalized ensembles with "flat" histograms can be used.

5.3 Parallel Tempering

Parallel tempering, introduced in Sect. 3.6 can be generalized in the same way. Using the generalized representation (54) we can write the combined weights of two configurations c_i and c_{i+1} with properties $\Lambda(c_i) = \lambda_i$ and $\Lambda(c_{i+1}) = \lambda_{i+1}$ simulated at parameters μ_i and μ_{i+1} as

$$\tilde{W}(c_i)\tilde{W}(c_{i+1})p(\mu_i, \lambda_i)p(\mu_{i+1}, \lambda_{i+1}) \tag{74}$$

before the swap and

$$\tilde{W}(c_{i+1})\tilde{W}(c_i)p(\mu_i, \lambda_{i+1})p(\mu_{i+1}, \lambda_i) \tag{75}$$

after the swap. Since the reduced weights $\tilde{W}(c)$ do not depend on the parameters μ, the Metropolis acceptance probability for the swap is

$$\min\left[1, \frac{p(\mu_i, \lambda_{i+1})p(\mu_{i+1}, \lambda_i)}{p(\mu_i, \lambda_i)p(\mu_{i+1}, \lambda_{i+1})}\right], \tag{76}$$

which reduces to (37) for the usual parallel tempering in temperature.

Applying parallel tempering to the magnetic field h in a classical Monte Carlo simulation and choosing a set of magnetic field strengths $\{h_i\}$, a swap between configurations at neighboring field strengths is then accepted with probability

$$\min\left[1, e^{-\beta(h_{i+1}-h_i)(M(c_{i+1})-M(c_i))}\right], \tag{77}$$

where $M(c)$ is the magnetization of the configuration c.

For a quantum system a parallel tempering swap in temperature is accepted with a probability

$$\min\left[1, e^{(\beta_{i+1}-\beta_i)(E_0(c_{i+1})-E_0(c_i))}\left(\frac{\beta_{i+1}}{\beta_i}\right)^{n_i-n_{i+1}}\right] \tag{78}$$

for continuous time path integrals and

$$\min\left[1, \left(\frac{\beta_{i+1}}{\beta_i}\right)^{n_i-n_{i+1}}\right] \tag{79}$$

in the SSE representation, where n_i and n_{i+1} refer to the order of the respective configuration.

The expressions for parallel tempering in coupling constants instead of temperatures can be derived in a similar fashion.

Optimal Temperature Sets

The algorithm to determine optimized temperature sets presented for classical systems in Sect. 3.6 can now be applied without modifications to the quantum case.

5.4 Wang-Landau Sampling and Optimized Ensembles

Just like histogram reweighting or parallel tempering, the multicanonical ensemble, Wang-Landau sampling and the optimized ensemble algorithms can not only be applied to the energy but to arbitrary observables Λ, by choosing the generalized multicanonical weight of a configuration c with $\Lambda(c) = \boldsymbol{\lambda}$:

$$p(c) = \frac{\tilde{W}(c)}{g(\boldsymbol{\lambda})} = \frac{W(c)}{p(\boldsymbol{\mu}, \boldsymbol{\lambda})g(\boldsymbol{\lambda})} \ . \tag{80}$$

The Wang-Landau algorithm can again be used to iteratively determine the generalized density of states $g(\boldsymbol{\lambda})$, and a flat histogram $H(\boldsymbol{\lambda})$ will be obtained. After the simulation, thermal averages $\langle A(\boldsymbol{\mu}) \rangle$ at arbitrary parameters $\boldsymbol{\mu}$ can be obtained using (57).

For example, to perform a multicanonical ensemble simulation in the magnetization M instead of the energy E, in an Ising model (58) we consider the density of states for the magnetization $g(M)$ of the magnetization $\boldsymbol{\lambda} = M = \sum_i \sigma_i$ and use a mixed weight

$$p(E_J, M) = e^{-\beta E_J} \frac{1}{g(M)} \tag{81}$$

where the exchange energy is defined as

$$E_J = J \sum_{\langle i,j \rangle} \sigma_i \sigma_j \ . \tag{82}$$

After the simulation, magnetic field dependent expectation values can be obtained at arbitrary values of the magnetic field $\boldsymbol{\mu} = h$:

$$\langle A(h) \rangle = \frac{\sum_M A(M)g(M)e^{\beta h M}}{\sum_M g(M)e^{\beta h M}} \tag{83}$$

Wang-Landau Sampling for Quantum Systems – High Temperature Expansion

It should now be obvious that Wang-Landau sampling and similar algorithms can be applied to quantum systems, by using the generalized density of states for quantum systems introduced in Sect. 5.1.

For thermal representations as a function of inverse temperature β, SSE offers a generalized density of states $g(n)$ in just the expansion order n, while the path integral representation requires a two-dimensional density of states $g(E_0, n)$ as a function of the non-interacting energy E_0 and the expansion order n. Since it is our experience that one-dimensional histograms perform, in general, better than higher dimensional histograms, we will focus only on the stochastic series expansion representation, for which the quantum version of Wang-Landau sampling was first introduced [35, 36].

In the following we discuss Wang-Landau sampling in the order n of a configuration, which is the equivalent of Wang-Landau sampling in energy space for classical systems. Following the generalized Wang-Landau algorithm, we replace the weight $W(c)$ of a configuration by the new weight (80) which here is

$$\frac{W(c)}{\beta^n g(n)} .$$

(84)

Normalization of $g(n)$ is simple, since $g(0)$ is just the total number of basis states (e.g. $(2+1)^N$ in a quantum spin model with N spins of size S.).

After the simulation, the partition function can easily be calculated as

$$Z = \sum_{n=0}^{\infty} g(n)\beta^n ,$$

(85)

and observables at arbitrary temperatures are calculated using (57), which here reads:

$$\langle A(\beta) \rangle = \frac{1}{Z} \sum_{n=0}^{\infty} A(n)g(n)\beta^n .$$

(86)

In any simulation the sums $\sum_{n=0}^{\infty}$ have to be truncated at some order K. What is the effect of this truncation? For canonical simulations we chose $K > O(N\beta)$, such that contributions from orders $n > K$ were negligibly small, and orders $n > K$ were never reached in the simulation. We could then ignore the cutoff. Using Wang-Landau sampling, the cutoff similarly restricts the validity of the results to temperatures where $g(K)\beta^K$, and hence contributions from terms $n > K$ are small. The cutoff K thus sets an upper bound for the accessible inverse temperatures β. In Fig. 17 results of calculations for the free energy F, entropy S and specific heat C of an $N = 10$ site antiferromagnetic Heisenberg chain, and compare to exact results. Using 10^8 sweeps, which can be performed in a few hours on a PC, the errors can be reduced down to the order of 10^{-4}. The cutoff was set to $K = 250$, restricting the accessible temperatures to $T \gtrsim 0.05J$. The sudden departure of the Monte Carlo data from the exact values below this temperature clearly shows this limit, which can be pushed lower by increasing K. The sudden deviation becomes even more pronounced in larger systems and provides a reliable indication for the range of validity of the results.

To illustrate the efficiency of the algorithm close to a thermal second order phase transition, we consider in our second example the Heisenberg antiferromagnet on a simple cubic lattice. From simulations of systems with L^3 sites,

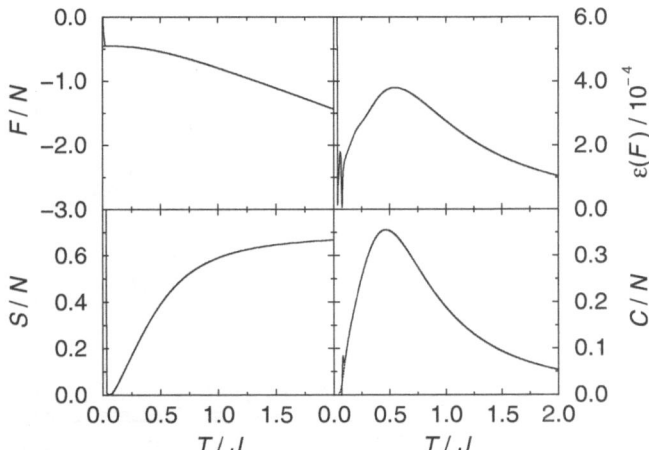

Fig. 17. Free energy (F), entropy (S) and specific heat (C) of an $N = 10$ site antiferromagnetic Heisenberg chain. Solid line correspond to the MC results, indistinguishable from the dotted lines for the exact results. Also shown is the relative error $\varepsilon(F)$ of F compared to the exact result

$L = 4, 6, 8, 12, 16$, we can calculate the staggered structure factor $S(\pi, \pi)$ for any value of the temperature using the measured histograms. Figure 18 shows the scaling plot of $S(\pi, \pi)/L^{2-\eta}$ with $\eta = 0.034$. The estimate for the critical temperature $T_c = 0.947J$, obtained in less than a day on a PC, compares well with earlier estimates [79].

Wang-Landau Sampling for Quantum Systems – Perturbation Expansion

Instead of performing a high temperature expansion, which is well suited to the investigation of finite temperature phase transitions, we can also apply Wang-Landau sampling to a perturbation expansion, better suited for quantum phase transitions. Instead of scanning a temperature range we vary one of the interactions at fixed temperature. Defining the Hamiltonian as $H = H_0 + \lambda V$ we can write the partition function equation as

$$Z = \sum_{n=0}^{\infty} \frac{\beta^n}{n!} \mathrm{Tr}(-H_0 - \lambda V)^n \equiv \sum_{n_\lambda=0}^{\infty} g(n_\lambda)\lambda^{n_\lambda} , \qquad (87)$$

where n_λ counts the powers of λ in the weight of a configuration. In this formulation of the algorithm, the cutoff K restricts the value of coupling parameter λ up to which the perturbation expansion is reliable.

The simplest case is when the parameter λ multiplies all terms on a subset of the bonds. An example is the bilayer Heisenberg quantum antiferromagnet with Hamiltonian

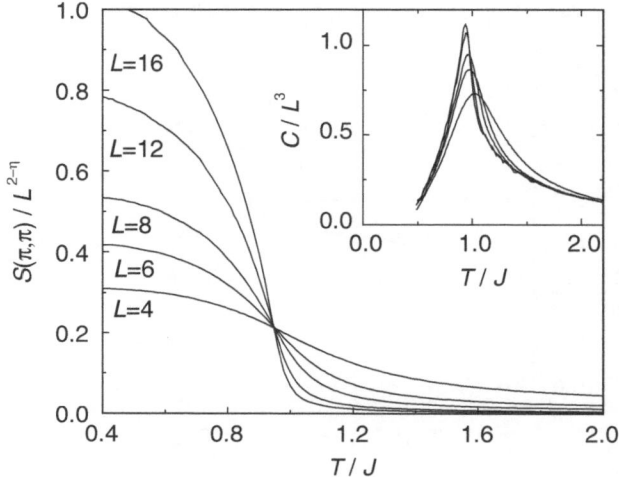

Fig. 18. Scaling plot of the staggered structure factor of a cubic antiferromagnet as a function of temperature, obtained from simulations at a fixed temperature for various lattice sizes. The inset shows the specific heat as a function of temperature. The cutoff $K = 500(L/4)^3$ restricts the accessible temperature range to $T \gtrsim 0.4J$

$$H_{\text{bilayer}} = J \sum_{l=1}^{2} \sum_{\langle i,j \rangle} \boldsymbol{S}_{i,l} \boldsymbol{S}_{j,l} + J' \sum_{\langle i,j \rangle} \boldsymbol{S}_{i,1} \boldsymbol{S}_{j,2} \,, \tag{88}$$

where $\boldsymbol{S}_{i,l}$ is the spin operator on site i in layer l. This model consists of two square lattices with coupling J between nearest neighbors inside each square lattice layer, and a coupling J' between adjacent spins in different layers, and we set $\lambda = J/J'$. For such models the algorithm remains very simple, and again only a few acceptance rates need to be changed in the code.

To normalize $g(n_\lambda)$ there are two options. If H_0 can be solved exactly, $g(0)$ can be determined directly. Otherwise, the normalization can be fixed using the high temperature expansion version of the algorithm to calculate $Z(\beta)$ at any fixed value of λ. Even without normalization we can still obtain entropy and energy differences.

We consider as an example the quantum phase transition in the bilayer Heisenberg antiferromagnet. Its ground state changes from quantum disordered to Néel ordered as the ratio $\lambda = J/J'$ of intra-plane (J) to inter-plane (J') coupling is increased [80]. From the histograms generated within *one* simulation we can calculate the staggered structure factor $S(\pi, \pi)$ of the system at *any value* of λ. In Fig. 19 we show a scaling plot of $S(\pi, \pi)/L^{2-z-\eta}$ as a function of λ. In short simulations, taking only a few days on a PC, we find the quantum critical point at $\lambda = 0.396$, which again compares well with earlier results.

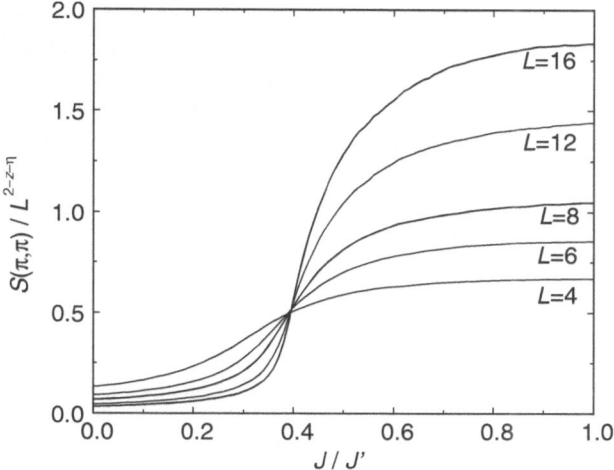

Fig. 19. Scaling plot of the staggered structure factor of a Heisenberg bilayer as a function of the coupling ratio $\lambda = J/J'$. Results are shown for various linear system sizes L. The temperature was chosen $\beta J' = 2L$, low enough to be in the scaling regime. The cutoff $K = 8L^3$ was chosen large enough to cover the coupling range $J/J' \lesssim 1$. The dynamical critical exponent of this model is $z = 1$ and $\eta = 0.034$

Optimized Ensembles for Quantum Systems

As in classical Monte Carlo simulations it turns out that a flat histogram $H(n)$ of the expansion orders n is not optimal, and again an optimized ensemble can be derived.

6 Summary

In this chapter we have reviewed two recent developments in the field of classical and quantum Monte Carlo simulations. In the first part we have presented a short review of extended ensemble techniques, including multi-canonical and parallel tempering simulations. Counter widespread assumptions "flat-histogram" multicanonical ensembles, or parallel tempering with constant, "flat" acceptance rates are not optimal. We have reviewed a recently developed iterative feedback algorithm to obtain an optimal multicanonical ensemble or an optimal choice of temperature set respectively that for a given model will maximize equilibration. We have shown examples ranging from classical spin systems, over dense liquids to protein folding.

In the second part we have given a short introduction and review of modern world line quantum Monte Carlo algorithms for quantum systems, which are free of any time discretization errors. We have highlighted the fact that the various world line representations map the quantum system to an effective classical system, and shown how efficient sampling schemes developed for classical simulations, such as histogram reweighting, parallel tempering, extended and optimized ensembles, can be applied also to quantum systems.

With examples ranging from quantum magnets to protein folding the ensemble optimization techniques have already been demonstrated to be useful for a wide range of models and scales, and will certainly help with the simulation of many other complex systems in the future.

References

1. F. Barahona (1982) On the computational complexity of Ising spin glass models. *J. Phys. A* **15**, p. 3241
2. S. Cook (1971) The complexity of theorem-proving procedures. *Conference Record of Third Annual ACM Symposium on Theory of Computing*, pp. 151–158
3. J. Kim and M. Troyer (1998) Low temperature behavior and crossovers of the square lattice quantum Heisenberg antiferromagnet. *Phys. Rev. Lett.* **80**, p. 2705
4. M. Troyer and U.-J. Wiese (2005) Computational complexity and fundamental limitations to fermionic quantum Monte Carlo simulations. *Phys. Rev. Lett.* **94**, p. 170201
5. N. Metropolis, A. R. Rosenbluth, M. N. Rosenbluth, A. H. Teller, and E. Teller (1953) Equation of state calculations on fast computing machines. *J. of Chem. Phys.* **21**, p. 1087
6. R. Swendsen and J.-S. Wang (1987) Nonuniversal critical dynamics in Monte Carlo simulations. *Phys. Rev. Lett.* **58**, p. 86
7. U. Wolff (1989) Collective Monte Carlo updating for spin systems. *Phys. Rev. Lett.* **62**, p. 361
8. O. Redner, J. Machta, and L. F. Chayes (1998) Graphical representations and cluster algorithms for critical points with fields. *Phys. Rev. E* **58**, p. 2749
9. H. Evertz, H. Erkinger, and W. von der Linden (2002) New cluster method for the Ising mode. In: *Computer Simulations in Condensed Matter Physics*, eds. D. Landau, S. P. Lewis, H.-B. Schüttler, vol. **XIV**, Springer, Berlin, p. 123
10. F. Alet, P. Dayal, A. Grzesik, A. Honecker, M. Körner, A. Läuchli, S. Manmana, I. McCulloch, F. Michel, R. Noack, G. Schmid, U. Schollwöck, F. Stöckli, S. Todo, S. Trebst, M. Troyer, P. Werner, and S. Wessel (2005) The ALPS project: open source software for strongly correlated systems. *J. Phys. Soc. Jpn. Suppl.* **74**, p. 30
11. B. A. Berg and T. Neuhaus (1991) Multicanonical algorithms for first order phase transitions. *Phys. Lett. B* **267**, p. 249
12. B. A. Berg and T. Neuhaus (1992) Multicanonical ensemble: A new approach to simulate first-order phase transitions. *Phys. Rev. Lett.* **68**, p. 9
13. F. Wang and D. P. Landau (2001) Efficient, multiple-range random walk algorithm to calculate the density of states. *Phys. Rev. Lett.* **86**, p. 2050
14. F. Wang and D. P. Landau (2001) Determining the density of states for classical statistical models: A random walk algorithm to produce a flat histogram. *Phys. Rev. E* **64**, p. 056101

15. C. Zhou and R. N. Bhatt (2005) *Phys. Rev. E* **72**, p. 025701(R)
16. H. K. Lee, Y. Okabe, and D. P. Landau (2006) Convergence and Refinement of the Wang-Landau Algorithm. *Comp. Phys. Comm.* **175**, p. 36
17. P. Dayal, S. Trebst, S. Wessel, D. Würtz, M. Troyer, S. Sabhapandit, and S. N. Coppersmith (2004) Performance limitations of flat-histogram methods. *Phys. Rev. Lett.* **92**, p. 097201
18. Y. Wu, M. Körner, L. Colonna-Romano, S. Trebst, H. Gould, J. Machta, and M. Troyer (2005) Overcoming the critical slowing down of flat-histogram Monte Carlo simulations: Cluster updates and optimized broad-histogram ensembles. *Phys. Rev. E* **72**, p. 046704
19. S. Alder, S. Trebst, A. K. Hartmann, and M. Troyer (2004) Dynamics of the Wang-Landau algorithm and Complexity of rare events for the three-dimensional bimodal Ising spin glass. *J. Stat. Mech.* P07008
20. S. Trebst, D. A. Huse, and M. Troyer (2004) Optimizing the ensemble for equilibration in broad-histogram Monte Carlo simulations. *Phys. Rev. E* **70**, p. 046701
21. S. Trebst, E. Gull, and M. Troyer (2005) Optimized ensemble Monte Carlo simulations of dense Lennard-Jones fluids. *J. Chem. Phys.* **123**, p. 204501
22. R. H. Swendsen and J. Wang (1986) Replica Monte Carlo Simulation of Spin-Glasses. *Phys. Rev. Lett.* **57**, p. 2607
23. E. Marinari and G. Parisi (1992) Simulated tempering: A new Monte Carlo scheme. *Europhys. Lett.* **19**, p. 451
24. A. P. Lyubartsev, A. A. Martsinovski, S. V. Shevkunov, and P. N. Vorontsov-Velyaminov (1992) *J. Chem. Phys.* **96**, p. 1776
25. K. Hukushima and Y. Nemoto (1996) Exchange Monte Carlo method and application to spin glass simulations. *J. Phys. Soc. Jpn.* **65**, p. 1604
26. H. G. Katzgraber, S. Trebst, D. A. Huse, and M. Troyer (2006) *J. Stat. Mech* p. P03018
27. S. Trebst, M. Troyer, and U. H. E. Hansmann (2006) Optimized parallel tempering simulations of proteins. *J. Chem. Phys.* **124** p. 174903
28. J. C. McKnight, D. S. Doering, P. T. Matsudaira, and P. S. Kim (1996) A thermostable 35-residue subdomain within villin headpiece. *J. Mol. Biol.* **260**, p. 126
29. Y. Duan and P. A. Kollman (1998) Pathways to a protein folding intermediate observed in a 1-microsecond simulation in aqueous solution. *Science* **282**, p. 740
30. B. Zagrovic, C. D. Snow, S. Khaliq, M. R. Shirts, and V. S. Pande (2002) Native-like mean structure in the unfolded ensemble of small proteins. *J. Mol. Biol.* **323**, p. 153
31. C.-Y. Liu, C.-K. Hu, and U. H. E. Hansmann (2003) Parallel tempering simulations of HP-36. *Proteins: Struct., Funct., Genet.* **52**, p. 436
32. U. H. E. Hansmann (2004) Simulations of a small protein in a specifically designed generalized ensemble. *Phys. Rev. E* **70**, p. 012902
33. M. J. Sippl, G. Némethy, and H. A. Sheraga (1984) Intermolecular potentials from crystal data. 6. Determination of empirical potentials for O-H...O=C hydrogen bonds from packing configurations. *J. Phys. Chem.* **88**, p. 6231
34. T. Ooi, M. Oobatake, G. Nemethy, and H. A. Scheraga (1987) Accessible surface-areas as a measure of the thermodynamic parameters of hydration of peptides. *Proc. Natl. Acad. Sci.* **84**, p. 3086
35. M. Troyer, S. Wessel, and F. Alet (2003) Flat histogram methods for quantum systems: algorithms to overcome tunneling problems and calculate the free energy. *Phys. Rev. Lett.* **90**, p. 120201

36. M. Troyer, F. Alet, and S. Wessel (2004) Histogram methods for quantum systems: from reweighting to Wang-Landau sampling. *Braz. J. of Physics* **34**, p. 377

37. R. Feynman (1953) Atomic theory of liquid helium near absolute zero. *Phys. Rev.* **91**, p. 1301

38. H. Trotter (1959) On the product of semi-groups of operators. *Proc. Am. Math. Soc.* **10**, p. 545

39. M. Suzuki (1976) Relationship between d-dimensional quantal spin systems and (d+1)-dimensional Ising systems – Equivalence, Critical Exponents and Systematic Approximants of the Partition Function and Spin Correlations. *Prog. Theor. Phys.* **56**, p. 1454

40. N. V. Prokofev, B. V. Svistunov, and I. S. Tupitsyn (1998) Exact, complete, and universal continuous-time worldline Monte Carlo approach to the statistics of discrete quantum systems. *JETP* **87**, p. 310

41. A. Sandvik and J. Kurkijärvi (1991) Quantum Monte Carlo simulation method for spin systems. *Phys. Rev. B* **43**, p. 5950

42. D. Handscomb (1962) The Monte Carlo method in quantum statistical mechanics. *Proc. Cambridge Philos. Soc.* **58**, p. 594

43. S. Sachdev, P. Werner, and M. Troyer (2004) Universal conductance of quantum wires near the superconductor-metal quantum transition. *Phys. Rev. Lett.* **92**, p. 237003

44. P. Werner, K. Völker, M. Troyer, and S. Chakravarty (2005) Phase diagram and critical exponents of a dissipative Ising spin chain in a transverse magnetic field. *Phys. Rev. Lett.* **94**, p. 047201

45. E. L. Pollock and D. M. Ceperley (1987) Path-integral computation of superfluid densities. *Phys. Rev. B* **36**, p. 8343

46. M. Jarrell and J. Gubernatis (1996) Bayesian inference and the analytic continuation of imaginary time Monte Carlo data. *Physics Reports* **269**, p. 133

47. W. von der Linden (1995) Maximum-entropy data analysis. *Applied Physics A* **60**, p. 155

48. K. S. D. Beach (2004) Identifying the maximum entropy method as a special limit of stochastic analytic continuation. *cond-mat/0403055*

49. M. Suzuki, S. Miyashita, and A. Kuroda (1977) Monte Carlo simulation of quantum spin systems. I. *Prog. Theor. Phys.* **58**, p. 1377

50. N. V. Prokofev, B. V. Svistunov, and I. S. Tupitsyn (1996) Exact quantum Monte Carlo process for the statistics of discrete systems. *JETP Lett.* **64**, p. 911

51. M. S. Makivić and H. Q. Ding (1991) Two-dimensional spin-1/2 Heisenberg antiferromagnet: A quantum Monte Carlo study. *Phys. Rev. B* **43**, p. 3562

52. H. G. Evertz, G. Lana, and M. Marcu (1993) Cluster algorithm for vertex models. *Phys. Rev. Lett.* **70**, p. 875

53. B. Beard and U. Wiese (1996) Simulations of discrete quantum systems in continuous Euclidean time. *Phys. Rev. Lett.* **77**, p. 5130

54. H. G. Evertz (2003) The loop algorithm. *Adv. in Physics* **52**, p. 1

55. N. Kawashima and K. Harada (2004) Recent developments of world-line Monte Carlo methods. *J. Phys. Soc. Jpn.* **73**, p. 1379

56. N. Kawashima and J. Gubernatis (1994) Loop algorithms for Monte Carlo simulations of quantum spin systems. *Phys. Rev. Lett.* **73**, p. 1295

57. N. Kawashima and J. Gubernatis (1995) Generalization of the Fortuin-Kasteleyn transformation and its application to quantum spin simulations. *J. Stat. Phys.* **80**, p. 169

58. K. Harada, M. Troyer and N. Kawashima (1998) The two-dimensional spin-1 quantum Heisenberg antiferromagnet at finite temperatures. *J. Phys. Soc. Jpn.* **67**, p. 1130

59. S. Todo and K. Kato (2001) Cluster algorithms for general-S quantum spin systems. *Phys. Rev. Lett.* **87**, p. 047203

60. N. Kawashima (1996) Cluster algorithms for anisotropic quantum spin models. *J. Stat. Phys.* **82**, p. 131

61. A. Sandvik (1999) Stochastic series expansion method with operator-loop update. *Phys. Rev. B* **59**, p. R14157

62. A. Dorneich and M. Troyer (2001) Accessing the dynamics of large many-particle systems using the stochastic series expansion. *Phys. Rev. E* **64**, p. 066701

63. O. Syljuasen and A. W. Sandvik (2002) Quantum Monte Carlo with directed loops. *Phys. Rev. E* **66**, p. 046701

64. U.-J. Wiese and H.-P. Ying (1992) Blockspin cluster algorithms for quantum spin systems. *Phys. Lett. A* **168**, p. 143

65. B. Frischmuth, B. Ammon, and M. Troyer (1996) Susceptibility and low-temperature thermodynamics of spin-1/2 Heisenberg ladders. *Phys. Rev. B* **54**, p. R3714

66. M. Greven, R. J. Birgeneau, and U. J. Wiese (1996) Monte Carlo study of correlations in quantum spin ladders. *Phys. Rev. Lett.* **77**, p. 1865

67. M. Troyer, M. Imada, and K. Ueda (1997) Critical exponents of the quantum phase transition in a planar antiferromagnet. *J. Phys. Soc. Jpn.* **66**, p. 2957

68. B. B. Beard, R. J. Birgeneau, M. Greven, and U.-J. Wiese (1998) Square-lattice Heisenberg antiferromagnet at very large correlation lengths. *Phys. Rev. Lett.* **80**, p. 1742

69. C. Yasuda, S. Todo, K. Hukushima, F. Alet, M. Keller, M. Troyer, and H. Takayama (2005) Néel temperature of quasi-low-dimensional Heisenberg antiferromagnets. *Phys. Rev. Lett.* **94**, p. 217201

70. D. C. Johnston, M. Troyer, S. Miyahara, D. Lidsky, K. Ueda, M. Azuma, Z. Hiroi, M. Takano, M. Isobe, Y. Ueda, M. A. Korotin, V. I. Anisimov, A. V. Mahajan, and L. L. Miller (2000) Magnetic susceptibilities of spin-1/2 antiferromagnetic Heisenberg ladders and applications to ladder oxide compounds. *cond-mat/0001147*

71. D. C. Johnston, R. K. Kremer, M. Troyer, X. Wang, A. Klümper, S. L. Budko, A. F. Panchula, and P. C. Canfield (2000) Thermodynamics of spin S=1/2 antiferromagnetic uniform and alternating-exchange Heisenberg chains. *Phys. Rev. B* **61**, p. 9558

72. R. Melzi, P. Carretta, A. Lascialfari, M. Mambrini, M. Troyer, P. Millet, and F. Mila (1999) $Li_2VO(Si,Ge)O_4$, a prototype of a two-dimensional frustrated quantum Heisenberg antiferromagnet. *Phys. Rev. Lett.* **85**, p. 1318

73. M. A. Korotin, I. S. Elfimov, V. I. Anisimov, M. Troyer, and D. I. Khomskii (1998) Exchange interactions and magnetic properties of the layered vanadates CaV_2O_5, MgV_2O_5, CaV_3O_7, and CaV_4O_9. *Phys. Rev. Lett.* **83**, p. 1387

74. F. Woodward, A. Albrecht, C. Wynn, C. P. Landee, and M. Turnbull (2002) Two-dimensional S= 1/2 Heisenberg antiferromagnets: Synthesis, structure, and magnetic properties. *Phys. Rev. B* **65**, p. 144412

75. G. Schmid, S. Todo, M. Troyer, and A. Dorneich (2002) Finite-temperature phase diagram of hard-core bosons in two dimensions. *Phys. Rev. Lett.* **88**, p. 167208

76. O. Nohadani, S. Wessel, B. Normand, and S. Haas (2004) Universal scaling at field-induced magnetic phase transitions. *Phys. Rev. B* **69**, p. 220402

77. A. Ferrenberg and R. Swendsen (1988) New Monte Carlo technique for studying phase transitions. *Phys. Rev. Lett.* **61**, p. 2635

78. A. Ferrenberg and R. Swendsen (1989) Optimized Monte Carlo data analysis. *Phys. Rev. Lett.* **63**, p. 1195

79. A. Sandvik (1998) Critical temperature and the transition from quantum to classical order parameter fluctuations in the three-dimensional Heisenberg antiferromagnet. *Phys. Rev. Lett.* **80**, p. 5196

80. A. Sandvik (1994) Order-disorder transition in a two-layer quantum antiferromagnet. *Phys. Rev. Lett.* **72**, p. 2777

The Coupled Electron-Ion
Monte Carlo Method

C. Pierleoni[1] and D.M. Ceperley[2]

[1] Department of Physics, University of L'Aquila, Polo di Coppito, Via Vetoio,
L'Aquila, 67010 Italy
carlo.pierleoni@aquila.infn.it

[2] Department of Physics and NCSA, University of Illinois at Urbana-Champaign,
Urbana, IL 61801, U.S.A.
david.ceperley@uiuc.edu

Carlo Pierleoni

C. Pierleoni and D.M. Ceperley: *The Coupled Electron-Ion Monte Carlo Method*, Lect. Notes
Phys. **703**, 641–683 (2006)
DOI 10.1007/3-540-35273-2_18

1 Introduction

Twenty years ago Car and Parrinello introduced an efficient method to perform Molecular Dynamics simulation for classical nuclei with forces computed on the "fly" by a Density Functional Theory (DFT) based electronic calculation [1]. Because the method allowed study of the statistical mechanics of classical nuclei with many-body electronic interactions, it opened the way for the use of simulation methods for realistic systems with an accuracy well beyond the limits of available effective force fields. In the last twenty years, the number of applications of the Car-Parrinello ab-initio molecular dynamics has ranged from simple covalent bonded solids, to high pressure physics, material science and biological systems. There have also been extensions of the original algorithm to simulate systems at constant temperature and constant pressure [2], finite temperature effects for the electrons [3], and quantum nuclei [4].

DFT is, in principle, an exact theory but the energy functional are treated approximately at the level of a self consistent mean field theory for practical purposes. Despite recent progress, DFT suffers from well-known limitations, for example, excited state properties such as optical gaps and spectra are generally unreliable. DFT shows serious deficiencies in describing van der Waals interactions, non-equilibrium geometries such as reaction barriers, systems with transition metals and/or cluster isomers with competing bonding patterns [5, 6]. As a consequence, current ab-initio predictions of metallization transitions at high pressures, or even the prediction of phase transitions are often only qualitative. Hydrogen is an extreme case [7–9] but even in silicon the diamond/β-tin transition pressure and the melting temperature are seriously underestimated [10].

An alternative route to the ground state properties of a system of many electrons in presence of nuclei is the Quantum Monte Carlo method (QMC) [6, 11]. QMC methods for bosons are "exact" meaning that all systematic errors are under control and can be reduced as much as desired with a computational cost growing as a power of the number of particles. However for fermions the "sign problem" makes a direct extension of QMC unstable and one has to resort to the "fixed node approximation" for practical calculations [6, 11]. Over the years, the level of accuracy of the fixed node approximation for simple homogeneous systems, such as ^{3}He and the electron gas, has been systematically improved [12–14]. In more complex, inhomogeneous situations such as atoms, molecules and extended systems of electrons and nuclei, progress have been somewhat slower. Nonetheless, in most cases, fixed-node QMC methods have proved to be more accurate than mean field methods (HF and DFT) [6]. Computing ionic forces with QMC to replace the DFT forces in the ab-initio MD, is more difficult and a general and efficient algorithm is still missing. Moreover, the computer time required for a QMC estimate of the electronic energy is, in general, more than for a corresponding DFT-LDA calculation. These problems have seriously limited the development of an ab-initio simulation method based on the QMC solution of the electronic problem "on the fly".

In recent years, we have developed a different strategy based entirely on the Monte Carlo method, both for solving the electronic problem, and for sampling the ionic configuration space [15, 16]. The new method, called the Coupled Electron-Ion Monte Carlo method (CEIMC) relies on the Born-Oppenheimer approximation for treating finite temperature ions coupled with ground state electrons. A Metropolis Monte Carlo simulation of the ionic degrees of freedom (represented either by classical point particles or by path integrals) at fixed temperature is performed based on the electronic energies computed during independent ground state Quantum Monte Carlo calculations. CEIMC has been applied, so far, to high pressure metallic hydrogen where it has found quite different effects of temperature than CPMD with Local Density Approximation (LDA) forces [17]. In this chapter, we present the theoretical basis of CEIMC. We start by describing, in some detail, the ground state QMC methods implemented in CEIMC, namely the Variational Monte Carlo and the Reptation Quantum Monte Carlo methods. We then describe the fixed node or restricted paths approximation necessary to treat fermions and the fixed phase method used to perform the average over boundary conditions needed for metallic system. In the subsequent section we describe how to implement the Metropolis algorithm when the energy difference has a statistical noise and discuss efficient strategies for energy differences within the CEIMC. The next section is devoted to describe the method to treat quantum mechanical protons and to integrate efficiently this new difficulty within CEIMC. Finally, we briefly review some CEIMC results for high pressure hydrogen and compare with existing CPMD results. Conclusions and perspectives for future developments are collected in the last section.

2 The Electronic Ground State Problem

Let us consider a system of N_p nuclei and N_e electrons in a volume V described by the non relativistic Hamiltonian

$$\hat{H} = -\sum_{i=1}^{N} \lambda_i \nabla_i^2 + \frac{e^2}{2} \sum_{i \neq j}^{N} \frac{z_i z_j}{|\hat{r}_i - \hat{r}_j|} \tag{1}$$

where $N = N_e + N_p$, and z_i, m_i, $\hat{r}_i$ represent the charge, mass and position operator of particle i respectively and $\lambda_i = \hbar^2/2m_i$. Let us denote with $R = (r_1, \ldots, r_{N_e})$ and $S = (r_{N_e+1}, \ldots, r_N)$ the set of coordinates of all electrons and nuclei respectively. We restrict the discussion to unpolarized systems, i.e. systems with a vanishing projection of the total spin along a given direction, say $S_z = 0$. Since the Hamiltonian does not flip spins, we can label electrons from 1 to $N_e/2$ as up spin ($\uparrow$) and electrons from $N_e/2 + 1$ to N_e as down spin ($\downarrow$).

Within the Born-Oppenheimer approximation, the energy of the system for a given nuclear state S is the expectation value of the Hamiltonian $\hat{H}$ over the corresponding exact ground state $|\Phi_0(S)\rangle$

$$E_{BO}(S) = \langle \Phi_0(S)|\hat{H}|\Phi_0(S)\rangle \tag{2}$$

which is a $3N_e$ dimensional integral over the electronic coordinates in configurational space

$$E_{BO}(S) = \int dR \, \Phi_0^*(R|S)\hat{H}(R,S)\Phi_0(R|S) = \int dR \, |\Phi_0(R|S)|^2 \, E_L(R|S) \tag{3}$$

with the *local energy* defined as

$$E_L(R|S) = \frac{\hat{H}(R,S)\Phi_0(R|S)}{\Phi_0(R|S)} \tag{4}$$

Since $\Phi_0(R|S)$ is normalized and $|\Phi_0(R|S)|^2 \geq 0$ everywhere, the $3N_e$ dimensional integral in (3) can be performed by standard Metropolis Monte Carlo by generating a Markov process which sample asymptotically $|\Phi_0(R|S)|^2$. Expectation values of any observable can be computed along the same Markov chain. In this respect, computing the properties of a many-body quantum system is similar to performing a MC calculation for a classical system. The square modulus of the ground state wave function plays the role of the classical Boltzmann distribution. An important quantity in what follows is the measure of the energy fluctuations for a given wave function. This can be defined by the variance of the local energy

$$\sigma^2(S) = \int dR \, |\Phi_0(R|S)|^2 \left(\frac{\hat{H}(R,S)\Phi_0(R|S)}{\Phi_0(R|S)} - E_{BO}(S) \right)^2$$

$$= \int dR \left(\hat{H}\Phi_0(R|S) \right)^2 - E_{BO}^2(S) \tag{5}$$

Note that for any exact eigenfunction of the hamiltonian, the local energy $E_L(R|S)$ does not depend on the electronic configuration R and is equal to the corresponding eigenvalue. This implies that the variance vanishes. This is known as the zero variance principle in Quantum Monte Carlo.

2.1 Variational QMC

The problem is to get a computable expression for the ground state wave function without solving the Schrödinger equation for the many body hamiltonian of (1), obviously an impossible task for any non trivial system. As usual in many body problems, we can resort to the variational principle which states that the energy of any proper trial state $|\Psi_T(S)\rangle$ will be greater or equal to the ground state energy

$$E_{BO}(S) \leq E_V(S) = \frac{\int dR \ \Psi_T^*(R|S)\hat{H}(R,S)\Psi_T(R|S)}{\int dR \ \Psi_T^*(R|S)\Psi_T(R|S)} \tag{6}$$

A proper trial wave function must satisfy the following requirements

- it has to have the right symmetry under particle permutation: $\Psi_T(\hat{P}R|S) = (-1)^P \Psi_T(R|S)$, where $\hat{P}$ is the permutation operator for electrons of same spin.
- the quantity $\hat{H}\Psi_T$ needs to be well defined everywhere which implies that both Ψ_T and $\nabla\Psi_T$ must be continuous whenever the potential is finite, including at the periodic boundaries.
- the integrals $\int dR|\Psi_T|^2$ and $\int dR\Psi_T^*\hat{H}\Psi_T$ must exist. Furthermore for a Monte Carlo evaluation of the variance σ^2 the integral $\int dR(\hat{H}\Psi_T)^2$ is also required to exist.

For a given trial function, it is essential to show analytically that these properties hold everywhere, in particular at the edge of the periodic box and when two particles approach each other (where generally the potential diverges). Otherwise, either the upper bound property is not guaranteed or the Monte Carlo error estimates are not valid.

The strategy in Variational Monte Carlo (VMC) is therefore to pick a proper form for a trial wave function based on physical insight for the particular system under study. In general, a number of parameters $(\alpha_1, \ldots, \alpha_k)$ will appear in the wave function to be treated as variational parameters. For any given set of $\{\alpha\}$ the Metropolis algorithm is used to sample the distribution

$$\Pi(R|S, \{\alpha\}) = \frac{\Psi_T(R|S, \{\alpha\})}{\int dR\Psi_T(R|S, \{\alpha\})} \tag{7}$$

and the electronic properties are then computed as averages over the generated Markov chain

$$E_V^T(S, \{\alpha\}) = \langle \ E_L(R|S, \{\alpha\}) \ \rangle \tag{8}$$

where the superscript T as been explicitly written to remember that the variational energy depends in general on the chosen analytical form and, for any given form, on the numerical values of the variational parameters $\{\alpha\}$.

Because of the zero variance principle stated above, the fluctuations in the local energy are entirely due to inaccuracies of the trial function for the particular configurations generated during the MC run. As the trial wave function approaches the exact eigenfunction (everywhere in configuration space!) the fluctuations decrease and the variational estimate of the energy converges more rapidly with the number of MC steps. At the same time, the estimate converges to the exact energy. This is at variance with classical Monte Carlo where fluctuations are induced by temperature.

The variational method is very powerful, and intuitively pleasing. One posits a form of the trial function and then obtains an upper bound for the energy. In contrast to other theoretical methods, no further approximations are made. The only restriction on the trial function is to be computable in a reasonable amount of time.

One of the problems with VMC is that it favors simple states over more complicated states. As an example, consider the liquid-solid transition in helium at zero temperature. The solid wave function is simpler than the liquid wave function because in the solid the particles are localized so that the phase space that the atoms explore is much reduced. This biases the difference between the liquid and solid variational energies for the same type of trial function, (e.g. a pair product form, see below) since the solid energy will be closer to the exact result than the liquid. Hence, the transition density will be systematically lower than the experimental value. Another illustration is the calculation of the polarization energy of liquid ^{3}He. The wave function for fully polarized helium is simpler than for unpolarized helium because antisymmetry requirements are higher in the polarized phase so that the spin susceptibility computed at the pair product level has the wrong sign!

The optimization of trial functions for many-body systems is time consuming, particularly for complex trial functions. The dimension of the parameter space increases rapidly with the complexity of the system and the optimization can become very cumbersome since it is, in general, a nonlinear optimization problem. Here we are not speaking of the computer time, but of the human time to decide which terms to add, to program them and their derivatives in the VMC code. This allows an element of human bias into VMC; the VMC optimization is more likely to be stopped when the expected result is obtained. The basis set problem is still plaguing quantum chemistry even at the SCF level where one only has 1-body orbitals. VMC shares this difficulty with basis sets as the problems get more complex.

Finally, the variational energy is insensitive to long range order. The energy is dominated by the local order (nearest neighbor correlation functions). If one is trying to compare the variational energy of a trial function with and without long range order, it is extremely important that both functions have the same short-range flexibility and both trial functions are equally optimized locally. Only if this is done, can one have any hope of saying anything about the long range order. The error in the variational energy is second order in the trial function, while any other property will be first order. Thus variational energies can be quite accurate while correlation functions are not very accurate.

As a consequence, the results typically reflect what was put into the trial function. Consider calculating the momentum distribution. Suppose the trial function has a Fermi surface. Then the momentum distribution will exhibits a discontinuity at k_f signaling the presence of a Fermi surface. This does not imply that the true wave function has a sharp Fermi surface.

2.2 Reptation Quantum Monte Carlo

It is possible to go beyond VMC by a number of related methods known as Projection Monte Carlo Methods. The general idea is to chose a trial function which has a non negligible overlap with the ground state wave function (in general, it is enough to require the right symmetry) and to apply a suitable projection operator which zeros out all the components of the trial wave function from the excited states in the Hilbert space of the system. We will limit the description to the method implemented in CEIMC, namely the Reptation Quantum Monte Carlo [18] or Variational Path Integral [19,20]. For the discussion of other projection QMC methods such as Diffusion Monte Carlo (DMC) and Green Function Monte Carlo (GFMC), we refer to the specialized literature [6,11].

Let us define $\{\Phi_i, E_i\}$ as the complete set of eigenfunctions and eigenvalues of the hamiltonian $\hat{H}$ in (1). Any trial state can be decomposed in the eigenstate basis:

$$|\Psi_T\rangle = \sum_i c_i |\Phi_i\rangle \tag{9}$$

where c_i is the overlap of the trial state with the ith eigenstate. Let us consider the application of the operator $e^{-t\hat{H}}$ onto this state

$$|\Psi(t)\rangle = e^{-t\hat{H}}|\Psi(0)\rangle = \sum_i c_i e^{-tE_i}|\Phi_i\rangle \tag{10}$$

with the initial state $|\Psi(0)\rangle = |\Psi_T\rangle$. Here t is a control parameter with dimension of inverse energy and we will call it "time" since it plays the role of imaginary time in the Bloch equation (see below). All excited states will be zeroed exponentially fast with increasing t, the rate of the convergence to the ground state depending on the energy gap between the ground state and the first excited state non-orthogonal to the trial function. The total energy as function of time is defined as

$$E(t) = \frac{\langle \Psi(t/2)|\hat{H}|\Psi(t/2)\rangle}{\langle \Psi(t/2)|\Psi(t/2)\rangle} = \frac{\langle \Psi_T|e^{-\frac{t}{2}\hat{H}}\hat{H}e^{-\frac{t}{2}\hat{H}}|\Psi_T\rangle}{\langle \Psi_T|e^{-t\hat{H}}|\Psi_T\rangle} \tag{11}$$

Similar to a thermal partition function, let us define the generating function of the moments of $\hat{H}$ as

$$Z(t) = \langle \Psi_T|e^{-t\hat{H}}|\Psi_T\rangle \ . \tag{12}$$

The total energy at time t is simply the derivative of the logarithm of $Z(t)$

$$E(t) = -\frac{\partial}{\partial t}lnZ(t) \tag{13}$$

and the variance of the energy is the second derivative

$$\sigma_E^2(t) = \langle (\hat{H} - E(t))^2 \rangle = -\frac{\partial}{\partial t}E(t) \geq 0 \tag{14}$$

which is non-negative by definition. This implies that the energy decreases monotonically with time. The ground state is reached at large time (much larger than the inverse gap) and

$$\lim_{t \to \infty} E(t) = E_0 \tag{15}$$

$$\lim_{t \to \infty} \sigma^2(t) = 0 \tag{16}$$

The last relation is the generalization of the zero variance principle in Projection Monte Carlo.

For observables $\hat{A}$ which do not commute with $\hat{H}$, for instance correlation functions, the average at "time" t, defined as in (11), takes the following form in configurational space[3]

$$A(t) = \langle \hat{A} \rangle_t = \frac{1}{Z(t)} \int dR_1 dR_2 dR_3 dR_4 \langle \Psi_T | R_1 \rangle \, \rho \left(R_1, R_2 \Big| \frac{t}{2} \right)$$

$$\langle R_2 | \hat{A} | R_3 \rangle \, \rho(R_3, R_4 | \frac{t}{2}) \, \langle R_4 | \Psi_T \rangle \tag{17}$$

where R_i represent the set of all electronic coordinates and $\rho(R, R', t)$ is the thermal density matrix of the system at inverse temperature t

$$\rho(R, R', t) = \langle R | e^{-t\hat{H}} | R' \rangle \tag{18}$$

Similarly, the expression of $Z(t)$ in configurational space is

$$Z(t) = \int dR_1 dR_2 \langle \Psi_T | R_1 \rangle \, \rho(R_1, R_2, t) \langle R_2 | \Psi_T \rangle \tag{19}$$

Thus, in order to compute any average over the ground state we need to know the thermal density matrix at large enough "time". Obviously, its analytic form for any non-trivial many-body system is unknown. However, at short time (or high temperature) the system approaches its classical limit and we can obtain approximations. Let us first decompose the time interval t in M smaller time intervals, $\tau = t/M$

$$\rho(R, R', t) = \langle R | e^{-(\tau \hat{H})^M} | R' \rangle = \int dR_1 \cdots dR_{M-1} \prod_{k=1}^{M-1} \langle R_{k-1} | e^{-\tau \hat{H}} | R_k \rangle \tag{20}$$

with the boundary conditions: $R_0 = R$ and $R_M = R'$ on the paths. For M large enough, we can apply the Trotter factorization to get an explicit form for the short time propagator. The simplest factorization, known as the "primitive" approximation, consists of ignoring the commutator of the kinetic and potential operators

$$\rho(R_{k-1}, R_k, \tau) = \langle R_{k-1} | e^{-\tau \hat{H}} | R_k \rangle \simeq \langle R_{k-1} | e^{-\tau \hat{K}} | R_k \rangle \, e^{-\frac{\tau}{2}[V(R_k) + V(R_{k-1})]} \tag{21}$$

---

[3] The expression gets slightly easier for observables diagonal in configurational space: $\langle R | A | R' \rangle = A(R) \delta(R - R')$.

A more accurate, but also more complex form, will be discussed later in the section on Path Integral Monte Carlo. Note that we have symmetrized the primitive form in order to reduce the systematic error of the factorization [19]. The explicit form of the kinetic propagator is the Green's function of the Bloch equation of a system of free particles [19, 21], i.e. a diffusion equation in configurational space

$$\langle R_{k-1}|e^{-\tau \hat{K}}|R_k\rangle = \left(\frac{1}{4\pi\lambda\tau}\right)^{\frac{3N}{2}} e^{-\frac{|R_k - R_{k-1}|^2}{4\lambda\tau}} \tag{22}$$

and therefore we get

$$\rho(R, R', t) = \int \prod_{k=1}^{M-1} dR_k \left[\prod_{k=1}^{M} \frac{e^{-\frac{|R_k - R_{k-1}|^2}{4\lambda\tau}}}{(4\pi\lambda\tau)^{3N/2}}\right] e^{-\tau\left[\frac{V(R_0)}{2} + \sum_{k=1}^{M-1} V(R_k) + \frac{V(R_M)}{2}\right]} \tag{23}$$

In the continuous limit ($M \to \infty$, $\tau \to 0$, $t = M\tau = $ const.) it becomes the Feynman-Kac formula [19]

$$\rho(R, R', t) = \left\langle \exp\left(-\int_0^t d\tau V(R(\tau))\right) \right\rangle_{RW} \tag{24}$$

where $\langle \cdots \rangle_{RW}$ indicate a path average over gaussian random walks $R(\tau)$ starting at $R(0) = R$ and ending at $R(t) = R'$ in a time t.

We have, in principle, developed a scheme for Monte Carlo calculations of ground state averages of a general quantum system. However, this scheme has a serious problem of efficiency which prevents its use for any non-trivial system. At the origin of the problem are the wild variations of the potential $V(R)$ in configuration space. There are cases like electron-proton systems where the potential is not bounded and therefore the primitive approximation of the propagator is not stable for any finite time step τ. However, even with well behaved effective potentials (like in Helium for instance) the large fluctuations of the potential energy would require a very small time step in order to observe convergence of the averages to their exact value. Moreover, the efficiency will degrade rapidly with more particles. The problem was recognized in the early days of QMC and the remedy introduced by Kalos in 1974. In the community of Ground State QMC it goes under the name of "importance sampling" (IS). A different strategy is applied in the PIMC community. We now describe importance sampling, not in the original form as introduced by Kalos in Green's Function Monte Carlo, but following a recent development by Baroni and Moroni in the framework of the Reptation QMC [18].

In VMC, a good trial function should have a local energy almost constant in configuration space. Let us assume to know such function Ψ_T. Let us then rewrite the hamiltonian $\hat{H}$ in terms of a new fictitious hamiltonian $\hat{\mathcal{H}}$

$$\hat{H} = \hat{\mathcal{H}} + E_L(R) \tag{25}$$

where

$$\hat{\mathcal{H}} = \lambda \left[-\nabla^2 + \frac{\nabla^2 \Psi_T}{\Psi_T} \right] \tag{26}$$

$$E_L(R) = V(R) - \lambda \frac{\nabla^2 \Psi_T}{\Psi_T} \tag{27}$$

We can now factorize the short time propagator in a different way (confr. (21))

$$\rho(R_{k-1}, R_k, \tau) \simeq \langle R_{k-1} | e^{-\tau \hat{\mathcal{H}}} | R_k \rangle \, e^{-\frac{\tau}{2} [E_L(R_k) + E_L(R_{k-1})]} \tag{28}$$

In this new form, the widely oscillating potential energy is replaced by the local energy which is much smoother for an accurate Ψ_T. We need to find the short time propagator of the importance sampling hamiltonian $\hat{\mathcal{H}}$ which is nothing but the solution of the corresponding Bloch equation [19, 21]

$$- \partial_t \rho_{IS}(R, R', t) = \hat{\mathcal{H}} \rho_{IS}(R, R', t) \tag{29}$$

$$\rho_{IS}(R, R', 0) = \delta(R - R') \tag{30}$$

It is not difficult to show by direct substitution, that the short time solution of this equation is

$$\rho_{IS}(R_{k-1}, R_k, \tau) = \frac{\Psi_T(R_{k-1})}{\Psi_T(R_k)} \left(\frac{1}{4\pi\lambda\tau} \right)^{\frac{3N}{2}} \exp\left\{ -\frac{(R_k - R_{k-1} - 2\lambda\tau F_{k-1})^2}{4\lambda\tau} \right\} \tag{31}$$

if we make the short time approximation $\left[1 + \tau \left(\nabla F + \frac{1}{F} \nabla^2 F \right) \simeq 1 \right]$. In these expressions, the drift force is defined as $F_k = F(R_k) = 2\nabla_{R_k} \ln \Psi_T(R_k)$.

This form of the short time propagator does not satisfy an important property of density matrices, namely the symmetry under exchange of the two legs, R and R'. We can remedy by taking the symmetrized density matrix as short time propagator

$$\rho_{IS}^s(R_{k-1}, R_k, \tau) = [\rho_{IS}(R_{k-1}, R_k, \tau)\rho_{IS}(R_k, R_{k-1}, \tau)]^{\frac{1}{2}} = \frac{e^{-L_s(R_{k-1}, R_k, \tau)}}{(4\pi\lambda\tau)^{\frac{3N}{2}}} \tag{32}$$

where the expression for the symmetrized link action is

$$L_s(R_{k-1}, R_k, \tau) = \frac{(R_k - R_{k-1})^2}{4\lambda\tau} + \frac{\lambda\tau}{2}(F_k^2 + F_{k-1}^2) + \frac{(R_k - R_{k-1}) \cdot (F_k - F_{k-1})}{2} \tag{33}$$

Using (32), (33) we obtain the propagator at any time t as

$$\rho(R, R', t) = \int \prod_{k=1}^{M-1} dR_k \left[\prod_{k=1}^{M} \frac{e^{-L_s(R_{k-1}, R_k, \tau)}}{(4\pi\lambda\tau)^{3N/2}} \right] e^{-\tau\left[\frac{E_L(R_0)}{2} + \sum_{k=1}^{M-1} E_L(R_k) + \frac{E_L(R_M)}{2} \right]} \tag{34}$$

In the continuum limit it is the generalized Feynman-Kac formula

$$\rho(R, R', t) = \left\langle \exp\left(-\int_0^t d\tau E_L(R(\tau))\right)\right\rangle_{DRW} \tag{35}$$

where $\langle \cdots \rangle_{DRW}$ indicate a path average over drifted random walks starting at $R(0) = R$ and ending at $R(t) = R'$ in a time t. With this form of the density matrix the generating function takes the form

$$Z(t) = \int dRdR' \Psi_T(R) \left\langle e^{-\int_0^t d\tau E_L(R(\tau))}\right\rangle_{DRW} \Psi_T(R') \tag{36}$$

and the average of a generic observable $\hat{A}$ becomes

$$A(t) = \frac{1}{Z(t)} \int \prod_{k=1}^4 dR_k \Psi_T(R_1) \left\langle e^{-\int_0^{\frac{t}{2}} d\tau E_L(R(\tau))}\right\rangle_{DRW} A(R_2, R_3)$$
$$\left\langle e^{-\int_0^{\frac{t}{2}} d\tau E_L(R(\tau))}\right\rangle_{DRW} \Psi_T(R_4) \tag{37}$$

with obvious boundary conditions on the path averages.

A special word on the calculation of the energy is in order. In the last equality in (11) the hamiltonian operator in the numerator can be pushed either to the left or to the right in such a way to operate directly on the trial state. Remembering the definition of the local energy, (4), we can write

$$E(t) = \frac{1}{Z(t)} \int dRdR' E_L(R)\Psi_T(R)\rho(R, R', t)\Psi(R')$$
$$= \frac{1}{Z(t)} \int dRdR' \Psi_T(R)\rho(R, R', t)\Psi(R')E_L(R')$$
$$= \frac{1}{2} \langle E_L(R) + E_L(R') \rangle \tag{38}$$

We use the last equality in order to improve the efficiency of the estimator. When computing the variance we push one $\hat{H}$ operator to the left and the other to the right to obtain

$$\sigma^2(t) = \int dRdR' E_L(R)\Psi_T(R)\rho(R, R', t)\Psi(R')E_L(R') - E^2(t) \tag{39}$$

Then reaching the ground state with vanishing variance means taking paths long enough (t large enough) for the correlation between the two ends to vanish. On the other hand the VMC method is obtained for $t = 0$ as $\rho(R, R', 0) = \delta(R - R')$.

2.3 Fermions

Up to now we have tacitly ignored the particle statistics and derived the formalism as if the particles were distinguishable. As far as the Hamiltonian does not depend explicitly on spin, the formalism remains valid for fermions if we consider states completely antisymmetric under particle permutation $|\hat{P}R_i\rangle = (-)^P|R_i\rangle$. The importance sampling hamiltonian $\hat{\mathcal{H}}$, defined in the same way, is symmetric under particle exchange even for antisymmetric trial states. The only place where we need care is in the initial condition of the Bloch equation, (30), which must be replaced by a completely antisymmetric delta function

$$\rho_{IS}(R, R', 0) = \mathcal{A}\delta(R - R') = \frac{1}{N!}\sum_P (-1)^P \delta(R - \hat{P}R') \tag{40}$$

Since $[\hat{H}, \hat{P}] = 0$, the imaginary time evolution preserves the symmetry and the fermion thermal density matrix takes the form

$$\rho_F(R, R', t) = \frac{1}{N!}\sum_P (-1)^P \rho_D(R, \hat{P}R', t) \tag{41}$$

where ρ_D is the density matrix of a system of distinguishable particles derived above (see (35)). The fermion density matrix between configurations R and R' at "time" t is the sum over permutations of the density matrix of distinguishable particles between the initial configurations R and the permutation of the final configuration $\hat{P}R'$, multiplied by the sign of the permutation. Each of those density matrices arises from the sum over all paths with given boundary conditions in time as expressed by the generalized Feynman-Kac formula. We can therefore think of a path in the configurational space of distinguishable particles as an object carrying not only a weight (given by the exponential of minus the integral of the local energy along the path) but also a sign fixed by its boundary conditions in time. The fermion density matrix is the algebraic sum over all those paths. While $\rho_D(R, R', t) \geq 0$ for any R' and t at given R, this property obviously does not apply to $\rho_F(R, R', t)$. This is at the origin of the "fermion sign problem" [26]. Briefly, the sign problem arises from the fact that the optimal probability to sample the electronic paths is the absolute value of the fermion density matrix which, however, is a bosonic density matrix (symmetric under particle permutation). With this sampling, the sign of the sampled paths will be left in the estimator for the averages. The normalization of any average will be given by the number of sampled positive paths minus the number of sampled negative paths. Since the sampling is bosonic, i.e. symmetric, these two numbers will eventually be equal and the noise on any average will blow up for a long enough sampling. The fundamental reason behind this pathology is that the Hilbert space of any time independent Hamiltonian (with local interactions) can be divided in the set of symmetric states, antisymmetric states and states of mixed symmetry. These sets are disjoint, e.g. any symmetric state is orthogonal to any antisymmetric one; in principle we cannot extract information for a fermionic state from a bosonic sampling [22].

A general solution of the fermion sign problem is still unavailable, although interesting algorithms have been proposed [11]. A class of methods try to build the antisymmetry constraint into the propagator, while other methods try to reformulate the problem of sampling in the space of antisymmetric wavefunctions (determinants) [23–25]. All these "fermions" methods are still at an early stage and their application has been limited so far to quite small numbers of fermions. The more robust and widely used, although approximate, method is the so-called restricted path or fixed node method [6, 26].

Within the fixed-node method, we need to consider the nodal surfaces of the fermion density matrix. For any given configuration R, these are defined by the implicit equation $\rho_F(R, R', t) = 0$, as the locations R' at which the density matrix at time t vanishes. The nodal surfaces of the initial configuration R divide the configurational space of R' in regions of positive ρ_F and regions of negative ρ_F. In terms of individual paths, the nodal locations are hypersurfaces in configurational space on which the sum of contributions of the positive and negative paths to the density matrix vanishes. Since the fermion density matrix satisfies the usual convolution relation [21]

$$\rho_F(R, R', t) = \int dR'' \rho_F(R, R'', \tau) \rho_F(R'', R', t - \tau) \qquad \forall \tau \in [0, t] \quad (42)$$

the configurations R'' belonging to the nodal surface of the initial point R at the arbitrary time τ will not contribute to the integral and therefore to the density matrix at any future time t. Therefore in constructing the fermion density matrix from R to R' at time t as sum over signed paths, we can safely disregard all those paths which have reached the nodal surface at any previous time $\tau \leq t$. If we define the reach of R at time t, $\Upsilon(R, t)$, as the set of points that can be reached from R in a time t without having crossed the nodal surfaces at previous times, the argument above can be formalized in the restricted paths identity[4]

$$\rho_F(R, R', t) = \frac{1}{N!} \sum_P (-)^P \left(\int_{Y(0)=R, Y(t)=PR'} \mathcal{D}Y \, e^{-S[Y]} \right)_{\Upsilon(R,t)} \quad (43)$$

where $S[Y]$ represent the action of the generic path Y (see (34) for its discretized form). Let us now consider the generating function $Z(t)$ of (19). Using the restricted paths identity we obtain

$$
\begin{aligned}
Z(t) &= \int dR dR' \Psi_T(R) \frac{1}{N!} \sum_P (-)^P \left(\int_{Y(0)=R, Y(t)=PR'} \mathcal{D}Y \, e^{-S[Y]} \right)_{\Upsilon(R,t)} \Psi_T(R') \\
&= \int dR dR'' \Psi_T(R) \left(\int_{Y(0)=R, Y(t)=R''} \mathcal{D}Y \, e^{-S[Y]} \right)_{\Upsilon(R,t)} \frac{1}{N!} \sum_P (-)^P \Psi_T(P^{-1} R'') \\
&= \int dR dR'' \Psi_T(R) \left(\int_{Y(0)=R, Y(t)=R''} \mathcal{D}Y \, e^{-S[Y]} \right)_{\Upsilon(R,t)} \Psi_T(R'')
\end{aligned}
$$

$$(44)$$

---

[4] In [27] an alternative proof based on the Bloch equation is provided.

where in the last equality we have used the fact that the trial wave function is antisymmetric under particle permutations: $(-)^P \Psi_T(\hat{P}^{-1}R) = \Psi_T(R)$. In this last form the integrand is always positive since for each R, the functional integral is restricted to paths inside its reach so that $\Psi_T(R)\Psi_T(R') \geq 0$ (if the nodes of the trial function are correct). Therefore using the restricted path identity we have proven that the generating function is a positive function at any time t and can be computed considering only positive paths which do not cross the nodal surfaces.

The restricted paths identity is by no means the solution of the sign problem since in order to know the nodal surfaces we have to know the density matrix itself. However rephrasing the problem in terms of spacial boundary conditions can lead to interesting approximate schemes. The nodal surface of the fermion density matrix for a system of N interacting particles is a highly non-trivial function in 6N dimensions and not much is known about it [27,28]. The approximate method, known as fixed node in ground state QMC [6] and as restricted paths in PIMC [26], consists in replacing the nodal surfaces of the exact fermion density matrix with the nodal surfaces of some trial density matrix. In ground state QMC, it is customary to restrict the class of possible nodal surfaces to time independent nodes and, within this class, the most reasonable choice is to assume the nodes of the trial wave function Ψ_T. In practice, this step requires a very minor modification of the algorithm: it is enough to ensure that $\Psi_T(R_{k-1})\Psi_T(R_k) > 0$ for any time interval along the sampled paths. In the continuous limit this restriction will enforce the restricted path identity.

The nodal surfaces of the trial wave function divide the configurational space in disconnected regions. In order to perform the configurational integral over R and R' in (44) it could appear necessary to sample all nodal regions. However, it can be proved that the nodal regions of the ground state of any Hamiltonian with a reasonable local potential are all equivalent by symmetry (Tiling theorem) [27]. This "tiling" theorem ensures that computing in a single nodal region is equivalent to a global calculation.

A further important property of the Fixed node method is the existence of a variational theorem: the FN-RQMC energy is an upper bound of the true ground state energy $E_T(\infty) \geq E_0$, and the equality holds if the trial nodes coincide with the nodes of the exact ground state [6]. Therefore for fermions, even projection methods such as RQMC are variational with respect to the nodal positions; the nodes are not optimized by the projection mechanism. The "quality" of the nodal location is important to obtain accurate results.

In some cases it is necessary to consider complex trial functions, for instance in the presence of a magnetic field [29] or in the twist average method to be discussed later [30]. In these cases we have to deal with a trial function of the form

$$\Psi_T(R) = |\Psi_T(R)|e^{i\varphi_T(R)} \tag{45}$$

where $\varphi_T(R)$, a real function, is the configuration dependent phase of the wave function. Obviously in VMC no differences arise in the use of complex trial functions other than in the estimators for the averages. For instance, the local energy is modified (confr. (27)) and contains an imaginary part

$$E_L(R) = V(R) - \lambda \frac{\nabla^2 |\Psi_T|}{|\Psi_T|} + \lambda (\nabla \varphi_T)^2 - i\lambda \left(2\nabla \varphi_T \frac{\nabla |\Psi_T|}{|\Psi_T|} + \nabla^2 \varphi_T \right) \quad (46)$$

Here, we limit the discussion to systems with time-reversal symmetry, i.e. zero magnetic fields. As Ψ_T approaches the exact ground state, the local energy approaches a real constant equal to the ground state energy while the imaginary part of the local energy vanishes. It is therefore natural to take the real part of (46) as energy estimator and, in addition to the variance, to monitor the deviation of the imaginary part from zero as an indicator of the quality of the trial wave function.

How do we have to modify the RQMC to work with complex wave functions? For general complex functions Φ, we can split the time independent Schrödinger equation into two coupled equations, one for the modulus $|\Phi|$ and one for the phase φ of the wave function

$$\left\{ -\lambda \nabla^2 + \left[V(R) + \lambda (\nabla \varphi)^2 \right] \right\} |\Phi| = E |\Phi| \quad (47)$$

$$\nabla^2 \varphi + 2\nabla \varphi \cdot \frac{\nabla |\Phi|}{|\Phi|} = 0 \quad (48)$$

In the "fixed phase" approximation [29, 31] one keeps the phase of the wave function $\varphi(R)$ fixed to some analytic form during the calculation, and solves the imaginary time dependent Schrödinger equation corresponding to the stationary problem of (47). Even for fermions this is a bosonic problem (since the modulus of the wave function must be symmetric under particle exchange) with a modified interaction $\left[V(R) + \lambda (\nabla \varphi)^2 \right]$. We can still perform the IS transformation and the formalism remains the same if the local energy is defined as the real part of (46). Note that the fixed node constraint for real trial functions can be recast into the fixed phase algorithm if we write $\varphi_T(R) = \pi[1 - \theta(\frac{\Psi_T(R)}{|\Psi_T(R)|})]$ so that the phase of the trial function changes by π across the nodes at it should. Since the phase is a step function, its gradient is a δ function and provides an infinite contribution in the action of paths crossing the nodes, i.e. a vanishing probability to cross the nodes.

In this section we have not explicitly indicated the dependence on the ionic state S. In the BO approximation, ions play the role of external fields for the electronic system so that their coordinates appear explicitly in the Hamiltonian, in the trial state, in the local energy and in drift force for the IS procedure.

2.4 Trial Wave Functions for Hydrogen

In this subsection, we describe some general properties of the trial wave functions for electronic systems. We will restrict our discussion to the case of a proton-electron system, i.e. hydrogen and refer to the literature for more complex systems (heavier elements) [6, 11]. In particular, we will not discuss the use of pseudopotentials in QMC and the related trial wave functions which, however, will be an important issue in future extensions of CEIMC.

The Pair Product Trial Function

The pair product trial wave function is the simplest extension of the Slater determinant of single particle orbitals used in mean field treatment of electronic systems (HF or DFT). This is also the ubiquitous form for trial functions in VMC

$$\Psi_{SJ}(R, \Sigma|S) = \exp\left(-\sum_{i<j} u_{ij}(r_{ij})\right) det[\theta_k(\boldsymbol{r}_i, \sigma_i|S)] \qquad (49)$$

where $\Sigma = \{\sigma_1, \ldots, \sigma_{N_e}\}$ is the set of spin variables of the electrons, $\theta_k(\boldsymbol{r}, \sigma|S)$ is the kth spin orbital for the given nuclear configuration and $\theta_k(\boldsymbol{r}_i, \sigma_i|S)$ is the Slater matrix. The additional term $u_{ij}(r_{ij})$ is the "pseudopotential" or pair correlation factor which introduces explicitly the two body correlations into the many body wave function. This term is of bosonic nature, i.e. is symmetric under particle exchange, while the antisymmetry is ensured by the determinant. Often the general form of both θ_k and u_{ij} are derived by some appropriate theory and then used in connection with some free variational parameters to be optimized.

Let us discuss first the appropriate form of the "pseudopotential" for Coulomb systems. There are important analytical properties for the "pseudopotential" that can be easily derived. Consider bringing two particles together and let us examine the dominant terms in the local energy. In a good trial action, the singularities in the kinetic energy must compensate those in the potential energy. The local energy for the two particle system is

$$e_L(r_{ij}) = e^{u_{ij}} \hat{h}_{ij}[e^{-u_{ij}}] = v_{ij}(r_{ij}) + \lambda_{ij}\left[\nabla^2 u_{ij} - (\nabla u_{ij})^2\right] \qquad (50)$$

where $\lambda_{ij} = \lambda_i + \lambda_j$, $\hat{h}_{ij}$ is the two body hamiltonian with the interaction potential v_{ij} and the trial wave function $\exp[-u_{ij}]$. Spin symmetry has been disregarded as well as the trivial term related to the center of mass motion. Therefore the short distance behavior of any good form of the pseudopotential should follow the solution of the two body Schrödinger equation. For the Coulomb potential the "cusp" condition derives from this constraint. Indeed substituting $v_{ij} = e^2 z_i z_j / r$ in (50) and zeroing the coefficient of the dominant power of r for $r \to 0$ provides

$$\left.\frac{du_{ij}}{dr}\right|_0 = -\frac{e^2 z_i z_j}{\lambda_{ij}(D-1)} \qquad (51)$$

It is also important to reproduce the correct behavior at large distances where a description in terms of collective coordinates is appropriate. The long-wavelength modes are important for the low energy response properties and are also the slowest modes to converge in QMC. It is possible to show that, within the Random Phase Approximation (RPA), the local energy is minimized by imposing

$$u_k^{ee} = -\frac{1}{2} + \sqrt{1 + a_k} \tag{52}$$

$$u_k^{ep} = \frac{-a_k}{\sqrt{1 + a_k}} \tag{53}$$

with $a_k = 12r_s/k^4$ and the electron sphere radius r_s is related to the volume per electron by $v = 4\pi r_s^3/3$ in atomic units [32]. Here $u_k^{\alpha\beta}$ indicates the Fourier Transform of $u_{\alpha\beta}(r)$. The obtained form of the "pseudopotential" is correct at short and long distances but not necessarily in between because of the approximation. One can improve slightly the quality of the VMC results considering the form

$$\tilde{u}_{ij}(r) = u_{ij}^{RPA}(r) - \alpha_{ij}e^{-r^2/w_{ij}^2} \tag{54}$$

with the variational parameter α_{ij}, w_{ij}. The additional term preserves the short and long distance behavior of the RPA function. This form of the pair trial function introduces four variational parameters, namely $\alpha_{ee}, w_{ee}, \alpha_{ep}, w_{ep}$.

We discuss now the choice of the spin orbitals. The spin-orbitals are conceptually more important than the pseudopotential because they provide the nodal structure of the trial function. With the fixed node approximation in RQMC, the projected ground state has the same nodal surfaces of the trial function, while the other details of the trial function are automatically "optimized" for increasing projection time. It is thus important that the nodes provided by given spin-orbitals be accurate. Moreover, the optimization of nodal parameters (see below) is, in general, more difficult and unstable than for the pseudopotential parameters [6].

The simplest form of spin-orbitals for a system with translational invariance are plane waves (PW) $\theta_k(r, \sigma) = \exp[\imath k \cdot r]$. This form was used in the first QMC study of metallic hydrogen [33]. It is particularly appealing for its simplicity and still qualitatively correct since electron-electron and electron-proton correlations are considered through the "pseudopotential". The plane waves orbitals are expected to reasonably describe the nodal structure for metallic atomic hydrogen, but no information about the presence of protons appears in the nodes with PW orbitals.

For insulating molecular hydrogen (i.e for $r_s \geq 1.5$), it is preferred to use localized gaussian orbitals. There are different possibilities: a single isotropic gaussian centered at the middle of the bond was used in the first QMC study [33], while a single multivariate gaussian was used in the first CEIMC attempt [15,16]. Another possibility is to form a molecular orbitals as linear combination of two atomic gaussian orbitals centered on each proton:

$$\theta_k(r, s_{k,1}, s_{k,2}) = \exp(-c|r - s_{k,1}|^2) + \exp(-c|r - s_{k,2}|^2) \tag{55}$$

where $s_{k,j}$ is the position of the jth proton of the kth molecule. This kind of orbitals have a single variational parameter c. At present we are experimenting using a trial molecular wave function with these orbitals multiplied by the corrected RPA Jastrow within the CEIMC.

The trouble with this strategy is that one should know which phase is stable before performing the calculation. This is typical of ground state studies. However in CEIMC we would rather let the system find its own state for given temperature and density (or pressure). In particular, this approach is not appropriate to address the interesting region of molecular dissociation and metallization. This problem can be solved by using orbitals obtained as solution of a single-electron problem as in band structure calculations or in self-consistent mean field methods. In previous works on ground state hydrogen, the single electron orbitals for a given protonic state S, were obtained from a DFT-LDA calculation [34–36]. One of this study [35] established that energies from plane-waves determinants in metallic hydrogen are higher than the more accurate estimates from DFT-LDA orbitals by 0.05 eV/atom at the density at which the transition between molecular and metallic hydrogen is expected ($r_s = 1.31$). Obtaining the orbitals from a DFT-LDA calculation has, however, several drawbacks in connection with the CEIMC. While for protons on a lattice we can solve the self-consistent theory for a primitive cell only, in a disordered configuration, we need to consider the entire simulation box. This is very expensive in CPU time and memory for large systems. Moreover, combining the LDA orbitals with Jastrow to improve the accuracy is not straightforward; substantial modification of the orbitals might be necessary requiring a reoptimization of the orbitals and the correlation factors, in principle, at each new ionic position.

Beyond the Pair Product Trial Function

Over the years there have been important progress in finding trial functions substantially more accurate then the pair product form for homogeneous systems [12,13]. Within the generalized Feynman-Kac formalism, it is possible to systematically improve a given trial function [13,14]. The first corrections to the pair product action with plane wave orbitals are a three-body correlation term which modifies the correlation part of the trial function (Jastrow) and a "backflow" transformation which changes the orbitals and therefore the nodal structure (or the phase) of the trial function [14]. The new trial function has the form

$$\Psi_T(R, \Sigma|S) = det[\theta_k(x_i, \sigma_i|S)]e^{-U_2-U_3} \tag{56}$$

where $U_2 = \sum_{i<j} \tilde{u}_{ij}$ is the two body "pseudopotential" discussed before, U_3 the three-body term of the form

$$U_3 = -\sum_{i=1}^{N_e} \left[\sum_{j=1}^{N} \xi_{ij}(r_{ij}) r_{ij} \right]^2 \tag{57}$$

and finally the "quasiparticle" coordinates appearing in the plane wave orbitals are given by

$$x_i = r_i + \sum_{j=1}^{N} \eta_{ij}(r_{ij})r_{ij}; \quad (i = 1, \ldots, N_e) \tag{58}$$

The functional form of the three-body term is that of a squared two-body force so its evaluation is not slower than a genuine 2-body term. In the homogeneous electron system, this term is particularly relevant at low density where correlation effects are dominant. On the other hand, the backflow transformation is more relevant at high density because in this limit the fermionic character of the system dominates. The same general framework should hold for hydrogen although a throughout study of the relative importance of those effects with density is still missing. The fundamental improvement of backflow orbitals for metallic hydrogen is that the nodal structure of the wave functions depends now on the proton positions. This provides better total energies for static protonic configurations [14] and improved energy differences and liquid structure in CEIMC [16,37]. However it is not clear how appropriate this kind of wave function will be when entering in the molecular phase of hydrogen.

The unknown functions, $\xi_{ij}(r), \eta_{ij}(r)$ in (57) and (58) need to be parameterized in some way. In a first attempt we have chosen gaussians with variance and amplitude as new variational parameters [16]. This form was shown to be suitable for homogeneous electron gas [13]. Approximate analytical forms for $\xi_{ij}(r)$ and $\eta_{ij}(r)$, as well as for the two-body pseudopotential, have been obtained later in the framework of the Bohm-Pines collective coordinates approach [14]. This form is particularly suitable for the CEIMC because there are no parameters to be optimized. This trial function is faster than the pair product trial function with the LDA orbitals, has no problems when protons move around and its nodal structure has the same quality as the corresponding one for the LDA Slater determinant [14]. We have extensively used this form of the trial wave function for CEIMC calculations of metallic atomic hydrogen.

Trial Function Optimization

For metallic hydrogen we have described a parameter-free trial function which does not need optimization. However, if we use the pair proton action both for molecular or LDA orbitals, we are left with free parameters in the Jastrow factor and with the width of the gaussians for molecular orbitals. Optimization of the parameters in a trial function is crucial for the success of VMC. Bad upper bounds do not give much physical information. Good trial functions will be needed in the Projector Monte Carlo method. First, we must decide on what to optimize and then how to perform the optimization. There are several possibilities for the quantity to optimize and depending on the physical system, one or other of the criteria may be best.

- The variational energy: E_V. If the object of the calculation is to find the least upper bound one should minimize E_V. There is a general argument suggesting that the trial function with the lowest variational energy will maximize the efficiency of Projector Monte Carlo [38].
- The variance of the local energy: $\sigma^2 = \int |\mathcal{H}\Psi|^2 - E_V^2$. If we assume that every step on a QMC calculation is statistically uncorrelated with the others, then the variance of the average energy will equal σ^2/p where p is the number of steps. The minimization of σ^2 is statistically more robust than the variational energy because it is a positive definite quantity with zero as minimum value. One can also minimize a linear combination of the variance and the variational energy.
- The overlap with the exact wave function: $\int \Psi\phi$. If we maximize the overlap, we find the trial function closest to the exact wave function in the least squares sense. This is the preferred quantity to optimize if you want to calculate correlation functions, not just ground state energies since, then, the VMC correlation functions will be closest to the true correlation functions. Optimization of the overlap will involve a Projector Monte Carlo calculation to determine the change of the overlap with respect to the trial function so it is more complicated and rarely used.

The most direct optimization method consists of running independent VMC calculations using different set of numerical values for the variational parameters. One can fit the energies to a polynomial, performing more calculations near the predicted minimum and iterating until convergence in parameter space is attained. The difficulty with this direct approach is that close to the minimum, the independent statistical errors will mask the variation with respect to the trial function parameters. This is because the derivative of the energy with respect to trial function parameters is very poorly calculated. Also, it is difficult to optimize, in this way, functions involving more than 3 variational parameters because so many independent runs are needed to cover the parameter space.

A correlated sampling method, known as reweighting [39,40] is much more efficient. One samples a set of configurations $\{R_j\}$ (usually several thousand points at least) according to some distribution function, usually taken to be the square of the wavefunction for some initial trial function: $|\Psi_T(R; \{\alpha\}_0)|^2$. Then, the variational energy (or the variance) for trial function nearby in parameter space can be calculated by using the same set of points:

$$E_v(a) = \frac{\sum_j w(R_j, \{\alpha\})E_L(R_j, \{\alpha\})}{\sum_j w(R_j, \{\alpha\})}, \qquad (59)$$

where the weight factor, $w(R) = |\Psi_T(R; \{\alpha\})/\Psi_T(R; \{\alpha\}_0)|^2$, takes into account that the distribution function changes as the variational parameters change. One then can use a minimizer to find the lowest variational energy or variance as a function of $\{\alpha\}$ keeping the configurations fixed. However, there is an instability: if the parameters move too far away, the weights span too large of a range and the error bars of the energy become large. The number of effective points of a weighted sum is:

$$N_{eff} = \left(\sum w_j\right)^2 / \sum w_j^2. \tag{60}$$

If this becomes much smaller than the number of points, one must resample and generate some new points. When minimizing the variance, one can also simply neglect the weight factors. Using the reweighting method one can find the optimal value of wavefunction containing tens of parameters.

2.5 Twist Average Boundary Conditions

Almost all QMC calculations in periodic boundary conditions have assumed that the phase of the wave function returns to the same value if a particle goes around the periodic boundaries and returns to its original position. However, with these boundary conditions, delocalized fermion systems converge slowly to the thermodynamic limit because of shell effects in the filling of single particle states. Indeed, with periodic boundary conditions the Fermi surface of a metal will be reduced to a discrete set of points in k-space. The number of k-points is equal to the number of electrons of same spin and therefore it is quite limited.

One can allow particles to pick up a phase when they wrap around the periodic boundaries,

$$\Psi_{\boldsymbol{\theta}}(\boldsymbol{r}_1 + L\boldsymbol{n}, \boldsymbol{r}_2, \cdots) = e^{i\boldsymbol{\theta}}\Psi_{\boldsymbol{\theta}}(\boldsymbol{r}_1, \boldsymbol{r}_2, \ldots) . \tag{61}$$

where we have assumed a cubic box of size L and $\boldsymbol{n}$ is a vector of integers. The boundary condition $\boldsymbol{\theta} = 0$ is periodic boundary conditions (PBC), and the general condition with $\boldsymbol{\theta} \neq 0$, twisted boundary conditions (TBC). If the periodic boundaries are used in all directions, each dimension can have an independent twist. Hence, in three dimension (3D), the twist angle is a three components vector. The free energy and therefore all equilibrium properties are periodic in the twist: $F(\boldsymbol{\theta} + 2\pi\boldsymbol{n}) = F(\boldsymbol{\theta})$ so that each component of the twist can be restricted to be in the range $-\pi < \theta_i \leq \pi$.

The use of twisted boundary conditions is commonplace for the solution of the band structure problem for a periodic solid, particularly for metals. In order to calculate properties of an infinite periodic solid, properties must be averaged by integrating over the first Brillouin zone.

For a degenerate Fermi liquid, finite-size shell effects are much reduced if the twist angle is averaged over: twist averaged boundary conditions (TABC) [30]. For any given property $\hat{A}$ the TABC is defined as

$$\langle \hat{A} \rangle = \int_{-\pi}^{\pi} \frac{d\boldsymbol{\theta}}{(2\pi)^d} \langle \Psi_{\boldsymbol{\theta}}|\hat{A}|\Psi_{\boldsymbol{\theta}} \rangle \tag{62}$$

TABC is particularly important in computing properties that are sensitive to the single particle energies such as the kinetic energy and the magnetic susceptibility. By reducing shell effects, accurate estimations of the thermodynamic limit for these properties can be obtained already with a limited number of electrons. What makes this very important is that the most accurate quantum methods have computational demands which increase rapidly with the number of fermions. Examples of such methods are exact diagonalization (exponential increase in CPU time with N), variational Monte Carlo (VMC) with wave functions having backflow and three-body terms [13] (increases as N^4), and transient-estimate and released-node Diffusion Monte Carlo methods [41] (exponential increase with N). Moreover, size extrapolation is impractical within CEIMC since it would have to be performed for any proposed ionic move prior to the acceptance test. Methods which can extrapolate more rapidly to the thermodynamic limit are crucial in obtaining high accuracy.

Twist averaging is especially advantageous in combination with stochastic methods (i.e. QMC) because the twist averaging does not necessarily slow down the evaluation of averages, except for the necessity of doing complex rather than real arithmetic. In a metallic system, such as hydrogen at very high pressure, results in the thermodynamic limit require careful integration near the Fermi surface because the occupation of states becomes discontinuous. Within LDA this requires "k–point" integration, which slows down the calculation linearly in the number of k-points required. Within QMC such k-point integration takes the form of an average over the (phase) twist of the boundary condition and can be done in parallel with the average over electronic configurations without significantly adding to the computational effort.

In CEIMC we can take advantage of the twist averaging to reduce the noise in the energy difference for the acceptance test of the penalty method (see below). In the electron gas, typically 1000 different twist angles are required to achieve convergence [30]. We have used the same number of twist angles in CEIMC calculations of metallic hydrogen. Different strategies can be used to implement the TABC [30]. We have used a fixed 3D grid in the twist angle space, at each grid point run independent QMC calculations and then averaged the resulting properties. This procedure can be easily and efficiently implemented on a parallel computer. Recently, we have devised a sampling procedure to randomize the grid points at each ionic step. We have limited experience, but, so far, we have evidence that good convergence in the electronic and ionic properties can already be reached for a number of twist angles as low as 30 [42].

2.6 Sampling Electronic States: The "Bounce" Algorithm

In this section we describe the way we have implemented electronic move in the CEIMC method. In particular we present an original algorithm for RQMC, particularly suitable for CEIMC, called the "bounce" algorithm.

First, how do the particles move in VMC? In the continuum it is usually more efficient to move the particles one at a time by adding a random vector to a particle's coordinate, where the vector is either uniform inside of a cube, or is a normally distributed random vector centered around the old position. Unfortunately, this procedure cannot be used with backflow orbitals. This is because the backflow transformation couples all the electronic coordinates in the orbitals so that once a single electron move is attempted the entire Slater determinant needs to be recomputed, an $O(N^3)$ operation. It is much more efficient to move all electrons at once, although global moves could become inefficient for large systems.

Next we describe the electronic sampling within RQMC. In the original work on RQMC [18], the electronic path space was sampled by a simple reptation algorithm, an algorithm introduced to sample the configurational space of linear polymer chains [43]. Remember that in RQMC the electronic configurational space is the space of $3N_e$-dimensional random paths of length "t". In practice, the imaginary time is discretized in M time slices $\tau = t/M$ and the paths become discrete linear chains of $M + 1$ beads. Let us indicate with Q the entire set of $3N(M + 1)$ coordinates $Q = \{R_0, \ldots, R_M\}$. According to (33), (34) and (36), the path distribution is

$$\Pi(Q) = |\Psi_T(R_0)\Psi_T(R_M)|e^{-\sum_{k=1}^{M} L_s(R_{k-1}, R_k, \tau)}$$
$$e^{-\tau\left[\frac{E_L(R_0)}{2} + \sum_{k=1}^{M-1} E_L(R_k) + \frac{E_L(R_M)}{2}\right]} \tag{63}$$

Given a path configuration Q, a move is done in two stages. First one of the two ends (either R_0 or R_M) is sampled with probability $1/2$ to be the growth end R_g. Then a new point near the growth end is sampled from a Gaussian distribution with center at $R_g + 2\lambda\tau F(R_g)$. In order to keep the number of links on the path constant, the old tail position is discarded in the trial move. The move is accepted or rejected with the Metropolis formula based on the probability of a reverse move. For use in the following, let us define the direction variable d as $d = +1$ for a head move ($R_g = R_M$), and $d = -1$ for a tail move ($R_g = R_0$). In standard reptation, the direction d is chosen randomly at each attempted step. The transition probability $P(Q \to Q')$ is the product of an attempt probability $T_d(Q \to Q')$ and an acceptance probability $a_d(Q \to Q')$. The paths distribution $\Pi(Q)$ does not depend on the direction d in which it was constructed. In the Metropolis algorithm, the acceptance probability for the attempted move is

$$a_d(Q \to Q') = \min\left[1, \frac{\Pi(Q')T_{-d}(Q' \to Q)}{\Pi(Q)T_d(Q \to Q')}\right] \tag{64}$$

which ensures that the transition probability $P_d(Q \to Q')$ satisfies detailed balance

$$\Pi(Q)P_d(Q \to Q') = \Pi(Q')P_{-d}(Q' \to Q) \tag{65}$$

The autocorrelation time of this algorithm, that is the number of MC steps between two uncorrelated configurations, scales as $[M^2/A]$, where A is the acceptance rate of path moves, an unfavorable scaling for large M (i.e. large projection time t). Moreover the occasional appearance of persistent configurations bouncing back and forth without really sampling the configuration space has been previously observed [44]. These are two very unfavorable features, particularly in CEIMC, where we need to perform many different electronic calculations. There is a premium for a reliable, efficient and robust algorithm.

We have found that a minimal modification of the reptation algorithm solves both of these problems. The idea is to chose randomly the growth direction at the beginning of the Markov chain, and reverse the direction upon rejection only, the "bounce" algorithm.

What follows is the proof that the bounce algorithm samples the correct probability distribution $\Pi(Q)$. The variable d is no longer randomly sampled, but, as before, the appropriate move is sampled from the same Gaussian distribution $T_d(Q \to Q')$ and accepted according to the (64). In order to use the techniques of Markov chains, we need to enlarge the state space with the direction variable d. In the enlarged configuration space $\{Q, d\}$, let us define the transition probability $P(Q, d \to Q', d')$ of the Markov chain. The algorithm is a Markov process in the extended path space, and assuming it is ergodic, it must converge to a unique stationary state, $\Upsilon(Q, d)$ satisfying the eigenvalue equation:

$$\sum_{Q,d} \Upsilon(Q, d) \, P(Q, d \to Q', d') = \Upsilon(Q', d') . \tag{66}$$

We show that our desired probability $\Pi(Q)$ is solution of this equation. Within the imposed rule, not all transitions are allowed, but $P(Q, d \to Q', d') \neq 0$ for $d = d'$ and $Q \neq Q'$ (accepted move), or $d' = -d$ and $Q = Q'$ (rejected move) only. Without loss of generality let us assume $d' = +1$ since we have symmetry between ± 1. Equation (66) with $\Upsilon(Q, d)$ replaced by $\Pi(Q)$ is

$$\Pi(Q')P(Q', -1 \to Q', 1) + \sum_{Q \neq Q'} \Pi(Q)P(Q, 1 \to Q', 1) = \Pi(Q') . \tag{67}$$

Because of detailed balance (65), we have

$$\Pi(Q)P(Q, 1 \to Q', 1) = \Pi(Q')P(Q', -1 \to Q, -1)$$

which, when substituted in this equation gives

$$\Pi(Q') \left[P(Q', -1 \to Q', 1) + \sum_{Q} P(Q', -1 \to Q, -1) \right] = \Pi(Q') . \tag{68}$$

Note that we have completed the sum over Q with the term $Q = Q'$ because its probability vanishes. The term in the bracket exhausts all possibilities for a move from the state $(Q', -1)$, thus it adds to one. Hence $\Pi(Q)$ is a solution of (66) and by the theory of Markov chains, it is the probability distribution of the stationary state.

3 Sampling Ionic States: The Penalty Method

In Metropolis Monte Carlo a Markov chain of ionic states S is generated according to the Boltzmann distribution $P(S) \propto e^{-\beta E_{BO}(S)}$ where $E_{BO}(S)$ is the Born-Oppenheimer energy for the ionic configuration S, and β the inverse temperature. From the state S a trial state S' is proposed with probability $T(S \rightarrow S')$ and the detailed balance condition is imposed by accepting the move with probability

$$A(S \rightarrow S') = \min \left[1, \frac{T(S' \rightarrow S)e^{-\beta E_{BO}(S')}}{T(S \rightarrow S')e^{-\beta E_{BO}(S)}} \right] \qquad (69)$$

Under quite general conditions on the system and on the a-priori transition probability $T(S \rightarrow S')$, after a finite number of MC steps the Markov chain so generated will visit the states of the configurational space with a frequency proportional to their Boltzmann's weight [43].

In CEIMC estimate of $E_{BO}(S)$ is affected by statistical noise. If we ignore the presence of noise and we use the standard Metropolis algorithm, the results will be biased, the amount of bias increasing for increasing noise level. A possible solution would be to run very long QMC calculations in order to get a negligibly small noise level resulting in a negligible bias. However, the noise level decreases as the number of independent samples to the power $1/2$, that is to decrease the noise level by one order of magnitude we should run 100 times longer, an unfavorable scaling if we realize that we have to repeat such calculation for any attempted move of the ions. The less obvious but far more efficient solution is to generalize the Metropolis algorithm to noisy energies. This is done by the Penalty Method [45]. The idea is to require the detailed balance to hold on average and not for any single energy calculation.

Let us consider two ionic states (S, S') and call $\delta(S, S')$ the "instantaneous" energy difference times the inverse temperature. Let us further assume that the average and the variance of $\delta(S, S')$ over the noise distribution $P(\delta|S \rightarrow S')$ exhist

$$\Delta(S, S') = \beta \,]E_{BO}(S') - E_{BO}(S)] = \langle \, \delta(S, S') \, \rangle \qquad (70)$$
$$\sigma^2(S, S') = \langle \, (\delta(S, S') - \Delta(S, S'))^2 \, \rangle \qquad (71)$$

We introduce the "instantaneous" acceptance probability, $a(\delta|S, S')$ and impose the detailed balance to hold for the average of $a(\delta|S, S')$ over the distribution of the noise

$$A(S \rightarrow S') = e^{-\Gamma(S,S')} A(S' \rightarrow S) \qquad (72)$$

where

$$A(S \rightarrow S') = \int_{-\infty}^{\infty} d\delta P(\delta|S \rightarrow S') a(\delta|S, S') \qquad (73)$$

and we have defined

$$\Gamma(S, S') = \Delta(S, S') - \ln\left[\frac{T(S' \to S)}{T(S \to S')}\right] \quad (74)$$

If we assume the quite general conditions $a(\delta|S, S') = a(\delta)$ and $P(\delta|S \to S') = P(-\delta|S' \to S)$ to hold, the detailed balance can be written

$$\int_{-\infty}^{\infty} d\delta \, P(\delta|S \to S') \left[a(\delta) - e^{-\Gamma} a(-\delta)\right] = 0 \quad (75)$$

which, supplemented by the condition $a(\delta) \geq 0$, is the equation to solve in order to obtain the acceptance probability $a(\delta)$. The difficulty is that during the MC calculation we do not know either $P(\delta|S \to S')$ nor $\Delta(S, S')$.

In order to make progress let us assume, as it happens in many interesting cases, that the noise of the energy difference is normally distributed so that

$$P(\delta|S \to S') = (2\pi\sigma^2)^{-1/2} \exp\left[-(\delta - \Delta)^2/2\sigma^2\right] \quad (76)$$

Let us, moreover, assume that we know the value of the variance σ. It is not difficult to check that the solution of (75) is

$$a_n(\delta|\sigma) = \min\left[1, \frac{T(S' \to S)}{T(S \to S')} \exp\left(-\delta - \frac{\sigma^2}{2}\right)\right] \quad (77)$$

The uncertainty in the energy difference just causes a reduction in the acceptance probability by an amount $\exp(-\sigma^2/2)$ for $\delta > -\sigma^2/2$. The integral of $a_n(\delta|\sigma)$ over the guassian measure provides

$$A(S \to S') = \tfrac{1}{2} \, \text{erfc}\left\{\left[\sigma^2/2 + \Gamma(S \to S')\right]/(\sqrt{2}\sigma)\right\} + $$
$$\tfrac{1}{2} \, \text{erfc}\left\{\left[\sigma^2/2 - \Gamma(S \to S')\right]/(\sqrt{2}\sigma)\right\} e^{-\Gamma(S \to S')} \quad (78)$$

which satisfies (72) since $\Gamma(S' \to S) = -\Gamma(S \to S')$. Note that (77) reduces to the standard Metropolis form for vanishing σ [43].

An important issue is to verify that the energy differences are normally distributed. Recall that if the moments of the energy difference are bounded, the central limit theorem implies that given enough samples, the distribution of the mean value will be Gaussian. Careful attention to the trial function to ensure that the local energies are well behaved may be needed.

In practice not only the energy difference Δ but also the variance σ is unknown and must be estimated from the data. Let us assume that, for a given pair of ionic states (S, S'), we generate n statistically uncorrelated estimates of the energy difference $\{y_1, \ldots, y_n\}$ each normally distributed with their first and second moments defined in the usual way

$$\Delta = \langle y_i \rangle \quad (79)$$
$$\sigma^2 = \langle (y_i - \Delta)^2 \rangle \quad (80)$$

Unbiased estimates of Δ and σ^2 are

$$\delta = \frac{1}{n} \sum_{i=1}^{n} y_i \tag{81}$$

$$\chi^2 = \frac{1}{n(n-1)} \sum_{i=1}^{n} (y_i - \delta)^2 \tag{82}$$

An extension of the derivation above provides [45]

$$a(\delta, \chi^2, n) = \min\left[1, \frac{T(S' \to S)}{T(S \to S')} \exp(-\delta - u_B)\right] \tag{83}$$

where

$$u_B = \frac{\chi^2}{2} + \frac{\chi^4}{4(n+1)} + \frac{\chi^6}{3(n+1)(n+3)} + \cdots \tag{84}$$

The first term in the r.h.s. of eq. (84) is the penalty in the case we know the variance. The error of the noise causes extra penalty which decreases as the number of independent samples n grows. In the limit of large n the first term dominates and we recover (77). Equation (84) must be supplemented by the condition $\chi^2/n \leq 1/4$ for the asymptotic expansion (84) to converge and the instantaneous acceptance probability to be positive [45].

The noise level of a system can be characterized by the relative noise parameter, $f = (\beta\sigma)^2 t/t_0$, where t is the computer time spent reducing the noise, and t_0 is the computer time spent on other pursuits, such as optimizing the VMC wave function or equilibrating the RQMC runs. A small f means little time is being spent on reducing noise, where a large f means much time is being spent reducing noise. For a double well potential, the noise level that gives the maximum efficiency is around $\beta\sigma \approx 1$, with the optimal noise level increasing as the relative noise parameter increases [45]. In CEIMC runs for hydrogen the noise level $\beta\sigma$ ranges between 0.3 and 3, the optimal value being around 1.

3.1 Efficient Strategies for Electronic Energy Differences

As explained above, we need to evaluate the energy difference and the noise between two protonic configurations (S, S'). The distance in configurational space between S and S' in an attempted move is however quite limited since we have to move all protons at once. Indeed, each backflow orbital depends on the position of all protons and single proton moves would require recomputing the entire determinant for each attempted move, a $O(N^4)$ operations for a global move. Instead, moving all protons together requires a single determinant calculation per ionic move, a $O(N^3)$ operation. In this case, performing two independent electronic calculations for S and S' to estimate the energy difference would be very inefficient since $\Delta E_{BO} \ll E_{BO}(S)$ and the noise on the energy difference would just be twice the noise on the single energy estimate. The strategy to adopt is to compute the energy difference from correlated sampling, i.e. sampling the electronic configurational space from a distribution which depends both on S and S' and estimating the energy difference and the other electronic properties of interest by reweighting the averages [16]. It is possible to show that the optimal sampling function, i.e. the sampling distribution for which the variance of the energy difference is minimal, takes the form

$$P(Q|S, S') \propto |\Pi(Q|S)(E_{BO}(S)$$
$$- \langle E_{BO}(S) \rangle) - \Pi(Q|S')(E_{BO}(S') - \langle E_{BO}(S') \rangle)| \qquad (85)$$

where $\langle\, E_{BO}\,\rangle$ is the estimate of the BO energy. In order to use this sampling probability we need to estimate the BO energies of the two states before performing the sampling. A simpler form which avoid this problem is

$$P(Q|S, S') \propto \Pi(Q|S) + \Pi(Q|S') \qquad (86)$$

We emphasize that the "reptile" space for the electron paths depends on the proton coordinates so that, because of the fixed-node restriction for fermions, legal paths for S may or may not be legal for S' and vice-versa. These two forms of importance sampling have the property that they sample regions of both configuration spaces (S and S') and make the energy difference rigorously correct with a bounded variance.

3.2 Pre-Rejection

We can use multi-level sampling to make CEIMC more efficient [19]. An empirical potential is used to "pre-reject" moves that would cause particles to overlap and be rejected anyway. A trial move is proposed and accepted or rejected based on a classical potential

$$A_1 = \min\left[1, \frac{T(S \to S')}{T(S' \to S)} \exp(-\beta \Delta V_{cl})\right] \qquad (87)$$

where $\Delta V_{cl} = V_{cl}(S') - V_{cl}(S)$ and T is the sampling probability for a move. If it is accepted at this first level, the QMC energy difference is computed and accepted with probability

$$A_2 = \min\left[1, \exp(-\beta \Delta E_{BO} - u_B) \exp(\beta \Delta V_{cl})\right] \qquad (88)$$

where u_B is the noise penalty. Compared to the cost of evaluating the QMC energy difference, computing the classical energy difference is much less expensive. Reducing the number of QMC energy difference evaluations reduces the overall computer time required.

For metallic hydrogen a single Yukawa potential was always found to be suitable for pre-rejection. For molecular hydrogen we use instead a Silvera-Goldman potential [46] riparametrized in such a way to have the center of interaction on any single proton in the molecule rather than on the molecular center of mass. In practice, we take a single Yukawa potential for intermolecular proton-proton interaction and the Kolos-Wolniewski potential [47] for the bonding interaction. The Yukawa screening length and the prefactor are optimized to reproduce the results of the Silvera-Goldman model. This new potential is suitable for pre-rejecting all types of moves that we attempt in the molecular hydrogen, namely molecular rotations, bond stretching and molecular translations. The original Silvera-Goldmann potential being spherically symmetric around the molecular center is not suitable to pre-reject the rotational moves.

4 Quantum Protons

By increasing pressure and/or decreasing temperature, ionic quantum effects can become relevant. Those effects are important for hydrogen at high pressure [7, 48]. Static properties of quantum systems at finite temperature can be obtained with the Path Integral Monte Carlo method (PIMC) [19]. We need to consider the ionic thermal density matrix rather than the classical Boltzmann distribution:

$$\rho_p(S, S'|\beta) = \langle S|e^{-\beta(\hat{K}_p + \hat{E}_{BO})}|S'\rangle \tag{89}$$

where $\hat{K}_p$ is the ionic kinetic energy operator and $\beta = (K_B T)^{-1}$ is the inverse physical temperature. Thermal averages of ionic operators $\hat{A}_p$ (diagonal in configurational space) are obtained as

$$A_p(\beta) = \frac{1}{Z(\beta)} \int dS A_p(S)\rho_p(S, S|\beta) \tag{90}$$

where Z is the partition function

$$Z(\beta) = \int dS \rho_p(S, S|\beta) = e^{-\beta F} \tag{91}$$

and F is the Helmholtz free energy of the system. As before, the thermal density matrix can be computed by a factorization of β in many (P) small intervals (time slices $\tau_p = \beta/P$) and by a suitable approximation for the "high temperature" (or short time) density matrix. According to the Feynman-Kac formula (see (24) and (35)) the diagonal part of the thermal density matrix is the sum over all closed paths, i.e. paths starting at S and returning to S after a "time" β. This is the famous "isomorphism" between quantum particles and ring polymers [19,21]. Considering particle statistics in the PIMC is more difficult than in RQMC. The reason is that in PIMC the state of the classical system, which has no symmetry built in, plays the role of the trial functions in RQMC and permutations need to be sampled explicitly with an additional level of difficulty in the method [19, 26]. However, protonic statistics become relevant when the quantum dispersion is comparable to the interionic distances $\Lambda_p = \sqrt{2\lambda_p\beta} \approx (N_p/V)^{-1/3} = n_p^{-1/3}$. This define a degeneracy temperature $k_B T_D(n_p) = 2\lambda_p n_p^{2/3}$ below which quantum statistics need to be considered. For hydrogen $T_D \simeq 66.2(K)/r_s^2$, where r_s is the usual ion sphere radius of coulomb systems $r_s = (3/4\pi n_p)^{1/3}$. Therefore proton statistics in metallic hydrogen ($r_s \leq 1.3$) becomes relevant below 50K depending on the density, a regime that we have not investigated yet. Proton statistics in molecular hydrogen is also quite important and results in the separation between ortho- and para-hydrogen [49]. Because this effect is relevant only at low temperature, we have disregarded it as well.

In order to implement the PIMC we need a suitable approximation for the high temperature density matrix $\rho_p(S, S'|\tau_p)$. We could use either the primitive approximation or the importance sampling approximation described earlier. However a better approximation, in particular for distinguishable particles, is the pair product action [19] which closely resembles the pair trial function. The idea is to build the many body density matrix as the product over all distinct pairs of a two-body density matrices obtained numerically for a pair of isolated particles. At high temperature the system approaches the classical Boltzmann distribution which is indeed of the pair product form. The method is described in detail in [19]. Here we just explain how we can take advantage of this methodology within the CEIMC scheme. In order to use the method of pair action, we need to have a pair potential between quantum particles. In CEIMC, however, the interaction among protons is provided by the many-body BO energy. Our strategy is to introduce an effective two-body potential between protons $\hat{V}_e$ and to recast the ionic density matrix as

$$
\begin{aligned}
\rho_P(S, S'|\tau_p) &= \langle\, S|e^{-\tau_p[\hat{H}_e + (\hat{E}_{BO} - \hat{V}_e)]}|S'\,\rangle \\
&\approx \langle\, S|e^{-\tau_p\hat{H}_e}|S'\,\rangle\, e^{-\frac{\tau_p}{2}[E_{BO}(S) - V_e(S)] + [E_{BO}(S') - V_e(S')]} \quad (92)
\end{aligned}
$$

where $\hat{H}_e = \hat{K}_p + \hat{V}_e$ and the corrections from the effective potential to the true BO energy are treated at the level of the primitive approximation. We can compute numerically the matrix elements of the effective pair density matrix $\hat{\rho}_e^{(2)}(\tau_p)$ as explained in [19]. The effective N_p-body density matrix is approximated by

$$
\langle S|e^{-\tau_p\hat{H}_e}|S'\rangle \approx \prod_{ij}^{N_p}\langle s_i, s_j|\hat{\rho}_e^{(2)}(\tau_p)|s_i', s_j'\rangle = \rho_0(S, S'|\tau_p)e^{-\sum_{ij} u_e(\boldsymbol{s}_{ij}, \boldsymbol{s}_{ij}'|\tau_p)}
$$

$$(93)$$

where ρ_0 is the free particle density matrix and $u_e(\boldsymbol{s}_{ij}, \boldsymbol{s}_{ij}'|\tau_p)$ is the effective pair action. The explicit form for the partition function is then

$$Z(\beta) = \int dS_1 \ldots dS_P \prod_{k=1}^{P} \exp \left\{ -\frac{(S_k - S_{k+1})^2}{4\lambda_p \tau_p} - \sum_{ij} u_e(s_{ij}^k, s_{ij}^{k+1}|\tau_p) \right\}$$

$$\exp \left\{ -\tau_p \sum_{k=1}^{P} [E_{BO}(S_k) - V_e(S_k)] \right\} \tag{94}$$

with the boundary condition: $S_{P+1} = S_1$. As for the pre-rejection step in the proton moves, we have used different effective potentials according to the system under consideration. In metallic hydrogen (a plasma) we have used a smooth screened coulomb form and found that it provides a fast convergence in τ_p. Convergence can be assessed by monitoring the various terms in the estimator for the proton kinetic energy [19]. A good effective potential should provide uniform convergence of the various orders in τ_p. With this effective potential, we have found convergence to the continuum limit ($\tau_p \to 0$) for $1/\tau_p \geq 3000 \, \text{K}$ which allows to simulate systems at room temperature with only $M \approx 10$ proton slices for $r_s \geq 1$. In molecular hydrogen we need to consider the extra contribution of the bonding potential. We have used the Kolos-Wolniewicz bonding potential in connection with the same smooth screened coulomb potential for the non bonding interactions. Convergence with τ_p is observed in a similar range.

A very nice feature of ionic PIMC in CEIMC is that considering ionic paths rather then classical point particles does not add any computational cost to the method. Let us suppose we run classical ions with a given level of noise $(\beta\sigma_{cl})^2$. Consider now representing the ions by P time slices. To have a comparable extra-rejection due to the noise we need a noise level per slice given by: $(\tau_p \sigma_k)^2 \approx (\beta\sigma_{cl})^2/P$ which provides $\sigma_k^2 \approx P\sigma_{cl}^2$. We can allow a noise per time slice P times larger which means considering P times less independent estimates of the energy difference per slice. However we need to run P different calculations, one for each different time slice, so that the amount of computing for a fixed global noise level is the same as for classical ions. In practice, however, because our orbitals depend on all proton positions, we are forced to move the proton positions at given imaginary time all together with a local (in imaginary time) update scheme. It is well known that the autocorrelation time of schemes with local updates rapidly increases with the chain length and this is the ultimate bottleneck of our present algorithm [19]. It is therefore essential to adopt the best factorization in order to minimize the number of time slices P needed and therefore to maximize the efficiency of the method.

When using TABC with quantum ions, for any proton time slice we should, in principle, perform a separate evaluation of the BO energy difference averaged over all twist angles. Instead at each protonic step, we randomly assign a subset of twist phases at each time slice and we compute the energy difference for that phase only. The TABC is then performed by adding up all the contributions from the different time slices. We have checked in few cases that this simplified procedure does not give detectable biases in the averages.

In practice we move all slices of all protons at the same time by a simple random move in a box and we pre-reject the moves with the effective pair action.

5 Summary of Results on High Pressure Hydrogen

In this section we briefly summarize some of the CEIMC results we have obtained for high pressure hydrogen. Figure 1 shows what is known and what has been predicted about the hydrogen phase diagram in a wide range of pressures and temperatures. The rectangular region in the right upper corner is the region where R-PIMC method can make reliable predictions [50–52]. There have been many studies of the ground state of hydrogen ($T = 0$) including some QMC investigations [33–36]. They have predicted a metallization density corresponding at $r_s = 1.31$ accompanied by a transition from a molecular m-hcp structure to a diamond lattice of protons. At intermediate temperatures a number of ab-initio Molecular Dynamics studies have been performed in the molecular and in the metallic phases, both in the crystal and in the liquid state [53–57].

Metallic Hydrogen

We first focus on the metallic system for pressure beyond the molecular dissociation threshold. In this region, hydrogen is a system of protons and delocalized electrons. At low enough temperature the protons order in a crystalline lattice which melts upon increasing temperature. The low temperature stable structure as a function of density is still under debate. The most accurate ground state QMC calculation [35], indicates that hydrogen at the edge of molecular dissociation will order in a diamond structure, and upon increasing density will undergo various structural transformations ultimately transforming to the bcc structure. However, these prediction are extrapolated from a single calculation at $r_s = 1.31$ and temperature effects are absent. With CEIMC we have investigated the density range $r_s \in [0.8, 1.2]$ and the temperature range $T \in [300\,\mathrm{K}, 5000\,\mathrm{K}]$ across the melting transition of the proton lattice and up to the lower limit of applicability of RPIMC. We limited the study to systems of $N_p = 32$ and $N_p = 54$ which, for cubic simulation boxes, form fcc and bcc lattices, respectively. We have observed the melting and refreezing of the protons and made a qualitative location of the melting line versus the density [16, 17] as shown by the green contour lines in Fig. 1.

A number of interesting question about the convergence and the efficiency of the CEIMC algorithm need to be answered before starting a systematic study. An important one is: how large must the electronic projection time be in order to get convergence in the energy difference to the ground state value and therefore obtain unbiased sampling in CEIMC? In order to answer such a question we have selected a pair of protonic configuration (S, S') at

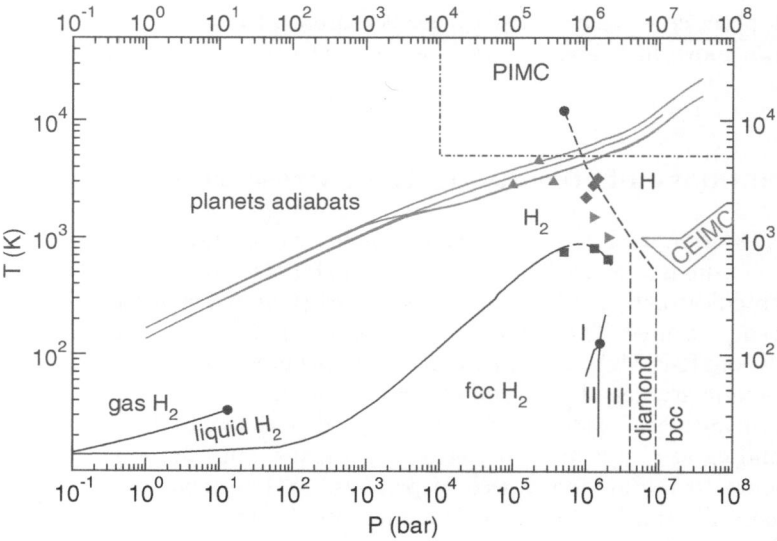

Fig. 1. Hydrogen phase diagram. Continuous transition lines are experimental results, *dashed lines* are theoretical prediction from various methods. *Squares* and
right-triangle are ab-initio MD predictions of molecular melting [57] and molecular
dissociation in the liquid phase [56]. The diamonds are shock-waves experimental
data through the liquid metalization [58]. The *triangles* are earlier CEIMC data for
the insulating molecular state [15, 16] while the light gray domain on the extreme
right indicates the CEIMC prediction for the melting [17]. Dark gray lines are model
adiabats for the interior of the giant planets of the solar system [59]

given density and computed the energy difference, together with total energy
and variance, versus the electronic projection time. The results reported in
Fig. 2 correspond to a system of $N_e = N_p = 16$ at $r_s = 1.31$ with the twist
phase $\theta = (0.4, 0.5, 0.6)\pi$. Results are obtained for an electronic time step
$\tau_e = 0.02\,\mathrm{H}^{-1}$, a compromise between accuracy and efficiency. In panel a)
we report the energy difference versus the projection time t to show that it
does not depend on the projection time when using the accurate trial function
with 3-body terms and backflow orbitals (3BF-A, black dots) discussed above.
This suggests that the proton configuration space can be sampled using VMC
for the electrons which is faster and more stable than RQMC. On the same
panel we have reported the VMC estimate obtained with a trial function with
a Slater determinant of simple plane waves and a two-body RPA Jastrow
(SJ, red triangle) and a RQMC result obtained for a trial function with a
2-body RPA-Jastrow and a Slater determinant of self consistent LDA orbitals
(LDA-J, blue squares). The simple SJ function at the variational level has a
much larger energy difference (in absolute value) and therefore will provide a
biased sampling of protonic configurations. It is not clear whether the RQMC
projection with such trial function will recover the correct energy difference

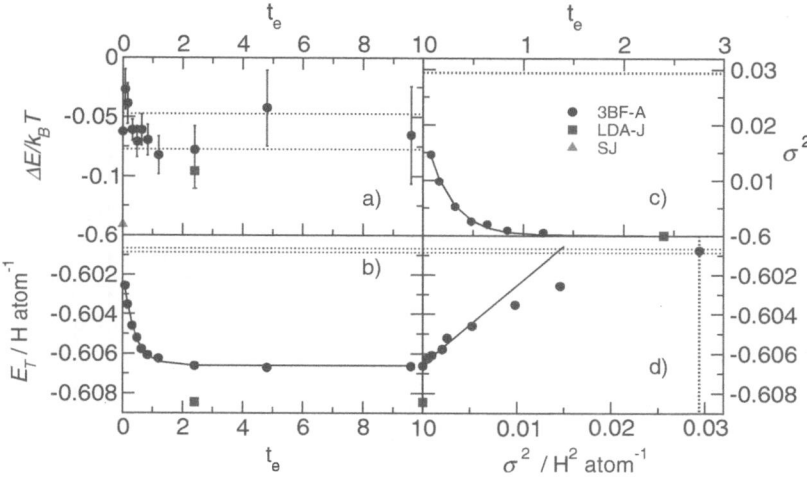

Fig. 2. $N_e = N_p = 16$, $r_s = 1.31$. Dependence of total energy, variance and energy difference for a pair of proton configurations (S, S') on the RQMC projection time. The study is performed for $\tau_e = 0.02\,H^{-1}$. *Dotted lines* represent the variational estimates with their error bars. In panel **b**) and **c**) the lines are exponential fits to data and in panel **d**) the continuous line is a linear fit in the region $\sigma^2 \leq 0.005$. *Black circles* (3BF-A) are results obtained with the analitical three-body and backflow trial wave functions discussed earlier, the red triangle is a variational result with a Slater-Jastrow trial function with simple plane wave orbitals and the blue squares are results from a trial function with LDA orbitals and an optmized two-body Jastrow

(remember that the two kind of trial functions have different nodal structure). On the other hand, no difference is detected between the RQMC results from the 3BF-A and the LDA-J trial functions at the same projection time. This gives an indication of the quality of the 3BF-A nodal surfaces. The total energy and the variance of configuration S versus t are reported in panels b) and c), respectively. The two panels nicely illustrate the limiting process toward the fixed-node ground state operated by the projection (see (11)). Within the 3BF-A trial function we observe a large gain in energy with the projection time, the difference being $E(\infty) - E(0) = 5.7\,mH/at = 1810\,K/at$. In both panels the results of the SJ trial function are off scale, while we have reported the result of the LDA-J wave function. According to the total energy, the quality of LDA-J function is superior to the one of 3BF-A function for this configuration. The same quality was instead detected for the bcc lattice configuration [14]. Finally, in panel d) we report the total energy versus the variance of the 3BF-A trial function to show that both quantities go linearly with the error in the wave function when we are close enough to the ground state (the fixed-node one) [13].

In order to check that we can indeed sample the proton space with VMC energy differences, we have performed two test runs for the systems of $N_p =$

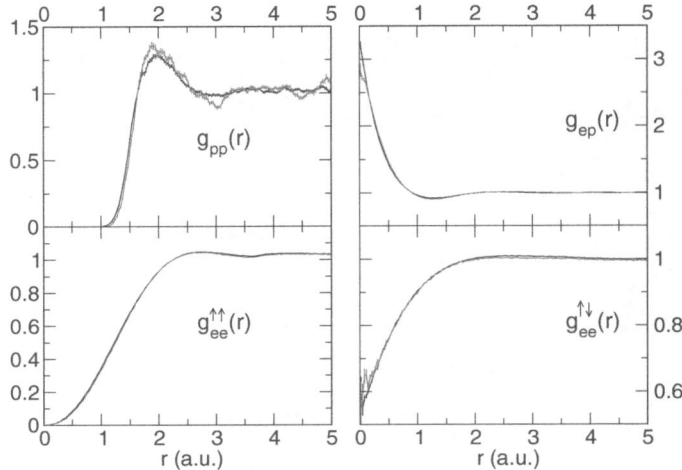

Fig. 3. $N_e = N_p = 54$, $r_s = 1.2$, $T = 5000$ K, Γ point. Comparison of VMC (*black*) and RQMC (*gray*) data for the pair correlation functions. The projection time in RQMC is $t_e = 0.68\,\mathrm{H}^{-1}$ with an electronic time step of $\tau_e = 0.01\,\mathrm{H}^{-1}$. Protons are considered as classical point particles

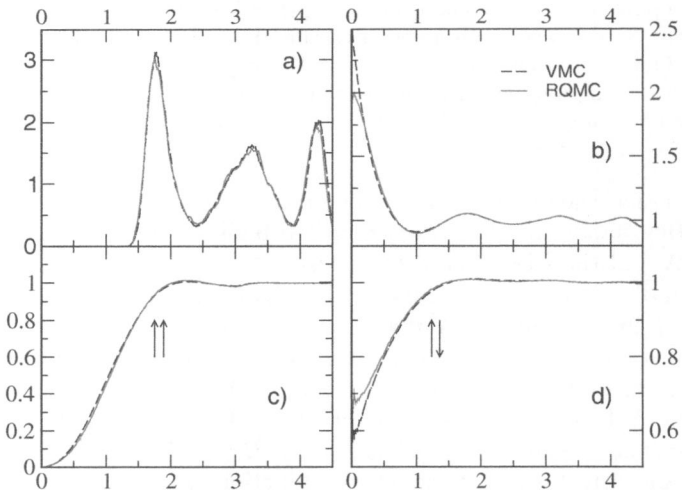

Fig. 4. $N_e = N_p = 54$, $r_s = 1.0$, $T = 1000$ K, Γ point. Comparison of VMC (*black*) and RQMC (*gray*) data for the pair correlation functions. The projection time in RQMC is $t_e = 1.0\,\mathrm{H}^{-1}$ with an electronic time step of $\tau_e = 0.02\,\mathrm{H}^{-1}$. Protons are considered as classical point particles. The proton-proton pair correlation functions exhibits a structure reminiscent of the bcc crystal structure. RQMC data for $g_{ep}(r)$ and $g_{ee}^{\uparrow\downarrow}(r)$ show a finite time step errors at short distances which however do not contribute significantly to the energy

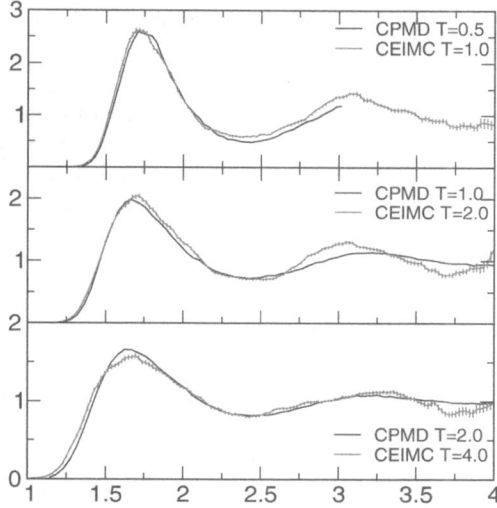

Fig. 5. Proton-proton pair correlation functions for classical protons at $r_s = 1$ and various temperatures. Comparison between CEIMC and CPMD-LDA data

$N_e = 54$ at the Γ point (zero twist angle), one at $r_s = 1.2$ and $T = 5000$ K and the other at $r_s = 1$ and $T = 1000$ K. The comparison for the pair correlation functions are shown in Figs. 3 and 4 respectively.

It is interesting to compare the predictions of CEIMC with other methods. In refs. [16, 17] we have compared with Restricted Path Integral Monte Carlo data and with Car-Parrinello Molecular Dynamics data with LDA forces (CPMD-LDA) [54]. In Fig. 5 we compare CEIMC and CPMD-LDA $g_{pp}(r)$'s for classical protons at $r_s = 1$ and various temperatures. Both calculations are done with PBC (zero twist angle) and CEIMC uses 54 protons while CPMD-LDA used 162 protons (both are closed shells in the reciprocal space, so that the electronic ground state is not degenerate). We see that in a wide range of temperatures, CPMD-LDA and CEIMC are off by a factor of 2 in temperature. CPMD-LDA predicts a less structured fluid and locates the melting transition at roughly $T_m \simeq 350$ K [54]. With CEIMC instead more structure is found and the melting is located between 1000 K and 1500 K [17]. Though several reasons could be at the origin of such unexpected discrepancy, we believe that the problem arises from a too smooth BO energy surface as provided by LDA. Evidences of this fact are also provided by previous ground state calculations for hydrogen crystal structures [35] which similarly found energy differences between various structures from LDA to be roughly half of the corresponding one from DMC. This will explain the factor of 2 in the temperature scale observed in Fig. 5.

In [17] we have reported data for the equation of state of metallic hydrogen. Proton quantum effects are quite important at such high density and need to be considered carefully to get accurate prediction of the equation of state. The importance of proton quantum effect is partially reported in [60]. Also we have shown that electronic VMC with 3BF-A trial function is accurate enough to sample the proton configuration space. However RQMC should be used in order to obtain accurate results for the energy and the pressure. A good strategy is to run CEIMC with VMC to sample efficiently the proton configurations and then run RQMC on fixed statistical independent protonic configurations previously generated.

Insulating Molecular Hydrogen

In this section we report very preliminary results in the insulating molecular liquid phase. We have investigated the state point at $r_s = 2.1$ and $T = 4530$ K because at this point, and other scattered points around it, experimental data for the equation of state are available from Gas-gun techniques [61]. Also the first CEIMC attempt [15, 16] focused on the same point although using a quite different trial functions. We used a pair product trial function with modified RPA Jastrows (electron-electron and electron-proton) and Slater determinants built with the molecular orbitals of (55). At variance with the metallic trial function, we have now 5 variational parameters to be optimized, 4 in the Jastrow factor and a single one in the orbitals. In particular, the nodal structure of the wave function will be affected by the width of the gaussian orbitals used to build the molecular orbitals. In the first CEIMC attempt [15, 16], a single multivariate gaussian, centered at the molecular center of mass and with a width different for each molecule was used. Thus the number of variational parameter was equal to three time the number of molecules. Also, the entire optimization procedure for each attempted protonic configuration was performed prior the Metropolis test. Clearly, the optimization step was a bottleneck of that scheme. In our present scheme, we have gained evidence that we can optimize the parameters on a single protonic configuration (even the parameters optimized on a lattice configuration are good) and use the same values for simulating the liquid phase. A similar conclusion was obtained in the metallic phase with numerically optimized orbitals [16]. In this way the optimization step needs to be performed only upon changing density and/or number of particles, while we can use the same set of values for the variational parameters to span the temperature axis at fixed density and number of particles. In Fig. 6 we compare the pair correlation functions as obtained by VMC and RQMC with $\tau_e = 0.01$ H^{-1} and $t_e = 1.6$ H^{-1} for a system of 27 hydrogen molecules. Since the system is insulating we do not average over the twist angle, but just use the Γ point. We observe that RQMC enhances slightly the stregth of the molecular bond and the electronic correlation inside molecules with respect to VMC. Good agreement with the early CEIMC results is observed not only for the correlation function, but also for the equation of state. In our present calculation we obtain $P(RQMC) = 0.224(5)$ Mbars to be compared with the previous estimate of 0.225(3) Mbars [15, 16] and with

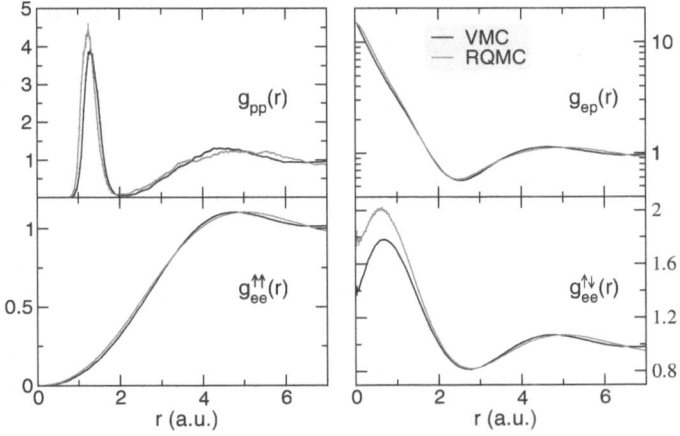

Fig. 6. Comparison between VMC and RQMC computed pair correlation functions for insulating molecular hydrogen at $r_s = 2.1, T = 4530\,\mathrm{K}$. $N_p = N_e = 54$

the experimental data $P = 0.234\,\mathrm{Mbars}$. The deviation from the experimental data is only roughly 5% and it is particularly encouraging if we consider that our data are for classical protons and quantum effects are expected to slightly increase the pressure.

6 Conclusions and Future Developments

In this paper we have described the principles of CEIMC and given some technical details on its practical implementation. The results for metallic and molecular hydrogen show that CEIMC is a practical strategy to couple ground state QMC methods for the electronic degrees of freedom with a finite temperature Monte Carlo simulation of the ionic degrees of freedom. To our knowledge, CEIMC is the only method available so far to perform ab-initio simulations based on QMC energies.

In a recent work, Grossman and Mitas have proposed a strategy to correct ab-initio Molecular Dynamics energies with QMC [62]. This is however different from CEIMC because the nuclear degrees of freedom are still sampled on the basis of DFT forces. Therefore, when applied to metallic hydrogen, that method would have found the same liquid structure as obtained by CPMD-LDA [54].

Very recently an interesting proposition on how to use noisy forces in ab-initio Molecular Dynamics has appeared [63]. The same strategy could be used with noisy QMC forces to simulate the dynamics of classical ions. A first attempt has already shown the feasibilty of this promising method [64] although the results are still very preliminary.

A crucial aspect of CEIMC is the choice of the electronic trial wave function. The ones we have discussed in this paper are either suitable for the metallic state or for the molecular state and their quality in describing the metalization-molecular dissociation transition is questionable. A current development within CEIMC is an efficient strategy to generate on the fly single electron orbitals depending on the instantaneous ionic configuration, in the spirit of the LDA orbitals previously used [35]. We have devised an efficient algorithm which is able to provide accurate orbitals in a reasonable computer time. We are at present using these orbitals to explore the molecular dissociation under pressure in the liquid state [65]. The same strategy can be used to obtain accurate prediction for the hydrogen equation of state in the temperature range between 50 K and 5000 K. Also application to helium and helium-hydrogen mixtures, very interesting systems for planetary physics, can be envisaged with the same methodology.

Tests for non-hydrogenic systems are needed to find the performance of the method on a broader spectrum of applications. The use of pseudopotentials within QMC to treat atoms with inner core is well tested [6]. What is not clear is how much time will be needed to generate trial functions, and to reduce the noise level to acceptable limits. Clearly, further work is needed to allow this next step in the development of microscopic simulation algorithms.

References

1. R. Car and M. Parrinello (1985) Unified Approach for Molecular Dynamics and Density-Functional Theory. *Phys. Rev. Letts.* **55**, p. 2471
2. M. Bernasconi, G. L. Chiarotti, P. Focher, S. Scandolo, E. Tosatti, and M. Parrinello (1995) First-principle-constant pressure molecular dynamics. *J. Phys. Chem. Solids* **56**, p. 501
3. A. Alavi, J. Kohanoff, M. Parrinello, and D. Frenkel (1994) Ab Initio Molecular Dynamics with Excited Electrons. *Phys. Rev. Letts.* **73**, pp. 2599–2602
4. D. Marx and M. Parrinello (1996) Ab initio path integral molecular dynamics: Basic ideas. *J. Chem. Phys.* **104**, p. 4077
5. R. M. Martin (2004) Electronic Structure. Basic Theory and Practical Methods. Cambridge University Press, Cambridge
6. M. W. C. Foulkes, L. Mitas, R. J. Needs, and G. Rajagopal (2001) Quantum Monte Carlo simulations of solids. *Rev. Mod. Phys.* **73**, p. 33
7. E. G. Maksimov and Y. I. Silov (1999) Hydrogen at high pressure. *Physics-Uspekhi* **42**, p. 1121
8. M. Stadele and R. M. Martin (2000) Metallization of Molecular Hydrogen: Predictions from Exact-Exchange Calculations. *Phys. Rev. Lett.* **84**, pp. 6070–6073
9. K. A. Johnson and N. W. Ashcroft (2000) Structure and bandgap closure in dense hydrogen. *Nature* **403**, p. 632
10. D. Alfé, M. Gillan, M. D. Towler, and R. J. Needs (2004) Efficient localized basis set for quantum Monte Carlo calculations on condensed matter. *Phys. Rev. B* **70**, p. 161101

11. B. L. Hammond, W. A. Lester Jr., and P. J. Reynolds (1994) Monte Carlo methods in Ab Initio Quantum Chemistry. World Scientific Singapore
12. R. M. Panoff and J. Carlson (1989) Fermion Monte Carlo algorithms and liquid 3He. *Phys. Rev. Letts.* **62**, p. 1130
13. Y. Kwon, D. M. Ceperley, and R. M. Martin (1994) Quantum Monte Carlo calculation of the Fermi-liquid parameters in the two-dimensional electron gas. *Phys. Rev. B* **50**, pp. 1684–1694
14. M. Holzmann, D. M. Ceperley, C. Pierleoni, and K. Esler (2003) Backflow correlations for the electron gas and metallic hydrogen. *Phys. Rev. E* **68**, p. 046707[1–15]
15. M. Dewing and D. M. Ceperley (2002) Methods in Coupled Electron-Ion Monte Carlo. In *Recent Advances in Quantum Monte Carlo Methods II* (Ed. S. Rothstein), World Scientific
16. D. M. Ceperley, M. Dewing, and C. Pierleoni (2002) The Coupled Electronic-Ionic Monte Carlo Simulation Method. *Lecture Notes in Physics* **605**, pp. 473–499, Springer-Verlag; physics/0207006
17. C. Pierleoni, D. M. Ceperley, and M. Holzmann (2004) Coupled Electron-Ion Monte Carlo Calculations of Dense Metallic Hydrogen. *Phys. Rev. Lett.* **93**, 146402[1–4]
18. S. Baroni and S. Moroni (1999) Reptation Quantum Monte Carlo: A Method for Unbiased Ground-State Averages and Imaginary-Time Correlations. *Phys. Rev. Letts.* **82**, pp. 4745–4748; S. Baroni, S. Moroni *Reptation quantum Monte Carlo* in "Quantum Monte Carlo Methods in Physics and Chemistry", eds. M. P. Nightingale and C. J. Umrigar (Kluwer, 1999), p. 313
19. D. M. Ceperley (1995) Path integrals in the theory of condensed helium. *Rev. Mod. Phys.* **67**, pp. 279–355
20. A. Sarsa, K. E. Schmidt, and W. R. Magro (2000) A path integral ground state method. *J. Chem. Phys.* **113**, p. 1366
21. R. P. Feynman (1998) Statistical Mechanics: a set of lectures. Westview Press
22. K. Huang (1988) Statistical Mechanics, John Wiley
23. S. Zhang and H. Krakauer (2003) Quantum Monte Carlo Method using Phase-Free Random Walks with Slater Determinants. *Phys. Rev. Lett.* **90**, p. 136401
24. R. W. Hall (2005) Simulation of electronic and geometric degrees of freedom using a kink-based path integral formulation: Application to molecular systems. *J. Chem. Phys.* **122**, p. 164112[1–8]
25. A. J. W. Thom and A. Alavi (2005) A combinatorial approach to the electron correlation problem. *J. Chem. Phys.* in print
26. D. M. Ceperley (1996) Path integral Monte Carlo methods for fermions. In *Monte Carlo and Molecular Dynamics of Condensed Matter Systems*, ed. by K. Binder and G. Ciccotti, Editrice Compositori, Bologna, Italy
27. D. M. Ceperley (1991) Fermion Nodes. *J. Stat. Phys.* **63**, p. 1237
28. D. Bressanini, D. M. Ceperley, and P. Reynolds (2001) What do we know about wave function nodes?. In *Recent Advances in Quantum Monte Carlo Methods II*, ed. S. Rothstein, World Scientfic
29. G. Ortiz, D. M. Ceperley, and R. M. Martin (1993) New stochastic method for systems with broken time-reversal symmetry: 2D fermions in a magnetic field. *Phys. Rev. Lett.* **71**, p. 2777
30. C. Lin, F. H. Zong, and D. M. Ceperley (2001) Twist-averaged boundary conditions in continuum quantum Monte Carlo algorithms. *Phys. Rev. E* **64**, 016702[1–12]

31. G. Ortiz and D. M. Ceperley (1995) Core Structure of a Vortex in Superfluid 4He. *Phys. Rev. Lett.* **75**, p. 4642
32. V. D. Natoli (1994) A Quantum Monte Carlo study of the high pressure phases of solid hydrogen, Ph.D. Theses, University of Illinois at Urbana-Champaign.
33. D. M. Ceperley and B. J. Alder (1987) Ground state of solid hydrogen at high pressures. *Phys. Rev. B* **36**, p. 2092
34. X. W. Wang, J. Zhu, S. G. Louie, and S. Fahy (1990) Magnetic structure and equation of state of bcc solid hydrogen: A variational quantum Monte Carlo study. *Phys. Rev. Lett.* **65**, p. 2414
35. V. Natoli, R. M. Martin, and D. M. Ceperley (1993) Crystal structure of atomic hydrogen. *Phys. Rev. Lett.* **70**, p. 1952
36. V. Natoli, R. M. Martin, and D. M. Ceperley (1995) Crystal Structure of Molecular Hydrogen at High Pressure. *Phys. Rev. Lett.* **74**, p. 1601
37. C. Pierleoni and D. M. Ceperley (2005) Computational methods in Coupled Electron-Ion Monte Carlo. *Chem. Phys. Chem.* **6**, p. 1872
38. D. Ceperley (1986) The Statistical Error of Green's Function Monte Carlo, in Proceedings of the Metropolis Symposium on. The Frontiers of Quantum Monte Carlo. *J. Stat. Phys.* **43**, p. 815
39. D. M. Ceperley and M. H. Kalos (1979) Monte Carlo Methods in Statistical Physics, ed. K. Binder, Springer-Verlag.
40. D. M. Ceperley, G. V. Chester, and M. H. Kalos (1977) Monte Carlo simulation of a many-fermion study. *Phys. Rev. B* **16**, p. 3081
41. D. M. Ceperley and B. J. Alder (1984) Quantum Monte Carlo for molecules: Green's function and nodal release. *J. Chem. Phys.* **81**, p. 5833
42. C. Pierleoni, K. Delaney, and D. M. Ceperley, to be published
43. D. Frenkel and B. Smit (2002) Understanding Molecular Simulations: From Algorithms to Applications, 2nd Ed., Academic Press, San Diego
44. S. Moroni, private communication
45. D. M. Ceperley and M. Dewing (1999) The penalty method for random walks with uncertain energies. *J. Chem. Phys.* **110**, p. 9812
46. I. F. Silvera and V. V. Goldman (1978) The isotropic intermolecular potential for H2 and D2 in the solid and gas phases. *J. Chem. Phys.* **69**, p. 4209
47. W. Kolos and L. Wolniewicz (1964) Accurate Computation of Vibronic Energies and of Some Expectation Values for H2, D2, and T2. *J. Chem. Phys.* **41**, p. 3674
48. E. Babaev, A. Sudbo, and N. W. Ashcroft (2004) A superconductor to superfluid phase transition in liquid metallic hydrogen. *Nature* **431**, p. 666
49. I. F. Silvera (1980) The solid molecular hydrogens in the condensed phase: Fundamentals and static properties. *Rev. Mod. Phys.* **52**, p. 393
50. C. Pierleoni, D. M. Ceperley, B. Bernu, and W. R. Magro (1994) Equation of State of the Hydrogen Plasma by Path Integral Monte Carlo Simulation. *Phys. Rev. Lett.* **73**, p. 2145; W. R. Magro, D. M. Ceperley, C. Pierleoni and B. Bernu (1996) Molecular Dissociation in Hot, Dense Hydrogen. *Phys. Rev. Lett.* **76**, p. 1240
51. B. Militzer and D. M. Ceperley (2000) Path Integral Monte Carlo Calculation of the Deuterium Hugoniot. *Phys. Rev. Lett.* **85**, p. 1890
52. B. Militzer and D. M. Ceperley (2001) Path integral Monte Carlo simulation of the low-density hydrogen plasma. *Phys. Rev. E* **63**, p. 066404
53. D. Hohl, V. Natoli, D. M. Ceperley, and R. M. Martin (1993) Molecular dynamics in dense hydrogen. *Phys. Rev. Lett.* **71**, p. 541

54. J. Kohanoff and J. P. Hansen (1995) Ab Initio Molecular Dynamics of Metallic Hydrogen at High Densities. *Phys. Rev. Lett.* **74**, pp. 626–629; *ibid.* (1996) Statistical properties of the dense hydrogen plasma: An ab initio molecular dynamics investigation. *Phys. Rev. E* **54**, pp. 768–781

55. J. Kohanoff, S. Scandolo, G. L. Chiarotti, and E. Tosatti (1997) Solid Molecular Hydrogen: The Broken Symmetry Phase. *Phys. Rev. Lett.* **78**, p. 2783

56. S. Scandolo (2003) Liquid-liquid phase transition in compressed hydrogen from first-principles simulations. *PNAS* **100**, p. 3051

57. S. A. Bonev, E. Schwegler, T. Ogitsu, and G. Galli (2004) A quantum fluid of metallic hydrogen suggested by first-principles calculations. *Nature* **431**, p. 669

58. S. T. Weir, A. C. Mitchell, and W. J. Nellis (1996) Metallization of Fluid Molecular Hydrogen at 140 GPa (1.4 Mbar). *Phys. Rev. Lett.* **76**, p. 1860

59. T. Guillot, G. Chabrier, P. Morel, and D. Gautier (1994) Nonadiabatic models of Jupiter and Saturn. *Icarus* **112**, p. 354; T. Guillot, P. Morel (1995) Coupled Electron Ion Monte Carlo Calculations of Atomic Hydrogen. *Astron. & Astrophys. Suppl.* **109**, p. 109

60. M. Holzmann, C. Pierleoni, and D. M. Ceperley (2005) Coupled Electron Ion Monte Carlo Calculations of Atomic Hydrogen. *Comput. Physics Commun.* **169**, p. 421

61. N. C. Holmes, M. Ross, and W. J. Nellis (1995) Temperature measurements and dissociation of shock-compressed liquid deuterium and hydrogen. *Phys. Rev. B* **52**, p. 15835

62. J. C. Grossman, L. Mitas (2005) Efficient Quantum Monte Carlo Energies for Molecular Dynamics Simulations. *Phys. Rev. Lett.* **94**, p. 056403

63. F. Krajewski and M. Parrinello (2005) Stochastic linear scaling for metals and nonmetals. *Phys. Rev. B* **71**, p. 233105; F. Krajewski, M. Parrinello, Linear scaling electronic structure calculations and accurate sampling with noisy forces. cond-mat/0508420

64. C. Attaccalite (2005) RVB phase of hydrogen at high pressure:towards the first ab-initio Molecular Dynamics by Quantum Monte Carlo, Ph.D. theses, SISSA-Trieste.

65. K. Delaney, C. Pierleoni and D.M. Ceperley (2006) Quantum Monte Carlo Simulation of the High-Pressure Molecular-Atomic Transition in Fluid Hydrogen. cond-mat/0603750, submitted to Phys. Rev. Letts.,

Path Resummations
and the Fermion Sign Problem

A. Alavi[1] and A.J.W. Thom[2]

[1] University of Cambridge, Chemistry Department, Lensfield Road, Cambridge
   CB2 1EW, U.K.
   asa10@cam.ac.uk
[2] University of Cambridge, Chemistry Department, Lensfield Road, Cambridge
   CB2 1EW, U.K.
   ajwt3@cam.ac.uk

Ali Alavi

A. Alavi and A.J.W. Thom: *Path Resummations and the Fermion Sign Problem*, Lect. Notes
Phys. **703**, 685–704 (2006)
DOI 10.1007/3-540-35273-2_19

686   A. Alavi and A.J.W. Thom

We review a recent method we have developed for Fermion quantum Monte Carlo. By using combinatorial arguments to perform resummations over paths, we reformulate the stochastic problem of sampling paths in terms of sampling "graphs", which are much better behaved with regards sign-cancellation problems encountered in path-integral simulations of Fermions. Detailed mathematical derivations of the new results are presented.

1 Introduction

In a recent paper [1], we have proposed a method to perform quantum Monte Carlo (QMC) simulations of Fermion systems based on the idea of sampling "graphs", rather than "paths". The latter, being rooted in Feynman's "real-space imaginary-time" path-integral (PI) theory [2–4], has been the paradigmatic approach to simulating quantum systems at finite temperature. For Fermion systems, however, there persists the infamous sign problem [5]. In the present contribution, we will review our new ideas, and in addition provide detailed mathematical derivations of theoretical results reported there, which we hope students may find useful. This paper is intended to be read in conjunction with [1].

Let us begin with a few words of motivation: why are we interested in developing QMC methods for treating electrons, when there exist highly accurate quantum chemical methods for computing electronic energies? The reason has to with the *scaling* of the algorithms with numbers of electrons. Wavefunction-based methods [6], such as Coupled-Cluster theory (CCSD(T)), which have emerged in the past few years as a "Rolls-Royce" method of obtaining accurate energies, scale as $\sim N^6 - N^7$ (N is the number of electrons), which is prohibitively expensive. On the other hand, the ever-popular density functional methods, which have a highly favourable $N^2 - N^3$ scaling, have important failings in treating strong-correlation systems, and non-local correlation phenomena such as van der Waals interactions. Quantum Monte Carlo methods, in principle, allow the way out, providing correlated energies with typically $\sim N^3$ scaling and, therefore, allow one to treat large systems whilst maintaining good accuracy. Currently the diffusion Monte Carlo method [7] within the fixed-node approximation is the most accurate of the QMC techniques. Attempts to go beyond the fixed-node approximation via the release-node method [8] fail as the number of electrons increase.

In PI theory, the sign problem arises from the alternating sign of the weights of paths which exchange the positions of identical Fermions: odd and even permutations lead to negative and positive weights, respectively. More specifically, let us denote the spatial and spin coordinates of N electrons with $\mathbf{X} = (\mathbf{x}_1, \ldots, \mathbf{x}_N)$. A given path can be described by a continuous function, $\mathbf{X}(\tau)$, with τ running from 0 to $\hbar\beta$. For equilibrium problems at finite temperature $kT = \beta^{-1}$, we are interested in paths which are closed, up to a permutation of the labels of the electrons:

$$\mathbf{X}(\hbar\beta) = \hat{P}\mathbf{X}(0) \tag{1}$$

According to the principles of equilibrium quantum statistical mechanics [3], we associate with each such path a Boltzmann weight determined by the Hamiltonian of the system, multiplied by the sign of the permutation:

$$w[\mathbf{X}(\tau); \mathbf{X}, \hat{P}\mathbf{X}] = \text{sign}(\hat{P}\mathbf{X}) \times$$
$$\left(\exp[-(1/\hbar) \int_0^{\hbar\beta} \left(m|\dot{\mathbf{X}}|^2/2 + V[\mathbf{X}(\tau)] \right) d\tau] \right)_{\mathbf{X}(\hbar\beta) = \hat{P}\mathbf{X}(0)} \tag{2}$$

The partition function is given by the path-integral of this weight over all such paths, for all permutations of end-points, and for all initial positions $\mathbf{X}$:

$$Q = \frac{1}{N!} \int d\mathbf{X} \sum_{\hat{P}} \int \mathcal{D}\mathbf{X}(\tau) w[\mathbf{X}(\tau); \mathbf{X}, \hat{P}\mathbf{X}] \tag{3}$$

The expectation value of the energy can be found through the relation:

$$E = -\frac{\partial \ln Q}{\partial \beta} \tag{4}$$

and takes the form

$$E = \frac{1}{Q} \frac{1}{N!} \int d\mathbf{X} \sum_{\hat{P}} \int \mathcal{D}\mathbf{X}(\tau) E[\mathbf{X}(\tau)] w[\mathbf{X}(\tau); \mathbf{X}, \hat{P}\mathbf{X}] \tag{5}$$

where $E[\mathbf{X}(\tau)]$ is an energy estimator derived from the Hamiltonian of the system [9]. In numerical implementation, the continuous path is discretized into a number P of time-slices:

$$\mathbf{X}(\tau) \rightarrow \mathbf{X}_1, \mathbf{X}_2 \ldots, \mathbf{X}_P, \hat{P}\mathbf{X}_1. \tag{6}$$

Since in a Metropolis Monte Carlo simulation one can generate paths only with a positive probability proportional to $|w|$, in order to compute expectation values of the type (5) one has to resort to a potentially problematic expression:

$$E = \frac{\langle \text{sign}(w[\mathbf{X}]) E[\mathbf{X}] \rangle_{|w|}}{\langle \text{sign}(w[\mathbf{X}]) \rangle_{|w|}} \tag{7}$$

Since $\text{sign}(w) = \pm 1$, the denominator can be very poorly behaved. As the paths get longer (i.e. P becomes large, which is necessary as β becomes large), or as N becomes large, one ends up sampling almost as many negative paths as positive ones, an exponential cancellation occurs in the sum over $\text{sign}(w)$ of the generated paths, making this denominator all but impossible to estimate.

Our proposal as a way forward is to reformulate the problem so that the entities which we end up sampling are not individual paths, but rather integrated entities in which exponentially large numbers of paths have been summed over. It is reasonable to expect that these entities (which we call "graphs") will, as a result, be much better behaved with regards sign fluctuations, and therefore enable one to sample them stochastically. In our paper [1] we presented evidence for this assertion in the specific case of a dissociating N_2 molecule. In this paper, we give another numerical example, a Hubbard model. In the next section, we first review the basic ideas of our approach.

2 Graphs versus Paths

In thinking about constructing paths in electron spaces, it is convenient from the outset to work in *antisymmetrized* spaces – conveniently done through the Slater Determinant formalism – and to construct paths in such spaces. Let us take a set of $2M$ one-electron orthonormal spin-orbitals, $u_1(\mathbf{x}), u_2(\mathbf{x}) \dots u_{2M}(\mathbf{x})$, and denote a point in our N-electron space with $D_{\mathbf{i}}(\mathbf{X})$ (the label $\mathbf{i}$ being an ordered N-tuple of integers):

$$D_{\mathbf{i}} \equiv D_{i_1 i_2 \dots i_N} = \frac{1}{\sqrt{N!}} \det[u_{i_1} \dots u_{i_N}] \,, \tag{8}$$

The orbitals can come from a variety of sources (e.g. they can be plane-waves, points on a regular 3D grid, or Hartree-Fock orbitals, Kohn-Sham orbitals etc). A given determinant describes an uncorrelated state of electrons, and in the limit of $M \to \infty$ this collection of $N_{det} = \binom{2M}{N}$ Slater determinants could be used to expand any correlated wavefunction (satisfying the boundary conditions of the problem). Obviously one wants this basis to provide a good starting point to describe the physics of N correlated electrons, and in common with wavefunction methods, the rate of convergence of our method will depend on the choice of this one-electron basis.

The advantage of using an antisymmetrized space is that one does not need to perform explicit averages over $N!$ permuted paths – as implied by the sum $(1/N!) \sum_{\hat{P}}$ in (3). Instead, traces over this space are traces over purely antisymmetric parts, as is appropriate for a Fermion system:

$$Q = \text{Tr}[e^{-\beta \hat{H}}] = \sum_{\mathbf{i}} \langle D_{\mathbf{i}} | e^{-\beta \hat{H}} | D_{\mathbf{i}} \rangle \tag{9}$$

$$= \sum_{\mathbf{i}_1} \sum_{\mathbf{i}_2} \cdots \sum_{\mathbf{i}_P} \langle D_{\mathbf{i}_1} | e^{-\beta \hat{H}/P} | D_{\mathbf{i}_2} \rangle \langle D_{\mathbf{i}_2} | e^{-\beta \hat{H}/P} | D_{\mathbf{i}_3} \rangle \dots \langle D_{\mathbf{i}_P} | e^{-\beta \hat{H}/P} | D_{\mathbf{i}_1} \rangle$$

A term in the above expansion corresponds to a closed path of P steps in Slater determinant space:

$$D_{\mathbf{i}_1} \to D_{\mathbf{i}_2} \to \dots \to D_{\mathbf{i}_P} \to D_{\mathbf{i}_1}$$

in a definite order; this is the analogue of a closed path as denoted in (6). The weight of such a path is:

$$w^{(P)}[\mathbf{i_1}, \mathbf{i_2}, \ldots, \mathbf{i_P}, \mathbf{i_1}] =$$

$$\langle D_{\mathbf{i_1}}|e^{-\beta\hat{H}/P}|D_{\mathbf{i_2}}\rangle\langle D_{\mathbf{i_2}}|e^{-\beta\hat{H}/P}|D_{\mathbf{i_3}}\rangle\ldots\langle D_{\mathbf{i_P}}|e^{-\beta\hat{H}/P}|D_{\mathbf{i_1}}\rangle \qquad (10)$$

which is analogous to (2). As regards the sign-problem we have still not made much progress, since $\text{sign}(w^{(P)})$ in (10) is not positive definite, and indeed a very poorly behaved quantity since it depends on the product of the signs of P factors, implying small variations in the path can produce wild fluctuations in the sign of the path.

Note there is no restriction that the elements along this path should be *distinct*. A given determinant can occur multiple times along a path. Suppose that this path consists of $n \leq P$ *distinct* determinants. This collection of determinants forms a n-vertex "graph" G:

$$G = \underbrace{\{D_{\mathbf{i}}, D_{\mathbf{j}}, D_{\mathbf{k}}, \ldots\}}_{n \text{ determinants}} \qquad (11)$$

The order in which the determinants appear in the above is of no consequence. The connectivity of the graph is determined by $\rho_{\mathbf{ij}} = \langle D_{\mathbf{i}}|e^{-\beta\hat{H}/P}|D_{\mathbf{j}}\rangle$. *G is an object on which one can represent the paths which visit exclusively all vertices in G.* The weight $w^{(n)}[G]$ of a given graph is obtained by summing over all such paths of length P:

$$w^{(n)}[G] = \sum_{\mathbf{i_1}\in G}\sum_{\mathbf{i_2}\in G}\sum_{\mathbf{i_3}\in G}\cdots\sideset{}{'}\sum_{\mathbf{i_P}\in G} w^{(P)}[\mathbf{i_1}, \mathbf{i_2}, \ldots, \mathbf{i_P}, \mathbf{i_1}] \qquad (12)$$

The $\prime$ indicates that the summation indices $\mathbf{i_1}, \mathbf{i_2}\ldots\mathbf{i_P}$ are chosen from the set G in such a way that each vertex in G is visited at least once. This condition ensures that the weights of two different graphs G_a and G_b (i.e. two graphs which differ in at least one vertex) do not include the same paths which visit only the vertices of the set $G_a \bigcap G_b$ (See Fig. 1 in [1]). Therefore, the sum $w[G_a] + w[G_b]$ will not double-count the contribution of such paths. As a result, the partition function can be represented not as a "sum-over-paths" but rather a "sum-over-graphs":

$$Q = \sum_n\sum_G w^{(n)}[G] \qquad (13)$$

which contains many fewer terms. It has been "contracted". In the above, the sum over n represents a sum over 1-vertex, 2-vertex, 3-vertex etc., graphs, and for each n, the sum over G represents the sum over all n-vertex graphs which can be constructed in the system. Since each graph represents a sum over exponentially large numbers of paths, it is reasonable to assume that the sign fluctuations of graphs will be better controlled than individual paths. There is, in addition, a more subtle reason why this contraction is beneficial, which has to do with the topology of graphs. Unlike paths, where there are equal number of odd and even permutations, certain graphs are positive definite and have no negative counterpart. Graphs which are trees (i.e. do not contain cycles) have positive weights. Any path on a tree which returns to its starting point must eventually retrace every outward step that it took, leading to a necessarily positive weight. Two-vertex graphs are the simplest and most important examples. If the terms with increasing n have a reasonable convergence, so as to enable a truncation of the vertex series at some small order of n, the distribution of the significant graphs will be strongly biased towards the positive graphs, and enable a sampling of graphs without encountering an exponential cancellation.

Given the expression (13) for the partition function, one can express E in a manner suitable for stochastic (Monte Carlo) method in which graphs are sampled with an appropriate probability and an energy estimator is averaged. Using (4), we find:

$$E = \frac{1}{Q} \sum_n \sum_G -\frac{\partial \ln w^{(n)}[G]}{\partial \beta} . w^{(n)}[G] \tag{14}$$

Therefore, if graphs can be sampled with an unnormalised probability given by $w^{(n)}[G]$ (using a Metropolis Monte Carlo method), then the energy estimator becomes:

$$\tilde{E}^{(n)}[G] = -\frac{\partial \ln w^{(n)}[G]}{\partial \beta} \tag{15}$$

i.e.

$$E = \langle \tilde{E}^{(n)}[G] \rangle_{w^{(n)}[G]} \tag{16}$$

In other words, if on step t of a Monte Carlo simulation consisting of K steps one is at graph G_t, then the energy is given by the running average of $\tilde{E}^{(n)}[G]$:

$$E = \lim_{K \to \infty} \frac{1}{K} \sum_t^K \tilde{E}^{(n)}[G_t] \tag{17}$$

In [1] we provide an algorithm to sample graphs generated according to a stochastic algorithm, and we will not elaborate further on this aspect.

Although (12)–(14) form the basis for a finite temperature method, it is possible (and indeed convenient) to modify the formalism to accelerate convergence towards the ground-state energy (the $\beta \to \infty$ limit), by ensuring that the graphs always contain a "good" determinant in them. For example, if we are using Hartree-Fock orbitals as the underlying one-particle basis, the Hartree-Fock determinant (composed out of the N lowest energy orbitals), is usually a good starting point. In this case, the formalism is modified as follows. Let us denote the HF determinant as D_0 and consider:

$$w_0 = \langle D_0 | e^{-\beta \hat{H}} | D_0 \rangle \tag{18}$$

and the function $\tilde{E}_0(\beta)$:

$$\tilde{E}_0(\beta) = -\frac{\partial \ln w_0}{\partial \beta} \tag{19}$$

In [1], we show that $\tilde{E}_0$ is, as a function of β, bounded from above by the Hartree-Fock energy E_{HF} and from below by the exact ground-state energy E_0:

$$\lim_{\beta \to 0} \tilde{E}_0 = E_{HF} \tag{20}$$

$$\lim_{\beta \to \infty} \tilde{E}_0 = E_0 \tag{21}$$

$$E_0 \le \tilde{E}_0 \le E_{HF} \tag{22}$$

w_0 here plays the role of the partition function Q, since it can itself be expanded as a path-integral, with the proviso that, instead of performing a trace over all initial positions, one computes the sum over all paths of length P which start and finish at D_0:

$$w_0 = \sum_{i_1} \sum_{i_2} \cdots \sum_{i_P} \langle D_0 | e^{-\beta \hat{H}/P} | D_{i_1} \rangle \langle D_{i_1} | e^{-\beta \hat{H}/P} | D_{i_2} \rangle \ldots \langle D_{i_P} | e^{-\beta \hat{H}/P} | D_0 \rangle \tag{23}$$

Analogously to (13), we now seek to write:

$$w_0 = \sum_n \sum_G w_0^{(n)}[G] \tag{24}$$

where $w_0^{(n)}[G]$ is defined analogously to (12), except for the fact that D_0 is considered the beginning and end point of paths:

$$w_0^{(n)}[G] = \sum_{i_2 \in G} \sum_{i_3 \in G} \cdots \sum_{i_P \in G} {}' w^{(P)}[\mathbf{0}, \mathbf{i_2}, \ldots, \mathbf{i_P}, \mathbf{0}] \tag{25}$$

In (25) it is required that G contains D_0 among its members. The ' has the same meaning as before: all vertices in G must be visited by the paths. Using (24) and (19) leads to an expression for $\tilde{E}_0$ suitable for a stochastic sampling of graphs, analogous to (15):

$$\tilde{E}_0 = \langle \tilde{E}_0^{(n)}[G] \rangle_{w_0^{(n)}[G]} \tag{26}$$

or in the case where $w_0^{(n)}[G]$ is not positive-definite:

$$\tilde{E}_0 = \frac{\langle \text{sign}(w_0^{(n)}[G]) \tilde{E}_0^{(n)}[G] \rangle_{|w|}}{\langle \text{sign}(w_0^{(n)}[G]) \rangle_{|w|}} \tag{27}$$

One might expect that the convergence of (24) with respect to the graph size n to be reasonably rapid. Starting from the Hartree-Fock determinant, 2-vertex graphs sum over paths which involve double-excitations, 3-vertex graphs include those with triple and quadruple excitations, and in general the n-vertex graph include paths with up to $2(n-1)$-fold excitations. From a physical view-point, it is reasonable to expect that a lot of correlation can be captured with fairly small graphs, in line with the experience from coupled-cluster and perturbation theories. Here the inverse-temperature-like parameter β plays a useful role. As we show for the N_2 molecule in [1], for small β, the convergence with respect to n is rapid, since at small β, the paths cannot move far from D_0. At large β the convergence at a given graph-size will be poor, and unphysical results will ensue. However, one get meaningful results at an intermediate β before convergence breakdown.

3 A Useful Combinatorial Identity

How can one compute terms in $w^{(n)}[G]$ in (13) (and $w_0^{(n)}[G]$ in (24)), which are necessary to do the Monte Carlo sampling? One can get an intuitive sense of this by rearranging the terms in the partition function to correspond with paths with increasing numbers of "hops":

$$
Q = \sum_i \Bigg[\rho_{ii}{}^P + \sum_j^{\prime i} \sum_{n=0}^{P-2} \sum_{m=0}^{P-2-n} \rho_{ii}{}^n \rho_{ij} \rho_{jj}{}^m \rho_{ji} \rho_{ii}{}^{P-2-m-n}
$$
$$
+ \sum_j^{\prime i} \sum_k^{\prime i,j} \sum_{n=0}^{P-3} \sum_{m=0}^{P-3-n} \sum_{l=0}^{P-3-n-m} \rho_{ii}{}^n \rho_{ij} \rho_{jj}{}^m \rho_{jk} \rho_{kk}{}^l \rho_{ki} \rho_{ii}{}^{P-3-n-m-l} + \dots \Bigg]
$$

$$(28)$$

where we have set $\rho_{ij} \equiv \langle D_i | e^{-\beta \hat{H}/P} | D_j \rangle$. Noting that we are getting nested sums, it proves useful to define a function $Z_h^{(P)}(x_1, x_2, \dots, x_h)$ as follows:

$$
Z_h^{(P)}(x_1, x_2, \dots, x_h) = \sum_{n_1=0}^{P-h} \sum_{n_2=0}^{P-h-n_1} \dots \sum_{n_h=0}^{P-h-\sum_{i=1}^{h-1} n_i} x_1^{n_1} x_2^{n_2} \dots x_h^{n_h} \quad (29)
$$

We will shortly discuss the properties of such nested sums. Let us for the moment note that we can cast each term in the square-brackets of (28) (which we define as w_i) as follows:

$$w_{\mathbf{i}} = \rho_{\mathbf{ii}}{}^P \left(1 + \sum_{\mathbf{j}}^{\prime\mathbf{i}} \frac{\rho_{\mathbf{ij}}\rho_{\mathbf{ji}}}{\rho_{\mathbf{ii}}{}^2} Z_2^{(P)} \left(1, \frac{\rho_{\mathbf{jj}}}{\rho_{\mathbf{ii}}} \right) \right.$$

$$+ \sum_{\mathbf{j}}^{\prime\mathbf{i}} \sum_{\mathbf{k}}^{\prime\mathbf{j},\mathbf{i}} \frac{\rho_{\mathbf{ij}}\rho_{\mathbf{jk}}\rho_{\mathbf{ki}}}{\rho_{\mathbf{ii}}{}^3} Z_3^{(P)} \left(1, \frac{\rho_{\mathbf{jj}}}{\rho_{\mathbf{ii}}}, \frac{\rho_{\mathbf{kk}}}{\rho_{\mathbf{ii}}} \right)$$

$$\left. + \sum_{\mathbf{j}}^{\prime\mathbf{i}} \sum_{\mathbf{k}}^{\prime\mathbf{j}} \sum_{\mathbf{l}}^{\prime\mathbf{k},\mathbf{i}} \frac{\rho_{\mathbf{ij}}\rho_{\mathbf{jk}}\rho_{\mathbf{kl}}\rho_{\mathbf{li}}}{\rho_{\mathbf{ii}}{}^4} Z_4^{(P)} \left(1, \frac{\rho_{\mathbf{jj}}}{\rho_{\mathbf{ii}}}, \frac{\rho_{\mathbf{kk}}}{\rho_{\mathbf{ii}}}, \frac{\rho_{\mathbf{ll}}}{\rho_{\mathbf{ii}}} \right) + \cdots \right) \quad (30)$$

In the sums above, each sum is prime with respect to (i.e. excludes) the index of the previous sum, except for the inner-most sum, which is also prime with respect to $\mathbf{i}$. For example, in the 4-hop term, $\mathbf{k}$ can equal $\mathbf{i}$, but not $\mathbf{j}$. Similarly, $\mathbf{l}$ can equal $\mathbf{j}$, but not $\mathbf{k}$ or $\mathbf{i}$.

Physically, $Z_h^{(P)}$ is a combinatorial factor which accounts for the number of ways the $P-h$ "stay-put" terms can be arranged for a given number of hops h. The rationale for this type of expansion is due to the diagonal dominance of the $\rho_{\mathbf{ij}}$ matrix. Every time we hop, we incur a penalty; therefore paths which involve many hops each carry an exponentially decreasing weight. On the other hand, the number of such paths increases exponentially. Therefore, the rate of convergence of the expansion above is difficult to predict. At high temperatures (or weak coupling) it will be fast (the off-diagonal $\rho_{\mathbf{ij}}$ are small), and otherwise slow. Our aim will be further summations in (30) to produce a rapidly convergent series.

In Appendix A, we prove the following identity:

$$Z_h^{(P)}(x_1, \ldots, x_h) = \frac{1}{2\pi i} \oint_C \frac{z^P - 1}{(z-1)\prod_\alpha (z - x_\alpha)} dz \quad (31)$$

where C is a contour which encloses the poles $(x_1, \ldots, x_h)$. This integral has some remarkable properties discussed in the Appendix. Crucially, it enables us *to sum over paths with different numbers of hops*. For example, consider a 3-cycle $\mathbf{i} \to \mathbf{j} \to \mathbf{k} \to \mathbf{i}$ involving the transition matrix elements $\rho_{\mathbf{ij}}\rho_{\mathbf{jk}}\rho_{\mathbf{ki}}$. This 3-cycle (whose weight can be negative), will be involved in the following 3-hop, 6-hop,..., paths which form a typical alternating series:

$$S = \frac{\rho_{\mathbf{ij}}\rho_{\mathbf{jk}}\rho_{\mathbf{ki}}}{\rho_{\mathbf{ii}}{}^3} Z_3 + \left(\frac{\rho_{\mathbf{ij}}\rho_{\mathbf{jk}}\rho_{\mathbf{ki}}}{\rho_{\mathbf{ii}}{}^3} \right)^2 Z_6 + \cdots$$

Defining the *primitive cycle function*:

$$A_{\mathbf{ijk}}(z) = \frac{\rho_{\mathbf{ij}}\rho_{\mathbf{jk}}\rho_{\mathbf{ki}}}{(z\rho_{\mathbf{ii}} - \rho_{\mathbf{ii}})(z\rho_{\mathbf{ii}} - \rho_{\mathbf{jj}})(z\rho_{\mathbf{ii}} - \rho_{\mathbf{kk}})} \quad (32)$$

we obtain:

$$S = \sum_n \frac{1}{2\pi i} \oint_C \frac{z^P - 1}{z - 1} [A_{\mathbf{ijk}}(z)]^n dz \quad (33)$$

Switching the order of summation and integration, we obtain:

$$S = \frac{1}{2\pi i} \oint_C \frac{z^P - 1}{z - 1} \frac{1}{1 - A_{ijk}(z)} dz \tag{34}$$

Since $1 - A_{ijk}(z) = 0$ is a cubic equation in z, it has three roots, which can be used to evaluate (34) in terms of three residues – see Appendix A.

The above is a simple example of the analytic summation over an alternating series. One can similarly treat more complex paths, which consist of sums of *combinations of cyclic paths*. In [1], we derive in detail sums over paths (which start and finish at **i**) constructed on two elemental types of graphs – star-graphs and chain-graphs. We quote below the results:

$$S_i^{star} = \frac{1}{2\pi i} \oint_C \frac{z^P - 1}{z - 1} \frac{1}{(1 - A_{G_1} - A_{G_2} - \cdots - A_{G_g})} dz \tag{35}$$

$$S_i^{chain} = \frac{1}{2\pi i} \oint_C \frac{z^P - 1}{z - 1} \frac{1}{1 - \frac{A_{G_1}}{1 - \frac{A_{G_2}}{1 - \cdots}}} dz \tag{36}$$

In the above, $A_{G_1}, A_{G_2}, \ldots$ are primitive cycle functions from which the star-graphs and chain-graphs are constructed.

Using these, we show that paths constructed on the general 3-vertex graph (which start and finish at **i**) contribute:

$$S_i^{(3)}[\{\mathbf{i, j, k}\}] = \frac{1}{2\pi i} \oint_C \frac{z^P - 1}{z - 1} \frac{1}{1 - \frac{A_{ij}}{1 - A_{jk}} - \frac{A_{ik}}{1 - A_{jk}} - 2\frac{A_{ijk}}{1 - A_{jk}}} dz$$

$$= \frac{1}{2\pi i} \oint_C \frac{z^P - 1}{z - 1} \frac{1 - A_{jk}}{1 - A_{ij} - A_{ik} - A_{jk} - 2A_{ijk}} dz \tag{37}$$

The denominator of the three-vertex term, $1 - A_{ij} - A_{ik} - A_{jk} - 2A_{ijk} = 0$, reduces to a cubic polynomial in z, which can be solved to yield three roots, which are in turn used to evaluated the contour integral using the residue theorem.

By a similar technique of enumerating the distinct circuits on a complete 4-vertex graph, one can construct the unfolded representation and hence evaluate $S_i^{(4)}[\{\mathbf{i, j, k, l}\}]$. This procedure is carried out in the Appendix B. Although the resulting expression is quite unwieldy, the final form simplifies remarkably to a quartic polynomial on the denominator of the contour-integral.

The final expression for the weights for graphs with 2, 3 and 4 vertices are given by (46-48) of [1], and they are not repeated here. These take into account a double-counting corrections as explained there. We are optimistic to be able to generalise our method to larger graphs.

4 An Example: The Hubbard Model

We demonstrate some of the ideas for a $\sqrt{18} \times \sqrt{18}$ Hubbard lattice containing 18 electrons, with periodic boundary conditions. This model has been extensively studied [11]. The Hubbard parameter was set to $U/t = 4$. The ground-state of the non-interacting system, being diagonalised by plane-waves, is a closed shell consisting of 1+4+4 doubly occupied levels. There are $2M = 36$ one particle orbtials, leading to $N_{det} = 9075135300$ determinants for an $N = 18$ system. There is a unique Fermi determinant. The matrix elements of the high-temperature density matrix ρ_{ij} were computed according to a Trotter decompostion formula ((30) of [1]). Corrections to this formula go as $(\beta/P)^2$, and are negligible for sufficiently small β/P (we used $\beta/P = 10^{-4}$). In addition, matrix elements with magnitude smaller 10^{-6} were set to zero, which represented a useful (factor of 2) but not essential saving in computer time. It is important to note that matrix elements between determinants which are more than double excitations away from each other, are zero. In addition, single excitations do not couple in this system due to translational invariance (momentum conservation).

We computed the weights of all graphs which include the Fermi determinant, up to 4-vertices. Graphs are classified according to whether they were trees or cyclic graphs. In Table 1, we give the total number of graphs which arise in these two catagories, and their integrated weight. In addition, the energies as measured by $\tilde{E}_0$ (19) are given for each vertex size truncation. The first observation is that there is an (expected) explosive growth in the number of graphs as the graph size n increases – by roughly 3 orders of magnitude from each increment in n. It is significant that the number of tree graphs remain overwhelmingly dominant even at the 4-vertex level, as do the relative weights of the tree graphs compared to cyclic graphs. Since the former are positive definite, this implies that a stochastic sampling of graphs would not encounter a serious sign-problem. The relative magnitude of the contributions of the 3 different graph sizes is β dependent. At $\beta = 1$, the 3-vertex graphs contribute the most to the overall weight, whereas the $\beta = 5$ the 4-vertex graphs are significant. This indicates that the convergence of the vertex series is strongly β dependent. This is also shown in the energies. Although at both β the energies decrease with increasing graph-size, the convergence is much faster at $\beta = 1$. The exact ground-state energy is known to be -17.25239, whilst the Hartree-Fock energy is -14. Therefore, at $\beta = 1$, the $\tilde{E}_0$ at the 4-vertex level captures some 65% of the ground-state correlation energy, whilst at $\beta = 5$ the result is somewhat worse (only 18%). This indicates that, for highly accurate calculations, we will need to go beyond the 4-vertex graphs.

5 Future Work

Although our method is very much in its infancy, we believe our ideas can be developed and usefully applied to many types of electronic problems. Current

Table 1. Number and the natural logarithm of the weights of graphs for a Hubbard model. The energies (in units of the Hubbard t) are also reported.

$\beta = 1$					
vertex	trees		cyclic		$\tilde{E}_0$
	number	$\ln w_0^{(n)}[G]$	number	$\ln w_0^{(n)}[G]$	
2	425	2.6174	0	-	−14.8593
3	248524	2.8557	4064	0.1497	−15.5963
4	166591656	1.7845	6222444	0.7217	−16.1097
$\beta = 5$					
vertex	trees		cyclic		$\tilde{E}_0$
	number	$\ln w_0^{(n)}[G]$	number	$\ln w_0^{(n)}[G]$	
2	425	15.3019	0	-	−14.1977
3	248484	108.0713	4064	5.3550	−14.3978
4	166591656	472.057	6222444	75.3802	−14.5974

work is to calibrate the method against the benchmark (Pople) set of molecules, which we hope to report on in the near future. It is an extremely exciting prospect to perform for Fermion systems a simulation which is stable, albeit approximate, through the truncation of of the graph-expansion of the partition function. We are currently working on the extension of the method to larger graphs. There are equally important issues of a more technical nature which will also need attention. A key property is the favourable scaling of the method: apart from an initialisation step of computing the 4-index integrals (itself an $N^4 M$ step), the calculation of the weights is an N^2 process. The main bottleneck very rapidly becomes the initialisation step if *all* 4-index integrals are calculated. A work-around this problem will become necessary if we are to tackle large systems consisting of hundreds of electrons, but we believe this issue will be resolvable through on-the-fly calculation and storage/retrieval of the integrals.

We are grateful for the support of the EPSRC through a Portfolio Award.

Appendix A: Proof of (31)

Consider the multiple nested-sum:

$$Z_h^{(P)}(x_1, \ldots, x_h) = \sum_{n_1=0}^{P-h} \sum_{n_2=0}^{P-h-n_1} \ldots \sum_{n_h=0}^{P-h-\sum_i^{h-1} n_i} x_1^{n_1} x_2^{n_2} \ldots x_h^{n_h} \quad (38)$$

This sum is non-zero only for $P \geq h$. For $P < h$, Z_h is identically zero for all $\{x_i\}$. We shall also assume that the $\{x_i\}$ can take arbitrary values, and in particular, that they can be "degenerate", i.e. some or all of them can be equal.

By induction, we would like to prove:

$$Z_h^{(P)}(x_1, \ldots, x_h) = \frac{1}{2\pi i} \oint_C \frac{z^P - 1}{(z-1)\prod_{j=1}^h (z - x_j)} dz \tag{39}$$

where the contour C encloses $x_1 \ldots, x_h$. The case $h = 1$ is immediately verified:

$$Z_1^{(P)}(x) = \frac{1}{2\pi i} \oint_C \frac{z^P - 1}{(z-1)(z-x)} dz$$

$$= \frac{x^P - 1}{x - 1} = \sum_{n=0}^{P-1} x^n \tag{40}$$

Now assume (39) holds for h, and consider $h + 1$.

$$Z_{h+1}^{(P)}(x_1, \ldots, x_h, x_{h+1}) = \sum_{n=0}^{P-h-1} x_{h+1}^n \sum_{n_1=0}^{P-h-1-n} \cdots \sum_{n_h=0}^{P-h-1-n-\sum_i n_i} x_1^{n_1} \ldots x_h^{n_h}$$

$$= \sum_{n=0}^{P-h-1} x_{h+1}^n Z_h^{(P-n-1)}(x_1, \ldots, x_h)$$

$$= \sum_{n=0}^{P-1} x_{h+1}^n Z_h^{(P-n-1)}(x_1, \ldots, x_h)$$

$$= \sum_{n=0}^{P-1} x_{h+1}^n \frac{1}{2\pi i} \oint_C \frac{z^{P-n-1} - 1}{(z-1)\prod_i^h (z-x_i)} dz$$

$$= \frac{1}{2\pi i} \oint_C \frac{\sum_{n=0}^{P-1}(x_{h+1}/z)^n z^{P-1} - x_{h+1}^n}{(z-1)\prod_i^h (z-x_i)} dz$$

$$= \frac{1}{2\pi i} \oint_C \frac{(z^P - x_{h+1}^P)/(z - x_{h+1}) - (1 - x_{h+1}^P)/(1 - x_{h+1})}{(z-1)\prod_i^h (z-x_i)} dz$$

$$= \frac{1}{2\pi i} \oint_C \frac{z^P - x_{h+1}^P}{(z-1)\prod_i^{h+1}(z-x_i)} dz - \frac{1}{2\pi i} \oint_C \frac{1 - x_{h+1}^P}{(z-1)(1-x_{h+1})\prod_i^h (z-x_i)} dz$$

$$= \frac{1}{2\pi i} \oint_C \frac{(z^P - x_{h+1}^P - x_{h+1}z^P + x_{h+1}^{P+1} - z + x_{h+1} + x_{h+1}^P z - x_{h+1}^{P+1} - 1 + 1}{(z-1)(1-x_{h+1})\prod_i^{h+1}(z-x_i)} dz$$

where in the last line we have added ± 1 to the numerator. Continuing:

$$= \frac{1}{2\pi i} \oint_C \frac{(z-1)(x_{h+1}^P - 1) - (z^P - 1)(x_{h+1} - 1)}{(z-1)(1-x_{h+1})\prod_i^{h+1}(z-x_i)} dz$$

$$= \frac{1}{2\pi i} \oint_C \frac{(x_{h+1}^P - 1)}{(1-x_{h+1})\prod_i^{h+1}(z-x_i)} dz + \frac{1}{2\pi i} \oint_C \frac{(z^P - 1)}{(z-1)\prod_i^{h+1}(z-x_i)} dz \tag{41}$$

Next, we use the identity [10]:

$$\frac{1}{2\pi i} \oint_C \frac{1}{\prod_i^{h+1}(z - x_i)} dz = 0 \text{ for } h \geq 1 \tag{42}$$

which can be verified by considering the Laurent expansion of the integrand:

$$\frac{1}{z^{h+1} \prod_i^{h+1}(1 - x_i/z)} = z^{-h-1} + (x_1 + \ldots + x_{h+1})z^{-h-2}$$

$$+ (x_1^2 + x_1 x_2 + \ldots)z^{-h-3} + \ldots. \tag{43}$$

Integrating term by term, everything vanishes if $h \geq 1$, which by hypothesis is the case. The first term in (41) is therefore zero, leaving:

$$Z_{h+1}^{(P)}(x_1, x_2, \ldots, x_{h+1}) = \frac{1}{2\pi i} \oint_C \frac{(z^P - 1)}{(z - 1) \prod_i^{h+1}(z - x_i)} dz \tag{44}$$

which is precisely the required result.

Incidently, the property that the contour-intergal formula (39) respects the identity $Z_h^{(P)} = 0$ for $h > P$, for any set of arguments $\{x_i\}$, follows from a generalisation of (42), namely:

$$\frac{1}{2\pi i} \oint_C \frac{z^P}{\prod_i^{h+1}(z - x_i)} dz = 0 \text{ for } P < h \tag{45}$$

which can be proven using the Laurent expansion of the integrand and integrating term-by-term, as above. This allows one to replace all summations over h to infinity, thereby considerably simplifying the manipulation of sums.

If the $x_1 \ldots, x_h$ are all distinct, the residue theorem gives:

$$Z_h^{(P)}(x_1, \ldots, x_h) = \sum_{i=1}^h \frac{x_i^P - 1}{(x_i - 1) \prod_{j \neq i}(x_i - x_j)}. \tag{46}$$

If there are multiple poles, then the appropriate version of the residue theorem needs to be applied, viz:

$$\frac{1}{2\pi i} \oint_C f(z) dz = (\text{sum of enclosed residues}) \tag{47}$$

where a residue for a pole of order m at z_0 is given by:

$$\frac{1}{(m-1)!} \frac{d^{m-1}}{dz^{m-1}} [(z - z_0)^m f(z)] \tag{48}$$

(46) has been derived by Hall [12] by an independent method.

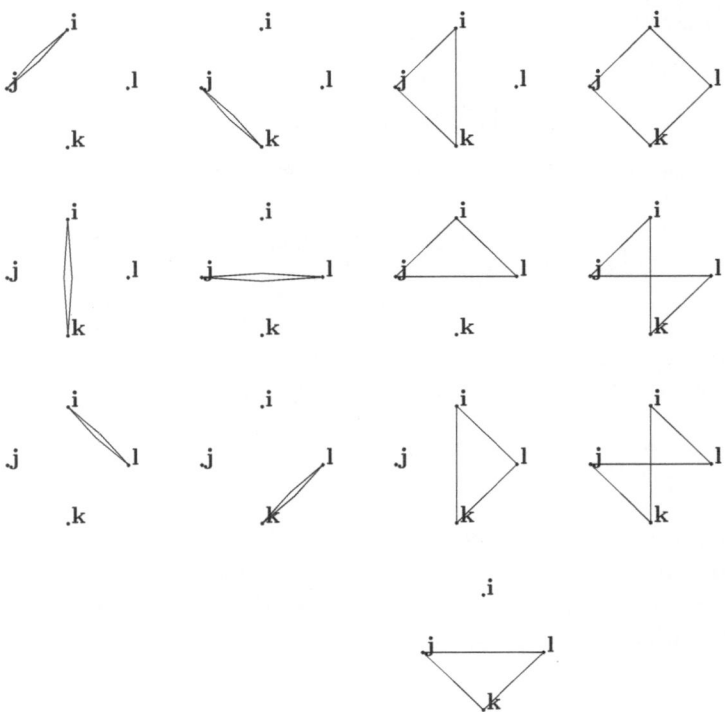

Fig. 1. All 13 distinct four-vertex primitive cycles

Appendix B: The Complete 4-vertex Graph

With the insight gained from constructing the complete three-vertex graph, we now develop an algorithm to create a graph constructed of primitive circuits linked in the form of chain and star graphs. Once such a graph has been constructed, it is a simple matter to write down the expression about which the contour must be taken to sum all possible graphs.

For a n-vertex graph, we must first list all possible primitive circuits. Here we give as an example the four-vertex graph, which has 20 primitive circuits (Fig. 1):

$$(ij), (ik), (il), (jk), (jl), (kl), (ijk), (ikj), (ijl), (ilj), (ikl), (ilk), (jkl), (jlk)$$
$$(ijkl), (ilkj), (ijlk), (iklj), (ikjl), (iljk)$$

Of these, thee are four which do not contain the pivot **i**, and the remaining 16 do.

The algorithm to construct the complete n-vertex graph is as follows:

1. Set the current pivot to **i**, the global pivot.

2. Take the current list of primitive graphs, and create a star graph consisting of all of the primitive graphs which contain the current pivot.
3. Recurse through each of the primitive graphs we have just added. For each vertex of each graph, repeat from step 2, having eliminated all graphs containing vertices between the current pivot and the vertex we are considering.

We give as an example the n-vertex algorithm applied to the four-vertex graph.

Setting **i** as our current pivot, the list of all graphs containing **i** is

$$(\mathbf{ij}), (\mathbf{ik}), (\mathbf{il}), (\mathbf{ijk}), (\mathbf{ikj}), (\mathbf{ijl}), (\mathbf{ilj}), (\mathbf{ikl}), (\mathbf{ilk})$$
$$(\mathbf{ijkl}), (\mathbf{ilkj}), (\mathbf{ijlk}), (\mathbf{iklj}), (\mathbf{ikjl}), (\mathbf{iljk})$$

Forming a star graph with all of these graphs pivoting and joining at **i** we result in Fig. 3. We now consider the two-, three- and four-vertex parts of this star graph.

Starting with (**ij**), we take **j** as the new pivot, we must now exclude all primitive graphs including **i**. From this, the list of graphs containing the new pivot **j** is,

$$(\mathbf{jk}), (\mathbf{jl}), (\mathbf{jkl}), (\mathbf{jlk})$$

A star graph is fomed by joining all these graphs at **j**. Now for each of these subgraphs, vertices **i** and **j** have been traversed, so the only remaining graph without these is (**kl**). Taking each of the graphs, we follow the same procedure, attaching a (**kl**) to the (**jk**) and an (**lk**) to the (**jl**). Looking at the (**jkl**) graph, we choose **k** as the next pivot. The only graph available to us is a (**kl**), which we attach. Similarly, we attach a (**lk**) to the l of (**jlk**). This is shown in Fig. 2(a). Recursing to the next level, there are no primitive graphs which do not contain **i, j**, and **k**, so the work on the (**ij**) graph is complete.

Returning to the pivot **i**, we now consider the (**ijk**) graph. At **j**, we follow the same procedure as above. At **k**, having excluded all graphs containing **i**, and **j**, only (**kl**) remains, which we attach to **k**. At l there are no graphs remaining to attach (Fig. 2(b)). The four-vertex graphs work in an analogous way, and are shown in Fig. 2(c).

The reasoning behind this algorithm is that at every point we must attempt to attach all possible graphs which would not lead to a graph which is accounted for elsewhere. At the **j** of the (**ij**) graph, if any graph containing **i** were attached, say (**jki**), traversing from our global pivot **i** would lead to a path **ijki**... which has already been accounted for under the (**ijk**) graph pivoted at **i**.

The result of this algorithm on the four-vertex graph yields the rather unweildy graph shown in Fig. 4 (which has been computer generated). This has the corresponding expression (also computer generated) associated with it:

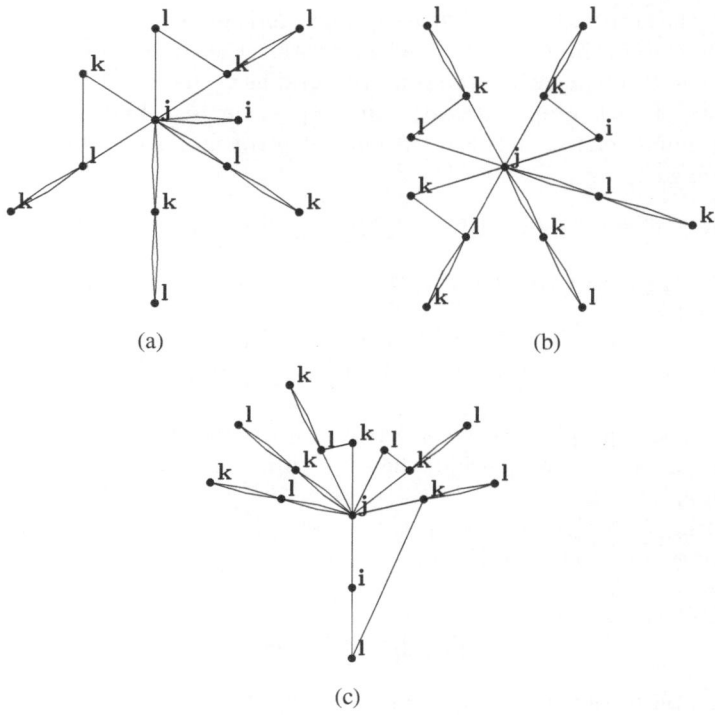

(a)   (b)

(c)

Fig. 2. The independent chains for a four-vertex graph beginning with (**a**) two-; (**b**) three- and (**c**) four-membered circuits

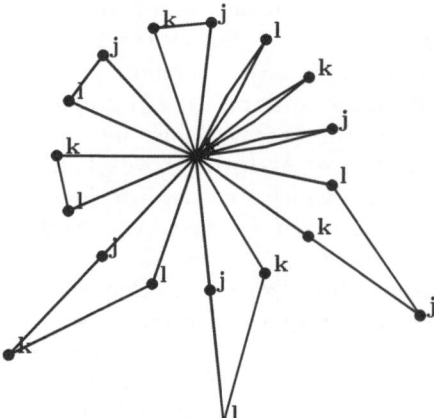

Fig. 3. All primitive circuits in the four-vertex graph which pivot at **i** joined in a star-graph

Fig. 4. The complete four-vertex graph of all independent circuits joined together and pivoted at **i**

$$S_{\mathbf{i}}^{(4)}[\{\mathbf{i},\mathbf{j},\mathbf{k},\mathbf{l}\}] =$$

$$\times \frac{\frac{1}{2\pi i}\oint_C \frac{z^P-1}{z-1}}{1-\dfrac{A_{\mathbf{ij}}}{1-\dfrac{A_{\mathbf{jk}}}{1-A_{\mathbf{kl}}}-\dfrac{A_{\mathbf{jl}}}{1-A_{\mathbf{kl}}}-\dfrac{2A_{\mathbf{jkl}}}{1-A_{\mathbf{kl}}}}-\dfrac{A_{\mathbf{ijk}}+A_{\mathbf{ijl}}+A_{\mathbf{ijkl}}+A_{\mathbf{ijlk}}}{\left(1-\dfrac{A_{\mathbf{ij}}}{1-A_{\mathbf{kl}}}-\dfrac{A_{\mathbf{jl}}}{1-A_{\mathbf{kl}}}-\dfrac{2A_{\mathbf{jkl}}}{1-A_{\mathbf{kl}}}\right)(1-A_{\mathbf{kl}})}\cdots} \qquad (49)$$

$$-\frac{A_{\mathbf{ik}}}{1-\dfrac{A_{\mathbf{jk}}}{1-A_{\mathbf{jl}}}-\dfrac{A_{\mathbf{kl}}}{1-A_{\mathbf{jl}}}-\dfrac{2A_{\mathbf{jkl}}}{1-A_{\mathbf{jl}}}}-\dfrac{A_{\mathbf{ijk}}+A_{\mathbf{ikl}}+A_{\mathbf{ikjl}}+A_{\mathbf{iklj}}}{\left(1-\dfrac{A_{\mathbf{jk}}}{1-A_{\mathbf{jl}}}-\dfrac{A_{\mathbf{kl}}}{1-A_{\mathbf{jl}}}-\dfrac{2A_{\mathbf{jkl}}}{1-A_{\mathbf{jl}}}\right)(1-A_{\mathbf{jl}})}\cdots$$

$$-\frac{A_{\mathbf{il}}}{1-\dfrac{A_{\mathbf{jl}}}{1-A_{\mathbf{jk}}}-\dfrac{A_{\mathbf{kl}}}{1-A_{\mathbf{jk}}}-\dfrac{2A_{\mathbf{jkl}}}{1-A_{\mathbf{jk}}}}-\dfrac{A_{\mathbf{ijl}}+A_{\mathbf{ikl}}+A_{\mathbf{ilkj}}+A_{\mathbf{iljk}}}{\left(1-\dfrac{A_{\mathbf{jl}}}{1-A_{\mathbf{jk}}}-\dfrac{A_{\mathbf{kl}}}{1-A_{\mathbf{jk}}}-\dfrac{2A_{\mathbf{jkl}}}{1-A_{\mathbf{jk}}}\right)(1-A_{\mathbf{jk}})}$$

which simplifies to

$$S_{\mathbf{i}}^{(4)}[\{\mathbf{i},\mathbf{j},\mathbf{k},\mathbf{l}\}]$$
$$= \frac{1}{2\pi i}\oint_C \frac{z^P-1}{z-1}\frac{1-A_{\mathbf{jk}}-A_{\mathbf{jl}}-A_{\mathbf{kl}}-2A_{\mathbf{jkl}}}{1-A_{\mathbf{ij}}(1-A_{\mathbf{kl}})-A_{\mathbf{ik}}(1-A_{\mathbf{jl}})-A_{\mathbf{il}}(1-A_{\mathbf{jk}})-A_{\mathbf{jk}}-A_{\mathbf{jl}}-A_{\mathbf{kl}}}$$
$$\qquad\qquad -2A_{\mathbf{ijk}}-2A_{\mathbf{ijl}}-2A_{\mathbf{ikl}}-2A_{\mathbf{jkl}}-2A_{\mathbf{ijkl}}-2A_{\mathbf{ijlk}}-2A_{\mathbf{ikjl}}$$

The denominator reduces to a quartic polynomial in z, whose roots lead to four residues.

References

1. A. J. W. Thom and A. Alavi (2005) A combinatorial approach to the electron correlation problem. *J. Chem. Phys.* **123**, pp. 204106
2. R. P. Feynman (1948) Space-Time Approach to Non-Relativistic Quantum Mechanics. *Rev. Mod. Phys.* **20**, pp. 367–387
3. R. P. Feynman, A. R. Hibbs (1965) Quantum Mechanics and Path Integrals, McGraw-Hill.
4. W. M. C. Foulkes, L. Mitas, R. J. Needs and G. Rajagopal (2001) Quantum Monte Carlo simulations of solids. *Rev. Mod. Phys.* **73**, pp. 33–83
5. D. M. Ceperley (1992) Path-integral calculations of normal liquid-He-3. *Phys. Rev. Lett.* **69**, pp. 331–334
6. For a comprehensive account of many quantum chemical methods see *Modern Electronic Structure Theory* by P. Jorgensen, J. Olsen and T. Helgaker (2000) Wiley, New York
7. J. B. Anderson (1975) Random-walk simulation of Schrodinger equation - H+3. *J. Chem. Phys.* **63**, pp. 1499–1503; *ibid.* (1976) Quantum chemistry by random-walk. **65**, pp. 4121–4127
8. A. Luchow and J. B. Anderson (1996) First-row hydrides: Dissociation and ground state energies using quantum Monte Carlo. *J. Chem. Phys.* **105**, pp. 7573–7578
9. M. E. Tuckerman and A. Hughes (1998) In "Classical and Quantum Dynamics in condensed systems", ed. B. J. Berne, G. Ciccotti, D. F. Coker, World Scientific
10. D. E. Knuth (1973) In "The Art of Computer Programming, Volume 1: Fundamental Algorithms", Addison Wesley
11. F. Becca, A. Parola, S. Sorella (2000) Ground-state properties of the Hubbard model by Lanczos diagonalizations. *Phys. Rev. B* **61**, pp. R16287–R16290
12. R. W. Hall (2002) An adaptive, kink-based approach to path integral calculations. *J. Chem. Phys.* **116**, pp. 1–7

Index

Lecture Notes in Physics

For information about earlier volumes
please contact your bookseller or Springer
LNP Online archive: springerlink.com